Protecting Groups

Protecting Groups

Strategies and Applications in Carbohydrate Chemistry

Edited by

Sébastien Vidal

WILEY-VCH
Verlag GmbH & Co. KGaA

Editor

Prof. Sébastien Vidal
Université Claude Bernard Lyon 1
CNRS
ICBMS - Equipe CO2 Glyco
43 Boulevard du 11 Nov. 1918
69622 Villeurbanne
France

Library of Congress Card No.: applied for

British Library Cataloguing-in-Publication Data
A catalogue record for this book is available from the British Library.

Bibliographic information published by the Deutsche Nationalbibliothek
The Deutsche Nationalbibliothek lists this publication in the Deutsche Nationalbibliografie; detailed bibliographic data are available on the Internet at <http://dnb.d-nb.de>.

Print ISBN: 978-3-527-34010-1
ePDF ISBN: 978-3-527-69700-7
ePub ISBN: 978-3-527-69702-1
oBook ISBN: 978-3-527-69701-4

Cover Design Tata Consulting Services
Typesetting SPi Global, Chennai, India
Printing and Binding C.O.S. Printers Pte Ltd, Singapore

Printed on acid-free paper

10 9 8 7 6 5 4 3 2 1

Contents

Foreword

There is no need to explain in the present monograph that *carbohydrates play important roles in biology*. The italicized part of this sentence has been used in myriad introductions to carbohydrate-related publications. Nevertheless, it is worth pointing out that glycosciences are a meeting point of many fields including chemistry, biochemistry, biology, medical sciences, and even physics, materials, and environmental sciences. The importance of carbohydrates and their derivatives in industry and society is growing continuously, and high-quality academic and industrial research is therefore a prerequisite for innovation and progress. When considering a monosaccharide as a starting material, the five hydroxyl groups display different reactivities, which have allowed chemists to design a multitude of protecting group strategies that allow for the precise control of structural transformations leading toward a dazzling variety of target structures for applications not only in glycosciences but also in every aspect of chemistry, biology, medicine, materials, and physics. This book is a collection of pioneering as well as recent developments for the protecting group strategies described in carbohydrate chemistry.

"Protective groups in organic synthesis" edited by Greene and Wuts represents one of the "must-have" books on any chemist's shelf. I was struck by the fact that although most of the protective groups compiled in this book could be or have been applied to carbohydrates, a reader wanting to make a selectively protected carbohydrate for the first time would probably have some difficulty in reaching its goal. We therefore embarked on a project a few years ago (it took 3 years to put the book together from all chapters' authors contributions) to provide a general tool not only for established carbohydrate chemists but also for chemists, biochemists, students, or any newcomer to the field to have the broadest possible panel of strategies to reach a suitably protected monosaccharide for their objectives.

The first series of chapters provide descriptions of the main protecting group strategies for each position of carbohydrates. The next series of chapters then focus on more specific families of carbohydrates (e.g. sialic acid, furanoside series, and sulfated oligosaccharides). The final chapters then illustrate the state-of-the-art applications of multiple protecting group strategies in areas such as automated oligosaccharide synthesis or in industrial production at the multi-kilogram scale.

The idea of this book has been in my mind for the past 10 years or so, and I am glad to finally reach this goal with the help of the numerous chapter authors. I would like to thank warmly and sincerely all the contributors who responded to my (phone) call(s) and agreed to devote a significant part of their time and expertise in order to provide the community with what I hope will be a useful and well-thumbed book.

Université Lyon 1 and CNRS
Editor
20 June 2018

Dr. Sébastien Vidal

Preface

The density of hydroxyl functional groups in carbohydrates is unmatched by any other class of natural products. As nature utilizes this tremendous possibility for structural diversity, particularly in oligosaccharide and glycoconjugate synthesis, regioselective access to these functional groups is of utmost importance. Nature solves this problem with enzymes that are functional group specific. Their generation requires a great synthetic effort: with the help of a synthesis machinery where amino acids are linked, thus generating an enzyme with a quaternary structure, that positions the carbohydrate in the active site, to enable only one specific reaction. This way, the enzyme formally also provides the functional group protection. Many carbohydrate-processing enzymes are known by now and some are even commercially available. However, so far, chemical synthesis via regioselective protection of carbohydrates dominates oligosaccharide and glycoconjugate synthesis as well as carbohydrate modification.

For regioselective carbohydrate protection, the distinctness of each of the functional groups has to be evaluated. Hence, the differentiation in (i) primary, secondary, and anomeric hydroxy groups; (ii) position of hydroxy groups; (iii) equatorial or axial orientation, particularly of hydroxy groups in pyranosides; and (iv) the presence of 1,2-*cis*-, 1,3-*cis*-diols, or diequatorial 1,2-*trans*-diols (for interactions with bifunctional protecting groups or chelating metal ions) provides already valuable information on reactivity differences. Additionally, the reactivity differences are strongly influenced by steric, electronic, and stereoelectronic effects (thermodynamic or kinetic anomeric effect, exoanomeric effect), thus permitting the planning of regioselective carbohydrate protection. To this state of the art, several generations of carbohydrate chemists have contributed and activities to improve the available methodology are still ongoing.

A huge arsenal of protecting groups has been studied and proposed over the years, that is confined in this book to the most frequently used protecting groups and their utility for regioselective protection. This development also led to the creation of new, quite useful terms for the classification of protecting groups, such as (i) permanent vs temporary, (ii) participating vs nonparticipating, (iii) arming vs disarming, (iv) cyclic vs acyclic, (v) orthogonal vs nonorthogonal, (vi) compatible vs noncompatible, and some more.

Different from most previous collections of protecting groups considering chemical properties, mode of attachment and cleavage, and eventually compatibility/orthogonality with other groups, in the present book, the individual character of the most important carbohydrates concerning regioselective protection is covered. This aspect is generally close to the problems and questions of practitioners of carbohydrate chemistry. As in most literature reports rather recipes for carbohydrate protection than real explanations of the experimental result are provided, the book authors were often forced to refrain from interpretations. Detailed mechanistic studies, elucidating the influence of ring-chain tautomerism, of steric, electronic, and stereoelectronic effects determining ring size and anomer ratio and this way the relative reactivities of functional groups via kinetic or thermodynamic product control, are often not reported in the literature.

The first chapters of this book discuss the methods for protecting group introduction, regioselectivity, cleavage method, and compatibility with and orthogonality to other protecting groups under special consideration of the most important carbohydrates (Chapters 2–5, 11). The frequent occurrence of amino sugars, particularly glucosamine, galactosamine, and neuraminic acid, requires specific methods for their protection that are compiled in Chapters 6 and 10. As regioselective protection is often a multistep, time-consuming endeavor, there were strong incentives to develop one-pot procedures. Indeed, such methods were successfully introduced for quite some protecting group arrays, as discussed in Chapter 7. The warnings in Chapter 8 on acyl migration are very important in the overall context of this book, as acyl groups play an important role in carbohydrate protection. Similar migratory aptitudes were also observed for *O*-silyl groups. The extensive discussion of pyranoside protection is extended to furanosides in Chapter 12. Only *O*-glycofuranoside protection is contained, as *N*-glycofuranoside protection is generally discussed in the context of nucleoside and nucleotide chemistry. The structural variation and synthesis of some naturally occurring oligosaccharide types, as for instance, cyclodextrins (Chapter 13) and glycosaminoglycans (Chapter 14), deserve a particular discussion of protecting group strategies. An important means to reduce the number of intermediate purification steps, particularly in oligosaccharide synthesis, is the use of tags (for instance, fluorous tags, Chapter 15) or to perform solid-phase oligosaccharide synthesis, eventually as "automated glycan assembly" (Chapter 16). As the linkage to the tag or to the polymer requires generally orthogonality to the other protecting groups, even a higher demand on the protecting group array for the building blocks is required. Thus, the overall efficiency is affected, as often excess of the building blocks is necessitated. Hence, it is impressive to learn how the existing methodologies for oligosaccharide synthesis can be optimized for the technical synthesis of heparin analogs (Chapter 17). To this end, standard oligosaccharide block synthesis was employed searching for an optimum of reaction yields under complete stereocontrol and a minimum of intermediate isolations and chromatographic purification steps.

Everyone working in carbohydrate chemistry will come across regioselective carbohydrate protection and protecting group cleavage. Hence, this book will be a great help in the synthesis planning. Therefore, this book should be within reach to everyone working in the field.

Richard R. Schmidt

Emeritus Professor
University of Konstanz, Germany
30 March 2018

1

Protecting Group Strategies in Carbohydrate Chemistry

Anne G. Volbeda, Gijs A. van der Marel, and Jeroen D. C. Codée

Leiden Institute of Chemistry, Leiden University, Einsteinweg 55, 2333 CC, Leiden, The Netherlands

Carbohydrates are the most densely functionalized class of biopolymers in nature. Every monosaccharide features multiple contiguous stereocenters and bears multiple hydroxyl functionalities. These can, in turn, be decorated with sulfate groups, acyl esters, lactic acid esters and ethers, or phosphate moieties. Amine and carboxylate functions can also be present. Most often, the amine groups are acetylated, but different amide functions are also found, as well as N-sulfates and alkylated amines. The discrimination of the functional groups on a carbohydrate ring has been and continues to be one of the great challenges in synthetic carbohydrate chemistry [1–3].

This chapter describes the differences in the reactivity of the various functional groups on a carbohydrate ring and how to exploit these in the design of effective protecting group strategies. The protecting groups on a carbohydrate dictate the reactivity of the (mono)saccharide, and this chapter will describe how protecting group effects can be used to control stereoselective transformations (most importantly, glycosylation reactions) and reactivity-controlled one-pot synthesis strategies. Applications and strategies in automated synthesis are also highlighted.

1.1 Discriminating Different Functionalities on a Carbohydrate Ring

The main challenge in the functionalization of a carbohydrate (mono)saccharide is the discrimination of the different hydroxyl functionalities. The – often subtle – differences in reactivity can be capitalized upon to formulate effective protecting group strategies (see Scheme 1.1A). The primary alcohol functionality is generally the most reactive of the hydroxyl groups because of steric reasons (see Chapter 2). It can be site selectively addressed using bulky protecting groups such as silyl or trityl ethers. The anomeric hydroxyl group discerns itself from the other secondary hydroxyl groups in that it is part of a hemiacetal functionality

Protecting Groups: Strategies and Applications in Carbohydrate Chemistry, First Edition.
Edited by Sébastien Vidal.

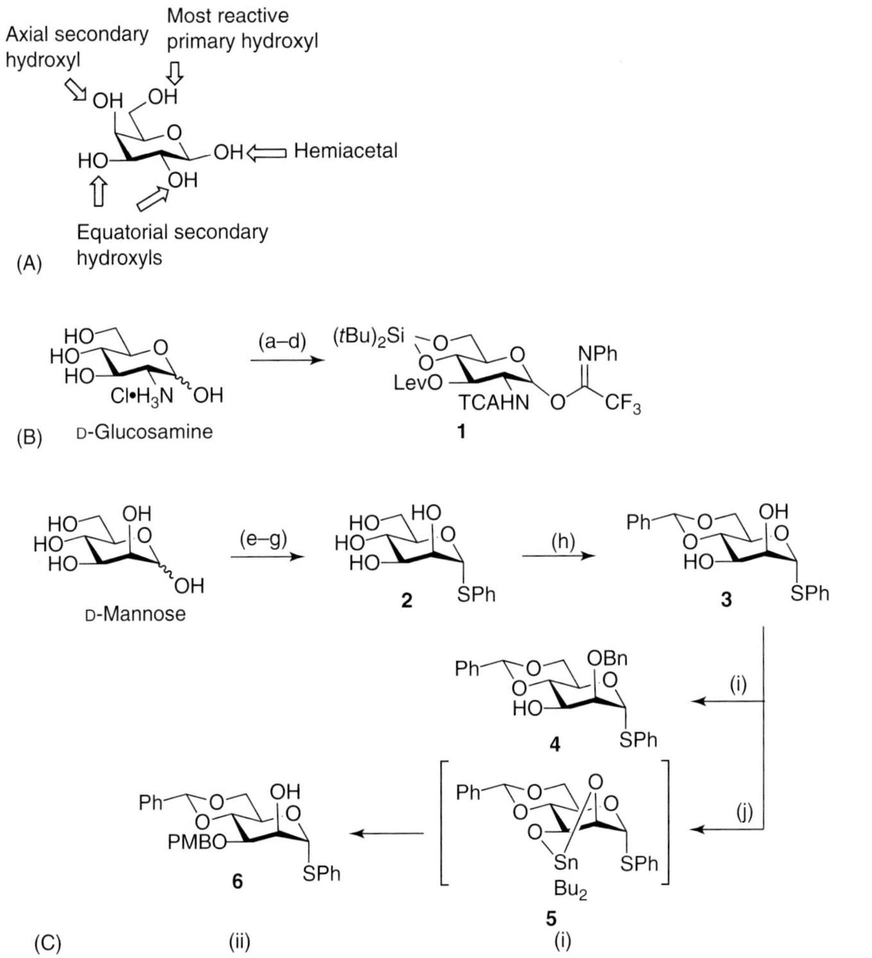

Scheme 1.1 (A) Relative reactivity of carbohydrate alcohols; (B) four-step reaction sequence to mask all functional groups in glucosamine; (a) Cl_3CCOCl, Et_3N, and MeOH; (b) $(tBu)_2Si(OTf)_2$, pyridine, and DMF, −40 °C (86% over 2 steps); (c) $CF_3C(=NPh)Cl$, Cs_2CO_3, and acetone (98%); (d) LevOH, DIC, DMAP, and DCM (82%). (C) Site-selective modification of mannosyl hydroxyl groups; (e) Ac_2O and pyridine; (f) PhSH, $BF_3{\cdot}OEt_2$, and DCM (75% over 2 steps); (g) NaOMe and MeOH (100%); (h) $HBF_4{\cdot}OEt_2$, $PhCH(OMe)_2$, and DMF (60%); (i) Bu_4NHSO_4, BnBr, NaOH, and DCM (75%); (j) (i) Bu_2SnO, toluene, and reflux; (ii) CsF, Bu_4NBr, PMBCl, toluene, and reflux (94%).

(see Chapter 5). It can, therefore, be selectively modified using acetal chemistry, and acid-catalyzed acetal and mixed thioacetal formations are among the most used methods to start a protecting group manipulation sequence. Because it is part of a hemiacetal functionality, the anomeric hydroxyl group is also the most acidic alcohol on a carbohydrate ring, and it can be chemoselectively modified under basic conditions. Conversely, it is less reactive than the other secondary alcohol groups under acidic conditions. Axial secondary alcohols are generally slightly less reactive than the equatorial ones on a carbohydrate ring, and these reactivity differences can often be exploited in designing an efficient protecting group scheme (see Chapters 3 and 4). Finally, the position of a hydroxyl group on the carbohydrate ring and the nature of its neighboring substituents affect its reactivity. In this regard, the use of cyclic protecting groups that engage two hydroxyl groups in a cyclic context (see Chapter 11) has proven to be a very powerful tool [4]. Benzylidene acetals and silylidene ketals can be used to mask C-4 C-6 diols, where isopropylidene groups and orthoesters are commonly employed to protect *cis*-hydroxyl groups in a five-membered ring constellation. Butane 2,3-bisacetals and the recently introduced *o*-xylylene groups can be used to protect vicinal diequatorial diols [5]. To illustrate how the reactivity of various alcohol groups can be exploited, two examples are given in Scheme 1.1B,C. The first example shows a four-step reaction sequence that has been used to site selectively mask all groups of a glucosamine synthon **1**. Thus, the nitrogen functionality in D-glucosamine can be chemoselectively protected with a trichloroacetyl group, by virtue of its higher nucleophilicity with respect to the alcohols present. Next, the primary alcohol at C-6 and the hydroxyl group at C-4 can be masked with a di-*tert*-butyl silylidene ketal. The selectivity of this transformation originates from the bulky nature of the protecting group and the fact that a stable *trans*-decalin system can be formed. Next, the anomeric hydroxyl group can be selectively addressed using basic conditions to install an imidate group. Finally, the remaining alcohol can be masked with a levulinoyl ester [6]. In the second example, the different hydroxyls of D-mannose are discriminated using the following steps (Scheme 1.1C). First, all hydroxyl groups are acetylated, concomitantly locking the mannose monosaccharide in a pyranoside ring. Next, the anomeric thioacetal is installed under Lewis acidic conditions. After saponification of the four remaining acetyl groups (**2**), the alcohol groups are diversified through the installation of a benzylidene acetal [7][1] (**3**) and selective benzylation of the C2-OH using phase transfer conditions (**4**) [8]. The selectivity in the latter transformation can be explained by taking into account the relative mild basic conditions (as opposed to the use of NaH in DMF) and the slightly higher acidity of the C2-OH because of its closer proximity to the anomeric center. Alternatively, the C3-OH can selectively be protected by exploiting the slightly higher nucleophilicity of this alcohol. Selective acylation is possible, as well as regioselective alkylation. To further enhance the reactivity difference between neighboring axial and equatorial hydroxyl groups, the use of stannylidene ketals presents a very effective approach [9]. Thus, diol **3** can be transformed into a dibutylstannylidene ketal (**5**) using dibutin oxide, after which the tin ketal can react with an appropriate electrophile, such as *para*-methoxybenzyl chloride under the *aegis* of cesium fluoride and tetrabutyl ammonium bromide (**6**).

Although the use of tin ketals, in stoichiometric and catalytic amounts, represents a very powerful means to discriminate alcohol functionalities, it requires the use of toxic tin species. To circumvent this drawback, Taylor and coworkers have introduced borinic acid catalysis to regioselectively protect glycosyl polyols [10, 11]. α-*O*-Methyl-fucopyranoside **7** can be regioselectively alkylated or

---

1 During this reaction, the formation of the double benzylidene acetal in which also the C2 and C3 hydroxyls react to form a second benzylidene acetal on the ring can be a major side reaction.

acylated using a catalytic amount of diphenylborinic ethylamine ester **8** and benzyl bromide or benzoyl chloride (Scheme 1.2). The reaction proceeds via borinate intermediate **9** that reacts in a highly regioselective manner to protect the equatorial alcohol at C-3.

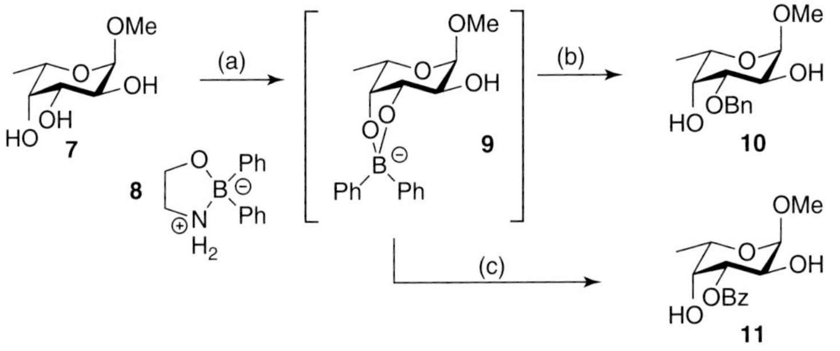

Scheme 1.2 Borinic acid catalysis to regioselectively protect alcohol functionalities: (a) **8**; (b) BnBr, Ag_2O, and MeCN, 40 °C, 48 h (94%); (c) BzCl, iPr_2NEt, and MeCN (92%).

To streamline the introduction of protecting groups, the groups Hung [12–15] and Beau [16–18] have devised a strategy to provide fully orthogonal protected building blocks in a one-pot manner (see Chapter 7). A key to the strategy is the transformation of all hydroxyl groups into trimethylsilyl (TMS) ethers, which renders the carbohydrate **12** well soluble in an organic solvent, such as dichloromethane, even at a low temperature. As shown in Scheme 1.3, the next steps in Hung's strategy involve the selective TMSOTf-mediated formation of a C4-C6

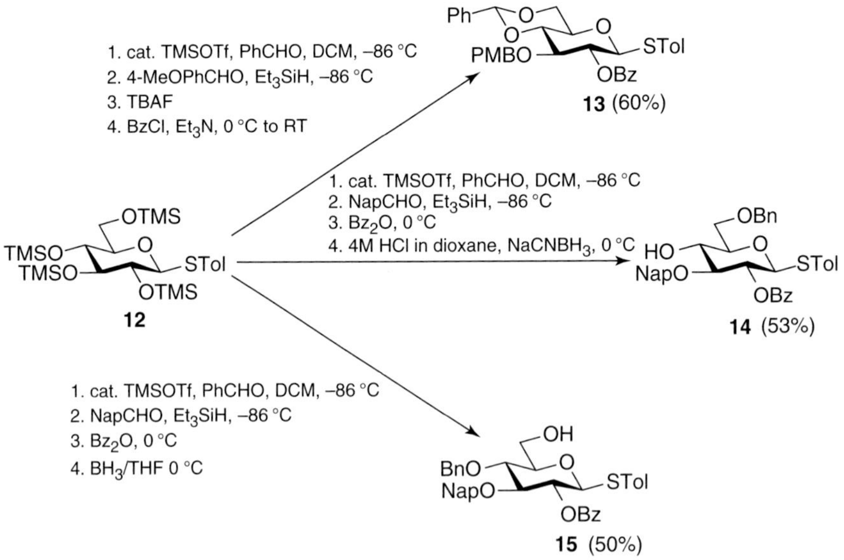

Scheme 1.3 One-pot protection of per-silylated thioglycoside to form different protected building blocks **13–15**.

acetal, ensuing the installation of a C2-C3 acetal and regioselective opening of the most reactive acetal (which is the acetal at C2-C3). This liberates the C2-O-TMS, which can be benzoylated to provide glucoside **13**. Regioselective, reductive opening of the C4-C6 acetal can then give access to either the C4 (**14**) or the C6 alcohol **15**. Using this strategy, the one-pot generation of a large variety of building blocks has been reported [12–15].

1.2 Strategies for an (Oligo)saccharide Synthesis Campaign

During an (oligo)saccharide synthesis campaign, different types of protecting groups can be discerned: those that will be removed during the assembly to allow for the manipulation of the unmasked alcohol, the temporary protecting groups; and those that are only to be removed at the very end of the assembly line, the permanent protecting groups. The latter groups should be stable to all reaction conditions used and be cleavable under mild conditions that do not jeopardize the integrity of the (oligo)saccharide target with all its functionalities. Benzyl ethers are by far the most used permanent protection used to date because they are stable to both acidic and basic conditions and can be removed using mild catalytic hydrogenation conditions. An impressive recent example of a synthesis, featuring benzyl groups for permanent protection, is presented in Scheme 1.4. Protected heparin eicosasaccharide **17** was built up from tetrasaccharide building block **16**. In the penultimate step 40, benzyl ethers and 10 azides were removed simultaneously to give the fully deprotected 20-mer **18** in 89% yield. In the final step, the 10 liberated amino groups were chemoselectively sulfated [19].

Also, dissolving metal reductions, such as the Birch reduction, has found much employment in global deprotection schemes. Permanent acyl protecting groups that are often employed (for example, to stereoselectively introduce glycosidic linkages, *vide infra*) are the pivaloyl and benzoyl esters. The former is more stable than the latter, representing an advantage during synthetic manipulations required during the assembly of the target compound. On the other hand, its stability necessitates harsh deprotection conditions that may affect other functionalities and linkages in the final product. Many types of protecting groups have been employed as temporary groups, including silyl ethers (substituted), acetyl esters, such as the levulinoyl and chloroacetyl esters, carbamates, carbonates, and allyl and substituted benzyl ethers.

The presence of double bonds precludes the use of catalytic hydrogenation for global deprotection of a target compound and therefore represents a synthetic challenge. Guo and coworkers have reported on the synthesis of a complex glycosyl phosphatidylinositol (GPI) anchor, bearing unsaturated lipids [20]. They selected PMB ethers to mask the hydroxyl functions throughout the synthesis. With monosaccharides **19**, **20**, and **21**, a trisaccharide was assembled, which was coupled to a disaccharide (constructed from **22**, **23**, and **24**) to form pentasaccharide **25**. Although PMB groups can be labile under Lewis acidic glycosylation conditions, no side reactions due to PMB cleavage occurred during the glycosylations. Deprotection commences with the reduction of the azide with

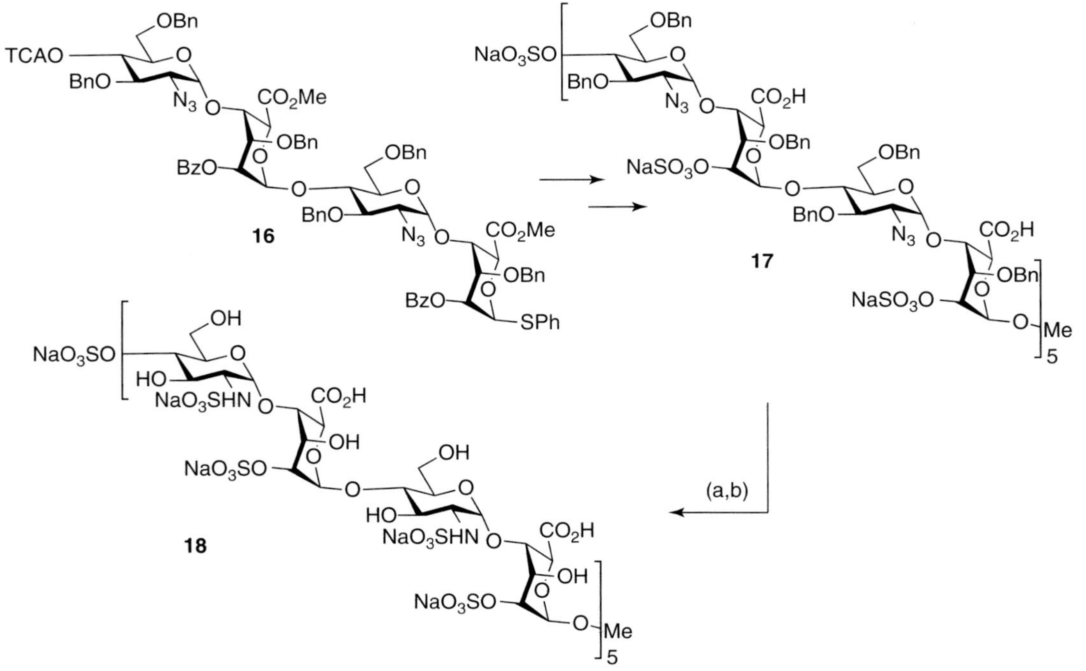

Scheme 1.4 Block coupling to heparin-like 20-mers. (a) $Pd(OH)_2/C$ and $EtOH/H_2O$ (89%); (b) SO_3·Pyridine and H_2O.

zinc in acetic acid, followed by base-catalyzed removal of the Fmoc and cyanoethyl groups. The last step is the removal of all PMB groups using trifluoroacetic acid. All PMB groups are removed without affecting the glycosidic linkages or the unsaturated lipid-bearing phosphatidylinositol.

Recently, Liu and coworkers described the use of TFA in toluene to remove substituted benzyl ethers for the global deprotection of oligosaccharides. They introduced PMB and 2-naphthylmethyl (Nap)-protected hydroxymethyl benzoates as acid-labile ester protecting groups for the same purpose [21]. Elongation of the reducing end terminus mannoside **27** with dibutyl phosphate donor **28** using stoichiometric amounts of TMSOTf provided dimer **29** (Scheme 1.5). Of note, under these Lewis acidic conditions, all protecting groups remained unaffected. Removal of the temporary tri-*iso*-propyl silyl ether (**30**) and ensuing coupling with another copy of **28** provided the target trisaccharide **31**. Global deprotection of this molecule by treatment with TFA in toluene gave the deprotected trisaccharide **32** in quantitative yield. Although it remains to be seen how general this methodology is, it can present a powerful alternative to the use of heterogeneous metal-catalyzed hydrogenolysis commonly used (Scheme 1.6).

1.3 Reactivity and Stereochemistry

Protecting groups have a major impact on the reactivity of a carbohydrate synthon. Electron-withdrawing protecting groups, such as acyl groups, deactivate a glycosyl donor because the electron-withdrawing effect of these groups destabilizes the buildup of (partial) positive charge at the anomeric center of the donor upon activation. This effect has been elegantly exploited and conceptualized by Fraser-Reid who introduced the armed–disarmed concept: benzyl ether carrying donors (so-called "armed" donors **33**) can be activated in the presence of acylated ones (termed "disarmed" donors **34**) allowing for the selective condensation of the armed donor with the disarmed building block (see Scheme 1.7A) [22]. Since the introduction of this seminal concept, an insight into glycosyl donor reactivity has tremendously increased, and it is now clear that, besides the nature of the protecting groups, the configuration and conformation of the donor glycoside, the orientation of the leaving group, and the exact position of the protecting groups all influence the reactivity of a donor building block [23]. The groups of Ley and Wong have developed reactivity scales, quantifying the relative reactivity of thioglycosides, setting the stage for effective one-pot assembly procedures involving multiple sequential glycosylation steps [24, 25].

The one-pot synthesis of tetrasaccharide **40** illustrates the use of relative reactivity values (RRVs) in oligosaccharide synthesis (Scheme 1.7B). The RRV values as determined by Wong and coworkers have been established with respect to the reactivity of tolyl 2,3,4,6-tetra-*O*-acetyl-1-thio-α-mannopyranoside (RRV = 1). The high RRV of thioglycoside **36** compared to thioglycoside **37** allows for the selective coupling of **36** to acceptor **37** in an NIS/TfOH-mediated glycosylation reaction. The obtained disaccharide donor is then treated with thioglycoside **38**, and an additional amount of NIS to form a trisaccharide. Tetrasaccharide **40** is obtained after addition of acceptor **39** and a third batch of NIS to the reaction mixture.

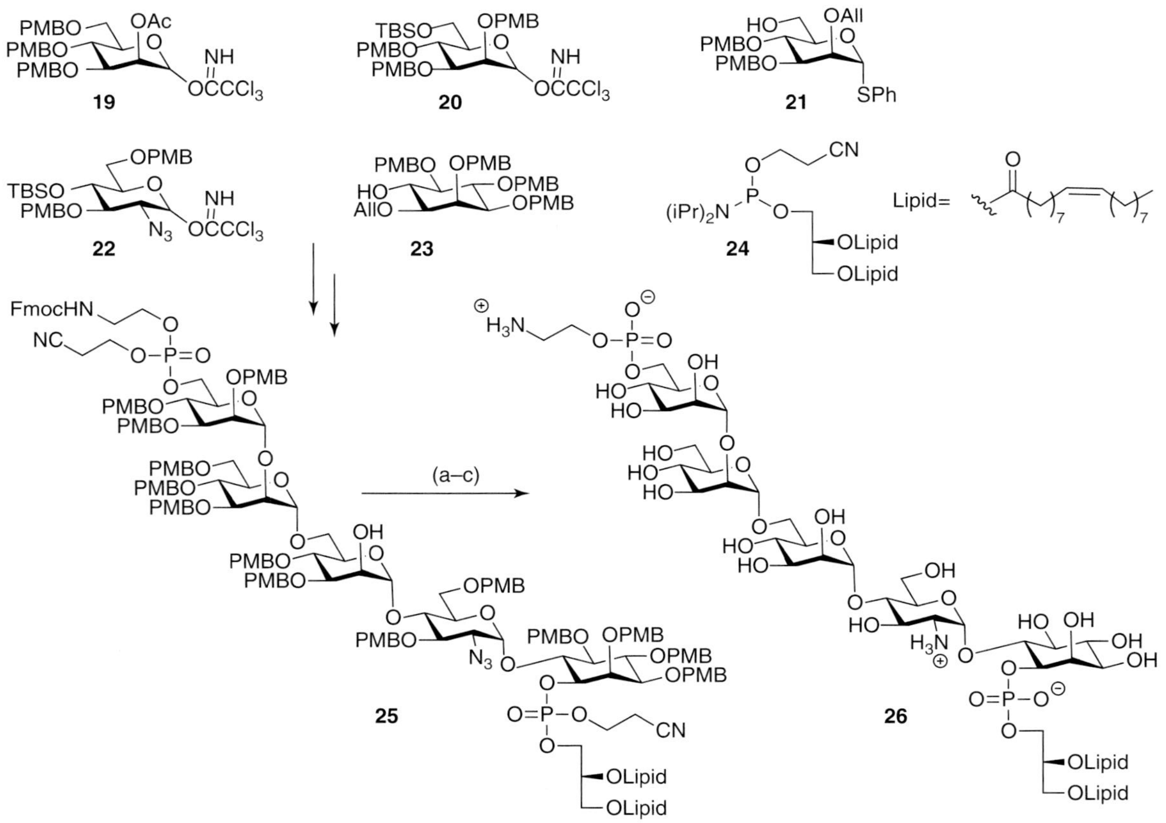

19 20 21 22 23 24 25 26

Scheme 1.5 GPI synthesis using a global deprotection strategy based on PMB protecting groups. (a) Zn, AcOH, and CH_2Cl_2, 2 h; (b) DBU and CH_2Cl_2, 1 h; (c) CH_2Cl_2-TFA (9 : 1), 1 h, 81% (3 steps).

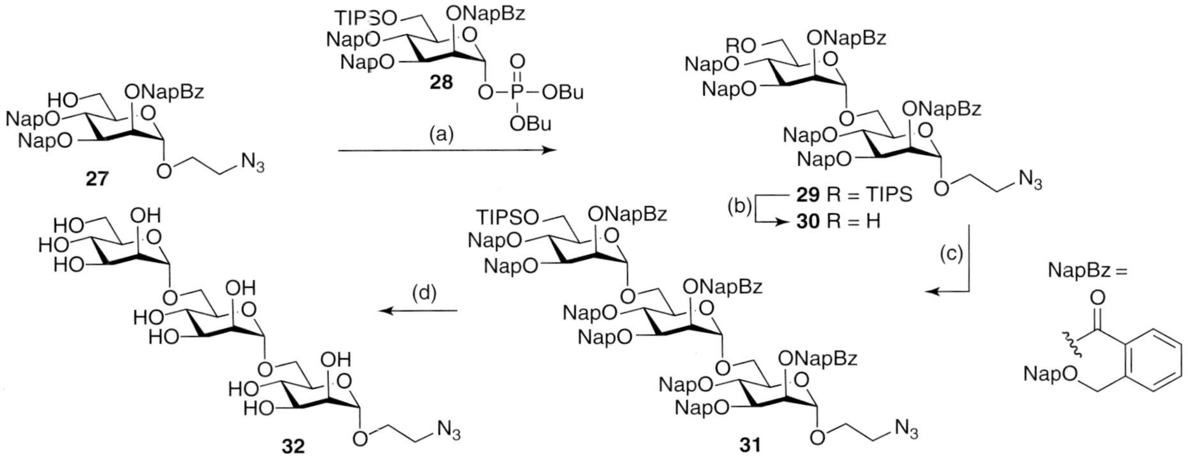

Scheme 1.6 Global deprotection using TFA in toluene: (a) TMSOTf and DCM, −20 °C (97%); (b) HF/pyridine and pyridine (91%); (c) **28**, TMSOTf, and DCM, −20 °C (94%); (d) TFA/toluene (10 : 1, v/v), 0 °C to RT (100%).

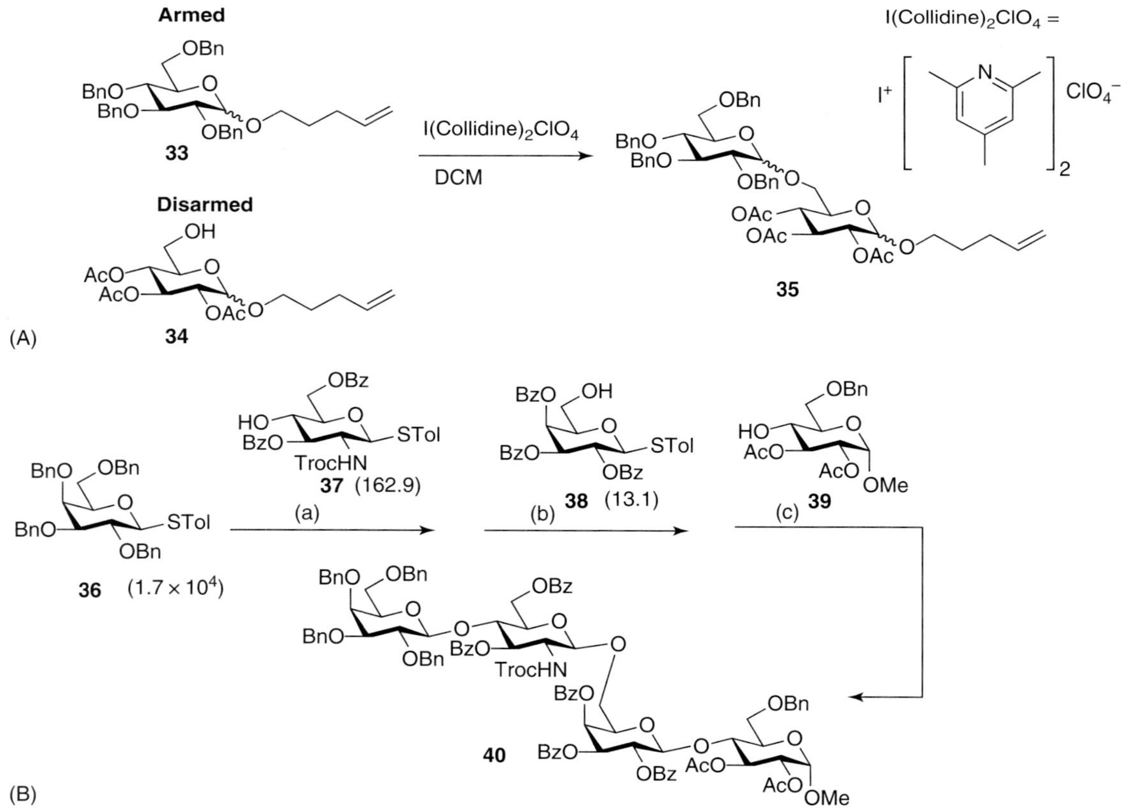

Scheme 1.7 (A) The armed–disarmed concept using *n*-pentenyl glycosides as conceptualized by Fraser-Reid. (B) Exploiting donor reactivity in a one-pot reaction sequence. (a) TfOH and NIS, −25 °C, DCM; (b) NIS, 0 °C, DCM; (c) NIS and DCM (40%).

The synthesis of this tetrasaccharide demonstrates the sophistication of the reactivity scales and their usefulness in the one-pot synthesis of oligosaccharides.

The impact of protecting groups on the stereochemical outcome of a glycosylation reaction is best illustrated by the anchimeric assistance that neighboring groups can provide during a glycosylation reaction. Glycosyl donors equipped with a C2-*O* or *N*-acyl group in general provide 1,2-*trans* products with great fidelity (exceptions occur because of stereochemical mismatch situations or overruling steric requirements) [26]. This can be explained by the formation of an intermediate dioxolenium ion that is formed by the attack of the C2-acyl group on the (developing) oxocarbenium ion. The dioxolenium ion bridge effectively shields one side of the carbohydrate ring, allowing the nucleophile only to approach from the opposite direction. Even though acyl groups are inherently more electron withdrawing than, for example, benzyl ethers, their presence can make a glycosyl donor more reactive because it can provide "active" anchimeric assistance. For example, disaccharide **41**, bearing three "disarming" benzoyl groups at C2, C3, and C4, could be selectively activated over building block **42**, carrying an arming benzyl group at C2, next to two disarming benzoates at C3 and C4, with the mild activator $Cu(OTf)_2$ [27]. Because of the limited reactivity of the activator, expulsion of the S-box aglycons only occurred when anchimeric assistance was provided by the neighboring C2-benzoate [28] (Scheme 1.8).

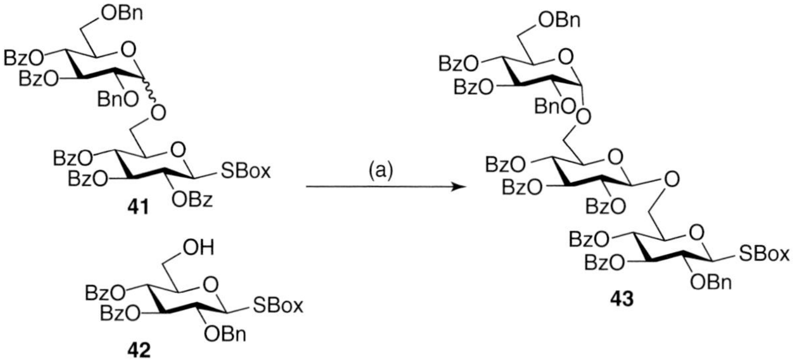

Scheme 1.8 Neighboring group participation-assisted selective activation: (a) $Cu(OTf)_2$, TfOH, and DCM (70%).

It has been proposed that acyl groups at positions other than C2 can also provide neighboring group participation, thereby influencing the stereochemical outcome of a glycosylation reaction [29, 30]. There are various examples describing the beneficial effect of C-6-acyl groups for the stereoselective synthesis of glucosyl, galactosyl, and mannosyl donors. Similarly, empirical evidence points to possible participation of ester groups at C4 of galactosyl and fucosyl donors. At the same time, studies with model compounds failed to convincingly demonstrate long-range participation leaving the subject open to further debate and showing that more sophisticated models and deeper insights into the effect of functional groups in glycosylation reactions are needed.

The stereoselective synthesis of 1,2-*cis*- and 2-deoxy glycosidic linkages is considerably more challenging than the construction of 1,2-*trans* bonds, but much progress has been made over the years in the stereoselective syntheses of these difficult linkages [31–34]. In all these syntheses, protecting groups play a key role in determining the overall shape and reactivity of the coupling partners. The overall reactivity of a glycosyl donor is decisive for the stereochemical outcome of a glycosylation reaction as it determines the stability of reactive intermediates that are formed upon activation. These include both covalent species [35, 36], such as anomeric triflates, and oxocarbenium ion intermediates, be it solvent separated or as part of a contact (or close) ion pair [37–39]. The equilibrium between these species, their stability, and the ease with which these are attached by an incoming nucleophile determine the overall stereochemical outcome of a glycosylation reaction. Because it is beyond the scope of this introductory chapter to provide an all-encompassing overview of these stereodirecting protecting group effects, only one – possibly the most prominent, but for sure the best studied one – example will be described here. Mannosyl donors, equipped with a benzylidene acetal spanning C4 and C6, can be used to effectively provide 1,2-*cis* mannosides. Crich and coworkers, who pioneered the method [40], have rationalized this stereochemical outcome through the intermediacy of the covalent α-triflate as the main product-forming intermediate (Scheme 1.9) [41]. The benzylidene acetal serves to limit the conformational freedom of the mannosyl ring, making it more difficult to adopt a flattened structure, which is required to accommodate the positive charge in an oxocarbenium ion intermediate. S_N2-type substitution on the anomeric triflate leads to the observed β-selectivity. This methodology has been applied in many different syntheses of complex

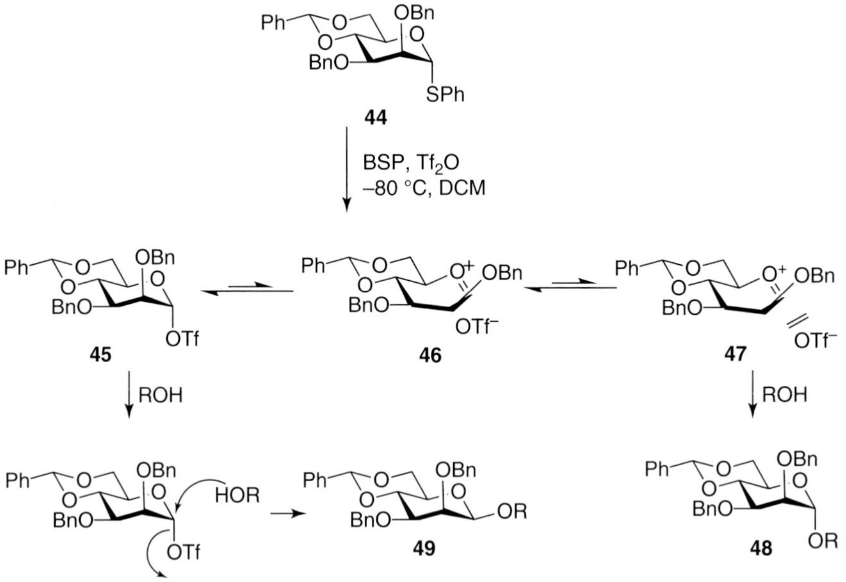

Scheme 1.9 Reaction mechanism manifold to account for the stereoselectivity in glycosylation reactions of benzylidene mannose donors.

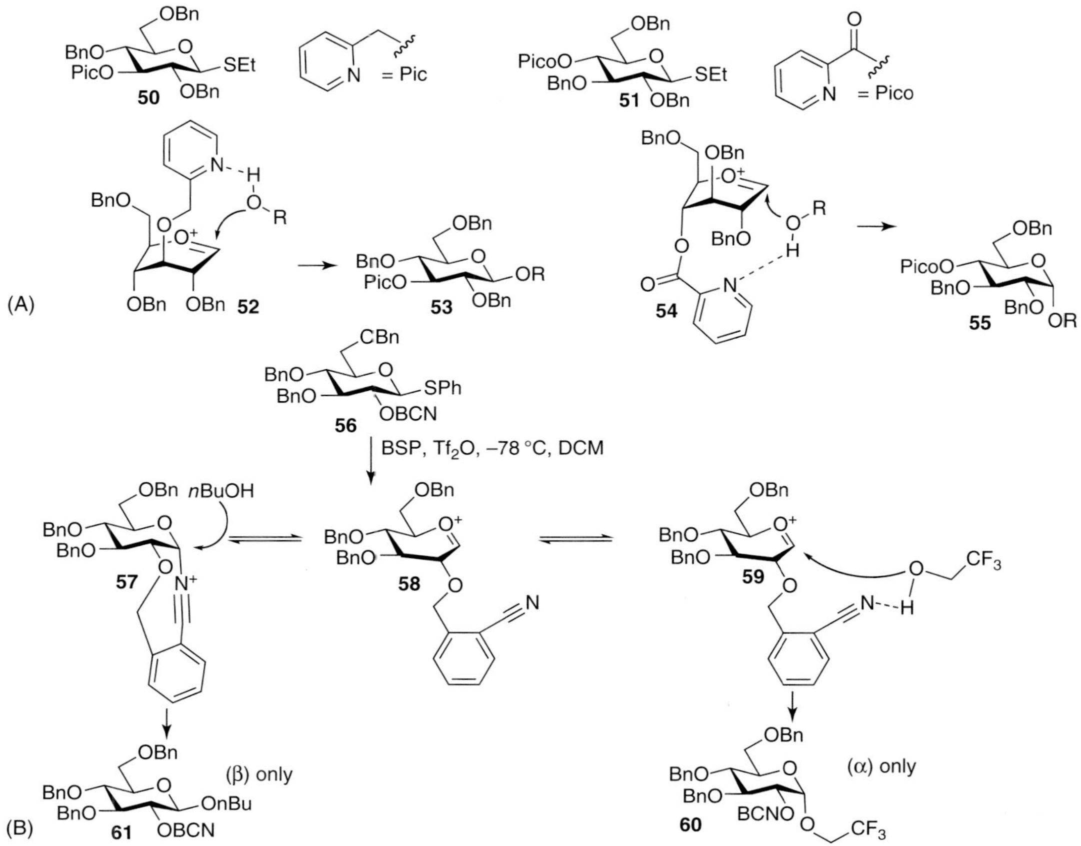

Scheme 1.10 (A) Hydrogen-bonding acceptor delivery by picolinyl and picolinoyl ether; (B) hydrogen-bonding acceptor delivery by cyanobenzyl ethers.

(bacterial) oligosaccharides and glycoconjugates, including the assembly of β-rhamnoside [32] and *cis*-linked heptose-containing oligomers [31]. To further investigate the origin of the striking selectivity, Crich and coworkers have conducted a number of seminal studies, including the determination of primary [42] and secondary [41] kinetic isotope effects and the development of "cation clock" methodology [43, 44] to discriminate between associative and dissociative product-forming pathways. Primary kinetic isotope effects indicated that the β-linked products are formed through an associative pathway, where the α-products in these reactions resulted from an attack of an oxocarbenium ion intermediate [45]. Secondary isotope effects measured in the glycosylation of between a benzylidene mannose donor and a methyl 2,3,6-tri-*O*-benzyl-α-D-glucopyranoside acceptor revealed that substantial oxocarbenium ion character developed in the transition state leading to the β-linked disaccharide, indicative of an S_N2-reaction with an exploded transition state. In contrast, *C*-glycosylation reactions of benzylidene mannose donors proceed through a dissociative pathway presumably via a $B_{2,5}$-oxocarbenium ion-like intermediate [46]. Overall, the benzylidene mannose system has not only developed to become the most direct and effective way to construct 1,2-*cis*-mannosidic linkages but it has also proven to be a rich breeding ground for the development of physical organic chemistry methods to investigate the principles underlying glycosylation stereochemistry.

Many different covalent-reactive species have been reported and characterized by spectroscopic techniques such as NMR [36]. However, in the majority of cases, the stereochemical outcome of glycosylation reactions involving these species cannot be simply traced back to the covalent-reactive intermediates. Clearly, other reactive intermediates have to be taken into account, and more insight is needed on how protecting and functional groups control the stability and reactivity of the different reactive intermediates.

Recently, several reports have appeared that make use of hydrogen bonding between donor and acceptor to direct glycosylation stereochemistry (Scheme 1.10). Demchenko and coworker have used picolinyl ethers (**50**) and picolinoyl esters (**51**) to direct the incoming nucleophile to the activated donor species with excellent facial selectivity [47, 48]. Hoang and Liu have described that glucosyl (**56**) and galactosyl donors bearing an *O*-cyanobenzyl ether at C-2 can provide either α- or β-linked products, depending on the reactivity of the acceptor and the solvent system used [49]. Reactive acceptors and the use of toluene lead to β-products, where unreactive alcohols and diethyl ether provide the opposite anomers. To account for the latter stereochemistry, the authors speculated that a hydrogen bond between the cyano group and the incoming acceptor could guide the nucleophile to the α-face of the donor molecule. How these new hydrogen-bonding protecting groups behave in the context of complex oligosaccharide synthesis will have to be shown in the near future.

1.4 Protecting Groups in Automated Synthesis

To streamline oligosaccharide assembly, much effort has been devoted to the development of automated synthesis techniques [50–52]. The automated solid-phase synthesis of peptides and nucleic acids is one of the major contributions of synthetic organic chemistry to the life sciences. However, solid-phase automated carbohydrate chemistry is significantly more challenging than the assembly of the other two biopolymers because one has to deal with all the different functionality present on the carbohydrate ring and the union of two carbohydrate building blocks involves the creation of a new stereocenter. Different strategies have been developed to automate oligosaccharide assembly based on either solution-phase synthesis or solid-phase techniques, and automated solid-phase synthesizers are now commercially available. Both techniques are based on the attachment of the growing oligosaccharide to a support. For the solution-phase approach, a light fluorous tag is used (see Chapter 15) [53], whereas the solid-phase methodology commonly employs a polystyrene-type resin (see Chapter 16) [54]. The support makes it possible to separate the target compound from the reagents used by filtration or a relatively simple fluorous solid-phase extraction step, thus allowing the use of excess reagents to drive reactions to completion. Other intermediate purification steps are not performed. Overall, this makes the process very efficient, but it also puts stringent constraints on the protecting groups used in the assembly. The use of excess reagent makes the reaction conditions employed harsher than the conditions that would be used in an equivalent solution-phase step. At the same time, cleavage of the temporary protecting groups has to proceed effectively because the buildup of deletion sequences leads to complex product mixtures necessitating a difficult, if not impossible, purification at the end of the assembly. Scheme 1.11 depicts the assembly of two oligomannuronic acid sequences through automated solid-phase [55] and automated fluorous-phase synthesis [56]. Both approaches rely on the use of mannuronic acid donor synthons because these enable the stereoselective formation of the 1,2-*cis* mannosidic linkage with great fidelity [57–59]. Obviously, the generation of epimeric mixtures is highly undesirable because it will generate very complex mixtures at the end of the assembly. Parallels between both approaches are the use of a double-bond-based linker system (cleavable by cross metathesis) and the use of imidate donors. Using the solid-phase approach, mannuronic acid tetramer **63**, octamer **64**, and dodecamer **65** were assembled (in 47%, 16%, and 11% over 8, 16, and 24 steps, respectively), whereas the latter approach was used to create hexasaccharide **70** (7% over 9 steps).

Two relevant protecting-group-related issues deserve mentioning here. Firstly, the methyl ester moieties can be used as precursors for the corresponding alcohol functionalities. It was shown that hexamannuronate **71** could be transformed into protected hexamannoside **73** through DIBAL reduction of the methyl esters in 82% yield. The second issue to note is that during the solid-phase assembly of the oligomers, deletion sequences were generated because of incomplete glycosylation steps (efficiency ~92% per step, no capping step was included). Saponification of the methyl esters allowed for the easy HPLC separation of the target stretches from their shorter counterparts. In designing automated oligosaccharide assemblies, it can be worthwhile to implement the possibility to purify semiprotected intermediates before the ultimate deprotection event because compounds featuring both hydrophilic and lipophilic groups allow for effective HPLC procedures, where fully protected compound can be too lipophilic and

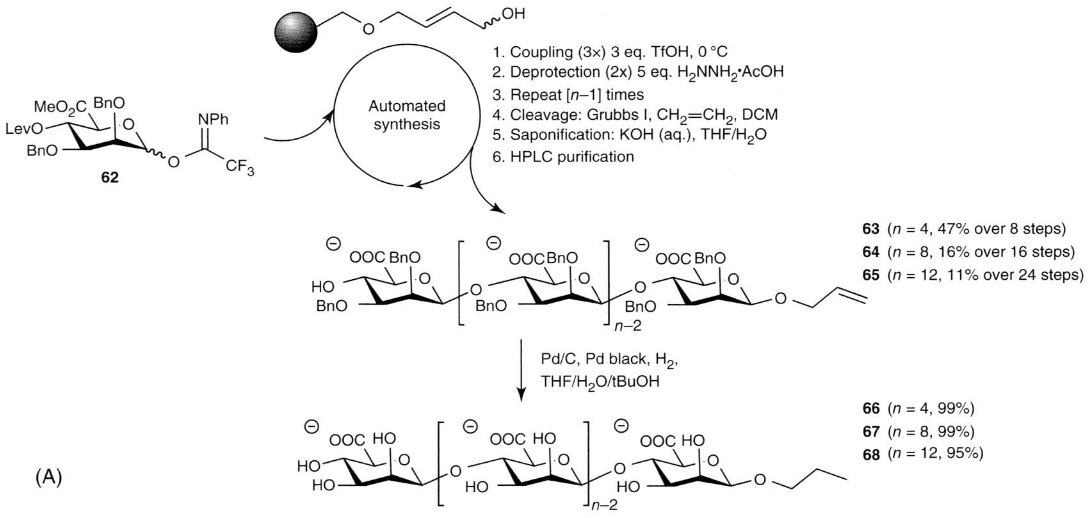

Scheme 1.11 Automated synthesis of oligomannuronic acids. (A) Solid-phase approach. (B) Fluorous-phase approach.

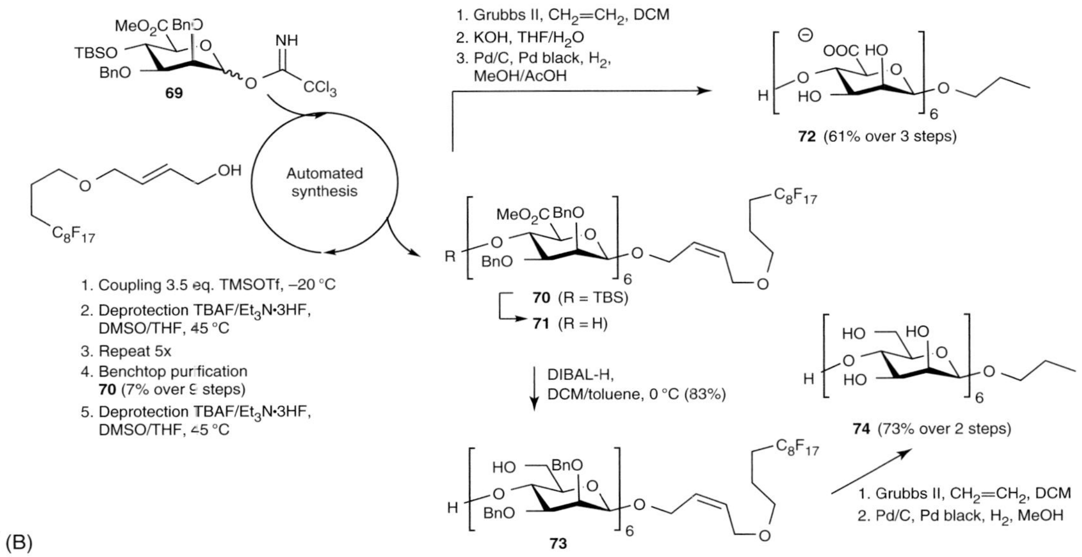

(B)

Scheme 1.11 (Continued).

fully deprotected compound too hydrophilic to efficiently purify. The latter strategy has also been applied in the automated solid-phase assembly of a set of hyaluronic acid (HA) oligomers [60]. HA-7-mer, 11-mer, and 15-mer were generated on a butanediol-functionalized polystyrene resin (Scheme 1.12) using

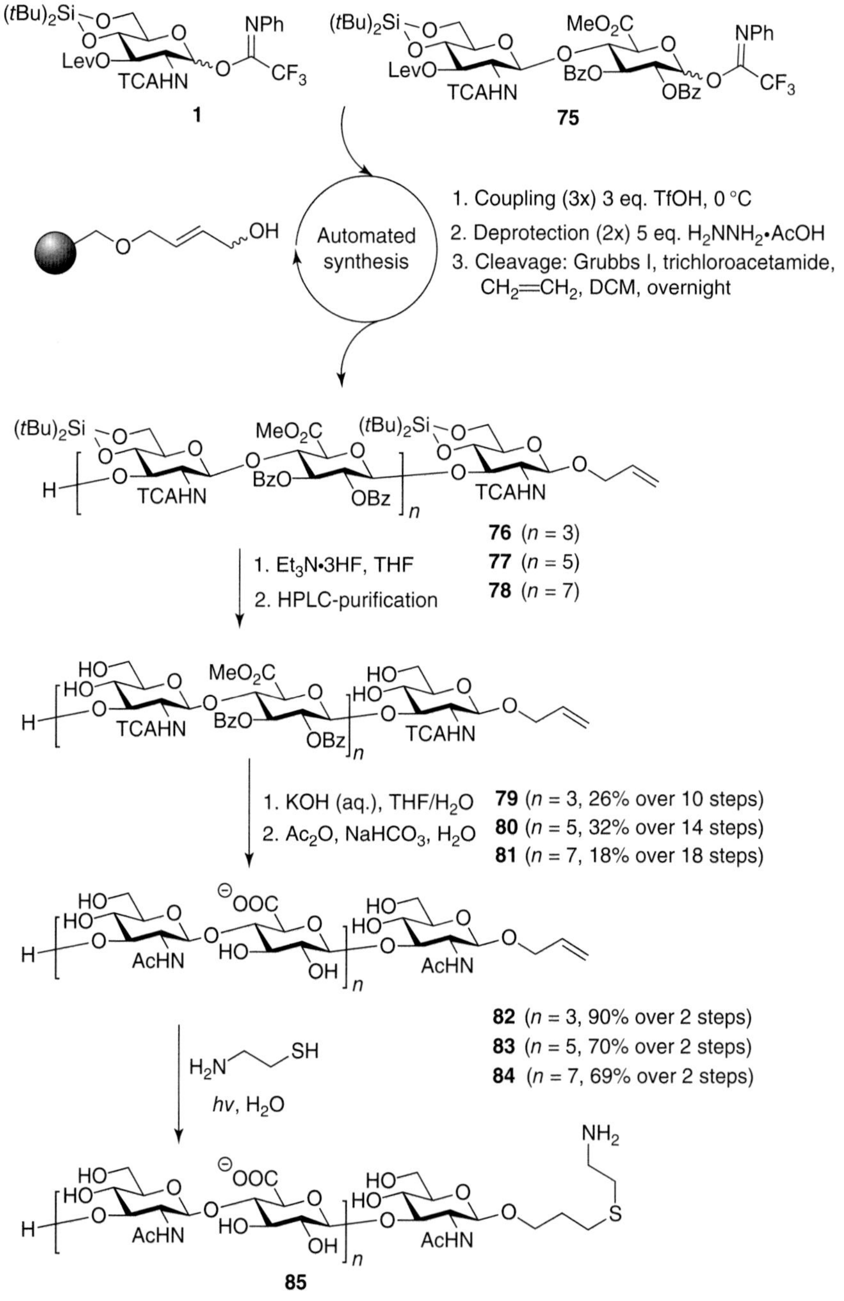

Scheme 1.12 Automated solid-phase assembly of hyaluronic acid oligosaccharides.

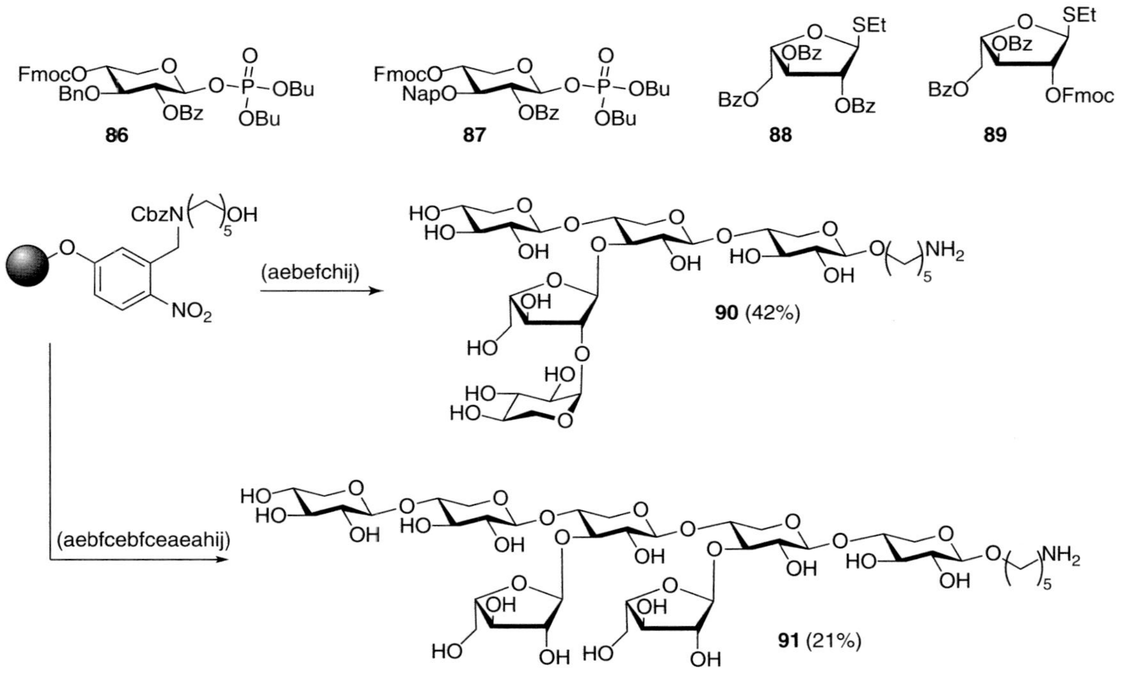

Scheme 1.13 Automated solid-phase assembly of plant cell wall arabinoxylan fragments; (a) donor **86**, TMSOTf, and DCM, −35 °C to −15 °C; (b) donor **87**, TMSOTf, and DCM, −35 °C to −15 °C; (c) donor **88**, NIS/TfOH, and DCM/dioxane, −40 °C to −20 °C; (d) donor **89**, NIS/TfOH, and DCM/dioxane, −40 °C to −20 °C; (e) 20% Et_3N in DMF, 25 °C; (f) 0.1 M DDQ in DCE/MeOH/H_2O (64 : 16 : 1); (g) Ac_2O and pyridine, 25 °C; (h) *hν* (305 nm); (i) NaOMe and THF/MeOH; (j) H_2, Pd/C, and EtOAc/MeOH/H_2O/AcOH.

monomeric building block **1** (Scheme 1.1) and disaccharide **75**. After cleavage of the resin by cross-metathesis, the fully protected oligomers **76–78** proved to be too lipophilic for purification, but removal of the silylidene ketals liberated two free alcohol groups per dimer repeat providing compounds **79–81** that were readily purified by HPLC. Of note, the silylidene group was employed in these syntheses because the corresponding benzylidene acetal proved to be too labile to withstand the acidic glycosylation conditions [6]. Global deprotection of the HA fragments was achieved by the saponification of all methyl and benzoyl esters and the trichloroacetyl amides. Selective *N*-acetylation gave the final compounds **82–84**. Because the protecting group strategy did not require the use of hydrogenation conditions, the reducing end anomeric allyl functionality could be retained. This in turn allowed the installment of a ligation handle through thiol–ene chemistry to give compound **85**.

Scheme 1.13 depicts the assembly of two plant arabinoxylans [61]. These syntheses nicely illustrate the use of the 9-fluorenylmethoxycarbonyl (Fmoc)-Nap couple as a set of orthogonal temporary protecting groups and the use of a UV-cleavable linker system. The former protecting group was used as a base-labile protecting group to mask the hydroxyl groups used for the elongation of the xylose backbone. Of note, cleavage of the Fmoc group generates a fulvene, the concentration of which can be measured spectroscopically providing an effective method to monitor the efficiency of the coupling events online. The Nap-ether was used at positions on the xylose building blocks where arabinofuranosyl branches were to be introduced. Cleavage of the Nap ethers was affected under oxidative conditions (DDQ) using a DCE/MeOH/H_2O solvent system. Although it is notable that aqueous solvent systems can be employed in combination with the polystyrene resin, the fact that the cleavage of the Nap ethers required seven repetitive reaction cycles illustrates the room for possible improvement. Cleavage of the arabinoxylan fragments from the solid support was affected by exposing the oligosaccharide-bearing resin to 305 nm UV light in a tailor-made continuous flow reactor [62].

1.5 Summary and Outlook

Protecting group chemistry can make or break any (oligo)saccharide synthesis effort. Much progress has been made over the years to understand and exploit reactivity differences between the functional groups on a carbohydrate and many efficient protecting group strategies, and schemes are now available. Even though these schemes may present multistep synthesis routes, they often involve optimized chemistry, assuring reliable synthetic outcomes. Nonetheless, there is a demand for ever-shorter synthetic routes, and the development of one-pot operations to introduce multiple protecting groups is therefore of high importance.

The demand for more efficiency can also be met by the development of better and more effective protecting groups. That is, protecting groups that are more robust during a synthesis campaign (for example, in a solid-phase setting) and/or can be removed more easily at the end of a synthesis. In this context, we have recently introduced two new pivaloyl-type groups that combine the advantages of the parent pivaloyl ester, i.e. stability and suppression of orthoester formation during glycosylation reactions, with ease of cleavage [63]. These two pivaloyl-based groups bear a reactive functionality appended to the pivaloyl core. The 2,2-dimethyl-4-(4-methoxy-phenoxy)-butanoate ester (MPDMB) and the 2,2-dimethyl-4-azido butanoate (AzDMB) are pivaloyl analogues that can be removed under either mild oxidative or reductive conditions, respectively (Scheme 1.14). An added advantage of the latter protecting group is found in the fact that it can be removed simultaneously with the commonly used permanent benzyl protecting groups using catalytic hydrogenation conditions.

There is continuous progress in the development of milder and more effective reaction conditions to affect protecting group manipulations. As shown in Scheme 1.13, the removal of the Nap ethers from the resin-bound arabinoxylans required seven repetitive cleavage cycles. To find more effective cleavage conditions for substituted benzyl ethers (PMB and Nap), we have recently used catalytic amounts of HCl in DCM/hexafluoro-*iso*-propanol (HFIP) [64]. These conditions were found to effectively cleave both PMB and Nap ethers while leaving other acid-labile functionalities (primary TBDPS ethers, glycosidic linkages) intact. In addition, the homogeneous conditions are amendable to a solid-phase setting [65, 66] and can therefore provide a more effective use of Nap ethers in solid-phase oligosaccharide synthesis.

Novel protecting groups and/or cleavage conditions are also required to mask amines on carbohydrate rings, especially functionalities that do not provide anchimeric assistance in glycosylation reactions. The only group that is now available for this purpose is the azide, and in cases where different orthogonally functionalized amine groups are required, the availability of more nonparticipating amine functionalities would be a valuable asset [67–69].[2]

Finally, it deserves mentioning that the last step(s) in the assembly of an oligosaccharide may be less trivial than they seem. Most oligosaccharide synthesis campaigns are based on a global deprotection event using a palladium-catalyzed hydrogenation as the key step to simultaneously remove a multitude of functional groups (benzyl ethers, benzyloxycarbonyl groups, benzylidene acetals, and azides).[3] Because many lipophilic groups are removed from the target compound to expose hydrophilic alcohols or amines, the polarity of the substrates increases tremendously leading to poorly soluble semiprotected intermediates, complicating the full deprotection of the target compounds. The presence of functional groups such as amines and thiols that can deactivate the palladium catalyst renders the final deprotection step(s) even more complicated. As an alternative to a catalytic hydrogenation, a dissolving metal (Birch) reduction can be employed. For these reductions, it also holds that the changing polarity of the substrate during the reaction can be a complication. Although impressive global

---

[2] Cyclic carbamates spanning the C2-N and C3-O have been used to create 1,2-*cis* glucosaminyl and galactosaminyl linkages. The stereochemistry in these glycosylation arises from a pathway in which initially formed β-linked products isomerize to the more stable α-products via an endocyclic ring-opening.

[3] Often the palladium is not used in a catalytic amount because the target compound is much more valuable than the precious metal catalyst.

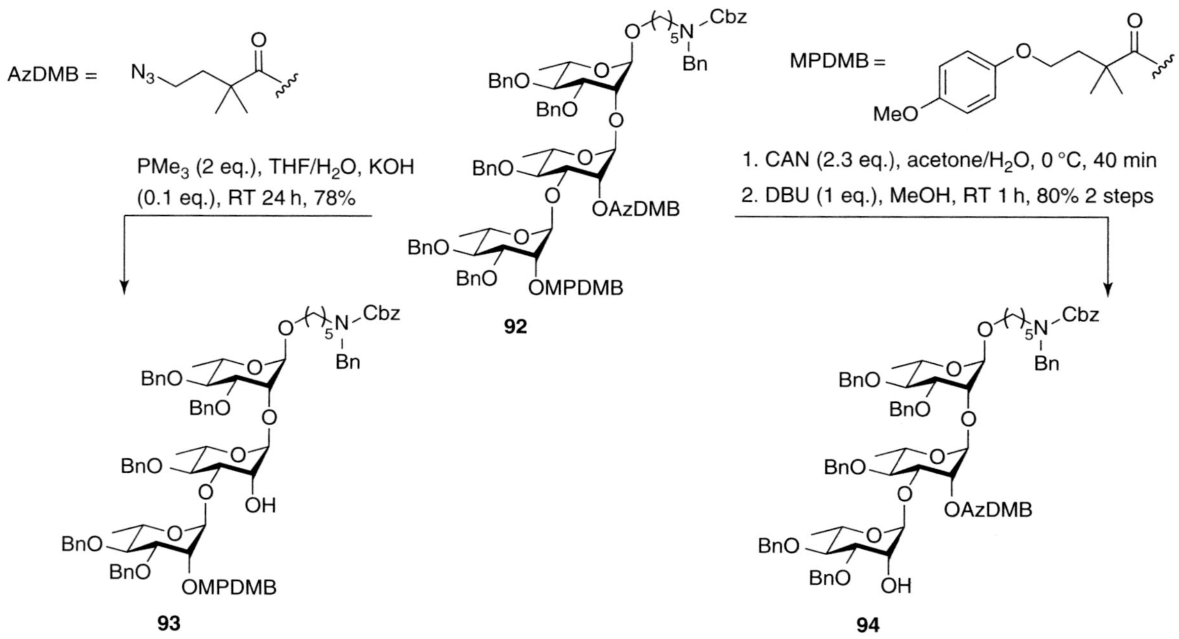

Scheme 1.14 Selective deprotection of AzDMB and MPDMB pivaloyl analogues.

deprotection events have been described using a Birch reduction, unexpected side reaction may occur. For example, in the final deprotection of *Micrococcus luteus* teichuronic acid stretches, composed of alternating *N*-acetyl mannosaminuronic acid and glucose residues, we encountered the unexpected cleavage of glycosidic linkages leading to fragmentation of the oligosaccharides (see Scheme 1.15) [70]. The cleavage occurred chemoselectively at the anomeric center of the mannosaminuronic acid residues, indicating that the cleavage was not the result of a β-elimination caused by the basic conditions of the Birch reduction.

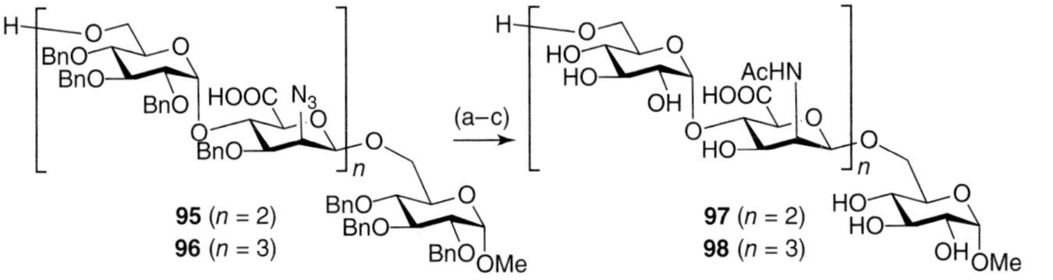

Scheme 1.15 Birch reduction of teichuronic acid oligosaccharides in which cleavage of the mannosaminuronic acid linkages was encountered; (a) Na (s), liquid NH_3, and THF, −60 °C; (b) HPLC purification; (c) Ac_2O, $NaHCO_3$, and THF/H_2O (**97**: 35% over 2 steps; **98**: 14% over 2 steps).

Unfortunately, often there is only a very limited amount of the final oligosaccharide available for deprotection and not much optimization can be done. Insight into why some global deprotection events proceed uneventfully, where others are accompanied by side reactions leading to complex reaction mixtures, and difficult purifications would be very valuable indeed. Innovative chromatography procedures to purify the highly polar target compounds, often lacking (UV)-chromophores for detection, would also represent a great addition to the oligosaccharide synthesis toolbox.

Abbreviations

Ac	acetyl
All	allyl
AzDMB	2,2,-dimethyl-4-azido-butanoate
BCN	2-cyanobenzyl
Bn	benzyl
Box	benzoxazolyl
BSP	1-benzenesulfinyl piperidine
Bu	butyl
Bz	benzoyl
Cbz	benzyloxycarbonyl
DBU	1,8-diazabicycloundec-7-ene
DCE	1,2-dichloroethane
DCM	dichloromethane

DDQ	2,3-dichloro-5,6-dicyano-1,4-benzoquinone
DIBAL	diisobutylaluminium hydride
DIC	*N,N*-diisopropylcarbodiimide
DMAP	4-(dimethylamino)pyridine
DMF	*N,N*-dimethylformamide
DMSO	dimethylsulfoxide
Fmoc	(9*H*-fluoren-9-yl)methoxycarbonyl
GPI	glycosyl phosphatidylinositol
HA	hyaluronic acid
HFIP	hexafluoro-*iso*-propanol
HPLC	high-performance liquid chromatography
hv	light
*i*Pr	*iso*-propyl
Lev	levulinoyl
Me	methyl
MPDMB	2,2,-dimethyl-4-(4-methoxy-phenoxy)-butanoate
Nap	2-naphthylmethyl
NIS	*N*-iodosuccinimide
NMR	nuclear magnetic resonance
Ph	phenyl
Pic	picolinyl
Pico	picolinoyl
PMB	*para*-methoxybenzyl
RRV	relative reactivity value
TBAF	tetrabutylammonium fluoride
TBS	*tert*-butyldimethylsilyl
TCA	trichloroacetyl
Tf	triflate
TFA	trifluoroacetic acid
THF	tetrahydrofuran
TMS	trimethylsilyl
Tol	tolyl
Troc	2,2,2-trichloroethoxycarbonyl
UV	ultraviolet

References

1 Codée, J.D.C., Ali, A., Overkleeft, H.S., and van der Marel, G.A. (2011). *C.R. Chim.* 14: 178–193.
2 Guo, J. and Ye, X.S. (2010). *Molecules* 15: 7235–7265.
3 Wittmann, V. (2006). *Angew. Chem. Int. Ed.* 45: 3399–3400.
4 Litjens, R.E.J.N., van den Bos, L.J., Codée, J.D.C. et al. (2007). *Carbohydr. Res.* 342: 419–429.
5 Balbuena, P., Gonçalves-Pereira, R., Jiménez Blanco, J.L. et al. (2013). *J. Org. Chem.* 78: 1390–1403.
6 Dinkelaar, J., Gold, H., Overkleeft, H.S. et al. (2009). *J. Org. Chem.* 74: 4208–4216.

7 Huang, M., Tran, H., Bohé, L., and Crich, D. (2014). *Proven Synthetic Methods, Carbohydrate Chemistry*, vol. 2, 175–182. CRC Press.
8 Garegg, P.J., Kvarnstrom, I., Niklasson, A. et al. (1993). *J. Carbohydr. Chem.* 12: 933–953.
9 Grindley, T.B. (1998). *Adv. Carbohydr. Chem. Biochem.* 53: 17–142.
10 Lee, D. and Taylor, M. (2012). *Synthesis* 44: 3421–3431.
11 McClary, C.A. and Taylor, M.S. (2013). *Carbohydr. Res.* 381: 112–122.
12 Hu, Y., Zhong, Y., Chen, Z.-G. et al. (2012). *J. Am. Chem. Soc.* 134: 20722–20727.
13 Huang, T.-Y., Zulueta, M.M.L., and Hung, S.-C. (2014). *Org. Biomol. Chem.* 12: 376–382.
14 Wang, C.-C., Lee, J.-C., Luo, S.-Y. et al. (2007). *Nature* 446: 896–899.
15 Huang, T.-Y., Zulueta, M.M.L., and Hung, S.-C. (2011). *Org. Lett.* 13: 1506–1509.
16 Français, A., Urban, D., and Beau, J.-M. (2007). *Angew. Chem. Int. Ed.* 46: 8662–8665.
17 Despras, G., Urban, D., Vauzeilles, B., and Beau, J.-M. (2014). *Chem. Commun.* 1067–1069.
18 Bourdreux, Y., Lemetais, A., Urban, D., and Beau, J.-M. (2011). *Chem. Commun.* 47: 2146–2148.
19 Hansen, S.U., Miller, G.J., Cliff, M.J. et al. (2015). *Chem. Sci.* 6: 6158–6164.
20 Swarts, B.M. and Guo, Z. (2010). *J. Am. Chem. Soc.* 132: 6648–6650.
21 Li, Y. and Liu, X. (2014). *Chem. Commun.* 50: 3155–3158.
22 Mootoo, D.R., Konradsson, P., Udodong, U., and Fraser-Reid, B. (1988). *J. Am. Chem. Soc.* 110: 5583–5584.
23 Fraser-Reid, B. and López, J.C. (2011). Armed–disarmed effects in carbohydrate chemistry: history, synthetic and mechanistic studies. In: *Reactivity Tuning in Oligosaccharide Assembly* (ed. B. Fraser-Reid and J. Cristóbal López), 1–29. Berlin Heidelberg: Springer.
24 Green, L., Hinzen, B., Ince, S.J. et al. (1998). *Synlett* 440–442.
25 Zhang, Z., Ollmann, I.R., Ye, X.S. et al. (1999). *J. Am. Chem. Soc.* 121: 734–753.
26 Spijker, N.M. and van Boeckel, C.A.A. (1991). *Angew. Chem. Int. Ed.* 30: 180–183.
27 Kamat, M.N. and Demchenko, A.V. (2005). *Org. Lett.* 7: 3215–3218.
28 Crich, D. and Li, M. (2007). *Org. Lett.* 9: 4115–4118.
29 Christina, A.E., van der Marel, G.A., and Codée, J.D.C. (2013). Recent developments in the construction of cis-glycosidic linkages. In: *Modern Synthetic Methods in Carbohydrate Chemistry*, 97–124. Weinheim: Wiley-VCH.
30 Komarova, B.S., Ustyuzhanina, N.E., Tsvetkov, Y.E., and Nifantiev, N.E. (2013). Stereocontrol of 1,2-cis-glycosylation by remote O-acyl protecting groups. In: *Modern Synthetic Methods in Carbohydrate Chemistry*, 125–159. Weinheim: Wiley-VCH.
31 Crich, D. and Li, M. (2008). *J. Org. Chem.* 73: 7003–7010.
32 Crich, D. and Li, L. (2009). *J. Org. Chem.* 74: 773–781.
33 Manabe, S. (2010). *Methods Enzymol.* 478: 413–435.
34 Nigudkar, S.S. and Demchenko, A.V. (2015). *Chem. Sci.* 6: 2687–2704.
35 Walvoort, M.T.C., van der Marel, G.A., Overkleeft, H.S., and Codée, J.D.C. (2013). *Chem. Sci.* 4: 897–906.
36 Frihed, T.G., Bols, M., and Pedersen, C.M. (2015). *Chem. Rev.* 115: 4963–5013.

37 Bohé, L. and Crich, D. (2011). *C.R. Chim.* 14: 3–16.
38 Bohé, L. and Crich, D. (2015). *Carbohydr. Res.* 403: 48–59.
39 Walvoort, M.T.C., Dinkelaar, J., van den Bos, L.J. et al. (2010). *Carbohydr. Res.* 345: 1252–1263.
40 Crich, D. and Sun, S. (1997). *J. Am. Chem. Soc.* 119: 11217–11223.
41 Crich, D. (2010). *Acc. Chem. Res.* 43: 1144–1153.
42 Huang, M., Garrett, G.E., Birlirakis, N. et al. (2012). *Nat. Chem.* 4: 663–667.
43 Adero, P.O., Furukawa, T., Huang, M. et al. (2015). *J. Am. Chem. Soc.* 137: 10336–10345.
44 Huang, M., Retailleau, P., Bohé, L., and Crich, D. (2012). *J. Am. Chem. Soc.* 134: 14746–14749.
45 See for the first observation by NMR of a glycosyl oxocarbenium ion in super acid media:Martin, A., Arda, A., Désiré, J. et al. (2015). *Nat. Chem.* 8: 186–191.
46 Moumé-Pymbock, M. and Crich, D. (2012). *J. Org. Chem.* 77: 8905–8912.
47 Yasomanee, J.P. and Demchenko, A.V. (2012). *J. Am. Chem. Soc.* 134: 20097–20102.
48 Yasomanee, J.P. and Demchenko, A.V. (2014). *Angew. Chem. Int. Ed.* 53: 10453–10456.
49 Le Mai Hoang, K. and Liu, X.-W. (2014). *Nat. Commun.* 5: 1–10.
50 Plante, O.J., Palmacci, E.R., and Seeberger, P.H. (2001). *Science* 291: 1523–1527.
51 Seeberger, P.H. (2015). *Acc. Chem. Res.* 48: 1450–1463.
52 Hsu, C.-H., Hung, S.-C., Wu, C.-Y., and Wong, C.-H. (2011). *Angew. Chem. Int. Ed.* 50: 11872–11923.
53 Roychoudhury, R. and Pohl, N.L.B. (2013). Light fluorous-tag-assisted synthesis of oligosaccharides. In: *Modern Synthetic Methods in Carbohydrate Chemistry*, 221–239. Weinheim: Wiley-VCH.
54 Seeberger, P.H. and Haase, W.C. (2000). *Chem. Rev.* 100: 4349–4393.
55 Walvoort, M.T.C., Van Den Elst, H., Plante, O.J. et al. (2012). *Angew. Chem. Int. Ed.* 1–5.
56 Tang, S.-L. and Pohl, N.L.B. (2015). *Org. Lett.* 17: 2642–2645.
57 van den Bos, L.J., Dinkelaar, J., Overkleeft, H.S., and van der Marel, G.A. (2006). *J. Am. Chem. Soc.* 128: 13066–13067.
58 Codée, J.D.C., van den Bos, L.J., de Jong, A.R. et al. (2009). *J. Org. Chem.* 74: 38–47.
59 Codée, J.D.C., Walvoort, M.T.C., de Jong, A.-R. et al. (2011). *J. Carbohydr. Chem.* 30: 438–457.
60 Walvoort, M.T.C., Volbeda, A.G., Reintjens, N.R.M. et al. (2012). *Org. Lett.* 14: 3776–3779.
61 Schmidt, D., Schuhmacher, F., Geissner, A. et al. (2015). *Chem. Eur. J.* 21: 1–6.
62 Eller, S., Collot, M., Yin, J. et al. (2013). *Angew. Chem. Int. Ed.* 52: 5858–5861.
63 Castelli, R., Overkleeft, H.S., van der Marel, G.A., and Codée, J.D.C. (2013). *Org. Lett.* 15: 2270–2273.
64 Volbeda, A.G., Kistemaker, H.A.V., Overkleeft, H.S. et al. (2015). *J. Org. Chem.* 80: 8796–8806.
65 Kistemaker, H.A.V., Lameijer, L.N., Meeuwenoord, N.J. et al. (2015). *Angew. Chem. Int. Ed.* 127: 4997–5000.
66 Palladino, P. and Stetsenko, D.A. (2012). *Org. Lett.* 14: 6346–6349.

67 Bindschädler, P., Noti, C., Castagnetti, E., and Seeberger, P.H. (2006). *Helv. Chim. Acta* 89: 2591–2610.
68 Hu, Y.-P., Lin, S.-Y., Huang, C.-Y. et al. (2011). *Nat. Chem.* 3: 557–563.
69 Lohman, G.J.S. and Seeberger, P.H. (2004). *J. Org. Chem.* 69: 4081–4093.
70 Walvoort, M.T.C., Lodder, G., Overkleeft, H.S. et al. (2010). *J. Org. Chem.* 75: 7990–8002.

2

Protecting Groups at the Primary Position of Carbohydrates

Marion Donnier-Maréchal, Sébastien Vidal, and Michele Fiore

CNRS and Université Claude Bernard Lyon 1, Institut de Chimie et Biochimie Moléculaires et Supramoléculaires (UMR 5246), Laboratoire de Chimie Organique 2-Glycochimie, 43 Boulevard du 11 November 1918, 69622 Villeurbanne, France

2.1 Introduction

Carbohydrates are multifunctional biomolecules with no less than five different hydroxyl groups displayed on cyclic pyranose or furanose scaffolds. Organic chemists have devoted much effort in selectively functionalizing each and every one of these alcohols through ingenious and efficient protecting group strategies. The primary position of (oligo)saccharides is particularly relevant because its differentiation from other anomeric or secondary alcohols will allow selective functionalization at this position. The synthetic strategies mainly rely on sterically hindered reagents that greatly prefer the more accessible primary position over secondary positions. In this chapter, protecting group strategies for only the α- or β-anomers of nonreducing carbohydrates are considered as nucleoside protecting groups would be required for furanosides (Figure 2.1).

Formation of ethers at the primary position requires bulky benzyl ethers such as the triphenylmethyl (also known as trityl and abbreviated as Tr). Similar hindered silyl ethers have also been applied successfully for the regioselective protection of the primary position. Access to ester functionalities has also been demonstrated using either chemical or chemoenzymatic methods, the latter being far more documented. Halogenation of the primary position can also be considered as a protecting group strategy because it masks the alcohol functionality; however, the corresponding halides (apart from the fluorides) can then be further converted into organic functional groups mainly through nucleophilic displacements (e.g. azides, amines, and sulfides). Finally, regioselective oxidation to the aldehyde or carboxylic acid functionalities can be performed chemically or chemoenzymatically.

Examples compiled in this chapter were selected based on access to a detailed experimental procedure useful for the reader. When a plethora of reports were identified for a protecting group, the selection was more difficult and focused on

Protecting Groups: Strategies and Applications in Carbohydrate Chemistry, First Edition.
Edited by Sébastien Vidal.

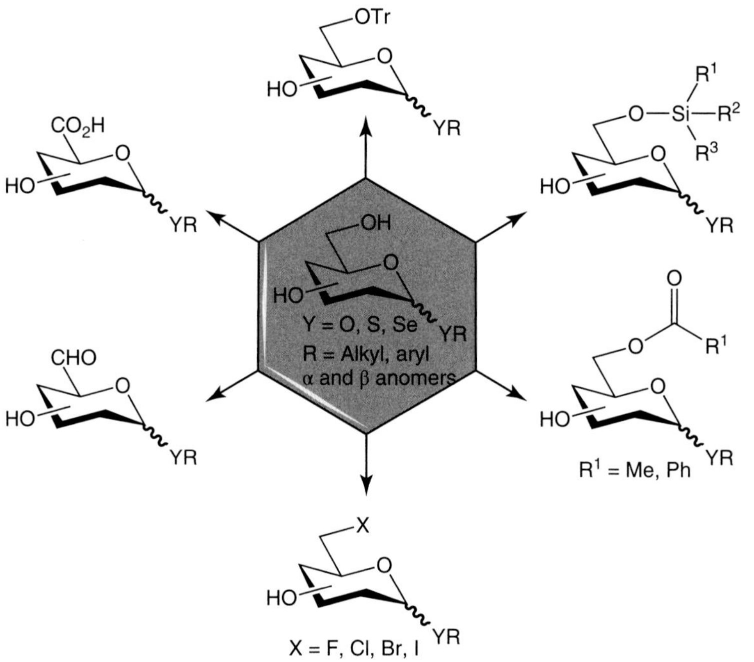

Figure 2.1 General outlook of the regioselective and/or chemoselective protections and chemical transformations possible at the primary position of saccharides.

the early and pioneering reports as well as modern synthetic adaptations providing improved results.

2.2 Selective Primary Hydroxyl Group Protection

2.2.1 Trityl Ethers

Triphenylmethyl chloride also known as trityl chloride (TrCl) is the most common reagent for the protection of primary alcohols of monosaccharides and disaccharides. The regioselectivity arises from the bulkiness of this tertiary alkyl halide with very few examples of ditritylated compounds [1]. The unprotected carbohydrate is usually dissolved in pyridine with an equimolar amount of trityl chloride. The procedure was applied successfully to nonreducing [1–4] (Table 2.1, entries 1–5), reducing [9] (Table 2.1 entry 7), modified monosaccharides [16] (Table 2.1, entry 6), disaccharides [12] (Table 2.1, entry 12), and glycals [7, 13, 14] (Table 2.1, entries 13 and 14). In some cases, the crude material was directly acetylated to improve the purification [3] followed by deacetylation [5] before applying further modifications (Table 2.1, entries 2 and 8). A catalytic amount of 4-dimethylaminopyridine (DMAP) [6, 11] (Table 2.1, entries 4 and 10), or 1,4-diazabicyclo[2.2.2]octane (DABCO) associated with DMAP [4] (Table 2.1, entries 9 and 10), is the other common base used to neutralize the hydrochloric acid generated during the etherification reaction. Nonconventional synthetic procedures were recently applied to

Table 2.1 Regioselective 6-*O*-tritylation.

Entry	Protection	Protected carbohydrate	Observations	References
1	TrCl, pyridine, 50–100 °C, 4–8 h, 50%	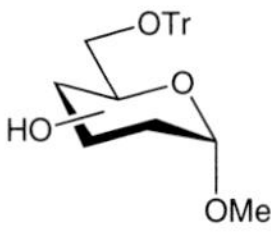 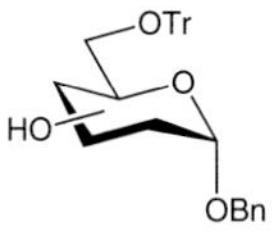	Glc, Man, Gal ditritylated by-products	[1]
2	1) TrCl, pyridine, RT then 50 °C, 5 h 2) Acetylation 3) Deacetylation 78% (3 steps)	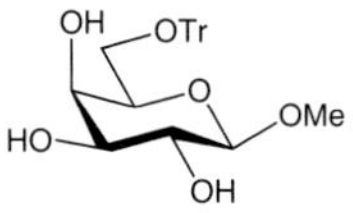	Purification of acetylated intermediate	[5]
3	TrCl, pyridine, 25 °C, 16 h, 95%	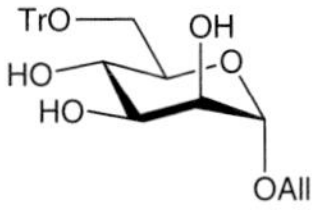	D-Mannan synthesis	[2]
4	TrCl, DMAP, pyridine, 80 °C, 16 h, 65–83%	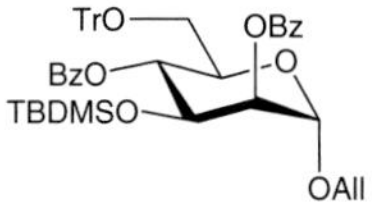	Glc, Man, Gal one-pot synthesis	[6]
5	TrCl, pyridine, 50 °C, 16 h 52%	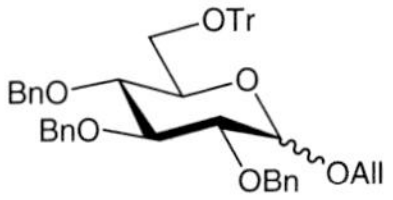	Benzylated intermediate isolated	[7]
6	TrCl, pyridine, 25 °C, 12 h, 87%	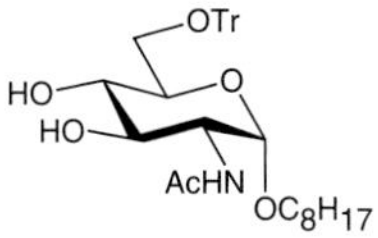	Glc, GlcNAc, Gal	[8]
7	TrCl, pyridine, 75 °C, 18 h, 94%	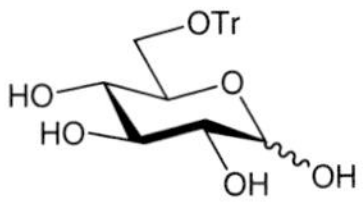	Glc	[9]
8	1) TrCl, pyridine, 75 °C, 18 h 2) Acetylation, 22% (2 steps)	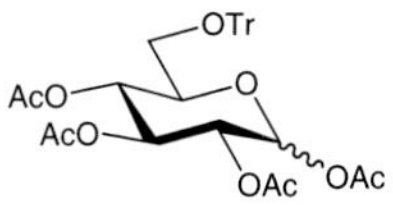	Poor isolated yield Anomeric mixture	[3]
9	TrCl, DABCO, Na_2CO_3, ball milling 600 rpm, 3–24 h, 87%	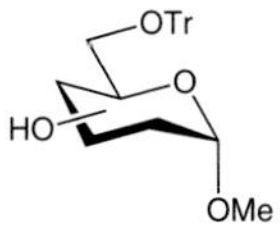	Glc, Gal, GlcNAc Ball milling *S*-glycosides	[4]
10	TrCl, DABCO, DMAP, CH_2Cl_2, 75 °C, 2–3.5 h, 98%	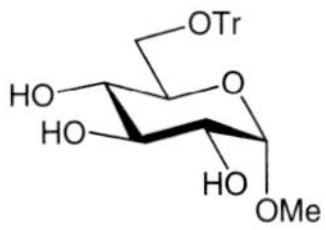	GlcNAc, Man, Gal, and protected carbohydrates	[10]
11	1) TrCl, pyridine, DMAP, 25 °C 2) Acetylation 60–84% (2 steps)	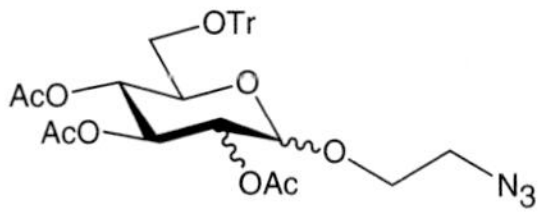	α-Glc, β-Glc, α-Man	[11]

(Continued)

Table 2.1 (Continued)

Entry	Protection	Protected carbohydrate	Observations	References
12	1) TrCl, pyridine, 90 °C, 7 h 2) Acetylation 62% (2 steps)	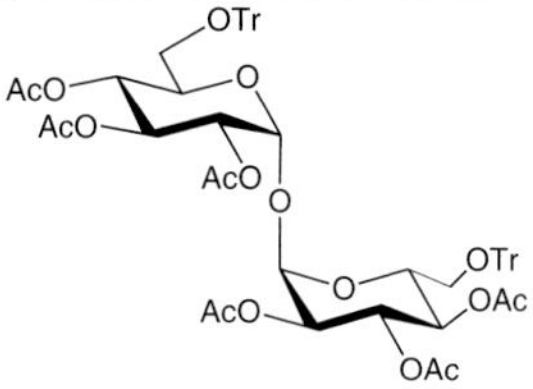	Acetylation of the crude product	[12]
13	TrCl, pyridine, CH_2Cl_2, RT, 16 h, 73%	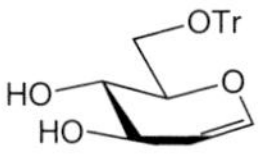	First full characterization of the product	[13]
14	TrCl, pyridine, RT, 18 h, 85%	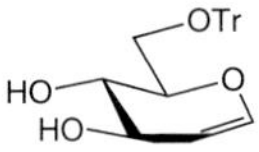		[14]
15	TrCl, DMAP, pyridine, RT, 18 h, 78%	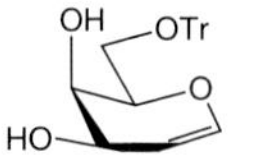	Glucal and galactal	[15]

improve yields such as ball milling [4], although with longer reactions. Association of DMAP was also used for the protection of galactals [15, 17] (Table 2.1, entry 15). Trityl chloride can also be extended to "one-pot synthesis" of building blocks in oligosaccharide synthesis [6, 7].

Deprotection of trityl ethers occurs in mild temperature and acidic conditions (Table 2.2). Classic deprotection is carried out in aqueous acetic acid or trifluoroacetic acid (TFA) for a few hours [8, 18] (Table 2.2 entry 16). This type of deprotection is not compatible with acid-sensitive protecting groups such as isopropylidene, although a chemoselective "one-pot" deprotection/purification can be achieved on a silica gel column [19] (Table 2.2, entry 17). Other deprotection methods are listed in Table 2.2; these include the use of Lewis acid [9], TFA [8], and pyridinium chloride [20] (Table 2.2, entries 18–20).

2.2.2 Silyl Ethers

Silyl ethers play an important role as protecting groups for primary hydroxyl functions and especially for glucal or galactal moieties in the synthesis of complex oligosaccharide or tumor antigen [21, 22]. They are widely relevant as protecting groups particularly when orthogonal strategies are required. Although trimethylsilyl (TMS) and triethylsilyl (TES) are typically not selective for the primary position, the required minimal bulkiness for 6-*O*-silylation can be accomplished with triisopropylsilyl (TIPS), *t*-butyldimethylsilyl (TBDMS), and *t*-butyldiphenylsilyl (TBDPS). Their introduction generally occurs in mild conditions (0 °C to RT, less than 12 h) with high yields (80–95%) using equimolar amounts of chlorosilanes in the presence of a base such as imidazole. Pyridine can also be used as both a base and a solvent, and DMAP is often used as a

Table 2.2 Selected conditions for the removal of trityl ethers.

Entry	Deprotection conditions	Deprotected carbohydrate	Observations	References
16	1) Tritylation 2) Benzoylation 3) AcOH, 85 °C, 6 h, 48% (3 steps)	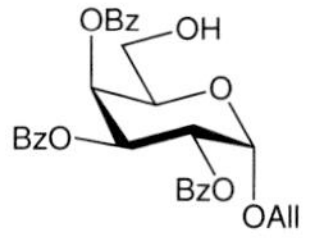	Gal	[18]
17	SiO_2 column, RT, 30 min, 79%	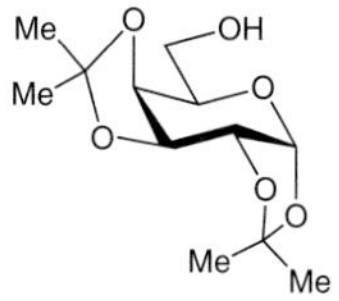	Glc, Gal, and trisaccharides Different protecting groups Compatible with benzyl and isopropylidene	[19]
18	$AlCl_3$, CH_2Cl_2/Et_2O (1/1, v/v), RT, 1 h, 58–73%	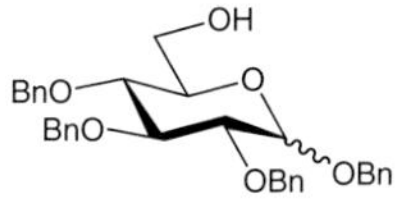	Glc	[9]
19	TFA/H_2O (9/1, v/v), CH_2Cl_2, RT, 30 min, 82%	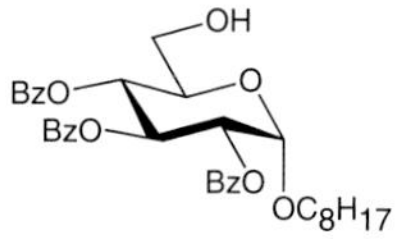	Glc	[8]
20	$C_5H_5N{\cdot}HCl$, EtOH, 80 °C, 2 h	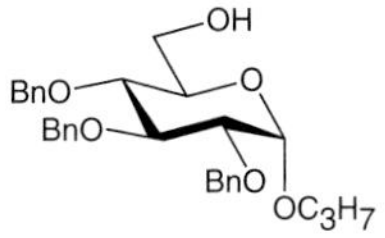	Glc	[20]

catalyst. Chromatography in polar solvents is necessary for the 6-*O*-silylated carbohydrates, which are obtained usually as colorless solids or pale-colored oils [23–30] (Table 2.3, entries 21–26, 28, 29, 32, and 33). The fastest reaction was reported to be less than 10 min for the large-scale 6-*O*-silylation of unprotected sugars [35] (Table 2.3, entry 31). Crude silylated intermediates were further protected in a "one-pot" procedure toward the preparation of multiprotected building blocks [29, 33, 34, 39] (Table 2.3, entries 27, 30, 32, and 37) and or disaccharide intermediates [44] (Table 2.3, entry 34). In general, it is accepted that bulky substituents such as TBDPS can be introduced at the primary positions in the order 6-OH > 6′-OH ≫ 1′-OH. Barros and coworkers recently showed that regioselective protection is possible by varying the amount of silyl chloride used for the protection of furanose rings in sucrose by using microwave irradiation [36]. Glucal and galactal were silylated at the primary alcohol [21, 22, 37–42, 45–47] (Table 2.3, entries 35–40), and the procedure was also extended to TES [43] with acceptable yields (Table 2.3, entry 41).

Silyl ethers are readily introduced through regioselective reagents, and they display good stability over a wide range of conditions. Their deprotection is also straightforward in 1 step using well-known techniques. The most representative

Table 2.3 Regioselective 6-*O*-silylation.

Entry	Protection	Protected carbohydrate	Observations	References
21	TIPSCl, imidazole, DMAP, DMF, RT, 29 h, 53%	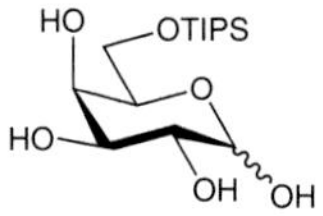	Gal	[23]
22	1) TIPSCl, imidazole, DMAP DMF, RT, 1 h 2) Benzoylation, 80% (2 steps)	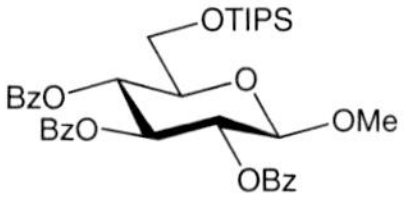	Glc Benzoylation performed	[24]
23	TIPSCl, imidazole, dry DMF, RT, 87%	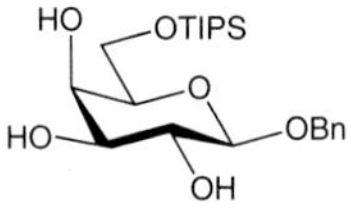	Gal	[25]
24	TBDMSCl, Et_3N, $DMAP_{cat.}$, dry DMF, −30 °C, 1 h, 87%	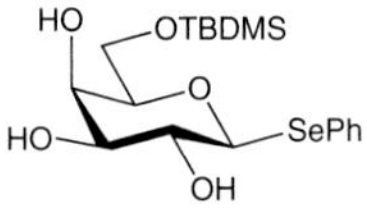	Gal	[31]
25	TBDMSCl, Et_3N, $DMAP_{cat.}$, dry DMF, RT, 72%	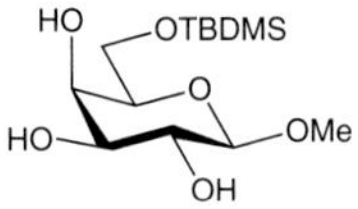	Gal	[26]
26	TBDMSCl, pyridine, DMAP, 0 °C to RT, 85%	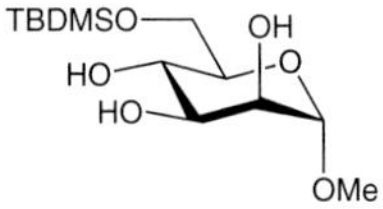	Man Large scale (10 g)	[32]
27	1) TBDMSCl, pyridine, DMAP 2) Acetylation, 95% (2 steps)	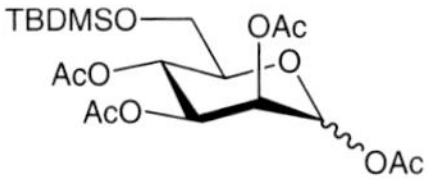		[33]
28	TBDMSCl, imidazole, DMF, 0–25 °C, 19 h, 92%	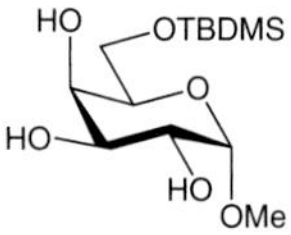	Gal	[27]
29	1) TBDMSCl, pyridine, DMF, RT, 12 h 2) Benzoylation 3) Debenzoylation, 56% (3 steps)	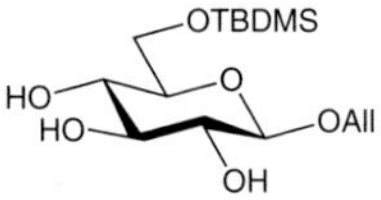	Glc	[28]
30	1) TBDMSCl, pyridine, DMF, RT, 4 h 2) Acetylation, 81% (2 steps)	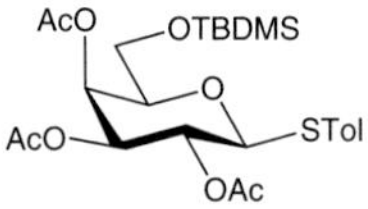	Gal	[34]
31	TBDMSCl, *N*-methylimidazole, iodine, CH_2Cl_2, RT, 10 min, 98%	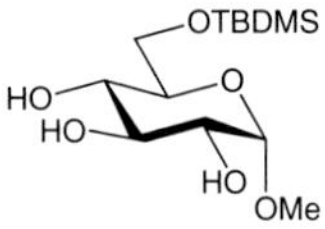	Improved and fast synthesis	[35]

Table 2.3 (Continued)

Entry	Protection	Protected carbohydrate	Observations	References
32	TBDPSCl, imidazole, DMF, 63%	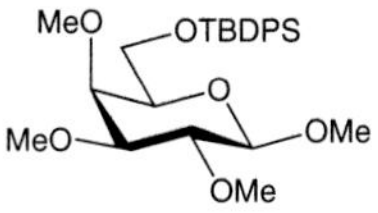	Gal	[29]
33	TBDPSCl, pyridine, RT, 4 h, 91%	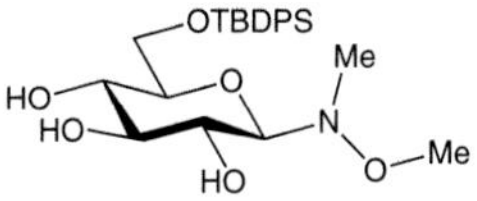	Only example in literature	[30]
34	TBDPSCl, DMAP, pyridine, μW (90 °C, 300 W), 5 min, 85%	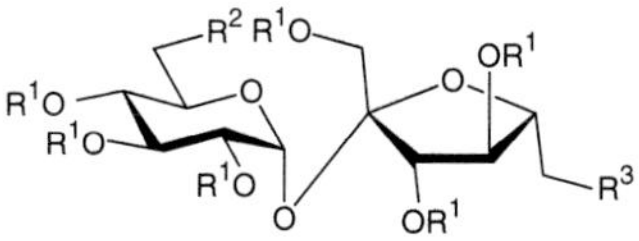 R^1 = H R^2 = OH, R^3 = OTBDPS 65% R^1 = H R^2 = R^3 = OTBDPS 85%	Selective conditions 1.1 eq. for the mono- and 2.2 eq. for the bis-ether	[36]
35	TBDMSCl, imidazole, DMF, 0–25 °C, 1 h, 82%	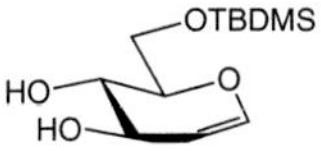	Gram-scale synthesis	[37]
36	TBDMSCl, TBAF, THF or NMP, RT, 24 h, 89–95%	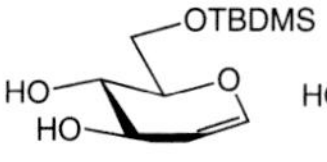 89% 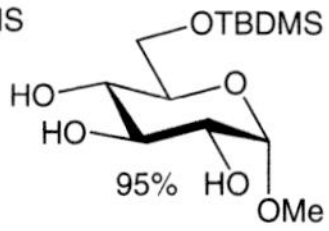 95%	Catalytic use of TBAF	[38]
37	1) TIPSCl, imidazole, DMF, RT 2) 3-*O*-benzylation 67% (2 steps)	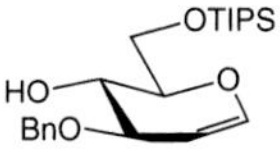	Glucal "one-pot" reaction	[39]
38	1) TIPSCl, imidazole, DMF, RT, 12 h 2) Acetylation 3) Deacetylation, 80% (3 steps)	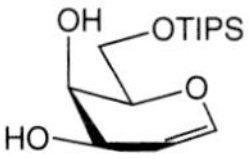	Galactal Detailed experimental conditions	[40]
39	TBDMSCl, imidazole, DMF, 0 °C, 40 min, 80–87%	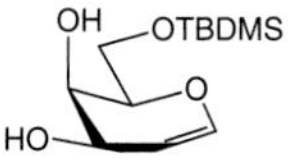		[41]
40	TBDPSCl, pyridine, DMAP, RT, 8 h, 90%		Large scale	[42]
41	RCl, imidazole, DMF, −18 °C, 14 h (R = TIPS) 89% (R = TBDPS) 95% (R = TBDMS) 74% (R = TES) 44%	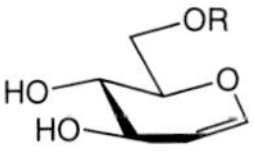	Glucal only In-depth methodology study	[43]

Table 2.4 Removal of TBDMS silyl ethers.

Entry	Deprotection conditions	Deprotected carbohydrate	Observations	References
42	IBr, MeOH, RT, 3 h, 85–90%	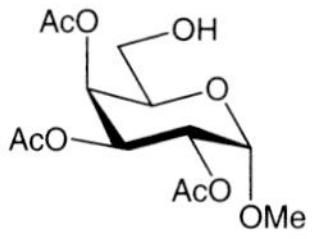	Crude reaction product virtually clean by NMR	[48]
43	HF-pyridine, 91%	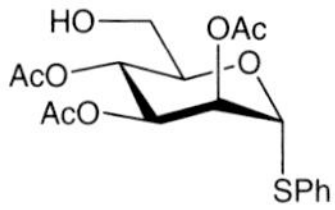		[49]
44	TBAF, THF, RT, 93%	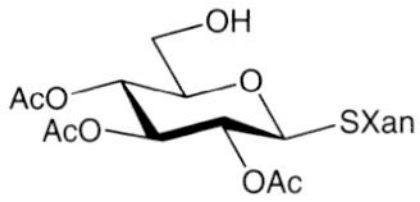	The crude mixture obtained was used for further reactions	[50]
45	TBAF, RT, 75%	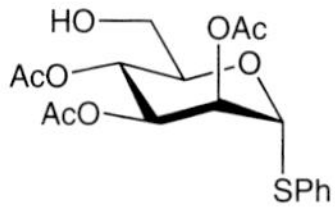	Man	[51]
46	TBAF, THF, RT, 9 h, 88%	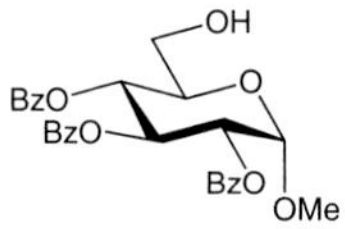	Scale-up does not compromise yield	[52]
47	Decaborane, THF/MeOH, (1/1 v/v), RT, 24 h, 90–99%	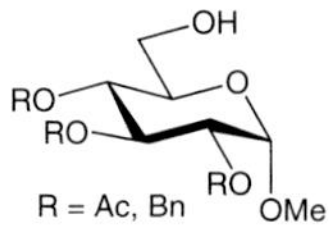 R = Ac, Bn	Glc	[53]
48	Phosphomolybdic acid supported on SiO_2	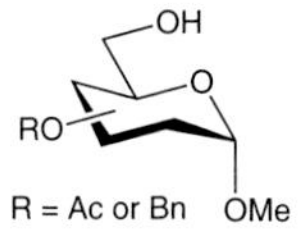 R = Ac or Bn	Glc, Gal, Man in presence of different protection groups	[54]
49	AcOH, 50 °C, 2 h, 86%	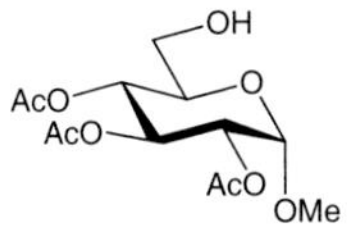	Glc	[55]
50	H_2SO_4, MeOH, RT, 93%	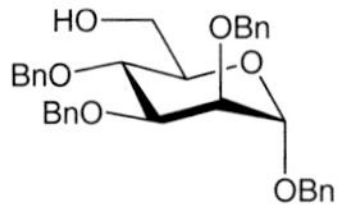	Man	[56]
51	SiO_2 column, RT, 30 min, 79%	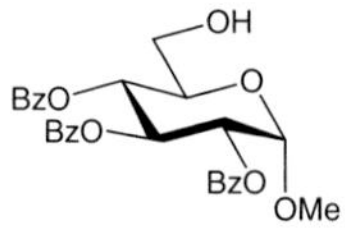	Glc	[19]
52	TfOH (triflic acid) SiO_2 optimized for MeCN, 50 °C, 30 min, 93%	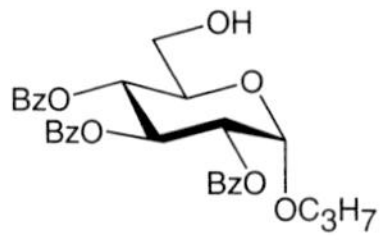	Glc Methodology paper	[57]

Table 2.4 (Continued)

Entry	Deprotection conditions	Deprotected carbohydrate	Observations	References
53	HCl, MeOH, RT, 88%	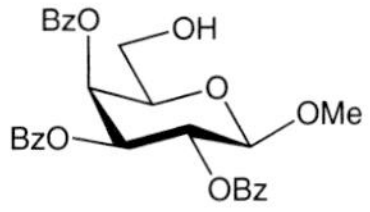	Gal TBDPS ether	[58]
54	1) Benzoylation 2) $BF_3{\cdot}OEt_2$, MeOH/CH_2Cl_2, (2/1, v/v), RT, 24 h, 79% (2 steps)	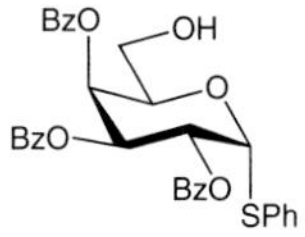	Gal TBDPS ether	[18]

examples of silyl ether deprotections under various mild conditions and high yields (85–99%) and purity are reported in Table 2.4. This selection includes fluoride-based reagents [49–52] (Table 2.4, entries 43–46), oxidizing/reducing reagents [48] (Table 2.4, entry 42), mild and strong acidic conditions [19, 52, 54–58] (Table 2.4, entries 46–53), and reducing agents such as decaborane [53] (Table 2.4, entry 47).

2.2.3 Regioselective 6-*O*-Sulfonylation

The regioselective 6-*O*-sulfonylation represents another approach for the protection of the primary positions and is mainly employed to access halides or other functional groups from the transiently activated alcohol. Reduction to the corresponding 6-deoxy derivatives (e.g. fucose or rhamnose) is another possibility. The typical regioselective sulfonylation of a carbohydrate at the primary position is achieved by dissolving the unprotected carbohydrate in dry pyridine and adding sulfonyl chlorides (RSO_2Cl) at low temperature (−10 to 0 °C). *p*-Toluenesulfonates (or tosylates) display the required bulkiness to achieve the 6-*O*-regioselective sulfonylation (Table 2.5, entries 55–57) using tosyl chloride (TsCl) [59–62, 71]. Acetylation of the crude mixture is often used to facilitate purifications of the 6-*O*-sulfonylated peracetylated products as described for mannose [63, 64] (Table 2.5, entry 58) and gulose [65] (Table 2.5, entry 59). However, this full protection is sometimes less efficient (30%), e.g. for glucose and galactosyl azide [66] (Table 2.5, entry 60). A similar strategy was used for the sulfonylation followed by benzoylation of mannose [67] (Table 2.5, entry 61) or glucose [72]. Moreover, regioselective 6-*O*-sulfonylation was reported with glucal [68] (Table 2.5, entry 62), thioglycosides [69] (Table 2.5, entry 63), and thiocolchicosides [70] (Table 2.5, entry 64). The sulfonylation at the primary position is typically followed by nucleophilic substitution with iodide [62, 72, 73] (Table 2.5, entry 57) or treatment with sodium azide [60, 74] (Table 2.5, entry 56) to give the corresponding 6-deoxy-6-iodo or 6-azido-6-deoxy-carbohydrates.

Table 2.5 Regioselective 6-*O*-tosylation.

Entry	Protection	Protected carbohydrate	Observations	References
55	TsCl, pyridine, −5 °C, 5 h, 96%	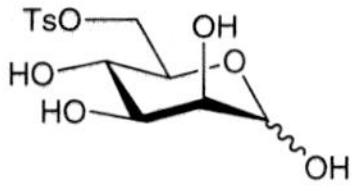	Man	[59]
56	TsCl, pyridine, 0 °C to RT, 17–24 h	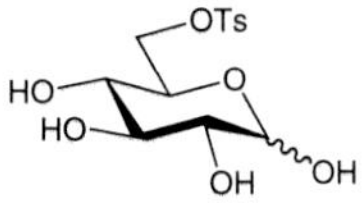	Glc Acetylation and azidation described	[60, 61]
57	TsCl, pyridine, −10 °C to RT, 15 h, 97%	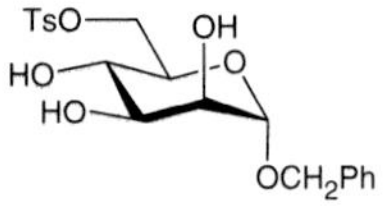	Man Conversion to 6-deoxy-6-iodo described	[62]
58	1) TsCl, pyridine, 20 °C, 2 h 2) Ac_2O, pyridine, 18 °C to RT, 2 h, 97% (2 steps)	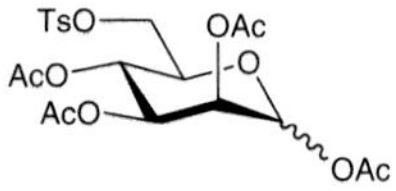	Man	[63, 64]
59	1) TsCl, pyridine, 20 °C, 2 h 2) Ac_2O, pyridine, 18 °C to RT, 2 h, 97% (2 steps)	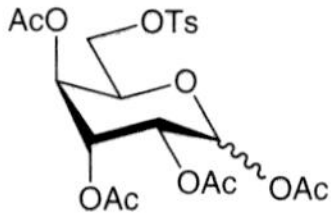	Gul	[65]
60	1) TsCl, pyridine, 0 °C, 20 h 2) Ac_2O, pyridine, 0 °C to RT, 1 h, 30–36% (2 steps)	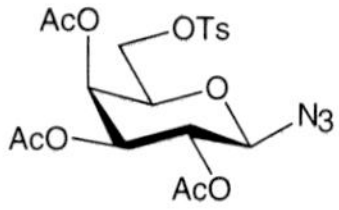	Glc, cellobiose	[66]
61	1) TsCl, pyridine, 0 °C to RT, 6 h 2) BzCl, pyridine, 0 °C to RT, 18 h, 38% (2 steps)	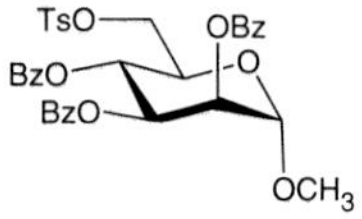	Man Conversion to 6-deoxy-6-iodo described	[67]
62	TsCl, pyridine, CH_2Cl_2, 3 h, 56%	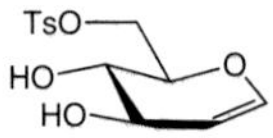	Glucal	[68]
63	TsCl, pyridine, 0 °C to RT, 5 h, 75%	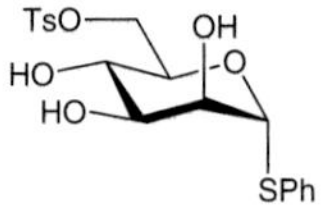	Thioglycoside	[69]
64	TsCl, Et_3N, DMAP, pyridine, RT, 20 h, 75%	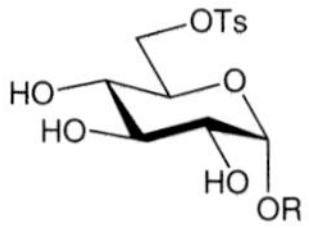	R = thiocolchicine derivatives	[70]

2.2.4 Regioselective 6-*O*-Acylation

2.2.4.1 Chemical Methods

The regioselective introduction of esters at the primary positions of hexopyranoses can be accomplished through two main strategies using either chemical (Table 2.6) or chemoenzymatic (Table 2.7) methodologies. In the chemical approaches, the amount of acylating reagent is crucial, and usually, only a slight excess is used to avoid diacylations. The introduction of a sterically hindered ester such as the pivaloate group was initially reported (Table 2.6, entry 65) in the total synthesis of glycosphingolipids [75] and then later in a study for the influence of the pyranose conformation in the glycosylation process [76] (Table 2.6, entry 66). Acetate, although less regioselective because of its smaller size in comparison to pivaloate, could also be introduced at the 6-position (Table 2.6, entries 67 and 68) at low temperature and using a slight excess of acetyl chloride [77, 78]. Other selective esterifications can be envisaged from the data reported, along with potential selective sulfonylations or silylations. Acetic acid activated with heterocycles (Table 2.6, entries 69 and 70) can also provide the regioselective 6-*O*-acetylation. 1-Acetyl imidazole was reported in combination with tetramethylammonium hydroxide as the best of eight bases screened [79] (Table 2.6, entry 69). 6-*O*-esterification with fatty acid chains (Table 2.6, entry 70) required the use of acylthiazolidine-2-thiones as the best of four acylating agents in lower amounts (0.5 eq.) [36, 80]. Benzoate derivatives were attainable through a sodium hydride-initiated multianion intermediate in the presence of the acyl chloride [81] (Table 2.6, entry 71). Interestingly, the same procedure with benzyl bromide provided the regioselective 6-*O*-benzylation. Tin-based esterification provided the desired 6-*O*-benzoates in good yields and regioselectivities (Table 2.6, entries 72 and 73). Dibutyltin(IV) oxide ($n\mathrm{Bu_2SnO}$) was perfectly regioselective (Table 2.6, entry 72) and the dibutylstannylene acetal intermediate could be accessed by heating 7 × 1 min under microwave activation [82]. Dimethyltin(IV) dichloride ($\mathrm{Me_2SnCl_2}$) was regioselective for β-glucosides (Table 2.6, entry 73) but led to 2-*O*-benzoylation for the α-anomer and 3-*O*-benzoylation for galacto and manno derivatives [83]. A general esterification method was reported for several alcohols (Table 2.6, entry 74), and glucopyranoside was used in this study using a carboxylic acid precursor with activating agents usually applied to peptide synthesis [84]. Very recently, γ-Keggin germanodecatungstate $[\gamma\text{-HGeW}_{10}\mathrm{O}_{36}]^{7-}$, a polyoxometalate, was applied in *trans*-esterification with vinyl acetate (Table 2.6, entry 75) leading to the desired 6-*O*-acetate in a very short reaction time (5 min), and this method is compatible with acid-sensitive alcohols [85]. The simplest method involves ethyl acetate as a solvent and *trans*-esterification agent catalyzed by concentrated sulfuric acid [86] (Table 2.6, entry 76). The regioselectivity was almost complete with 0–5% of diacetylated carbohydrates.

2.2.4.2 Chemoenzymatic Methods

A general approach for the regioselective 6-*O*-acylation of carbohydrates based on chemoenzymatic approaches uses both a variety of acyl donors (typically a vinyl ester) and several different enzymes [111].

Table 2.6 Chemical regioselective 6-*O*-acylation of saccharides.

Entry	Acylation conditions	Acylated carbohydrate	Observations	References
65	PivCl (1.3 eq.), pyridine, 0–25 °C, 14 h, 65%	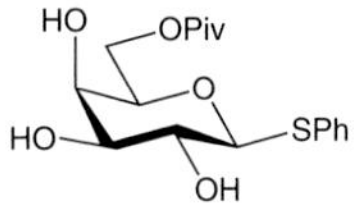	Single example	[75]
66	PivCl (2.2 eq.), DMAP (0.1 eq.), pyridine, 0 °C, 1 h, 42%	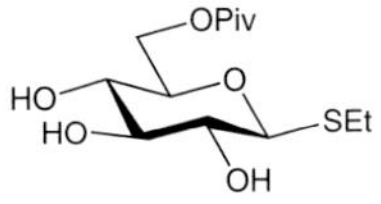	Single example	[76]
67	AcCl (1.2 eq.), collidine (2 eq.), CH_2Cl_2, −40 to 25 °C, 4 h, 75%	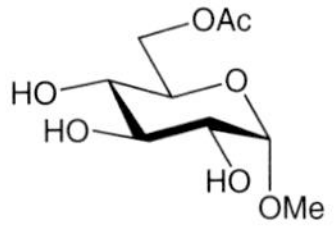	Ribose also reported	[77]
68	AcCl (1.2 eq.), collidine (2 eq.), CH_2Cl_2, −40 °C, 30 min, 60%	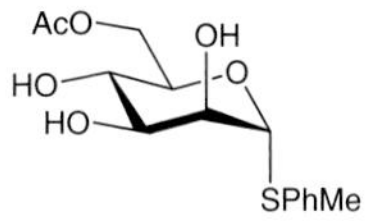	Single example	[78]
69	1-Acetylimidazole (3 eq.), Me_4NOH (1.2 eq.), H_2O, 60 °C, 16 h, 70%	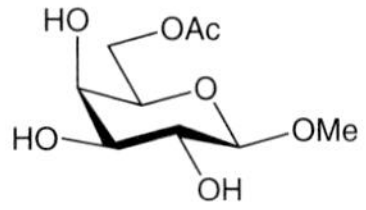	Glc, Man, Gal	[79]
70	Acylating agent (0.5 eq.), DMAP (0.25 eq.), NaH (0.08 eq.), pyridine, RT, 2 h, 88%	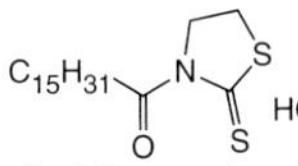 Acylating agent 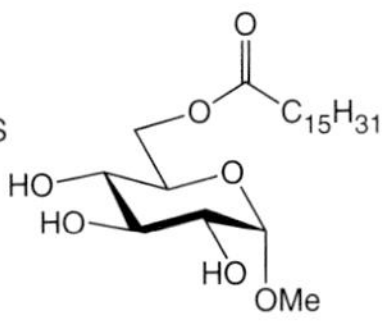	Man, Gal Reducing carbohydrates or methyl glycosides	[80]
71	NaH (4 eq.), MePhCOCl (1.7 eq.), DMF, 20 °C, 2 h, 57% (R = OPh) 61% (R = OMe) 76% (R = OPhOMe) 81% (R = SPh)	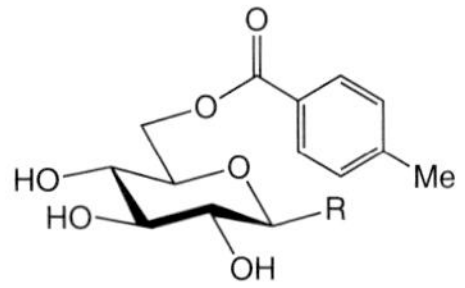	Mono-benzylation also reported >60% yield for α-anomers	[81]
72	nBu$_2$SnO (1 eq.), BzCl (1 eq.), PhMe, RT, 30 min, 76%	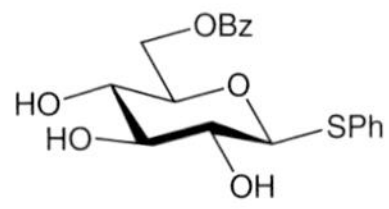	40% yield for α-OMe anomer	[82]
73	Me_2SnCl_2 (0.05 eq.), BzCl (1.2 eq.), iPr$_2$NEt (2 eq.), THF, RT, 79%	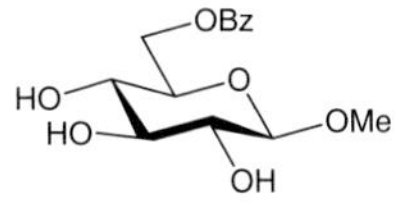	2-*O*-Benzoylation for α-anomer 3-*O*-Benzoylation for Gal and Man	[83]

Table 2.6 (Continued)

Entry	Acylation conditions	Acylated carbohydrate	Observations	References
74	RCO_2H (1 eq.), TBTU (1 eq.), *i*Pr_2NEt (2 eq.), DMF, RT, 5 h, 81% (R = Ph) 78% (R = $(BnOCH_2)_3C$)	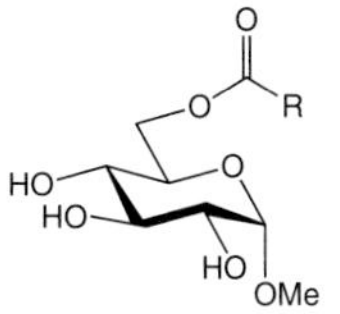	General esterification method	[84]
75	Vinyl acetate (0.5 eq.), [γ-$HGeW_{10}O_{36}]^{7-}$ (5 ppm), DMSO, RT, 5 min, 63%	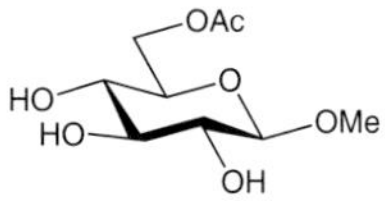	Single example	[85]
76	EtOAc, $H_2SO_{4\,conc.}$, RT, 12 h, 90%	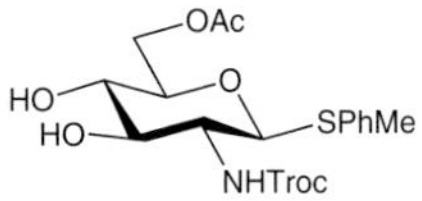	Glc, Man, Gal, GlcNAc	[86]

The early report of a porcine pancreatic lipase as a selective 6-*O*-acetylation system using a 2,2,2-trichloroethyl ester as the acyl donor (Table 2.7, entry 77) involved using pyridine with gentle heating to provide reasonable solubility of the starting material [87] (i.e. native hexoses). The monochloroethyl or dichloroethyl esters did not afford good yields or selectivities. The same approach was used for acetylation of methyl glucopyranoside (Table 2.7, entry 78) using lipase from *Candida cylindracea* and vinyl acetate [88]. Even though the yield was not reported in this study, more than 90% regioselectivity was observed for the 6-*O*-acetate. Vinyl esters provide acetaldehyde as a by-product that is compatible with the chemoenzymatic reaction conditions. Acyl acetoximes were used in combination with lipase PS from *Pseudomonas cepacia* (Table 2.7, entry 79), providing the desired 6-*O*-acylated glucopyranosides in good yields [89]. The oxime by-product generated does not participate in the reverse deacylation reaction [112]. Selective 6-*O*-acetylation of *N*-acetylmannosamine (ManNAc) was accomplished with protease N from *Bacillus subtilis* using isoprenyl acetate as the acyl donor [90] (Table 2.7, entry 80). The enzyme was solubilized in a phosphate buffer, then freeze-dried, and the preparation was used for the chemoenzymatic reaction in dimethylformamide (DMF). The same ManNAc building block was obtained from subtilisin 8399 with vinyl acetate (Table 2.7, entry 81) in the synthesis of modified sialic acid residues [91] using conditions based on a previously reported general study [113]. Galactal required small amount of water and short reaction time (i.e. 45 min) to provide the 6-*O*-acetyl galactal derivative [92] (Table 2.7, entry 82). Galactolipids with bifurcated lipid chain aglycons were also regioselectively 6-*O*-acetylated (Table 2.7, entry 83) toward the preparation of GM_4 ganglioside analogs [93].

Acylation with fatty acid esters was performed with lipase Novozym 435 (Novozym 435 and *Candida antarctica* lipase B also known as CAL-B are the same enzymes) and a carboxylic acid as the acyl donor [94] (Table 2.7, entry 84).

Table 2.7 Chemoenzymatic regioselective 6-*O*-acylation of saccharides.

Entry	Acylation conditions	Acylated carbohydrate	Observations	References
77	Porcine pancreatic lipase, 2,2,2-trichloroethyl laurate, pyridine, 45 °C, 48 h, 91%	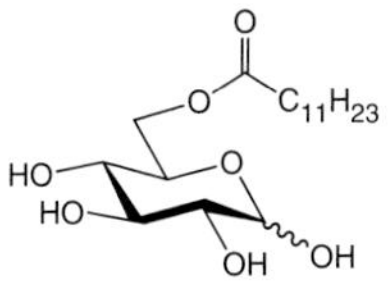	Butyryl, acetyl, capryl also reported Gal and Man	[87]
78	Lipase from *Candida cylindracea*, vinyl acetate, benzene/pyridine (2/1 v/v), 28 °C, 120 h	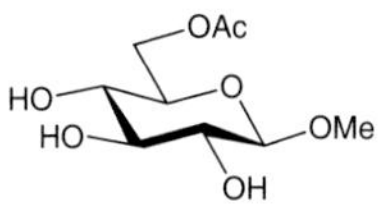	Man, ManNAc yield not reported	[88]
79	Lipase PS from *Pseudomonas cepacia*, acetyl acetoxime, pyridine, 25 °C, 72 h, 76%	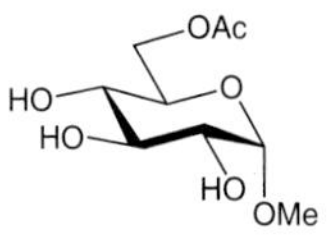	Glucose also reported four different esters	[89]
80	Protease N from *Bacillus subtilis*, isoprenyl acetate, DMF, 45 °C, 44 h, 76%	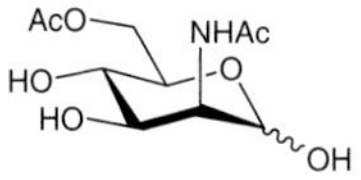	Specific enzyme preparation	[90]
81	Subtilisin 8399, vinyl acetate, DMF, 45 °C, 120 h, 92%	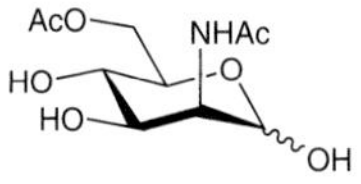	Lactate also reported	[91]
82	Lipase from *Candida cylindracea*, vinyl acetate, EtOAc, RT, 24 h, 90%	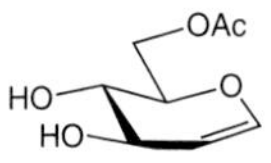	Galactal also reported	[92]
83	Lipase PS from *Pseudomonas cepacia*, vinyl acetate, 45 °C, 48 h, 90%	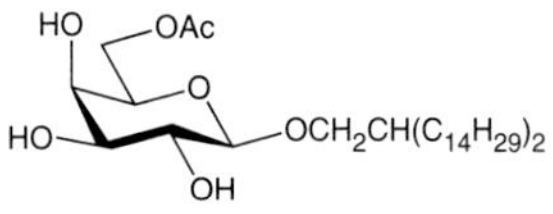	Galactolipids	[93]
84	Lipase Novozym 435, $C_{11}H_{23}CO_2H$, Et_2O, 110 °C (60 μW–120 W), 2 h, 97%	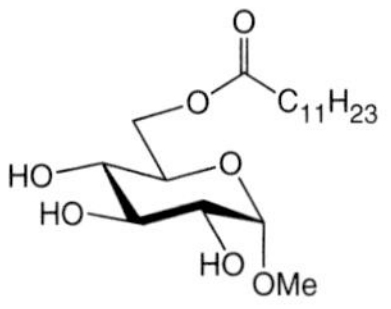	Trehalose and Glc also reported microwaves	[94]
85	Lipase B from *Candida antarctica*, $C_7H_{15}CO_2H$, MeCN, 70 °C, 24 h, 72%	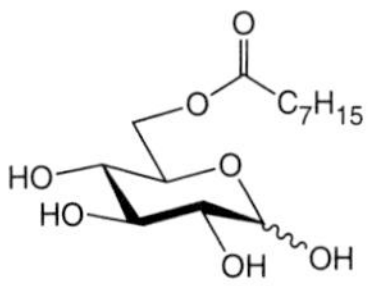	Single example	[95]
86	Lipase B from *Candida antarctica*, $C_{14}H_{29}CO_2H$, *t*-amyl alcohol, 80 °C, 24 h, 31%	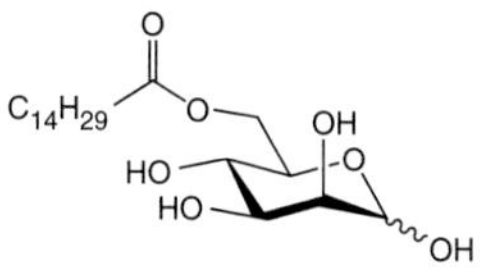	Fluorinated esters also reported moderate yields	[96]

Table 2.7 (Continued)

Entry	Acylation conditions	Acylated carbohydrate	Observations	References
87	Lipase B from *Candida antarctica*, $Ph(CH_2)_3CO_2H$, *t*BuOH, 60 °C, 24 h, 80%	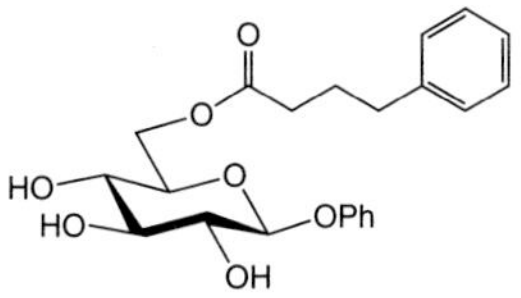	Five aglycons and six esters tested	[97]
88	Lipase B from *Candida antarctica*, $C_4H_9CO_2H$, 70 °C, 28 h, 72%	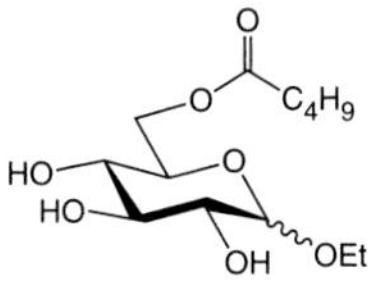	20 kg-scale reported six fatty acids	[98]
89	Lipase Novozym 435, $H_2C{=}COCOC_{13}H_{27}$, THF/pyridine (4/1 v/v), 40 °C, 48 h, 94%	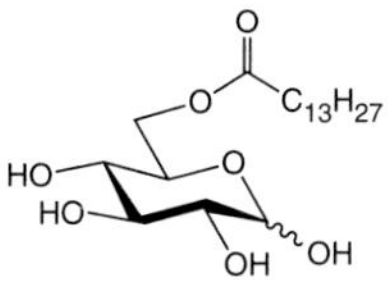	Limited amount of pyridine	[99]
90	Lipase Novozym 435, $H_2C{=}COCOR$, MeCN, 45 °C, 48 h 90% ($R = C_7H_{15}$), 73% ($R = C_9H_{19}$), 56% ($R = C_{11}H_{21}$)	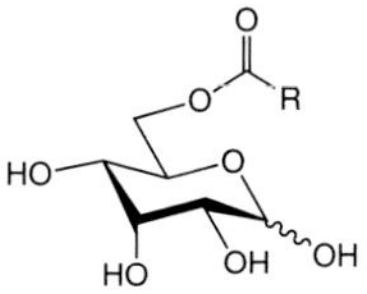	Glucose also reported	[100]
91	Lipase from *Humicola lanuginosa*, vinyl laurate, *t*-amyl alcohol/ DMSO (4/1 v/v), 40 °C, 10 h, 80%	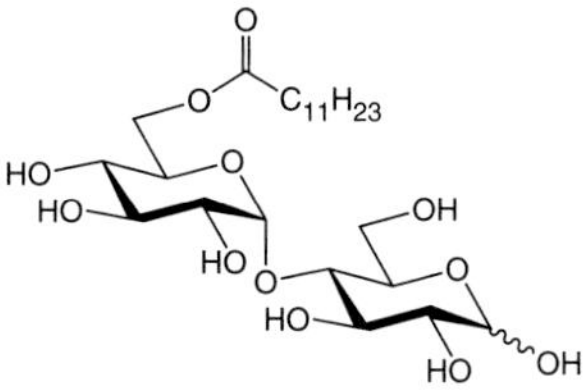	Maltose and leucrose also reported Several esters	[101]
92	Lipase Novozym 435, vinyl methacrylate, MeCN, 50 °C, 120 h, 70%	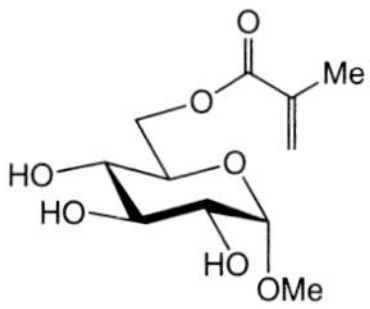	8 g scale	[102]
93	Lipase Novozym 435, vinyl methacrylate, *t*BuOH, 50 °C, 24 h, 80%	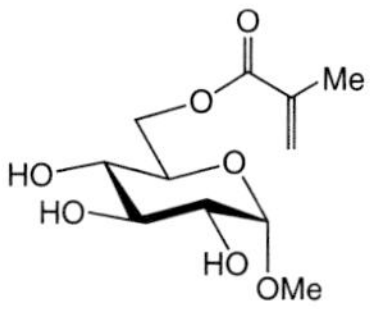	Also with *p*-bromothio-phenyl aglycon (78%)	[103]
94	Lipase Novozym 435, methyl acrylate, *t*BuOH, 55 °C, 24 h, 35%	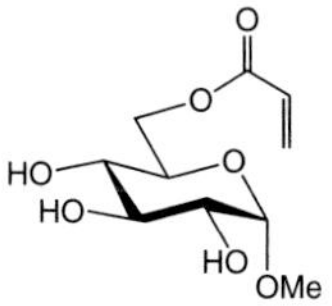	Methyl ester as the acyl donor	[104]

(Continued)

Table 2.7 (Continued)

Entry	Acylation conditions	Acylated carbohydrate	Observations	References
95	Lipase B from *Candida antarctica*, $HS(CH_2)_2CO_2H$, *t*-amyl alcohol, 80 °C, 24 h, 48%	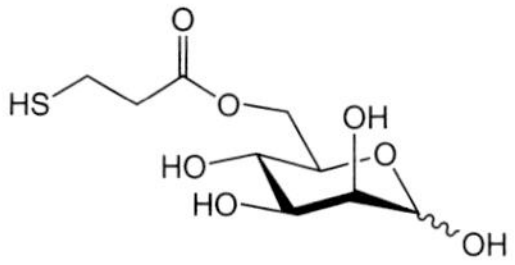		[105]
96	Subtilisin, dimethyl malonate, pyridine, 45 °C, 16 h, 69%	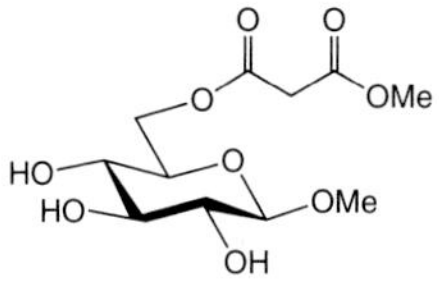	Single example	[106]
97	Alkaline protease from *Bacillus subtilis*, divinyl dicarboxylate, pyridine, 55 °C, 72 h, 30% (*n* = 2), 53% (*n* = 4), 35% (*n* = 6)	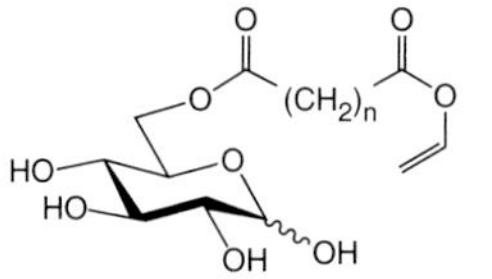	Glucose only	[107]
98	Subtilisin (ChiroCLEC-BL), vinyl laurate, pyridine, 40 °C, 72 h, 60%	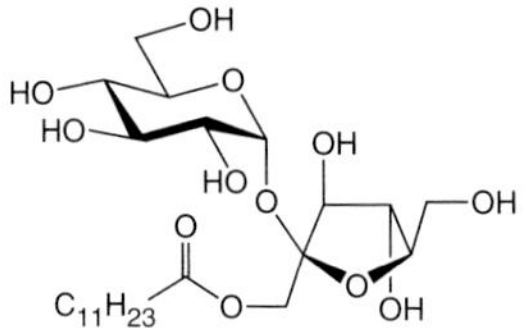	Myristate and stearate reported	[108]
99	Lipase from *Humicola lanuginosa*, vinyl laurate, *t*-amyl alcohol/DMSO (4/1 v/v), 40 °C, 24 h, 70%	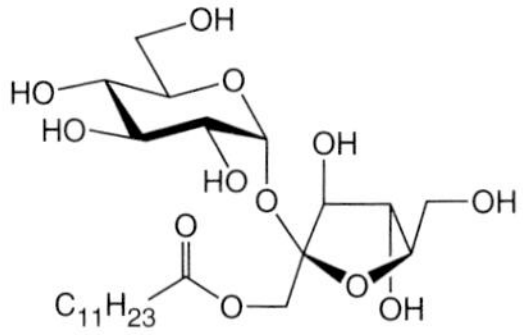	Palmitate also reported	[109]
100	Protease N from *Bacillus subtilis*, $CF_3CH_2OCOC_3H_7$, DMF/H_2O (85/15 v/v), 45 °C, 24 h, 76%	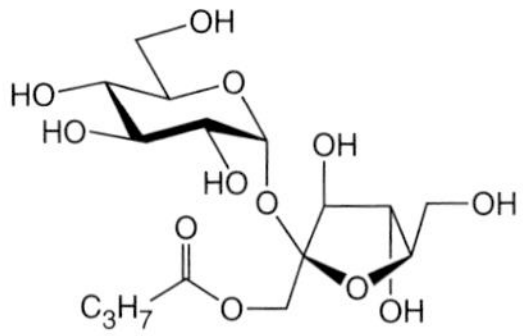	Other short esters	[110]

Similar acylation was also accomplished with CAL-B (Table 2.7, entry 85) using 10% of octanoic acid in acetonitrile for the optimal solubility of glucose [95]. *t*-Amyl alcohol was also used as a solvent (Table 2.7, entry 86) for carboxylic acids bearing fluorinated chains [96] while *t*-butanol was reported for six different aryl aliphatic esters (Table 2.7, entry 87) with high regioselectivity but moderate to good isolated yields from 14% to 67% [97]. Further applications toward carbohydrate-based surfactants were also reported using the same regioselective acylation strategy [114]. CAL-B was applied to the large-scale (up to 20 kg) synthesis of 6-*O*-acyl glucose [98] (Table 2.7, entry 88) based on a systematic study reported earlier [115]. Vinyl esters were also used as acyl donors, and the influence of the solvent [116] or enzyme [117] was investigated. A small portion of pyridine (i.e. 20%) was used in tetrahydrofuran (THF) to ensure the optimal solubility of glucose (Table 2.7, entry 89), which was pre-dried in a vacuum oven for 48 h to remove water for the reaction to proceed efficiently [99]. 6-*O*-acylation of allose was studied with Novozym 435 (Table 2.7, entry 90) with a vinyl ester in acetonitrile [100] while maltose and leucrose regioselective acylation (Table 2.7, entry 91) was accomplished with a lipase from *Humicola lanuginosa* [101].

Introduction of acrylates at the 6-position leads to polymerizable building blocks for the design of glycosylated polymers. Novozym 435 was applied to the 8 g-scale synthesis of the 6-*O*-methacrylate of glucose (Table 2.7, entries 92–94) from vinyl methacrylate in acetonitrile [102] (Table 2.7, entry 92), but switching to *t*-butanol (Table 2.7, entry 93) resulted in much shorter reaction times [103]. The same esterification could be accomplished from the methyl ester (Table 2.7, entry 94), although in moderate 35% isolated yield [104]. CAL-B was used for the introduction of a 3-thiopropanoate (Table 2.7, entry 95) from the corresponding acid in good isolated yield, demonstrating the compatibility with thiolated species [105]. Dimethyl malonate was also used as an acyl donor to provide the corresponding desymmetrized malonate (Table 2.7, entry 96) in a reaction with subtilisin [106] while divinyl diesters of several diacids (Table 2.7, entry 97) required an alkaline phosphatase from *B. subtilis* to provide the desired esters in moderate isolated yields (30–53%) [107]. The influence of water content was also studied for this particular esterification process [118].

The mono-acylation of sucrose has been studied [119] with emphasis on the effect of salt hydrate pair [120], and more general studies for several disaccharides have been reported [121]. The regioselective 6-*O*-acylation of sucrose could be readily achieved in good isolated yields using either subtilisin [108] (Table 2.7, entry 98) or lipase from *H. lanuginosa* [109] (Table 2.7, entry 99) from the vinyl esters. 2,2,2-Trifluoroethanol esters were also reported with protease N from *B. subtilis* [110] (Table 2.7, entry 100) and the synthetic strategy applied in further studies [122] including the self-organization properties of the resulting amphiphiles [123].

Novozym 435 and CAL-B were also investigated using ionic liquids that improve the solubility of glucose, thus leading to lower amounts of diacylated products [124–126].

2.3 Selective Primary Hydroxyl Group Deprotection

2.3.1 Regioselective 6-*O*-Desilylation

Selective 6-*O*-deprotection of silyl ethers (TMS, TBDMS, and TIPS) occur in mild basic or acid conditions at low temperature (e.g. 0 °C) from the corresponding persilylated carbohydrates. The process is typically under kinetic control, and further desilylation is usually observed when longer reaction

Table 2.8 Regioselective desilylation at the primary position of saccharides.

Entry	Cleavage	Deprotected carbohydrate	Observations	References
101	K_2CO_3, MeOH 0 °C, 20 min, 65–90%	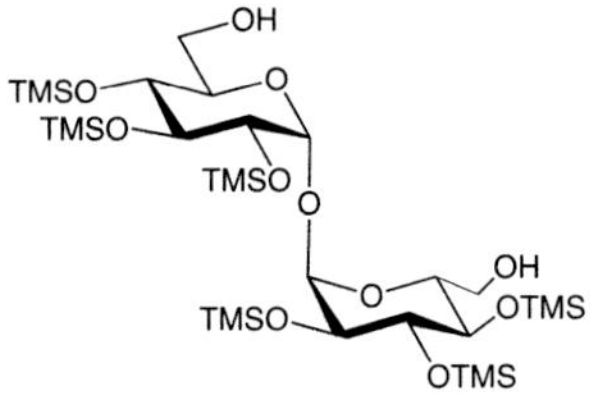	Trehalose mono-desilylation	[128]
102	K_2CO_3, MeOH, 0 °C, 5 h, 82%	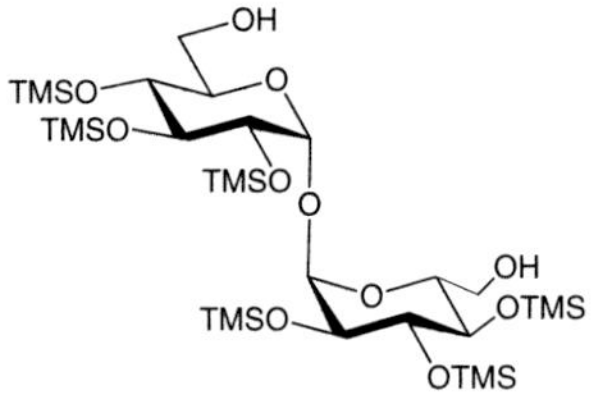	Improved synthetic procedure from Datta et al. [128]	[129]
103	AcOH, MeOH/Acetone (3/4 v/v), 0 °C, 79%	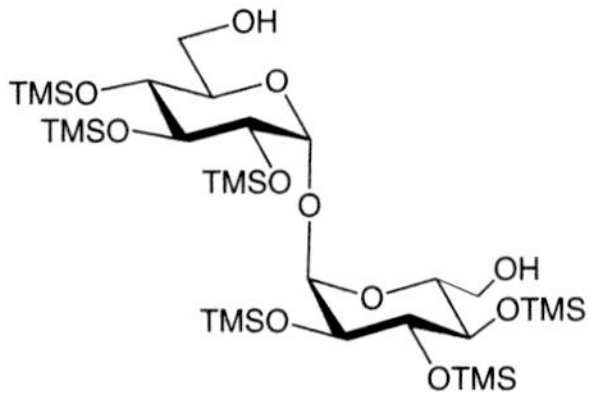		[130]
104	1) Silylation 2) NH_4OAc, CH_2Cl_2, RT, 14 h, 77% (2 steps)	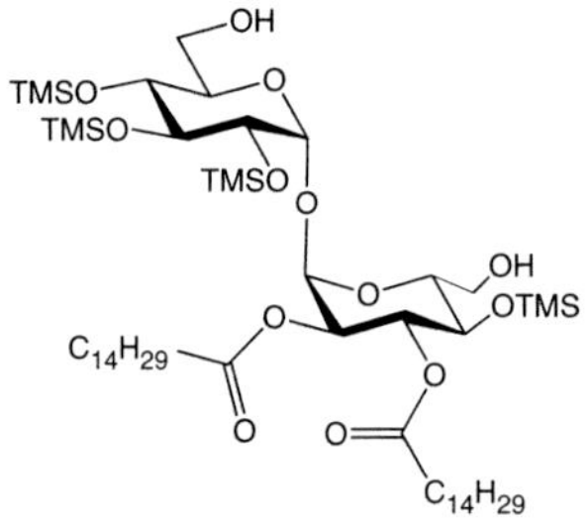	Compatible with esters	[131]
105	NH_4OAc, CH_2Cl_2/MeOH (1/3 v/v), RT, 63%	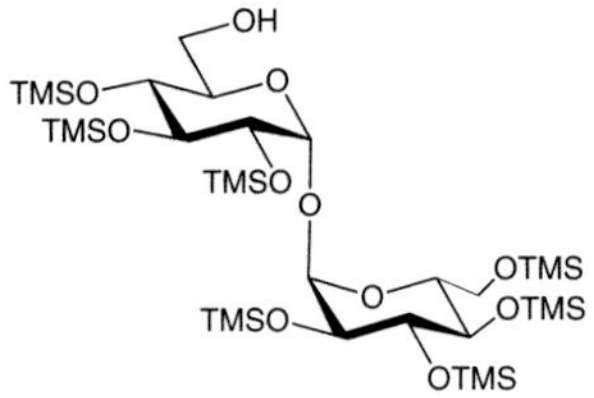	mono-desilylation	[132]
106	K_2CO_3, MeOH, 20 °C, 2 h, 67%	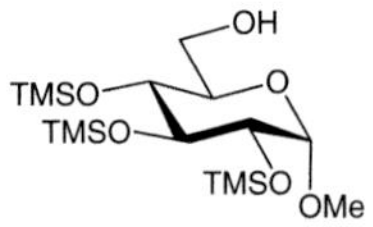	Glc (67%) Man (37%)	[133]
107	5% AcOH, MeOH, 0 °C to RT, 2 h, 75%	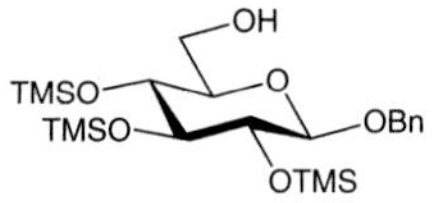	Glc	[134]

Table 2.8 (Continued)

Entry	Cleavage	Deprotected carbohydrate	Observations	References
108	AcOH, MeOH/acetone (3/4 v/v), RT, 8 h, 78%	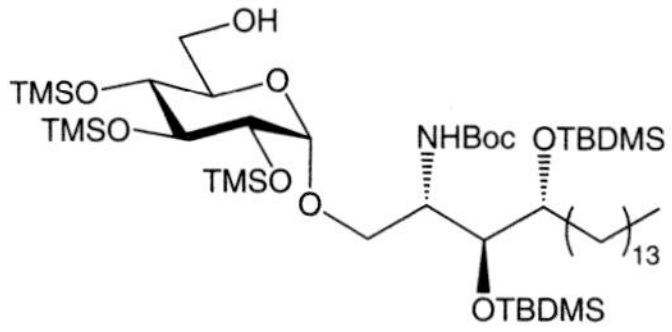	Orthogonal with TBDMS and NHBoc-protecting groups	[135]
109	AcOH, MeOH/acetone (3/4 v/v), RT, 48 h, 58%	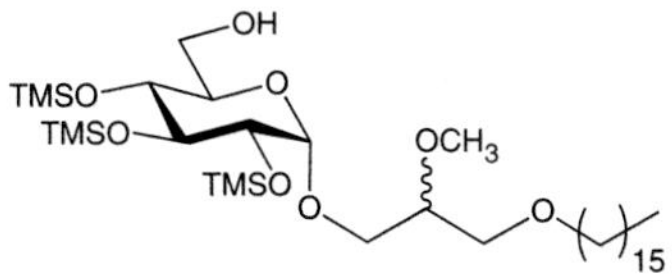	Racemic mixture	[136]
110	NH_4OAc, CH_2Cl_2/MeOH (1/1, v/v), RT, 9–12 h, 65–90%	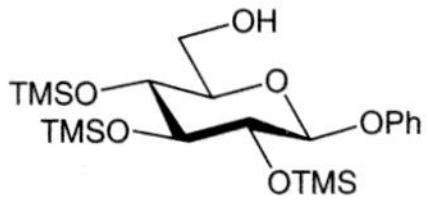	Glc, Gal Trehalose	[137]
111	K_2CO_3, MeOH, 0 °C, 20 min, 50%	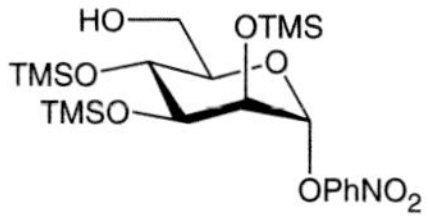	Man	[127]
112	*hν* (245 nm, 8 W), CBr_4, MeOH, RT, 1–39 h, 71–94%	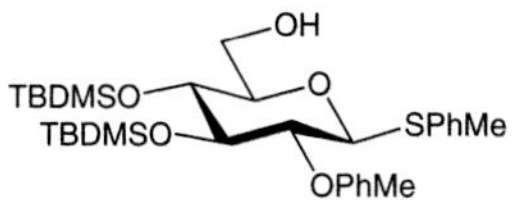	Glc, Gal, Man, Neu5Ac Selective and compatible with esters	[138]
113	BCl_3, $ClCH_2CH_2OH$, THF, RT, 12 min, 81%	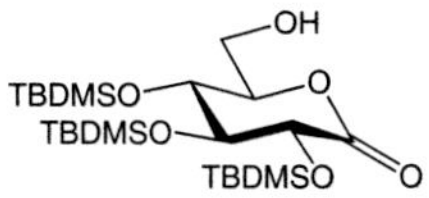	γ- and δ-Lactones	[139]
114	TFA/H_2O/THF (1 : 1 : 2.5), RT, 6 h, 70%	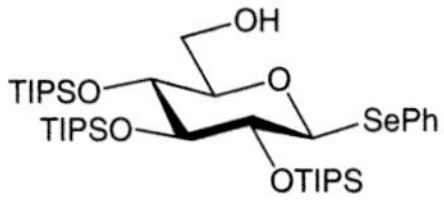	Glc	[140]
115	1) Deacetylation 2) Persilylation 3) HF/pyridine, THF, 3 h, RT, 63% (3 steps)	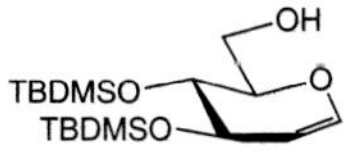	"One-pot" from peracetylated glucal	[141]

times are used. Deprotection of per-trimethylsilylated mannose or glucose can be performed in methanolic solution of potassium carbonate at low temperature and are relatively rapid from several minutes [127] (Table 2.8, entry 111) to a few hours [133, 134] (Table 2.8, entries 106–107). Other conditions use a source of acetate ions (NH_4OAc or acetic acid) in MeOH–acetone or MeOH–CH_2Cl_2 [134–137] (Table 2.8, entries 107–110). Similar procedures were applied to per-trimethylsilylated trehalose [128–130, 142] (Table 2.8, entries 101–103) or modified trehalose derivatives [131] (Table 2.8, entry 104) with high regioselectivity of the mono-desilylation distinguishing between the two primary positions of the disaccharide with acceptable isolated yields [132, 142] (Table 2.8, entry 105).

The regioselective photochemical deprotection of TBDMS ether was carried out on different protected substrates including protected Glc, Gal, 2-azido-2-deoxy-Gal, Man, and Neu5Ac with several aglycons (e.g. OAll, OMe, and STol) [138, 143] (Table 2.8, entry 112). The reactions were carried out using a UV lamp at 245 nm in the presence of CBr_4. The reactions were relatively long, but the wide variety of deprotected molecules and high isolated yields (71–94%) made this an interesting method. Strong Lewis acid (i.e. BCl_3) also selectively removed the 6-*O*-TBDMS ether in 12 min in carbohydrate-based protected δ- and γ-lactones [139] (Table 2.8, entry 113). Tri-*O*-acetyl-D-glucal was first deacylated and then the 6-position was selectively unmasked from per-TBDMS glucal by using commercial HF/pyridine solution. In this case, the procedure for silylation/desilylation was performed "one-pot" and isolated yields were reported after 3 steps [141] (Table 2.8, entry 115).

A single example was reported for the 6-*O*-regioselective removal of TIPS ether from per-triisopropylsilylated phenylselenyl glucopyranose [140] using a mixture of TFA/H_2O/THF at room temperature with good yields (Table 2.8, entry 114).

2.3.2 Regioselective 6-*O*-Deacylation

2.3.2.1 Chemical Methods

Peracylated carbohydrates are among the most common intermediates used because they are readily prepared by acylation of the native and natural saccharides but also because of their good shelf stability allowing long-term storage and their commercial availability. Selective removal of an ester moiety at the primary position of a carbohydrate is a challenging task for organic chemists because of the poor difference in reactivities between primary and secondary esters. Although only two chemical methods have been reported (Table 2.9), a wealth of procedures is available taking advantage of chemoenzymatic methods (Table 2.10). A small number of benzoylated carbohydrates have been used while the vast majority of esters reported are acetates, which are more compatible with esterases or lipases for enzymatic reactions.

Selective removal of the 6-*O*-acetate can be readily achieved with iodine in methanol (Table 2.9, entry 116), a method initially applied to nucleosides but also to glucopyranoses [144]. The $[t\text{BuSnOH(Cl)}]_2$ catalyst was more recently introduced and is easily accessible from *t*-butyl chloride and tin(IV) chloride [163]. The catalyst was investigated toward saccharides among a large family of acetylated alcohols [145] (Table 2.9, entry 117), and the scope of the reaction is very broad. A later report took advantage of this selective deacetylation to access fluorinated ketoheptoses [146] (Table 2.9, entry 118). The same tin-based catalyst led to the total synthesis of capuramycin [147], which proved its versatility in the presence of other protecting groups (Table 2.9, entry 119).

Table 2.9 Chemical regioselective 6-*O*-deacylation of peracylated saccharides.

Entry	Deacylation conditions	Deacylated carbohydrate	Observations	References
116	I_2, MeOH, 70 °C, 5.5 h, 38% (R = OAc) 41% (R = N_3) 43% (R = NPhth)	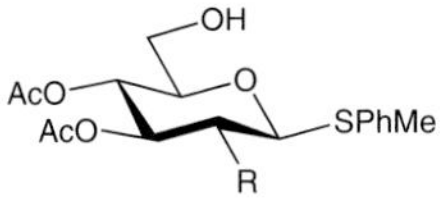	Also applied to nucleosides	[144]
117	[*t*BuSnOH(Cl)]$_2$, THF/MeOH (1/1 v/v), 0–30 °C, 2–24 h, 81–91%	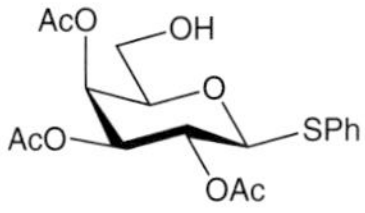	Eight monosaccharides	[145]
118	[*t*BuSnOH(Cl)]$_2$, MeOH, 30 °C, 4 h, 91%	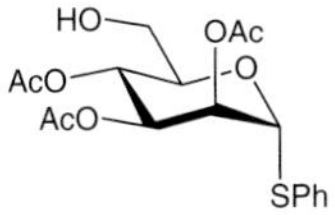	2.5 g scale	[146]
119	[*t*BuSnOH(Cl)]$_2$, MeOH, RT, 12 h, 70%	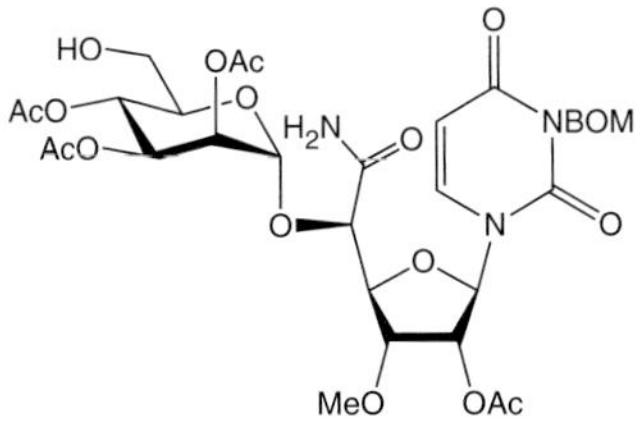	Compatible with other protecting groups	[147]

2.3.2.2 Chemoenzymatic Methods

Chemoenzymatic reactions are the most common and successful methods for the regioselective deacylation of peracylated carbohydrates. A few enzymes have been described in this context, and some recent studies have discussed the importance and influence of the support used for the immobilization of the enzyme [164–166].

The first report was for the 6-*O*-deacylation of methyl glucopyranosides protected with acetates, pentanoates, and octanoates [148] using a lipase from *C. cylindracea* (Table 2.10, entry 120). The reaction was highly dependent on the type of ester used because acetates did not provide any transformation while octanoates led to emulsions under the reaction conditions. Pentanoates were the most suitable esters and provided 90% yield of the desired glucopyranoside (Table 2.10, entry 120). Mannosides and galactosides were regioselectively 6-*O*-deacylated in only 33% and 29% yield, respectively, showing the influence of the hexose configuration on the reaction. Addition of a cosolvent [149] helped improving the reaction with shorter reaction times and acetates being selectively deprotected under these conditions (Table 2.10, entry 121). Octanoates were then made compatible by using DMF [150] (Table 2.10, entry 122). The introduction of 50% [BMIM]PF$_6$ (1-butyl-3-methylimidazolium) ionic liquid as a cosolvent improved the solubility and therefore the isolated yields [151]. But if the ionic liquid is used at 100%, then the regioselectivity is changed in favor of the

Table 2.10 Chemoenzymatic regioselective 6-*O*-deacylation of peracylated saccharides.

Entry	Deacylation conditions	Deacylated carbohydrate	Observations	References
120	Lipase from *Candida cylindracea*, acetone, phosphate buffer (100 mM, pH 7), $CaCl_2$ (3 mM), NaCl (200 mM), RT, 72 h, 90%	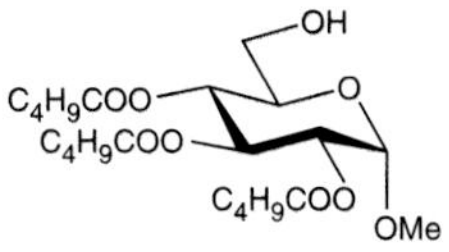	Substrate-dependent pentanoates	[148]
121	Lipase from *Candida cylindracea*, $n\text{Bu}_2\text{O}$/phosphate buffer (100 mM, pH 7) (1/9 v/v), RT, 3 h, 100%	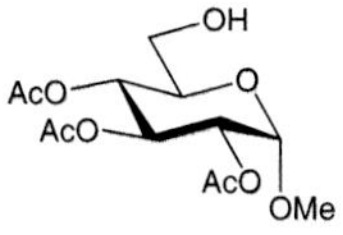	Adding $n\text{Bu}_2\text{O}$ improves rate and acetates are compatible	[149]
122	Lipase from *Candida cylindracea*, DMF/phosphate buffer (50 mM, pH 7) (1/9 v/v), RT, 16 h, 78%	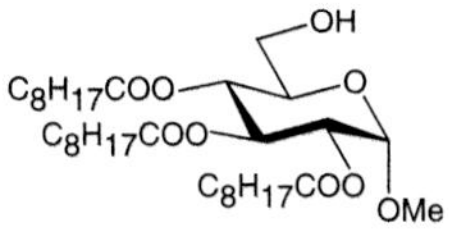	Substrate-dependent pentanoates and octanoates	[150]
123	Lipase from *Candida cylindracea*, [BMIM] PF_6/phosphate buffer (50 mM, pH 7) (1/1 v/v), RT, 24 h, 100%	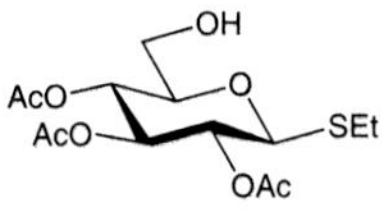	Influence of ionic liquid	[151]
124	Lipase from *Candida cylindracea*, acetone/$i\text{Pr}_2\text{O}$/phosphate buffer (pH 7) (1/1.6/10 v/v), RT, 16 h, 92%	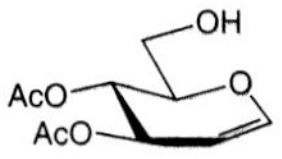	Glucal	[152]
125	Lipase from *Candida rugosa*, phosphate buffer (100 mM, pH 7), RT, 2.5 h, 94%	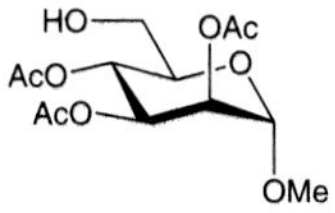	Glc and Man acetates and propanoates Influence of EtOH	[153]
126	Lipase from *Candida rugosa*, MeCN/phosphate buffer (25 mM, pH 5) (1/3 v/v), RT, 19 h, 94%	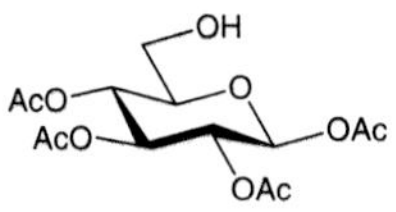	Acyl migration at higher pH > 7 Glc and Gal	[154]
127	Lipase from *Candida rugosa*, MeCN/phosphate buffer (25 mM, pH 5) (15/85 v/v), 25 °C, 2.5 h, 75%	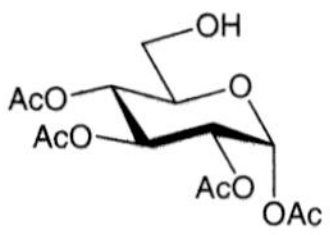	Also selective for ribose	[155]
128	Lipase from *Candida rugosa*, MeCN/phosphate buffer (50 mM, pH 5) (1/4 v/v), 25 °C, 5 h, 98%	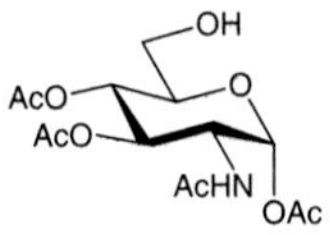	GlcNAc, Gal, glucal and galactal	[156]

Table 2.10 (Continued)

Entry	Deacylation conditions	Deacylated carbohydrate	Observations	References
129	Lipase from *Candida rugosa*, MeCN/sodium acetate (50 mM, pH 5) (1/4 v/v), 25 °C, 96 h, 95%	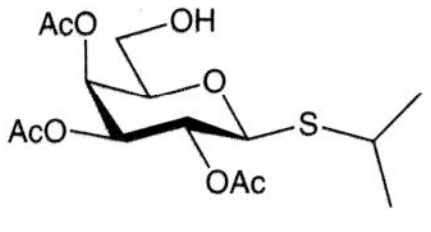	Galactal	[157]
130	Lipase from *Candida rugosa*, dioxane/ phosphate buffer (25 mM, pH 4) (1/4 v/v), 30 °C, 48 h, 95%	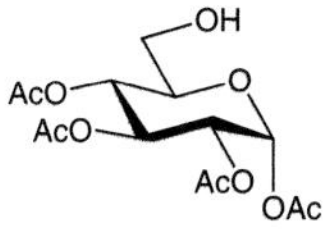	β-OAc unchanged	[158]
131	Lipase from *Candida rugosa*, dioxane/acetate buffer (200 mM, pH 4.8) (1/4 v/v), 40 °C, 40 h, 88%	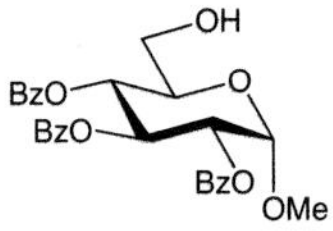	Glc, Gal, Man	[159]
132	Lipase from *Thermomyces lanuginosus*, *i*Pr_2O/ MeOH (98/2 v/v), 48 °C, 24 h, 84%	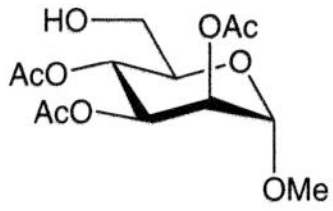	Glc, Gal, Man	[160]
133	Esterase from *Rhodosporidium toruloides*, citrate phosphate buffer (pH 5), 30 °C, 24 h, 80%	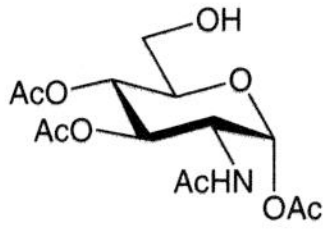	Large scale (8 g)	[161]
134	Esterase from *Rhodosporidium toruloides*, citrate phosphate buffer (pH 5), 30 °C, 20 h, 88%	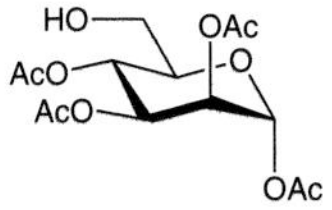	Glc, GlcNAc, Man, Gal, GalNAc	[162]

4-*O*-deacetylated glucopyranoside (Table 2.10, entry 123). This methodology was applied to the synthesis of (1→4)-*C*-disaccharides [167]. Successful adaptation to glucal (Table 2.10, entry 124) was also reported using two cosolvents [152] (i.e. acetone and diisopropyl ether) with further developments toward synthesis of 1,6-oligosaccharides [168] or carbohydrate-based chiral ligands in asymmetric synthesis [169].

Lipase from *Candida rugosa* is the new name for lipase from *C. cylindracea* after 1995 [170], and this enzyme was initially described [153] in an overview of 6-*O*-selective esterification with acetates or pentanoates that also investigated the influence of ethanol in the reaction outcome (Table 2.10, entry 125). A large set of carbohydrates can be regioselectively unmasked at the 6-position by using acetonitrile as a cosolvent (Table 2.10, entries 126–129). The optimal pH for this enzyme is 5 because acyl migration from the 4- to the 6-position was observed at pH > 7 [154]. Some examples in the ribofuranose series [155] or glycals [156, 157] are also reported. Dioxane was also used as a cosolvent [158] (Table 2.10, entry 130), leading to applications in the synthesis of glucose–nucleoside conjugates as anti-HIV prodrugs [171] or glycopolymers [172]. Benzoates could be removed using this enzyme but required the use of dioxane as a cosolvent [159] (Table 2.10, entry 131).

A single example using a lipase from *Thermomyces lanuginosus* was reported (Table 2.10, entry 132), providing good isolated yields of monosaccharides in the gluco, manno, and galacto series [160]. Another esterase from *Rhodosporidium toruloides* gave access to almost all common hexasaccharides [162] with a large-scale (8 g) synthesis described for GlcNAc [161] (Table 2.10, entries 133 and 134).

2.3.3 Regioselective 6-*O*-Debenzylation

The selective 6-*O*-debenzylation of perbenzylated saccharides represents a rapid access method to partially protected derivatives with a primary hydroxyl group available for further chemical transformation with the remaining benzyl ethers acting as stable long-term protecting groups. Benzyl ethers are traditionally removed by hydrogenolysis, but the selective 6-*O*-deprotection requires milder

Table 2.11 Regioselective 6-*O*-debenzylation of perbenzylated saccharides.

Entry	Debenzylation conditions	Debenzylated carbohydrate	Observations	References
135	$CrCl_2$ or $CrCl_3$, LiI, EtOAc/H_2O (9/0.005 v/v), 60 °C, 8 h, 79%	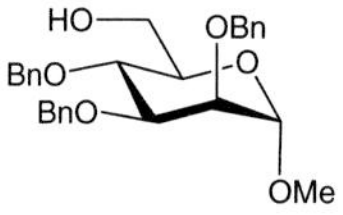	Substrate dependent	[173]
136	1) AcOH, Ac_2O, $ZnCl_2$, RT, 1.5 h 2) NaOMe, MeOH, RT, 5 h, 78% (2 steps) (R^1 = OBn, R^2 = H) or 1) Ac_2O, TMSOTf, CH_2Cl_2, −78 °C, 1 h 2) NaOMe, MeOH, RT, 5 h, 75% (2 steps) (R^1 = H, R^2 = OBn)	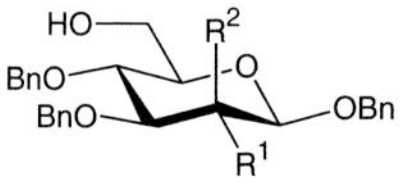	Acetolysis then methanolysis	[174]
137	1) Ac_2O, TMSOTf, CH_2Cl_2, −78 °C, 9 h 2) NaOMe, MeOH, RT, 30 min, 50% (2 steps)	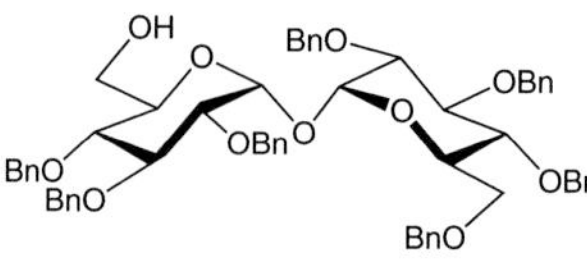	Mono-debenzylation, 34% Recovery of perbenzylated trehalose	[175]
138	DIBAL-H, PhMe, 50 °C, 5 h, 80%	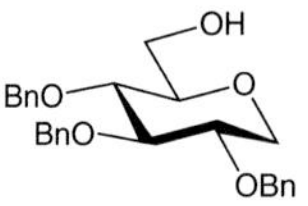	Specific of 1,5-anhydro-glucitol	[176]

conditions to leave the secondary ethers unchanged. $CrCl_2$/LiI was applied for the regioselective removal of benzyl ethers on carbohydrate scaffolds [173] (Table 2.11, entry 135). The method was highly substrate dependent as perbenzylated methyl mannopyranoside was regioselectively deprotected at the primary position while the corresponding methyl glucopyranosides (α or β) were deprotected at the 6- and 3-positions in a mixture of mono-alcohols. This difference in reactivities was attributed to the formation of a three-point coordination with Cr(II) or Cr(III) species, which is possible with manno- but not with gluco-configured carbohydrates. Later on, acetolysis of the primary benzyl ether was achieved with acetic anhydride in acidic conditions with either a protic acid (Table 2.11, entry 136) or a Lewis acid (Table 2.11, entries 136 and 137). The reaction performed on perbenzylated glucose or mannose provided the desired primary alcohol in good yields after removal of the transient acetate [174]. The same reaction with perbenzylated trehalose led to the mono-debenzylated disaccharide in good yield (50%) with substantial recovery of starting material (34%) [175]. Lithium aluminum hydrides have been extensively applied to the regioselective deprotection of cyclodextrins (see Chapter 13) and an example of this use (Table 2.11, entry 138) led to the regioselective 6-*O*-debenzylation of perbenzylated 1,5-anhydro-glucitol in high yield [176]. The difficulty in such selective debenzylations is reflected by the limited number of methods available.

2.4 Regioselective Transformations at the Primary Position

2.4.1 Halogenation

Conversion of alcohols into the corresponding alkyl halides (F, Cl, Br, or I) is of great potential in glycochemistry because halogenated derivatives are useful precursors in the preparation of further functionalized intermediates. Fluorination can be readily achieved by direct methods using diethylaminosulfur trifluoride (DAST) at low temperature and with short reaction time to avoid decomposition [177, 178] (Table 2.12, entries 139 and 140). Chlorination or iodination at the primary position of monosaccharide, disaccharide, and trisaccharide is carried out in the presence of a halogen source such as CCl_4 [179], CBr_4 [181, 183] (Table 2.12, entries 143 and 145), *N*-bromosuccinimide (NBS) [180] (Table 2.12, entry 142), or triphenylphosphine with Br_2 or I_2 [182] (Table 2.12, entries 143 and 144). The same synthetic strategy was extended to disaccharides and trisaccharides such as sucrose [184, 185, 187] (Table 2.12, entries 147–149), trehalose [175, 183, 188, 189, 191, 192], and maltotriose [190]. The use of microwave-assisted halogenations was successfully applied to glucose [179] and sucrose [36, 186] with high yields and reduced reaction times (10–15 min).

2.4.2 Oxidation to the Carboxylic Acid

The selective oxidation of primary alcohol to uronic acids can be achieved by heterogeneous catalysis over platinum [193] (Table 2.13, entry 154). However,

Table 2.12 Regioselective halogenation at the primary position of saccharides.

Entry	Condition	Resulting carbohydrate	Observations	References
139	DAST, CH_2Cl_2, −10 °C, 45 min, 31%	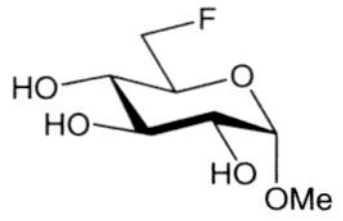	Glc, Man Poor regioselectivity Difluorinated derivatives	[177]
140	DAST, CH_2Cl_2, −40 °C to RT, 45 min, 31%	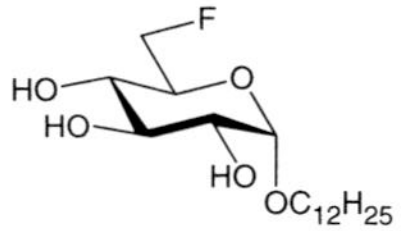	Glc, Man, trehalose Poor selectivity	[178]
141	PPh_3, KCl, pyridine, CCl_4, 10 min, 100 °C, 68–92%	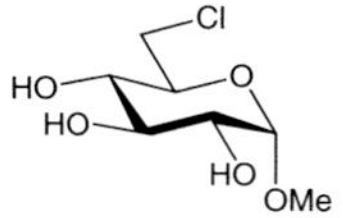	Gram-scale synthesis	[179]
142	NBS, PPh_3, DMF, RT, 48 h	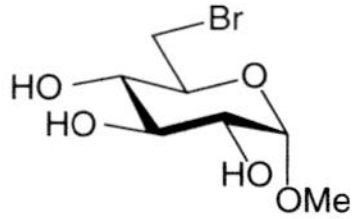	Glc, Man	[180]
143	PPh_3, CBr_4 MeCN, RT, 6 h, 73%	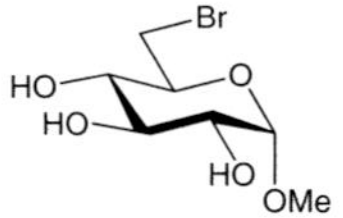	Glc	[181]
144	PPh_3, Br_2 or I_2 in CH_2Cl_2, DMF, RT, 1–2 h, 85–90%	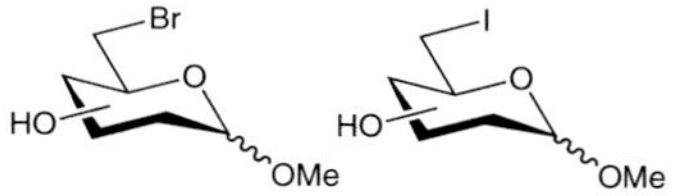	Glc, Gal, Man	[182]

145	1) CBr_4, (chloro-phenylthio-methylene) dimethylammonium chloride (CPMA), DMF, RT, 48 h 2) Acetylation, 80–95% (2 steps)	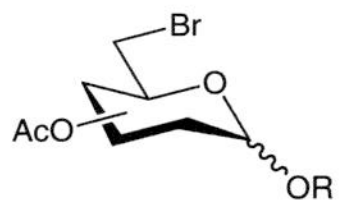	Glc, Man, Gal, lactose, trehalose, cyclodextrins	[183]
146	PPh_3, CCl_4, pyridine, CCl_4, 10 min, μW (70 °C, 300 W), 73–76%	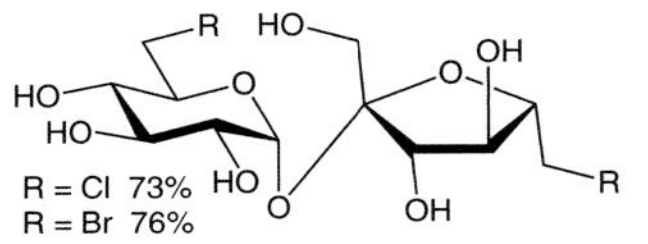	Compatible with silyl ethers	[36]
147	PPh_3, pyridine, KCl, CCl_4, 30 min, 65–70 °C, 71%	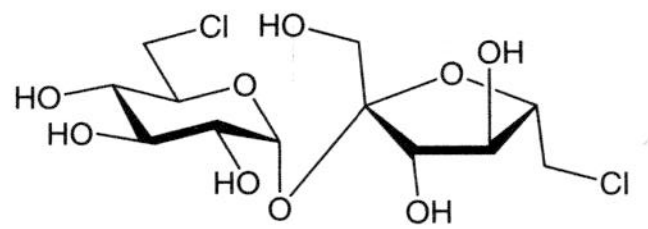	Selective for primary alcohols	[184]
148	PPh_3, I_2, imidazole (R = I) or CBr_4, PhMe (R = Br), 65–70 °C, 6 h, 40–91%	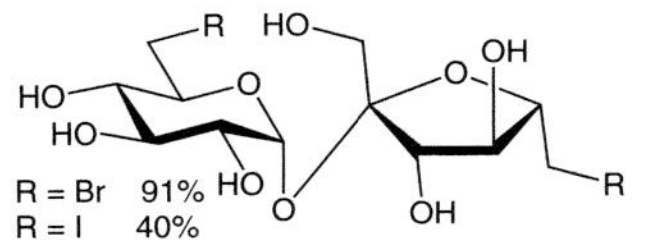	Sucrose	[185]
149	PPh_3, CBr_4, pyridine, μW (90 °C, 400 W), 10 min, 76%	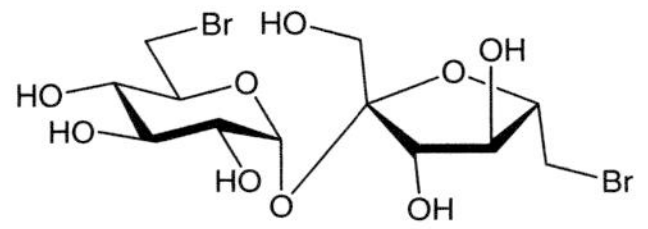	Microwave-assisted halogenation	[186]
150	PPh_3, I_2, imidazole, DMF, 80 °C, 1.5 h, 52%	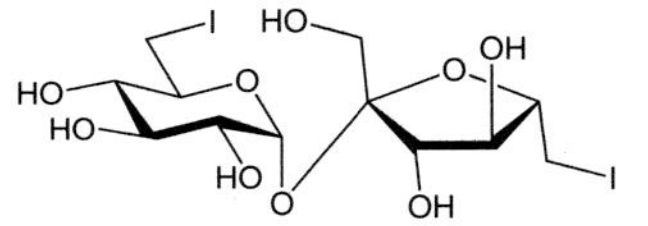	Sucrose	[187]

(Continued)

Table 2.12 (Continued)

Entry	Condition	Resulting carbohydrate	Observations	References
151	1) NBS, DMF, RT, 48 h 2) Acetylation, 58% (2 steps)	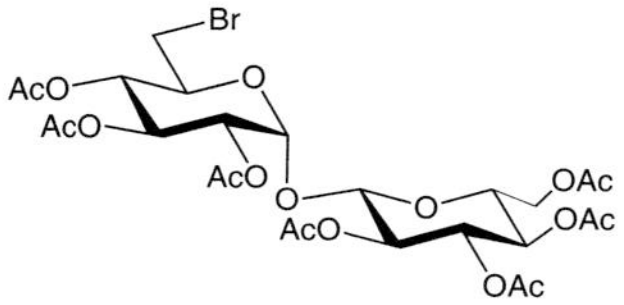	Trehalose Large-scale (10 g) mono-halogenation	[188]
152	PPh_3, I_2, imidazole, DMF, 80–90 °C, 1.5–3.5 h, 71–77%	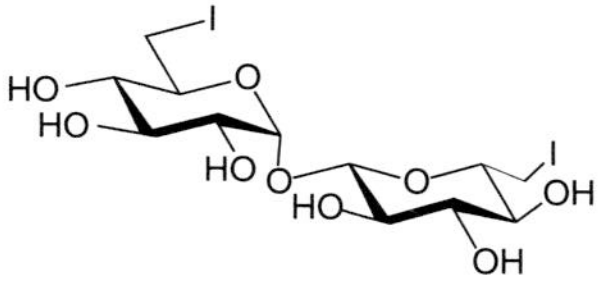	Trehalose Large scale (3 g)	[189]
153	PPh_3, pyridine, CCl_4, 40 °C, 30 min, 95%	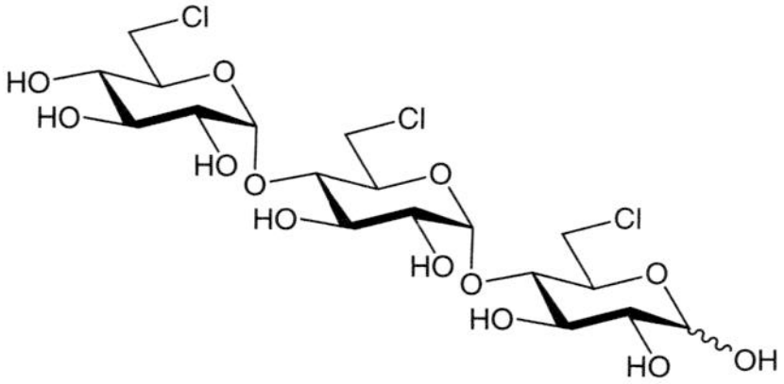	Maltotriose	[190]

the most popular method is the use of the NaOCl/TEMPO ((2,2,6,6-tetramethylpiperidin-1-yl)oxy) system. Indeed, its efficiency was demonstrated not only for unprotected carbohydrates such as galactose in both L- and D- series [194] (Table 2.13, entry 155) or mannose [195] (Table 2.13, entry 156) but also for

Table 2.13 Regioselective oxidation at the primary position of saccharides to carboxylic acids.

Entry	Protection	Protected carbohydrate	Observations	References
154	Pt/O_2, $NaHCO_3$, 70 °C, 50 h, 94%	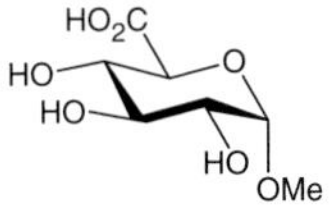	Glc [^{2}H] labeled	[193]
155	TEMPO, NaBr, 5% NaOCl, H_2O, 0 °C, 2–7 h, 50–70%	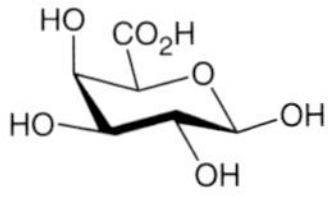	Gal Reaction with α and β in L and D series	[194]
156	1) TEMPO, KBr, 15% NaOCl, H_2O, 0 °C, 1.5 h, 87% 2) Dowex-H^+ resin, H_2O, 100 °C, 16 h, 65%	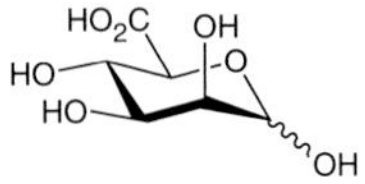	Man Procedure described in supporting information	[195]
157	TEMPO, KBr, NaOCl, 0–3 °C, pH = 10, 5 h, 70%	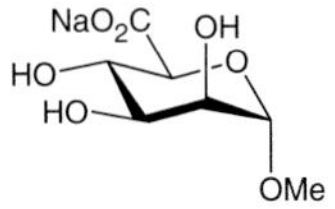	Man Crystal structure	[196]
158	TEMPO, KBr, 12% NaOCl, H_2O, 5 °C, pH = 10.5, 82%	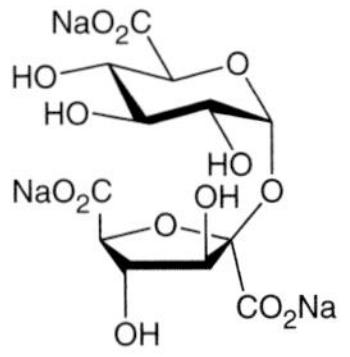	Sucrose Sonication effect on the yield	[197]
159	Polymer-immobilized TEMPO, NaOCl, RT, pH 9.5, 90%	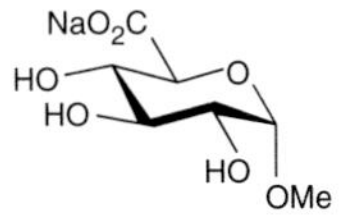	Glc Review on TEMPO reaction	[198]
160	RVC anode/carbon cathode, 20% TEMPO, Na_2CO_3 (400 mM), $NaHCO_3$ (300 mM), 30 mA, 5.5 F mol^{-1} undivided cell, 98%	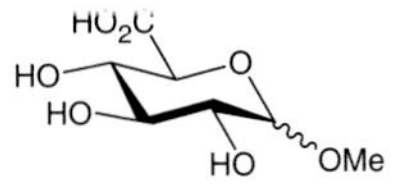	Glc Photovoltaic cell Mixture of anomers	[199]
161	Platinum foil anode/ graphite cathode, TEMPO, carbonate buffer, 0.53 V, 20 °C, 45–96%	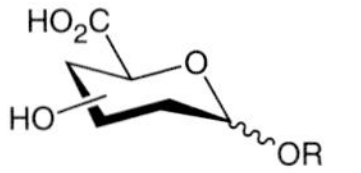 R = Me, C_8H_{17}, $C_{12}H_{25}$, PO_3Na_2	Glc, Gal, Man, salicin Reaction described also with disaccharide, oligosaccharide, and polysaccharide	[200]

nonreducing [196, 201] (Table 2.13, entry 157) and modified carbohydrates [197] (Table 2.13, entry 158). The NaOCl oxidizing agent can also be replaced with *t*-BuOCl in order to avoid the presence of substantial amounts of inorganic salts, which usually hamper purification [202]. The recyclability of TEMPO is also a challenge for sustainable chemistry, and this is why a polymer-immobilized TEMPO has been developed as a recyclable heterogeneous catalyst. In contrast to previously reported systems, bromide is not necessary to afford the oxidized compounds with good yield [198, 203] (Table 2.13, entry 159). Finally, although direct oxidation reactions can be powerful synthetic tools, electrochemical reactions are typically more selective, easily worked up, and scaled up. Different systems have been developed such as reticulated vitreous carbon (RVC) anode/carbon cathode [199] (Table 2.13, entry 160) or platinum foil anode/graphite cathode [200] (Table 2.13, entry 161) both in the presence of TEMPO. These systems are compatible not only with simple monosaccharides (e.g. Glc, Gal, and Man) but also with disaccharides, oligosaccharides, and polysaccharides.

2.4.3 Oxidation to the Aldehyde

The chemoselective oxidation of the primary alcohol of carbohydrates can be achieved with TEMPO and a nonaqueous solvent (e.g. dichloromethane) in order to stop the reaction at the aldehyde stage. However, these solvents are

Table 2.14 Regioselective oxidation at the primary position of saccharides to aldehydes.

Entry	Protection	Protected carbohydrate	Observations	References
162	TEMPO, trichloroisocyanuric acid (TCC), $NaHCO_3$, DMF, 0 °C, 4–9 h, 65–84%		Glc, Gal, Man	[204]
163	$TEMPO^+/BF_4^-$, 2,6-lutidine, DMF, 25 °C, 60%	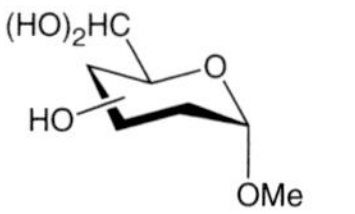	Man, Glc, Gal	[205]
164	Binuclear copper (II) catalyst derived from Cu_2(bpdpo)-TEMPO, O_2, 60 °C, 24 h, 35%	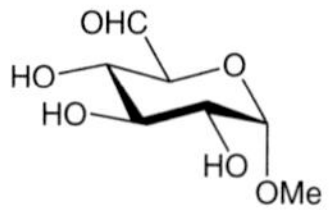	Glc	[206]
165	GO, catalase, O_2, sodium phosphate buffer, pH 7, 4 °C, 14–24 h, >95%	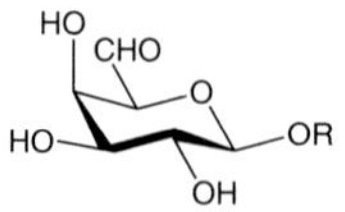 R = Me, CH_2Ph	Limited to Gal	[207]
166	GO, HRP, catalase, H_2O, RT, 24 h	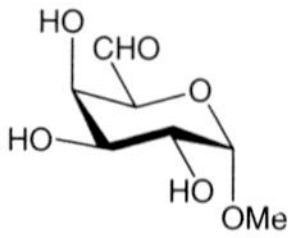	Limited to Gal	[208]

generally incompatible with unprotected carbohydrates because of solubility issues. Alternative solvents were studied, and the most efficient system for this transformation appeared to be DMF with TEMPO, NaOCl as the base and trichloroisocyanuric acid (TCC) as a secondary oxidant [204] (Table 2.14, entry 162). It was also demonstrated that the use of $TEMPO^+/BF_4^-$ salt in DMF afforded aldehyde derivatives in their hydrated forms [205] (Table 2.14, entry 163). Moreover, the undesired oxidation of primary alcohols to the carboxylic acid was overcome by the development of a binuclear copper(II)/TEMPO catalyst, leading to the desired aldehyde [206] (Table 2.14, entry 164). The bioconversion of galactose into aldehyde derivatives is also possible using galacto-oxidase (GO) in the presence of peroxidase and catalase [207, 208] (Table 2.14, entries 165 and 166), and this approach was used for labeled carbohydrates [209]. However, this reaction is restricted to galactose.

2.5 Summary and Conclusions

The preparation of a few milligrams of key synthetic building blocks is crucial for the preparation of therapeutic oligosaccharide targets. Avoiding several protection and deprotection steps can result in a reduction of time, effort, money, materials, and energy. In this survey, we have described both chemical and chemoenzymatic approaches allowing the selective regioselective protection, deprotection, and functionalization of the 6-position of both simple carbohydrates and complex oligosaccharides. Particular emphasis has been placed on classical approaches to providing the protected monosaccharide or disaccharide at the primary position in the absence of other protecting groups at the secondary positions (except for the anomeric position). Additionally, activation of the primary alcohols with hindered sulfonates (e.g. tosylates) was discussed as an approach to offering access to further chemical modifications at this position. A description of the extensive use of enzymatic reactions for the regioselective introduction or removal of a protecting group at the 6-position has been covered. Finally, oxidation to the aldehyde or carboxylic acid was also discussed.

2.6 Experimental Section

2.6.1 Typical Experimental Procedure for Regioselective Silylation [23] (Table 2.3, Entry 21)

Triisopropyl chlorosilane (TIPSCl, 4 ml, 18.32 mmol, 1.1 eq.) was added dropwise to a stirred suspension of D-galactose (3 g, 16.65 mmol, 1 eq.), imidazole (2.5 g, 36.63 mmol, 2.2 eq.), and DMAP (0.2 g, 1.66 mmol, 0.1 eq.) in anhydrous DMF (16.7 ml). After stirring at RT during 29 h, the reaction mixture was diluted with methanol (40 ml) and concentrated under reduced pressure to give a pale yellow syrup. Silica gel column chromatography (5% MeOH in EtOAc) gave 6-*O*-TIPS-D-galactopyranose (3.23 g, 58%) as a colorless solid.

2.6.2 Typical Experimental Procedure for Desilylation [52] (Table 2.4, Entry 46)

To a solution of silyl ether (523 mg, 0.67 mmol) in THF (5.5 ml) at 0 °C was added dropwise a solution of tetrabutylammonium fluoride (0.739 mmol, 1 M in THF). After stirring for 9 h at room temperature, the reaction was quenched with saturated aqueous $NaHCO_3$ and extracted with EtOAc. The combined organic layers were dried ($MgSO_4$), filtered, and evaporated. Flash chromatography (EtOAc/hexane 1/3) afforded the desilylated product (321 mg, 88%) as a white solid.

Abbreviations

Ac	acetyl
All	allyl
BMIM	1-butyl-3-methylimidazolium
Boc	*t*-butoxycarbonyl
Bz	benzoyl
CAL-B	Novozym 435, *Candida antarctica* lipase B
Cat.	catalytic
Conc.	concentrated
CPMA	(chloro-phenylthio-methylene)dimethylammonium chloride
DABCO	1,4-diazabicyclo[2.2.2]octane
DAST	diethylaminosulfur trifluoride
DIBAL-H	diisobutylaluminium hydride
DMAP	4-dimethylaminopyridine
DMF	dimethylformamide
DMSO	dimethylsulfoxide
eq.	equivalent
Gal	galactose
Glc	glucose
GlcNAc	*N*-acetylglucosamine
GO	galacto-oxidase
Gul	gulose
HRP	horseradish peroxidase
Man	mannose
ManNAc	*N*-acetylmannosamine
NBS	*N*-bromosuccinimide
Neu5Ac	*N*-acetyl neuraminic acid
Piv	pivaloyl
RT	room temperature
RVC	reticulated vitreous carbon
TBAF	tetrabutylammonium fluoride
TBDMS	*t*-butyldimethylsilyl
TBDPS	*t*-butyldiphenylsilyl
TBTU	*O*-(benzotriazol-1-yl)-*N,N,N′,N′*-tetramethyluronium tetrafluoroborate

TCC	trichloroisocyanuric acid
TEMPO	(2,2,6,6-tetramethylpiperidin-1-yl)oxy
TES	triethylsilyl
Tf	trifluoromethanesulfonyl
TFA	trifluoroacetic acid
THF	tetrahydrofuran
TIPS	triisopropylsilyl
TMS	trimethylsilyl
Tol	toluyl
Tr	trityl, triphenylmethyl
Ts	tosyl, *p*-toluenesulfonyl

References

1 Koto, S., Morishima, N., Yoshida, T. et al. (1983). *Bull. Chem. Soc. Jpn.* 56: 1171–1175.
2 Ogawa, T. and Yamamoto, H. (1985). *Carbohydr. Res.* 137: 79–88.
3 Liu, L.-D. and Liu, H.-W. (1989). *Tetrahedron Lett.* 30: 35–38.
4 Patil, P.R. and Ravindranathan Kartha, K.P. (2008). *J. Carbohydr. Chem.* 27: 279–293.
5 Kováč, P., Sokoloski, E.A., and Glaudemans, C.P.J. (1984). *Carbohydr. Res.* 128: 101–109.
6 Du, Y., Zhang, M., and Kong, F. (2000). *Org. Lett.* 2: 3797–3800.
7 Deranieh, R.M., Greenberg, M.L., Le Calvez, P.-B. et al. (2012). *Org. Biomol. Chem.* 10: 9601–9619.
8 Wing, C., Errey, J.C., Mukhopadhyay, B. et al. (2006). *Org. Biomol. Chem.* 4: 3945–3950.
9 Amigues, E.J., Greenberg, M.L., Ju, S. et al. (2007). *Tetrahedron* 63: 10042–10053.
10 Gadakh, B.K., Patil, P.R., Malik, S., and Kartha, K.P.R. (2009). *Synth. Commun.* 39: 2430–2438.
11 Lamandé-Langle, S., Collet, C., Hensienne, R. et al. (2014). *Bioorg. Med. Chem.* 22: 6672–6683.
12 Kanemaru, M., Yamamoto, K., and Kadokawa, J.-I. (2012). *Carbohydr. Res.* 357: 32–40.
13 Esswein, A., Rembold, H., and Schmidt, R.R. (1990). *Carbohydr. Res.* 200: 287–305.
14 Di Bussolo, V., Caselli, M., Pineschi, M., and Crotti, P. (2002). *Org. Lett.* 4: 3695–3698.
15 Lee, S., Lim, D., Lee, E. et al. (2014). *J. Med. Chem.* 57: 7425–7434.
16 Yang, Z., Lai-Ming Wong, E., Yuen-Ting Shum, T. et al. (2005). *Org. Lett.* 7: 669–672.
17 Kim, Y., Kim, J., Oh, K. et al. (2012). *ACS Med. Chem. Lett.* 3: 151–154.
18 Ziegler, T., Eckhardt, E., and Keller, D. (1997). *J. Carbohydr. Chem.* 16: 719–738.
19 Roy, B., Verma, P., and Mukhopadhyay, B. (2009). *Carbohydr. Res.* 344: 145–148.

20 Castro, M.J.L., Kovensky, J., and Cirelli, A.F. (1997). *Tetrahedron Lett.* 38: 3995–3998.
21 Gordon, D.M. and Danishefsky, S.J. (1992). *J. Am. Chem. Soc.* 114: 659–663.
22 Park, T.K., Kim, I.J., Hu, S. et al. (1996). *J. Am. Chem. Soc.* 118: 11488–11500.
23 Craig, D., Munasinghe, V.R.N., Tierney, J.P. et al. (1999). *Tetrahedron* 55: 15025–15044.
24 Pei, Z., Dong, H., Caraballo, R., and Ramström, O. (2007). *Eur. J. Org. Chem.* 4927–4934.
25 Parhi, A.K., Mootoo, D.R., and Franck, R.W. (2008). *Tetrahedron* 64: 9821–9827.
26 Villalobos, A. and Danishefsky, S.J. (1989). *J. Org. Chem.* 54: 12–15.
27 Chambers, D.J., Evans, G.R., and Fairbanks, A.J. (2005). *Tetrahedron* 61: 7184–7192.
28 Kishida, M. and Akita, H. (2005). *Tetrahedron: Asymmetry* 16: 2625–2630.
29 Boto, A., Hernández, D., Hernández, R., and Suárez, E. (2006). *J. Org. Chem.* 71: 1938–1948.
30 Dasgupta, S. and Nitz, M. (2011). *J. Org. Chem.* 76: 1918–1921.
31 Mallet, A., Mallet, J.-M., and Sinay, P. (1994). *Tetrahedron: Asymmetry* 5: 2593–2608.
32 Belakhov, V., Dovgolevsky, E., Rabkin, E. et al. (2004). *Carbohydr. Res.* 339: 385–392.
33 Kantchev, E.A.B. and Parquette, J.R. (2005). *Synlett* 1567–1570.
34 Dasgupta, S., Pramanik, K., and Mukhopadhyay, B. (2007). *Tetrahedron* 63: 12310–12316.
35 Bartoszewicz, A., Kalek, M., and Stawinski, J. (2008). *Tetrahedron* 64: 8843–8850.
36 Barros, M.T., Petrova, K.T., Correia-da-Silva, P., and Potewar, T.M. (2011). *Green Chem.* 13: 1897–1906.
37 Nicolaou, K.C., Mitchell, H.J., Jain, N.F. et al. (1999). *Chem. Eur. J.* 5: 2648–2667.
38 Johnson, D.A. and Taubner, L.M. (1996). *Tetrahedron Lett.* 37: 605–608.
39 Schell, P., Orgueira, H.A., Roehrig, S., and Seeberger, P.H. (2001). *Tetrahedron Lett.* 42: 3811–3814.
40 Costantino, V., Fattorusso, E., Imperatore, C., and Mangoni, A. (2002). *Tetrahedron* 58: 369–375.
41 Jacquinet, J.-C. (1990). *Carbohydr. Res.* 199: 153–181.
42 Danishefsky, S.J., Gervay, J., Peterson, J.M. et al. (1995). *J. Am. Chem. Soc.* 117: 1940–1953.
43 Lellouche, J.-P. and Koeller, S. (2001). *J. Org. Chem.* 66: 693–696.
44 Potopnyk, M.A., Cmoch, P., and Jarosz, S. (2012). *Org. Lett.* 14: 4258–4261.
45 Aurrecoechea, J.M., Gil, J.H., and López, B. (2003). *Tetrahedron* 59: 7111–7121.
46 Baba, T., Huang, G., and Isobe, M. (2003). *Tetrahedron* 59: 6851–6872.
47 Sabitha, G., Reddy, S.S.S., and Yadav, J.S. (2010). *Tetrahedron Lett.* 51: 6259–6261.
48 Ravindranathan Kartha, K.P. and Field, R.A. (1999). *Synlett* 311–312.
49 Valverde, S., Garcia, M., Gomez, A.M., and Lopez, J.C. (2000). *Chem. Commun.* 813–814.
50 Falconer, R.A. (2002). *Tetrahedron Lett.* 43: 8503–8505.
51 Mouffouk, F., Morère, A., Vidal, S. et al. (2004). *Synth. Commun.* 34: 303–307.

52 Berkowitz, D.B., Bose, M., Pfannenstiel, T.J., and Doukov, T. (2000). *J. Org. Chem.* 65: 4498–4508.
53 Jeong, Y.J., Lee, J.H., Park, E.S., and Yoon, C.M. (2002). *J. Chem. Soc., Perkin Trans.* 1: 1223–1225.
54 Kishore Kumar, G.D. and Baskaran, S. (2005). *J. Org. Chem.* 70: 4520–4523.
55 Ellis, D., Norman, S.E., and Osborn, H.M.I. (2008). *Tetrahedron* 64: 2832–2854.
56 Fokt, I., Skora, S., Conrad, C. et al. (2013). *Carbohydr. Res.* 368: 111–119.
57 Yan, S., Ding, N., Zhang, W. et al. (2012). *Carbohydr. Res.* 354: 6–20.
58 Nashed, E.M. and Glaudemans, C.P.J. (1987). *J. Org. Chem.* 52: 5255–5260.
59 Sivapriya, K., Suguna, P., and Chandrasekaran, S. (2007). *Tetrahedron Lett.* 48: 2091–2095.
60 Almant, M., Moreau, V., Kovensky, J. et al. (2011). *Chem. Eur. J.* 17: 10029–10038.
61 Gabrielli, L., Calloni, I., Donvito, G. et al. (2014). *Eur. J. Org. Chem.* 2014: 5962–5967.
62 Branchaud, B.P. and Meier, M.S. (1989). *J. Org. Chem.* 54: 1320–1326.
63 Jeong, L.S., Schinazi, R.F., Beach, J.W. et al. (1993). *J. Med. Chem.* 36: 2627–2638.
64 Chu, C.K., Beach, J.W., Jeong, L.S. et al. (1991). *J. Org. Chem.* 56: 6503–6505.
65 Beach, J.W., Jeong, L.S., Alves, A.J. et al. (1992). *J. Org. Chem.* 57: 2217–2219.
66 Maschauer, S., Haubner, R., Kuwert, T., and Prante, O. (2014). *Mol. Pharmaceutics* 11: 505–515.
67 Haskins, W.T., Hann, R.M., and Hudson, C.S. (1946). *J. Am. Chem. Soc.* 68: 628–632.
68 Basava, V., Gorun, S.M., and Marzabadi, C.H. (2014). *Carbohydr. Res.* 391: 106–111.
69 Gomez, A.M., Casillas, M., Rodriguez, B. et al. (2010). *Arkivoc* (iii): 288–302.
70 Gelmi, M.L., Fontana, G., Pocar, D. et al. (2007). *J. Med. Chem.* 50: 2245–2248.
71 Georges, M. and Fraser-Reid, B. (1984). *Carbohydr. Res.* 127: 162–164.
72 Lopez, O.L., Fernandez-Bolanos, J.G., Lillelund, V.H., and Bols, M. (2003). *Org. Biomol. Chem.* 1: 478–482.
73 Blattner, R., Furneaux, R.H., Kemmitt, T. et al. (1994). *J. Chem. Soc., Perkin Trans. 1* 3411–3421.
74 Ueda, M., Sawai, Y., Wada, Y., and Yamamura, S. (2000). *Tetrahedron* 56: 5123–5130.
75 Nicolaou, K.C., Caulfield, T.J., Kataoka, H., and Stylianides, N.A. (1990). *J. Am. Chem. Soc.* 112: 3693–3695.
76 Okada, Y., Mukae, T., Okajima, K. et al. (2007). *Org. Lett.* 9: 1573–1576.
77 Ishihara, K., Kurihara, H., and Yamamoto, H. (1993). *J. Org. Chem.* 58: 3791–3793.
78 Thompson, A.J., Williams, R.J., Hakki, Z. et al. (2012). *Proc. Natl. Acad. Sci. U.S.A.* 109: 781–786.
79 Lu, Y., Wei, P., Pei, Y. et al. (2014). *Green Chem.* 16: 4510–4514.
80 Baczko, K. and Plusquellec, D. (1991). *Tetrahedron* 47: 3817–3828.
81 Sawada, Y., Nanboku, N., Yanase, E., and Nakatsuka, S.-I. (2010). *Heterocycl. Commun.* 16: 21–24.
82 Morcuende, A., Valverde, S., and Herradón, B. (1994). *Synlett* 89–91.
83 Demizu, Y., Kubo, Y., Miyoshi, H. et al. (2008). *Org. Lett.* 10: 5075–5077.

84 Twibanire, J.-D.A.K. and Grindley, T.B. (2011). *Org. Lett.* 13: 2988–2991.
85 Sugahara, K., Satake, N., Kamata, K. et al. (2014). *Angew. Chem. Int. Ed.* 53: 13248–13252.
86 Liang, P.-H., Lu, Y.-J., and Tang, T.-H. (2010). *Tetrahedron Lett.* 51: 6928–6931.
87 Therisod, M. and Klibanov, A.M. (1986). *J. Am. Chem. Soc.* 108: 5638–5640.
88 Wang, Y.F., Lalonde, J.J., Momongan, M. et al. (1988). *J. Am. Chem. Soc.* 110: 7200–7205.
89 Gotor, V. and Pulido, R. (1991). *J. Chem. Soc., Perkin Trans. 1* 491–492.
90 Kim, M.J., Hennen, W.J., Sweers, H.M., and Wong, C.H. (1988). *J. Am. Chem. Soc.* 110: 6481–6486.
91 Liu, J.L.C., Shen, G.J., Ichikawa, Y. et al. (1992). *J. Am. Chem. Soc.* 114: 3901–3910.
92 Holla, E.W. (1989). *Angew. Chem. Int. Ed.* 28: 220–221.
93 Zhang, X., Kamiya, T., Otsubo, N. et al. (1999). *J. Carbohydr. Chem.* 18: 225–239.
94 Gelo-Pujic, M., Guibe-Jampel, E., Loupy, A. et al. (1996). *J. Chem. Soc., Perkin Trans. 1* 2777–2780.
95 Ljunger, G., Adlercreutz, P., and Mattiasson, B. (1994). *Biotechnol. Lett* 16: 1167–1172.
96 Favrelle, A., Boyère, C., Laurent, P. et al. (2011). *Carbohydr. Res.* 346: 1161–1164.
97 Otto, R.T., Bornscheuer, U.T., Syldatk, C., and Schmid, R.D. (1998). *J. Biotechnol.* 64: 231–237.
98 Adelhorst, K., Björkling, F., Godtfredsen, S.E., and Kirk, O. (1990). *Synthesis* 112–115.
99 Davis, R.A., Lin, C.-H., and Gervay-Hague, J. (2012). *Chem. Commun.* 9083–9085.
100 Afach, G., Kawanami, Y., and Izumori, K. (2005). *Biosci. Biotechnol., Biochem.* 69: 833–835.
101 Ferrer, M., Angeles Cruces, M., Plou, F.J. et al. (2000). *Tetrahedron* 56: 4053–4061.
102 Albertin, L., Stenzel, M., Barner-Kowollik, C. et al. (2004). *Macromolecules* 37: 7530–7537.
103 Petrova, K.T., Dey, S.S., and Barros, M.T. (2015). *Carbohydr. Polym.* 125: 281–287.
104 Mann, D., Chattopadhyay, S., Pargen, S. et al. (2014). *RSC Adv.* 4: 62878–62881.
105 Boyère, C., Favrelle, A., Broze, G. et al. (2011). *Carbohydr. Res.* 346: 2121–2125.
106 Danieli, B., Bertario, A., Carrea, G. et al. (1993). *Helv. Chim. Acta* 76: 2981–2991.
107 Wu, Q., Lu, D., Cai, Y. et al. (2001). *Biotechnol. Lett* 23: 1981–1985.
108 Polat, T., Bazin, H.G., and Linhardt, R.J. (1997). *J. Carbohydr. Chem.* 16: 1319–1325.
109 Ferrer, M., Cruces, M.A., Bernabé, M. et al. (1999). *Biotechnol. Bioeng.* 65: 10–16.
110 Potier, P., Bouchu, A., Descotes, G., and Queneau, Y. (2000). *Tetrahedron Lett.* 41: 3597–3600.
111 Plou, F.J., Cruces, M.A., Ferrer, M. et al. (2002). *J. Biotechnol.* 96: 55–66.

112 Pulido, R., Ortiz, F.L., and Gotor, V. (1992). *J. Chem. Soc., Perkin Trans. 1* 2891–2898.

113 Riva, S., Chopineau, J., Kieboom, A.P.G., and Klibanov, A.M. (1988). *J. Am. Chem. Soc.* 110: 584–589.

114 Boyere, C., Favrelle, A., Leonard, A.F. et al. (2013). *J. Mater. Chem. A* 1: 8479–8487.

115 Bjorkling, F., Godtfredsen, S.E., and Kirk, O. (1989). *Chem. Commun.* 934–935.

116 Danieli, B., Luisetti, M., Sampognaro, G. et al. (1997). *J. Mol. Catal. B* 3: 193–201.

117 Cao, L., Bornscheuer, U.T., and Schmid, R.D. (1999). *J. Mol. Catal. B* 6: 279–285.

118 Kitagawa, M., Raku, T., Shimakawa, H. et al. (2002). *Macromol. Biosci.* 2: 233–237.

119 Rich, J.O., Bedell, B.A., and Dordick, J.S. (1995). *Biotechnol. Bioeng.* 45: 426–434.

120 Kim, J.E., Han, J.J., Yoon, J.H., and Rhee, J.S. (1998). *Biotechnol. Bioeng.* 57: 121–125.

121 Riva, S., Nonini, M., Ottolina, G., and Danieli, B. (1998). *Carbohydr. Res.* 314: 259–266.

122 Potier, P., Bouchu, A., Gagnaire, J., and Queneau, Y. (2001). *Tetrahedron: Asymmetry* 12: 2409–2419.

123 Molinier, V., Kouwer, P.J.J., Fitremann, J. et al. (2006). *Chem. Eur. J.* 12: 3547–3557.

124 Park, S. and Kazlauskas, R.J. (2001). *J. Org. Chem.* 66: 8395–8401.

125 Li, L., Ji, F., Wang, J. et al. (2015). *Carbohydr. Res.* 416: 51–58.

126 Ganske, F. and Bornscheuer, U.T. (2005). *Org. Lett.* 7: 3097–3098.

127 Vaillant, O., Cheikh, K.E., Warther, D. et al. (2015). *Angew. Chem. Int. Ed.* 54: 5952–5956.

128 Datta, A.K., Takayama, K., Nashed, M.A., and Anderson, L. (1991). *Carbohydr. Res.* 218: 95–109.

129 Schiefelbein, L., Keller, M., Weissmann, F. et al. (2010). *Eur. J. Pharm. Biopharm.* 76: 342–350.

130 Zeng, X., Smith, R., and Zhu, X. (2013). *J. Org. Chem.* 78: 4165–4170.

131 Sarpe, V.A. and Kulkarni, S.S. (2014). *Org. Lett.* 16: 5732–5735.

132 Liu, C. and Mariano, P.S. (2015). *Tetrahedron Lett.* 56: 3008–3010.

133 Meldal, M., Christensen, M.K., and Bock, K. (1992). *Carbohydr. Res.* 235: 115–127.

134 Fernández, C., Nieto, O., Rivas, E. et al. (2000). *Carbohydr. Res.* 327: 353–365.

135 Jervis, P.J., Cox, L.R., and Besra, G.S. (2011). *J. Org. Chem.* 76: 320–323.

136 Sevrain, C.M., Haelters, J.-P., Chantome, A. et al. (2013). *Org. Biomol. Chem.* 11: 4479–4487.

137 Cui, Y., Cheng, Z., Mao, J., and Yu, Y. (2013). *Tetrahedron Lett.* 54: 3831–3833.

138 Chen, M., Lu, K., Shih-Yuan Lee, A., and Lin, C.C. (2002). *Tetrahedron Lett.* 43: 2777–2780.

139 Yang, Y.-Y., Yang, W.-B., Teo, C.-F., and Lin, C.-H. (2000). *Synlett* 1634–1636.

140 Abe, H., Terauchi, M., Matsuda, A., and Shuto, S. (2003). *J. Org. Chem.* 68: 7439–7447.
141 Anquetin, G., Rawe, S.L., McMahon, K. et al. (2008). *Chem. Eur. J.* 14: 1592–1600.
142 Narouz, M.R., Soliman, S.E., Bassily, R.W. et al. (2013). *Synlett* 2271–2273.
143 Chen, M.-Y., Patkar, L.N., Lu, K.-C. et al. (2004). *Tetrahedron* 60: 11465–11475.
144 Ren, B., Cai, L., Zhang, L.-R. et al. (2005). *Tetrahedron Lett.* 46: 8083–8086.
145 Orita, A., Hamada, Y., Nakano, T. et al. (2001). *Chem. Eur. J.* 7: 3321–3327.
146 Waschke, D., Leshch, Y., Thimm, J. et al. (2012). *Eur. J. Org. Chem.* 948–959.
147 Kurosu, M., Li, K., and Crick, D.C. (2009). *Org. Lett.* 11: 2393–2396.
148 Sweers, H.M. and Wong, C.H. (1986). *J. Am. Chem. Soc.* 108: 6421–6422.
149 Kloosterman, M., Mosmuller, E.W.J., Schoemaker, H.E., and Meijer, E.M. (1987). *Tetrahedron Lett.* 28: 2989–2992.
150 Hennen, W.J., Sweers, H.M., Wang, Y.F., and Wong, C.H. (1988). *J. Org. Chem.* 53: 4939–4945.
151 Gervaise, C., Daniellou, R., Nugier-Chauvin, C., and Ferrières, V. (2009). *Tetrahedron Lett.* 50: 2083–2085.
152 Crotti, P., Di Bussolo, V., Favero, L. et al. (2002). *Tetrahedron* 58: 6069–6091.
153 Hsiao, K.-F., Yang, F.-L., Wu, S.-H., and Wang, K.-T. (1995). *Biotechnol. Lett* 17: 963–968.
154 Terreni, M., Salvetti, R., Linati, L. et al. (2002). *Carbohydr. Res.* 337: 1615–1621.
155 Fernandez-Lorente, G., Palomo, J.M., Cocca, J. et al. (2003). *Tetrahedron* 59: 5705–5711.
156 Filice, M., Palomo, J.M., Bonomi, P. et al. (2008). *Tetrahedron* 64: 9286–9292.
157 Rodrigues, D.S., Mendes, A.A., Filice, M. et al. (2009). *J. Mol. Catal. B* 58: 36–40.
158 Rodríguez-Pérez, T., Lavandera, I., Fernández, S. et al. (2007). *Eur. J. Org. Chem.* 2769–2778.
159 Esmurziev, A., Sundby, E., and Hoff, B.H. (2009). *Eur. J. Org. Chem.* 1592–1597.
160 Hietanen, A. and Kanerva, L.T. (2012). *Eur. J. Org. Chem.* 2729–2737.
161 Chaplin, D., Crout, D.H.G., Bornemann, S. et al. (1992). *J. Chem. Soc., Perkin Trans. 1* 235–237.
162 Horrobin, T., Hao Tran, C., and Crout, D. (1998). *J. Chem. Soc., Perkin Trans. 1* 1069–1080.
163 Puff, H., Hevendehl, H., Höfer, K. et al. (1985). *J. Organomet. Chem.* 287: 163–178.
164 Mendes, A.A., Rodrigues, D.S., Filice, M. et al. (2008). *Tetrahedron* 64: 10721–10727.
165 Palomo, J.M., Filice, M., Fernandez-Lafuente, R. et al. (2007). *Adv. Synth. Catal.* 349: 1969–1976.
166 Fernandez-Lorente, G., Filice, M., Terreni, M. et al. (2008). *J. Mol. Catal. B* 51: 110–117.
167 Levoirier, E., Canac, Y., Norsikian, S., and Lubineau, A. (2004). *Carbohydr. Res.* 339: 2737–2747.
168 Di Bussolo, V., Checchia, L., Romano, M.R. et al. (2008). *Org. Lett.* 10: 2493–2496.

169 Grugel, H., Albrecht, F., and Boysen, M.M.K. (2014). *Adv. Synth. Catal.* 356: 3289–3294.

170 Bashir, N.B., Phythian, S.J., Reason, A.J., and Roberts, S.M. (1995). *J. Chem. Soc., Perkin Trans. 1* 2203–2222.

171 Rodríguez-Pérez, T., Fernández, S., Sanghvi, Y.S. et al. (2010). *Bioconjugate Chem.* 21: 2239–2249.

172 Pearson, S., Lu, H., and Stenzel, M.H. (2015). *Macromolecules* 48: 1065–1076.

173 Falck, J.R., Barma, D.K., Venkataraman, S.K. et al. (2002). *Tetrahedron Lett.* 43: 963–966.

174 Lu, W., Navidpour, L., and Taylor, S.D. (2005). *Carbohydr. Res.* 340: 1213–1217.

175 Wang, M., Xu, Z., Tu, P. et al. (2004). *Bioorg. Med. Chem. Lett.* 14: 2585–2588.

176 Lecourt, T., Herault, A., Pearce, A.J. et al. (2004). *Chem. Eur. J.* 10: 2960–2971.

177 Somawardhana, C.W. and Brunngraber, E.G. (1983). *Carbohydr. Res.* 121: 51–60.

178 Miethchen, R. and Zur, C. (1995). *J. Prakt. Chem.* 337: 189–195.

179 Cléophax, J., Liagre, M., Loupy, A., and Petit, A. (2000). *Org. Process Res. Dev.* 4: 498–504.

180 Nakane, M., Hutchinson, C.R., and Gollman, H. (1980). *Tetrahedron Lett.* 21: 1213–1216.

181 Cicero, D., Varela, O., and De Lederkremer, R.M. (1990). *Tetrahedron* 46: 1131–1144.

182 Leon-Ruaud, P. and Plusquellec, D. (1991). *Tetrahedron* 47: 5185–5192.

183 Xue, W. and Zhang, L. (2012). *Chin. J. Chem.* 30: 1429–1432.

184 Chen, C.-C., Whistler, R.L., and Daniel, J.R. (1983). *Carbohydr. Res.* 117: 318–321.

185 Hough, L., Sinchareonkul, L.V., Richardson, A.C. et al. (1988). *Carbohydr. Res.* 174: 145–160.

186 Potewar, T.M., Petrova, K.T., and Barros, M.T. (2013). *Carbohydr. Res.* 379: 60–67.

187 García Fernández, J.M., Gadelle, A., and Defaye, J. (1994). *Carbohydr. Res.* 265: 249–269.

188 Wang, M., Tu, P.-F., Xu, Z.-D. et al. (2003). *Helv. Chim. Acta* 86: 2637–2644.

189 García Fernández, J., Ortiz Mellet, C., Jiménez Blanco, J. et al. (1995). *Carbohydr. Res.* 268: 57–71.

190 Thiebault, N., Lesur, D., Godé, P. et al. (2008). *Carbohydr. Res.* 343: 2719–2728.

191 Koga, K., Ishida, K., Yamada, T. et al. (1999). *Tetrahedron Lett.* 40: 923–926.

192 Srinivasachari, S., Liu, Y., Zhang, G. et al. (2006). *J. Am. Chem. Soc.* 128: 8176–8184.

193 Heinkele, G., Geditz, M.C.K., Ganchev, B. et al. (2014). *J. Labelled Compd. Radiopharm.* 57: 699–703.

194 Baisch, G. and Öhrlein, R. (1998). *Carbohydr. Res.* 312: 61–72.

195 Muthana, M.M., Qu, J., Xue, M. et al. (2015). *Chem. Commun.* 4595–4598.

196 Xu, P., Dauter, Z., and Kováč, P. (2014). *Synthesis* 1073–1078.

197 Lemoine, S., Thomazeau, C., Joannard, D. et al. (2000). *Carbohydr. Res.* 326: 176–184.

198 Dijksman, A., Arends, I.W.C.E., and Sheldon, R.A. (2000). *Chem. Commun.* 271–272.

199 Nguyen, B.H., Perkins, R.J., Smith, J.A., and Moeller, K.D. (2015). *Beilstein J. Org. Chem.* 11: 280–287.
200 Schnatbaum, K. and Schäfer, H.J. (1999). *Synthesis* 864–872.
201 Gabrielli, L., Merlo, S., Airoldi, C. et al. (2014). *Bioorg. Med. Chem.* 22: 2576–2583.
202 Melvin, F., McNeill, A., Henderson, P.J.F., and Herbert, R.B. (1999). *Tetrahedron Lett.* 40: 1201–1202.
203 Sheldon, R.A. and Arends, I.W.C.E. (2004). *Adv. Synth. Catal.* 346: 1051–1071.
204 Angelin, M., Hermansson, M., Dong, H., and Ramström, O. (2006). *Eur. J. Org. Chem.* 4323–4326.
205 Breton, T., Bashiardes, G., Léger, J.-M., and Kokoh, K.B. (2007). *Eur. J. Org. Chem.* 1567–1570.
206 Striegler, S., Dunaway, N.A., Gichinga, M.G., and Milton, L.K. (2010). *Tetrahedron* 66: 7927–7932.
207 Mazur, A.W. and Hiler, G.D. (1997). *J. Org. Chem.* 62: 4471–4475.
208 Leppänen, A.-S., Niittymäki, O., Parikka, K. et al. (2010). *Carbohydr. Res.* 345: 2610–2615.
209 Schoevaart, R. and Kieboom, T. (2002). *Carbohydr. Res.* 337: 899–904.

3

Protecting Groups at the Secondary Positions of Carbohydrates

Sébastien Vidal and Peter G. Goekjian

Université de Lyon, CNRS and Université Claude Bernard Lyon 1, Institut de Chimie et Biochimie Moléculaires et Supramoléculaires (UMR 5246), Laboratoire de Chimie Organique 2-Glycochimie, 43 Boulevard du 11 Novembre 1918, 69622 Villeurbanne, France

3.1 Introduction

The use of protecting groups in carbohydrate chemistry is in some respects very similar, and in other respects quite different, from the use of protecting groups in any other organic synthesis. Thus, much, if not all, of what has been written on protecting groups is fully applicable here, and the reader is referred to a number of excellent writings on this topic [1–3]. In this chapter, we will focus on some of the specific concerns met in carbohydrate chemistry. Also, one might reasonably expect that after all the work that has been published to date, protecting group chemistry would be a solved problem; far from it. An inordinate number of chemical syntheses are still shipwrecked on the reefs of protecting group chemistry. This chapter may therefore be of interest to a larger audience, as it highlights some of the difficulties encountered with protecting group chemistry and some solutions that have been brought to bear.

A comprehensive study of protecting groups, even within carbohydrates, is an impossibly broad undertaking. Our objective is to provide the reader a more concise overview with some key principles and insights. This chapter is limited to oxygen protecting groups (for aminosugars, see Chapter 6). We have arbitrarily limited ourselves to *O*- and *S*-glycosides. The literature on *N*-glycosides includes nucleosides and nucleotides, an exceptionally rich source of protecting group chemistry, although more oriented toward ribofuranoses [4, 5]. The vast chemistry of protecting groups within the polyether *C*-glycosyl compounds is also overlooked here, although the use of carbohydrate starting materials allows us to benefit from some of this work.

3.1.1 Regioselectivity

The first peculiarity of protecting groups in carbohydrate chemistry is the issue of regioselectivity. If the starting material is the parent sugar (although *de novo*

Protecting Groups: Strategies and Applications in Carbohydrate Chemistry, First Edition.
Edited by Sébastien Vidal.

approaches should not be overlooked, see Chapter 9), then the five hydroxyl moieties of a hexose need to be differentiated. There is generally one (aldose) or two (ketose) primary hydroxyl groups (see Chapter 2) and one hemiacetal hydroxyl group, which can be differentiated based on either its acetal chemistry or its differential hydroxyl reactivity (see Chapter 5). However, this leaves either two or three alcohol functions that need to be differentiated, and the body of literature on selective protections in carbohydrate chemistry provides an outstanding case study of the kinetic and thermodynamic control of regioselectivity in alcohol protection. The regioselectivity will depend on the protecting group, the steric and electronic environment of each hydroxyl, and their relative stereochemistry. Exquisite regiocontrol has been achieved in many of the carbohydrate stereoisomers in both pyranose and furanose forms. The regioselective protection of carbohydrates is covered in Chapters 4, 14, 16, and 17.

3.1.2 Temporary or Persistent Protecting Groups: Orthogonality

The synthesis of oligosaccharides requires several types of protecting groups. "Terminal" protecting groups will be unmasked in the final step to reveal the free hydroxyl groups of the desired oligosaccharide. As free sugars are relatively tedious to purify, groups such as acetyl and benzyl, which can be deprotected to volatile by-products, are particularly popular. In certain cases, the permethylated carbohydrate is as active as the polyhydroxyl compound [6, 7], which avoids the need for final deprotection altogether. Among the remaining protecting groups, we can differentiate between persistent, "late-stage" functionalization groups, which will be deprotected in the penultimate stages to allow polyfunctionalization by sulfates, phosphates, or other carbohydrates, and temporary "iterative" protecting groups used in reductive-to-nonreductive sense oligosaccharide synthesis. In the latter case, an activated carbohydrate bearing an iterative protecting group at the appropriate position will be coupled to a glycoside acceptor, the iterative protecting group will be removed, and the process will be repeated (Scheme 3.1). In the case of branched oligosaccharides, several different orthogonal iterative protecting groups will be needed. More broadly, the full control of the synthesis of carbohydrate derivatives requires the ability to place different, mutually orthogonal protecting groups at each of the positions of the carbohydrate. A recent review of orthogonal protecting groups provides an excellent overview of the topic [8].

3.1.3 Anchimeric Participation in the Control of the Anomeric Configuration

One final constraint in the choice of protecting groups is the issue of anchimeric assistance in controlling the anomeric stereochemistry in oligosaccharide synthesis. If the protecting group at the 2-position (and to various extents other positions) carries a Lewis basic site that can interact with the oxycarbenium ion intermediate, such as an ester, it will direct the incoming nucleophile to provide the 1,2-*trans* glycoside under kinetic conditions. Anchimeric C-2 participation is perhaps the most reliable stereocontrol element in oligosaccharide synthesis.

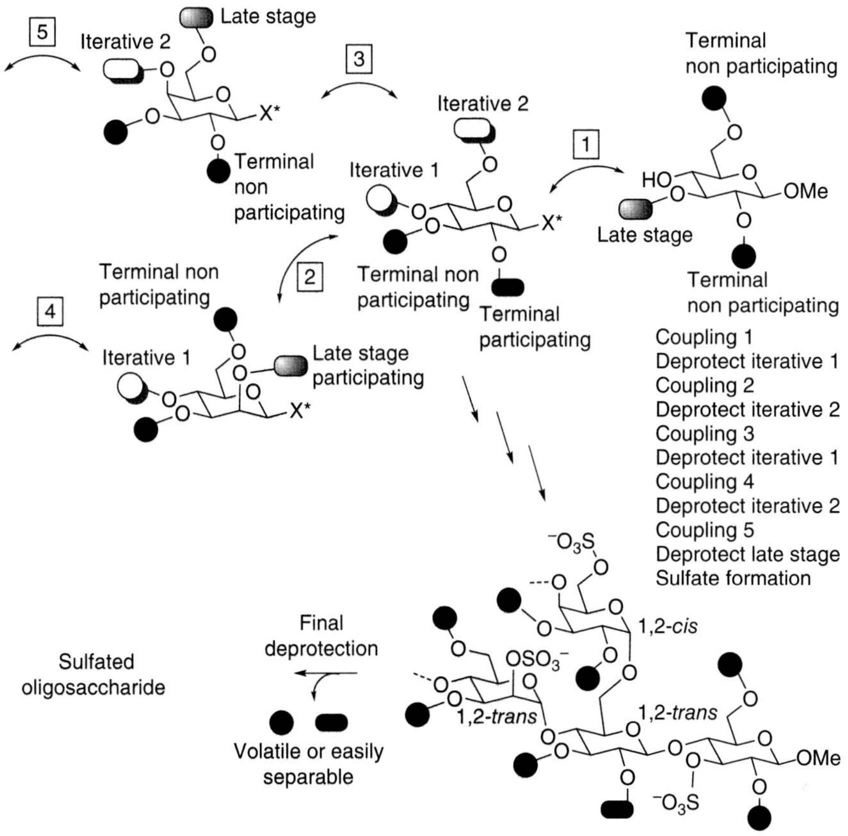

Scheme 3.1 Different protecting group roles in a hypothetical oligosaccharide synthesis.

The presence of an electron-withdrawing protecting group on O-2 or the presence of cyclic acetals also significantly modulates the reactivity of the adjacent anomeric center, resulting in "armed" vs "disarmed" activated glycoside donors. As a result, it is possible to do one-pot, sequential glycosylations by programming in the relative reactivity of the glycoside donors [9–11]. Silyl protecting groups can affect the conformation of the pyranose ring, thus affecting the stereoselectivity and rate of the glycosylation [12, 13].

3.1.4 Protecting Group Migration

If chemical orthogonality were the sole issue, carbohydrate protecting group chemistry would indeed be a solved problem. However, as carbohydrates systematically have vicinal hydroxyl groups, protecting group migration becomes a recurrent problem. Indeed, silyl and acyl protecting groups, in particular, readily undergo 1,2-migration under non-neutral conditions. The migration is faster in acyclic or 1,2-*cis* systems but can also occur in 1,2-*trans* cases. In some cases, useful selectivities can be obtained, but in most cases, protecting group migration will result in mixtures. More troublesome, protecting group migration during the reaction can result in the synthesis of a regioisomeric product, which could be problematic if the migration was not properly anticipated in the synthetic design [14–17]. The issue of functional group migration will be treated in depth in Chapter 8.

3.2 The Major Protecting Group Motifs

There is a very large panel of protecting groups available today, which we have organized according to the three most widely used classes: benzyl-type protecting groups, silyl protecting groups, and esters. A listing of protecting groups is provided within each class, with a detailed discussion of one or two examples. The purpose here is to give a broad overview of the concepts and challenges, and many interesting new protecting groups, as well as some classical ones, are not discussed here [8].

3.2.1 Benzyl and Substituted Benzyl Ethers

A wide variety of benzyl and substituted benzyl groups have been developed as protecting groups, all of which are applicable to carbohydrate chemistry (Table 3.1). Benzyl and *p*-methoxybenzyl (PMB) ethers are discussed in more detail below. The (2-naphthyl)methyl (Nap) group has generally similar reactivity to the PMB group and can be removed under similar mild conditions. The *p*-iodo, *p*-bromo, and *p*-chlorobenzyl groups can be removed selectively under acidic conditions, or sequentially under Hartwig–Buchwald conditions via the labile *p*-aminobenzyl group. The *o*- and *p*-nitrobenzyl protecting groups can be deprotected by a similar strategy after reduction of the nitro group; the *o*-nitrobenzyl group can also be removed under photolysis conditions, thus providing highly orthogonal deprotection conditions. Both the *o*-nitrobenzyl and the nicotinyl protecting groups act as anchimeric participating groups, favoring 1,2-*trans* glycosylation.

3.2.1.1 Benzyl Ethers

The benzyl group is as close to an ideal terminal protecting group as can be found and has been the overwhelming choice in this role. It is usually formed under Williamson conditions (Table 3.2), is stable to the protection and deprotection conditions for the majority of protecting groups as well as to most coupling and functional group manipulation conditions, and can be removed quantitatively by hydrogenolysis over Pd or $Pd(OH)_2$ on carbon or over Raney nickel (Table 3.3), leaving only toluene as a by-product after filtration of the heterogeneous catalyst.

The most commonly used Williamson conditions for benzyl protection require strong bases and are therefore not compatible with certain protecting groups and functionalities, in particular esters and silyl ethers, even though there are reports showing that even these can be tolerated in certain cases (Table 3.2). However, in general, either hydrolysis or migration will be observed, and nonadjacent esters are thus more easily tolerated. Methyl esters can be compatible, providing

Table 3.1 General overview of substituted benzyl groups.

Protecting group	Protection conditions	References	Deprotection conditions	References	Role
Benzyl (Bn)	BnBr, NaH, *n*Bu_4NI, THF/DMF BnCl, KOH, dioxane	[18] [19]	H_2, Pd/C, MeOH Na, NH_3(l), THF	[19] [20]	Nonparticipating terminal
p-Methoxybenzyl (PMB)	PMBCl, NaH, *n*Bu_4NI, THF/DMF PMBCl, *i*Pr_2NEt, 150 °C PMBOC(=NH)CCl_3, TfOH, CH_2Cl_2	[21] [22] [23]	DDQ or CAN, CH_2Cl_2/H_2O TFA, CH_2Cl_2, RT H_2, Pd/C, MeOH/H_2O	[24] [25] [26]	Nonparticipating, late stage or iterative
p-Bromo- or *p*-chlorobenzyl	$ArCH_2Cl$, NaH, *n*Bu_4NI, DMF	[27, 28]	$FeCl_3$, CH_2Cl_2 R_2NH, $PdCl_2$, X-Phos, then HCl, dioxane–water	[27] [28]	Nonparticipating, late stage
p-Nitrobenzyl (PNB)	*p*-NO_2PhCH_2Br, 2,4,6-collidine, AgOTf, 35 °C	[29]	In, NH_4Cl, MeOH/*i*PrOH, 85 °C	[29]	Nonparticipating, late stage
o-Nitrobenzyl	BF_3 or $PhBCl_2$, Et_3SiH (from *o*-nitrobenzylidene) $ArCH_2Br$, NaH, DMF	[30] [31]	hν, MeOH or MeCN H_2, Pd/C, MeOH	[30] [31]	C-2 participating, late stage or iterative
(2-Naphthyl)methyl (Nap)	$ArCH_2Br$, NaH, DMF	[32]	DDQ, CH_2Cl_2/H_2O H_2, Pd/C, EtOH-THF H_2, $Pd(OH)_2$/C, MeOH, AcOH	[33, 34] [35] [36]	Nonparticipating late stage or iterative
Nicotinyl	$PyrCH_2Br$, NaH, DMF	[37]	H_2, Pd/C, MeOH	[37]	C-2 participating, late stage

Table 3.2 Benzyl protection conditions and functional group compatibility.

Conditions	Compatible FG	References
BnBr, NaH, $n\mathrm{Bu_4NI}$, THF/DMF	Benzoyl, levulinoyl (non adjacent)	[38]
	Pivaloyl	[39]
	Methyl ester	[40, 41]
	Silylidene	[42]
	TBDMS (1,2-*trans*)	[43]
	TBDPS (non-adjacent)	[44, 45]
	TIPS	[46]
	Epoxide	[47]
	Acetonide	[47]
	Benzylidene	[48]
	Glycal (enol ether)	[49–51]
	Orthoester	[52, 53]
	Trityl	[54]
	NPhth	[45, 55]
	Azide	[46, 56]
	S-glycoside	[57, 58]
	Silyl group migration	[16]
BnCl, KOH, $n\mathrm{Bu_4NI}$, PhMe, reflux	Methyl ester	[59]
BnBr, NaOH, $n\mathrm{Bu_4NHSO_4}$, H_2O/PhMe	Piv (1,2-*trans*)	[60]
	OMs	[60]
BnBr, Ag_2CO_3, PhMe, 60 °C	Acetyl (1,2-*cis*)	[61, 62]
BnBr, Ag_2O	OTs (1,2-*cis*)	[63]
BnBr, BaO, $Ba(OH)_2$, DMF, 0 °C to RT	OTs (1,2-*trans*)	[64]
BnOC(=NH)CCl_3, TfOH, CH_2Cl_2/heptane, RT	Acetyl (1,2-*cis*)	[65, 66]
	Benzoyl (1,2-*trans*)	[66]
	Pivaloyl (1,2-*cis*)	[67]
	Levulinoyl (1,2-*trans*)	[68]
	Methyl ester	[69]
	Trityl	[70]
	Benzylidene	[66]
	Alkene and alkyne	[71]
	Alkyl iodide	[72]
	TBDPS	[68]

that any benzyl ester formed in the reaction is transesterified back to the methyl ester.

Benzyl groups can be installed early in the sequence to circumvent such problems, yet the order of introduction is typically governed by the regioselectivity of the first protection steps (see Chapter 4). A variety of milder or acid-catalyzed conditions have thus been reported for reactions with more sensitive functionalities. The use of silver carbonate or barium hydroxide/barium oxide mixtures has given good results in the presence of esters and other reactive functional groups, yet the use of benzyl trichloroacetimidate in the presence of triflic acid is generally the method of choice, particularly in the presence of adjacent ester groups.

Table 3.3 Benzyl deprotection conditions and group compatibility.

Deprotection	Conditions	Compatible FG	References	Reactive FG	References
Hydrogenolysis	H_2, Pd/C	Acetyl (1,2-*cis*)	[73, 74]	Ester (migration 1,2-*trans*)	[91–93]
	EtOH, or MeOH	Benzoyl (1,2-*cis*)	[75]	Ester (hydrolysis)	[94]
		Ester (1,2-*cis*)	[76]	Alkene, alkyne (reduction)	[90, 95, 96]
		Chloroacetyl (non-adjacent)	[77]	Benzylidene, PMB	[26, 76]
		TBDMS (1,2-*trans* OH)	[77]	*N*-Cbz	[81]
		TBDPS (non-adjacent)	[78]	Silyl ethers	[97–99]
		Lev	[79]	THP, EE	[100]
		Benzylidene	[80]	Azide (reduction)	[96, 100]
		N-Cbz	[81]	NHTCA (reduction)	[91, 95]
		N-Fmoc	[82]		
		NPhth	[83]		
		Carbamates, ureas	[82]		
		S-glycoside	[84, 85]		
		SO_2R glycoside	[86]		
		Alkene	[87]		
		Sulfate	[88, 89]		
		Phosphate	[90]		
	H_2, Pd/C	Acetyl (1,2-*cis*)	[102, 103]		
	EtOAc, dioxane, or THF	Ester (1,2-*cis*)	[104]		
	(aprotic solvent)	TES (1,2-*cis*)	[105]		
		TBDMS (non-adjacent)	[102, 103]		
	H_2, Pd/C, MeOH	Acetyl (1,2-*cis*)	[106]	Acetyl migration	[110]
	Acid	Benzoyl (1,2-*cis*)	[107, 108]	TBDPS	[111]
	(AcOH, TFA or HCl)	NHAc	[107]		
		S-glycoside	[109]		
		NPhth	[109]		

(*Continued*)

Table 3.3 (Continued)

Deprotection	Conditions	Compatible FG	References	Reactive FG	References
	H_2, $Pd(OH)_2/C$	Ester (1,2-*cis*)	[112]		
	EtOAc/DMF or EtOH	Chloroacetyl	[113]		
		NPhth	[113]		
		TBDMS (1,2-*cis*)	[114, 115]		
		Benzylidene	[116]		
		S-glycoside	[117]		
		N-Cbz	[116]		
		OMs (1,2-*trans* OH)	[105]		
		Phosphate	[112]		
		Phosphonate	[118]		
	$NH_4^+ \cdot HCO_2^-$, Pd/C	Silylidene	[119]		
	MeOH, reflux	Benzylidene	[120]		
		Sulfate	[121]		
	Cyclohexene, $Pd(OH)_2/C$	PMB	[26]		
	MeOH, reflux				
	Raney Ni	Acetyl (non-adjacent)	[122]	Thioether	[122]
	EtOH or MeOH	Benzoyl (non-adjacent)	[122]	Azide	[124]
		Acetonide	[122]	Benzylidene	[123]
		Phosphoramidate	[123]		
Birch	Na, NH_3(l)	Orthoester	[125]	Trityl, acetyl	[137]
		Alkene (non-allylic)	[126]	Bz, Ac (cleavage)	[129, 134]
		Ester	[126]	Methyl ester to 1° amide	[134]
		Allylic alcohol and amine	[127–129]	NHTCA (reduction)	[138]
		Phosphate	[130, 131]	*S*-phenyl glycoside	[20]
		S-alkyl glycosides	[85, 132–134]	*N*-sulfonamide (aminolysis)	[139]
		Carboxylic acid	[135]	Propargyl alcohol	[127, 140]
		Nitrile, amide	[134]	Azide (reduction)	[135]
		Amines	[134]	Alkyne	[141]
		PMB	[136]		

Oxidative	$NaBrO_3$, $Na_2S_2O_4$	Acetyl (1,2-*cis*)	[142]		
	H_2O/AcOEt	Azide	[143]		
		NHTFA	[143]		
		Succinamide ester	[144]		
		N-Cbz	[145]		
	NBS, hν, $CaCO_3$	Acetyl, Pivaloyl (1,2-*cis*)	[146]		
	CCl_4/H_2O	Ethyl ester	[146]		
		TBDMS	[146]		
		NPhth	[146]		
		SPh glycoside	[146]		
	DDQ	Benzoyl	[147]	Acetyl (1,2-*cis*) migration	[148]
	H_2O/CH_2Cl_2, reflux	TBDPS	[147]		
		SMe glycoside	[147]		
		SPh glycoside	[148]		
Hard acid soft nucleophile	$BF_3{\cdot}OEt_2$, NaI	Benzoyl (1,2-*cis*)	[149]		
		Acetyl (non-adjacent)	[149]		
		SAr glycoside	[149]		
	m-MeC_6H_4OH	Acetamide	[150]	Benzylidene	[150]
	Me_2S, TFA, TfOH, −15 °C	*N*-Fmoc	[150]	*t*-Butyl ether	[150]
		Peptide	[150]	*t*-Butyl ester	[150]
				Trityl	[150]
	TMSBr, TFA, PhSMe	Acetamide	[151]	Benzylidene	[151]
		Fmoc	[151]	Partial glycolysis	[151]
		Sulfate	[152]		
		Sulfide	[152]		
		Amine (1°)	[152]		

(Continued)

Table 3.3 (Continued)

Deprotection	Conditions	Compatible FG	References	Reactive FG	References
	TMSI	Hemiketal	[153]	Phosphonate deprotection	[153, 155]
		SAr glycoside	[154]		
		Acetyl (non-adjacent)	[154]		
		Acetamide	[154]		
	$FeCl_3$	Ester (1,2-*trans*)	[156]	Benzylidene	[156]
		SPh glycoside	[157]		
		Sulfate	[156]		
		Unsaturated ester	[156]		
		Vinyl phosphonate	[158]		
	BCl_3 or BBr_3	Bz (1,2-*trans*)	[159]		
	CH_2Cl_2, −78 °C	TBDPS (non-adjacent)	[160]		
		TMS alkyne	[161]		
		Alkyne	[161]		
		SMe glycoside	[162]		
		SPyr glycoside	[162]		
		Azide	[163]		
		Alkene	[163]		
Acetolysis	Ac_2O, TMSOTf	Ester	[164]	OMe glycoside	[164, 166]
		ONs	[164]	TIPS	[165]
		N-vinylimidazole	[164]		
		Alkyne and TMS alkyne	[165]	Benzyloxymethyl	[167]
		Alkene	[166]	p-Methoxybenzyl	[167]
	Ac_2O, $FeCl_3$	(2-trimethylsilyl)ethyl glycoside	[167]		

Benzyl groups are most commonly removed by hydrogenolysis, which is compatible with a wide range of carbohydrate structures, including those containing sulfates, phosphates, *S*-glycosides, acetonides, and carboxylic and carbonic acid derivatives (Table 3.3). Notable exceptions are unsaturated derivatives and azides, which will generally be reduced. Benzylidene, *N*-carboxybenzyl, *N*-Fmoc, and substituted benzyl ethers will generally be removed under Pd-catalyzed hydrogenolysis conditions, although controlled hydrogenolysis of benzyl ethers in the presence of each of these groups have been reported. However, these may be difficult to monitor in practice, and more convenient conditions may include Birch reduction with LiDBB for debenzylation in the presence of PMB, transfer hydrogenation with ammonium formate or oxidative debenzylation with 2,3-dichloro-5,6-dicyano-1,4-benzoquinone (DDQ) for debenzylation in the presence of benzylidenes, and oxidative cleavage with $NaBrO_3$ in the case of *N*-carboxybenzyl groups (Table 3.2).

The use of catalytic hydrogenation should render the deprotection orthogonal to protecting groups removed under nontransition-metal-catalyzed conditions. However, the situation is not as clear-cut as one might expect. Concurrent deprotection or migration of silyl and ester protecting groups is commonly observed, either because of the residual acidity resulting from the activation of the catalyst or to hydrogenolysis in the case of silyl groups. Thus, deprotection of a benzyl group adjacent to a silyl or acyl moiety requires special care, as both migration and hydroxyl-assisted hydrolysis occur under even mildly acidic or mildly basic conditions. The deprotection of nonadjacent ester groups seems less troublesome. Considerable caution must be exercised in the presence of adjacent acyl and silyl groups, as migration may or may not occur under very similar conditions, and reactions in which migration occurs are not always reported.

Despite these limitations, hydrogenolysis remains the method of choice for debenzylation in the presence of silyl groups. Alternatively, the debenzylation in the presence of *t*-butyldimethylsilyl (TBDMS) groups can be achieved with lithium aluminum hydride [168]. *t*-Butyldimethylsilyl groups are the most sensitive to hydrogenolysis, so triisopropyl, triethylsilyl (TES), or *t*-butyldiphenylsilyl (TBDPS) groups are generally preferable. In the presence of esters, debenzylation with $FeCl_3$ or $NaBrO_3$ are the most useful methods (Table 3.3). A reliable alternative may be to use a PMB instead of a benzyl group, which can be deprotected with DDQ in buffered biphasic solution (see Table 3.4).

Benzyl groups can be removed in the presence of alkenes under Birch conditions, although the hydrogenolysis of benzyl groups in the presence of cyclic alkenes has been reported (Table 3.3). Alkyne and azide groups will usually be reduced under both hydrogenolysis and Birch conditions, and silyl ethers are typically removed before Birch reduction. BCl_3 is compatible with both alkynes and azides as well as various heterocycles and carboxylic acid derivatives. Oxidative debenzylation with $NaBrO_3$ or DDQ is also effective in the presence of an azide group. Acetolysis of benzyl groups provides another approach to debenzylation in the presence of alkenes, alkynes, and other reactive groups, although it is not particularly useful in the presence of other ester protecting groups if orthogonal deprotection is still needed. Sulfates, phosphates, and other functionalities are generally compatible with hydrogenolysis conditions, in particular,

Table 3.4 *p*-Methoxybenzyl ether deprotection conditions and group compatibility.

Deprotection	Conditions	Compatible FG	References	Reactive FG	References
DDQ	CH_2Cl_2/H_2O (buffered)	Acetyl (1,2-*cis*)	[169]		
		Benzoyl (1,2-*cis*)	[170, 171]		
		Pivaloyl (1,2-*cis*)	[172]		
		Levulinoyl (1,2-*trans*)	[173]		
		Methyl ester	[171]		
		TBDMS (1,2-*cis*)	[172, 174]		
		SEM (1,2-*cis*)	[173]		
		TBDPS	[173]		
		Benzylidene, benzyl	[169]		
		NPhth, NHAc	[169, 171]		
		N-Cbz, *N*-Troc	[170, 175]		
		Acetonide	[176]		
		Azide	[169]		
		Alkene	[173, 176]		
$Ce(NH_4)_2(NO_3)_6$	H_2O/MeCN, 0 °C	Acetyl (1,2-*cis*)	[177]		
		Benzoyl (1,2-*cis*)	[178]		
		Ester (1,2-*cis*)	[179, 180]		
		TBDMS (1,2-*trans*)	[180]		
		N(Ac)-allyl glycoside	[177]		
		(2-Trimethylsilyl)ethoxy	[181, 182]		
		Benzylidene	[181]		
		Azide	[183]		
		Nap	[33]		

TFA	CH_2Cl_2, RT	Acetyl (1,2-*cis*)	[25, 184]
		Benzoyl (1,2-*cis*)	[25, 184]
		TBDPS	[25, 184]
		NPhth, azide	[25, 184]
		Sulfate	[185]
HCl	$(CF_3)_2CHOH$, RT	Acetyl (1,2-*trans*)	[186]
		Pivaloyl (1,2-*cis*)	[186]
		Lev, Methyl ester	[186]
		TBDMS, TBDPS	[186]
		N-Cbz	[186]
		S-Ar glycoside	[186]
Hard acid soft nucleophile	PhSH, $SnCl_4$, CH_2Cl_2	Acetyl (1,2-*cis*)	[187]
Hydrogenolysis	(see Table 3.3)		

by catalytic transfer hydrogenation, but also with many of the alternative debenzylation conditions.

3.2.1.2 *p*-Methoxybenzyl Ethers (PMB)

The PMB group is a common late-stage or iterative protecting group. It is stable to both the protection and deprotection conditions for esters and silyl ethers. The most common protecting conditions are under Williamson conditions, using either NaH or Hunig's base (Table 3.1), but acid-catalyzed conditions with PMB trichloroacetimidate, or regioselective reductive opening of *p*-methoxybenzylidenes (see Chapter 11), are also particularly efficient. Most, if not all, of the protection conditions described for the benzyl group can be extended to PMB, Nap, and dimethoxybenzyl groups and are therefore not duplicated here.

The PMB group can be removed efficiently in the presence of benzyl groups, esters, silyl groups, and a variety of functional groups with DDQ or ceric ammonium nitrate (CAN) in buffered biphasic solution (Table 3.4). These are conditions of choice to avoid 1,2-migration. The PMB group can also be removed by hydrogenolysis, along with benzyl, benzylidene, and other substituted benzyl groups. In addition, it can be removed either under Bronsted acid conditions or with a hard Lewis acid and a soft nucleophile. Naphthylmethyl (Nap) and dimethoxybenzyl groups can be protected and deprotected under similar conditions. Some protecting group compatibility can be achieved, as, for example, the PMB and Nap protecting groups can be removed sequentially with CAN an then DDQ [33], catalytic hydrogenation [35], or HCl/HFIP (1,1,1,3,3,3-hexafluoroisopropanol) [186].

3.2.2 Silyl Ethers

Silyl protecting groups are an important addition to the carbohydrate protecting group arsenal and are excellent iterative or late-stage protecting groups. They can also be used as terminal protecting groups, as while it is necessary to separate the silane derivatives produced in the deprotection, their very low polarity makes this step relatively trivial. In addition, their steric bulk induces long-range interactions that significantly affect carbohydrate conformation and stereoelectronic interactions, and hence their reactivity during glycosylations [12, 13].

The most classical silyl protecting group for secondary alcohols is the TBDMS. The TBDPS group is used mostly for primary alcohols but can also be used on secondary alcohols. A number of more stable protecting groups, such as the TES, diethylisopropylsilyl (DEIPS), triisopropylsilyl (TIPS), and thexyldimethylsilyl (TDS) groups, have been used, and some orthogonality exists between these groups. The simplest silyl protecting group, trimethylsilyl (TMS), is too labile to be carried through chemical steps but has been used *in situ* as a blocking group in a one-pot tandem process. Conditions for both protection and deprotection are similar for the different silyl groups but may require slightly more or less forcing conditions (Table 3.5). Deprotection conditions would typically be tetra-*n*-butylammonium fluoride, pyridine/HF, or trifluoroacetic acid (TFA).

The common protection conditions are compatible with most other protecting groups and functionalities, including esters [198, 199], *S*-glycosides [198, 199], other silyl groups [198, 199], benzyl and trityl [198, 199] ethers, benzylidenes

Table 3.5 General overview of silyl ethers.

Protecting group	Protection conditions	References
tBuMe$_2$Si (TBDMS)	TBDMSCl, imidazole, DMF	[188]
	TBDMSCl, pyr, $AgNO_3$	[189]
	TBDMSOTf, lutidine, CH_2Cl_2	[190]
Et_3Si (TES)	TESCl, DMAP, DMF	[191]
	TESOTf, iPr$_2$NEt, CH_2Cl_2	[192]
iPr$_3$Si (TIPS)	TIPSOTf, lutidine, CH_2Cl_2	[193]
	TIPSCl, imidazole, DMF	[194]
tBuPh$_2$Si (TBDPS)	TBDPSCl, imidazole, DMF	[195]
	TBDPSOTf, lutidine, CH_2Cl_2	[196]
iPrEt$_2$Si (DEIPS)	DEIPSCl, imidazole, DMF	[197]
$Me_2CHC(Me_2)SiMe_2$ (TDS)	TDSCl, imidazole, −10 °C	[16]

[200], acetonides [193], tetrahydropyrans [201], and carbamates [192], as well as reactive functionalities such as anomeric sulfones [202], alkyl iodides and toluenesulfonates [188, 190, 201], and enol ethers [191]. The choice of conditions turns more on the steric hindrance of the alcohol to be protected. The use of chlorosilane with imidazole as a nucleophilic catalyst and base are usually adequate for single protections but will generally lead to partial protection of carbohydrate polyols (see Chapter 4). Persilylation or silylation of sterically hindered alcohols typically requires the silyl triflate with lutidine or the chlorosilane in the presence of silver salts.

The most common deprotection conditions are with tetra-n-butylammonium fluoride, which is compatible with many protecting groups and functionalities, but these conditions are nonetheless quite basic, and hydrolysis of esters, in particular acetates, can be observed. Aqueous hydrogen fluoride is very effective but can present some safety concerns. Various superstoichiometric complexes, such as 10% HF in pyridine or $Et_3N{\cdot}3HF$, have been used as safer forms of HF, and tetrafluorosilane provides a mild source of fluoride. Silyl groups can also be removed under acidic conditions with TFA or HCl in ethanol, or with Lewis acids. They can undergo 1,2-migration under mild conditions, which represents a potential difficulty while deprotecting adjacent functional groups but can also represent interesting opportunities [203–206] (see Chapter 8). There is some degree of protecting group compatibility among silyl groups, as, for example, the TES can be removed in the presence of TBDMS with $PdCl_2(MeCN)_2$ in air [207], and the TBDMS can be removed in the presence of TBDPS in methanol in the presence of pyridinium p-toluenesulfonate (PPTS) [208] or camphorsulfonic acid (CSA) [209] (Table 3.6).

3.2.3 Esters

Esters are probably the most important protecting groups in carbohydrate chemistry, as they have been used in just about every role in oligosaccharide synthesis.

Table 3.6 Silyl ether deprotection conditions and group compatibility.

Conditions	Compatible FG	References	Reactive FG	References
TBAF, THF	Benzoyl (1,2-*trans*)	[210]	Acetyl 1,2-*cis*-migration	[213]
	NPhth	[211]	Isobutyryl 1,3-migration	[214]
	Acetonide	[211]	Ac partial hydrolysis	[48]
	Acyclic carbonate	[211]	Chloroacetyl hydrolysis	[210]
	Azide	[211]		
	Trimethylsilylethoxy	[212]		
	Alkyne	[212]		
TBAF, AcOH, DMF/THF				
DEIPS	Benzoyl (1,2-*trans*)	[215]		
	Acetyl (non-adjacent)	[215]		
	N-Troc, *N*-Cbz (non-adjacent)	[215]		
TBDPS (RT to 65 °C)	Benzoyl (1,2-*cis*)	[216]		
	p-Chlorobenzoyl (1,2-*trans*)	[217]	Acetyl (1,2-*cis*) migration	[213]
	(2-Trimethylsilyl)ethoxy	[216]		
	NPhth	[217]		
	Benzylidene	[217]		
	Amide, urea, carbamate	[217]		
10% HF/pyr, DMF				
TBDPS (2°), 1 GPa pressure	Acetyl (1,2-*cis*)	[218–220]		
	Benzoyl (1,2-*cis*)	[218]		
	Acetonide	[218–220]		
	NPhth	[218–220]		

TBDMS, TES	Methyl ester	[218]
	p-Methoxybenzoyl (1,2-*trans*)	[221]
	Acetyl (non-adjacent)	[221]
	Chloroacetyl (non-adjacent)	[222]
	Methyl ester	[223]
	Benzoyl	[223]
	Tris-unsaturated ester	[222]
	p-Methoxybenzylidene	[223]
	NHTCA	[223]
	1,3-Diene	[222]
	Ketal methyl glycoside	[222]
	Sulfate	[223]
$Et_3N{\cdot}3HF$	NHTCA	[224]
	Hemiacetal	[224]
SiF_4, MeCN, RT	Chloroacetyl (1,2-*trans*)	[77]
	Ester (1,2-*trans*)	[77]
	Acetyl, lactone, ketone, (non-adjacent)	[77]
Protic acids		
90% aq. TFA or TFA, $CHCl_3$, RT	Benzoyl (1,2-*cis*)	[225, 226]
	Chloroacetyl (non-adjacent)	[225, 226]
	NHTFA	[225, 226]
HCl, EtOH, RT (for TES)	Chloroacetyl (non-adjacent)	[227]
	Glucuronate ester	[225, 226]
PPTS, EtOH, or MeOH	TBDPS	[208]

(Continued)

Table 3.6 (Continued)

Conditions	Compatible FG	References	Reactive FG	References
(for TBDMS, TES)	Benzoyl (1,2-*cis*)	[208, 209]		
	SEM	[228]		
	Bis(*t*-butyl)silylene	[228]		
CSA, MeOH (for TES)	TBDPS	[209]		
HBF_4	*N*-Cbz	[188]	Trityl, benzylidene	[188]
Lewis acids				
$BF_3{\cdot}OEt_2$, MeCN, 0 °C (for TBDMS)	Acetyl (1,2-*trans*)	[229]		
	Chloroacetyl (non-adjacent)	[229]		

Esters are the prototypical participating group and have been used at each of the positions of carbohydrates to help control the stereochemistry at the anomeric position under kinetic conditions [230]. Esters can be removed cleanly under Zemplén conditions, and thus serve as terminal protecting groups. They can be both formed and hydrolyzed under orthogonal conditions to other protecting groups and to each other and thus have served extensively as transient protecting groups. Finally, the reactivity of the ester carbonyl can be fine-tuned using steric and electronic effects.

Another important aspect of ester protecting groups is their propensity for 1,2-migration. Ester groups with adjacent free hydroxyl groups migrate under very mild conditions [231, 232] via a readily accessible hemiorthoester intermediate. Conditions to suppress migration are rather difficult to maintain, and robust conditions under which esters do not migrate are difficult to define. Finally, during glycosylation reactions, the nucleophile may intercept the dioxolenium ion intermediate to give rise to either a relatively stable orthoester or to transesterification of the C-2 ester protecting group onto the glycosyl acceptor [233, 234].

Ester protecting groups can be divided into two groups: simple esters and triggered esters (Tables 3.7 and 3.9), in which a second functional group on the protecting group can participate in the hydrolysis of the ester, allowing for deprotection under conditions that are compatible with other protecting groups, including other esters.

3.2.3.1 Simple Ester Protecting Groups

Of the simple ester protecting groups above, acetate is the most ubiquitous and the most versatile. Acetyl protecting groups are used in every role, as a terminal protecting group, because methanolysis leaves only volatile methyl acetate as a by-product, as a late-stage or persistent protecting group, as they are stable enough to carry through a synthesis, and for directing the stereochemistry of the glycosylation by C-2 participation. They can be used as iterative protecting groups, providing they are not required in one of the other roles.

The benzoyl group is a more stable alternative to the acetyl group and also presents the advantage of a chromophore for UV detection (ultraviolet). It plays all the same roles as the acetyl group, including as a terminal protecting group, because separating the apolar methyl benzoate from the carbohydrate is rarely troublesome. It is stable to some deacetylation conditions (see Table 3.8), although the chloroacetyl or levulinate esters are more common partners in combination with benzoate groups.

The pivaloyl group is a sterically hindered variant of the acetyl group, making it far more stable and also particularly suitable for regioselective acylations (see Chapters 2 and 4). It is compatible with the acetyl group, as it is stable to either mildly basic or acidic deacetylation conditions. It is also widely used as an internal nucleophile for stereospecific inversion of neighboring alcohols [292]. Additional alkyl esters, such as *n*-butyl, isobutyryl, and isovaleryl esters, can be used in place of the acetyl group to provide better solubility.

The reactivity of the acetyl and benzoyl groups can be fine-tuned by substitution with an electron-withdrawing or electron-donating group. Thus,

Table 3.7 Simple ester protecting groups.

Protecting group	Protection conditions	References	Deprotection conditions	References	Role
Acetyl (Ac)	Ac_2O, pyr, DMAP AcCl, pyr, DMAP	[235] [236]	MeOH, Na MeOH, HCl MeOH, NH_3	[237] [238] [239]	Participating terminal, late stage or iterative
Benzoyl (Bz)	BzCl, pyr	[240]	MeOH, Na	[240]	Participating terminal, late stage or iterative
Pivaloyl (Piv)	PivCl, pyr, DMAP	[241]	MeOH, NaOMe	[233]	Participating, late stage or persistent
Isobutyryl	RCOCl, pyr	[242]	NaOMe, MeOH LiOH, H_2O/MeOH, RT Na_2CO_3, H_2O/MeOH	[243] [242] [244]	
Methoxyacetyl Phenoxyacetyl	$MeOCH_2COCl$, pyr $(PhOCH_2CO)_2O$, pyr, DMAP	[245] [246]	*t*$BuNH_2$, MeOH/$CHCl_3$, 10 °C NH_3, CH_2Cl_2/MeOH NaOMe, MeOH, 5 min, 0 °C	[247] [248] [249]	Participating, iterative, late stage or terminal
Dichloroacetyl	$(Cl_2CHCO)_2O$, pyr	[250]	NH_4OH, H_2O/MeOH $Zn(OAc)_2$, MeOH	[250] [251]	Participating iterative
2,4,6-Trimethylbenzoyl	ArCOCl, pyr, DMAP	[252]	0.5 M NaOH, MeOH/THF	[253]	Participating, late stage or terminal
o-Toluoyl	*o*-MePhCOCl, pyr	[250]	NH_4OH, MeOH	[250]	Participating, late stage or terminal
2,6-Dimethoxybenzoyl	ArCOCl, pyr	[252]	Na, MeOH	[252]	Participating, late stage or terminal

Table 3.8 Acetyl deprotection conditions and group compatibility.

Conditions	Compatible FG	References	Reactive FG	References
NaOMe, MeOH (Zemplén)	Benzoyl (1,2-*cis*)	[254]	1,2-TBDMS migration	[267]
	Lev (1,2-*cis*)	[255]	1,2-TDS migration	[16]
	Ester (1,2-*cis*)	[256]	Benzoyl (non-adjacent)	[264]
	Methyl ester	[257]	Ester (non-adjacent hydrolysis)	[256]
	TDS (1,2-*trans*, −15 °C)	[16]	Ester (adjacent hydrolysis)	[256]
	TBDMS (1,2-*trans*)	[206]		
	TBDMS (non-adjacent)	[258]		
	ArOTBDMS (non-adjacent)	[259]		
	TBDPS (non-adjacent)	[45, 260]		
	TIPS (non-adjacent)	[261]		
	N-Cbz	[260]		
	N-Troc	[262]		
	NPhth	[45]		
	N-dimethylmaleoyl	[263]		
	NHAc	[260]		
	NHTCA	[264]		
	Azide	[265]		
	Sulfate	[266]		
K_2CO_3, MeOH	TBDMS (1,2-*trans*)	[268]		
	TIPS (non-adjacent)	[269]		
	TBDPS (non-adjacent)	[270]		
	Methyl ester	[271]		
	NPhth	[270]		
LiOH, H_2O_2	TDS (non-adjacent)	[272]	Methyl ester	[272]

(Continued)

Table 3.8 (Continued)

Conditions	Compatible FG	References	Reactive FG	References
$Mg(OMe)_2$, MeOH	TIPS (non-adjacent)	[273]		
	Benzoyl (non-adjacent)	[273]		
	Methyl ester	[274]		
	Lev ester (non-adjacent)	[274]		
	TBDPS (non-adjacent)	[274]		
	SEM (non-adjacent)	[274]		
	Azide	[274]		
Bu_2SnO, MeOH, reflux	Benzoyl (1,2-*trans*)	[275]	Ester migration (1,2-*trans*)	[276]
NH_3, MeOH	Ester (1,2-*cis*)	[277]	Ester migration (1,2-*cis*)	[278]
	Methyl ester	[277]	Ester (non-adjacent, aminolysis)	[256]
	Enol ether	[277]		
N_2H_4-H_2O, MeOH/$CHCl_3$, RT	TBDMS (1,2-*trans*)	[279]		
or EtOH, reflux			TBDMS migration (1,2-*trans*)	[203]
AlH(*i*Bu)$_2$, CH_2Cl_2, 0 °C	TBDMS (1,2-*trans*)	[280]	*p*-Methoxybenzylidene	[280]
	TBDPS (1,2-*cis*)	[281]		
	TMS acetylene	[281]		
HCl, MeOH or AcCl in MeOH or EtOH	Benzoyl (1,2-*cis*)	[282, 283]	*O*-Aryl glycoside	[256]
	S-glycoside	[283]	*O*-CMe_2R glycoside	[289]
	Pivaloyl (1,2-*trans*)	[284]	Ester migration (1,2-*cis* and	[289]
	Methyl ester	[285]	1,2-*trans*)	[280]
	NPhth	[286]	TBDMS (adjacent)	[290]
	O-Troc	[287]	TMS	
	Azide	[288]		
CuCl, MeOH, H_2O	Benzoyl (1,2-*cis*)	[282]		
HBF_4, THF, RT	Benzoyl (1,2-*trans*)	[291]		
	Benzylidene	[291]		

2-methoxyacetyl, 2-phenoxyacetyl, and 2,2-dichloroacetyl groups are more electrophilic and have been used as cleavable protecting groups in the presence of other esters, including acetyl and benzoyl groups. The reactivity and the solubility of the benzoyl group can be modulated by *para* substituents, for example, *p*-toluoyl [293], *p*-bromobenzoyl [294], *p*-methoxybenzoyl [295], and *p*-nitrobenzoyl groups [296]. In addition, *ortho*-substituted benzoyl groups are stabilized by steric hindrance of the carbonyl group and are less prone to migration [252], and *o*-toluoyl, 2,6-dimethoxybenzoyl, and 2,4,6-trimethylbenzoyl groups have found some use as stabilized versions of the benzoyl protecting group.

Relatively few protection conditions are required for acyl protecting groups, as the most common conditions shown in Table 3.7 are compatible with most other protecting groups and functionalities, including ethers, acetals, esters, silyl ethers, nitrogen protecting groups, azides, phosphates, and more reactive functional groups, such as alkyl halides [297], sulfonates [298], and sulfates [299]. Other conditions were developed for regioselective acylation and are discussed in Chapter 4. These conditions may be useful if particularly mild conditions are needed for a nonregioselective acylation. It is in the deprotection step that compatibility with other protecting groups and functionalities can become an issue. The most effective conditions are clearly the Zemplén conditions, although many other protecting groups may not be compatible. In the case of neighboring ester or silyl groups, methanolysis [256] or migration [278] can be either observed or not under similar conditions, generally depending on more rigorous control of the temperature and reaction time. A variety of conditions with milder bases, such as $Mg(OMe)_2$, dibutyltin oxide, or ammonia, can be used, particularly in the presence of acid-sensitive functionalities. Although potassium carbonate is a milder base than sodium methoxide, the reactive species, potassium methoxide, has similar reactivity, so there is actually relatively little difference in reaction profiles with the Zemplén conditions, simply that the stoichiometry of the base is less critical. Additional conditions, such as LiOH, H_2O_2, or aqueous sodium hydroxide in methanol, can be used when hydrolysis to a carboxylic acid is desired, for example, in uronic acids.

The overwhelmingly favored condition for the hydrolysis of acetyl groups in the presence of neighboring benzoate or other esters is methanolysis under acidic conditions (Table 3.8). However, even under acidic conditions, acyl migration or silyl hydrolysis can be observed [289, 300]. Reduction with diisobutylaluminum hydride appears to be the best conditions to avoid silyl migration, as the covalent complex to the aluminum presumably blocks the free hydroxyl group.

3.2.3.2 Triggered Ester Protecting Groups

A second group of esters contains a secondary functional group capable of triggering the hydrolysis of the ester under specific conditions (Table 3.9). These can be either electrophilic groups, which form a heterocyclic ring when reacted with a bis-nucleophile, or masked nucleophiles, which, when revealed, react with the ester group in an intramolecular manner to release the carbohydrate hydroxyl group.

Table 3.9 Triggered ester protecting groups.

Protecting group	Protection conditions	Deprotection conditions	Role	References
Chloroacetyl (ClAc)	$ClCH_2COCl$, pyr, DMAP	Thiourea, pyr, EtOH	Participating, iterative	[301]
	$ClCH_2CO_2H$, DCC, DMAP	NH_2NH_2, AcOH, MeOH	Participating, iterative	[302]
Bromoacetyl (BrAc)	$BrCH_2COBr$, tetramethylurea	Thiourea, collidine, MeOH		[303, 304]
Levulinoyl (Lev)	LevOH, EDC, DMAP, 0 °C	$N_2H_4 \cdot AcOH$, EtOH	Participating, Iterative or persistent	[305]
2-(Allyloxy)phenylacetyl	$ArCH_2CO_2H$, DCC, DMAP	Proton sponge, $Pd(PPh_3)_4$, $EtOH/H_2O$	Participating, iterative	[306]
2-(Chloroacetoxymethyl)-benzoyl (CAMB)	CAMBCl, pyr (5 eq.), CH_2Cl_2	thiourea, $MeOH/CH_2Cl_2$	Participating, iterative	[234]
3-(2′-Benzyloxyphenyl)-3,3-dimethylpropanoyl	$ArCMe_2CH_2CO_2H$, CDI, DBU	H_2, $Pd(OH)_2/C$, MeOH/EtOAc	Participating, terminal or late stage	[307]
4-Azidobutanoyl	$N_3(CH_2)_3COCl$, pyr, CH_2Cl_2 derived from cyclic orthoester [308]	PPh_3, dioxane, H_2O	Participating iterative or late stage	[309]
2-(Azidomethyl)benzoyl		H_2, Pd/C EtOH		[310]
(2-Nitrophenyl)acetyl	$(NO_2PhCH_2CO)_2O$, pyr	Zn, NH_4Cl, MeOH/THF, RT	Participating, iterative or late stage	[311]
	$NO_2PhCH_2CO_2H$, DCC, DMAP	LiOH, THF/MeOH, RT		[312]

EDC, *N*-(3-dimethylaminopropyl)-*N*′-ethyl carbodiimide.

In the electrophilic category, the most widely used protecting groups are the chloroacetyl and levulinate esters. In the case of the chloroacetyl or bromoacetyl group, substitution with thiourea will release the ester by the formation of a 2-aminothiazolinone. The keto group of the levulinate ester reacts with hydrazine to form a diazinone. The masked nucleophile category includes protecting groups with an allyl, chloroacetyl, or benzyl-protected γ or δ-hydroxyester, or an γ-azide or nitro group that can be reduced to an amine. These functional groups are typically orthogonal with other triggered esters and acetyl groups [306].

Protection conditions are similar to those of the simple esters and are compatible with most protecting groups and functionalities. As to the deprotection conditions, there are typically two categories: specific conditions for deprotection of iterative and late-stage protecting groups in the presence of various compatible protecting groups and conditions used for global deprotection when used as a terminal or late-stage protecting group. In the case of protected nucleophiles, the conditions are those for the deprotection of the allyl, trichloroacetyl, and benzyl groups, respectively, which are generally compatible with other groups within the molecule. In the case of the azide and nitrophenyl-triggered esters, they can be removed either under Staudinger or single-electron-transfer conditions, which are compatible with the majority of protecting groups [311], or under hydrogenolysis, which may be used during a global deprotection. In the case of levulinic esters, deprotection in the presence of hydrazinium acetate is compatible with acetyl, chloroacetyl, benzoyl, and silyl groups with little or no migration [305, 313]. The levulinate ester is orthogonal to the acetyl group, as the latter can be removed in the presence of levulinate [255]. In the case of the chloroacetyl group, the most common conditions are with thiourea, particularly in the presence of other esters (Table 3.10). The use of collidine or lutidine instead of pyridine reduces acyl group migration [303, 304]. Like other esters, the chloroacetyl group is compatible with base-stable protecting groups and functionalities (benzyl and substituted benzyl ethers, acetals, *S*-glycosides, amides, carbamates, etc.) but can also be removed cleanly in the presence of other esters, silyl groups, phthalimide protecting groups, and various functional potentially reactive groups such as azides and alkyl halides. There are to our knowledge no specific report for the deprotection of chloroacetyl groups in the presence of adjacent silyl groups, so it is unclear whether silyl migration will be an issue or not. It can almost certainly be suppressed, although it may be preferable to use a levulinate ester in this case [313]. The chloroacetyl group can also be removed under mild methanolysis conditions by taking advantage of the inductive effect of the chlorine. The chloroacetyl can thus be removed in the presence of acetyl, levulinate, and other esters by trans-esterification in the presence of mild bases.

3.2.4 Other Protecting Groups

There remains a wealth of other protecting groups that we have not treated here. These generally fall into three categories: specialized orthogonal protecting

Table 3.10 Chloroacetyl deprotection conditions and group compatibility.

Conditions	Compatible FG	References
Thiourea, *s*-collidine, EtOH, or MeOH	Acetyl (1,2-*cis*)	[314]
	Benzoyl (1,2-*cis*)	[315]
	Methyl ester	[316]
	Pivaloyl (non-adjacent)	[317]
	TBDPS (non-adjacent)	[316]
	TBDMS (non-adjacent)	[318]
	N-Troc	[319]
	NHTFA	[320]
	NPhth	[321]
	Azide	[314]
	Alkyl chloride	[322]
1-Selenocarbamoylpiperidine, THF, 40 °C	Benzoyl (non-adjacent)	[323]
	Pivaloyl (1,3-*trans*)	[323]
	TIPS (non-adjacent)	[323]
	(2-Trimethylsilyl)ethoxy	[323]
	2,3-Dimethoxybutadiyl	[323]
	O-Fmoc	[323]
Hydrazine dithiocarbonate, lutidine/AcOH, MeOH/CH_2Cl_2	Methyl ester	[324]
	Acetyl (non-adjacent)	[324]
	Benzoyl (1,2-*trans*)	[325]
	N-Cbz	[324]
	Alkyl bromide (1°, 2°)	[325]
NH_2NH_2, AcOH, MeOH	Acetyl (1,2-*trans*)	[326]
	TBDMS (non-adjacent)	[326]
	Glycosyl fluoride	[327]
	Lev (cleavage)	[328]
DABCO, PhMe/EtOH, 55 °C	Acetyl	[328]
	Lev	[328]
	Benzylidene	[328]
	NHTCA	[328]
K_2CO_3, THF/MeOH	TBDMS (non-adjacent)	[193, 329]
	TIPS (non-adjacent)	[193]
	Methyl ester	[329]
	N-Cbz	[329]
$NaHCO_3$, H_2O/MeOH/THF, 60 °C	Lev (non-adjacent)	[330]
	TBDPS (non-adjacent)	[330]

groups that have been developed particularly for library generation and oligosaccharide synthesis; transient protecting groups, such as TMS ethers [12, 13] or borinate esters [331] (Chapter 7); or classical protecting groups, such as allyl ethers or acyclic acetals such as methoxymethyl (MOM), benzyloxymethyl (BOM), (2-trimethylsilyl)ethoxymethyl (SEM), and many others [2, 3]. However, we hope that the discussion above will provide a general framework that will allow the reader to assimilate the multitude of protecting groups available to the carbohydrate chemist.

3.3 Conclusion

As can be seen from the above discussion, the choice of protecting groups is not a solved problem in carbohydrate chemistry. In particular, the issue of functional group compatibility and protecting group migration remains a significant issue that must be carefully considered both in the synthetic design and during the execution of the synthesis. A simple analysis based on functional group chemistry is not sufficient for predicting functional group behavior in such functionally dense structures, and even a straightforward prediction like "an ester will not react with hydrogen over Pd on carbon" requires a considerable caveat. A detailed look into functional group compatibility as a function of reaction conditions sheds remarkably little light: a given group may or may not react under seemingly similar conditions; in fact, one can almost systematically find examples of both results in the literature. We have tried to give some subjective guidance based on our own experience, but in the end, like in any synthesis, controlling the selectivity of the reaction requires careful control of the reaction and purification conditions, rather than a simple recipe. We hope that this chapter will provide the reader with some of the issues to consider and the possible solutions that can be brought to bear.

3.4 Experimental Section

3.4.1 Phenyl 2,3-Di-*O*-benzyl-4,6-*O*-benzylidene-α-D-glucopyranosyl-[(1→4)-2,3,6-tri-*O*-benzyl-α-D-glucopyranosyl]-(1→4)-2,3,6-tri-*O*-benzyl-1-thio-β-D-glucopyranoside [27]

A solution of the polyol precursor (5.03 g, 4.29 mmol) in *N*,*N*-dimethylformamide (DMF 80 ml) was cooled to 0 °C; treated sequentially with NaH (11.96 g), benzyl bromide (25.51 ml, 214.55 mmol), tetraethylammonium iodide (141 mg, 0.055 mmol), and imidazole (37.3 mg, 0.055 mmol); warmed to 23 °C; and stirred for 20 h. The mixture was cooled to 0 °C and treated dropwise with MeOH (50 ml) and then H_2O (50 ml). The mixture was diluted with CH_2Cl_2 (200 ml). The organic layer was washed with saturated aqueous $NaHCO_3$ (2 ml × 80 ml), H_2O (2 ml × 80 ml), and brine (80 ml) and dried ($MgSO_4$). Evaporation and flash chromatography (PhMe/EtOAc 40 : 1–20 : 1) gave the title compound as a white solid (9.06 g).

3.4.2 5-Azidopentyl 2-*O*-acetyl-4,6-*O*-benzylidene-β-D-mannopyranosyl-(1→4)-*O*-(3,6-di-*O*-benzyl-2-deoxy-2-phthalimido-β-D-glucopyranosyl)-(1→4)-*O*-3,6-di-*O*-benzyl-2-deoxy-2-phthalimido-β-D-glucopyranoside [169]

Toasolutionof5-azidopentyl-2-*O*-acetyl-4,6-*O*-benzylidene-3-*O*-*p*-methoxybenzyl-β-D-mannopyranosyl-(1→4)-*O*-(3,6-di-*O*-benzyl-2-deoxy-2-phthalimido-β-D-glucopyranosyl)-(1→4)-*O*-3,6-di-*O*-benzyl-2-deoxy-2-phthalimido-β-D-glucopyranoside (1.0 g, 0.673 mmol) in CH_2Cl_2/H_2O (10 ml, 10/1) was added DDQ

(0.183 g, 0.808 mmol) at 0 °C and the resulting reaction mixture was stirred at room temperature and then filtered. The organic layer was washed with water (2 ml × 30 ml). The aqueous layer was further extracted with CH_2Cl_2 (2 ml × 50 ml). The combined organic layers were washed with brine (40 ml), dried (Na_2SO_4), and concentrated *in vacuo*. The residue was purified by flash column chromatography to afford the title compound (0.652 g, 70%) as colorless foam.

Abbreviations

Bn	benzyl
Boc	*t*-butoxycarbonyl
BrAc	bromoacetyl
BrBn	*p*-bromobenzyl
Bz	benzoyl
CAMB	2-(chloroacetoxymethyl)benzoyl
CAN	ceric ammonium nitrate
Cbz	benzyloxycarbonyl
CDI	1,1′-carbonyldiimidazole
ClAc	chloroacetyl
CSA	camphorsulfonic acid
DABCO	1,4-diazabicyclo[2.2.2]octane
DBU	1,8-diazabicycloundec-7-ene
DCC	*N,N*′-dicyclohexylcarbodiimide
DDQ	2,3-dichloro-5,6-dicyano-1,4-benzoquinone
DEIPS	diethylisopropylsilyl
DMAP	*N,N*-dimethyl-4-aminopyridine
DME	1,2-dimethoxyethane
DMF	*N,N*-dimethylformamide
EDC	*N*-(3-dimethylaminopropyl)-*N*′-ethyl carbodiimide
EE	1-ethoxyethyl
FG	functional group
Fmoc	9-fluorenylmethyloxycarbonyl
HFIP	1,1,1,3,3,3-hexafluoroisopropanol
Lev	levulinoyl
LiDBB	4,4′-di-*tert*-butylbiphenylide
MOM	methoxymethyl
Ms	methanesulfonyl
Nap	(2-naphthyl)methyl
NBS	*N*-bromosuccinimide
Ns	2-nitrotoluenesulfonate
Phth	phthaloyl
Piv	pivaloyl
PMB	*p*-methoxybenzyl
PNB	*p*-nitrobenzyl
PPTS	pyridinium *p*-toluenesulfonate
pyr	pyridine

RT	room temperature
SEM	(2-trimethylsilyl)ethoxymethyl
TBAF	tetra-*n*-butylammonium fluoride
TBDMS	*t*-butyldimethylsilyl
TBDPS	*t*-butyldiphenylsilyl
TCA	trichloroacetyl
TDS	thexyldimethylsilyl
TES	triethylsilyl
Tf	trifluoromethanesulfonyl
TFA	trifluoroacetic acid or trifluoroacetyl
THF	tetrahydrofuran
TIPS	triisopropylsilyl
TMS	trimethylsilyl
Troc	2,2,2-trichloroethoxycarbonyl
Ts	*p*-toluenesulfonyl
UV	ultraviolet
X-Phos	2-dicyclohexylphosphino-2′,4′,6′-triisopropylbiphenyl

References

1 Hung, S.-C. and Wang, C.-C. (2016). Protecting group strategies in carbohydrate synthesis. In: *Glycochemical Synthesis: Strategies and Applications*, Chapter 2 (ed. S.-C. Hung and M.M.L. Zulueta), 35–68.

2 Wuts, P.G.M. (2014). *Greene's Protective Groups in Organic Synthesis*, 5e. Hoboken, NJ: Wiley.

3 Kocienski, P.J. (2005). *Protecting Groups*. Stuttgart: Georg Verlag Thieme.

4 Muller, S., Wolf, J., and Ivanov, S.A. (2004). *Curr. Org. Synth.* 1: 293–307.

5 Roy, B., Depaix, A., Périgaud, C., and Peyrottes, S. (2016). *Chem. Rev.* 116: 7854–7897.

6 Buller, H.R., Cohen, A.T., Davidson, B. et al. (2007). *N. Engl. J. Med.* 357: 1094–1104.

7 Herbert, J.M., Hérault, J.P., Bernat, A. et al. (1998). *Blood* 91: 4197–4205.

8 Ágoston, K., Streicher, H., and Fügedi, P. (2016). *Tetrahedron: Asymmetry* 27: 707–728.

9 Mootoo, D.R., Konradsson, P., Udodong, U., and Fraser-Reid, B. (1988). *J. Am. Chem. Soc.* 110: 5583–5584.

10 Zhang, Z., Ollmann, I.R., Ye, X.-S. et al. (1999). *J. Am. Chem. Soc.* 121: 734–753.

11 Fraser-Reid, B., Wu, Z., Andrews, C.W. et al. (1991). *J. Am. Chem. Soc.* 113: 1434–1435.

12 Pedersen, C.M., Nordstrom, L.U., and Bols, M. (2007). *J. Am. Chem. Soc.* 129: 9222–9235.

13 Bols, M. and Pedersen, C.M. (2017). *Beilstein J. Org. Chem.* 13: 93–105.

14 Bourne, E.J., Huggard, A.J., and Tatlow, J.C. (1953). *J. Chem. Soc.* 735–741.

15 Bonner, W. (1959). *J. Org. Chem.* 24: 1388–1390.

16 Lassaletta, J.M., Meichle, M., Weiler, S., and Schmidt, R.R. (1996). *J. Carbohydr. Chem.* 15: 241–254.

17 Phanumartwiwath, A., Hornsby, T.W., Jamalis, J. et al. (2013). *Org. Lett.* 15: 5734–5737.
18 Czernecki, S., Georgoulis, C., and Provelenghiou, C. (1976). *Tetrahedron Lett.* 17: 3535–3536.
19 Sundberg, R.L., McCloskey, C.M., Rees, D.E., and Coleman, G.H. (1945). *J. Am. Chem. Soc.* 67: 1080–1084.
20 Pfaeffli, P.J., Hixson, S.H., and Anderson, L. (1972). *Carbohydr. Res.* 23: 195–206.
21 Takaku, H. and Kamaike, K. (1982). *Chem. Lett.* 11: 189–192.
22 Gathirwa, J.W. and Maki, T. (2012). *Tetrahedron* 68: 370–375.
23 Takahashi, H., Kittaka, H., and Ikegami, S.J. (2001). *Org. Chem.* 66: 2705–2716.
24 Gigg, J., Gigg, R., Payne, S., and Conant, R. (1987). *J. Chem. Soc., Perkin Trans. 1* 1165–1170.
25 Reina, J.J. and Rojo, J. (2006). *Tetrahedron Lett.* 47: 2475–2478.
26 Lubineau, A., Augé, C., Le Goff, N., and Le Narvor, C. (1997). *Carbohydr. Res.* 305: 501–509.
27 Hoffmann, B., Zanini, D., Ripoche, I. et al. (2001). *Helv. Chim. Acta* 84: 1862–1888.
28 Plante, O.J., Buchwald, S.L., and Seeberger, P.H. (2000). *J. Am. Chem. Soc.* 122: 7148–7149.
29 Nakatsuka, T., Tomimori, Y., Fukuda, Y., and Nukaya, H. (2004). *Bioorg. Med. Chem. Lett.* 14: 3201–3203.
30 Zhu, C.-J., Yi, H., Chen, G.-R., and Xie, J. (2008). *Tetrahedron* 64: 10687–10693.
31 Buda, S., Gołębiowska, P., and Mlynarski, J. (2013). *Eur. J. Org. Chem.* 3988–3991.
32 Crich, D., Hutton, T.K., Banerjee, A. et al. (2005). *Tetrahedron: Asymmetry* 16: 105–119.
33 Wright, J.A., Yu, J., and Spencer, J.B. (2001). *Tetrahedron Lett.* 42: 4033–4036.
34 Borbás, A., Szabó, Z.B., Szilágyi, L. et al. (2002). *Tetrahedron* 58: 5723–5732.
35 Lazar, L., Janossy, L., Csavas, M. et al. (2012). *Arkivoc* 5: 312–325.
36 Lee, Y.J., Ishiwata, A., and Ito, Y. (2008). *J. Am. Chem. Soc.* 130: 6330–6331.
37 Smoot, J.T., Pornsuriyasak, P., and Demchenko, A.V. (2005). *Angew. Chem. Int. Ed.* 44: 7123–7126.
38 Fulse, D.B., Jeon, H.B., and Kim, K.S. (2007). *J. Org. Chem.* 72: 9963–9972.
39 Lloyd, D., Bylsma, M., Bright, D.K. et al. (2017). *J. Org. Chem.* 82: 3926–3934.
40 Suhara, Y., Yamaguchi, Y., Collins, B. et al. (2002). *Bioorg. Med. Chem.* 10: 1999–2013.
41 Ott, A.-J., Brackhagen, M., Davtyan, A. et al. (2001). *J. Carbohydr. Chem.* 20: 611–636.
42 Kumagai, D., Miyazaki, M., and Nishimura, S.I. (2001). *Tetrahedron Lett.* 42: 1953–1956.
43 Chen, G., Xie, J., Li, J. et al. (2009). Application of amino acid saccharide-series compounds as protein tyrosine phosphatase-1B inhibitor. CN101475615A.
44 Ratcliffe, A.J., Konradsson, P., and Fraser-Reid, B. (1990). *J. Am. Chem. Soc.* 112: 5665–5667.
45 Watabe, J., Singh, L., Nakahara, Y. et al. (2002). *Biosci. Biotechnol. Biochem.* 66: 1904–1914.
46 Santana, A.G., Bastida, Á., del Campo, T.M. et al. (2011). *Synlett* 219–222.

47 van Boeckel, C.A.A., Beetz, T., Vos, J.N. et al. (1985). *J. Carbohydr. Chem.* 4: 293–321.

48 Ichikawa, Y., Manaka, A., and Kuzuhara, H. (1985). *Carbohydr. Res.* 138: 55–64.

49 Upreti, M., Ruhela, D., and Vishwakarma, R.A. (2000). *Tetrahedron* 56: 6577–6584.

50 Spassova, M.K., Bornmann, W.G., Ragupathi, G. et al. (2005). *J. Org. Chem.* 70: 3383–3395.

51 Li, W., Silipo, A., Molinaro, A., and Yu, B. (2015). *Chem. Commun.* 51: 6964–6967.

52 Hori, M. and Nakatsubo, F. (2001). *Carbohydr. Res.* 332: 405–414.

53 Jayaprakash, K.N. and Fraser-Reid, B. (2004). *Org. Lett.* 6: 4211–4214.

54 Clausen, M.H., Jorgensen, M.R., Thorsen, J., and Madsen, R. (2001). *J. Chem. Soc., Perkin Trans. 1* 543–551.

55 Rabuka, D. and Hindsgaul, O. (2002). *Carbohydr. Res.* 337: 2127–2151.

56 Joosten, J.A.F., Loimaranta, V., Appeldoorn, C.C.M. et al. (2004). *J. Med. Chem.* 47: 6499–6508.

57 Marmuse, L., Nepogodiev, S.A., and Field, R.A. (2005). *Tetrahedron: Asymmetry* 16: 477–485.

58 Damager, I., Olsen, C.E., Møller, B.L., and Motawia, M.S. (2002). *Synthesis* 418–426.

59 Zelenka, K., Trnka, T., Tislerova, I. et al. (2004). *Collect. Czech. Chem. Commun.* 69: 1149–1160.

60 Sakairi, N., Hayashida, M., and Kuzuhara, H. (1987). *Tetrahedron Lett.* 28: 2871–2874.

61 Malik, S., Dixit, V.A., Bharatam, P.V., and Kartha, K.P.R. (2010). *Carbohydr. Res.* 345: 559–564.

62 Mangion, I.K. and MacMillan, D.W.C. (2005). *J. Am. Chem. Soc.* 127: 3696–3697.

63 Haradahira, T., Maeda, M., Kai, Y., and Kojima, M. (1985). *Chem. Commun.* 364–365.

64 Bauder, C. (2008). *Org. Biomol. Chem.* 6: 2952–2960.

65 Nemati, N., Karapetyan, G., Nolting, B. et al. (2008). *Carbohydr. Res.* 343: 1730–1742.

66 Wessel, H.P., Iversen, T., and Bundle, D.R. (1985). *J. Chem. Soc., Perkin Trans. 1* 2247–2250.

67 Stanca-Kaposta, E.C., Gamblin, D.P., Cocinero, E.J. et al. (2008). *J. Am. Chem. Soc.* 130: 10691–10696.

68 Jaipuri, F.A. and Pohl, N.L. (2008). *Org. Biomol. Chem.* 6: 2686–2691.

69 Voss, A., Nemati, N., Poghosyan, H. et al. (2012). *Carbohydrate Chemistry: Proven Synthetic Methods*, vol. 1, 303–326. CRC Press, Taylor & Francis.

70 Arimitsu, K., Kimura, H., Kajimoto, T. et al. (2013). *J. Labelled Compd. Radiopharm.* 56: 562–572.

71 Chai, Y., Zu, Y., Sheng, J. et al. (2015). Preparation of 1,5-enynol compounds as sugar terminal protecting groups. CN104513137A.

72 Jensen, T., Mikkelsen, M., Lauritsen, A. et al. (2009). *J. Org. Chem.* 74: 8886–8889.

73 Lu, Y.P., Li, H., Cai, M.S., and Li, Z.J. (2001). *Carbohydr. Res.* 334: 289–294.

74 Ruiz Contreras, R., Kamerling, J.P., Breg, J., and Vliegenthart, J.F.G. (1988). *Carbohydr. Res.* 179: 411–418.
75 Kovac, P., Taylor, R.B., and Glaudemans, C.P.J. (1985). *J. Org. Chem.* 50: 5323–5333.
76 Fuerstner, A., Jeanjean, F., Razon, P. et al. (2003). *Chem. Eur. J.* 9: 320–326.
77 Postema, M.H.D., TenDyke, K., Cutter, J. et al. (2009). *Org. Lett.* 11: 1417–1420.
78 Haller, M.F. and Boons, G.-J. (2002). *Eur. J. Org. Chem.* 2033–2038.
79 Yamamura, T., Hada, N., Kaburaki, A. et al. (2004). *Carbohydr. Res.* 339: 2749–2759.
80 Elsaidi, H.R.H., Paszkiewicz, E., and Bundle, D.R. (2015). *Carbohydr. Res.* 408: 96–106.
81 Kinzy, W. and Schmidt, R.R. (1987). *Carbohydr. Res.* 164: 265–276.
82 Prosperi, D., Ronchi, S., Lay, L. et al. (2004). *Eur. J. Org. Chem.* 395–405.
83 Paulsen, H. and Tietz, H. (1985). *Carbohydr. Res.* 144: 205–229.
84 Kong, L., Almond, A., Bayley, H., and Davis, B.G. (2016). *Nat. Chem.* 8: 461–469.
85 Wu, X., Lipinski, T., Paszkiewicz, E., and Bundle, D.R. (2008). *Chem. Eur. J.* 14: 6474–6482.
86 Lattova, E., Svetlikova, Z., Mikusova, K. et al. (2013). *RSC Adv.* 3: 17784–17792.
87 Cheng, K., Liu, J., Liu, X. et al. (2009). *Carbohydr. Res.* 344: 841–850.
88 Petitou, M., Duchaussoy, P., Lederman, I. et al. (1986). *Carbohydr. Res.* 147: 221–236.
89 Jacquinet, J.-C. and Sinaÿ, P. (1987). *Carbohydr. Res.* 159: 229–253.
90 Wu, X., Shen, Z., Zeng, X. et al. (2008). *Carbohydr. Res.* 343: 1718–1729.
91 Chassagne, P., Raibaut, L., Guerreiro, C., and Mulard, L.A. (2013). *Tetrahedron* 69: 10337–10350.
92 Greimel, P., Jabs, S., Storch, S. et al. (2006). *Carbohydr. Res.* 341: 918–924.
93 Hargreaves, J.M., Le Guen, Y., Guerreiro, C. et al. (2014). *Org. Biomol. Chem.* 12: 7728–7749.
94 Narouz, M., Soliman, S., Bassily, R. et al. (2013). *Lett. Org. Chem.* 10: 502–509.
95 Boutet, J. and Mulard, L.A. (2008). *Eur. J. Org. Chem.* 5526–5542.
96 Barbaud, C., Bols, M., Lundt, I., and Sierks, M.R. (1995). *Tetrahedron* 51: 9063–9078.
97 Toshima, K., Yanagawa, K., Mukaiyama, S., and Tatsuta, K. (1990). *Tetrahedron Lett.* 31: 6697–6698.
98 Cormier, J.F., Isaac, M.B., and Li-Feng, C. (1993). *Tetrahedron Lett.* 34: 243–246.
99 Ikawa, T., Hattori, K., Sajiki, H., and Hirota, K. (2004). *Tetrahedron* 60: 6901–6911.
100 Kaisalo, L.H. and Hase, T.A. (2001). *Tetrahedron Lett.* 42: 7699–7701.
101 Gadikota, R.R., Callam, C.S., Appelmelk, B.J., and Lowary, T.L. (2003). *J. Carbohydr. Chem.* 22: 149–170.
102 Tanaka, M., Nashida, J., Takahashi, D., and Toshima, K. (2016). *Org. Lett.* 18: 2288–2291.
103 Finan, P.A. and Warren, C.D. (1962). *J. Chem. Soc.* 3089–3092.
104 Tsutsui, N., Tanabe, G., Kita, A. et al. (2013). *Tetrahedron Lett.* 54: 451–453.
105 Sun, X., Lee, H., Lee, S., and Tan, K.L. (2013). *Nat. Chem.* 5: 790–795.

106 Tarling, C.A. and Withers, S.G. (2004). *Carbohydr. Res.* 339: 2487–2497.
107 Pozsgay, V. (1995). *J. Am. Chem. Soc.* 117: 6673–6681.
108 Lichtenthaler, F.W., Klares, U., Szurmai, Z., and Werner, B. (1998). *Carbohydr. Res.* 305: 293–303.
109 Kaesbeck, L. and Kessler, H. (1997). *Liebigs Ann. Recl.* 165–167.
110 Ohara, K., Lin, C.-C., Yang, P.-J. et al. (2013). *J. Org. Chem.* 78: 6390–6411.
111 Tanabe, G., Matsuda, Y., Oka, M. et al. (2016). *J. Org. Chem.* 81: 3407–3415.
112 Kleeb, S., Jiang, X., Frei, P. et al. (2016). *J. Med. Chem.* 59: 3163–3182.
113 Nicolaou, K.C., Caulfield, T.J., Kataoka, H., and Stylianides, N.A. (1990). *J. Am. Chem. Soc.* 112: 3693–3695.
114 Lewis, J.C., Bastian, S., Bennett, C.S. et al. (2009). *Proc. Natl. Acad. Sci. U.S.A.* 106: 16550–16555.
115 Arbues, A., Malaga, W., Constant, P. et al. (2016). *ACS Chem. Biol.* 11: 2865–2875.
116 Si, A. and Misra, A.K. (2016). *Tetrahedron* 72: 4435–4441.
117 Crich, D. and Li, H. (2000). *J. Org. Chem.* 65: 801–805.
118 Veleti, S.K., Lindenberger, J.J., Ronning, D.R., and Sucheck, S.J. (2014). *Bioorg. Med. Chem.* 22: 1404–1411.
119 Lee, A., Farrand, K.J., Dickgreber, N. et al. (2006). *Carbohydr. Res.* 341: 2785–2798.
120 Bieg, T. and Szeja, W. (1985). *Synthesis* 76–77.
121 Kozioł, A., Lendzion-Paluch, A., and Manikowski, A. (2013). *Org. Process Res. Dev.* 17: 869–875.
122 Crich, D. and Li, L. (2009). *J. Org. Chem.* 74: 773–781.
123 Ashmus, R.A. and Lowary, T.L. (2014). *Org. Lett.* 16: 2518–2521.
124 Liang, F.-S., Wang, S.-K., Nakatani, T., and Wong, C.-H. (2004). *Angew. Chem. Int. Ed.* 43: 6496–6500.
125 Paulsen, H., Bien, L.N., Sinnwell, V. et al. (1985). *Liebigs Ann. Chem.* 1513–1536.
126 Ali, H.M., Koza, G., Rwoa'a, T.H. et al. (2016). *Tetrahedron* 72: 7143–7158.
127 Matsui, R., Seto, K., Sato, Y. et al. (2011). *Angew. Chem. Int. Ed.* 50: 680–683.
128 Ogawa, S., Nose, T., Ogawa, T. et al. (1985). *J. Chem. Soc., Perkin Trans. 1* 2369–2374.
129 Cheng, J.M.H., Dangerfield, E.M., Timmer, M.S.M., and Stocker, B.L. (2014). *Org. Biomol. Chem.* 12: 2729–2736.
130 Jaworek, C.H., Iacobucci, S., Calias, P., and d'Alarcao, M. (2001). *Carbohydr. Res.* 331: 375–391.
131 Pallanca, J.E. and Turner, N.J. (1993). *J. Chem. Soc., Perkin Trans. 1* 3017–3022.
132 Knapp, S. and Kukkola, P.J. (1990). *J. Org. Chem.* 55: 1632–1636.
133 Zhu, X., Dere, R.T., and Jiang, J. (2011). *Tetrahedron Lett.* 52: 4971–4974.
134 Collin, M.-P., Hobbie, S.N., Bottger, E.C., and Vasella, A. (2009). *Helv. Chim. Acta* 92: 230–266.
135 Walvoort, M.T.C., Lodder, G., Overkleeft, H.S. et al. (2010). *J. Org. Chem.* 75: 7990–8002.
136 Li, X., Saleh, Z., Egri, B. et al. (2015). *Tetrahedron Lett.* 56: 1420–1422.
137 Kovac, P. and Bauer, S. (1972). *Tetrahedron Lett.* 13: 2349–2350.
138 Wang, A. and Auzanneau, F.-I. (2010). *Carbohydr. Res.* 345: 1216–1221.

139 Akhtar, T. and Cumpstey, I. (2007). *Tetrahedron Lett.* 48: 8673–8677.
140 Crich, D. and Karatholuvhu, M.S. (2008). *J. Org. Chem.* 73: 5173–5176.
141 Mukai, C., Kataoka, O., and Hanaoka, M. (1995). *J. Org. Chem.* 60: 5910–5918.
142 Cheng, L., Chen, Q., Liu, J., and Du, Y. (2007). *Carbohydr. Res.* 342: 975–981.
143 Yao, W., Yan, J., Chen, X. et al. (2015). *Carbohydr. Res.* 401: 5–10.
144 Adinolfi, M., Barone, G., Iadonisi, A., and Schiattarella, M. (2001). *Tetrahedron Lett.* 42: 5971–5972.
145 Emmadi, M. and Kulkarni, S.S. (2014). *Carbohydr. Res.* 399: 57–63.
146 Riley, J.G. and Grindley, T.B. (2001). *J. Carbohydr. Chem.* 20: 159–169.
147 Manabe, S., Yamaguchi, M., and Ito, Y. (2013). *Chem. Commun.* 49: 8332–8334.
148 Crich, D. and Vinogradova, O. (2007). *J. Org. Chem.* 72: 3581–3584.
149 Dhenin, S.G.Y., Moreau, V., Nevers, M.-C. et al. (2009). *Org. Biomol. Chem.* 7: 5184–5199.
150 Nakahara, Y., Ozawa, C., Tanaka, E. et al. (2007). *Tetrahedron* 63: 2161–2169.
151 Ichiyanagi, T., Takatani, M., Sakamoto, K. et al. (2002). *Tetrahedron Lett.* 43: 3297–3300.
152 Yoshida, T., Chiba, T., Yokochi, T. et al. (2001). *Carbohydr. Res.* 335: 167–180.
153 Forget, S.M., Bhattasali, D., Hart, V.C. et al. (2012). *Chem. Sci.* 3: 1866–1878.
154 Arimitsu, K., Kajimoto, T., Kimura, H. et al. (2011). *Heterocycles* 83: 2779–2802.
155 Moravcova, J., Heissigerova, H., Kocalka, P. et al. (2003). *Tetrahedron Lett.* 44: 8797–8800.
156 Gau, B., Lemétais, A., Lepore, M. et al. (2013). *ChemBioChem* 14: 2413–2417.
157 Bourgeaux, E. and Combret, J.C. (2000). *Tetrahedron: Asymmetry* 11: 4189–4205.
158 Downey, A.M. and Cairo, C.W. (2013). *Carbohydr. Res.* 381: 123–132.
159 Nakagawa, A., Tanaka, M., Hanamura, S. et al. (2015). *Angew. Chem. Int. Ed.* 54: 10935–10939.
160 Ueda, Y., Furuta, T., and Kawabata, T. (2015). *Angew. Chem. Int. Ed.* 54: 11966–11970.
161 Wei, Y., Wen, J., Zhu, G. et al. (2015). Preparation of glucopyranosyl derivatives and their uses as SGLT2 antidiabetic agents. WO2015043511A1.
162 Li, Y., Shi, Z., Chen, L. et al. (2017). *J. Med. Chem.* 60: 4173–4184.
163 Xie, J., Menand, M., and Valery, J.-M. (2005). *Carbohydr. Res.* 340: 481–487.
164 Kumar, P., Shustov, G., Liang, H. et al. (2012). *J. Med. Chem.* 55: 6033–6046.
165 Alzeer, J. and Vasella, A. (1995). *Helv. Chim. Acta* 78: 177–193.
166 Stick, R.V. and Stubbs, K.A. (2005). *J. Carbohydr. Chem.* 24: 529–547.
167 Kartha, K.P.R., Kiso, M., and Hasegawa, A. (1989). *J. Carbohydr. Chem.* 8: 675–679.
168 Binder, T.P. and Robyt, J.F. (1986). *Carbohydr. Res.* 147: 149–154.
169 Shivatare, S.S., Chang, S.-H., Tsai, T.-I. et al. (2013). *J. Am. Chem. Soc.* 135: 15382–15391.
170 Zegelaar-Jaarsveld, K., van der Plas, S.C., van der Marel, G.A., and van Boom, J.H. (1996). *J. Carbohydr. Chem.* 15: 591–610.
171 Kaitsubata, Y., Aramaki, R., Nishioka, K. et al. (2016). *Tetrahedron Lett.* 57: 1154–1157.
172 Kwon, Y.T., Lee, Y.J., Lee, K., and Kim, K.S. (2004). *Org. Lett.* 6: 3901–3904.

173 Compostella, F., Ronchi, S., Panza, L. et al. (2006). *Chem. Eur. J.* 12: 5587–5595.

174 Nicolaou, K.C., Mitchell, H.J., Rodriguez, R.M. et al. (2000). *Chem. Eur. J.* 6: 3149–3165.

175 Mong, K.-K.T. and Wong, C.-H. (2002). *Angew. Chem. Int. Ed.* 41: 4087–4090.

176 Shiozaki, M., Tashiro, T., Koshino, H. et al. (2010). *Carbohydr. Res.* 345: 1663–1684.

177 Ramos, D., Rollin, P., and Klaffke, W. (2001). *J. Org. Chem.* 66: 2948–2956.

178 Gu, G., Wei, G., and Du, Y. (2004). *Carbohydr. Res.* 339: 1155–1162.

179 Chassagne, P., Fontana, C., Guerreiro, C. et al. (2013). *Eur. J. Org. Chem.* 4085–4106.

180 Zhu, S.-Y., Huang, J.-S., Zheng, S.-S. et al. (2013). *Org. Lett.* 15: 4154–4157.

181 Rich, J.R. and Bundle, D.R. (2004). *Org. Lett.* 6: 897–900.

182 Hori, K., Sawada, N., Ando, H. et al. (2003). *Eur. J. Org. Chem.* 3752–3760.

183 He, Y., Yang, Y., and Iyer, S.S. (2016). *Bioconjugate Chem.* 27: 1509–1517.

184 Chen, W.A., Mautino, M.R., and Martin, B. (2014). Preparation of aminodeoxy disaccharide-modified glycoproteins and uses thereof as vaccines, immunostimulants, and antiviral agents. WO2014151423A1.

185 Veerapen, N., Yuan, Y., Sanders, D.A.R., and Pinto, B.M. (2004). *Carbohydr. Res.* 339: 2205–2217.

186 Volbeda, A.G., Kistemaker, H.A.V., Overkleeft, H.S. et al. (2015). *J. Org. Chem.* 80: 8796–8806.

187 Yu, W., Su, M., Gao, X. et al. (2000). *Tetrahedron Lett.* 41: 4015–4017.

188 Albert, R., Dax, K., Pleschko, R., and Stütz, A.E. (1985). *Carbohydr. Res.* 137: 282–290.

189 Uryu, T., Yamanaka, M., Henmi, M. et al. (1986). *Carbohydr. Res.* 157: 157–169.

190 Faghih, R. and Fraser-Reid, B. (1987). *Carbohydr. Res.* 169: 247–251.

191 Smith, A.B. III and Rivero, R.A. (1987). *J. Am. Chem. Soc.* 109: 1272–1274.

192 Nicolaou, K.C., Hummel, C.W., Pitsinos, E.N. et al. (1992). *J. Am. Chem. Soc.* 114: 10082–10084.

193 Nicolaou, K.C., Mitchell, H.J., Fylaktakidou, K.C. et al. (2000). *Chem. Eur. J.* 6: 3116–3148.

194 Ishiwata, A., Munemura, Y., and Ito, Y. (2008). *Tetrahedron* 64: 92–102.

195 Noya, B., Paredes, M.D., Ozores, L., and Alonso, R. (2000). *J. Org. Chem.* 65: 5960–5968.

196 Ikeda, Y., Furukawa, K., and Yamada, H. (2002). *Carbohydr. Res.* 337: 1499–1501.

197 Boltje, T.J., Li, C., and Boons, G.-J. (2010). *Org. Lett.* 12: 4636–4639.

198 Du, Y., Zhang, M., and Kong, F. (2000). *Org. Lett.* 2: 3797–3800.

199 Grayson, E.J., Bernardes, G.J.L., Chalker, J.M. et al. (2011). *Angew. Chem. Int. Ed.* 50: 4127–4132.

200 Yang, Z., Lin, W., and Yu, B. (2000). *Carbohydr. Res.* 329: 879–884.

201 Nicolaou, K.C., Daines, R.A., Ogawa, Y., and Chakraborty, T.K. (1988). *J. Am. Chem. Soc.* 110: 4696–4705.

202 Cassidy, J.F. and Williams, J.M. (1986). *Tetrahedron Lett.* 27: 4355–4358.

203 Akashi, Y., Takao, K.-I., and Tadano, K.-I. (2009). *Tetrahedron Lett.* 50: 1139–1142.

204 Arias-Perez, M.S. and Santos, M.J. (1996). *Tetrahedron* 52: 10785–10798.

205 Halmos, T., Montserret, R., Filippi, J., and Antonakis, K. (1987). *Carbohydr. Res.* 170: 57–69.

206 Yoon, S. (2002). *J. Korean Chem. Soc.* 46: 590–594.

207 Tschamber, T., Adam, S., Matsuya, Y. et al. (2007). *Bioorg. Med. Chem. Lett.* 17: 5101–5106.

208 Du, Y., Zhang, M., and Kong, F. (2001). One-pot three-step synthesis of 6-O-trityl-3(or 2)-O-silyl monoglycoside and its application. CN1324801A.

209 Kuczynska, K., Cmoch, P., Rarova, L. et al. (2016). *Carbohydr. Res.* 423: 49–69.

210 Gu, G., Fang, M., and Du, Y. (2011). *Carbohydr. Res.* 346: 2801–2804.

211 Martin-Lomas, M., Flores-Mosquera, M., and Chiara, J.L. (2000). *Eur. J. Org. Chem.* 1547–1562.

212 Wang, B., McClosky, D.D., Anderson, C.T., and Chen, G. (2016). *Carbohydr. Res.* 433: 54–62.

213 Matsuo, I., Wada, M., Manabe, S. et al. (2003). *J. Am. Chem. Soc.* 125: 3402–3403.

214 Brown, R.T., Mayalarp, S.P., and Watts, J. (1997). *J. Chem. Soc., Perkin Trans. 1* 1633–1638.

215 Buskas, T., Li, Y., and Boons, G.-J. (2005). *Chem. Eur. J.* 11: 5457–5467.

216 Ohtsuka, I., Hada, N., Sugita, M., and Takeda, T. (2002). *Carbohydr. Res.* 337: 2037–2047.

217 Meutermans, W., West, M.L., and Adamson, G. et al. (2003). Combinatorial library solid phase synthesis of disaccharides for drug discovery. WO2003093286A1.

218 Takatani, M. and Ito, Y. (2006). *Chem. Asian J.* 1: 64–75.

219 Nakano, J., Ishiwata, A., Ohta, H., and Ito, Y. (2007). *Carbohydr. Res.* 342: 675–695.

220 Matsuo, I., Wada, M., and Ito, Y. (2002). *Tetrahedron Lett.* 43: 3273–3275.

221 Peng, W., Tang, P., Hu, X. et al. (2007). *Bioorg. Med. Chem. Lett.* 17: 5506–5509.

222 Nicolaou, K.C., Li, Y., Fylaktakidou, K.C. et al. (2001). *Angew. Chem. Int. Ed.* 40: 3854–3857.

223 Tully, S.E., Mabon, R., Gama, C.I. et al. (2004). *J. Am. Chem. Soc.* 126: 7736–7737.

224 Salamone, S., Guerreiro, C., Cambon, E. et al. (2015). *J. Org. Chem.* 80: 11237–11257.

225 Argunov, D.A., Karelin, A.A., Grachev, A.A. et al. (2011). *Russ. Chem. Bull.* 60: 1004–1011.

226 Karelin, A.A., Tsvetkov, Y.E., Paulovicova, L. et al. (2010). *Carbohydr. Res.* 345: 1283–1290.

227 Li, K. and Helm, R.F. (1995). *J. Agric. Food Chem.* 43: 2098–2103.

228 Denmark, S.E., Kobayashi, T., and Regens, C.S. (2010). *Tetrahedron* 66: 4745–4759.

229 Miyagawa, A., Matsuda, T., and Yamamura, H. (2015). *J. Carbohydr. Chem.* 34: 215–246.

230 Crich, D., Hu, T., and Cai, F. (2008). *J. Org. Chem.* 73: 8942–8953.

231 Hixson, J.L., Hayasaka, Y., Curtin, C.D. et al. (2016). *J. Agric. Food Chem.* 64: 9401–9411.

232 Zhu, Y. and Ralph, J. (2011). *Tetrahedron Lett.* 52: 3729–3731.

233 Kunz, H. and Harreus, A. (1982). *Liebigs Ann. Chem.* 1982: 41–48.
234 Ziegler, T. and Pantkowski, G. (1994). *Liebigs Ann. Chem.* 1994: 659–664.
235 Reynolds, D.D. and Evans, W.L. (1942). *Org. Synth.* 22: 56–58.
236 Klausener, A., Beyer, G., Leismann, H. et al. (1989). *Tetrahedron* 45: 4989–5002.
237 Braun, G. (1937). *Org. Synth.* 17: 34–35.
238 Staněk, J. and Jitka Černá, M. (1963). *Tetrahedron Lett.* 4: 35–37.
239 Wolfrom, M.L., McWain, P., Shafizadeh, F., and Thompson, A. (1959). *J. Am. Chem. Soc.* 81: 6080–6082.
240 Robertson, G.J. and Griffith, C.F. (1935). *J. Chem. Soc.* 1193–1201.
241 Hough, L., Chowdhary, M.S., and Richardson, A.C. (1978). *Chem. Commun.* 664–665.
242 Heinkele, G., Geditz, M.C.K., Ganchev, B. et al. (2014). *J. Labelled Compd. Radiopharm.* 57: 699–703.
243 Josien-Lefebvre, D., Desmares, G., and Le Drian, C. (2001). *Helv. Chim. Acta* 84: 890–897.
244 O'Neill, P.M., Scheinmann, F., Stachulski, A.V. et al. (2001). *J. Med. Chem.* 44: 1467–1470.
245 Shah, S., Nagata, M., Yamasaki, S., and Williams, S.J. (2016). *Chem. Commun.* 52: 10902–10905.
246 Bauer, J. and Rademann, J. (2005). *J. Am. Chem. Soc.* 127: 7296–7297.
247 Doz, E., Rose, S., Court, N. et al. (2009). *J. Biol. Chem.* 284: 23187–23196.
248 Zhulin, V.M., Makarova, Z.G., Klimov, E.M. et al. (1989). *Russ. Chem. Bull.* 38: 2568–2573.
249 Barroca, N. and Schmidt, R.R. (2004). *Org. Lett.* 6: 1551–1554.
250 Veeneman, G.H., Gomes, L.J.F., and van Boom, J.H. (1989). *Tetrahedron* 45: 7433–7448.
251 Allevi, P., Anastasia, M., Ciuffreda, P. et al. (1993). *J. Org. Chem.* 58: 4175–4178.
252 Abbas, S. and Haines, A.H. (1975). *Carbohydr. Res.* 41: 298–303.
253 Nunomura, S. and Ogawa, T. (1988). *Tetrahedron Lett.* 29: 5681–5684.
254 Zhu, Y. and Kong, F. (2000). *Synlett* 1783–1787.
255 Bourke, J., Brereton, C.F., Gordon, S.V. et al. (2014). *Org. Biomol. Chem.* 12: 1114–1123.
256 Ueda, Y., Chuang, J.M., Crast, L.B., and Partyka, R.A. (1988). *J. Org. Chem.* 53: 5107–5113.
257 Srivastava, O.P. and Hindsgaul, O. (1987). *Carbohydr. Res.* 161: 195–210.
258 Descotes, G., Mentech, J., and Roques, N. (1989). *Carbohydr. Res.* 188: 63–70.
259 Duynstee, H.I., de Koning, M.C., Ovaa, H. et al. (1999). *Eur. J. Org. Chem.* 2623–2632.
260 Hanessian, S., Saavedra, O.M., Mascitti, V. et al. (2001). *Tetrahedron* 57: 3267–3280.
261 Chery, F., Cronin, L., O'Brien, J.L., and Murphy, P.V. (2004). *Tetrahedron* 60: 6597–6608.
262 Koeller, K.M. and Wong, C.-H. (2000). *Chem. Eur. J.* 6: 1243–1251.
263 Aly, M.R.E., Ibrahim, E.-S.I., El-Ashry, E.-S.H.E., and Schmidt, R.R. (2000). *Eur. J. Org. Chem.* 319–326.

264 Pekari, K., Tailler, D., Weingart, R., and Schmidt, R.R. (2001). *J. Org. Chem.* 66: 7432–7442.
265 Jung, K.-H., Hoch, M., and Schmidt, R.R. (1989). *Liebigs Ann. Chem.* 1099–1106.
266 Meyer, B. and Stuike-Prill, R. (1990). *J. Org. Chem.* 55: 902–906.
267 Adak, S., Emmadi, M., and Kulkarni, S.S. (2014). *RSC Adv.* 4: 7611–7616.
268 Nicolaou, K.C., Rodríguez, R.M., Mitchell, H.J. et al. (2000). *Chem. Eur. J.* 6: 3095–3115.
269 Wang, P., Kim, Y.-J., Navarro-Villalobos, M. et al. (2005). *J. Am. Chem. Soc.* 127: 3256–3257.
270 Wei, A., Boy, K.M., and Kishi, Y. (1995). *J. Am. Chem. Soc.* 117: 9432–9436.
271 Düffels, A., Green, L.G., Ley, S.V., and Miller, A.D. (2000). *Chem. Eur. J.* 6: 1416–1430.
272 Dhamale, O.P., Lawrence, R., Wiegmann, E.M. et al. (2017). *ACS Chem. Biol.* 12: 367–373.
273 Kwon, Y.-U., Soucy, R.L., Snyder, D.A., and Seeberger, P.H. (2005). *Chem. Eur. J.* 11: 2493–2504.
274 Fan, R.-H., Achkar, J., Hernández-Torres, J.M., and Wei, A. (2005). *Org. Lett.* 7: 5095–5098.
275 Wang, S.-M., Li, J., Li, H.-Y. et al. (2008). *Chin. J. Org. Chem.* 28: 120–122.
276 Mastihubová, M. and Biely, P. (2010). *Carbohydr. Res.* 345: 1094–1098.
277 Ayers, S. and Sneden, A.T. (2004). *J. Nat. Prod.* 67: 221–224.
278 Ueda, Y., Chuang, J.M., Crast, L.B., and Partyka, R.A. (1989). *J. Antibiotics* 42: 1379–1392.
279 Terauchi, T., Koyama, Y., Machida, S. et al. (2011). *J. Appl. Glycosci.* 59: 11–19.
280 Roen, A., Mayato, C., Padron, J.I., and Vazquez, J.T. (2008). *J. Org. Chem.* 73: 7266–7279.
281 Bording, S. and Bach, T. (2014). *Chem. Commun.* 50: 4901–4903.
282 Aqeel, A., Sato, K.-I., Hashimoto, H., and Yoshimura, J. (1989). *J. Carbohydr. Chem.* 8: 405–412.
283 Rich, J.R., McGavin, R.S., and Reimer, K.B. (2001). *Carbohydr. Res.* 330: 517–521.
284 Peng, W., Sun, J., Lin, F. et al. (2004). *Synlett* 259–262.
285 Abronina, P.I., Galkin, K.I., Backinowsky, L.V., and Grachev, A.A. (2009). *Russ. Chem. Bull.* 58: 457–467.
286 Tsvetkov, Y.E., Backinowsky, L.V., and Kochetkov, N.K. (1989). *Carbohydr. Res.* 193: 75–90.
287 Sanapala, S.R. and Kulkarni, S.S. (2016). *Org. Lett.* 18: 3790–3793.
288 Sherman, A.A., Yudina, O.N., Mironov, Y.V. et al. (2001). *Carbohydr. Res.* 336: 13–46.
289 Grkovic, T., Pouwer, R.H., Wang, Q. et al. (2015). *J. Nat. Prod.* 78: 1215–1220.
290 Smiljanic, N., Halila, S., Moreau, V., and Djedaïni-Pilard, F. (2003). *Tetrahedron Lett.* 44: 8999–9002.
291 Huang, G.-L., Mei, X.-Y., Liu, M.-X., and Liu, T.-C. (2004). *Bioorg. Med. Chem. Lett.* 14: 6027–6029.
292 Laurent, N., Lafont, D., Boullanger, P., and Mallet, J.M. (2005). *Carbohydr. Res.* 340: 1885–1892.

293 Paulsen, H., Rauwald, W., and Weichert, U. (1988). *Liebigs Ann. Chem.* 75–86.

294 Padrón, J.I. and Vázquez, J.T. (1995). *Tetrahedron: Asymmetry* 6: 857–858.

295 Suzuki, K., Nonaka, H., and Yamaura, M. (2004). *J. Carbohydr. Chem.* 23: 253–259.

296 Tatibouët, A., Lefoix, M., Nadolny, J. et al. (2001). *Carbohydr. Res.* 333: 327–334.

297 Bhattacharyya, S., Magnusson, B.G., Wellmar, U., and Nilsson, U.J. (2001). *J. Chem. Soc., Perkin Trans. 1* 886–890.

298 Spohr, U. and Lemieux, R.U. (1988). *Carbohydr. Res.* 174: 211–237.

299 Abdel-Malik, M.M. and Perlin, A.S. (1989). *Carbohydr. Res.* 190: 39–52.

300 Nakahara, Y. and Ogawa, T. (1988). *Carbohydr. Res.* 173: 306–315.

301 Bertolini, M. and Glaudemans, C.P.J. (1970). *Carbohydr. Res.* 15: 263–270.

302 Kanie, O., Takeda, T., and Ogihara, Y. (1990). *J. Carbohydr. Chem.* 9: 159–165.

303 Kováč, P. and Taylor, R.B. (1987). *Carbohydr. Res.* 167: 153–173.

304 Soliman, S.E. and Kováč, P. (2015). *J. Org. Chem.* 80: 11227–11232.

305 Yoshida, M., Kawakami, Y., Ishida, H. et al. (1996). *J. Carbohydr. Chem.* 15: 399–418.

306 Arranz, E. and Boons, G.-J. (2001). *Tetrahedron Lett.* 42: 6469–6471.

307 Crich, D. and Cai, F. (2007). *Org. Lett.* 9: 1613–1615.

308 Arasappan, A. and Fuchs, P.L. (1995). *J. Am. Chem. Soc.* 117: 177–183.

309 Thompson, C., Ge, M., and Kahne, D. (1999). *J. Am. Chem. Soc.* 121: 1237–1244.

310 Velarde, S., Urbina, J., and Peña, M.R. (1996). *J. Org. Chem.* 61: 9541–9545.

311 Daragics, K. and Fügedi, P. (2010). *Org. Lett.* 12: 2076–2079.

312 Wang, Y.-H., Yeh, H.-W., Wang, H.-W. et al. (2013). *Carbohydr. Res.* 375: 118–135.

313 Manthorpe, J.M., Szpilman, A.M., and Carreira, E.M. (2005). *Synthesis* 3380–3388.

314 Li, Q., Li, H., Li, Q. et al. (2002). *Carbohydr. Res.* 337: 1929–1934.

315 Pozsgay, V. (2000). *Tetrahedron: Asymmetry* 11: 151–172.

316 Behera, A., Emmadi, M., and Kulkarni, S.S. (2014). *RSC Adv.* 4: 58573–58580.

317 Barroca, N. and Jacquinet, J.-C. (2002). *Carbohydr. Res.* 337: 673–689.

318 Zhang, X., Zhou, Y., Zuo, J., and Yu, B. (2015). *Nat. Commun.* 6: 5879.

319 Pozsgay, V. (1998). *Angew. Chem. Int. Ed.* 37: 138–142.

320 Pearson, A.G., Peak, I.R., Wilson, J.C., and Grice, I.D. (2011). *Carbohydr. Res.* 346: 2805–2811.

321 Nilsson, M. and Norberg, T. (1988). *Carbohydr. Res.* 183: 71–82.

322 Guillemineau, M. and Auzanneau, F.-I. (2012). *J. Org. Chem.* 77: 8864–8878.

323 Sogabe, S., Ando, H., Koketsu, M., and Ishihara, H. (2006). *Tetrahedron Lett.* 47: 6603–6606.

324 Petitou, M., Duchaussoy, P., Lederman, I. et al. (1987). *Carbohydr. Res.* 167: 67–75.

325 Allen, J.G. and Fraser-Reid, B. (1999). *J. Am. Chem. Soc.* 121: 468–469.

326 Seki, M. and Mori, K. (2001). *Eur. J. Org. Chem.* 3797–3809.

327 Ohtsuka, I., Ako, T., Kato, R. et al. (2006). *Carbohydr. Res.* 341: 1476–1487.

328 Shakti, U., Ashwani, S., Anuradha, S. et al. (2009). Carbohydrate based toll-like receptor (TLR) antagonists. WO2009047792A2.

329 Nicolaou, K.C., Mitchell, H.J., Jain, N.F. et al. (1999). *Chem. Eur. J.* 5: 2648–2667.

330 Wong, C.-H., Zhang, Z., Ollmann, I. et al. (2000). Programmable one-pot oligosaccharide synthesis and structural effects of monosaccharides on the anomeric glycosylation reactivity. WO2000009527A1.

331 Lee, D. and Taylor, M.S. (2011). *J. Am. Chem. Soc.* 133: 3724–3727.

4

Regioselective Protection at the Secondary Positions of Carbohydrates with Acyclic Protecting Groups

Peter G. Goekjian and Sébastien Vidal

Université de Lyon, Institut de Chimie et Biochimie Moléculaires et Supramoléculaires (UMR 5246), Laboratoire de Chimie Organique 2-Glycochimie, CNRS and Université Claude Bernard Lyon 1, 43 Boulevard du 11 Novembre 1918, 69622, Villeurbanne, France

4.1 Introduction

The synthesis of carbohydrate derivatives, be it for oligosaccharide synthesis, medicinal chemistry, or total synthesis, generally requires the ability to install different protecting groups on each position of a monosaccharide. The anomeric position can be differentiated by exploiting its acetal-based reactivity (Chapter 5), and the primary alcohol based on sterics (Chapter 2), but it remains necessary to differentiate among the three secondary alcohols. The use of cyclic protecting groups contributes significantly to this effort (Chapter 11), but regioselective protection of secondary alcohols with acyclic protecting groups is usually required at some point. We will limit ourselves here to reactions that differentiate pyranoside secondary alcohols and thus not treat aminosugars (Chapter 6) or furanosides (Chapter 12) in this chapter. Some exceptionally efficient, one-pot reaction schemes to differentially protected sugars are presented in Chapter 7. We will not focus here on specific orthogonal protecting schemes but rather on installing silyl-, ether-, or ester-type protecting groups to allow for the greatest flexibility in designing a protection manifold.

The regioselective protection of secondary alcohols of carbohydrates generally relies on a few broad concepts: steric interactions, electronic effects, metal complexes, and kinetic vs thermodynamic regioselectivity. The inductive effect of anomeric acetal oxygens makes the C-2 hydroxyl group both more acidic and less nucleophilic, both aspects of which can be exploited under kinetic control. For sterics, gauche interactions normally render axial hydroxyl groups more hindered than the equatorial ones, but longer range interactions of sterically hindered protecting groups can easily overcome this effect under both kinetic and thermodynamic control. The nature of the protection reaction (silylation, etherification, or acylation) will also influence the nature of the transition state and hence the selectivity under kinetic conditions. In the case of esters and silyl ethers, migration under mild conditions [1, 2] makes it difficult to differentiate

Protecting Groups: Strategies and Applications in Carbohydrate Chemistry, First Edition.
Edited by Sébastien Vidal.

between kinetic and thermodynamic conditions. A variety of stable or transient metal complexes, in particular those involving tin, copper, or iron, can be used to direct the reaction to one or the other position. A certain number of specific chiral, hydrogen-bonding, or peptide catalysts have also been brought to bear on this problem. As a result of the interplay between all of these factors, there are not always straightforward rules to predict the regioselectivity in the protection of the secondary alcohols of monosaccharides. Furthermore, seemingly similar conditions can give different results.

We have, therefore, chosen to present an overview of empirical results in this area. The results will be presented based on the positions being selectively protected, the stereochemistry of the carbohydrate, the number of free hydroxyl groups in the starting material, and the protecting group being installed. Regioselective *bis*-protections of free pyranoside tetraols will be discussed in Section 4.5. An alternative approach is the regioselective mono-deprotection of a fully protected monosaccharide, which will be discussed in Section 4.6. We will discuss here regioselective protections with some of the major protecting groups (benzyl, substituted benzyl, 2-naphthylmethyl, and allyl ethers, silyl ethers, esters, and carbonates). We hope that these will give enough of a perspective to allow reasonable predictions for other protecting groups.

4.2 Regioselective Protections at the 2-Position

The regioselective monoprotection of the secondary alcohol at the 2-position of a sugar has been widely reported. The 2-hydroxyl group is generally the most reactive, yet careful control of the conditions is necessary to achieve good selectivity. As the 4- and 6-hydroxyl groups can be selectively protected using a cyclic acetal protecting group (Chapter 11), the most common regioselective protection thus differentiates between the alcohols at the 2- and 3-positions.

4.2.1 Glucose

4.2.1.1 Monoprotection at O-2 of Glucopyranoside 2,3-Diols

Monoprotection of the 2-hydroxyl group of 4,6-*O*-benzylidene and 4,6-*O*-isopropylidene glucopyranosides has been extensively studied (Table 4.1). Both nonparticipating (benzyl, substituted benzyl, allyl, and silyl) and C-2 participating groups (esters and carbonates) have been installed regioselectively at this position, leaving the 3-hydroxyl group free. In some situations, nonselective monoprotection and separation of the two regioisomers can be an efficient solution, in particular if both regioisomers can eventually be exploited, but we will focus here on the more selective examples.

Silyl Ethers There are relatively few examples of regioselective silylation of the 2-position of 4,6-*O*-benzylidene glucopyranoside sugars, as monosilylation generally favors the 3-position. As a general rule, the 2-*O*-silyl ether is the kinetic product, and the 3-*O*-silyl ether is the thermodynamic product in the benzylidene glucose series. Selective protection of the 2-position requires using either the

Table 4.1 Regioselective 2-*O*-functionalization of glucopyranoside 4,6-*O*-acetals.

Entry	Product	Conditions	Observations	References
1	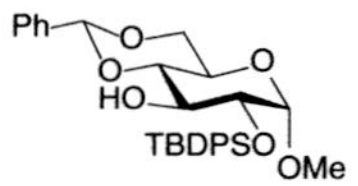	TBDPSCl, pyridine, imidazole (cat.), 94%	No experimental section provided	[3]
2	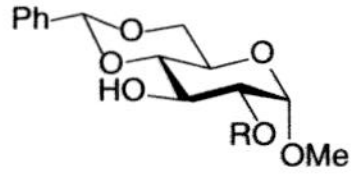	Trialkylsilyl-2-methylprop-2-ene-1-sulfinate (1 eq.), DMF, RT R = Me_3Si (83%), Et_3Si (97%), *i*Pr_3Si (95%), *t*$BuMe_2Si$ (96%)	Silylating reagent required but easily accessible	[4]
3	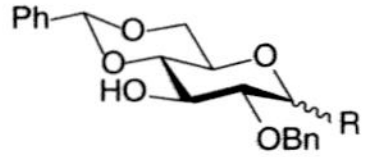	BnBr (1.23 eq.), *n*Bu_4NI (0.33 eq.), NaOH (1.6 eq.), CH_2Cl_2/H_2O (10/1), RT, 24 h R = αOMe (50%), βOMe (56%), βOBn (55%), βOPh (72%), βO(2-NitroBn) (83%)	Mixture of 2-*O*- and 3-*O*-regioisomers separated by chromatography	[5]
4	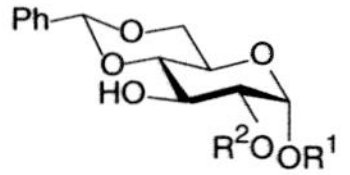	R^2Br (1.25–1.4 eq.), *n*Bu_4NX (0.12–0.25 eq.), 5–10% NaOH, CH_2Cl_2/H_2O (10/1), RT or reflux, 8–24 h $R^1 = OCH_2CCl_3$/X = Br R^2 = Bn (59%), All (72%) [6] R^1 = OMe/X = HSO_4 R^2 = Nap (51%) [7]	Mixture of 2-*O*- and 3-*O*-regioisomers separated by chromatography	[6, 7]
5	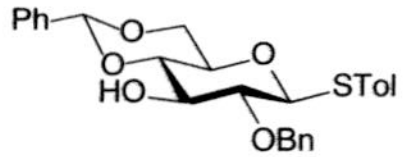	BnBr (1.3 eq.), NaH (1.3 eq.), DMF, RT, 30 min, 65% (3-*O*-Bn = 15%)	Mixture of 2-*O*- and 3-*O*-regioisomers separated by chromatography	[8]
6	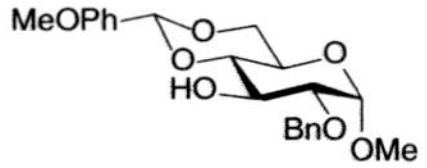	BnBr (3 eq.), *n*Bu_2SnO (1.1 eq.), *n*Bu_4NBr (1.1 eq.), MeCN, reflux, 18 h, 54% (3-*O*-Bn = 15%)	Mixture of 2-*O*- and 3-*O*-regioisomers separated by chromatography	[9]

(Continued)

Table 4.1 (Continued)

Entry	Product	Conditions	Observations	References
7	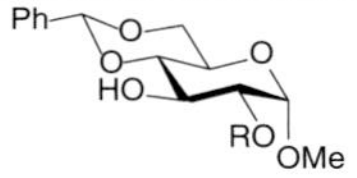	BnBr (20 eq.) or AllBr (8 eq.), $(nBu_3Sn)_2O$ (2 eq.), PhMe, 16–48 h R = Bn (89%), All (75%)	Mixture of 2-*O*- and 3-*O*-regioisomers separated by chromatography	[10]
8	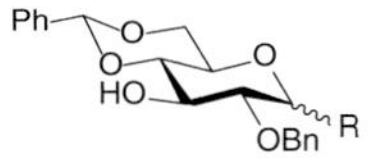	BnBr (1.1 eq.), $Fe(dibm)_3$ (0.05 eq.), K_2CO_3 (1.5 eq.), MeCN, 80 °C, 8–12 h R = αOMe (74%), βOMe (89%), βSPh (92%)		[11]
9	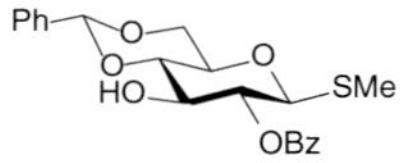	BzCl (1.2 eq.), nBu_4NI (1.5 eq.), K_2CO_3 (5 eq.), $CHCl_3$, RT, 48 h, 60%		[12]
10	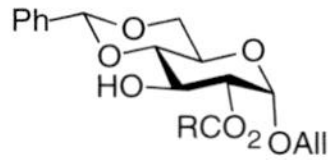	RCOCl (1–1.2 eq.), pyridine, −17 °C to RT, 3–16 h RCO = Bz (80%) [13, 14], Piv (80%) [15]	β-OAll (75%)	[13–15]
11	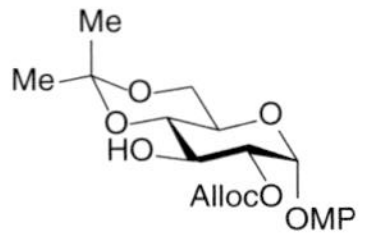	AllocCl (1.1 eq.), DMAP (0.1 eq.), pyridine (4 eq.), CH_2Cl_2, −35–0 °C, 1 h, 86%		[16]
12	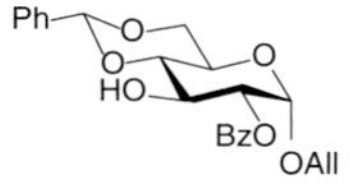	BzCl (1 eq.), imidazole (2 eq.), $CHCl_3$, reflux, 7 h, 76%		[17]
13	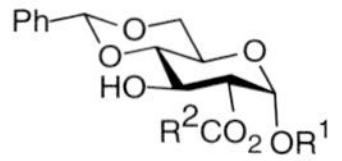	$(R^2CO)_2O$ (1.4 eq.), Et_3N (9 eq.), CH_2Cl_2, RT, 16 h R^1 = Me/R^2CO = Ac (80%) R^1 = Me/R^2CO = Bz (93%) R^1 = All/R^2CO = Ac (64%) R^1 = All/R^2CO = Bz (78%)		[18]

14	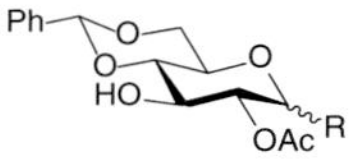	Ac_2O (1.1 eq.), $n$$Bu_4NOAc$ (0.6 eq.), MeCN, 40 °C, 8–12 h R = αOMe (88%), βOMe (85%)	3-*O*-acetylation with galactoside	[19]
15	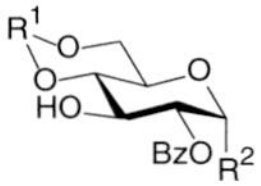	BzCN (1–1.2 eq.), Et_3N or DMAP (0.1 eq.), MeCN or CH_2Cl_2, −78 °C or RT, 0.5–4 h R^1 = PhCH, R^2 = OMe (89%) [20] R^1 = CMe_2, R^2 = OBn (76%) [21]		[20, 21]
16	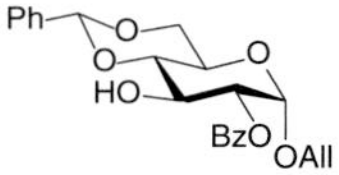	1-Benzoyloxy-benzotriazole (1.1 eq.), Et_3N (1.1 eq.), CH_2Cl_2, RT, 16 h, 82%		[22]
17	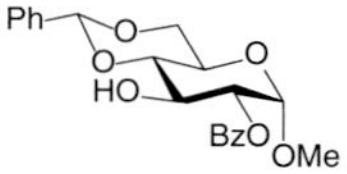	2-Benzoyliminomalononitrile (1.4 eq.), Et_3N (1.2 eq.), CH_2Cl_2, 0 °C to RT, 16 h, 89–93%	3-*O*-benzoylation with galactoside	[23]
18	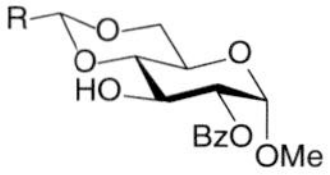	$PhCO_2H$ (0.75 eq.), PPh_3 (3 eq.), DEAD (0.5 eq.), THF, RT, 20 h then 60 °C, 20 h R = Ph (80%), 4-Me-Ph (75%)	Difficult to separate from reduced DEAD and phosphine oxide	[24]
19	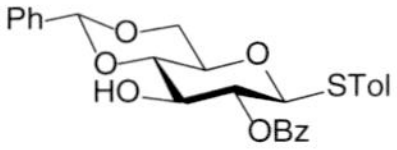	BzCl (1.25 eq.), Ag_2O (1.5 eq.), KI (0.2 eq.), CH_2Cl_2, RT, 8 h, 86%	Also with Gal and Man derivatives and several aglycons	[25]
20	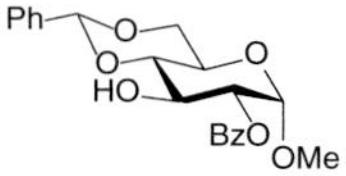	BzCl (1.1 eq.), $NiCl_2$ (1 eq.) NaH (2 eq.), THF, RT, 4–16 h, 85%	3-*O*-benzoylation with galactoside	[26]
21	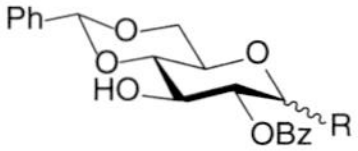	1. $n$$Bu_2SnO$ (1–1.05 eq.), PhMe or MeOH, reflux, 1 h 2. BzCl (1.1 eq.), PhMe or dioxane, RT, 1 h R = β$O(CH_2)_3NHCOCF_3$ (74%) [27] R = αOEt (90%) [28]		[27, 28]

(Continued)

Table 4.1 (Continued)

Entry	Product	Conditions	Observations	References
22	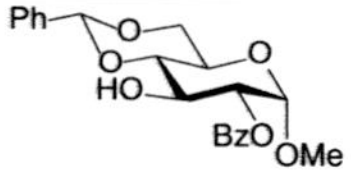	1. $Me_2Sn(OMe)_2$ (1.1 eq.), PhMe, reflux 2. BzCl (1.1 eq.), Et_3N (0.05 eq.), PhMe, 0 °C to RT, 1 h, 92%		[29]
23	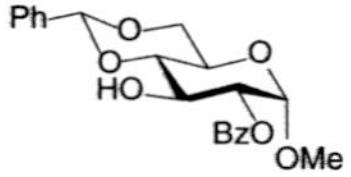	BzCl (1.2 eq.), Me_2SnCl_2 (0.01 eq.), K_2CO_3 (2 eq.), THF, RT, 12 h, 92%		[30]
24	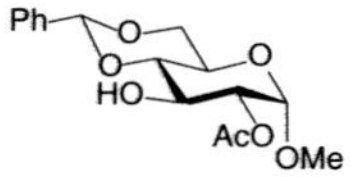	1. $Me_2Si(OMe)_2$ (2 eq.) or $MeSi(OMe)_3$ (1.2 eq.), MeCN, 80 °C 2. Ac_2O or AcCl (1.1 eq.), *n*Bu_4NOAc (0.3 eq.), MeCN, 40 °C, 63–65%	3-*O*-acetylation with galactoside	[31]
25	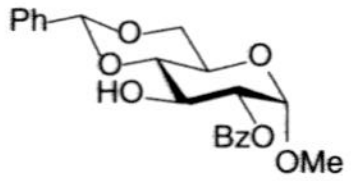	Bz_2O (1.3 eq.), $(CF_3CO_2)_2Cu$ (1.3 eq.), sym-collidine (1.3 eq.), MeCN, RT, 6 h, 82%	3-*O*-benzoylation with galactoside	[32]
26	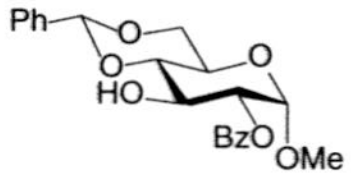	BzCl (1 eq.), *i*Pr_2NEt (1 eq.), $CuCl_2$(*R*)-PhBOX (0.1 eq.), $CHCl_3$, −40 °C, 85%	3-*O*-benzoylation with mannoside	[33]
27	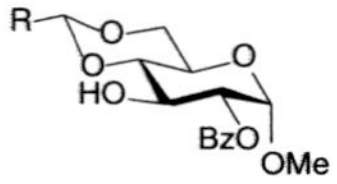	Bz_2O (10 eq.), peptide-DMAP (0.1 eq.), Et_3N (20 eq.), $CHCl_3$, −22 °C, 36 h R = Ph (91%), 4-O_2N-Ph (90%)	3-*O*-benzoylation with galactoside	[34]

bulky *t*-butyldiphenylsilyl group (entry 1) or an original sulfinate-based silylating agent for less-bulky silyl ethers (entry 2).

Alkyl Ethers Monoetherification of the 2-position of glucose 4,6-*O*-acetals with benzyl, substituted benzyl, 2-naphthylmethyl, or allyl protecting groups can be performed under traditional biphasic conditions (entries 3 and 4) or with sodium hydride in DMF (entry 5). The higher reactivity can be attributed to stabilization of a partial negative charge on the C2-oxygen in the transition state by the inductive effect of the ring oxygen. Reasonable selectivities can be achieved regardless of the nature or stereochemistry of the glycoside.

Metal complexes can be used to improve selectivity; stable cyclic or acyclic tin complexes are most popular (entries 6 and 7). However, these conditions give mixtures that need to be separated by chromatography and thus only present a moderate advantage over the metal-free conditions above. The use of iron complexes in catalytic amounts gives a more appealing approach, involving transient iron complexation with the sugar (entry 8).

Esters and Carbonates Esters can be installed regioselectively on the C-2 position in much the same way as alkyl ethers, both with and without metals. However, there is the added issue of acyl group migration (Chapter 8), which means that the selectivity can be under either kinetic or thermodynamic control under similar conditions. Monoacylation without metal complexes is generally achieved with acid chlorides (entries 9–12) or anhydrides (entries 13 and 14), using potassium carbonate (entry 9), DMAP-pyridine (entries 10 and 11), imidazole (entry 12), triethylamine (entry 13), or tetrabutylammonium acetate (entry 14) as base and/or nucleophilic catalyst. Very interesting results have been reported with more exotic reagents such as benzoyl cyanides (entry 15), hydroxybenzotriazole esters (entry 16), *O*-benzoyloximes (entry 17), or under Mitsunobu conditions (entry 18). The selectivity can be further improved by the use of metal complexes, include silver (entry 19), nickel (entry 20), tin (entries 21–23), silicon (entry 24), and copper (entry 25). In the presence of chiral catalysts (entries 26 and 27), particularly in the case of α-glucopyranosides, the C-2 hydroxyl is in a more asymmetric environment than the C-3 hydroxyl and can thus be differentiated on a basis similar to a kinetic resolution.

4.2.1.2 Monoprotection at O-2 of Glucopyranoside 2,3,4-Triols and 2,3,4,6-Tetraols

The primary position of a glucopyranoside can easily be differentiated based on sterics, leaving the three secondary alcohols free. The C-2 hydroxyl group can then be selectively silylated, alkylated, or acylated (Table 4.2).

Regioselective silylation of methyl 6-*O*-trityl-α-D-glucopyranoside, prepared *in situ*, gives the 2-*O*-*t*-butyldimethylsilyl-glucoside in excellent yield after acetylation (entry 28). However, the regioselectivity depends strongly on the configuration of the glycoside, as the β-octyl and the β-thiophenyl glucoside both give the 3-*O*-isomer (see Section 3.1).

In alkylation reactions, the Fe-based methodology as used for the diol was equally effective in the regioselective 2-*O*-benzylation of primary protected glucose triols as well as tetraols (entries 29 and 30).

Table 4.2 Regioselective 2-*O*-functionalization of glucopyranoside 2,3,4-triols and 2,3,4,6-tetraols.

Entry	Product	Conditions	Observations	References
28	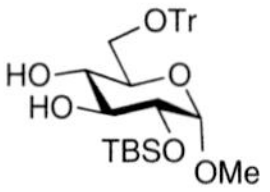	TBSCl (1.1 eq.), imidazole (2 eq.), pyridine, 86% (3 steps, one pot)	β-Glc, α-Man, and β-Gal give 3-*O*-silyl	[35]
29	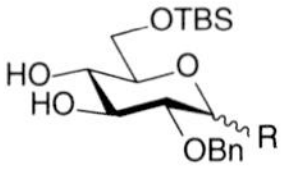	BnBr (1.5 eq.), $Fe(dibm)_3$ (0.2 eq.), K_2CO_3 (1.5 eq.), MeCN, 80 °C, 24 h R = αOMe (78%), βOMe (50%)		[11]
30	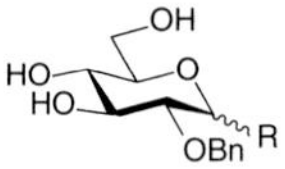	BnBr (1.5 eq.), $Fe(dibm)_3$ (0.2 eq.), K_2CO_3 (1.5 eq.), MeCN/DMF (9/1), 80 °C, 24 h R = αOMe (75%), βOMe (70%)		[11]
31	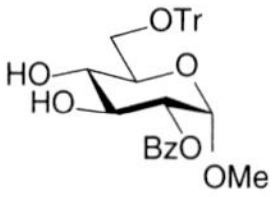	BzOH (1.2 eq.), BOPCl (2.4 eq.), DMAP (1.2 eq.), pyridine, RT, 15 h, 73% (3-*O*-Bz = 14%)	Mixture of 2-*O*- and 3-*O*-regioisomers separated by chromatography	[36]
32	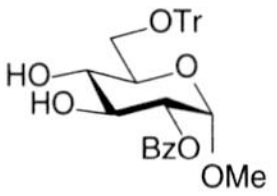	Bz_2O (1.3 eq.), $(CF_3CO_2)_2Cu$ (1.3 eq.), sym-collidine (1.3 eq.), MeCN, RT, 6 h, 85%	3-*O*-Bz with β-Gal	[37]
33	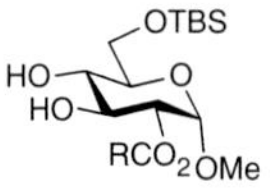	RCOCl (1 eq.), $Cu(OTf)_2$ (0.1 eq.), Ligand (0.1 eq.), iPr$_2$NEt (1.5 eq.), CH_2Cl_2, RT, 16 h RCO = Bz (84%), Piv (74%)	Ligand = (*S*,*S*)-Ph-PyBox	[38]
34	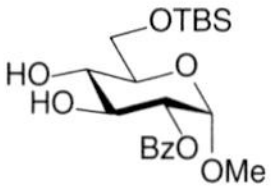	BzCl (1 eq.), Oriyama's catalyst (1 eq.), Et_3N (1 eq.), CH_2Cl_2, −60 °C, 1 h, 82%	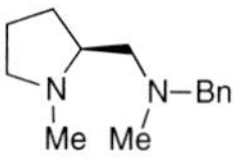	[39]
35	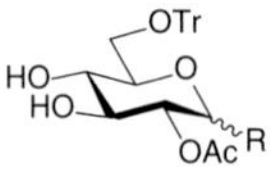	Vinyl acetate, CRL, RT, 36 h R = αOMe (98%), βOMe (90%)	High selectivity	[40]
36	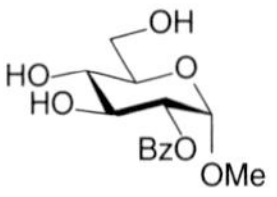	1. Resin-Tr-Cl, nBu$_4$NI, sym-collidine, DMF, RT, 72 h 2. nBu$_2$SnO, 4 Å MS, DMF, 90 °C, 12 h 3. BzCl, CH_2Cl_2, RT, 1 h 4. TFA, iPr$_3$SiH, CH_2Cl_2, 94%		[41]
37	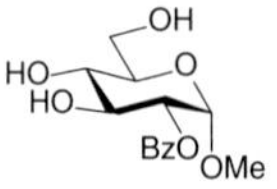	BzCl (1.2–1.3 eq.), THF, RT, 9 h With nBu$_2$SnCl$_2$ (0.1 eq.), PEMP (1.3 eq.), and nBu$_4$NI (0.5 eq.) = 93% 6-*O*-acylation possible with Me_2SnCl_2 [42] With Me_2SnCl_2 (0.05 eq.), iPr$_2$NEt (2 eq.), 82% [43]		[42, 43]

Table 4.2 (Continued)

Entry	Product	Conditions	Observations	References
38	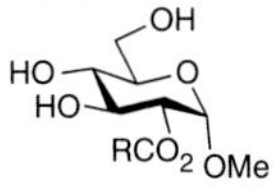	RCOCl (1.2 eq.), LiCl (3.5 eq.), Et_3N (2.5 eq.), DMA, 0 °C, 4–24 h RCO = Bz (87%), PhNHCO (70%), Piv (75%), $nC_7H_{15}CO$ (81%), EtOCO (93%)		[44]

There are more reports of regioselective 2-*O*-acylation than of 2-*O*-alkylation in the triol and tetraol series. In the triol case, the three primary alcohols could be differentiated without metal complexation by BOP-mediated esterification (entry 31). The copper-mediated conditions used for the diol above gave equally good results in the triol case (entry 32). Chiral copper bis-oxazoline complexes under conditions that are similar to those used for the diol gave equally impressive selectivity in the triol (entry 33). Other asymmetric conditions, using Oriyama's diamine catalyst with benzoyl chloride (entry 34), or enzymatic conditions with a lipase (entry 35), were also effective in selectively acylating the C2 position. Immobilization of the tetraol, presumably at the primary position, followed by tin-mediated acylation of the triol also gave access to the 2-*O*-acylated carbohydrate (entry 36).

Acylation in the tetraol series provided interesting selectivities for the 2-position over even the primary position with both tin and lithium complexes. In the case of catalytic tin complexes in the presence of nBu_4NI, the dibutyltin complex gave the 2-*O*-benzoyl sugar, whereas the dimethyltin complex gave the 6-*O*-benzoyl glucoside [42] (entry 37). However, the latter complex also gave the 2-*O*-acylated sugar in the absence of iodide [43]. Stoichiometric lithium complexes have also demonstrated their effectiveness with a wider set of protecting groups, including pivaloyl, benzoyl, carbamates, carbonates, and aliphatic esters (entry 38).

4.2.2 Mannose

Regioselective protection of the C-2 hydroxyl group of D-mannosides can be achieved in both the diol and triol series under a variety of conditions, despite the fact that the hydroxyl is in an axial position (Table 4.3).

4.2.2.1 Benzyl and Substituted Benzyl Ethers

Benzyl, 2-naphthylmethyl, and *p*-methoxybenzyl ethers can be selectively introduced by simple alkylation, under either biphasic or homogeneous conditions (entries 39–42). The inductive effect of the anomeric oxygens thus dominates over the steric effects tied to the axial position of the oxygen.

4.2.2.2 Esters

In the case of esters, selective 2-*O*-protection is somewhat more challenging in the mannose series as it is necessary to avoid acyl migration. Benzoylation under phase transfer catalysis gives satisfactory results (entry 43). Silver oxide–potassium

Table 4.3 Regioselective 2-*O*-functionalization of mannopyranosides.

Entry	Product	Conditions	Observations	References
39	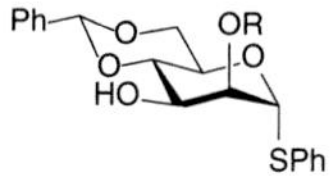	BnBr (1.1 eq.) or PMBCl (1.2 eq.), nBu_4NHSO_4 (0.2 eq.), NaOH (5 eq.), CH_2Cl_2/H_2O (3/1), reflux, 20 h R = Bn (75%) [45], PMB (70%) [46]		[45, 46]
40	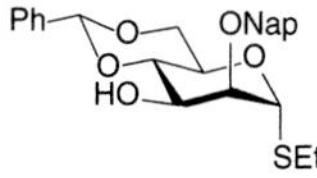	NapBr (1.5 eq.), nBu_4NHSO_4 (0.2 eq.), 5% NaOH, CH_2Cl_2/H_2O (10/3), reflux, 16 h, 87%		[47]
41	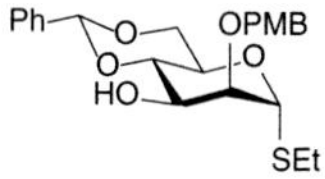	PMBCl (0.5 eq.), nBu_4NBr (0.3 eq.), 5% NaOH, CH_2Cl_2/H_2O (10/1), reflux, 30 h, 54%		[48]
42	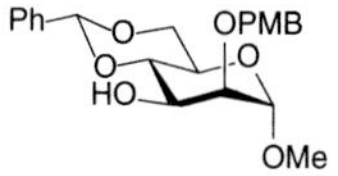	PMBCl (1 eq.), NaH (1.02 eq.), DMF, 0 °C, 1 h, 43%		[49]
43	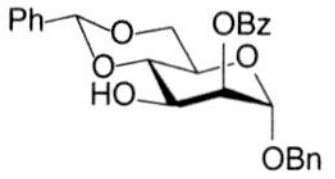	BzCl (1.5 eq.), nBu_4NHSO_4 (0.2 eq.), 5% NaOH, CH_2Cl_2/H_2O (9/1), −5 °C, 1.5 h, 58%		[50]
44	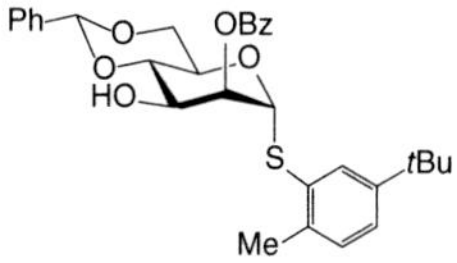	1. $PhC(OMe)_3$ (12 eq.), CSA (0.2 eq.), RT, 1 h 2. AcOH, 80 °C, 1 h, 69%		[51]
45	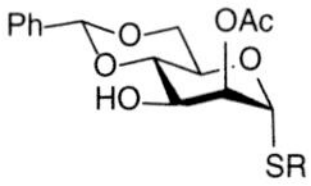	R = Ph [52] 1. $MeC(OEt)_3$, CSA, RT, 30 min 2. AcOH, RT, 1 h (75%) R = Et [53] 1. $MeC(OEt)_3$, *p*TsOH, MeCN, RT 2. AcOH, RT (73%)		[52, 53]
46	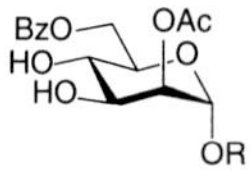	1. $MeC(OMe)_3$ (3 eq.), CSA (0.2 eq.), MeCN, RT, 1 h 2. 1 M HCl, EtOAc R = Me (85%), $CH_2CH_2N_3$ (70%)		[54]
47	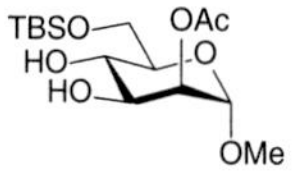	AcCl (1.2 eq.), chiral catalyst (0.2 eq.), $iPr_2NEt{\cdot}HCl$ (0.03 eq.), iPr_2NEt (1.2 eq.), *t*-amyl alcohol, 4 °C, 2 h, 74%	Catalyst 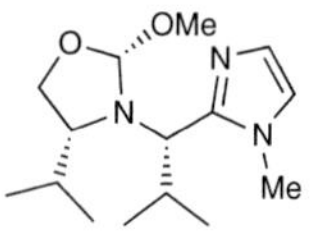	[55]
48	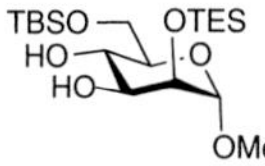	TESCl (1.2 eq.), chiral catalyst (0.2 eq.), $iPr_2NEt{\cdot}HCl$ (0.03 eq.), iPr_2NEt (1.2 eq.), *t*-amyl alcohol, 4 °C, 2 h, 84%	Catalyst 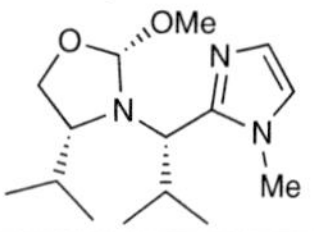	[55]

iodide-mediated benzoylation also gave good results in the mannose series (Table 4.1, entry 19). However, the best results have been achieved with the regioselective hydrolysis of orthoester intermediates in both the diol (entries 44 and 45) and triol (entry 46) series. In the latter case, selectivity for the axial 2-position over the two equatorial positions was achieved presumably by selective formation of the *cis*-2,3-*O*-orthoester. Direct regioselective 2-*O*-acylation of a 6-*O*-silyl-D-mannoside 2,3,4-triol was also reported using a chiral imidazole catalyst (entry 47).

4.2.2.3 Silyl Ethers

Regioselective 2-*O*-silylation of the 6-*O*-*t*-butyldimethylsilyl-D-mannoside 2,3,4-triol was achieved using a chiral imidazole catalyst under the same conditions as used for the acetate above (entry 48). It should be noted that the 2,6-*O*-di-*t*-butyldiphenylsilyl-D-mannoside can also be obtained by *bis*-silylation of the tetraol (see Section 4.5).

4.2.3 Galactose

Regioselective protection of the 2-hydroxyl group has been achieved in the galactose 4,6-*O*-benzylidene series with silyl, benzyl, and ester groups (Table 4.4). A regioselective enzymatic acetylation of 6-*O*-benzyl-galactoside was also reported.

4.2.3.1 Silyl Ethers

In the galactoside 4,6-*O*-acetal series, regioselective 2-*O*-silylation was reported using TIPSCl and imidazole under standard conditions (entry 49). Once again, silylation with *t*-butyldimethylsilyl chloride led to the 3-*O*-silyl compound under similar conditions (see Section 3.3). Silylation of 6-*O*-*t*-butyldimethylsilyl-α-D-galactoside with TESCl and *N*-methylimidazole gave an 86 : 14 selectivity in favor of the 2-*O*-triethylsilyl product in 77% combined yield by NMR [55]. It should also be noted that *bis*-silylation of unprotected galactosides usually leads to the 2,6-*O*-disilyl products under thermodynamic control [62] (see Section 4.5).

4.2.3.2 Benzyl Ethers

Alkylation with benzyl bromide or *p*-methoxybenzyl chloride under phase transfer conditions gives the 2-*O*-protected galactoside in acceptable yields (entries 50 and 51).

4.2.3.3 Esters

For regioselective esterification reactions between O-2 and O-3, galactose might be expected to be similar to glucose, yet many conditions that give selective 2-*O*-acylation in the glucose series gave the 3-*O*-acyl product in the galactose series (Table 4.1, entries 14, 17, 20, 24, 25, and 27).

Direct acylation at the 2-position was achieved under biphasic phase transfer conditions (Table 4.4, entry 52), or under the silver oxide-potassium iodide-mediated benzoylation conditions that gave good results in the glucose and mannose series (entry 53 and Table 4.1, entry 19). Dibutyltin oxide conditions in the absence of halide salts provided 2-*O*-acylation in high yields (entry 54). Finally,

Table 4.4 Regioselective 2-*O*-functionalization of galactopyranosides.

Entry	Product	Conditions	Observations	References
49	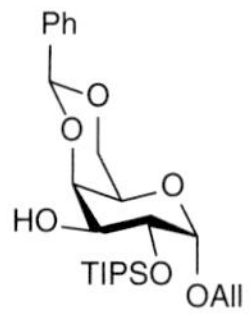	TIPSCl, imidazole, DMF, 0 °C to RT, 8 h, 83% over 2 steps	Yield reported after acetylation	[56]
50	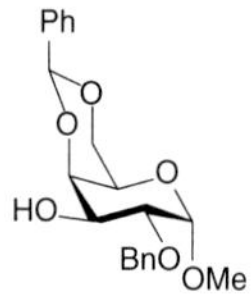	BnBr (1.1 eq.), Aliquat 336 (0.1 eq.), KOH (1.5 eq.), CH_2Cl_2, RT, 12 h, 60% (3-*O*-Bn = 20%)	Mixture of 2-*O*- and 3-*O*-regioisomers separated by chromatography	[57]
51	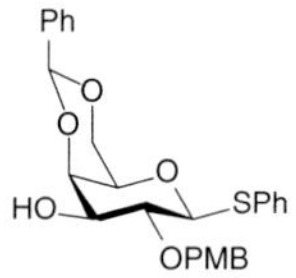	PMBCl (1.15 eq.), *n*Bu_4NBr (1 eq.), 10% NaOH, CH_2Cl_2/H_2O (9/1), RT, 10 h, 67%		[58]
52	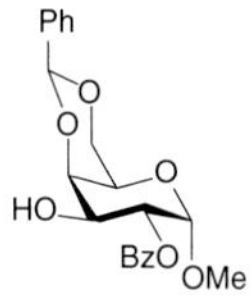	BzCl (1.05 eq.), *n*Bu_4NCl (0.05 eq.), 40% NaOH, CH_2Cl_2/H_2O (6/1), RT, 20 min, 40%		[59]
53	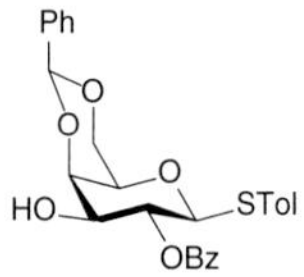	BzCl, Ag_2O, KI, CH_2Cl_2, 70%		[25]
54	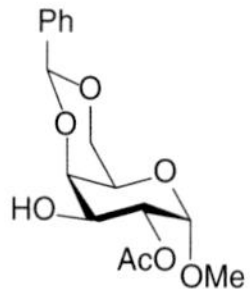	1. *n*Bu_2SnO (1 eq.), MeOH, reflux, 2 h 2. AcCl (1.5 eq.), Et_3N (1.5 eq.), dioxane, RT, 16 h, 78%		[60]
55	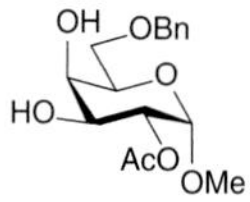	Vinyl acetate, PCL, 37 °C, 10 h, 95%	High selectivity	[61]

good selectivity for the 2-position was achieved in the triol under enzymatic conditions using *Pseudomonas cepacia* lipase (entry 55).

4.2.4 Other Monosaccharides

Regioselective monoprotection at the 2-position has been reported in less-common monosaccharides including allose, gulose, talose, and altrose (Table 4.5). Selective 2-*O*-allylation of 4,6-*O*-benzylidene-α-D-allopyranoside has been

Table 4.5 Regioselective 2-*O*-functionalization of other hexopyranosides.

Entry	Product	Conditions	Observations	References
56	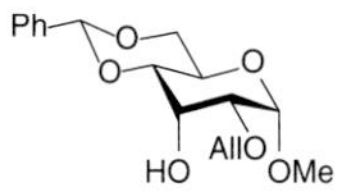	1. nBu$_2$SnO (1.03 eq.), MeOH, reflux, 1.5 h 2. All-I (1.8 eq.), DMF, 100 °C, 1 h 68% (3-*O*-All = 7%)	Mixture of 2-*O*- and 3-*O*-regioisomers separated by chromatography	[63]
57	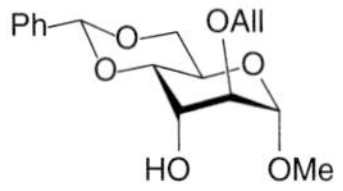	AllBr, nBu$_4$NBr, 10% NaOH/ CH$_2$Cl$_2$, 48 h, 76%	nBu$_4$NI, 20 h, 79% [64]	[65]
58	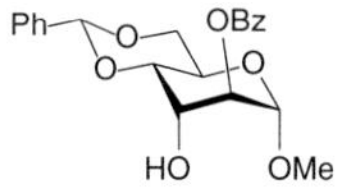	BzCl, nBu$_4$NCl, NaI, 40% NaOH/CH$_2$Cl$_2$, 5 min, 91% [66] BzOBt, CH$_2$Cl$_2$, Et$_3$N, RT, 72 h, 90% [67]	Alloside (89%) Migration without NaI	[66, 67]
59	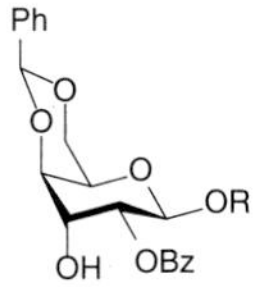	R = All [68] BzCl (1.1 eq.), CH$_2$Cl$_2$/pyridine (4/1), 0 °C, 40 min, 94% R = p-O$_2$N-Ph [69] BzCl (1.05 eq.), CH$_2$Cl$_2$/ pyridine (5/1), 0 °C, 1.5 h, 79%		[68, 69]
60	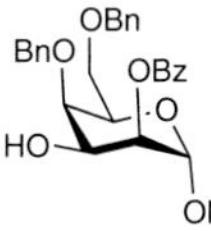	1. PhC(OMe)$_3$ (10 eq.), pTsOH (cat.), 5 h 2. AcOH, 30 min, 81%		[70]
61	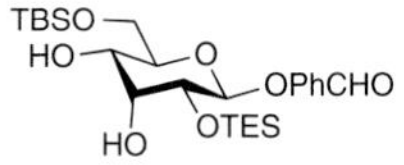	TESCl (1.2 eq.), chiral catalyst (0.2 eq.), iPr$_2$NEt·HCl (0.03 eq.), iPr$_2$NEt (1.2 eq.), THF, 4 °C, 2 h, 90%	Catalyst 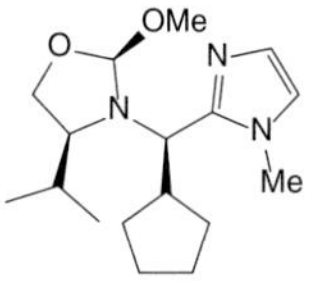	[55]

reported using a tin complex (entry 56). Methyl 2-*O*-allyl-4,6-*O*-benzylidene-α-D-altropyranoside was similarly obtained under phase transfer conditions (entry 57). In the remaining series, direct monobenzylation of 4,6-*O*-benzylidene-L-idose gave moderate selectivity at low conversion [71]. A similar monobenzylation of methyl 4,6-di-*O*-benzyl-α-D-talopyranoside with benzyl bromide, nBu$_4$NHSO$_4$, and sodium hydroxide gave somewhat better selectivity (4 : 1) at low conversion, in 51% yield based on recovered starting material [72].

The corresponding benzoylation in both the allose and altrose series was reported under phase transfer conditions using benzoyl chloride (entry 58). The 2-*O*-benzoyl-4,6-*O*-benzylidene-α-D-alloside was also obtained under homogenous conditions using the HOBt ester. It was shown in the allose series that the presence of sodium iodide prevented acyl migration under phase transfer conditions.

In the gulose series, the 4,6-*O*-benzylidene was selectively benzoylated by direct acylation with benzoyl chloride under homogeneous conditions (entry 59).

Selective 2-*O*-benzoylation of methyl 4,6-di-*O*-benzyl-α-D-taloside was achieved via the orthoester in good yields (entry 60).

In an interesting regioselective reaction of a 2,3,4-triol, an aryl 6-*O*-*t*-butyldimethylsilyl-D-alloside was regioselectively silylated at the 2-position with a chiral imidazole catalyst (entry 61).

4.3 Regioselective Protections at the 3-Position

Regioselective monoprotection of carbohydrate diols, triols, and tetraols at the 3-position can usually be achieved with silyl protecting groups and/or the use of tin acetals in the presence of halide ions. However, a fair number of other interesting approaches have been reported.

4.3.1 Glucose

4.3.1.1 Silyl Ethers

Selective silylation at the 3-position of 4,6-*O*-benzylidene glucosides has been achieved in the glucose series under relatively standard conditions (TBSCl, imidazole) at low temperature (Table 4.6, entry 62). It should be noted that the bulkier TBDPS group gave selectively the 2-substituted compound (Table 4.1, entry 1) [1]. In the 2,3,4-triol series, silylation of 6-*O*-trityl-β-glucosides or thioglucosides similarly provided the 3-*O*-*t*-butyldimethylsilyl-D-glucoside (entry 63), although the corresponding α-methyl glycoside provided the 2-*O*-TBS isomer (Table 4.2, entry 28).

4.3.1.2 Other Ethers

Selective alkylation at the 3-position of 4,6-*O*-benzylidene-D-glycopyranosides has been achieved with MOMCl in the presence of an amine base (entry 64). However, cyclic tin acetals have been the most widely used with benzyl and allyl-type ethers. Opening of a tin acetal in the presence of iodide (entry 65) or fluoride salts (entry 66 and Table 4.8, entry 90) leads to the formation of the 3-*O*-benzyl, allyl, or 2-naphthylmethyl glucopyranosides. Selective alkylation of the 3-position of 4,6-*O*-benzylidene-D-glucopyranosides has also been reported using a catalytic iron complex (Table 4.7, entry 73).

4.3.1.3 Esters

Benzoylation at the 3-position of glucopyranoside 4,6-*O*-acetals can be achieved under standard conditions with BzCl in pyridine, under similar conditions to 2-*O*-benzoylation (Table 4.1, entry 10), but at very low temperature (−50 °C, entry 67). Otherwise, acetylation at the 3-position has been reported using tin acetals (entry 68), or by enzymatic methods, starting either from the 2,3-diol with several enzymes (entry 69) or from the 2,3,4-triol with *Candida rugosa* lipase (CRL) in supercritical CO_2 (entry 70). Regioselective 3-*O*-acylation of 6-*O*-*t*-butyldimethylsilyl-D-glucopyranoside was also reported with benzoic anhydride and tetrabutylammonium benzoate or acetic anhydride-tetrabutylammonium acetate (Table 4.7, entry 81 and Table 4.8, entry 106, respectively).

Table 4.6 Regioselective 3-*O*-functionalization of glucopyranosides.

Entry	Product	Conditions	Observations	References
62	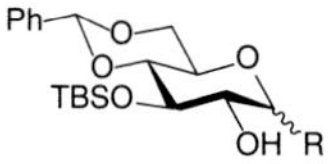	TBSCl (1.1–1.2 eq.), imidazole (1.5 eq.), DMF, 0 °C, 16 h R = βSMe (quantitative) [73] R = βSEt (86%) [74] R = αOAll (47%) with DMAP (0.2 eq.) [75]		[73–75]
63	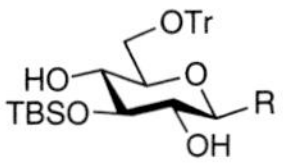	TBSCl (1.1 eq.), imidazole (2 eq.), pyridine, RT, 12 h R = SPh (75%), R = OC_8H_{17} (65%) (3 steps, one pot)	Also βS*i*Pr Gal (83%), αOMe Man (79%)	[35]
64	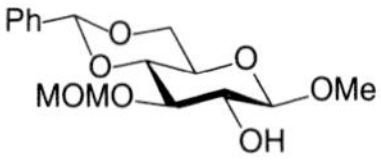	MOMCl (1.5 eq.), *i*Pr_2NEt (1.2 eq.), CH_2Cl_2, 0 °C, 12 h, 84%		[76]
65	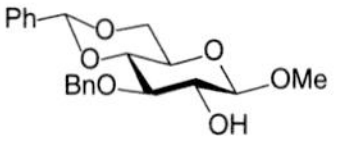	1. *n*Bu_2SnO (1.2 eq.), benzene, reflux, 16 h 2. BnBr (1.07 eq.), *n*Bu_4NI (1.13 eq.), benzene, reflux, 16 h, 85%	Also reported with αOMe [77], βOMe [78], βOMP [79, 80], βSPh [81], βO$(CH_2)_2$NHCBz [82], or Gal with PMB [83]	[84]
66	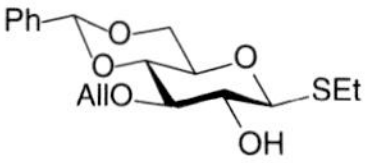	1. *n*Bu_2SnO (1.19 eq.), MeOH, reflux, 1 h 2. AllBr (1.2 eq.), CsF (1.3 eq.), DMF, RT, 16 h, 67%	Also reported with Gal [85, 86]	[87]
67	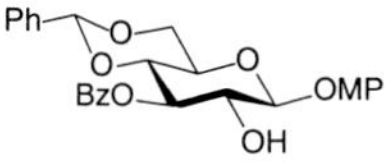	BzCl (1.1 eq.), pyridine (2 eq.), CH_2Cl_2, −50 °C, 3 h, 55%		[88]
68	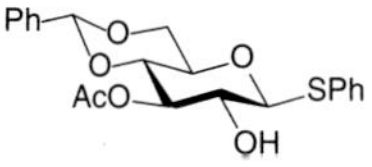	1. *n*Bu_2SnO (1.1 eq.), benzene, reflux, 6 h 2. AcCl (1.25 eq.), CH_2Cl_2, 0 °C, 2 h, 83%	2-*O*-Ac also isolated	[89]
69	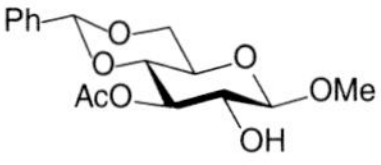	Vinyl acetate, PFL, RT, 24 h, 92% [90, 91] or Vinyl acetate, CAB, 45 °C, 20 h, 79% [92]	Similar results reported with Lipase P from *Pseudomonas cepacia* with chloroacetate and pivaloate [93, 94]	[90–92]
70	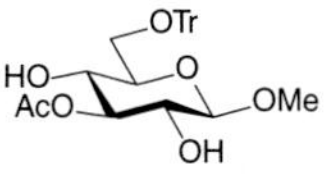	Vinyl acetate, CRL, $scCO_2$, 20 MPa, 50 °C, 20 h, quantitative	Also reported with PCL [61]	[95]

Table 4.7 Regioselective 3-*O*-functionalization of mannopyranosides.

Entry	Product	Conditions	Observations	References
71	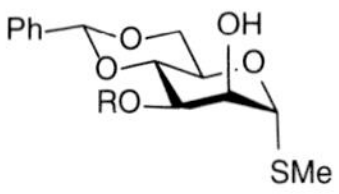	R = TBDPS [96] TBDPSCl (1.5 eq.), imidazole (2 eq.), CH_2Cl_2, RT, 14 h, 84% R = TIPS [97] TIPSCl (1.1 eq.), imidazole (2.5 eq.), DMF, 0 °C to RT, 18 h, 70%		[96, 97]
72	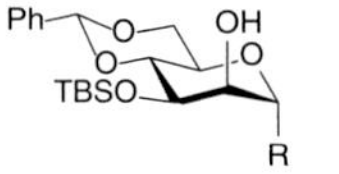	TBSOTf (1.1 eq.), Et_3N (1.1 eq.), DMF, −10 °C to RT, 2 h R = SPh (79%) [98] R = OAll (61%) [99]	Yield over two steps is reported and indicated here	[98, 99]
73	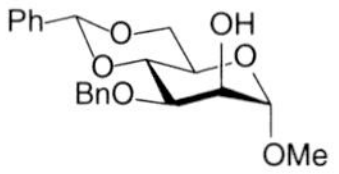	BnBr (1.1 eq.), $Fe(dibm)_3$ (0.05 eq.), K_2CO_3 (1.5 eq.), MeCN, 80 °C, 8 h, 91%	Several examples with Glc, Gal and Man and PMB, allyl, BrBn, CNBn	[11]
74	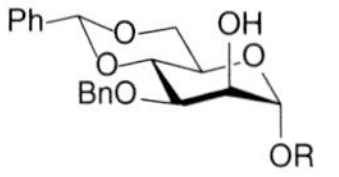	1. $n$$Bu_2SnO$ (1.02 eq.), PhMe, reflux, 3 h 2. BnBr (1.05 eq.), $n$$Bu_4NBr$ (1.06 eq.), CsF (1.02 eq.), PhMe, reflux, 3 h R = Bn (88%), All (88%), Me (89%)	Also reported with Nap [100] or PMB [83]	[101]
75	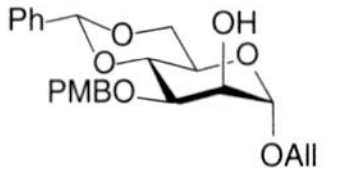	1. $n$$Bu_2SnO$ (1 eq.), PhMe, reflux, 3 h 2) PMBCl (1.05 eq.), CsF (1.07 eq.), $n$$Bu_4NBr$ (1.07 eq.), PhMe, RT, 48 h, 98%	Also reported with α$OPhNO_2$ [102], αSPh [98, 103–106], αSTr [107], α$S(CH_2)_2SiMe_3$ [108], α$O(CH_2)_2N_3$ [109]	[99]
76	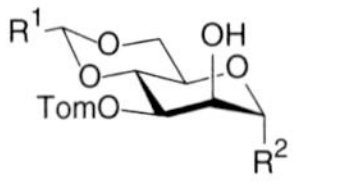	1. $n$$Bu_2SnCl_2$ (1 eq.), $i$$Pr_2NEt$ (3.6 eq.), DCE, RT, 1.5 h 2. TomCl (1.3 eq.), 80 °C, 15 min R^1 = Ph/R^2 = OAll (91%) [110] R^2 = SPh/R^1 = Ph (78%), MeOPh (68%) [47]		[47, 110]
77	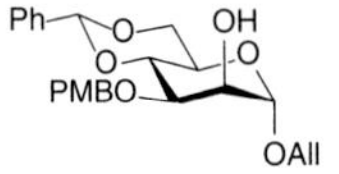	PMBST (1.6 eq.), DTBMP (1.25 eq.), AgOTf (1.6 eq.), CH_2Cl_2, RT, 48 h, 0 °C to RT, 16 h, 57%	PMBST 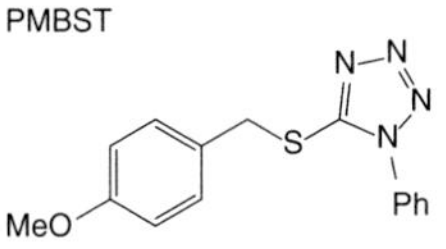	[111]
78	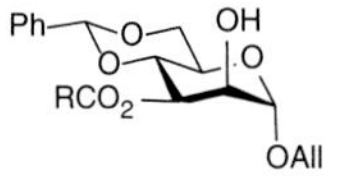	RCOCl (1.5 eq.), pyridine (10 eq.), CH_2Cl_2, 0 °C to RT, 2 h RCO = Ac (91%) [112] RCO = Bz (91%) [113]		[112, 113]

Table 4.7 (Continued)

Entry	Product	Conditions	Observations	References
79	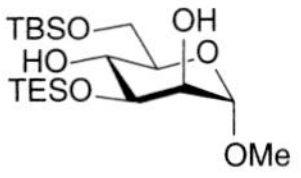	TESCl (1.2 eq.), iPr$_2$NEt (1.2 eq.), catalyst (0.05 eq.), iPr$_2$NEt·HCl (0.03 eq.), t-amyl alcohol, −15 °C to RT, 2–4 h, 98%	Catalyst 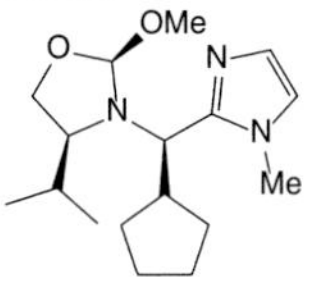	[114]
80	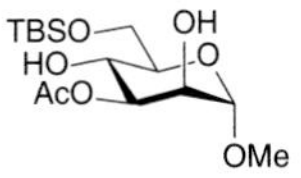	AcCl (1.2 eq.), iPr$_2$NEt (1.2 eq.), catalyst (0.05 eq.), iPr$_2$NEt·HCl (0.03 eq.), t-amyl alcohol, −15 °C to RT, 2–4 h, 96%	Catalyst 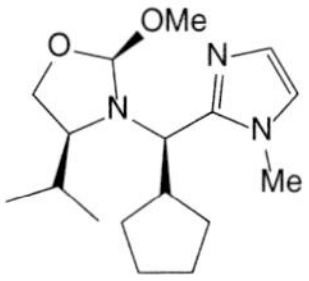	[114]
81	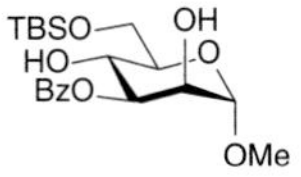	Bz$_2$O (1.1 eq.), nBu$_4$NOBz (0.2 eq.), MeCN, 40 °C, 8–12 h, 84%	Five examples with Glc, Gal, and Man	[115]
82	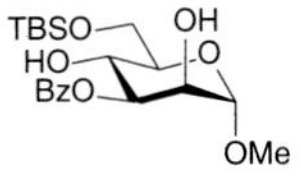	BzCl (1.5 eq.), iPr$_2$NEt (1.5 eq.), catalyst (0.1 eq.), MeCN, RT, 4 h, 86%	Also with Gal and for 3-O-Fmoc catalyst 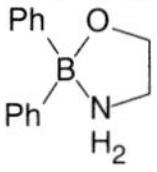	[116]
83	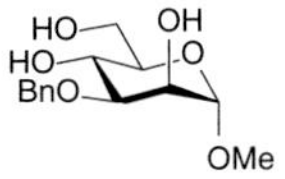	BnBr (4 eq.), nBu$_2$SnO (0.1 eq.), nBu$_4$NBr (0.3 eq.), iPr$_2$NEt (2.5 eq.), 70 °C, 4.5 h, 64%	Sealed tube, solvent-free, catalytic (Sn) Several examples with Gal, Man, and allyl, Bn [117, 118]	[118]
84	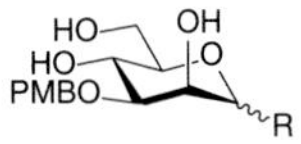	1. nBu$_2$SnO (1.1 eq.), PhMe, reflux, 3 h 2. PMBCl (1.5 eq.), nBu$_4$NI (0.2 eq.), PhMe, RT to reflux, 2 h R = αSPh (83%) [83] R = βSTol (65%) [119]	Also reported with Bn [120] or allyl [121] Also with Gal or αSEt [122] or with Bn [123]	[83, 119]
85	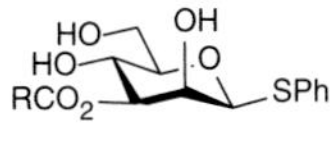	1. Me$_2$SnCl$_2$ (0.05 eq.), iPr$_2$NEt (2 eq.), THF/H$_2$O (19/1), RT, 5 min 2. RCOCl (1.1 eq.), RT, 2.5 h RCO = Ac (90%), Bz (92%)	Examples with Gal and Man also reported for Bz [43]	[124]

4.3.2 Mannose

4.3.2.1 Monoprotection at O-3 of Mannopyranoside 2,3-Diols

Silylation of the 3-position of mannopyranoside 4,6-O-acetals with t-butyldiphenylsilyl, t-butyldimethylsilyl, and triisopropylsilyl groups can be achieved under standard conditions (entries 71 and 72). A similar preparation of methyl 3-O-t-butyldiphenylsilyl-4,6-O-cyclohexylidene-α-D-thiomannoside was also reported [161].

Table 4.8 Regioselective 3-*O*-functionalization of galactopyranosides.

Entry	Product	Conditions	Observations	References
86	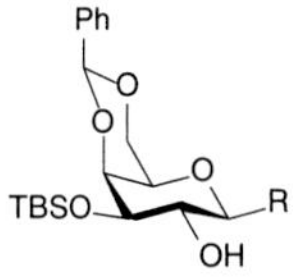	TBSCl (1.3 eq.), imidazole (2.5 eq.), DMF, 40 °C, 16 h R = OMe (65%) [125] R = STol (80%) in CH_2Cl_2 [126]		[125, 126]
87	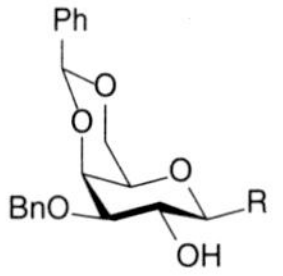	BnBr (1.1 eq.), pyridine or NaH, DCE or THF, RT, 4–16 h R = OMe (61%), SPh (64%)		[26]
88	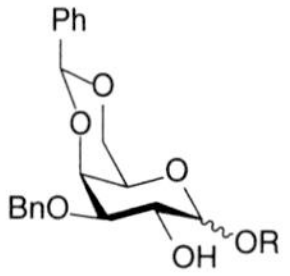	1. *n*Bu_2SnO (1.5 eq.), MeOH, reflux, 3 h 2. BnBr (6–12 eq.), 1,4-dioxane, 50 °C, 5–6 h OR = αOMe (71%), βOMe (95%)	Also reported with βOBn [127]	[128]
89	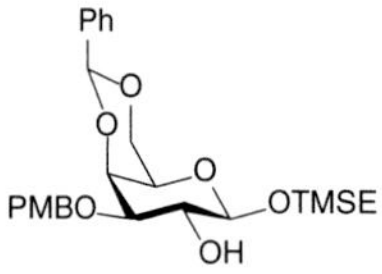	1. *n*Bu_2SnO (0.6 eq.), MeOH, reflux, 1 h 2. PMBCl (0.8 eq.), *n*Bu_4NI, MeCN, 100 °C, 2 h, 70%	Also reported with βOPent [129], βOAll [130]	[131]
90	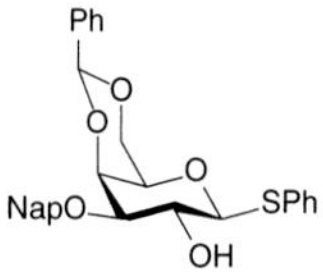	1. *n*Bu_2SnO (1.8 eq.), PhMe, reflux, 4 h 2. NapBr (2 eq.), CsF (2 eq.), *n*Bu_4NI (2 eq.), DMF, reflux, 16 h, 60%	Also reported with Man [47, 132–135] or Glc [136]	[137]
91	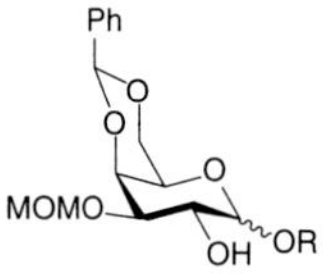	1. *n*Bu_2SnO (1.5 eq.), MeOH, reflux, 3 h 2. MOMCl (1.5 eq.), 1,4-dioxane, 50 °C, 1.5 h OR = αOMe (90%) [128], βOMe (93%) [128], βOPent (93%) [138]		[128, 138]
92	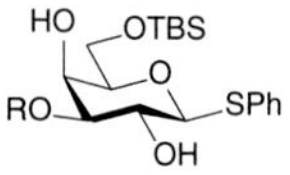	1. RX (1.5 eq.), Me_2SnCl_2 (0.1 eq.), Ag_2O (1.1 eq.), MeCN, RT, 24 h R = Bn (84%), All (81%), PMB (78%), Nap (89%)	Examples with 4,6-*O*-benzylidene and 6-*O*-Bz	[139]
93	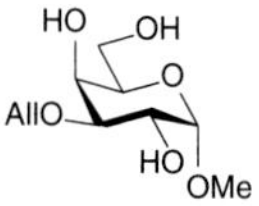	1. *n*Bu_2SnO (0.9 eq.), PhMe, reflux, 2 h 2. AllBr (10 eq.), *n*Bu_4NI (0.9 eq.), PhMe, 60 °C, 18 h, 76%	Also with βOMP [140], βS*i*Pr [141], Nap [142], Bn [143], Man [144]	[145]
94	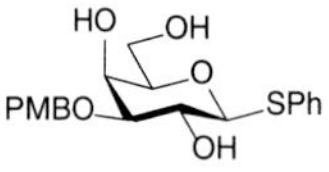	1. *n*Bu_2SnO (1.05 eq.), MeOH, reflux, 4 h 2. PMBCl (3 eq.), CsF (1.2 eq.), DMF, 30 °C, 48 h, 66%	Also reported with allyl and Bn [146, 147]	[148]

Table 4.8 (Continued)

Entry	Product	Conditions	Observations	References
95	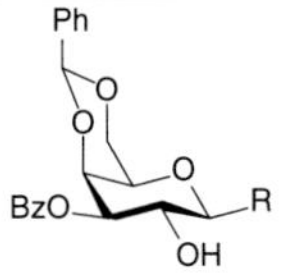	BzCl (1.1 eq.), pyridine or NaH, DCE or THF, RT, 4–16 h R = OMe (70%), SPh (67%)		[26]
96	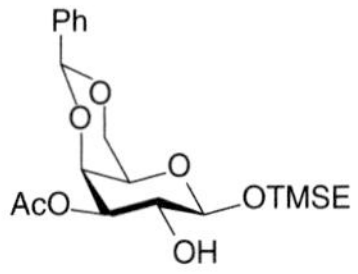	AcCl (0.98 eq.), pyridine/ CH_2Cl_2 (1/4), RT, 1 h, 80%	Also with βSMe/ ClAc [149]	[131]
97	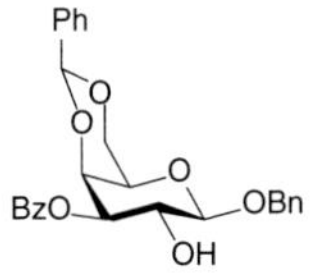	BzCl (1 eq.), imidazole (2 eq.), $CHCl_3$, reflux, 18 h, 91%		[150]
98	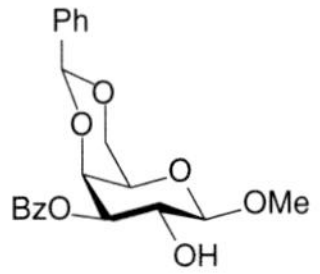	BzCl (1.1 eq.), HOBt (1.1 eq.), Et_3N (1.1 eq.), CH_2Cl_2, RT, 18 h, 99%	1-Benzoyloxy-1*H*-benzotriazole (BBTZ) generated *in situ*	[151]
99	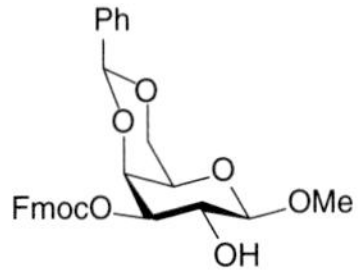	FmocCl (1.1 eq.), DMAP (0.1 eq.), pyridine/MeCN (2/1), RT, 16 h, 68%		[152]
100	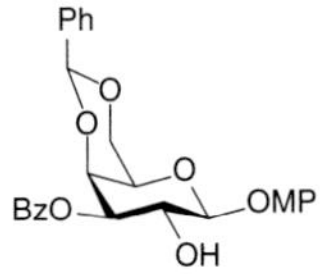	BzCN (1.3 eq.), Et_3N, MeCN, RT, 30 min, 75%	Also reported with a galactolipid [153]	[154]
101	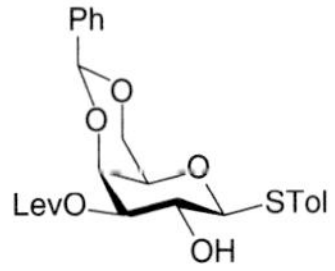	LevOH (1.2 eq.), DCC (1.5 eq.), DMAP (0.5 eq.), CH_2Cl_2, RT, 3 h, 56%		[155]
102	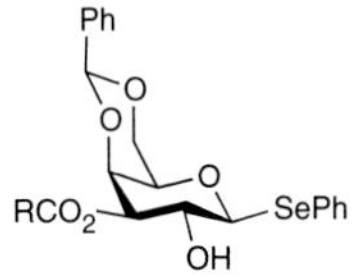	RCOCl (1.2 eq.), $CuCl_2$ (1 eq.), NaH (2 eq.), THF, RT, 5 h RCO = Bz (34%), Piv (35%)		[156]
103	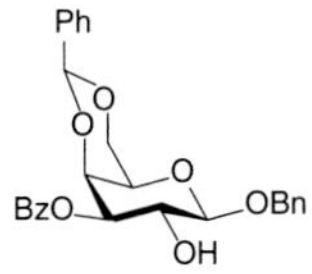	Bz_2O (1.25 eq.), $(CF_3CO_2)_2Cu$ (1.25 eq.), pyridine (1.25 eq.), MeCN, RT, 16 h, 77%	$MoO_2(acac)_2$ was also used for other Gal and Man derivatives [157]	[32]

(Continued)

Table 4.8 (Continued)

Entry	Product	Conditions	Observations	References
104	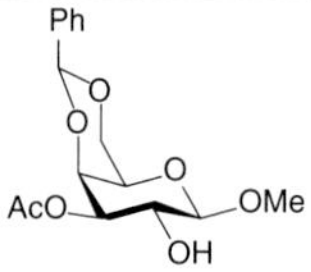	1. $Me_2Si(OMe)_2$ (2 eq.), MeCN, 80 °C 2. AcCl (1.18 eq.), MeCN, 40 °C, 76%		[31]
105	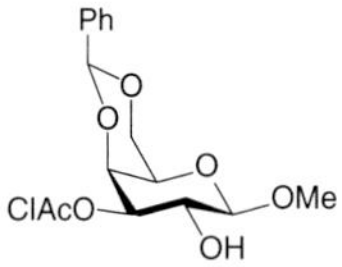	1. nBu$_2$SnO (1.1 eq.), MeOH, reflux, 2 h 2. ClAcCl (0.8 eq.), MS 4 Å, PhMe, 0 °C, 30 min, 79%	Also reported with βSPh [158]	[159]
106	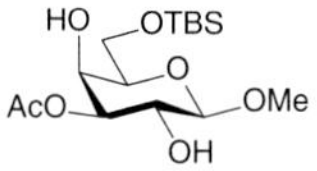	Ac_2O (1.1 eq.), nBu$_4$NOAc (0.3 eq.), MeCN, 40 °C, 8–12 h, 81%	Also with Glc, Man	[160]
107	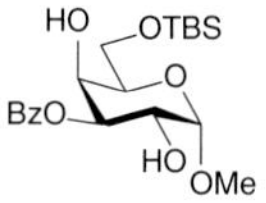	BzCl (1 eq.), $Cu(OTf)_2$ (0.1 eq.), catalyst (0.1 eq.), iPr$_2$NEt (1.5 eq.), MeCN, −5 °C, 16 h, 83%	Also with Piv, Cbz and for Man (*R,R*)- or (*S,S*)-catalyst 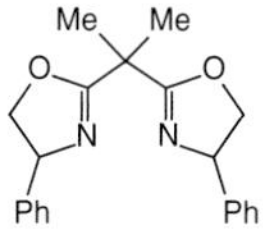	[38]

Selective alkylation of the 3-position of 4,6-*O*-benzylidene mannosides has been reported using a catalytic iron complex (entry 73) leading to allyl, benzyl, *p*-methoxybenzyl, and *p*-cyanobenzyl ethers. Tin acetals in the presence of halide salts give the corresponding 3-*O*-benzyl, allyl, 2-naphthylmethyl, *p*-methoxybenzyl, and triisopropylsilyloxymethyl (Tom) ethers in good yields (entries 74–76 and Table 4.8, entry 90). Regioselective alkylation was also reported using 5-(*p*-methoxylbenzylthio)-1-phenyl-1*H*-tetrazole and a silver catalyst (entry 77).

Selective acylation at the 3-position of the mannopyranoside 2,3-diols can be achieved under standard conditions (RCOCl, pyridine, entry 78). The regioselective 3-*O*-benzoylation of 4,6-*O*-benzylidene-D-mannopyranosides has also been reported with a chiral copper catalyst [33] (Table 4.1, entry 26) and a molybdenum oxide complex (Table 4.8, entry 103).

4.3.2.2 Monoprotection at O-3 of Mannopyranoside 2,3,4-Triols and 2,3,4,6-Tetraols

Regioselective 3-*O*-silylation of 6-*O*-trityl-α-D-mannopyranoside under relatively standard conditions was reported in good yields (Table 4.6, entry 63). An interesting selective silylation of the 3-position of the 2,3,4-triol with a triethylsilyl group has been reported using a chiral imidazole-based nucleophilic catalyst (entry 79). Regioselective 3,6-di-*O*-silylation of mannopyranoside tetraols gives the 3,6-isomer under kinetic conditions [162] (Section 4.5).

Selective acylation of mannoside 2,3,4-triols at the 3-position can be achieved with the chiral imidazole-based nucleophilic catalyst described above for silylation (entry 80), or using benzoic anhydride in the presence of tetrabutylammonium benzoate (entry 81). Similar phase transfer conditions were reported with acetic anhydride and tetrabutylammonium acetate (Table 4.8, entry 106). The use of borinic esters also gives access to 3-*O*-acylated mannosides under mild conditions (entry 82). The use of chiral copper catalysts discussed below for galactose (Table 4.8, entry 107) also has been reported for producing 3-*O*-benzoyl-mannopyranosides.

Remarkable selectivity has also been achieved in the 2,3,4,6-tetraols, using bistin acetals. 3-*O*-Allyl, benzyl, *p*-methoxybenzyl ethers have thus been obtained in a single step from the corresponding simple *O*- and *S*-mannopyranosides (entries 83–85 and Table 4.8, entry 93).

4.3.3 Galactose

4.3.3.1 Silyl and Substituted Benzyl Ethers

In the galactose series, silylation if the 2,3-diol under standard conditions usually gives the 3-*O*-silylated galactoside (entry 86). Regioselective silylation of 6-*O*-trityl-β-D-galactopyranosides also gave the 3-*O*-*t*-butyldimethylsilyl product, although the same conditions in the α-galactopyranoside gave a nearly 1 : 1 mixture of 2-*O*- and 3-*O*-silyl products (Table 4.6, entry 63).

Benzylation of 4,6-*O*-benzylidene 2,3-diols under standard, non-phase transfer conditions gives the 3-*O*-benzyl galactoside (entry 87). The more common approach uses once again the tin acetals to prepare benzyl, *p*-methoxybenzyl, 2-naphthylmethyl, and methoxymethyl ethers at the 3-position (entries 88–91 and Table 4.6, entries 65 and 66). Selective 3-*O*-alkylation of 4,6-*O*-benzylidene-D-galactopyranosides has been reported using a catalytic iron complex (Table 4.7, entry 73).

In the case of the 2,3,4-triols, allyl, benzyl, *p*-methoxybenzyl, and 2-naphthylmethyl ethers have been prepared using dimethyltin acetals, with either a silyl or a benzoyl protecting group at the 6-position (entry 92).

In the case of the 2,3,4,6-tetraols, selective introduction of allyl, benzyl, *p*-methoxybenzyl, or 2-naphthylmethyl can be achieved using tin acetals (entries 93 and 94 and Table 4.7, entry 84).

4.3.3.2 Esters

Esterification of galactoside 2,3-diols and 2,3,4-triols typically provide the 3-*O*-benzoyl product under conditions that provided the 2-*O*-benzoyl-D-glucoside: 3-*O*-acyl galactopyranosides can thus be prepared using phase transfer conditions, an *O*-benzoyl oxime, silyl acetals, copper complexes, or chiral catalysts (Table 4.1, entries 14, 17, 24, 25, and 27 and Table 4.2, entry 32). Selective esterification of the 3-position of galactose 2,3-diols can also be achieved under typ ical, non-phase transfer conditions, using various nucleophilic catalysts (pyridine, imidazole, HOBt, entries 95–98). The use of FmocCl in the presence of DMAP also gives regioselectively the 3-*O*-carbonate (entry 99). Using benzoyl cyanide as an electrophile also gives selective acylation of the 3-position (entry 100), as does levulinic acid in the presence of DCC (entry 101). The use of copper(II) complexes directs the acylation to the 3-position using either the acid chloride (entry 102) or anhydride (entry 103). Cyclic silyl (entry 104) and tin acetals (entry 105 and Table 4.7, entry 85) also provide access to the 3-*O*-acylated galactosides.

In the case of the 2,3,4-triols, acetic anhydride in the presence of tetrabutylammonium acetate gives the 3-*O*-acetyl galactoside (entry 106). Similar conditions led to the 3-*O*-benzoyl-6-*O*-*t*-butyldimethylsilyl-D-galactopyranoside, as did benzoyl chloride in the presence of borinate esters (Table 4.7 entries 81 and 82). Chiral *bis*-oxazoline-based copper catalysts also provide access to the 3-*O*-Bz, Piv, Cbz-protected galactosides (entry 107).

4.3.4 Other Sugars

Methyl 4,6-*O*-benzylidene-D-taloside was benzoylated at the equatorial 3-position with benzoyl chloride and pyridine in 70% yield, along with 30% of the 2,3-di-*O*-benzoate [163].

4.4 Regioselective Protections at the 4-Position

For the regioselective protection of the 4-position, the most common route to selectively 4-*O*-protected sugars is the regioselective opening of 4,6-*O*-cyclic acetals, as described in Chapter 11.

In the case of regioselective acylation of the 4-position, the route to a diol precursor is less straightforward, except in the galactose series, via the 3,4-*O*-acetonide. Regioselective 4-*O*-acylation of 2,6-di-*O*-benzyl- and 2,6-di-*O*-acyl-D-galactopyranosides have been reported using trimethyl orthoacetate or trimethylorthobenzoate (entries 108 and 109). Direct 4-*O*-benzoylation was reported with benzoic anhydride in the presence of tetrabutylammonium benzoate (entry 110), or more generally with acyl cyanides (entry 111).

In the case of pyranoside 2,3,4-triols and tetraols, a remarkable regioselective protection of a glucopyranoside tetraol uses a chiral, peptide-substituted DMAP catalyst, which affords high yields of octyl 4-*O*-acetyl-β-D-glucopyranoside from the tetraol (entry 112). Similar results were reported in the thioglycoside and mannoside series, the latter with the isobutyryl ester.

The regioselective monosilylation at the 4-position of 6-*O*-*t*-butyldimethylsilyl-β-D-altroside was performed in the presence of a chiral imidazole catalyst (entry 113). In the L-mannose series, enzymatic esterification of 6-*O*-isobutyryl-L-mannoside with trifluoroethyl butyrate and porcine pancreatic lipase gave the 4,6-di-*O*-isobutyryl-L-mannoside, although these conditions are not applicable to the D-series [164] (Table 4.9).

4.5 Regioselective *bis*-Protection of the 2,6-, 3,6-, and 4,6-Positions of Hexopyranoside Tetraols

In the case of simple glucopyranoside 2,3,4,6-tetraols, there are useful reports of the simultaneous, regioselective protection of two positions, providing synthetically useful routes to the 2,6-3,6- and 4,6-di-*O*-protected monosaccharides (Table 4.10).

Table 4.9 Regioselective 4-*O*-functionalization of hexopyranosides.

Entry	Product	Conditions	Observations	References
108	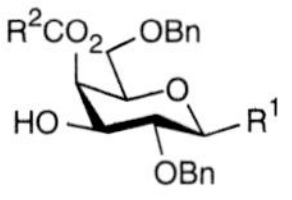	R^1 = βOBn or OMe/ R^2CO = Ac [165] 1. $MeC(OEt)_3$ (10 eq.), TsOH, PhMe 2. AcOH, H_2O, 89% R^1 = αOMe/R^2CO = Bz (with 2-*O*-Me) [166] 1. $PhC(OMe)_3$ 2. AcOH, H_2O, 74%	Many other glycosides reported since	[165, 166]
109	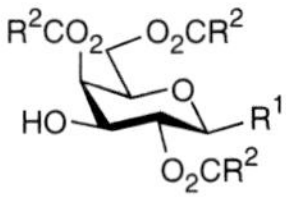	R^2CO = Ac/R^1 = MeOPh or $O(CH_2)_8N_3$ [167] $MeC(OEt)_3$ (5 eq.), TsOH (0.3 eq.), PhMe, RT (85%) R^2CO = Bz [168] $PhC(OMe)_3$ (3 eq.), CSA, C_6H_6, RT (99%)		[167, 168]
110	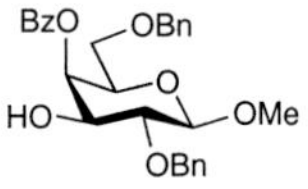	Bz_2O, $n\text{Bu}_4NOBz$		[115]
111	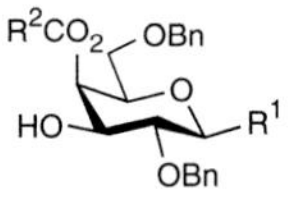	R^2COCN (1.0 eq.), DMAP (0.1 eq.), CH_2Cl_2/MeCN, −78 °C R^2CO = Bz (92%), R^2CO = Ac (81%), R^1 = OMe, STol (Bz)	PhCOCN also 2,6-di-*O*-Bz and 2,6-di-*O*-Ac; PhCOF or Bz_2O gives 3-*O*-Bz	[20]
112	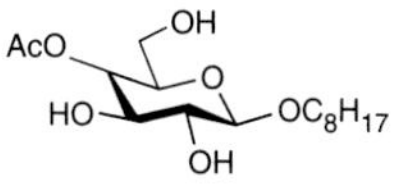	Ac_2O (1.1 eq.), catalyst (0.1 eq.), collidine (1.5 eq.), $CHCl_3$, −20 °C, 24 h, 96%	Peptide-based catalyst Also reported with βSC_8H_{17}	[169]
113	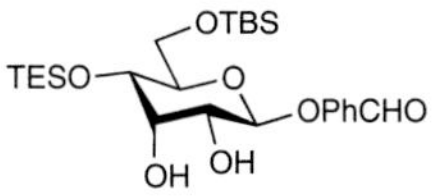	TESCl (1.2 eq.), chiral catalyst (0.05 eq.), $i\text{Pr}_2NEt{\cdot}HCl$ (0.03 eq.), $i\text{Pr}_2NEt$ (1.2 eq.), THF, 4 °C, 2 h, 63%	Catalyst 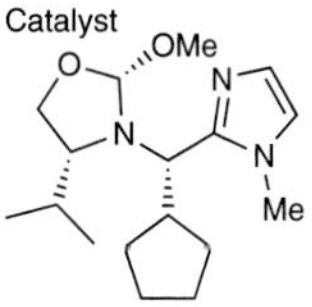	[55]

4.5.1 Glucose

The regioselective *bis*-benzylation at the 2- and 6-positions of methyl-α-D-glucopyranoside was reported using dibutyltin oxide and tetrabutylammonium bromide, providing the 2,6-di-*O*-benzyl glucoside in a single step in 85% yield (entry 114).

Persilylation of sugars requires forcing conditions, and partial silylation can yield synthetically useful partially protected products. In the glucose series, partial silylation leads selectively to the 3,6-di-*O*-TBS-protected glucose under

Table 4.10 Regioselective bis-functionalization of hexopyranosides.

Entry	Product	Conditions	Observations	References
114	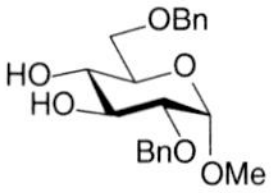	*n*Bu_2SnO (2.1 eq.), *n*Bu_4NBr (0.5 eq.), BnBr (4 eq.), PhMe, 100 °C, 10 h, 85%	Also reported with βGlcOMe (72%) or in two-step process [170]	[171]
115	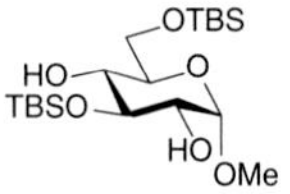	TBSCl (2.5 eq.), *n*Bu_4NBr (0.3 eq.), pyridine (5 eq.), 50 °C, 4.5 h, 60%	Solvent-free Also reported with Man or glucal	[172]
116	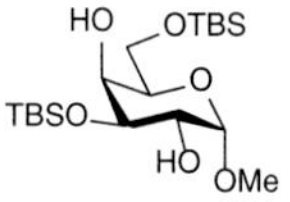	TBSCl (5 eq.), 3,5-di-CF_3-$PhB(OH)_2$ (0.2 eq.), *n*$Bu_3P{=}O$ (0.2 eq.), *i*Pr_2NEt (5 eq.), MeCN, 60 °C, 16 h, 81%	Also reported with Man (86%) [173] Also reported with $\alpha ManO(CH_2)_3N_3$ [174]	[173]
117	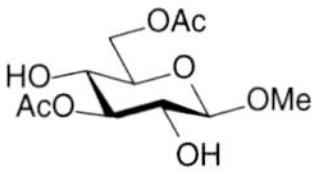	1. *n*Bu_2SnO (2.2 eq.), MeOH, 70 °C, 2 h 2. Ac_2O, DMF, 0 °C to RT, 6 h, 70%		[175]
118	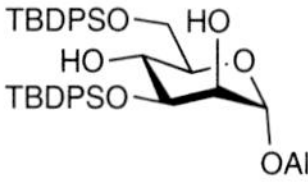	TBDPSCl (2.2 eq.), imidazole (5 eq.), DMF, 20 °C; 8 h, 68% (21% 2,6-di-*O*-TBDPS)		[162]
119	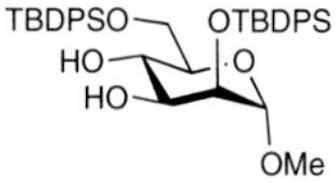	TBDPSCl (2.2 eq.), imidazole (5 eq.), DMF, 20 °C; 48 h, 62% (24% 3,6-di-*O*-TBDPS)		[162]
120	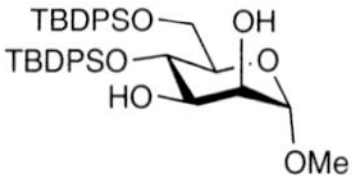	From 3,6-di-*O*-TBDPS isomer: *n*BuLi, −76 °C, THF (rigorously anhydrous), 1 h, 68%		[162]
121	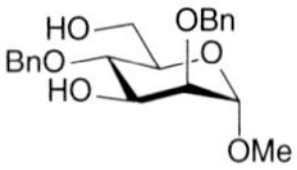	1. $PhCH(OMe)_2$, CSA, DMF, 70 °C, 250 mbar 2. BH_3·THF (10 eq.), $Cu(OTf)_2$ (0.3 eq.), 0 °C to RT, 1.5 h, 61% from the *bis*-benzylidene	Reductive opening of a *bis*-benzylidene acetal intermediate	[176]
122	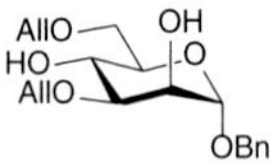	1. *n*Bu_2SnO (3 eq.), *n*Bu_4NBr (0.5 eq.), MeOH, reflux, 4 h 2. AllBr (10 eq.), CsF (3 eq.), DMF, 0 °C to RT, 4 days, 58%		[117]
123	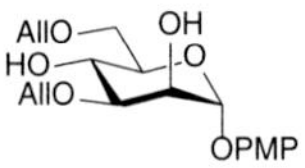	1. *n*Bu_2SnO (1.5 eq.), PhMe, reflux, 4 h 2. AllBr (10 eq.), *n*Bu_4NBr (0.3 eq.), PhMe, 80 °C, 7 days, 51%		[177]
124	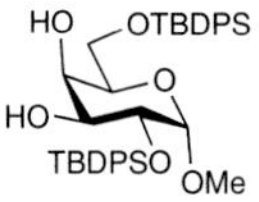	TBDPSCl, imidazole, DMF, 20 °C; 48 h, 65% (14% 3,6-di-*O*-TBDPS)		[62]

Table 4.10 (Continued)

Entry	Product	Conditions	Observations	References
125	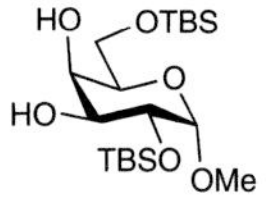	TBDMSCl (3.2 eq.), imidazole (6.4 eq.), DMF, RT, 24 h, 84% (14% 2,3,6-tri-*O*-TBS)	αGlcOMe 70% (2.2 eq. TBSCl) Poor results with βGalOMe	[2]
126	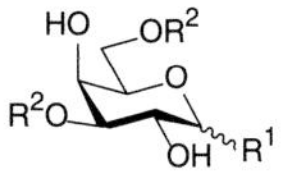	1. nBu_2SnCl_2 (0.1 eq.), nBu_4NBr (0.1 eq.), K_2CO_3 (1.5 eq.), R^2Br (2 eq.), MeCN/DMF (10/1), 80 °C, 3 h 2. R^2Br (1.5 eq.), NaOH (1.5 eq.), 80 °C, 3 h Nine examples/70–88%	R^1 = αOMe, βOMe, βSPh R^2 = All, Bn, PMB Very specific to Gal and poorly selective for Man	[178]
127	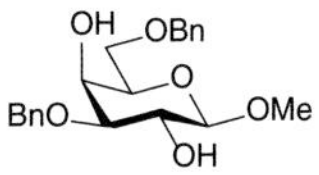	nBu_2SnO (2.1 eq.), nBu_4NBr (0.5 eq.), BnBr (5 eq.), PhMe, 100 °C, 12 h, 80%	Also reported with αManOMe (70%)	[171]
128	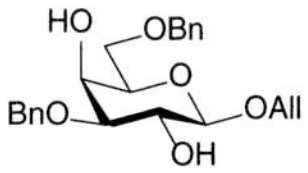	1. $(nBu_3Sn)_2O$ (1.5 eq.), PhMe, reflux, 6 h 2. BnBr (2 eq.), nBu_4NI (3 eq.), PhMe, 100 °C, 16 h, 70%		[179, 180]

solvent-free conditions (entry 115). However, *bis*-silylation under relatively standard conditions was reported to give the 2,6-*O*-*t*-butyldimethylsilyl-α-D-glucopyranoside under the same conditions as used in the galactose series (entry 125). The 3,6-bis-*O*-silyl glucopyranoside can also be obtained using a transient boronate esters (entry 116). Using dibutyltin oxide, this time in the absence of halide salts, yields selectively the 3,6-di-*O*-acetyl sugar in acceptable yield (entry 117).

4.5.2 Mannose

In the mannose series, bis-*t*-butyldimethylsilylation at the 3- and 6-positions was reported under the same conditions as in the glucose series, using tetrabutylammonium bromide and pyridine (entry 115) or boronic acid catalysts in high yield (entry 116). The 3,6-di-*O*-*t*-butyldiphenylsilyl-D-mannoside could also be obtained in moderate yield by direct silylation at short reaction times (entry 118). At longer reaction times, imidazole-catalyzed silyl group migration led to the 2,6-di-*O*-TBDPS isomer (entry 119), whereas migration with *n*-butyllithium led preferentially to the 4,6-di-*O*-TBDPS-D-mannopyranoside (entry 120).

The 3,6-di-*O*-benzyl-mannopyranoside was accessed via double regioselective reduction of the *bis*-benzylidene intermediate (entry 121). The use of dibutyltin oxide in the presence of halides provides the 3,6-di-*O*-allyl-mannopyranoside (entries 122 and 123) or the corresponding 3,6-di-*O*-benzyl derivative (entry 127).

4.5.3 Galactose

Finally, in the galactose series, *bis*-silylation under relatively standard conditions yields the 2,6-di-*O*-silyl-α-D-galactopyranosides (entries 124 and 125). Again, a strong dependence on the anomeric stereochemistry was reported, as the β-galactoside gave poor selectivity in the latter case.

Alkylation via tin acetals in the presence of tetrabutylammonium bromide provides the 3,6-di-*O*-benzyl, allyl, or PMB galactopyranoside in a two-step, one-pot process (entry 126). Otherwise, conditions that are generally similar to those used for the *bis*-allylation in the mannose series provide the di-*O*-benzyl galactopyranoside in high yields, based on David's pioneering work (entry 127). Finally, using transient monovalent tributyltin ethers similarly gives 3,6-di-*O*-benzyl galactopyranoside in good yield (entry 128).

4.6 Regioselective Mono-deprotection of Peracetyl and Perbenzyl Monosaccharides

Regioselective mono-deprotection provides the most direct access to a carbohydrate with a single free position (Table 4.11). The most widely reported reaction is the selective deprotection of the 2-position of a tetraprotected monosaccharide.

The synthesis of the 1,3,4,6-tetra-*O*-acetylglucopyranose from tetracetyl-D-glucopyranosyl chloride in acetone–DMF–water (entry 129) occurred presumably via the formation of an intermediate hemiorthoester. The alternative approach is the selective deprotection of peracetylated glucopyranosides, thanks to the electron-withdrawing effect of the endocyclic pyranose oxygen on the C2-subsitutent. The C2-acetoxy group is therefore more electrophilic and can be hydrolyzed selectively using either trifluoroacetic acid for the pentaacetate or hydrazine with the methyl glucoside (entries 130 and 131). A similar regioselective 2-*O*-debenzoylation in the allose series with hydrazine hydrate and pyridine at room temperature gave methyl 3,4,6-tri-*O*-benzoyl-D-alloside in 77% yield [191].

Enzymatic hydrolysis of acetyl groups offers access to other positions. *Aspergillus niger* lipase (ANL) deprotects selectively the 2-position of tetraacetyl methyl α-D- galactopyranoside (entry 132) and the 3-position of tetracetyl methyl α-D-glucopyranoside (entry 133). On the latter substrate, the CRL selectively deprotects the 6-position at pH 4, which after 4→6-*O*-acyl migration at pH 8 provides the 2,3,6-triacetate with a free alcohol at the 4-position (entry 134).

Sollogoub has reported a series of regioselective monodeprotections of tetra-*O*-benzyl glycosides using trialkylaluminum reagents. In the galactose and glucose series, selective deprotection of the 2-*O*-benzyl group is observed in high yields (entries 135 and 136). In the mannose series, the selectivity between the 2- and 3-positions is poor. On the other hand, the regioselective deprotection of the 4-position of allyl 2,3,4,6-tetra-*O*-benzyl-α-D-galactopyranoside was reported, in the presence of triethylsilane and iodine (entry 137). In the same study, they report the sequential regioselective deprotection of the 2-*O*-benzyl group after acylation of the 4-hydroxyl group.

Table 4.11 Regioselective mono-deprotection of hexopyranosides.

Entry	Product	Conditions	Observations	References
129	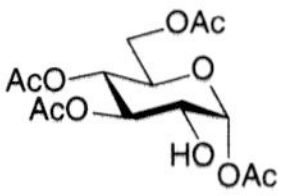	$Me_2CO/DMF/H_2O$ (100/4/10), 2.5 h, RT, 60%	From tetraacetyl glucopyranosyl chloride	[181]
130	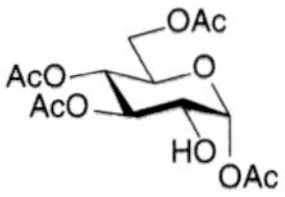	TFA/H_2O (10/1), 5 h, RT, 70%	From the peracetyl α-glucopyranose	[182]
131	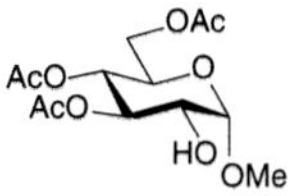	$N_2H_4{\cdot}H_2O$, THF, 10 h, 5 °C, 49%	Also reported with αOAll (47%) and βOAll (46%)	[183]
132	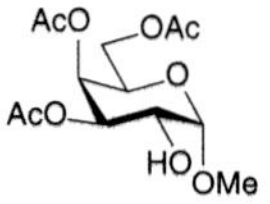	ANL, phosphate buffer/ Me_2CO (9/1), pH 7, 68%		[184]
133	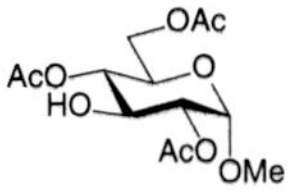	ANL, phosphate buffer/ Me_2CO (9/1), pH 7, 72%		[184]
134	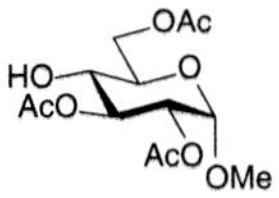	CRL, phosphate buffer/ MeCN (4/1), pH 4 then pH 8 68%	Acyl migration through change in pH Similar migration strategy reported for αOAc [185]	[186]
135	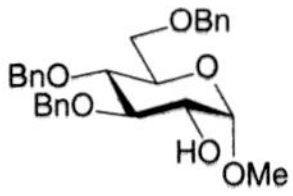	*i*Bu_3Al, PhMe, 3 h, 50 °C, 98% [187] or $AlMe_3$, PhMe, 4 days, 60 °C, 61% [188]	$AlMe_3$ is much slower than *i*Bu_3Al Also reported with Gal [187] (98%)	[187, 188]
136	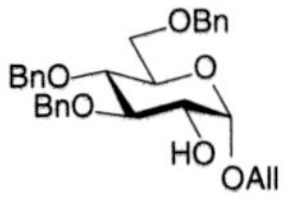	*i*Bu_3Al, PhMe, 16 h, 50 °C, 55%	Also reported with Gal (62%)	[189]
137	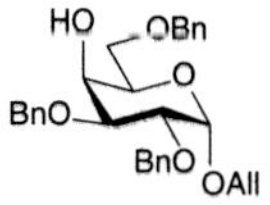	Et_3SiH, I_2, CH_2Cl_2, −40 to −20 °C, 50 min, 67%	*In situ* generated HI	[190]

4.7 Summary and Conclusions

The issue of protecting groups in carbohydrate chemistry represents an exceptionally rich example of the control of regioselectivity and chemoselectivity in organic chemistry. A rich body of literature has been published, part of which has been summarized in this chapter. Ultimately, any arrangement of protecting groups can be achieved in the three major carbohydrate configurations (Glc, Gal, and Man), and these results provide enough guidance to make reasonable predictions for other configurations and other protecting groups. Current progress focuses on designing particularly efficient reaction schemes, and on the regioselective protection of disaccharide building blocks, which are not addressed here.

A subtle interplay between various mechanistic factors, such as electronic effects, sterics, the reactivity of metal complexes, kinetic differentiation by chiral catalysts, and acyl and silyl group migration, renders simple predictive rules for synthetic design elusive. However, an awareness of these issues becomes important in the execution of a targeted protection scheme, as similar conditions can give different and sometimes unexpected outcomes, and a simple reliance on recipes may prove both frustrating and unproductive. Nonetheless, the basic premise that an effective protection scheme can be achieved, more or less efficiently, holds essentially true and that few syntheses will fail on the inability to achieve orthogonal protection of a carbohydrate precursor.

4.8 Experimental Section

4.8.1 Typical Experimental Procedure for Tin-acetal-mediated Regioselective 3-*O*-Benzylation [101] (Table 4.7, Entry 74)

To a solution containing the corresponding methyl 4,6-*O*-benzylidene-α-D-mannopyranoside (1 eq.) in dry toluene (7 ml $mmol^{-1}$) was added $n\text{Bu}_2\text{SnO}$ (1.02 eq.). The reaction mixture was stirred under reflux under an argon atmosphere for 3 h, cooled down to room temperature, followed by the addition of $n\text{Bu}_4\text{NBr}$ (1.06 eq.), CsF (1.02 eq.), and BnBr (1.05 eq.). The reaction mixture was stirred under reflux for 3 h, cooled down to room temperature, and diluted with EtOAc (40 ml) and saturated $NaHCO_3$ (30 ml). The organic layer was separated, and the water layer was extracted with EtOAc (3 × 40 ml). The combined organic layers were washed with water (30 ml) and brine (30 ml), dried, and concentrated. The crude product was purified by column chromatography (hexane/EtOAc, 4 : 1) to give methyl 3-*O*-benzyl-4,6-*O*-benzylidene-α-D-mannopyranoside in 89% yield.

4.8.2 Typical Experimental Procedure for Regioselective 2-*O*-Debenzylation of Perbenzyl Methyl Glucopyranoside [187] (Table 4.11, Entry 135)

A 1 M solution of triisobutylaluminium (5.4 ml, 5.4 mmol) in toluene was added to a solution of methyl 2,3,4,6-tetra-*O*-benzyl-α-D-glucopyranoside (500 mg, 0.90 mmol) in dry toluene (5 ml) at 0 °C under argon. The mixture was stirred at 50 °C for 3 h when TLC (EtOAc/C_6H_{12}, 1 : 2) indicated complete conversion of the starting material (R_f 0.5) into a single product (R_f 0.3). Ice-cold water (10 ml) was added, and the mixture was filtered, the organic layer was separated, and the aqueous layer was extracted with EtOAc (3 × 20 ml). The combined organic layers were dried ($MgSO_4$), concentrated, and the residue was purified by flash chromatography to afford methyl 3,4,6-tri-*O*-benzyl-α-D-glucopyranoside as a white crystalline solid (409 mg, 98%).

Abbreviations

Ac	acetyl
acac	acetylacetonate
All	allyl
ANL	*Aspergillus niger* lipase
BBTZ	1-benzoyloxy-1*H*-benzotriazole
Bn	benzyl
BOP	benzotriazole-1-yl-oxy-tris-(dimethylamino)-phosphonium hexafluorophosphate
Bz	benzoyl
CAB	*Candida antarctica* lipase B
CBz	benzyloxycarbonyl
ClAc	2-chloroacetyl
CRL	*Candida rugosa* lipase
CSA	camphorsulfonic acid
DCC	*N,N'*-dicyclohexylcarbodiimide
DCE	1,2-dichloroethane
dibm	diisobutyrylmethane
DMAP	4-dimethylaminopyridine
DMF	*N,N*-dimethylformamide
DTBMP	2,6-di-*t*-butyl-4-methylpyridine
eq.	equivalent
Fmoc	9-fluorenylmethyloxycarbonyl
HOBt	1-hydroxybenzotriazole
Lev	levulinoyl
MOM	methoxymethyl
MP	*p*-methoxyphenyl
MS	molecular sieves
Nap	2-naphthylmethyl
PCL	*Pseudomonas cepacia* lipase
PEMP	1,2,2,6,6-pentamethylpiperidine
Pent	pentenyl
PFL	*Pseudomonas fluorescens* lipase
Piv	pivaloyl
PMB	*p*-methoxybenzyl
RT	room temperature
TBS	*t*-butyldimethylsilyl
TBDPS	*t*-butyldiphenylsilyl
TES	triethylsilyl
TFA	trifluoroacetic acid
THF	tetrahydrofuran
TIPS	triisopropylsilyl
TLC	thin layer chromatography
TMS	trimethylsilyl
TMSE	2-trimethylsilylethyl
Tol	toluyl

Tom	triisopropylsilyloxymethyl
Tr	trityl, triphenylmethyl
Ts	tosyl, *p*-toluenesulfonyl

References

1 Silva, S., Simão, A.C., Tatibouët, A. et al. (2008). *Tetrahedron Lett.* 49: 682–686.
2 Halmos, T., Montserret, R., Filippi, J., and Antonakis, K. (1987). *Carbohydr. Res.* 170: 57–69.
3 Wood, W.W. and Rashid, A. (1987). *Tetrahedron Lett.* 28: 1933–1936.
4 Marković, D., Tchawou, W.A., Novosjolova, I. et al. (2016). *Chem. Eur. J.* 22: 4196–4205.
5 Khanbabaee, K., Lötzerich, K., Borges, M., and Großer, M. (1999). *J. Prakt. Chem.* 341: 159–166.
6 Doboszewski, B. and Zamojski, A. (1987). *Carbohydr. Res.* 164: 470–476.
7 Lázár, L., Jánossy, L., Csávás, M. et al. (2012). *Arkivoc* v: 312–325.
8 Fan, Q.-H., Li, Q., Zhang, L.-H., and Ye, X.-S. (2006). *Synlett* 1217–1220.
9 Keddie, N.S., Bultynck, G., Luyten, T. et al. (2009). *Tetrahedron: Asymmetry* 20: 857–866.
10 Dasgupta, F. and Garegg, P.J. (1994). *Synthesis* 1121–1123.
11 Ren, B., Ramström, O., Zhang, Q. et al. (2016). *Chem. Eur. J.* 22: 2481–2486.
12 Fügedi, P., Birberg, W., Garegg, P.J., and Pilotti, Å. (1987). *Carbohydr. Res.* 164: 297–312.
13 Huang, G.-L., Mei, X.-Y., and Liu, M.-X. (2005). *Carbohydr. Res.* 340: 603–608.
14 Lefeber, D.J., Kamerling, J.P., and Vliegenthart, J.F.G. (2001). *Chem. Eur. J.* 7: 4411–4421.
15 Yang, Z., Lin, W., and Yu, B. (2000). *Carbohydr. Res.* 329: 879–884.
16 Zong, G., Barber, E., Aljewari, H. et al. (2015). *J. Org. Chem.* 80: 9279–9291.
17 Li, Z.-J., Huang, H.-Q., and Cai, M.-S. (1994). *Carbohydr. Res.* 265: 227–236.
18 Lu, X.-A., Chou, C.-H., Wang, C.-C., and Hung, S.-C. (2003). *Synlett* 1364–1366.
19 Zhou, Y., Rahm, M., Wu, B. et al. (2013). *J. Org. Chem.* 78: 11618–11622.
20 Peng, P., Linseis, M., Winter, R.F., and Schmidt, R.R. (2016). *J. Am. Chem. Soc.* 138: 6002–6009.
21 Ross, A.J., Ivanova, I.A., Ferguson, M.A.J., and Nikolaev, A.V. (2001). *J. Chem. Soc., Perkin Trans. 1* 72–81.
22 Pelyvás, I., Lindhorst, T., and Thiem, J. (1990). *Liebigs Ann. Chem.* 761–769.
23 Burugupalli, S., Shah, S., van der Peet, P.L. et al. (2016). *Org. Biomol. Chem.* 14: 97–104.
24 Wang, G., Ella-Menye, J.-R., St. Martin, M. et al. (2008). *Org. Lett.* 10: 4203–4206.
25 Wang, H., She, J., Zhang, L.-H., and Ye, X.-S. (2004). *J. Org. Chem.* 69: 5774–5777.
26 Gangadharmath, U.B. and Demchenko, A.V. (2004). *Synlett* 2191–2193.
27 Krylov, V., Ustyuzhanina, N., Grachev, A. et al. (2007). *Synthesis* 3147–3154.
28 Norsikian, S. and Lubineau, A. (2005). *Org. Biomol. Chem.* 3: 4089–4094.

29 Boons, G.-J., Castle, G.H., Clase, J.A. et al. (1993). *Synlett* 913–914.
30 Iwasaki, F., Maki, T., Onomura, O. et al. (2000). *J. Org. Chem.* 65: 996–1002.
31 Zhou, Y., Ramstrom, O., and Dong, H. (2012). *Chem. Commun.* 48: 5370–5372.
32 Evtushenko, E.V. (2015). *J. Carbohydr. Chem.* 34: 41–54.
33 Allen, C.L. and Miller, S.J. (2013). *Org. Lett.* 15: 6178–6181.
34 Huber, F. and Kirsch, S.F. (2016). *Chem. Eur. J.* 22: 5914–5918.
35 Du, Y., Zhang, M., and Kong, F. (2000). *Org. Lett.* 2: 3797–3800.
36 Dalpozzo, R., De Nino, A., Maiuolo, L. et al. (2004). *Synth. Commun.* 34: 4207–4217.
37 Evtushenko, E.V. (2012). *Carbohydr. Res.* 359: 111–119.
38 Chen, I.H., Kou, K.G.M., Le, D.N. et al. (2014). *Chem. Eur. J.* 20: 5013–5018.
39 Hu, G. and Vasella, A. (2002). *Helv. Chim. Acta* 85: 4369–4391.
40 Cipolla, L., Lotti, M., De Gioia, L., and Nicotra, F. (2003). *J. Carbohydr. Chem.* 22: 631–644.
41 Peri, F., Cipolla, L., and Nicotra, F. (2000). *Tetrahedron Lett.* 41: 8587–8590.
42 Muramatsu, W. and Takemoto, Y. (2013). *J. Org. Chem.* 78: 2336–2345.
43 Demizu, Y., Kubo, Y., Miyoshi, H. et al. (2008). *Org. Lett.* 10: 5075–5077.
44 Watanabe, Y., Kawamoto, M., Shintaku, H. et al. (2013). *Asian J. Org. Chem.* 2: 927–930.
45 Crich, D., Li, W., and Li, H. (2004). *J. Am. Chem. Soc.* 126: 15081–15086.
46 Crich, D., Li, L., and Shirai, M. (2009). *J. Org. Chem.* 74: 2486–2493.
47 Boltje, T.J., Li, C., and Boons, G.-J. (2010). *Org. Lett.* 12: 4636–4639.
48 Ohtsuka, I., Hada, N., Sugita, M., and Takeda, T. (2002). *Carbohydr. Res.* 337: 2037–2047.
49 Kwon, Y.T., Lee, Y.J., Lee, K., and Kim, K.S. (2004). *Org. Lett.* 6: 3901–3904.
50 Lemanski, G. and Ziegler, T. (2000). *Helv. Chim. Acta* 83: 2676–2697.
51 Despras, G., Robert, R., Sendid, B. et al. (2012). *Bioorg. Med. Chem.* 20: 1817–1831.
52 Chevalier, R., Esnault, J., Vandewalle, P. et al. (2005). *Tetrahedron* 61: 7669–7677.
53 Mukhopadhyay, B. and Roy, N. (2003). *Carbohydr. Res.* 338: 589–596.
54 Reina, J.J., Di Maio, A., Ramos-Soriano, J. et al. (2016). *Org. Biomol. Chem.* 14: 2873–2882.
55 Sun, X., Lee, H., Lee, S., and Tan, K.L. (2013). *Nat. Chem.* 5: 790–795.
56 Ban, L. and Mrksich, M. (2008). *Angew. Chem. Int. Ed.* 47: 3396–3399.
57 Dubreuil, D., Cleophax, J., and Loupy, A. (1994). *Carbohydr. Res.* 252: 149–157.
58 Sarkar, V. and Mukhopadhyay, B. (2015). *Carbohydr. Res.* 406: 65–70.
59 Hakamata, W., Nishio, T., and Oku, T. (2000). *Carbohydr. Res.* 324: 107–115.
60 Rashid, A., Mackie, W., Colquhoun, I.J., and Lamba, D. (1990). *Can. J. Chem.* 68: 1122–1127.
61 MacManus, D.A. and Vulfson, E.N. (1995). *Carbohydr. Res.* 279: 281–291.
62 Arias-Perez, M.S., Lopez, M.S., and Santos, M.J. (2002). *J. Chem. Soc., Perkin Trans. 2* 1549–1552.
63 Vatele, J.M. (1986). *Tetrahedron* 42: 4443–4450.
64 Du, Y., Mao, W., and Kong, F. (1996). *Carbohydr. Res.* 282: 315–323.
65 Yoshimura, J., Kawauchi, N., Yasumori, T. et al. (1984). *Carbohydr. Res.* 133: 255–274.
66 Szeja, W. (1983). *Carbohydr. Res.* 115: 240–242.

67 Kim, S., Chang, H., and Kim, W.J. (1985). *J. Org. Chem.* 50: 1751–1752.
68 Rich, J.R., Wakarchuk, W.W., and Bundle, D.R. (2006). *Chem. Eur. J.* 12: 845–858.
69 Chen, H.-M. and Withers, S.G. (2010). *Carbohydr. Res.* 345: 2596–2604.
70 Lowary, T.L., Eichler, E., and Bundle, D.R. (2002). *Can. J. Chem.* 80: 1112–1130.
71 Davis, N.J. and Flitsch, S.L. (1994). *J. Chem. Soc., Perkin Trans. 1* 359–368.
72 Liu, J., Kong, F., and Cao, L. (1993). *Carbohydr. Res.* 240: 295–300.
73 Manabe, S. and Ito, Y. (2002). *J. Am. Chem. Soc.* 124: 12638–12639.
74 Mo, K.-F., Li, H., Mague, J.T., and Ensley, H.E. (2009). *Carbohydr. Res.* 344: 439–447.
75 Watson, A.J.A., Alexander, S.R., Cox, D.J., and Fairbanks, A.J. (2016). *Eur. J. Org. Chem.* 1520–1532.
76 Sato, K.-I., Sakai, K., Tsushima, K., and Akai, S. (2007). *Tetrahedron Lett.* 48: 3745–3748.
77 Mandal, P.K. and Misra, A.K. (2007). *Synthesis* 2660–2666.
78 Murphy, P.V., O'Brien, J.L., and Smith, A.B. (2001). *Carbohydr. Res.* 334: 327–335.
79 Verma, P., Kabra, V.K., and Mukhopadhyay, B. (2011). *Carbohydr. Res.* 346: 2342–2347.
80 Cox, D.J., Singh, G.P., Watson, A.J.A., and Fairbanks, A.J. (2014). *Eur. J. Org. Chem.* 4624–4642.
81 Pedretti, V., Veyrières, A., and Sinaÿ, P. (1990). *Tetrahedron* 46: 77–88.
82 Dhara, D. and Misra, A.K. (2013). *Tetrahedron: Asymmetry* 24: 1488–1494.
83 Nicolaou, K.C., Mitchell, H.J., Rodriguez, R.M. et al. (2000). *Chem. Eur. J.* 6: 3149–3165.
84 Murphy, P.V., O'Brien, J.L., Gorey-Feret, L.J., and Smith, A.B. (2003). *Tetrahedron* 59: 2259–2271.
85 Ghosh, S. and Misra, A.K. (2009). *J. Carbohydr. Chem.* 28: 447–462.
86 Csávás, M., Herczeg, M., and Borbás, A. (2013). *Synlett* 719–722.
87 Alpe, M. and Oscarson, S. (2002). *Carbohydr. Res.* 337: 1715–1722.
88 Roy, B., Pramanik, K., and Mukhopadhyay, B. (2007). *Glycoconjugate J.* 25: 157.
89 Thompson, C., Ge, M., and Kahne, D. (1999). *J. Am. Chem. Soc.* 121: 1237–1244.
90 Iacazio, G. and Roberts, S.M. (1993). *J. Chem. Soc., Perkin Trans. 1* 1099–1101.
91 Matsuo, I., Isomura, M., Walton, R., and Ajisaka, K. (1996). *Tetrahedron Lett.* 37: 8795–8798.
92 Danieli, B., Luisetti, M., Sampognaro, G. et al. (1997). *J. Mol. Catal. B* 3: 193–201.
93 Panza, L., Brasca, S., Riva, S., and Russo, G. (1993). *Tetrahedron: Asymmetry* 4: 931–932.
94 Gridley, J.J., Hacking, A.J., Osborn, H.M.I., and Spackman, D.G. (1998). *Tetrahedron* 54: 14925–14946.
95 Palocci, C., Falconi, M., Chronopoulou, L., and Cernia, E. (2008). *J. Supercrit. Fluids* 45: 88–93.
96 Attolino, E., Rising, T.W.D.F., Heidecke, C.D., and Fairbanks, A.J. (2007). *Tetrahedron: Asymmetry* 18: 1721–1734.
97 Ishiwata, A., Sakurai, A., Nishimiya, Y. et al. (2011). *J. Am. Chem. Soc.* 133: 19524–19535.
98 Tamigney Kenfack, M., Blériot, Y., and Gauthier, C. (2014). *J. Org. Chem.* 79: 4615–4634.

99 Codée, J.D.C., Hossain, L.H., and Seeberger, P.H. (2005). *Org. Lett.* 7: 3251–3254.
100 Seth, P.P., Yu, J., Jazayeri, A. et al. (2012). *J. Org. Chem.* 77: 5074–5085.
101 Poláková, M., Roslund, M.U., Ekholm, F.S. et al. (2009). *Eur. J. Org. Chem.* 870–888.
102 Müller, A., Kobarg, H., Chandrasekaran, V. et al. (2015). *Chem. Eur. J.* 21: 13723–13731.
103 Oshitari, T. and Kobayashi, S. (1995). *Tetrahedron Lett.* 36: 1089–1092.
104 Adero, P.O., Furukawa, T., Huang, M. et al. (2015). *J. Am. Chem. Soc.* 137: 10336–10345.
105 Lee, B.-Y., Baek, J.-Y., Jeon, H.-B., and Kim, K.-S. (2007). *Bull. Korean Chem. Soc.* 28: 257–262.
106 Tsutsui, N., Tanabe, G., Gotoh, G. et al. (2013). *Tetrahedron* 69: 9917–9930.
107 Sajtos, F., Lázár, L., Borbás, A. et al. (2005). *Tetrahedron Lett.* 46: 5191–5194.
108 Lázár, L., Bajza, I., Jakab, Z., and Lipták, A. (2005). *Synlett* 2242–2244.
109 Alpe, M., Oscarson, S., and Svahnberg, P. (2003). *J. Carbohydr. Chem.* 22: 565–577.
110 Codée, J.D.C., Kröck, L., Castagner, B., and Seeberger, P.H. (2008). *Chem. Eur. J.* 14: 3987–3994.
111 Kotturi, S.R., Tan, J.S., and Lear, M.J. (2009). *Tetrahedron Lett.* 50: 5267–5269.
112 Zhang, J. and Kong, F. (2003). *Carbohydr. Res.* 338: 1719–1725.
113 Ma, Z., Zhang, J., and Kong, F. (2004). *Carbohydr. Res.* 339: 1761–1771.
114 Sun, X., Lee, H., Lee, S., and Tan, K.L. (2013). *Nat. Chem.* 5: 790–795.
115 Zhang, X., Ren, B., Ge, J. et al. (2016). *Tetrahedron* 72: 1005–1010.
116 Lee, D. and Taylor, M.S. (2011). *J. Am. Chem. Soc.* 133: 3724–3727.
117 Unverzagt, C., Gundel, G., Eller, S. et al. (2009). *Chem. Eur. J.* 15: 12292–12302.
118 Giordano, M. and Iadonisi, A. (2014). *J. Org. Chem.* 79: 213–222.
119 Yu, H.N., Furukawa, J.-I., Ikeda, T., and Wong, C.-H. (2004). *Org. Lett.* 6: 723–726.
120 Polat, T. and Wong, C.-H. (2007). *J. Am. Chem. Soc.* 129: 12795–12800.
121 Zhang, X., Gu, G., and Guo, Z. (2015). *Eur. J. Org. Chem.* 7075–7085.
122 Cherif, S., Clavel, J.-M., and Monneret, C. (2002). *J. Carbohydr. Chem.* 21: 123–130.
123 Rising, T.W.D.F., Claridge, T.D.W., Davies, N. et al. (2006). *Carbohydr. Res.* 341: 1574–1596.
124 Emmadi, M. and Kulkarni, S.S. (2011). *J. Org. Chem.* 76: 4703–4709.
125 Kihlberg, J., Frejd, T., Jansson, K., and Magnusson, G. (1986). *Carbohydr. Res.* 152: 113–130.
126 Bartetzko, M.P., Schuhmacher, F., Hahm, H.S. et al. (2015). *Org. Lett.* 17: 4344–4347.
127 David, S., Thieffry, A., and Veyrières, A. (1981). *J. Chem. Soc., Perkin Trans. 1* 1796–1801.
128 Haque, M.E., Kikuchi, T., Yoshimoto, K., and Tsuda, Y. (1985). *Chem. Pharm. Bull.* 33: 2243–2255.
129 Yoshida, T., Chiba, T., Yokochi, T. et al. (2001). *Carbohydr. Res.* 335: 167–180.
130 Amer, H., Hofinger, A., Puchberger, M., and Kosma, P. (2001). *J. Carbohydr. Chem.* 20: 719–731.
131 Oscarson, S. and Sehgelmeble, F.W. (2005). *Tetrahedron: Asymmetry* 16: 121–125.

132 Crich, D., Wu, B., and Jayalath, P. (2007). *J. Org. Chem.* 72: 6806–6815.
133 Sanapala, S.R. and Kulkarni, S.S. (2014). *Chem. Eur. J.* 20: 3578–3583.
134 Tony Mong, K.-K., Shiau, K.-S., Lin, Y.H. et al. (2015). *Org. Biomol. Chem.* 13: 11550–11560.
135 Ohira, S., Yamaguchi, Y., Takahashi, T., and Tanaka, H. (2015). *Tetrahedron* 71: 6602–6611.
136 Liao, G., Zhou, Z., Burgula, S. et al. (2015). *Bioconjugate Chem.* 26: 466–476.
137 Crich, D. and Vinogradova, O. (2007). *J. Org. Chem.* 72: 6513–6520.
138 Ley, S.V., Abad-Somovilla, A., Anderson, J.C. et al. (2008). *Chem. Eur. J.* 14: 10683–10704.
139 Saikam, V., Dara, S., Yadav, M. et al. (2015). *J. Org. Chem.* 80: 11916–11925.
140 Williams, R.J., McGill, N.W., White, J.M., and Williams, S.J. (2010). *J. Carbohydr. Chem.* 29: 236–263.
141 Li, A. and Kong, F. (2005). *Bioorg. Med. Chem.* 13: 839–853.
142 Kumar, V., Locke, R.D., and Matta, K.L. (2009). *Synlett* 2633–2636.
143 Geng, X., Wang, L., Gu, G., and Guo, Z. (2016). *Carbohydr. Res.* 427: 13–20.
144 Valerio, S., Pastore, A., Adinolfi, M., and Iadonisi, A. (2008). *J. Org. Chem.* 73: 4496–4503.
145 Kalikanda, J. and Li, Z. (2011). *Carbohydr. Res.* 346: 2380–2383.
146 Oßwald, M., Lang, U., Friedrich-Bochnitschek, S. et al. (2003). *Z. Naturforsch. B* 58: 764–774.
147 Wang, Z., Zhou, L., El-Boubbou, K. et al. (2007). *J. Org. Chem.* 72: 6409–6420.
148 Mehta, S. and Whitfield, D.M. (2000). *Tetrahedron* 56: 6415–6425.
149 Nilsson, M. and Norberg, T. (1988). *Carbohydr. Res.* 183: 71–82.
150 Chittenden, G.J.F. (1971). *Carbohydr. Res.* 16: 495–496.
151 Pelyvás, I.F., Lindhorst, T.K., Streicher, H., and Thiem, J. (1991). *Synthesis* 1015–1018.
152 Angles d'Ortoli, T. and Widmalm, G. (2016). *Tetrahedron* 72: 912–927.
153 Jakas, A., Perc, M., Suć, J. et al. (2016). *J. Carbohydr. Chem.* 35: 69–85.
154 Jacquinet, J.-C. (2004). *Carbohydr. Res.* 339: 349–359.
155 Yang, B., Yoshida, K., Yin, Z. et al. (2012). *Angew. Chem. Int. Ed.* 51: 10185–10189.
156 Osborn, H.M.I., Brome, V.A., Harwood, L.M., and Suthers, W.G. (2001). *Carbohydr. Res.* 332: 157–166.
157 Evtushenko, E.V. (2010). *J. Carbohydr. Chem.* 29: 369–378.
158 Kanaya, T., Schweizer, F., Takeda, T. et al. (2012). *Carbohydr. Res.* 361: 55–72.
159 Liakatos, A., Kiefel, M.J., Fleming, F. et al. (2006). *Bioorg. Med. Chem.* 14: 739–757.
160 Ren, B., Rahm, M., Zhang, X. et al. (2014). *J. Org. Chem.* 79: 8134–8142.
161 Ito, Y., Ando, H., Wada, M. et al. (2001). *Tetrahedron* 57: 4123–4132.
162 Arias-Perez, M.S. and Santos, M.J. (1996). *Tetrahedron* 52: 10785–10798.
163 Öberg, C.T., Blanchard, H., Leffler, H., and Nilsson, U.J. (2008). *Bioorg. Med. Chem. Lett.* 18: 3691–3694.
164 Colombo, D., Ronchetti, F., and Toma, L. (1991). *Tetrahedron* 47: 103–110.
165 Paulsen, H., Hasenkamp, T., and Paal, M. (1985). *Carbohydr. Res.* 144: 45–55.
166 Aspinall, G.O., Chatterjee, D., and Khondo, L. (1984). *Can. J. Chem.* 62: 2728–2735.

167 McGill, N.W. and Williams, S.J. (2009). *J. Org. Chem.* 74: 9388–9398.
168 Sherman, A.A., Yudina, O.N., Mironov, Y.V. et al. (2001). *Carbohydr. Res.* 336: 13–46.
169 Kawabata, T., Muramatsu, W., Nishio, T. et al. (2007). *J. Am. Chem. Soc.* 129: 12890–12895.
170 Simas, A.B.C., da Silva, A.A.T., dos Santos Filho, T.J., and Barroso, P.T.W. (2009). *Tetrahedron Lett.* 50: 2744–2746.
171 Zhou, Y., Li, J., Zhan, Y. et al. (2013). *Tetrahedron* 69: 2693–2700.
172 Traboni, S., Bedini, E., and Iadonisi, A. (2016). *Beilstein J. Org. Chem.* 12: 2748–2756.
173 Lee, D. and Taylor, M.S. (2013). *Org. Biomol. Chem.* 11: 5409–5412.
174 Teumelsan, N. and Huang, X. (2007). *J. Org. Chem.* 72: 8976–8979.
175 Dong, H., Pei, Z., Byström, S., and Ramström, O. (2007). *J. Org. Chem.* 72: 1499–1502.
176 Pastore, A., Adinolfi, M., Iadonisi, A., and Valerio, S. (2010). *Carbohydr. Res.* 345: 1316–1323.
177 Heidecke, C.D. and Lindhorst, T.K. (2007). *Chem. Eur. J.* 13: 9056–9067.
178 Xu, H., Ren, B., Zhao, W. et al. (2016). *Tetrahedron* 72: 3490–3499.
179 Argunov, D.A., Krylov, V.B., and Nifantiev, N.E. (2015). *Org. Biomol. Chem.* 13: 3255–3267.
180 David, S. and Fernandez-Mayoralas, A. (1987). *Carbohydr. Res.* 165: c11–c13.
181 Dick, W.E. (1972). *Carbohydr. Res.* 21: 255–268.
182 Chittenden, G.J.F. (1988). *Carbohydr. Res.* 183: 140–143.
183 Li, J. and Wang, Y. (2004). *Synth. Commun.* 34: 211–217.
184 Hsiao, K.-F., Wu, S.-H., and Wang, K.-T. (1993). *Bioorg. Med. Chem. Lett.* 3: 2125–2128.
185 Filice, M., Guisan, J.M., Terreni, M., and Palomo, J.M. (2012). *Nat. Protoc.* 7: 1783.
186 Terreni, M., Salvetti, R., Linati, L. et al. (2002). *Carbohydr. Res.* 337: 1615–1621.
187 Sollogoub, M., Das, S.K., Mallet, J.-M., and Sinaÿ, P. (1999). *C. R. Acad. Sci.* 2: 441–448.
188 Jia, C., Pearce, A.J., Blériot, Y. et al. (2004). *Tetrahedron: Asymmetry* 15: 699–703.
189 Listkowski, A., Ing, P., Cheaib, R. et al. (2007). *Tetrahedron: Asymmetry* 18: 2201–2210.
190 Pastore, A., Valerio, S., Adinolfi, M., and Iadonisi, A. (2011). *Chem. Eur. J.* 17: 5881–5889.
191 Ishido, H., Sakairi, N., Sekiya, M., and Nakazaki, N. (1981). *Carbohydr. Res.* 97: 51–79.

5

Protecting Groups at the Anomeric Position of Carbohydrates

Chadamas Sakonsinsiri[1] *and W. Bruce Turnbull*[2]

[1]*Department of Biochemistry, Faculty of Medicine, Khon Kaen University, Khon Kaen 40002, Thailand*
[2]*School of Chemistry and Astbury Centre for Structural Molecular Biology, University of Leeds, Leeds LS2 9JT, UK*

5.1 Introduction

The chemistry of carbohydrates is dominated by the chemistry of the anomeric center. Pentose, hexose, and higher sugars have a high propensity to cyclize to form five-atom furanose and six-atom pyranose rings in which a hemiacetal or hemiketal group is formed, thus creating the anomeric center. This new stereocenter can have two possible configurations, referred to as α and β, the relative proportions of which tend to vary from sugar to sugar [1]. When monosaccharides are attached to other molecules, e.g. other sugars, lipids, or proteins, the glycosidic linkages are most commonly acetals/ketals involving the anomeric center. Formation of glycosidic linkages occurs through the process of glycosylation, which typically involves the loss of a substituent from the anomeric center, thus generating an oxacarbenium ion that may then react with a nucleophile, which is often an alcohol.

The purpose of this chapter is not to review glycosylation but rather to consider protecting groups at the anomeric position of carbohydrates. The term protecting group implies the temporary blocking of a reactive group while other chemical reactions are performed on the molecule. The protecting group should then be removable to reveal the original functionality. In the case of the anomeric center, that functionality is the hydroxyl group of a hemiacetal or hemiketal. Therefore, it is not surprising that all common classes of alcohol protecting groups have found use at the anomeric center. However, the reactivity of acetals and ketals presents a second strategy for anomeric protection in which the anomeric hydroxyl group is replaced by another element (e.g. sulfur) to create what may also be considered a latent hydroxyl group. Examples of this type of anomeric protection include thioglycosides which are commonly used as glycosyl donors [2]. Therefore, should all glycosyl donors leaving groups be considered to be protecting groups? No – many of the common anomeric leaving groups, e.g. Br and Cl, are not stable enough to allow other chemical manipulations on the sugar before their removal and so they do not meet the standard definition for a

Protecting Groups: Strategies and Applications in Carbohydrate Chemistry, First Edition.
Edited by Sébastien Vidal.

protecting group. Furthermore, glycosylation of an acceptor with a glycosyl donor does not constitute deprotection of the anomeric center if the reaction leads to a new glycosidic linkage. In such cases, the anomeric functional group is being used as a leaving group and not as a protecting group. Therefore, within the scope of this review, while glycosylation may sometimes constitute the process by which an anomeric protecting group is introduced, it will not be considered to be a deprotection step unless the acceptor alcohol is water, thus forming a hemiacetal/hemiketal. However, where a protecting group is more commonly converted into a glycosyl donor, e.g. a halide, then selected examples will also be included.

Nevertheless, glycosylation is still the primary driving force for chemists wanting to selectively protect/deprotect the anomeric center. For example, it is often convenient to introduce aglycons such as ceramide lipids late in a glycolipid synthesis [3]. The use of temporary anomeric protection in such cases allows the installation of popular leaving groups such as Schmidt's trichloroacetimidate for the final glycosylation reaction [4]. Furthermore, cyclic acetal protecting groups can provide convenient access to furanose versions of some sugars (e.g. glucose and xylose; see Chapter 12), and following other protecting group manipulations to fix the sugar in the furanose ring form, the anomeric center needs to be deprotected to allow conversion into a glycosyl donor and then into the desired glycoside.

In this chapter, we will first consider oxygen protecting groups including alkyl and aryl glycosides, esters, cyclic acetals/ketals and orthoesters (see Chapter 11), and silyl ethers. We will then consider alternative protection strategies in which a non-oxygen substituent is attached to the anomeric carbon atom, including thioglycosides and glycosyl amine derivatives. Although the literature is full of exciting new innovations that often promise more than they ultimately deliver, here we will focus on tried and tested methods that have an established history of widespread use, while only briefly mentioning some of the more exotic protecting groups that may demonstrate their worth in the years to come.

5.2 *O*-alkyl and *O*-aryl Glycosides

The most simple of anomeric oxygen protecting groups is the methyl group. Methyl glycosides are most conveniently prepared by Fischer glycosylation in which a strong acid is used to catalyze the reaction in methanol as the solvent. Although such reactions are typically performed under thermodynamic control leading to a mixture of α- and β-pyranosides [1], in some cases, e.g. xylose, it is possible to isolate furanosides as a kinetic product earlier in the reaction (Table 5.1, entry 1) [5, 6]. Following protection of the other hydroxyl groups, for example, by benzylation, it is possible to selectively hydrolyze the methyl glycoside to regain a hemiacetal (Table 5.1, entry 2) that may then be activated as a good leaving group for a glycosylation reaction [6]. Alternatively, deprotection can be achieved in a two-step process involving acetolysis followed by transesterification under Zemplén conditions (Table 5.1, entry 3) [7, 20].

Table 5.1 Installation and removal of alkyl and aryl anomeric protecting groups.

Entry	Example conditions	References
1	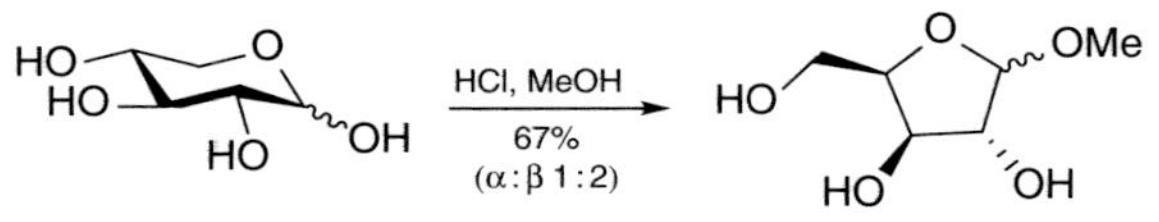	[5]
2	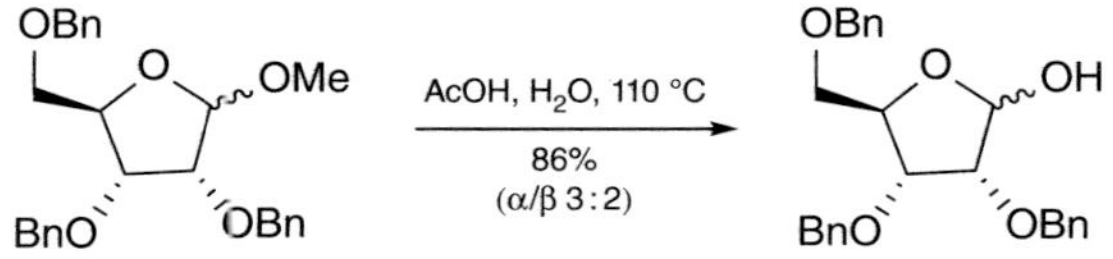	[6]
3	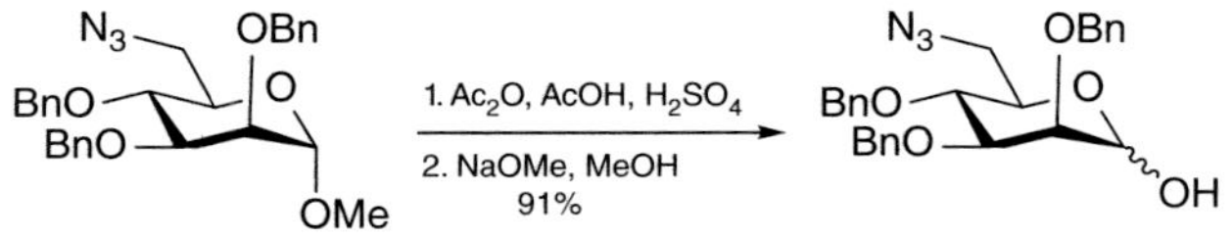	[7]
4	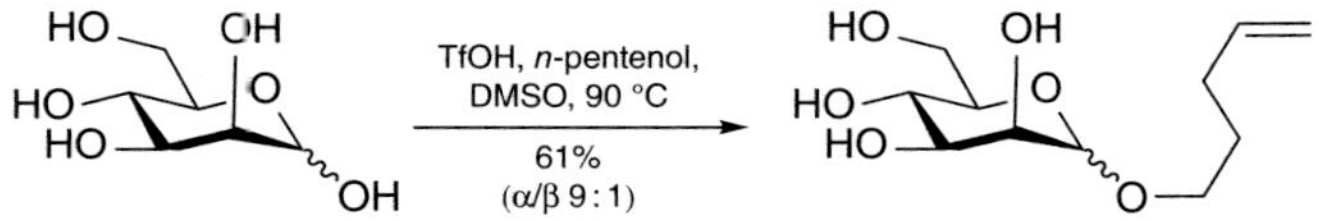	[8]
5	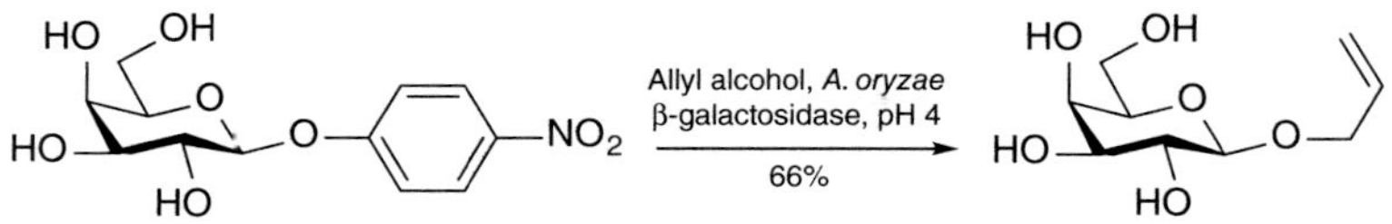	[9]

(Continued)

Table 5.1 (Continued)

Entry	Example conditions	References
6	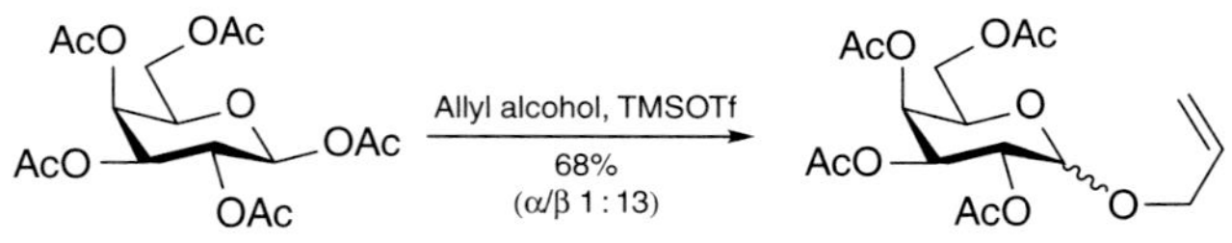	[10]
7	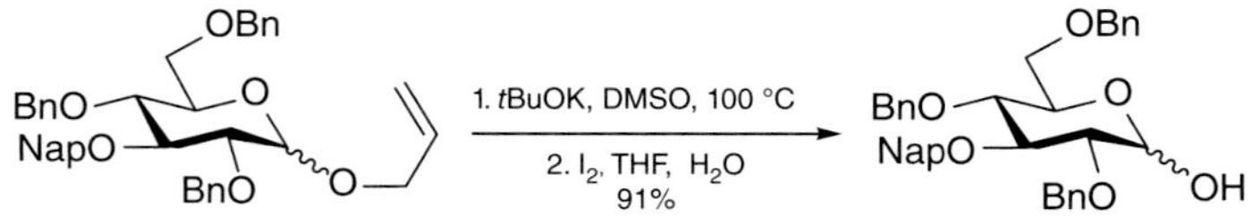	[11]
8	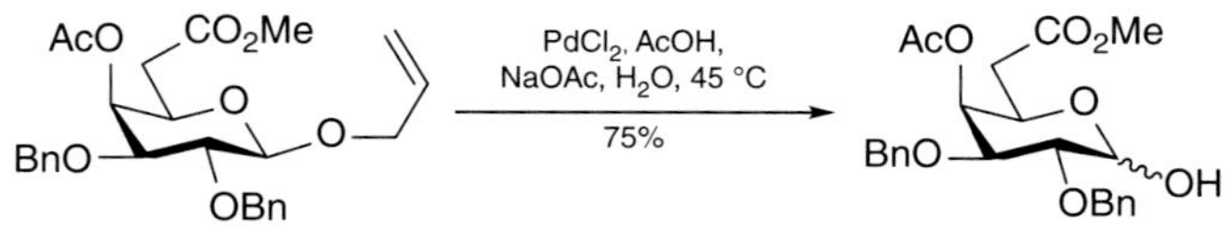	[12]
9	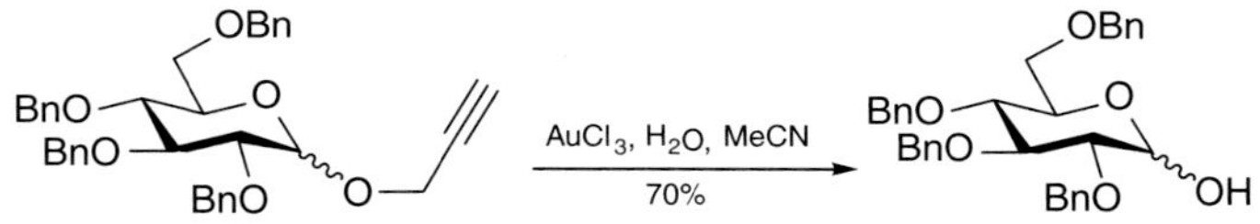	[13]
10	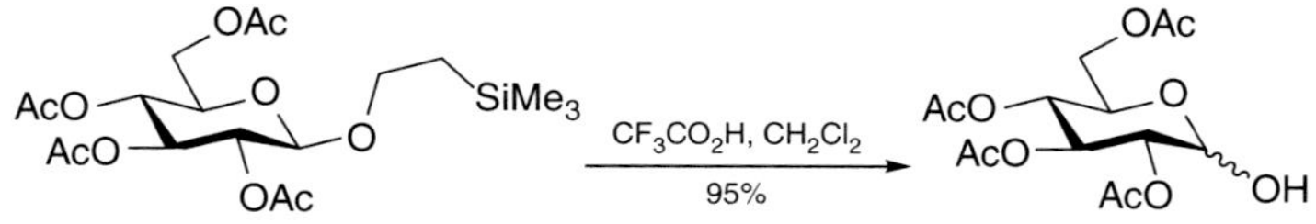	[14]

11	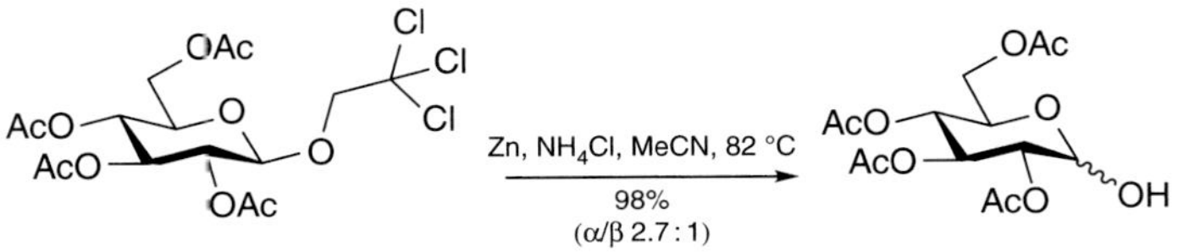	[15]
12	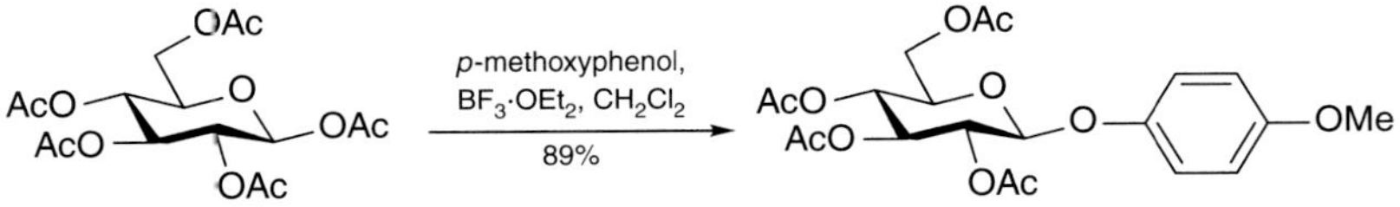	[16]
13	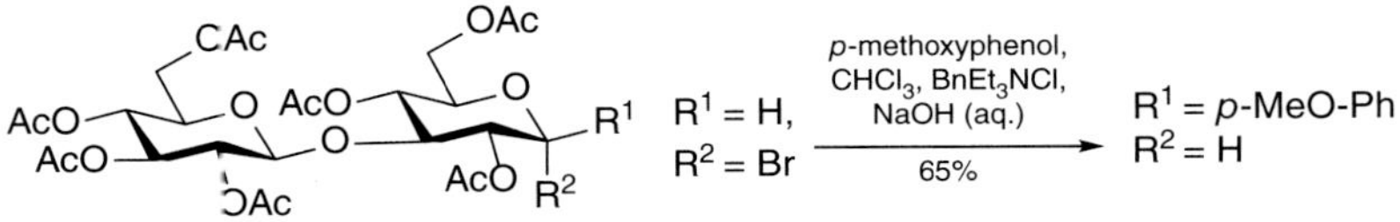	[17]
14	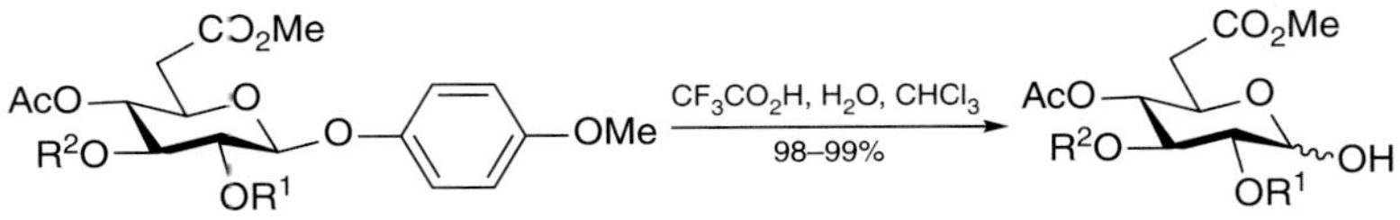 R^1, R^2 = other protected sugar residues	[18]
15	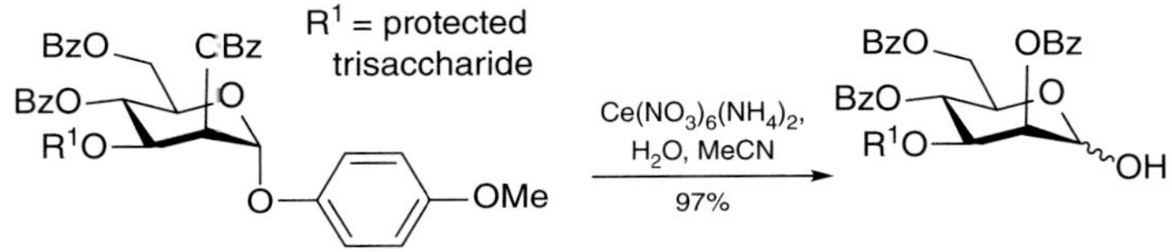	[19]

Other, relatively simple alkyl glycoside protecting groups can also be introduced by Fischer glycosylation (e.g. allyl, propargyl, or *n*-pentenyl) [8, 12, 13, 21]. However, this approach has some disadvantages: little control of anomeric stereochemistry can make analysis more complex, and solubility of the sugar in less polar alcohols can become limiting; however, this point can be alleviated by use of only a few equivalents of the alcohol in a polar solvent such as DMSO (Table 5.1, entry 4) [8]. An alternative glycosylation approach for introduction of such protecting groups is to use a glycosidase enzyme (Table 5.1, entry 5) [9], but more common is to treat a peracetylated sugar with, typically, 1–4 eq. of a Lewis acid to promote departure of the anomeric acetate group and generate an oxacarbenium ion intermediate (Table 5.1, entry 6) [10]. In this case, neighboring group participation by the 2-*O*-acetyl group usually provides only the β-glycoside product.

Aglycons that have additional chemical functionality are often more useful as protecting groups than simple alkyl groups, if that functional group can aid their release under mild and selective conditions. For example, an allyl group can be isomerized to an acid-labile enol ether by use of a strong base such as potassium *tert*-butoxide (Table 5.1, entry 7) [11], or a palladium catalyst (Table 5.1, entry 8) [12, 22], but other transition metals can also be used [23, 24]. Although such a transformation is not possible for a *n*-pentenyl group, activation with a source of I^+ or Br^+ leads to a cationic intermediate that may cyclize onto the anomeric oxygen atom and release the protecting group. This approach has been applied both for direct glycosylation of alcohols [25], but also for hydrolysis of the *n*-pentenyl glycosides [26]. An analogous reaction can be applied to propargyl glycosides, but in this case with activation by $AuCl_3$ (Table 5.1, entry 9) [13].

The 2-(trimethylsilyl)ethyl group has been applied extensively in the synthesis of glycolipids. It may be released from complex protected oligosaccharides under relatively mild conditions using aqueous trifluoroacetic acid (TFA; Table 5.1, entry 10) [14]. Alternatively, they can be converted directly into glycosyl esters in the presence of a Lewis acid and acid anhydride [14] or they can be converted to glycosyl chlorides using 1,1-dichloromethyl methyl ether and zinc chloride [27].

2,2,2-Trichloroethyl glycosides, prepared via Lewis acid activation of peracetylated sugars [28], are amenable to a wide variety of protecting group manipulations before anomeric deprotection by use of zinc powder and ammonium chloride (Table 5.1, entry 11) [15].

The introduction of aryl *O*-glycoside as protecting groups is also frequently accomplished by the treatment of acetylated sugars with the appropriate phenol in the presence of a Lewis acid (Table 5.1, entry 12) [16]. Alternatively, the greater acidity of phenols also permits their introduction through an S_N2-type of displacement of a glycosyl halide (Table 5.1, entry 13) [17]. The *p*-methoxyphenyl group has become a popular anomeric protecting group in recent years. As with the 2-(trimethylsilyl)ethyl group, it can be hydrolyzed in the presence of aqueous TFA (Table 5.1, entry 14) [18], but it can also be removed under oxidative conditions by use of ceric ammonium nitrate (CAN) (Table 5.1, entry 15) [19, 29]. Alternatively, it can be converted directly into a variety of glycosyl donors, e.g. chlorides, bromides, and phenyl thioglycosides [16, 30].

5.3 Glycosyl Esters

Esters form arguably the most important class of protecting groups used in carbohydrate chemistry. The first step of many oligosaccharide syntheses involves complete esterification of the sugar starting material. As such, regioselective protection is not a general consideration in the most common procedures for introducing anomeric ester protecting groups. There are many effective methods for acetylation of carbohydrates, most of which employ acetic anhydride as the acetylating agent. The classic method involves nucleophilic catalysis by pyridine (Table 5.2, entry 1) [31], sometimes supplemented with the more nucleophilic dimethylaminopyridine [41]. As such reactions are typically conducted under kinetic control, they usually lead to mixtures of α- and β-anomeric acetates in proportions that reflect the α/β ratio of the starting sugar; however, pure anomers can sometimes be obtained upon crystallization. It is also common to use Brønsted or Lewis acid catalysis to activate the acetic anhydride, for example, perchloric acid or zinc chloride (Table 5.2, entry 2) [31]. HBr in acetic acid can be used to promote acetylation of many sugars as a first step toward making glycosyl bromides [44]. Iodine-mediated catalysis can provide a very mild method of acetylation (Table 5.2, entry 3) [32]. Although these methods also tend to lead to mixtures of anomers, the use of hot acetic anhydride in the presence of sodium acetate provides an excellent method for preparing only the β-anomer of both mono- and disaccharides (Table 5.2, entry 4) [31, 33]. Although the acetylation step in this reaction is presumably also under kinetic control, the rate of hemiacetal anomerization is faster than the rate of acetylation of the α-anomer; therefore, the less-hindered β-anomer reacts selectively.

Acetolysis of oligo- and polysaccharides employing acetic anhydride, acetic acid, and sulfuric acid is another classical method for preparing anomeric acetates [7, 20]. This reaction is especially useful for preparing maltoheptaose derivatives via selective ring opening and acetylation of β-cyclodextrin derivatives (Table 5.2, entry 5) [34].

Both anhydrides and acid chlorides are used for installation of simple ester groups such as acetyl or chloroacetyl [45, 46]. However, in the case of more complex esters, e.g. benzoate and pivalate, it is much more common to use acid chlorides as their anhydrides are significantly less reactive (Table 5.2, entry 6; see Chapter 17) [35, 42, 47, 48]. Levulinyl groups have been introduced by activation of the acid with *N,N′*-diisopropylcarbodiimide (DIPC; Table 5.2, entry 7) [36]. Alternatively, a carboxylic acid can be alkylated with either a glycosyl halide (Table 5.2, entry 8) [37] or via a Mitsunobu reaction starting from the hemiacetal (Table 5.2, entry 9) [38].

Selective anomeric de-esterification is a very common step before the formation of imidate glycosyl donors, among other transformations. In the case of anomeric acetate groups, this transformation is most commonly achieved using hydrazinium acetate (Table 5.2, entry 10) [39], but a variety of other reagent systems including heavy metal salts, other amines, and perchloric acid on silica have also been investigated (Table 5.2, entry 11) [40]. In some cases, enzymatic hydrolysis of anomeric acetates can be very high yielding (Table 5.2, entry 12) [41]; however, predicting which ester(s) will be removed by any given enzyme is

Table 5.2 Installation and removal of ester anomeric protecting groups.

Entry	Example conditions	References
1	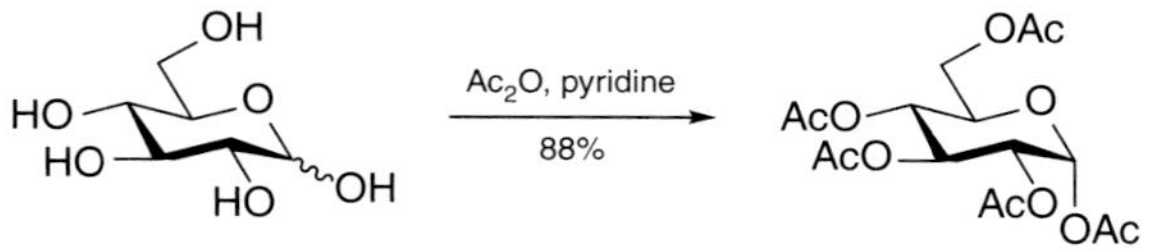	[31]
2	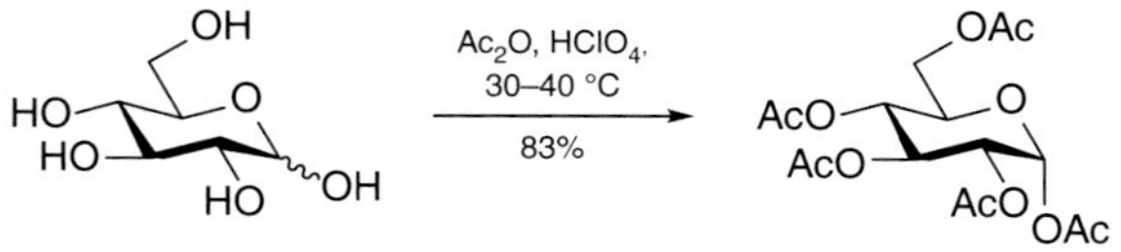	[31]
3	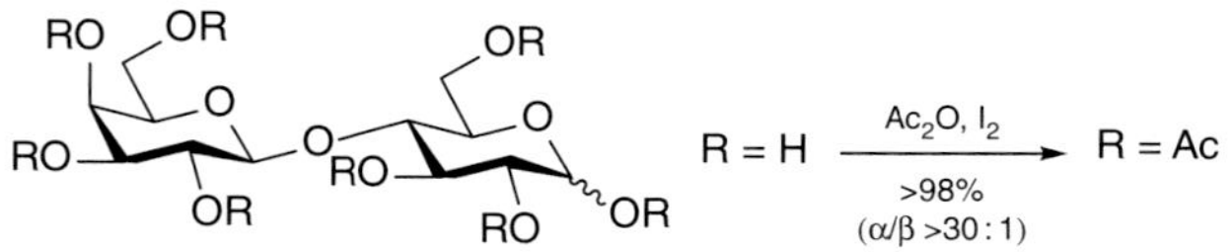	[32]
4	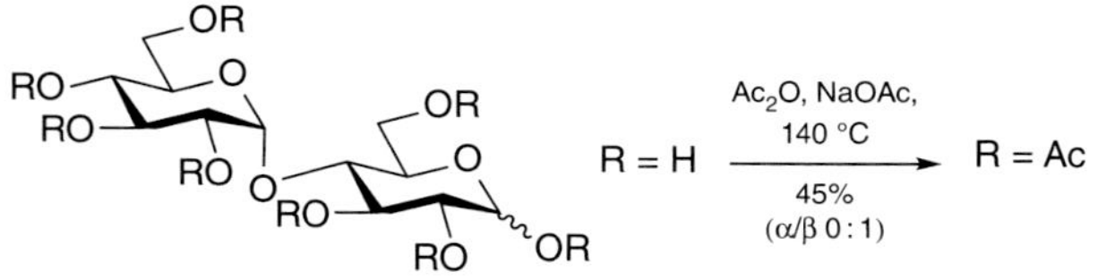	[33]
5	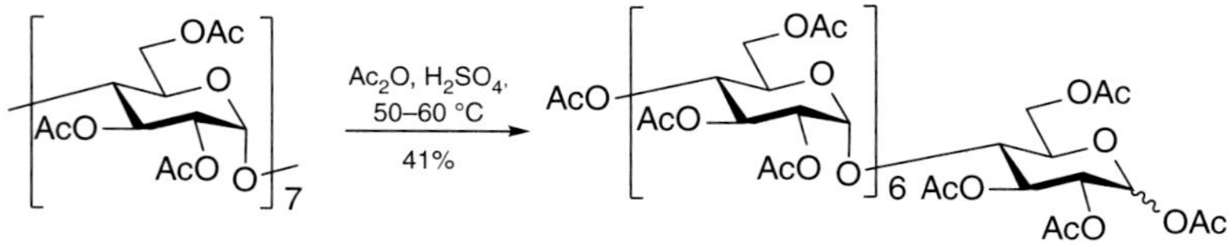	[34]

6 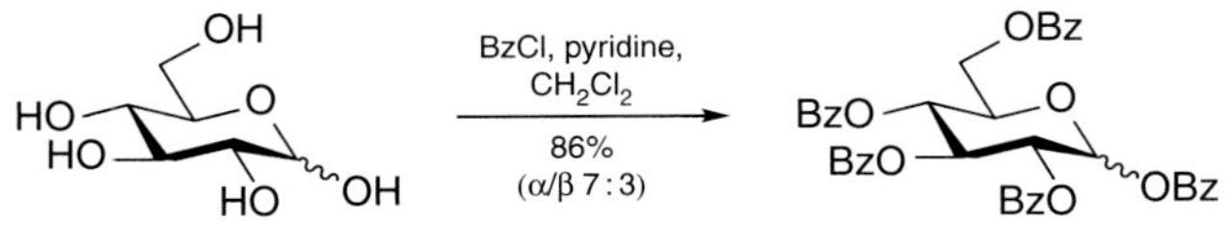

[35]

7 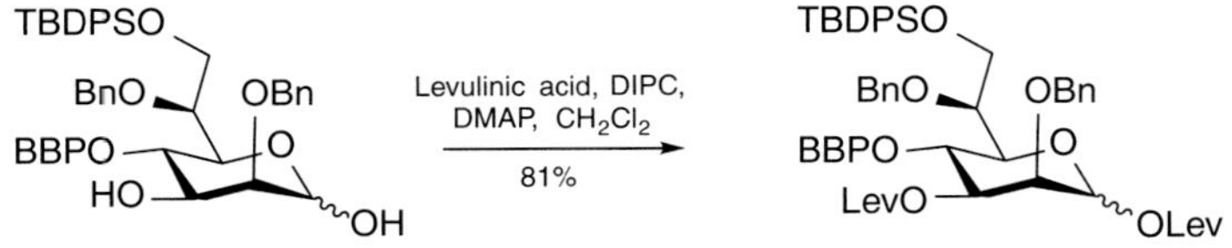

[36]

8 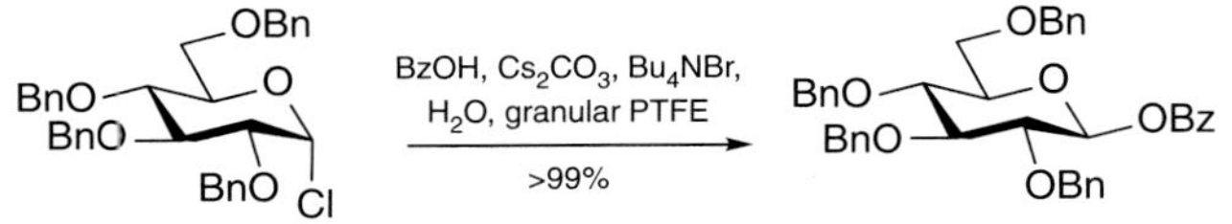

[37]

9 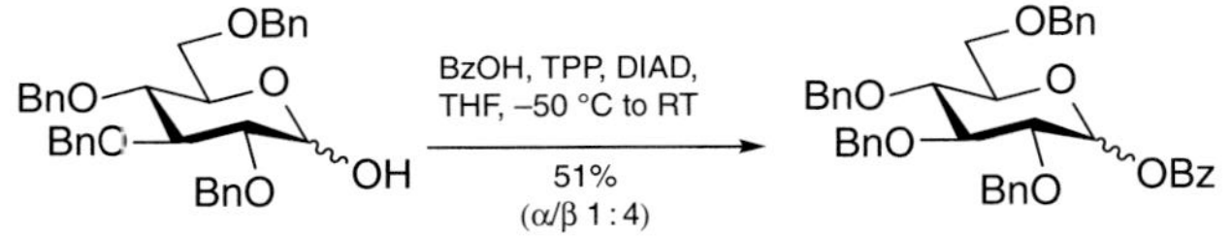

[38]

10 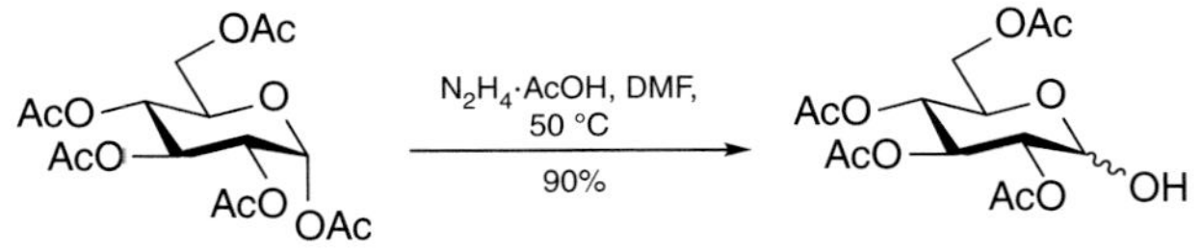

[39]

(Continued)

Table 5.2 (Continued)

Entry	Example conditions	References
11	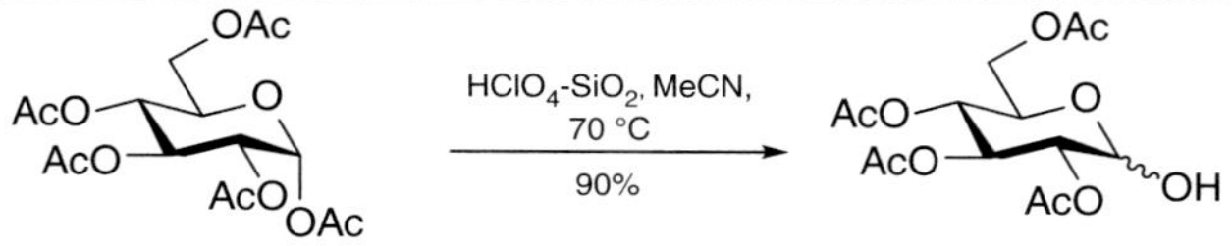	[40]
12	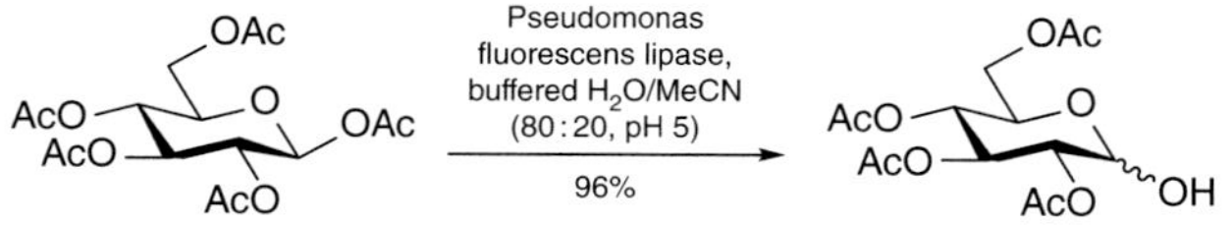	[41]
13	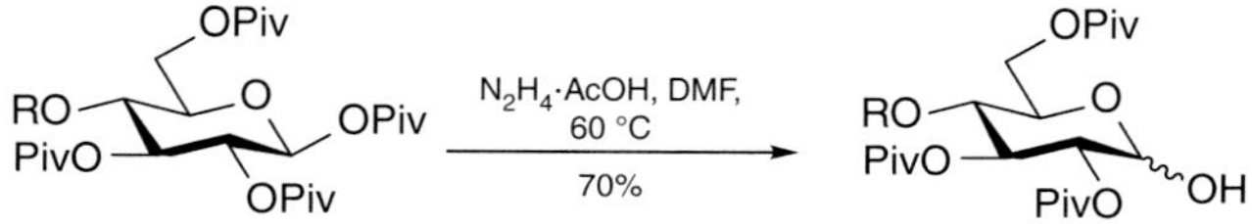 R = a fully pivalylated disaccharide	[42]
14	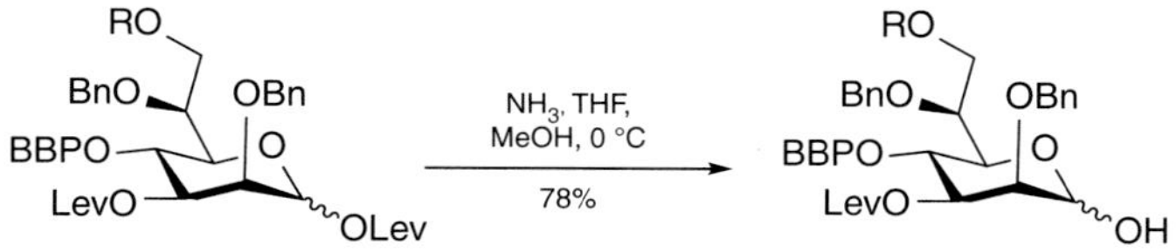 R = a protected monosaccharide	[36]
15	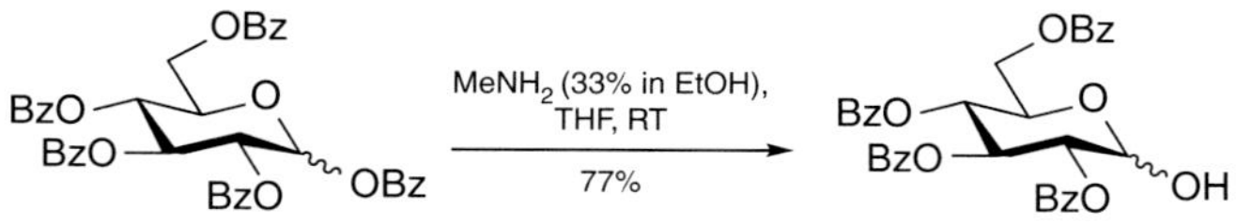	[43]

an empirical process. Hydrazinium acetate has also been used with other anomeric esters, e.g. pivalates (Table 5.2, entry 13) [42]. Although hydrazine-based reagents provide the standard methods for removing levulinyl groups, where there are multiple levulinates in a carbohydrate derivative, ammonia in a mixture of tetrahydrofuran and methanol has been shown to remove the anomeric levulinyl ester selectively (Table 5.2, entry 14) [36]. This reagent mixture is also suitable for selective removal of anomeric benzoates [47]; however, methyl amine is possibly used most widely for this purpose (Table 5.2, entry 15) [42, 43, 49].

5.4 Cyclic Acetals, Ketals, and Orthoesters

Although protecting group strategies for 1,2-diols are covered in depth in Chapter 11, the importance of cyclic acetals, ketals, and orthoesters to selective protection of the anomeric center also warrants mention in this chapter. The classic example of cyclic ketal protection of anomeric center is the use of the acetonide (isopropylidene ketal) of glucose. Many synthetic methods for diacetone glucose have been reported, but typically, they employ anhydrous acetone as solvent and catalysis with either a Brønsted and/or Lewis acid (e.g. sulfuric acid (Table 5.3, entry 1) or zinc chloride and phosphoric acid) [50]. The anomeric acetonide protecting group can withstand a broad range of other protecting group manipulations, including selective hydrolysis of the 5,6-acetonide group [50], making it a very useful entry to the synthesis of glucofuranose and xylofuranose derivatives. Hydrolysis of the 1,2-acetonide group requires more forcing conditions, e.g. H_2SO_4(aq.)/1,4-dioxane under reflux (Table 5.3, entry 2) [51], H_2SO_4/acetic acid/water at 70 °C (Table 5.3, entry 3) [52], and 90% aqueous TFA at room temperature (Table 5.3, entry 4) [53].

1,2-*O*-benzylidene acetals can be made in an analogous manner employing benzaldehyde as the solvent in the presence of a Lewis acid, e.g. zinc chloride [54]; however, the reaction is very inefficient and excess benzaldehyde is inconvenient to remove. It is preferable to use benzaldehyde dimethyl acetal in *N,N*-dimethylformamide (DMF) in the presence of a Brønsted acid, e.g. Amberlite IR-120 (H^+) resin; however, even this method is not very efficient (Table 5.3, entry 5) [54]. An alternative synthetic strategy is to perform a reductive cyclization reaction on a 2-*O*-benzoyl-1-bromo-sugar (Table 5.3, entry 6) [55]. Under catalysis by potassium iodide, a 2-*O*-benzoyl (or 2-*O*-(*p*-methoxybenzoyl)) group can cyclize onto the anomeric carbon, with reduction by sodium borohydride to yield the (*p*-methoxy)benzylidene acetal. Regardless of the synthetic route chosen, formation of a 1,2-acetal creates a new stereogenic center giving rise to mixtures of *endo*- and *exo*-isomers of the benzylidene acetal.

It is possible to deprotect the anomeric center selectively by reductive ring opening of benzylidene acetals to install a 2-*O*-benzyl, or 2-*O*-(*p*-methoxybenzyl), ether [54, 56, 57]. The optimal choice of reductant is dependent on the configuration of the sugar: mannosyl 1,2-*O*-benzylidene acetals can be converted efficiently to the corresponding 2-*O*-ether using either borane-tetrahydrofuran complex and trimethylsilyl triflate in tetrahydrofuran or sodium cyanoborohydride and trimethylsilyl chloride in acetonitrile (Table 5.3,

Table 5.3 Installation and removal of cyclic acetals, ketals, and orthoesters protecting groups.

Entry	Example conditions	References
1	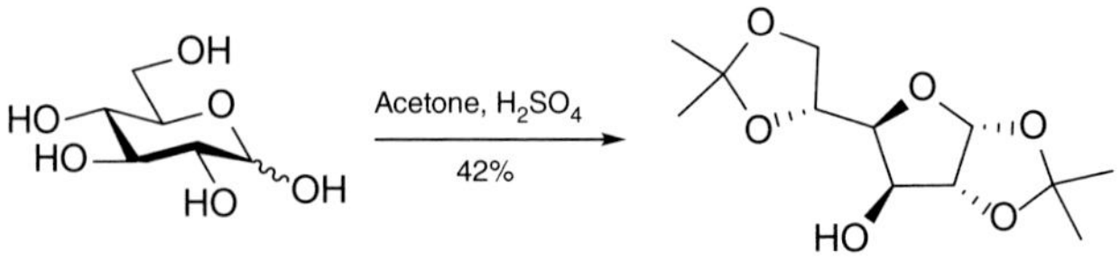	[50]
2	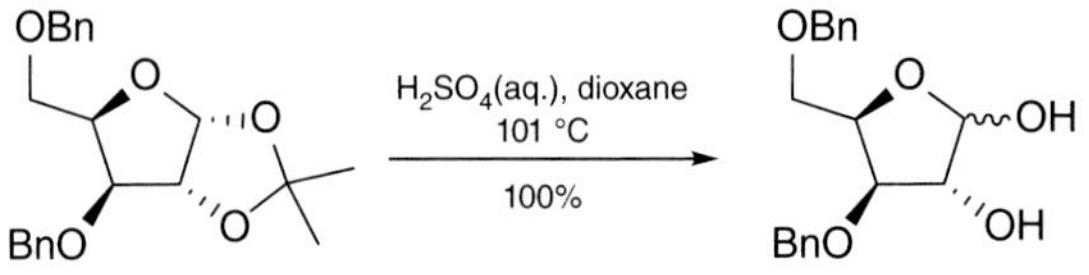	[51]
3	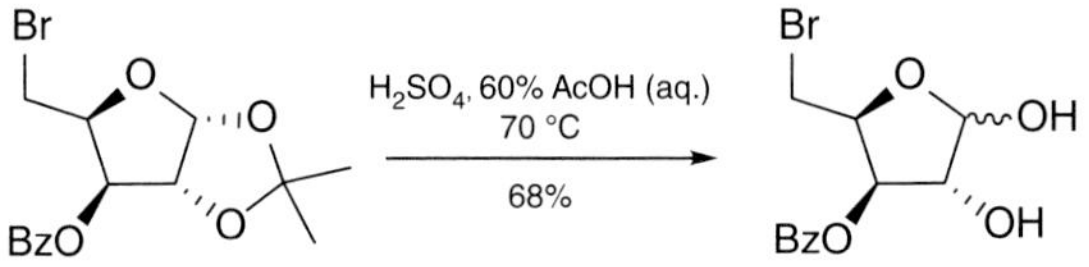	[52]
4	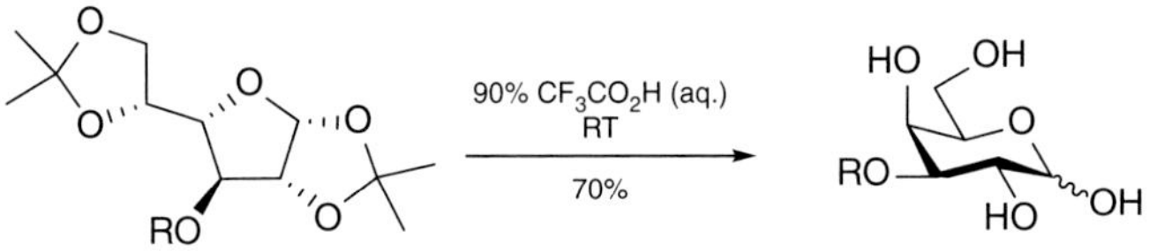 R = a protected monosaccharide	[53]
5	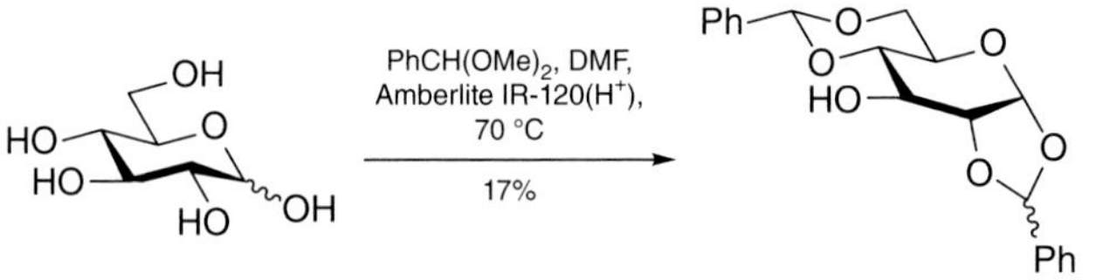	[54]
6	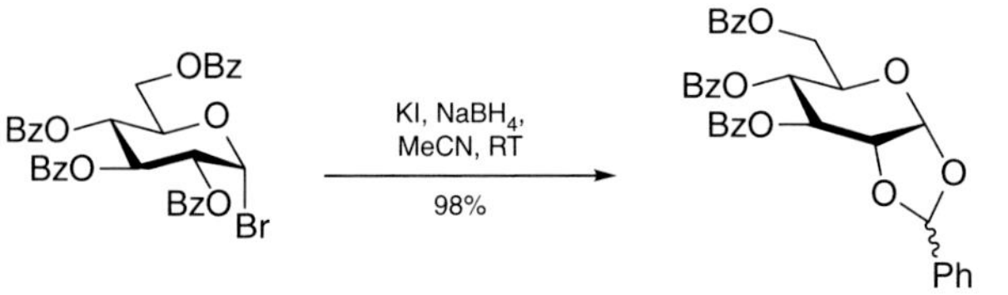	[55]
7	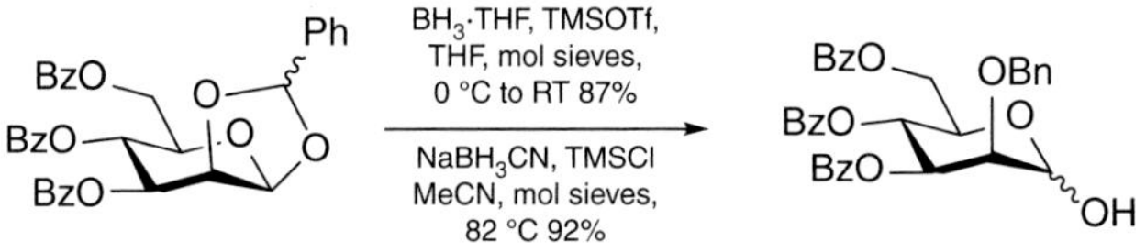	[56]
8	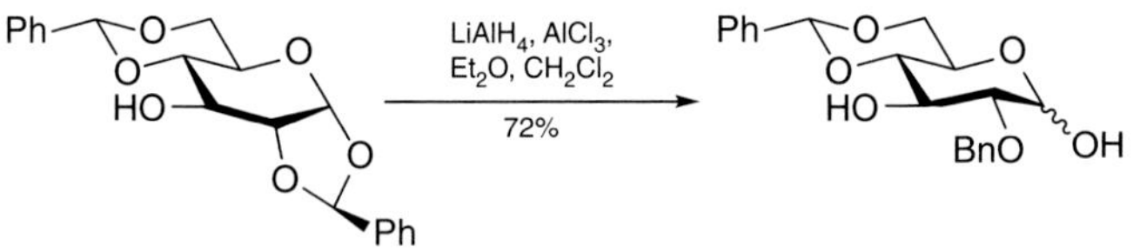	[54]

Table 5.3 (Continued)

Entry	Example conditions	References
9	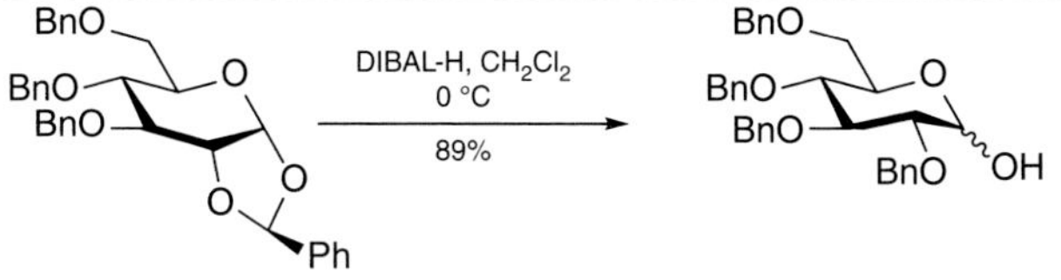	[57]
10	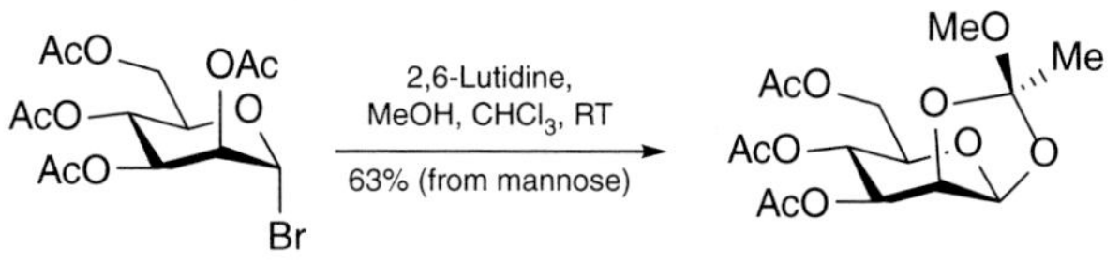	[58]
11	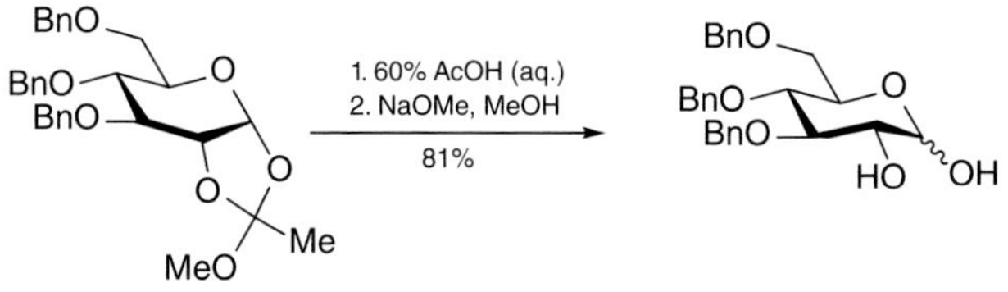	[59]

entry 7) [56]. Exposure of the corresponding gluco-configured sugars to these reagents tends to lead to mixtures of 1-*O*- and 2-*O*-benzyl ethers. However, use of a combination of lithium aluminum hydride with aluminum chloride (Table 5.3, entry 8) [54], or diisobutylaluminium hydride (Table 5.3, entry 9) [57], allows reductive ring opening of gluco-configured 1,2-*O*-benzylidene acetals to free the anomeric center; however, the former reaction only works for the *exo*-isomer of the 1,2-*O*-benzylidene acetal.

1,2-Orthoesters are also useful anomeric protecting groups for situations in which it is necessary to have O-3, O-4, and O-6 differentially protected from O-1 and O-2. They can be prepared conveniently by activation of acetylated glycosyl bromides in the presence of an alcohol and the hindered base 2,6-lutidine (Table 5.3, entry 10) [58]. As the orthoester is resistant to basic conditions, other protecting group manipulations can be performed on the sugar before ring opening of the orthoester under acidic conditions [59, 60]. Although hydrolysis restores the ester function, this reaction is not always completely regioselective [60]; nevertheless, subsequent treatment with sodium methoxide in methanol allows efficient deprotection of O-1 and O-2 (Table 5.3, entry 11) [59].

5.5 Silyl Ethers

Bulky silyl ethers are most frequently employed for selective protection of primary positions on carbohydrates (see Chapter 2). However, their advantage of high stability to diverse reaction conditions has also led to their widespread use as temporary anomeric protecting groups. The silyl ethers that appear most frequently at the anomeric center are the thexyldimethylsilyl (TDS) and *tert*-butyldimethylsilyl (TBS or TBDMS) groups. The TDS group is usually introduced using TDSCl and imidazole as catalyst in either DMF or dichloromethane (Table 5.4, entry 1) [61]. The anomeric TDS ether is stable during a wide range of protecting group manipulations; however, it is known to undergo migration between O-1 and O-2 under Zemplén deacetylation conditions [67]. It can be removed to regenerate a hemiacetal upon treatment with either tetrabutylammonium fluoride (Table 5.4, entry 2) [62], or HF-pyridine complex (Table 5.4, entry 3) [61].

TBS protecting groups are also often attached using the silyl chloride and imidazole in dichloromethane or DMF solution (Table 5.4, entry 4) [63, 64]. An alternative procedure employs TBSCl in pyridine solution (Table 5.4, entry 5) [53]. An important feature of these silylation reactions is that they tend to favor formation of the β-isomer for D-glucose and D-galactose derivatives as the equatorial β-configuration is less sterically hindered. Silylation can also influence the balance between furanosyl and pyranosyl forms of a sugar; for example, a thio-linked sialyl galactose disaccharide that existed as a mixture of furanose and pyranose isomers in solution was converted almost completely to the β-pyranose ring form upon introduction of TBS groups at O-1 and O-6 [53]. In contrast to this reaction performed in pyridine, it has been found that when free D-galactose reacts with TBSCl in DMF in the presence of imidazole, the β-furanose is the sole product formed (Table 5.4, entry 4) [63]. Anomeric TBS groups are relatively resistant to TFA hydrolysis, which is often used to remove primary TBS ethers [53], but can be removed by acetolysis (Table 5.4, entry 6) [53], or TBAF with TFA (Table 5.4, entry 7) [64].

The more bulky *tert*-butyldiphenylsilyl (TBDPS) and triisopropylsilyl (TIPS) groups have also found occasional use as anomeric protecting groups. TBDPS is again introduced using the silyl chloride in DMF in the presence of imidazole (Table 5.4, entry 8) and is removed using HF-pyridine complex (Table 5.4, entry 9) [65]. TIPS can also be introduced under these conditions [63], but for enhanced reactivity, it can be preferable to use triisopropylsilyl triflate (TIPSOTf) in the presence of a hindered base such as 2,6-lutidine (Table 5.4, entry 10) [66]. The additional steric hindrance surrounding the silicon atom in TIPS allows it to remain intact while a TBS group is removed using TBAF from a primary position in the same molecule; however, treatment with HF-pyridine complex will also remove the anomeric TIPS group (Table 5.4, entry 11) [66].

5.6 *S*-glycosyl and *N*-glycosyl Derivatives

All examples of anomeric protecting groups discussed thus far have been substituents attached to the anomeric oxygen atom. However, it is evident that in some cases, the anomeric oxygen atom is replaced during the deprotection step, for example, upon hydrolysis of methyl or *p*-methoxyphenyl glycosides. Therefore, protection of the anomeric center need not involve an oxygen substituent at all if a hydroxyl group can be installed during deprotection. In this regard, thioglycosides constitute one of the most widely used anomeric protection

Table 5.4 Installation and removal of silyl ether anomeric protecting groups.

Entry	Example conditions	References
1	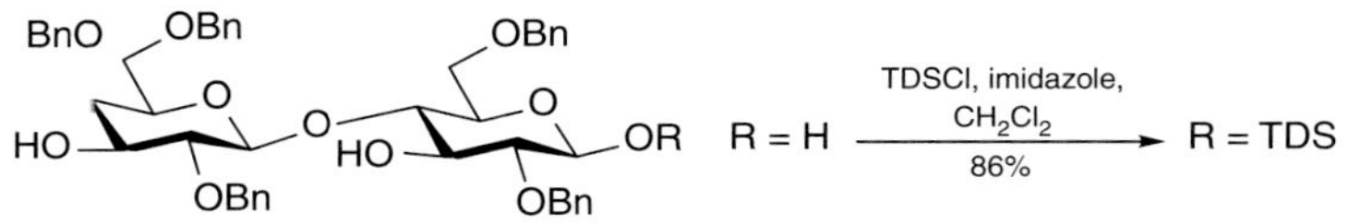	[61]
2	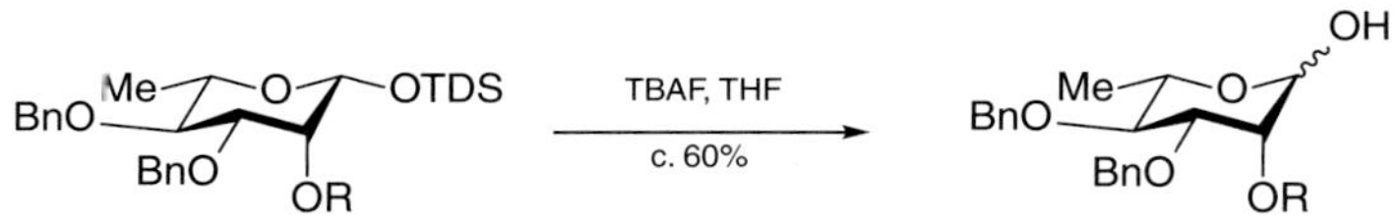 R = a protected pseudo-disaccharide	[62]
3	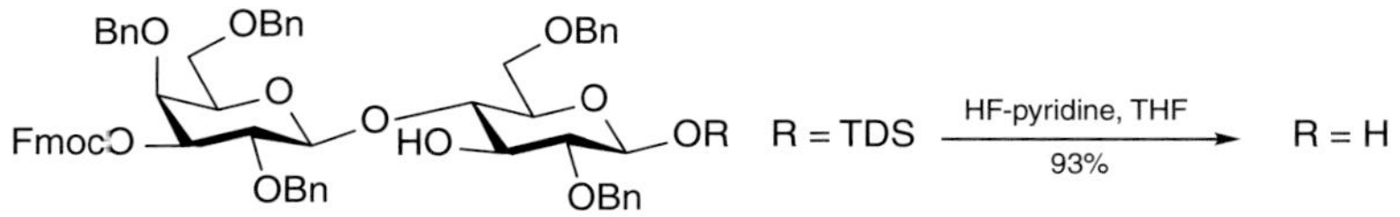	[61]
4	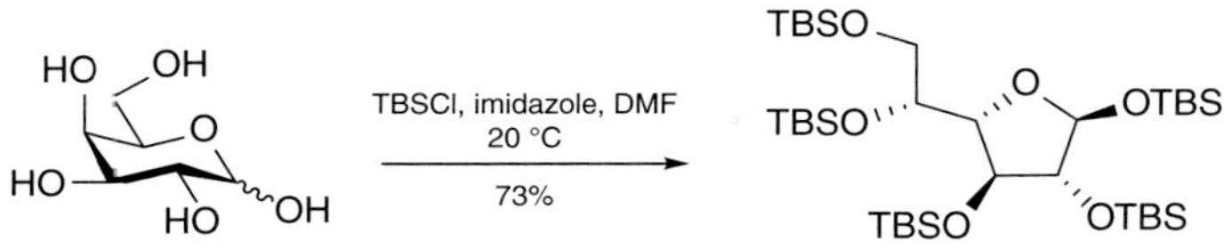	[63]

(*Continued*)

Table 5.4 (Continued)

Entry	Example conditions	References
5	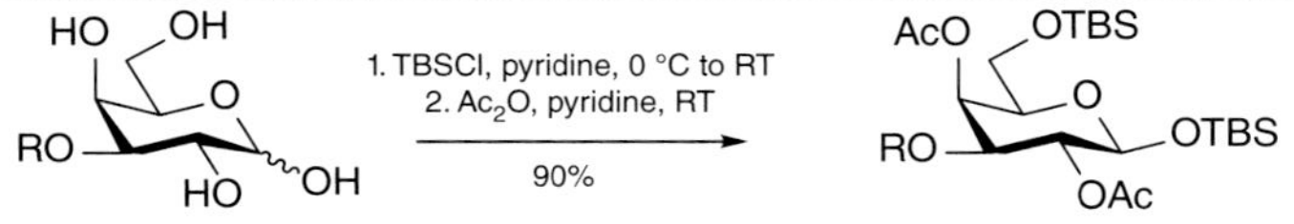 R = a protected monosaccharide	[53]
6	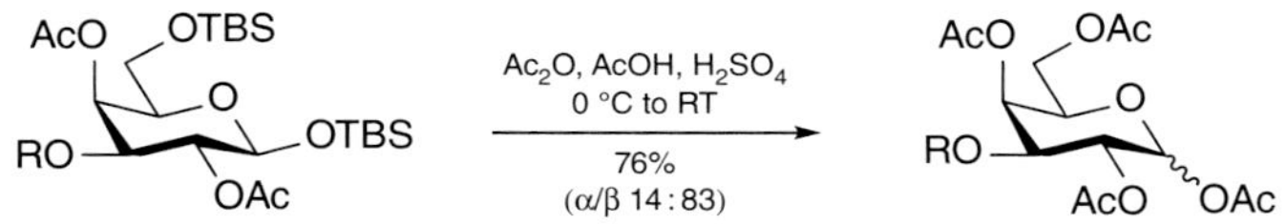 R = a protected monosaccharide	[53]
7	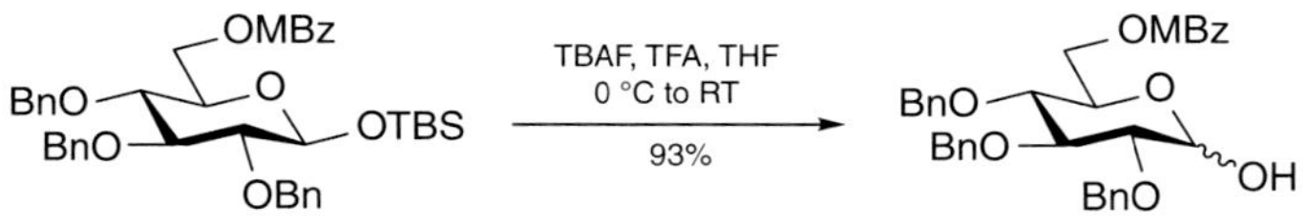	[64]
8	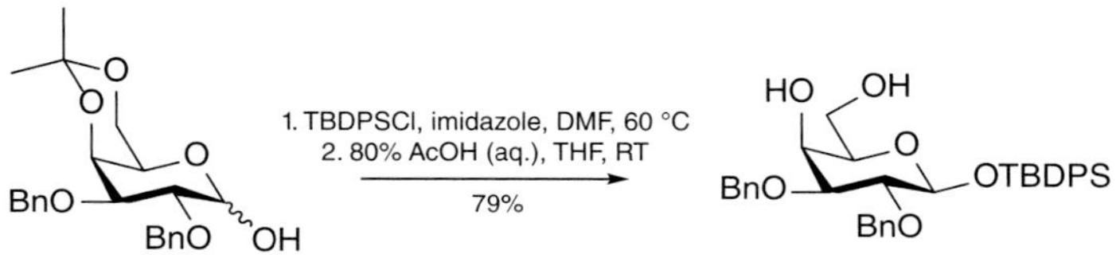	[65]

9	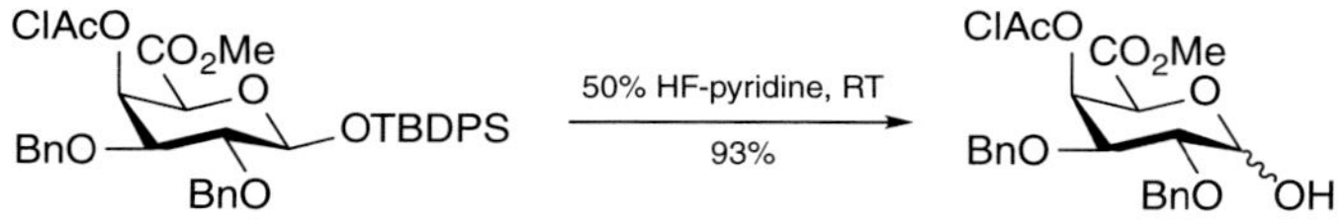	[65]
10	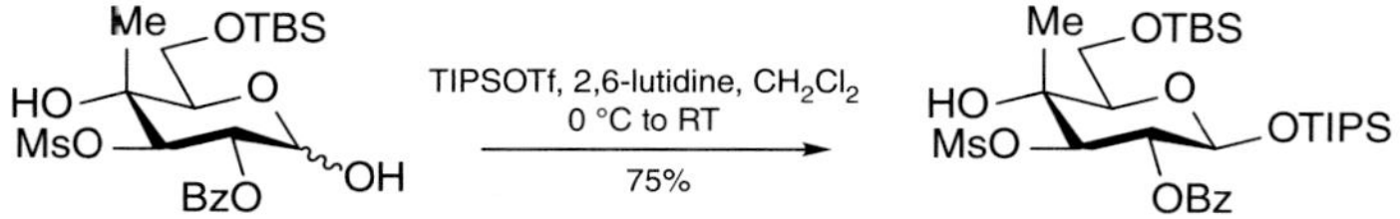	[66]
11	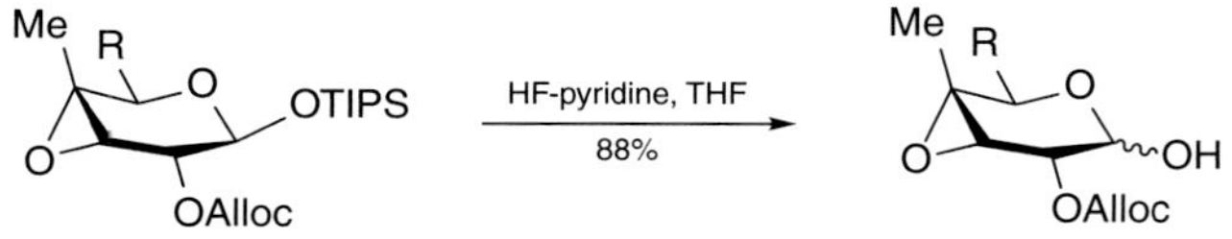 R = a polyunsaturated hydrocarbon	[66]

strategies. Despite their widespread utility as glycosyl donors, many researchers have found advantages in converting thioglycosides to other types of glycosyl donors, often via hemiacetal intermediates [68–71], but deprotection to generate a hemiacetal intermediate can enable other synthetic methods such as homologation to higher sugars [72]. Thioglycosides are stable to a very broad range of protecting group manipulations, yet can be hydrolyzed under conditions that are orthogonal to those used for most other protecting groups, typically by use of *N*-bromosuccinimide in aqueous acetone [73].

Although thiophenyl and thiotolyl glycosides are probably the most frequently used as temporary protecting groups [68, 69, 72, 74–76], others include thioethyl glycosides [70] and *S*-thiazolinyl (STaz) glycosides [77]. There are many routes to the synthesis of thioglycosides [78]; however, the most common strategies involve either activation of a glycosyl acetate with a Lewis acid such as $BF_3{\cdot}OEt_2$ (Table 5.5, entries 1 and 2) [79, 80] or reaction of a glycosyl bromide with the appropriate thiolate (Table 5.5, entries 3 and 4) [77, 78, 81]. In addition to the *N*-bromosuccinimide-promoted hydrolysis already mentioned (Table 5.5, entry 5) [68, 69, 72, 73, 75–77], other activation conditions for hydrolysis typically involve *N*-iodosuccinimide in combination with an aqueous acid, e.g. (Table 5.5, entries 6 and 7) sulfuric acid coated on silica [74], TFA [70]. The STaz group can be hydrolyzed in aqueous acetone upon addition of 0.2–1 eq. of a strong acid (Table 5.5, entry 8) [77].

N-glycosyl derivatives have been described much less frequently in the literature, but the growing familiarity of this class of compounds for chemical biology studies warrants their inclusion in this section. For example, *N*,*O*-dialkoxyamine-*N*-glycosides that are widely used to form glycoconjugates [84], have also been used as anomeric protecting groups [82, 83]. *N*,*O*-dimethylhydroxylamine reacts with reducing sugars in sodium acetate buffer to provide the β-*N*-glycoside (Table 5.5, entry 9) [82]. Such compounds are stable to a range of protecting group manipulations and glycosylation reactions before hydrolysis of the *N*-glycoside protecting group using *N*-chlorosuccinimide in aqueous THF (Table 5.5, entry 10) [82], or trimethylsilyl chloride in THF/methanol (9 : 1) (Table 5.5, entry 11) [83].

5.7 Concluding Remarks

There is a vast array of anomeric protecting groups available to the carbohydrate chemist. As with many things in both life and science, the choice of anomeric protecting group is subject to the manner of the time. Data from Scifinder Scholar show that the number of publications reporting the use of 2-(trimethylsilyl) ethyl glycosides rose from one publication in 1981 to peak at 44 papers in 1996 and has now fallen to typically fewer than 10 reports each year. In contrast, *p*-methoxyphenyl glycosides have seen an exponential increase in publications since the late 1980s and are probably only now reaching their zenith (over 100 reports per annum since 2013). The protecting group of preference for a particular application is very dependent on the overall protecting group strategy,

Table 5.5 Installation and removal of *S*-glycosyl and *N*-glycosyl anomeric protecting groups.

Entry	Example conditions	References
1	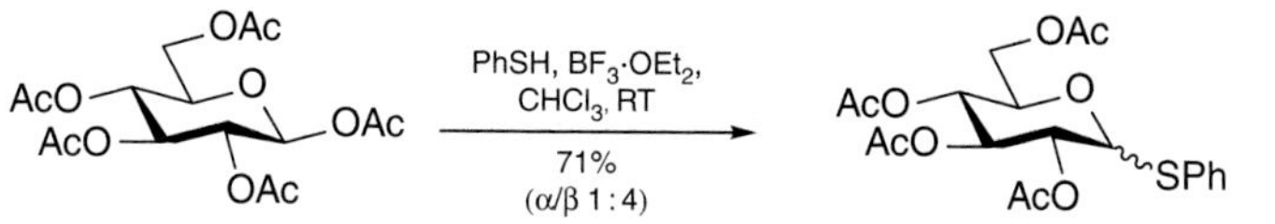	[79]
2	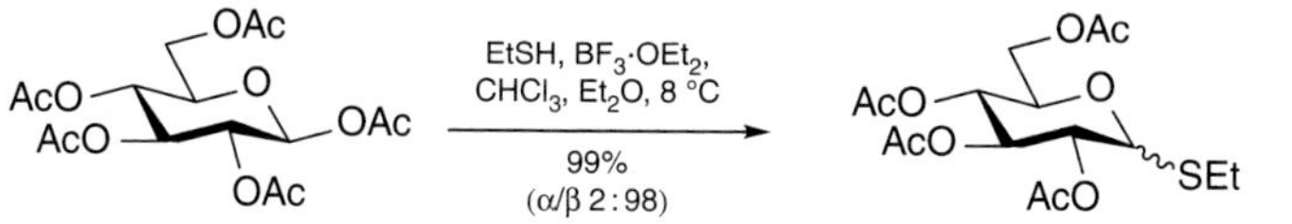	[80]
3	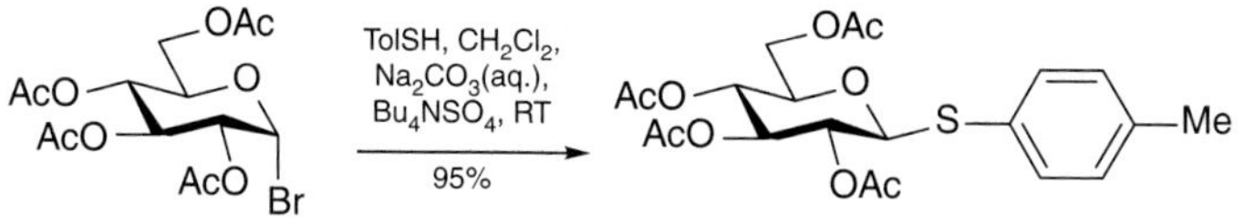	[81]
4	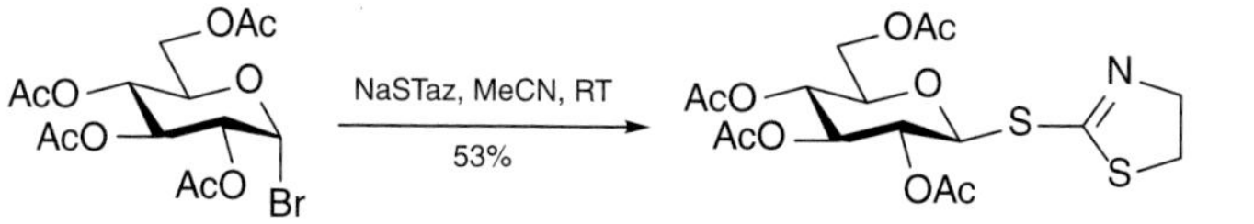	[77]
5	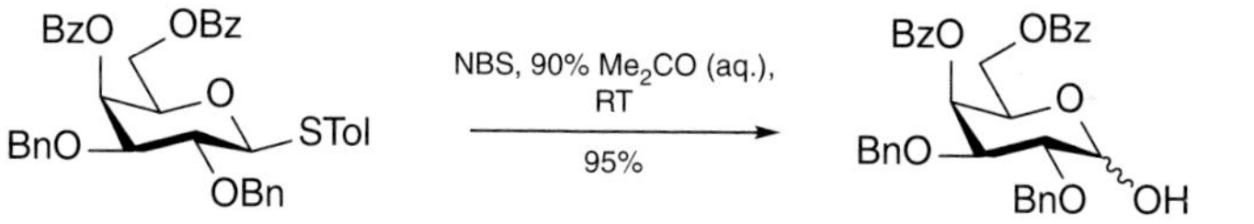	[68]
6	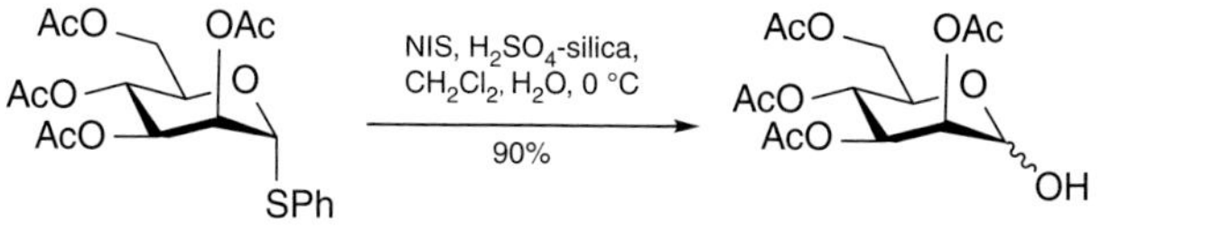	[74]
7	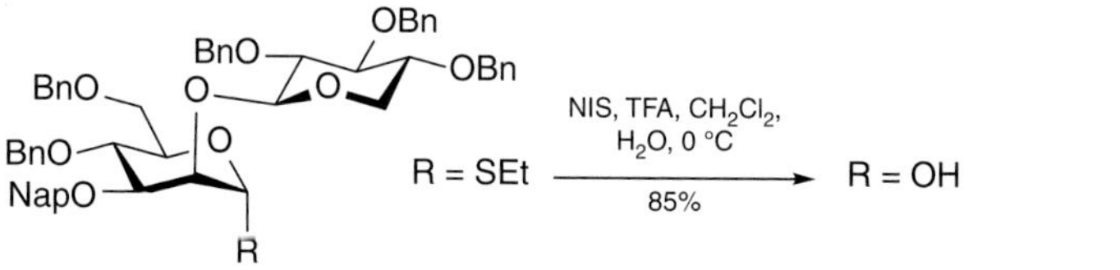	[70]
8	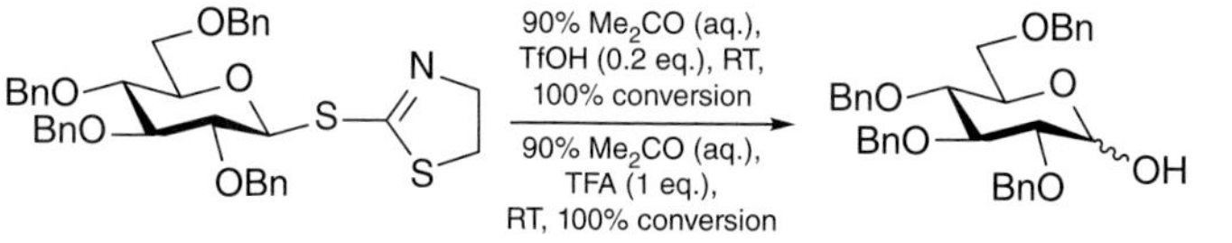	[77]
9	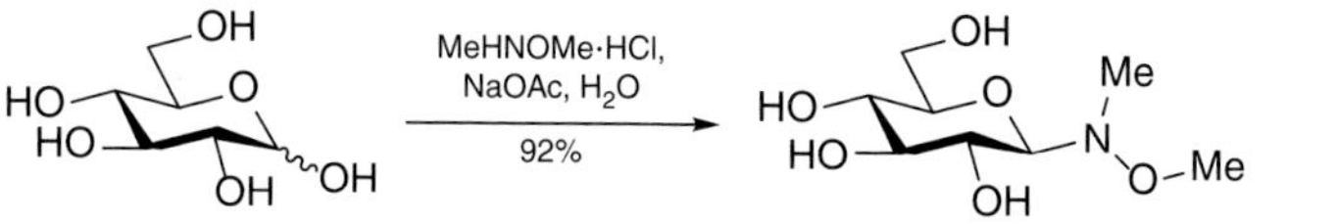	[82]

(Continued)

Table 5.5 (Continued)

Entry	Example conditions	References
10	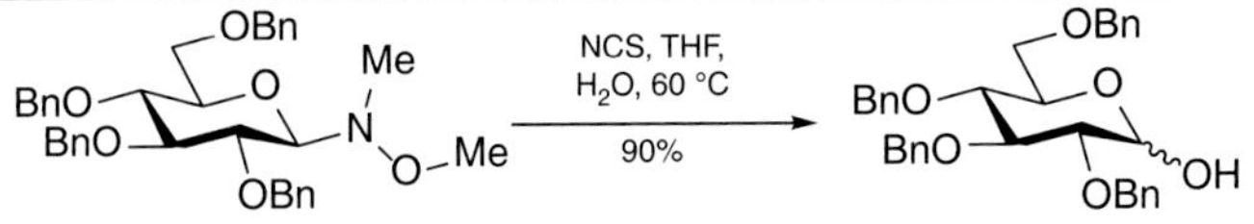	[82]
11	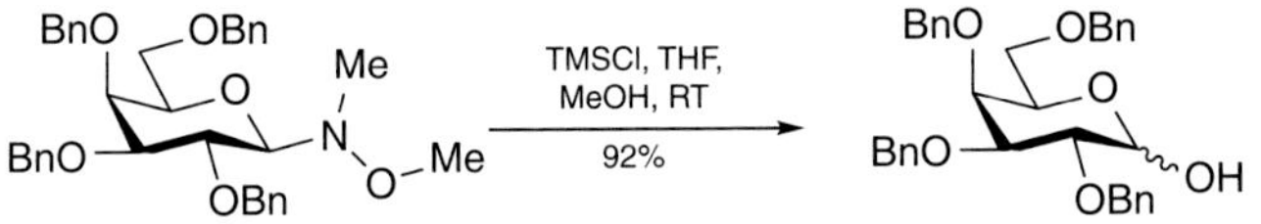	[83]

which, in turn, is dependent on the location and nature of the glycosidic bonds to be synthesized. Those groups that can also allow use of their derivatives as glycosyl donors (e.g. the thioglycosides) will be perennial favorites, but there is room in the chemists' armoury for more novel groups such as the *N,O*-dimethyloxyamine-*N*-glycosides, which can be introduced into unprotected sugars in high yield, without first recourse to peracetylation.

5.8 Example Experimental Procedures

5.8.1 Synthesis of a *p*-Methoxyphenyl Glycoside [16]

1,2,3,4,6-Penta-*O*-acetyl-β-D-glucose (10.0 g, 25.6 mmol, 1 eq.) and 4-methoxyphenol (4.31 g, 38.4 mmol, 1.5 eq.) in dry dichloromethane (100 ml) were treated with $BF_3{\cdot}OEt_2$ (5.3 ml, 42.1 mmol, 1.64 eq.) under argon. After 7 h, the mixture was diluted with dichloromethane (250 ml) and washed with water (200 ml), saturated aq. $NaHCO_3$, (200 ml), and water (2 × 150 ml). The organic phase was dried, filtered, and concentrated. The crude product was crystallized (heptane/ethyl acetate, −20 °C) to give *p*-methoxyphenyl 2,3,4,6-tetra-*O*-acetyl-β-D-glucopyranoside (7.80 g). The mother liquor was concentrated and crystallized to give a second crop of the same product (2.5 g; total yield 89%).

5.8.2 Synthesis of a *N,O*-dimethyloxyamine-*N*-glycoside [82]

N,O-dimethylhydroxylamine hydrochloride (1.19 g, 12.2 mmol, 1.1 eq.) and sodium acetate (1.00 g, 12.2 mmol, 1.1 eq.) were dissolved in approximately 1 ml of water, and the solution was added slowly to a solution of glucose (2.00 g, 11.1 mmol, 1 eq.) in water (15 ml) at 0 °C. The reaction was allowed to proceed for 20 h at room temperature at which time thin-layer chromatography (TLC) analysis showed conversion of the starting material to a faster moving product (dichloromethane/methanol 5 : 1, R_f 0.4). The water was evaporated *in vacuo*, and the product was purified by flash silica gel column chromatography eluting with (dichloromethane/methanol, 6 : 1) to give *N,O*-dimethyl-*N*-(β-D-glucopyranosyl) hydroxylamine (2.27 g, 92%) as a white solid.

5.8.3 Synthesis of a Thexyldimethylsilyl Glycoside [61]

Chloro(dimethyl)thexylsilane (TDSCl; 0.90 ml, 4.55 mmol, 1.15 eq.) and imidazole (0.54 g, 7.92 mmol, 2 eq.) were added under argon to a solution of *O*-(2,4,6-tri-*O*-benzyl-β-D-galactopyranosyl)-(1→4)-2,3,6-tri-*O*-benzyl-D-glucopyranose (3.50 g, 3.96 mmol, 1 eq.) in dichloromethane. The reaction mixture was stirred for 16 h and the solvent was then evaporated *in vacuo*. The residue was adsorbed onto silica gel in toluene and purified by flash silica gel column chromatography (petroleum ether/ethyl acetate, 8 : 2) to afford thexyldimethylsilyl (2,4,6-tri-*O*-benzyl-β-D-galactopyranosyl)-(1→4)-2,3,6-tri-*O*-benzyl-β-D-glucopyranoside (3.49 g, 86%) as a gum.

Abbreviations

Ac	acetyl
aq.	aqueous
Bn	benzyl
Bz	benzoyl
CAN	ceric ammonium nitrate
ClAc	2-chloroacetyl
D	dextrorotatory
DIAD	diisopropyl azodicarboxylate
DIBAL-H	diisobutylaluminium hydride
DIPC	*N,N'*-diisopropylcarbodiimide
DMAP	4-dimethylaminopyridine
DMF	*N,N*-dimethylformamide
DMSO	dimethyl sulfoxide
eq.	equivalent
Fmoc	fluorenylmethyloxycarbonyl
Lev	levulinyl
MBz	4-methoxybenzoyl
mol sieves	molecular sieves
Nap	2-naphthylmethyl
NIS	*N*-iodosuccinimide
p	para
Tf	trifluoromethanesulfonyl (triflyl)
Ph	phenyl
Piv	pivaloyl
RT	room temperature
STaz	*S*-thiazolinyl
TBAF	tetrabutylammonium fluoride
TBDMS	*tert*-butyldimethylsilyl
TBDPS	*tert*-butyldiphenylsilyl
TBS	*tert*-butyldimethylsilyl
TCS	chloro(dimethyl)thexylsilyl
TDS	thexyldimethylsilyl
TFA	trifluoroacetic acid

Tf	trifluoromethanesulfonyl
THF	tetrahydrofuran
TIPS	triisopropylsilyl
TLC	thin layer chromatography
TMS	trimethylsilyl
TPP	triphenylphosphine

References

1 Collins, P. and Ferrier, R. (1995). *Monosaccharides–Their Chemistry and Their Roles in Natural Products*. New York: Wiley.
2 van der Marel, G.A., van den Bos, L.J., Overkleeft, H.S. et al. (2007). Sequential glycosylation strategies: a focus on thioglycosides as donors and acceptors. In: *Frontiers in Modern Carbohydrate Chemistry*, Chapter 12, vol. 960 (ed. A.V. Demchenko), 190–208. American Chemical Society.
3 Vankar, Y.D. and Schmidt, R.R. (2000). *Chem. Soc. Rev.* 29: 201–216.
4 Schmidt, R.R. and Kinzy, W. (1994). *Adv. Carbohydr. Chem. Biochem.* 50: 21–123.
5 Augestad, I. and Berner, E. (1954). *Acta Chem. Scand.* 8: 251–256.
6 van Rijssel, E.R., van Delft, P., Lodder, G. et al. (2014). *Angew. Chem. Int. Ed.* 53: 10381–10385.
7 Hiruma, K., Kanie, O., and Wong, C.-H. (1998). *Tetrahedron* 54: 15781–15792.
8 Konradsson, P., Roberts, C., and Fraser-Reid, B. (1991). *Rec. Trav. Chim. Pays-Bas* 110: 23–24.
9 Vetere, A., Medeot, M., Campa, C. et al. (2003). *J. Mol. Catal. B: Enzym.* 21: 153–156.
10 Khamsi, J., Ashmus, R.A., Schocker, N.S., and Michael, K. (2012). *Carbohydr. Res.* 357: 147–150.
11 Ouairy, C., Cresteil, T., Delpech, B., and Crich, D. (2013). *Carbohydr. Res.* 377: 35–43.
12 Pogosyan, A., Gottwald, A., Michalik, D. et al. (2013). *Carbohydr. Res.* 380: 9–15.
13 Hotha, S. and Kashyap, S. (2006). *J. Am. Chem. Soc.* 128: 9620–9621.
14 Jansson, K., Ahlfors, S., Frejd, T. et al. (1988). *J. Org. Chem.* 53: 5629–5647.
15 Zhang, J., Fu, J., Si, W. et al. (2011). *Carbohydr. Res.* 346: 2290–2293.
16 Zhang, Z. and Magnusson, G. (1996). *Carbohydr. Res.* 295: 41–55.
17 Montel, E., Hrmova, M., Fincher, G.B. et al. (2009). *Aust. J. Chem.* 62: 575–584.
18 Kim, Y.-J., Wang, P., Navarro-Villalobos, M. et al. (2006). *J. Am. Chem. Soc.* 128: 11906–11915.
19 Hou, D., Skogman, F., and Lowary, T.L. (2008). *Carbohydr. Res.* 343: 1778–1789.
20 Tashiro, T., Nakagawa, R., Inoue, S. et al. (2008). *Tetrahedron Lett.* 49: 6827–6830.
21 Bornaghi, L.F. and Poulsen, S.-A. (2005). *Tetrahedron Lett.* 46: 3485–3488.
22 Boltje, T.J., Li, C., and Boons, G.-J. (2010). *Org. Lett.* 12: 4636–4639.
23 Hu, Y.-J., Dominique, R., Das, S.K., and Roy, R. (2000). *Can. J. Chem.* 78: 838–845.
24 Jambal, I., Kefurt, K., Hlaváčková, M., and Moravcová, J. (2012). *Carbohydr. Res.* 360: 31–39.

25 Arasappan, A. and Fraser-Reid, B. (1996). *J. Org. Chem.* 61: 2401–2406.
26 Wilson, B.G. and Fraser-Reid, B. (1995). *J. Org. Chem.* 60: 317–320.
27 Jansson, K., Noori, G., and Magnusson, G. (1990). *J. Org. Chem.* 55: 3181–3185.
28 Doboszewski, B. and Zamojski, A. (1987). *Carbohydr. Res.* 164: 470–476.
29 Bodine, K.D., Gin, D.Y., and Gin, M.S. (2004). *J. Am. Chem. Soc.* 126: 1638–1639.
30 Emmadi, M. and Kulkarni, S.S. (2013). *Org. Biomol. Chem.* 11: 3098–3102.
31 Wolfrom, M.L. and Thompson, A. (1963). Acetylation. In: *Methods in Carbohydrate Chemistry*, vol. 2 (ed. R.L. Whistler and M.L. Wolfrom), 211–215. New York: Academic Press.
32 Kartha, K.P.R. and Field, R.A. (1997). *Tetrahedron* 53: 11753–11766.
33 Wolfrom, M.L. and Thompson, A. (1962). β-Maltose monohydrate: enzymic preparation of maltose from starch. In: *Methods in Carbohydrate Chemistry*, vol. 1 (ed. R.L. Whistler and M.L. Wolfrom), 334–335. New York: Academic Press.
34 Sakairi, N., Wang, L.-X., and Kuzuhara, H. (1995). *J. Chem. Soc., Perkin Trans. 1* 437–443.
35 Barros, M.T., Maycock, C.D., Rodrigues, P., and Thomassigny, C. (2004). *Carbohydr. Res.* 339: 1373–1376.
36 Yang, Y., Oishi, S., Martin, C.E., and Seeberger, P.H. (2013). *J. Am. Chem. Soc.* 135: 6262–6271.
37 Feng, S. and Li, C. (2015). *J. Agric. Food. Chem.* 63: 5732–5739.
38 Smith, A.B., Rivero, R.A., Hale, K.J., and Vaccaro, H.A. (1991). *J. Am. Chem. Soc.* 113: 2092–2112.
39 Excoffier, G., Gagnaire, D., and Utille, J.-P. (1975). *Carbohydr. Res.* 39: 368–373.
40 Tiwari, P. and Misra, A.K. (2006). *Tetrahedron Lett.* 47: 3573–3576.
41 Filice, M., Guisan, J.M., Terreni, M., and Palomo, J.M. (2012). *Nat. Protoc* 7: 1783–1796.
42 Yao, Q., Song, J., Xia, C. et al. (2006). *Org. Lett.* 8: 911–914.
43 Al-Mughaid, H. and Grindley, T.B. (2006). *J. Org. Chem.* 71: 1390–1398.
44 Kartha, K.P.R. and Jennings, H.J. (1990). *J. Carbohydr. Chem.* 9: 777–781.
45 Fürstner, A., Jeanjean, F., Razon, P. et al. (2003). *Chem. Eur. J.* 9: 307–319.
46 Zhu, Y. and Ralph, J. (2011). *Tetrahedron Lett.* 52: 3729–3731.
47 Martin, R., Schäfer, T., Theumer, G. et al. (2009). *Synthesis* 3488–3492.
48 Gauthier, C., Legault, J., Lebrun, M. et al. (2006). *Bioorg. Med. Chem.* 14: 6713–6725.
49 Guo, T., Liu, Q., Wang, P. et al. (2009). *Carbohydr. Res.* 344: 1167–1174.
50 Schmidt, O.T. (1963). Isopropylidene derivatives. In: *Methods in Carbohydrate Chemistry*, vol. 2 (ed. R.L. Whistler and M.L. Wolfrom), 318–325. New York: Academic Press Inc.
51 Tanabe, G., Matsuoka, K., Yoshinaga, M. et al. (2012). *Bioorg. Med. Chem.* 20: 6321–6334.
52 Stalford, S.A., Kilner, C.A., Leach, A.G., and Turnbull, W.B. (2009). *Org. Biomol. Chem.* 7: 4842–4852.
53 Turnbull, W.B. and Field, R.A. (2000). *J. Chem. Soc., Perkin Trans. 1* 1859–1866.
54 Lipták, A., Imre, J., Harangi, J., and Nánási, P. (1983). *Carbohydr. Res.* 116: 217–225.
55 Suzuki, K., Mizuta, T., and Yamaura, M. (2003). *J. Carbohydr. Chem.* 22: 143–147.

56 Suzuki, K., Nonaka, H., and Yamaura, M. (2003). *Tetrahedron Lett.* 44: 1975–1977.
57 Suzuki, K., Nonaka, H., and Yamaura, M. (2004). *J. Carbohydr. Chem.* 23: 253–259.
58 Beignet, J., Tiernan, J., Woo, C.H. et al. (2004). *J. Org. Chem.* 69: 6341–6356.
59 Carpintero, M., Nieto, I., and Fernández-Mayoralas, A. (2001). *J. Org. Chem.* 66: 1768–1774.
60 Zhu, C., Peng, W., Li, Y. et al. (2006). *Carbohydr. Res.* 341: 1047–1051.
61 Roussel, F., Knerr, L., and Schmidt, R.R. (2001). *Eur. J. Org. Chem.* 2067–2073.
62 Ronchi, P., Scarponi, C., Salvi, M. et al. (2013). *J. Org. Chem.* 78: 5172–5183.
63 Dureau, R., Legentil, L., Daniellou, R., and Ferrières, V. (2012). *J. Org. Chem.* 77: 1301–1307.
64 Tang, Y., Li, J., Zhu, Y. et al. (2013). *J. Am. Chem. Soc.* 135: 18396–18405.
65 Nakahara, Y. and Ogawa, T. (1988). *Carbohydr. Res.* 173: 306–315.
66 Takano, D., Nagamitsu, T., Ui, H. et al. (2001). *Org. Lett.* 3: 2289–2291.
67 Lassaletta, J.M. and Schmidt, R.R. (1995). *Synlett* 925–927.
68 Fan, G.-T., Pan, Y.-S., Lu, K.-C. et al. (2005). *Tetrahedron* 61: 1855–1862.
69 Shirahata, T., Kojima, A., Teruya, S. et al. (2011). *Tetrahedron* 67: 6482–6496.
70 Guazzelli, L., Ulc, R., Rydner, L., and Oscarson, S. (2015). *Org. Biomol. Chem.* 13: 6598–6610.
71 Shiozaki, M., Doi, H., Tanaka, D. et al. (2006). *Tetrahedron* 62: 205–225.
72 Waschke, D., Thimm, J., and Thiem, J. (2011). *Org. Lett.* 13: 3628–3631.
73 Motawia, M.S., Marcussen, J., and Møller, B.L. (1995). *J. Carbohydr. Chem.* 14: 1279–1294.
74 Dasgupta, S., Roy, B., and Mukhopadhyay, B. (2006). *Carbohydr. Res.* 341: 2708–2713.
75 Sharma, I., Bohé, L., and Crich, D. (2012). *Carbohydr. Res.* 357: 126–131.
76 Shiozaki, M., Tashiro, T., Koshino, H. et al. (2013). *Carbohydr. Res.* 370: 46–66.
77 Pornsuriyasak, P. and Demchenko, A.V. (2006). *Chem. Eur. J.* 12: 6630–6646.
78 Turnbull, W.B., Fascione, M.A., and Stalford, S.A. (2007). Product class 14: glycosyl sulfur, selenium, and tellurium compounds. In: *Science of Synthesis*, vol. 29 (ed. S.L. Warriner), 923–969. Stuttgart: Georg Thieme Verlag.
79 Ferrier, R.J. and Furneaux, R.H. (1976). *Carbohydr. Res.* 52: 63–68.
80 Saibal Kumar, D. and Roy, N. (1996). *Carbohydr. Res.* 296: 275–277.
81 Kumar, R., Tiwari, P., Maulik, P.R., and Misra, A.K. (2006). *Eur. J. Org. Chem.* 74–79.
82 Dasgupta, S. and Nitz, M. (2011). *J. Org. Chem.* 76: 1918–1921.
83 Huang, S., Liao, J., Zhao, Q. et al. (2013). *J. Carbohydr. Chem.* 32: 158–168.
84 Cló, E., Blixt, O., and Jensen, K.J. (2010). *Eur. J. Org. Chem.* 540–554.

6

N-protecting Groups for 2-Amino-2-deoxy-glycosides

Sébastien Vidal

CNRS and Université Claude Bernard Lyon 1, Institut de Chimie et Biochimie Moléculaires et Supramoléculaires (UMR 5246), Laboratoire de Chimie Organique 2-Glycochimie, 43 Boulevard du 11 Novembre 1918, 69622 Villeurbanne, France

6.1 Introduction

One of the many findings in glycobiology was the determination of the structures of the ABO blood group antigens (Figure 6.1). These oligosaccharides are conjugated to proteins and are constituted of a core trisaccharide unit for the O-antigen (Figure 6.1, R = H) and with either an additional α-D-GalNAc residue for the A-antigen of an α-D-Gal unit for the B-antigen. The variation from A- to B-antigen is on a single position of the nonreducing end galactoside with the acetamido or hydroxy group, respectively. This subtle change is controlled by a specific gene for each individual coding for either glycosyltransferase capable of adding the glycoside unit with complete regiocontrol and stereocontrol of the created glycosidic bond.

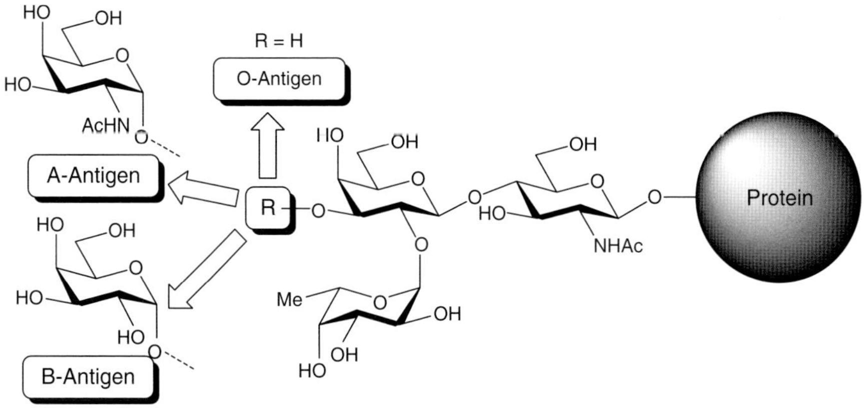

Figure 6.1 Structure of the three main blood group antigens O, A, and B conjugated to their protein.

Protecting Groups: Strategies and Applications in Carbohydrate Chemistry, First Edition.
Edited by Sébastien Vidal.

Chitin is one of the most abundant sources of saccharide on earth as the polymeric form of GlcNAc linked through β-(1→4) glycosidic bonds (Figure 6.2). Cellulose is the corresponding polysaccharide with glucose units and major constituent of plants. The simple change of the 2-position from OH to NHAc moiety largely influences the properties of these chemically very similar polymers. Similarly, the properties of chitin and chitosan are modulated by varying degrees of *N*-acetylation.

Glycosaminoglycans (GAGs) are a large family of polysaccharides containing GlcNAc or GalNAc residues with a wide range of roles in biological processes (Figure 6.2). The sulfation pattern of these GAGs is highly variable and critical for the exact biological implication of such glycopolymers. For instance, dermatan is found in skin and fibrous connective tissues, keratan in cornea, hyaluronic acid in loose connective tissues and eyes, and chondroitin sulfate in cartilage and intervertebral disc (see Chapter 14). It is interesting to note that hyaluronic acid involves a GlcNAc residue, and a subtle change to a GalNAc residue leads to chondroitin with very different biological properties. Heparin is a highly sulfated

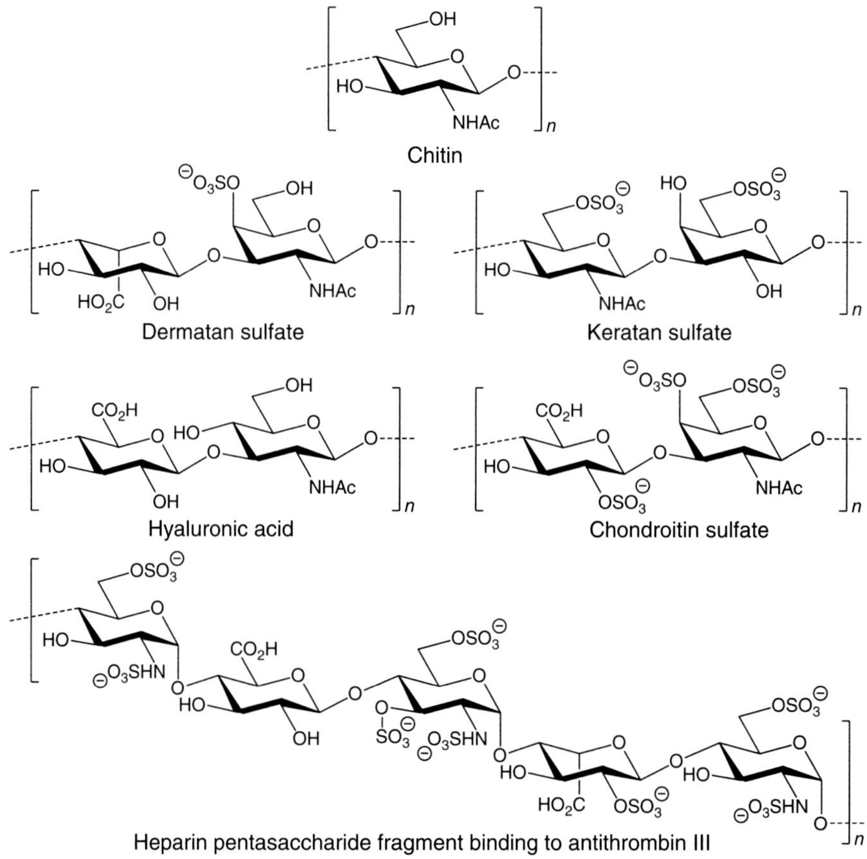

Figure 6.2 General structures of glycosaminoglycans and their sulfated species.

GAG influencing blood coagulation through complexation with antithrombin III (see Chapter 17).

These polysaccharides are interesting examples of how the variations of stereochemistry or functionalities can largely influence their properties. Other aspects of 2-amino-2-deoxy-glycosides are important in biological signaling such as *N*-acetyl-neuraminic acid (Neu5Ac) involved in infectious diseases or cancer, for instance, for which ManNAc is the precursor for its biosynthesis. The design of suitably functionalized ManNAc analogs with reporter groups at specific positions could lead to intense and fruitful investigations for the better understanding of post-translational protein glycosylation patterns. For that matter, protecting group strategies were required for the efficient synthesis of such ManNAc analogs.

The *O*-GlcNAc-transferase (OGT) is an enzyme responsible for the specific *O*-glycosylation of Ser or Thr residues of proteins in concert with *O*-phosphorylations by protein kinases as a key element of signal transduction. Recent developments in this field demonstrated the cross talk between *O*-GlcNAcylation and phosphorylation in numerous diseases [1–5]. General approaches toward GlcNAc analogs were also applied efficiently in this context and provided several molecules leading to important findings on the roles and consequences of *O*-GlcNAcylation.

The careful choice of protecting group strategy for the 2-amino-2-deoxy-glycoside framework appears as a crucial aspect of carbohydrate chemistry. This chapter will discuss the different categories of protecting groups generally used for 2-amino-2-deoxy-glycosides, but the choice of glycosylation conditions for these saccharides will not be discussed in detail because this aspect has been reviewed elsewhere [6–8]. 2-Amino-2,6-dideoxy-glycosides (e.g. FucNAc and RhaNAc) have not been covered herein. The amide protecting groups will be discussed along with imides, carbamates, and other protecting groups of amines. Another aspect will be considered when the amine functionality will be hidden in an azido or a glycal moiety and then the NHAc functionality installed at the late stage of the synthesis. The reader can also refer to the following review on the protecting group topic for further information [9].

6.2 *N*-acyl-based Protecting Groups

6.2.1 Acetyl Protecting Group

This 2-acetamido group is not a real protecting group *per se*, but the desired final functionality to be installed at the 2-position of such glycosides. Nevertheless, the chemical stability of the acetamide moiety is rather limited but sometimes allows for the proper synthesis of 2-acetamido-2-deoxy-glycosides in good yields through a rather short synthetic sequence. As no specific introduction of removal of this protecting group is necessary, the implication of 2-acetamido moiety is mainly noticeable in the glycosylation process (Figure 6.3). A carbocation is readily generated at the anomeric carbon atom after removal of the anomeric

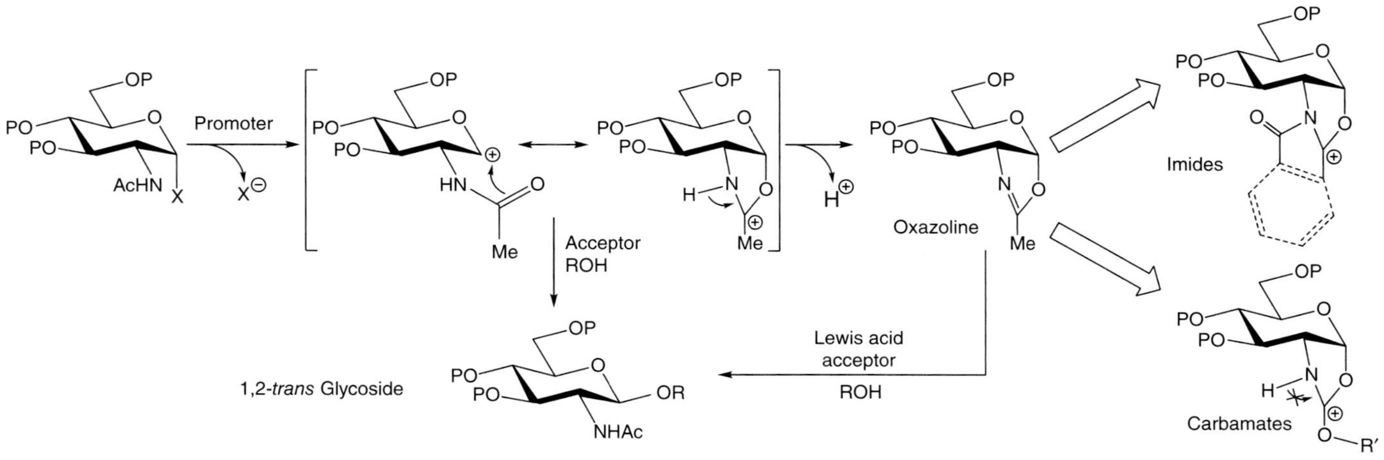

Figure 6.3 General aspects of glycosylation with 2-acetamido-2-deoxy-glycosides.

substituent by activation of the promoter. This carbocation can then be stabilized by anchimeric assistance of the 2-acetamido group, and the resulting five-membered ring can then evolve quite rapidly to the corresponding bicyclic oxazoline by deprotonation. The carbocation intermediate can react with the acceptor (ROH) to provide the 1,2-*trans* glycoside. The oxazoline can be isolated and stored as a stable intermediate and then re-engaged in a glycosylation reaction requiring a Lewis acid to promote the ring opening by the acceptor.

The formation of oxazoline is observed for most of the amide functional groups at the 2-position and more specifically with the less nucleophilic secondary alcohols. Several other halogenated acetamido groups have, therefore, been introduced to reduce this propensity to provide oxazoline. In addition, the introduction of imides or carbamates is preventing this undesired deprotonation and leads to the desired 1,2-*trans* glycosides in good yields and stereoselectivities.

6.2.2 Trichloroacetamido Protecting Group (TCA)

The trichloroacetyl group provides an increased electrophilicity to the anomeric carbon during the glycosylation process, while limiting the deprotonation to the oxazoline and leading to a good 1,2-*trans* stereoselectivity. The same comment applies to the trifluoroacetyl group (see Section 6.2.3).

The trichloroacetyl group is readily introduced (Scheme 6.1) on the peracetylated glucosamine hydrochloride **1** with trichloroacetyl chloride [10, 11] to provide the amide **2**. Triethylamine was used both to quench the hydrochloric acid formed in the reaction and to deprotonate the hydrochloride salt of glucosamine. The same amidation can be obtained from the nonprotected glucosamine hydrochloride **3** in water [12] with the same acyl chloride to yield the hydroxylated intermediate **4**, which could be further acetylated to compound **2**, although as a mixture of anomers under the reported conditions (Ac_2O, pyridine, 16 h, room temperature (RT)).

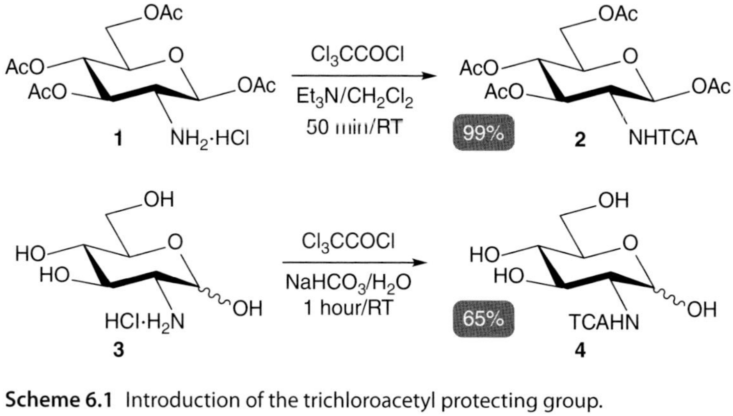

Scheme 6.1 Introduction of the trichloroacetyl protecting group.

The trichloroacetyl group was not deprotected but rather reduced under various conditions (Scheme 6.2) to the corresponding acetamido moiety required for

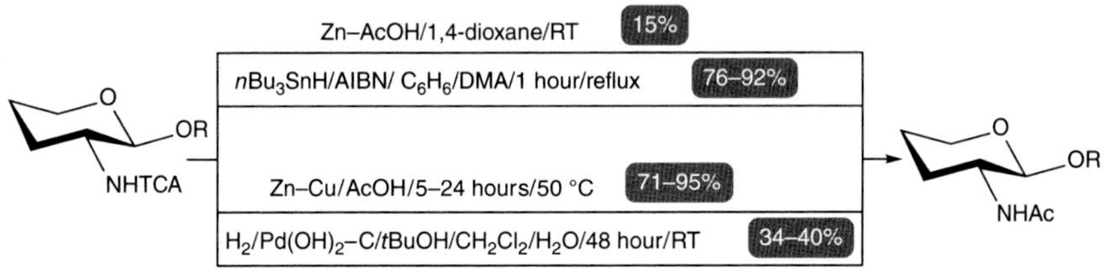

Scheme 6.2 Reduction of the trichloroacetyl protecting group to the acetyl moiety.

the target oligosaccharides. This process was usually performed at the late stage of the synthesis. It is worth pointing out that the perfect control of the total triple reduction is required to avoid mixtures contaminated with residual mono-, di-, or tri-halogenated species [13, 14].

The standard reduction with zinc and acetic acid provides poor yields [13]. The group of professor Jacquinet has intensively investigated this reduction process and identified tin-based hydride reduction as a quite reliable reduction system under radical conditions (with α,α′-azoisobutyronitrile (AIBN)) and providing the desired acetamides on complex oligosaccharide analogs of chondroitins [15–17]. Nevertheless, difficulties in purifying the tin species [16] or to reach complete reduction [14, 18] and sometimes difficulties in reaching reproducible yields prompted the same group to investigate an alternative reduction protocol under smoother conditions with Zn—Cu couple [18]. Finally, reduction by hydrogenolysis also afforded the desired 2-acetamido-oligosacchairdes in modest yields [19]. The deprotection of trichloroacetamide to the free amine group would require basic or harsher reductive conditions and the present reduction protocols are compatible with ester groups such as acetates.

6.2.3 Trifluoroacetyl Protecting Group (TFA)

The trifluoroacetyl group can be readily introduced by using either trifluoroacetic anhydride (TFAA) [20–24] from the corresponding 2-deoxy-2-amino-glycoside **1** to the amide **5** (Scheme 6.3). Chemoselective trifluoroacetamidation

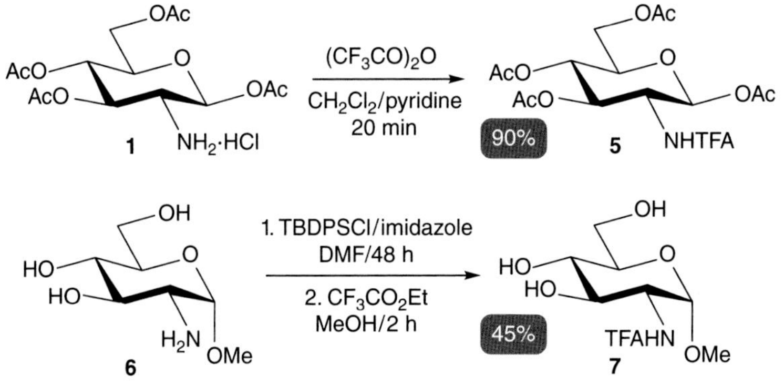

Scheme 6.3 Introduction of the trifluoroacetyl protecting group.

can be achieved in the presence of alcohols using ethyl trifluoroacetate from amine **6** to the diol **7** [25, 26].

The trifluoroacetamide can be hydrolyzed to the corresponding amine (Scheme 6.4) under either acidic conditions [21, 22] or with lithium [24], potassium [26], or sodium hydroxide [27]. *O*-acetates can be chemoselectively saponified using sodium methoxide and leaving the trifluoroacetamide unchanged [28].

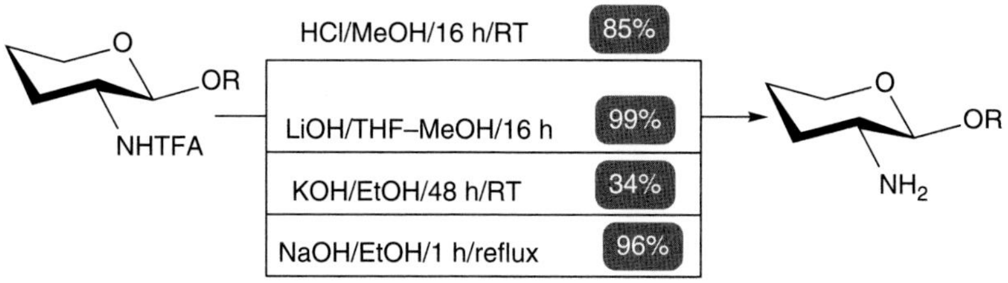

Scheme 6.4 Hydrolysis of the trifluoroacetyl protecting group.

6.3 Imido-based Protecting Groups

Imido-based protecting groups have been investigated [29] for their tendency to prevent the formation of oxazolines during glycosylation as typically observed with 2-amido-2-deoxy-glycosides (Figure 6.3). Also sometimes, the rotation around the amide C—N bond can be partially hampered by the bulky halogenated acetamides. The resulting rotamer species can then be differentiated by nuclear magnetic resonance (NMR) spectroscopy and their structural assignments becoming more complex. The C_2-symmetry of imides will not provide rotamers, thus allowing an easier characterization of the protected intermediates.

6.3.1 Phthaloyl Protecting Group (Phth)

The pioneering report for the preparation of 2-deoxy-2-phthalimido-glycoside [30] described the 2-two-step synthesis from glucosamine **8** with phthalic anhydride providing the 2-(2-carboxybenzamido)-2-deoxy-glycoside **9** in high yield (Scheme 6.5). The cyclization to the imide **10** was then achieved during the

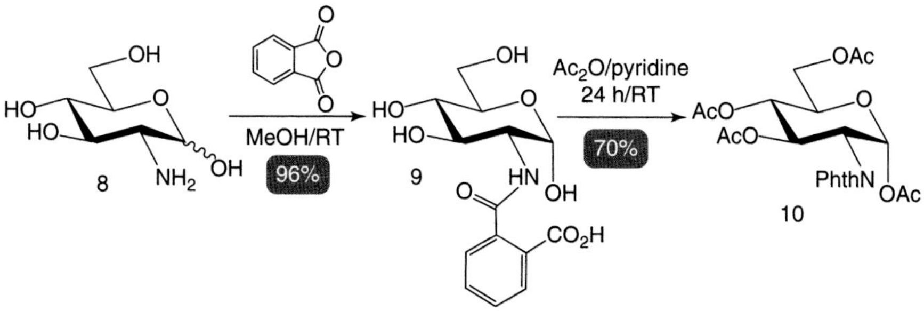

Scheme 6.5 Installation of the phthaloyl protecting group.

subsequent peracetylation of the intermediate **9**. The formation of the phthalimide moiety requires the initial opening of the cyclic phthalic anhydride by the amine and then the intramolecular cyclization into the imide. Nevertheless, the carboxylic acid intermediate formed is not electrophilic enough to allow such cyclization, and activation with acetic [23, 31] or pivalic anhydride [32] will generate a mixed anhydride suitable for this cyclization.

Removal of phthalimide can be performed (Scheme 6.6) under aminolysis conditions using hydrazine [25], ethylenediamine [23], or hydroxylamine [33]. Direct transformation into the 2-acetamido-2-deoxy-glycoside can also be readily achieved using hydrazine [34], ethylenediamine [35], or butylamine [36] followed by acetylation of the crude material.

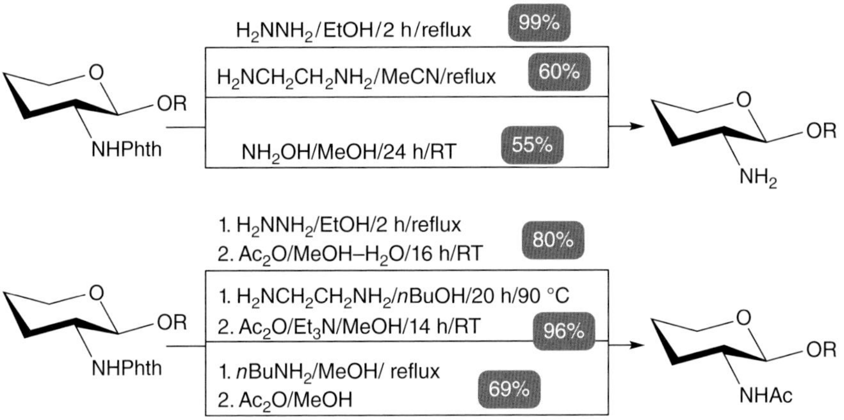

Scheme 6.6 Aminolysis of the phthaloyl protecting group.

6.3.2 Dichlorophthaloyl Protecting Group (DCP)

Aminolysis of the phthalimido moiety is sometimes difficult because of the extended reaction time or high temperature, which can be particularly troublesome when performed on large and complex oligosaccharides. Halogenated phthalimides such as dichlorophthalimide (DCP) or tetrachlorophthalimide (TCP, see Section 6.3.3) have been investigated and are more reactive toward ring opening by amines [7]. The DCP will be less stable than Phth toward aminolysis but yet more stable than TCP. The DCP moiety is readily introduced from the corresponding anhydride (Scheme 6.7) and subsequent acetylation [37].

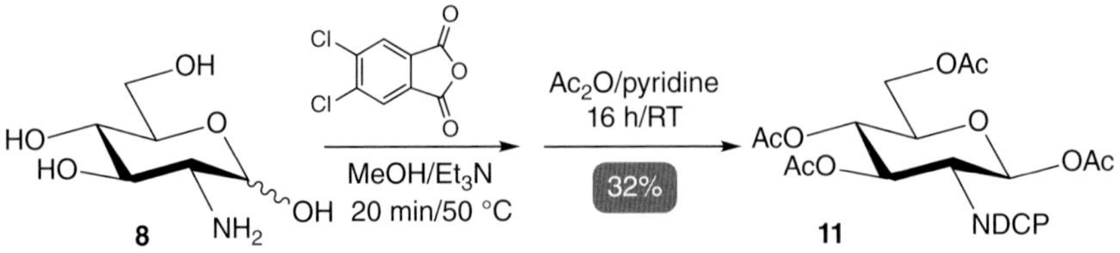

Scheme 6.7 Installation of the dichlorophthaloyl protecting group.

Deprotection of DCP is achieved with ethylenediamine, and acetylation of the crude material affords the acetamido derivatives [37, 38].

6.3.3 Tetrachlorophthaloyl Protecting Group (TCP)

The tetrachlorophthaloyl group has been used as an amine protecting group in carbohydrate chemistry because of its smoother removal under aminolysis conditions in comparison to DCP and Phth. *N*-differentiation of linear glucosamines could even be achieved using a TCP imide, which was removed chemoselectively in the presence of a Phth imide using ethylenediamine [39–41].

The introduction of TCP is performed (Scheme 6.8) from glucosamine **8** using the anhydride and subsequent acetylation of the crude material to the corresponding imide **12** [22, 41, 42]. Removal was initially proposed by reduction with sodium borohydride [42] generating phthalide as a by-product. Nevertheless, more standard and compatible conditions with multifunctional oligosaccharide structures were described using ethylenediamine [22, 41, 43] or hydrazine [22].

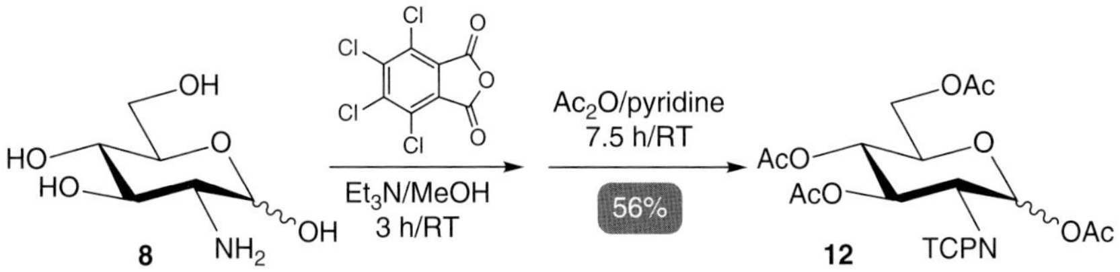

Scheme 6.8 Installation of the tetrachlorophthaloyl protecting group.

6.3.4 Other Imido-based Protecting Groups (Dts, DTG, DPM, and DMM)

A series of rarely used imido-based protecting groups have also been reported usually for their ability to provide 1,2-*trans* glycosides and also for a highly orthogonal deprotection protocol leaving most of other protecting groups unchanged in complex oligosaccharides.

The dithiasuccinyl group (Dts) can be introduced in two steps (Scheme 6.9) from the glucosamine **8** by condensation with *S*-carboxymethyl *O*-ethyl xanthate to provide the thiocarbamate **13**, which was converted to dithiasuccinimide with chlorocarbonyl sulfenyl chloride [44]. The same two-step synthesis was later revisited and allowed higher yields for both steps (80% for each step) using

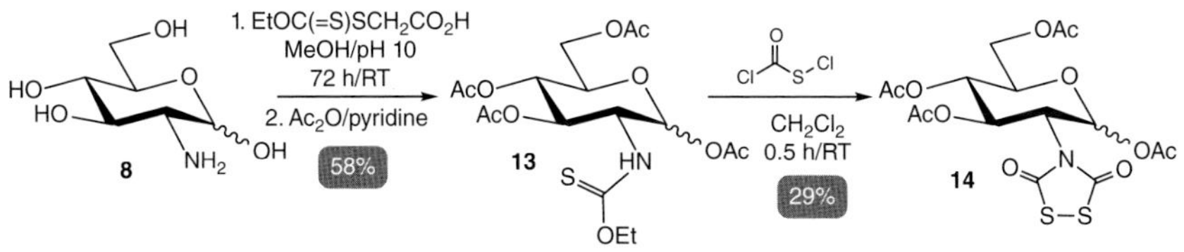

Scheme 6.9 Installation of the dithiasuccinyl protecting group.

bis(ethoxythiocarbonyl)sulfide to install the thiocarbamate **13** with the second step modified only in the protocol used but not the reagents [45]. The Dts group is also very interesting because it can be deprotected using very chemoselective conditions such as with dithiothreitol (DTT) [44, 46], sodium borohydride [44], or zinc in acetic acid [45].

The dithioglycolyl group (DTG) gives access to 1,2-*trans* glycosides without the risk of oxazoline formation because of its imide character. It is introduced by the reaction of the acetylated glucosamine **1** with thiodiglycolic anhydride and subsequent cyclization to the amide with acetic anhydride [42] to the thiodiglycolyl-protected glucosamine **15** (Scheme 6.10). If a direct reductive cleavage is performed with only tributyltin hydride, the corresponding succinimide is formed in 95% yield through an intramolecular C—C bond formation between the two radical intermediates generated. Therefore, the ring opening of the imide with sodium methoxide and then treatment with tributyltin hydride followed by reacetylation with acetic anhydride afforded the corresponding 2-acetamido-2-deoxy-glycoside in high yield (84–89%).

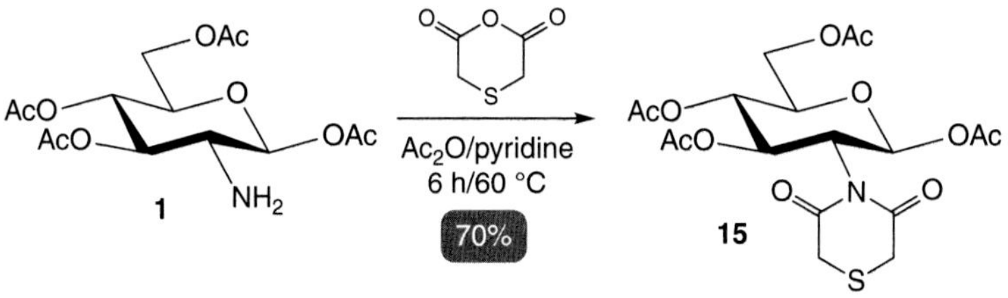

Scheme 6.10 Installation of the thiodiglycolyl protecting group.

The diphenylmaleoyl (DPM) group is structurally analogous to the phthalimides but has a yellow color and also has fluorescence properties that can be useful for characterization of detection aspects. The DPM group is introduced on glucosamine **8** using diphenylmaleic anhydride (Scheme 6.11) followed by acetylation with acetic anhydride to the DPM-protected intermediate **16**. DPM can be readily removed with hydrazine [47].

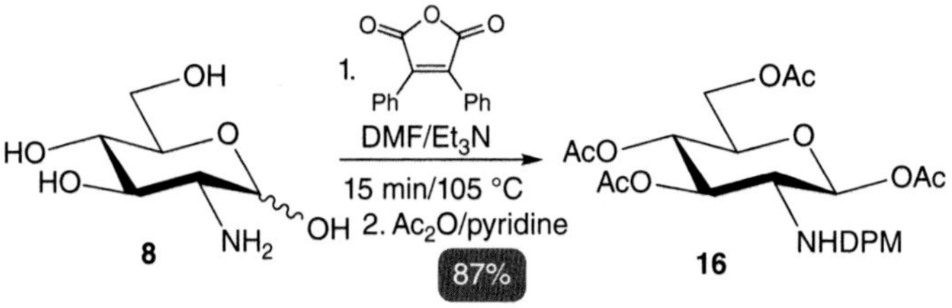

Scheme 6.11 Installation of the diphenylmaleoyl protecting group.

The dimethylmaleoyl (DMM) group is introduced [48–50] using dimethylmaleic anhydride with the subsequent acetylation to cyclize the imide to compound

17 (Scheme 6.12). Removal was accomplished in two steps by first ring opening of the imide with sodium hydroxide to provide an open-chain amide/carboxylic acid. The carboxylic acid then recyclizes at the amide carbonyl to generate a butenolide, which is then hydrolyzed with HCl to the corresponding 2-amino-2-deoxy-glycoside. The wide applicability of the DMM protecting group can be exemplified by numerous glycosylations reported for glycolipids [51], GAGs [52], oligochitoses [53], peptidoglycans [48, 49], *N*-glycans [54], human milk oligosaccharides [55], or other oligosaccharides [56, 57] and is also compatible with solid-phase synthesis of oligosaccharides [58].

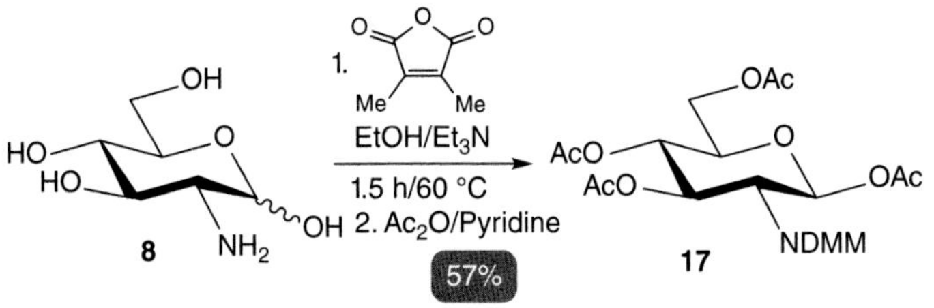

Scheme 6.12 Installation of the dimethylmaleoyl protecting group.

6.4 Carbamate-based Protecting Groups

Acyl- and imide-based protecting groups have been reported for the convenient 1,2-*trans* glycosylation. Nevertheless, the amides suffer from formation of oxazolines or sometimes limited stability under various reaction conditions while the imides are more stable but sometimes suffer from difficult removal under harsh conditions. Carbamates display a good chemical stability leading to a complementary set of orthogonal functions with most of the other protecting groups typically used in carbohydrate chemistry. They also provide the 1,2-*trans* glycosides through anchimeric assistance in high yields and stereoselectivities [59, 60].

6.4.1 2,2,2-Trichloroethoxycarbonyl Protecting Group (Troc)

The highly electron-withdrawing 2,2,2-trichloroethoxycarbonyl protecting group increases the electrophilicity of the anomeric carbon atom, thus exalting the reactivity during the glycosylation process. *N*-carbamates are typically obtained from 2,2,2-trichloroethyl chloroformate (TrocCl) and the corresponding glycosylamine (Scheme 6.13) with or without protecting groups on the saccharide [60–71].

Removal of the Troc group is typically performed by reduction with Zn/AcOH [63, 70, 72] and subsequent acetylation of the crude mixture to provide the corresponding 2-acetamido-2-deoxy-glycosides (Scheme 6.14). A similar reduction can be achieved with cadmium [68] while treatment with sodium hydroxide would also generate the 2-amino-2-deoxy-glycoside intermediate [65]. The free amine can also be obtained by simply treating with Zn dust in acetic acid without

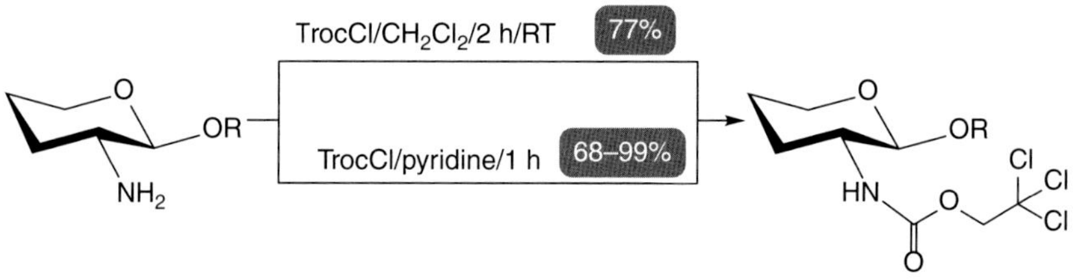

Scheme 6.13 Installation of the 2,2,2-trichloroethoxycarbonyl protecting group.

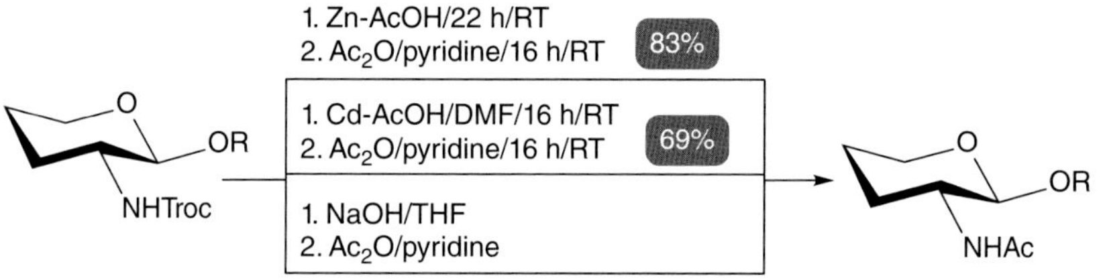

Scheme 6.14 Removal of the 2,2,2-trichloroethoxycarbonyl protecting group.

reacetylation of the crude material [67, 73]. In addition, deprotection with tetrabutylammonium fluoride (TBAF) was also reported and is useful when other functionalities are sensitive to strong basic or reductive conditions [74]. Deprotection could also be achieved with $(Bu_3Sn)_2$ in good yields and even on solid support [75]. The Troc group can also be conserved when *O*-deacetylation is accomplished with guanidine/guanidinium nitrate [76].

6.4.2 Benzyloxycarbonyl Protecting Group (Cbz or Z)

Benzyl chloroformate is typically introduced as a protecting group of 2-amino-2-deoxy-glycosides (Scheme 6.15) in the presence of an inorganic base ($NaHCO_3$, $KHCO_3$) using a biphasic system with water and dichloromethane [77–79], chloroform [80, 81], or diethyl ether [71].

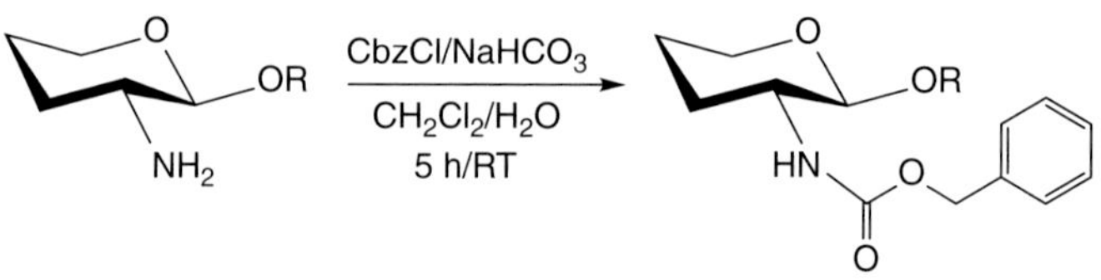

Scheme 6.15 Installation of the benzyloxycarbonyl protecting group.

Removal of the Cbz carbamate has been systematically reported under hydrogenolysis conditions with Pd—C as a catalyst and in water/organic solvent mixtures for solubility aspects. The Cbz group is largely orthogonal to most of the reaction conditions used in these syntheses and compatible with multifunctional oligosaccharides bearing a wide range of other protecting groups such as in the multistep synthesis of sialyl Lewis X analogs [77], glycosylphosphatidylinositols [82], aminoglycoside antibiotics [83], chitodisaccharides [84], teichoic acid [85], heparan sulfates [86, 87], or for the design of sulfatase inhibitors [88].

The *p*-nitrobenzyloxycarbonyl (PNZ) protecting group was also reported as an alternative to Cbz and is introduced with the corresponding chloroformate and removed under either hydrogenolysis or with sodium dithionite under neutral conditions [89, 90].

6.4.3 Allyloxycarbonyl Protecting Group (Alloc)

The allyloxycarbonyl protecting group is introduced (Scheme 6.16) as a carbamate using the 2-amino-2-deoxy-glycosides with allylchloroformate with triethylamine as the base [91, 92]. Later on, biphasic systems have been reported for the same reaction [93, 94].

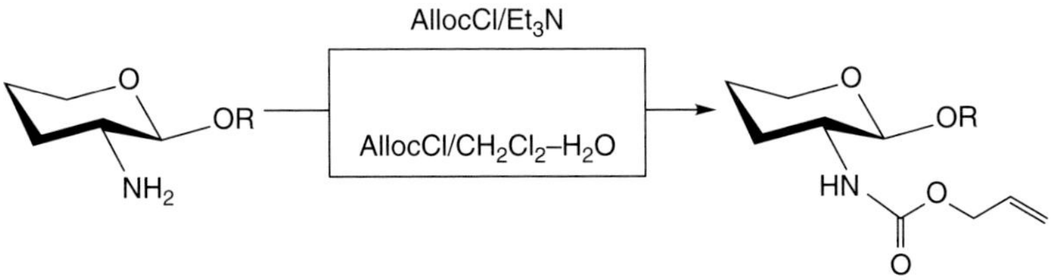

Scheme 6.16 Installation of the allyloxycarbonyl protecting group.

The removal of the allyloxycarbonyl group to the free amine requires the formation of a π-allyl complex with Pd(0) with an allyl acceptor such as dimethyl [91, 92] or diethyl malonate [94, 95] (Scheme 6.17), both providing volatile byproducts leading to easier purification of the resulting amines. The similar conditions using tributyltin hydride also led to the desired amine [93, 96], along with dimethylaminotrimethylsilane (TMSDMA) [97, 98] or polysiloxanes [77, 99].

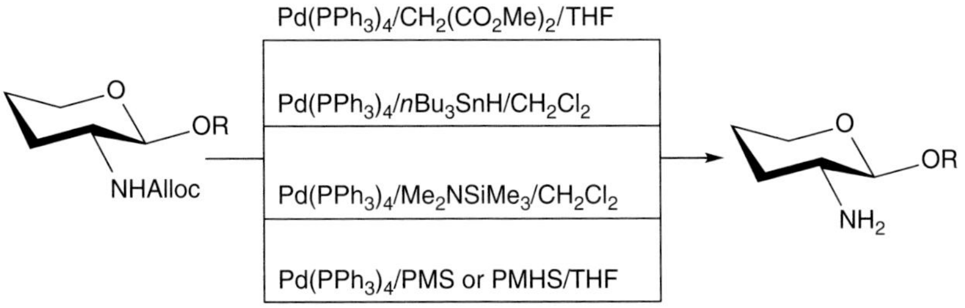

Scheme 6.17 Removal of the 2,2,2-trichloroethoxycarbonyl protecting group.

6.4.4 *t*-Butoxycarbonyl Protecting Group (Boc)

The *t*-butoxycarbonyl (Boc) protecting group has been largely applied in peptide synthesis and can also be used as a convenient alternative to other carbamates for an optimal deprotection under acidic conditions. Introduction of *t*-butoxycarbamate is readily achieved (Scheme 6.18) from the corresponding amine using di-*t*-butyl dicarbonate (Boc2O) in organic or aqueous solvents [100–103] or with 2-(*t*-butoxycarbonyloxyimino)-2-phenyl-acetonitrile (Boc-ON) [104, 105]. The same Boc-protected 2-amino-2-deoxy-glycosides can be obtained in a two-step sequence (Scheme 6.18) from the corresponding 2-azido-2-deoxy-glycosides by reduction with DTT [106], triphenylphosphine [107, 108], hydrogen sulfide [109], propanedithiol [110], or ethanedithiol [111].

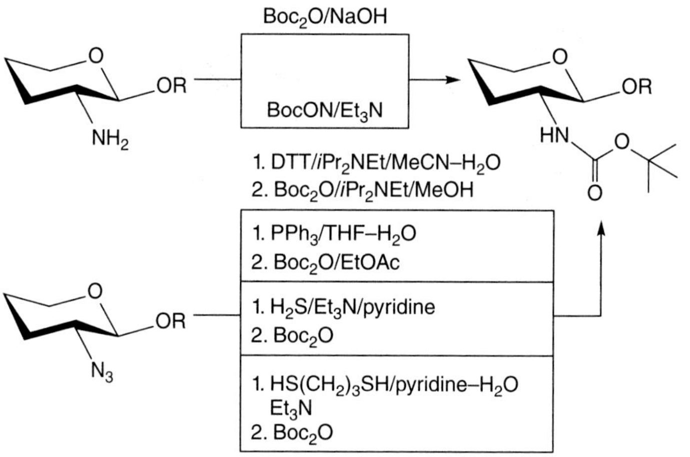

Scheme 6.18 Installation of the *t*-butoxycarbonyl protecting group from 2-amino- or 2-azido-2-deoxy-glycosides.

If the access to the free amine functionality is difficult, the Boc can be introduced directly from the 2-acetamido-2-deoxy-glycosides through the *N*-acetyl-*N*-Boc-protected intermediates (Scheme 6.19) and then selective hydrolysis of the acetyl group with sodium methoxide [112] or hydrazine [113–115].

Unmasking of the Boc carbamate was unequivocally performed under acid conditions using trifluoroacetic acid [100, 101, 103, 107, 108, 110, 111, 113–115] or with *p*-toluenesulfonic acid [112].

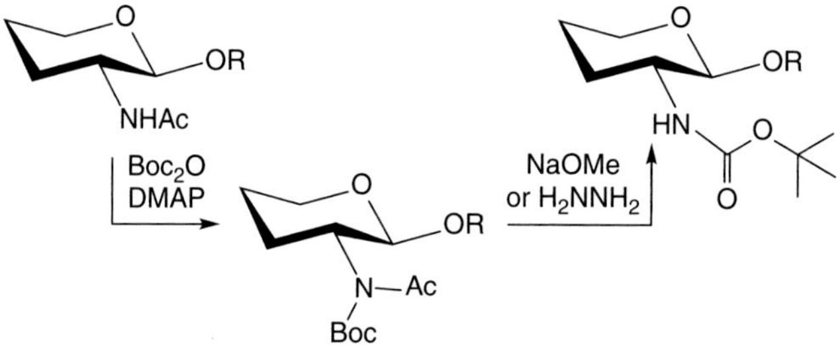

Scheme 6.19 Installation of the *t*-butoxycarbonyl protecting group from 2-acetamido-2-deoxy-glycosides.

6.4.5 9-Fluorenylmethoxycarbonyl Protecting Group (Fmoc)

The base-sensitive 9-fluorenylmethoxycarbamate (Fmoc) is a good alternative to the acid-labile Boc group and other carbamates. Fmoc can be introduced at the primary amine position (Scheme 6.20) either with the FmocCl chloroformate [104, 116–120] or with the FmocOSu activated carbonate [121, 122].

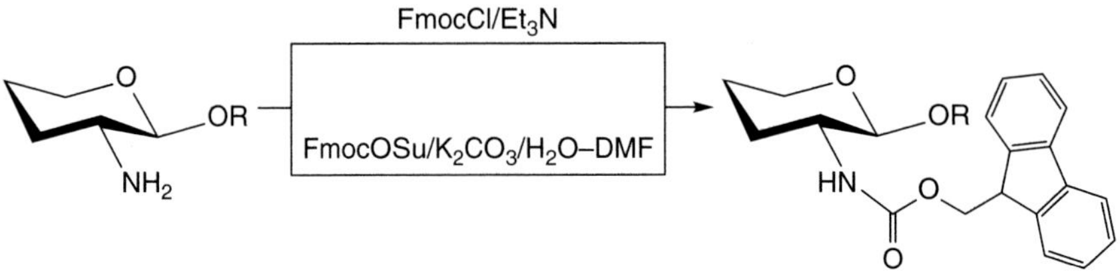

Scheme 6.20 Installation of the 9-fluorenylmethoxycarbonyl protecting group.

Unmasking was typically accomplished with a tertiary amine (piperidine [117, 122], DBU [116, 118–121], or morpholine [105]) as a base to generate dibenzofulvene and releasing CO_2 to afford the desired 2-amino-2-deoxy-glycosides. Sodium methoxide can also be used with the concomitant *O*-deacetylation of the carbohydrate allowing the removal of both *N*-Fmoc and *O*-Ac in a single step [123]. TBAF can also be used in combination with 1-octanethiol as a dibenzofulvene scavenger [124]. Similarly, *N*-(2-mercaptoethyl)aminomethyl-polystyrene was also used in combination with DBU as another dibenzofulvene scavenger [105].

6.4.6 2,3-Oxazolidinone Protecting Group

2,3-Oxazolidinone-protected 2-amino-2-deoxy-glycosides are presenting an increased reactivity at the 4-position toward glycosylation [125, 126], although this 4-position has been described as among the most difficult to glycosylate [127]. The very first example of such a 2,3-oxazolidinone (Scheme 6.21) has been

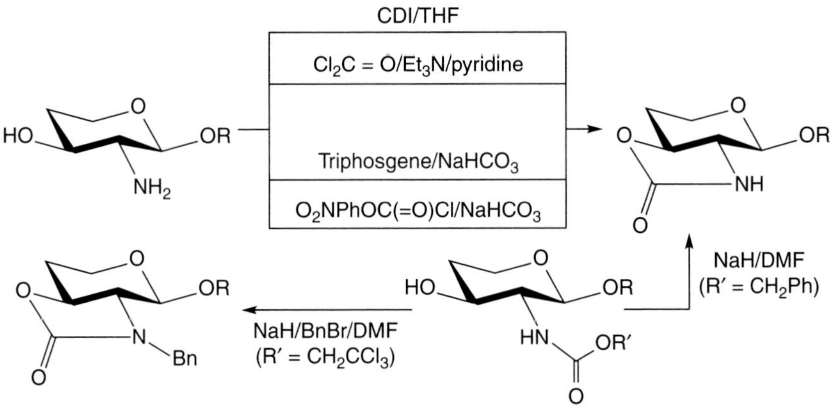

Scheme 6.21 Installation of the 2,3-oxazolidinone protecting group.

reported from benzyl 2-amino-4,6-*O*-benzylidene-2-deoxy-β-D-glucopyranoside using *N,N'*-carbonyldiimidazole in tetrahydrofuran (THF) [128]. The following procedures for this protection involved triphosgene as a less toxic and less hazardous alternative to phosgene [126, 129–133], but also *p*-nitrophenylchloroformate [125, 134–137]. The 2,3-oxazolidinone can also be obtained from a *N*-carbamate precursor, which is deprotonated with sodium hydroxide at the 3-position. The resulting alkoxide will then cyclize on the carbonyl group of the carbamate to the corresponding 2,3-oxazolidinone releasing the carbamate's alcohol (R'OH). This cyclization can proceed from the Cbz-protected intermediate to the corresponding 2,3-oxazolidinone [138, 139], but the Troc-protected derivatives would include an additional 2-*N*-benzyl group when treated with sodium hydride and benzyl bromide in *N,N*-dimethylformamide (DMF) [140–142]. Highly α-stereoselective glycosylation could be achieved with these *N*-benzyl-2,3-oxazolidinone derivatives [140–142].

The removal of the 2,3-oxazolidinone protecting group can be accomplished through three main synthetic strategies from the N-H, N-Ac, or N-Bn derivatives under basic conditions (Scheme 6.22). The N-Ac substituent was sometimes introduced to improve the stability of the 2,3-oxazolidinone and also to provide a 1,2-*trans* anchimeric assistance during the glycosylation with this additional acetyl group. On the contrary, the N-Bn substituent was introduced to improve the stability of the 2,3-oxazolidinone but leading preferably to 1,2-*cis* glycosides as a nonparticipating group. The *N*-substituent can be conserved or removed by a fine-tuning of the reaction conditions. The N-H 2,3-oxazolidinone can be hydrolyzed with sodium hydroxide to the corresponding 2-amino-2-deoxy-glycoside [135, 143]. When an additional acetyl group is introduced at the nitrogen atom, the resulting N-Ac 2,3-oxazolidinone can then be converted to the

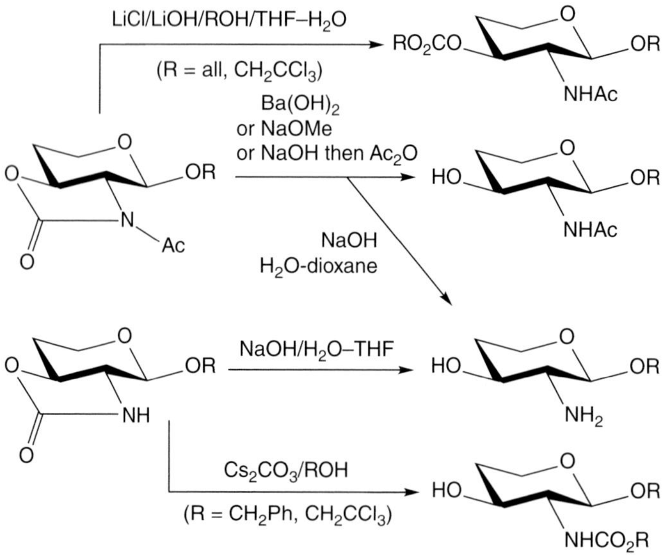

Scheme 6.22 Removal of the 2,3-oxazolidinone protecting group from N-H or N-Ac precursors.

amines using sodium hydroxide [144, 145] but treatment with barium hydroxide [125] or sodium methoxide [146] hydrolyzes the 2,3-oxazolidinone and leaves the acetamide unchanged leading to the 2-acetamido-2-deoxy-glycosides. The same result can be obtained by complete removal of both 2,3-oxazolidinone and acetamide with sodium hydroxide and reacetylation of the crude mixture with acetic anhydride [139]. Finally, ring opening of the N-Ac 2,3-oxazolidinone with lithium chloride and lithium hydroxide in the presence of an alcohol leads to the 2-acetamido-3-*O*-carbonate [134].

The N-Bn-substituted 2,3-oxazolidinone can be converted to the 2-acetamido-2-deoxy-glycoside (Scheme 6.23) in three synthetic steps with saponification of the cyclic carbamate, followed by hydrogenolysis of the *N*-benzyl group and acetylation of the resulting amine [140, 147]. The *N*-benzyl group can also be conserved by treatment with potassium *t*-butoxide [141, 148]. It is worth pointing out that the use of *N*-benzylated derivatives as a protecting group has been scarcely investigated and gave moderate results [149].

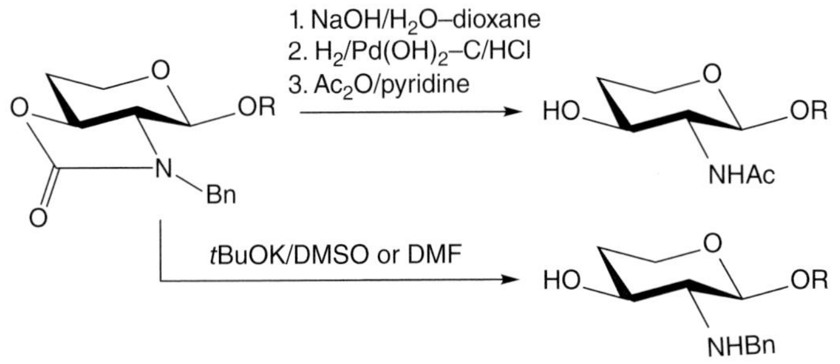

Scheme 6.23 Removal of the 2,3-oxazolidinone protecting group from N-Bn precursors.

6.5 Imine- or Enamine-based Protecting Groups

6.5.1 *N*-arylidene Protecting Groups

2-Amino-2-deoxy-glycosides can be readily converted to the *N*-arylidene glycosides with the corresponding aldehydes. The generated imine appears as a temporary protecting groups because of its moderate stability to standard carbohydrate chemistry conditions and more importantly to acidic media typically used for glycosylation. Salicylaldehyde [150, 151] was first used and then anisaldehyde [152, 153] was applied to the synthesis of the peracetylated 2-amino-2-deoxy-glycoside with a highly reliable procedure [154] (Scheme 6.24). Hydrolysis

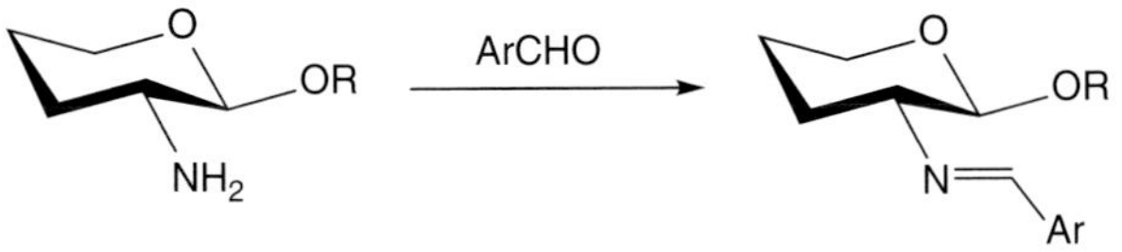

Scheme 6.24 Synthesis of *N*-arylidene glycosides.

of the imine with hydrochloric acid provides the 2-amino-2-deoxy-glycoside. A larger series of aryl substituents have also been used for the design of chiral iminophosphinite ligands [155].

6.5.2 4,4-Dimethyl-2,6-dioxocyclohexylidenemethyl (Ddm) and 4,4-Dimethyl-2,6-dioxocyclohexylideneethyl (Dde) Protecting Groups

Reaction of 5,5-dimethyl-2-(phenylaminomethylene)-1,3-cyclohexanedione with 2-amino-2-deoxy-glycosides affords the 4,4-Dimethyl-2,6-dioxocyclohexylidene-methyl (Ddm)-protected intermediate [156] (Scheme 6.25). Further activation with acetyl bromide led to the concomitant *O*-acetylation and bromination of the anomeric position. This protecting group provides 1,2-*trans* glycosides because of the participation of the carbonyl groups to the stabilization and shielding of the α-face of the anomeric carbocation intermediate.

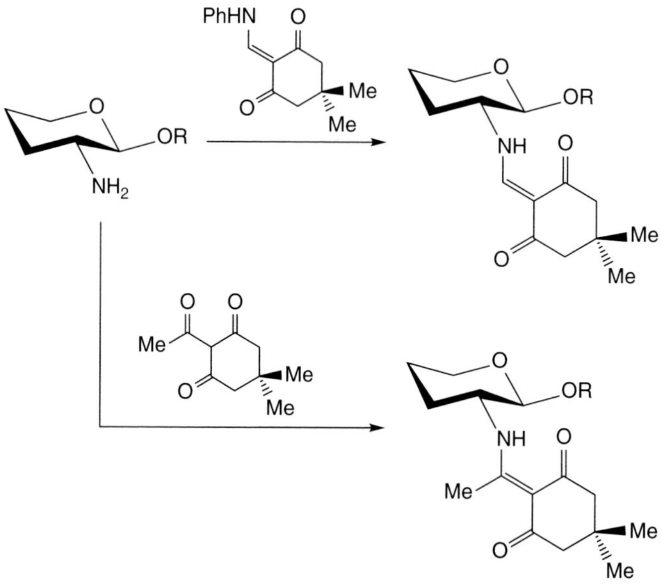

Scheme 6.25 Synthesis of Ddm- and Dde-protected glycosides.

Similarly, the condensation of 2-amino-2-deoxy-glycosides with acetyldimedone provides the Dde-protected intermediates that can be converted to several activated glycosyl donors [157], which provide 1,2-*cis* stereoselectivity in a glycosylation, although the nonparticipating capabilities of this protecting group has not been further investigated.

The Ddm protecting group can be removed with chlorine toward the amine hydrochloride salt in high yields (85–98%). The Dde protecting group is readily removed using ammonia, hydrazine, or primary amines and is quite unstable toward aqueous hydroxides.

6.5.3 (1,3-Dimethyl-2,4,6-trioxopyrimidine-5-ylidene)methyl (DTPM) Protecting Group

The (1,3-dimethyl-2,4,6-trioxopyrimidine-5-ylidene)methyl protecting group (DTPM) was developed as an analog to Ddm and Dde as a cheap and efficient protecting group compatible with several reaction conditions used in carbohydrate chemistry. It is introduced on 2-amino-2-deoxy-glycosides using 1,3-dimethyl-5-[(dimethylamino)methylene]-2,4,6-trioxopyrimidine with completion in typically 1 min with precipitation of the DTPM-protected carbohydrates [158] (Scheme 6.26). Further functional group transformations can afford a series of glycosyl donors with preference toward the 1,2-*trans* glycosylation [159]. The DTPM-protecting group can be removed with ammonia, hydrazine, or primary amines.

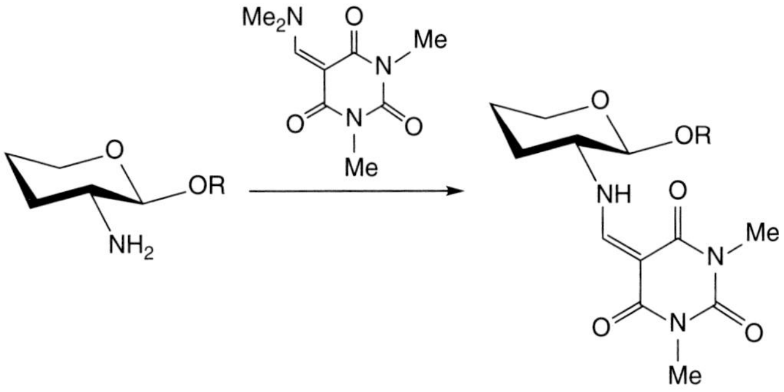

Scheme 6.26 Synthesis of DTPM-protected glycosides.

6.6 2-Deoxy-2-azido Derivatives as a Protecting Group

Masking the 2-amino functionality can be readily achieved using the 2-azido derivatives predominantly leading to α-glycosides when performing glycosylations. The 2-azido moiety is rather stable under a large series of conditions apart from reductive media that would provide the corresponding amines. Hence, this 2-azido moiety is not a "protecting group" *per se* but provides an easy approach to masked 2-amino functionalities in an elegant manner.

The introduction of the 2-azido group is readily achieved by diazo transfer [160, 161] from the corresponding glucosamine hydrochloride in the presence of an inorganic base (e.g. K_2CO_3) and a catalytic amount of copper sulfate [160, 162–164] (Scheme 6.27). The reaction proceeds in water, methanol, or toluene

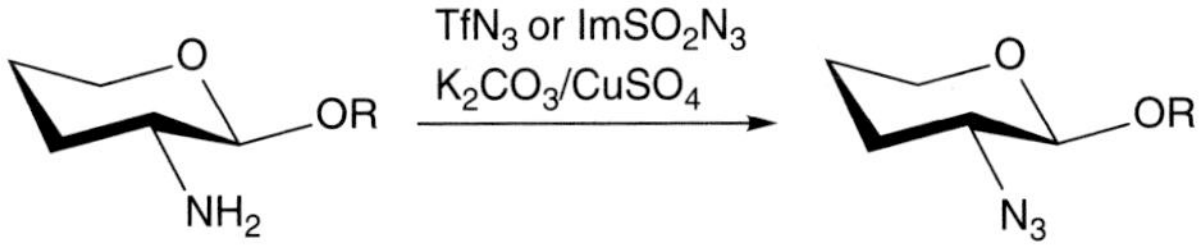

Scheme 6.27 Conversion of 2-amino-glycosides into 2-azido-glycosides.

or mixtures of these solvents and does not affect the stereochemistry at the C-2 carbon as the C—N bond is not affected. Triflic azide was freshly prepared from triflic anhydride and sodium azide, and this reagent is hazardous because of both the presence of azide and acids potentially generating the toxic hydrazoic acid and for potential explosion under concentrated solutions such as during evaporation of the solvents during workup of the reactions. Similar conditions were reported using triethylamine [113, 137, 165–167] or DMAP [166, 168–171] instead of inorganic carbonates. Imidazole-1-sulfonyl azide hydrochloride was later introduced as an efficient, shelf-stable, and inexpensive diazo transfer reagent [161] and can be readily applied for the preparation of 2-azido-glycosides [172–174]. Finally, nonafluorobutanesulfonyl azide was also reported as a shelf-stable alternative to triflic azide [175].

Conversion of the 2-azido moiety into the 2-acetamido group is achieved by reaction with thioacetic acid [137, 176] or by reduction with $NiCl_2/NaBH_4$ [177], zinc dust [62], or hydrogenolysis [178] followed by acetylation of the crude mixtures with acetic anhydride (Scheme 6.28). The recovery of the 2-amino functionality usually required in the final oligosaccharide targets can also be performed using several reducing conditions associated (or not) with an additional functionalizing agent to provide the free amine [179], the *N*-sulfate [180], or the 2,2,2-trichloroethoxycarbamate [181].

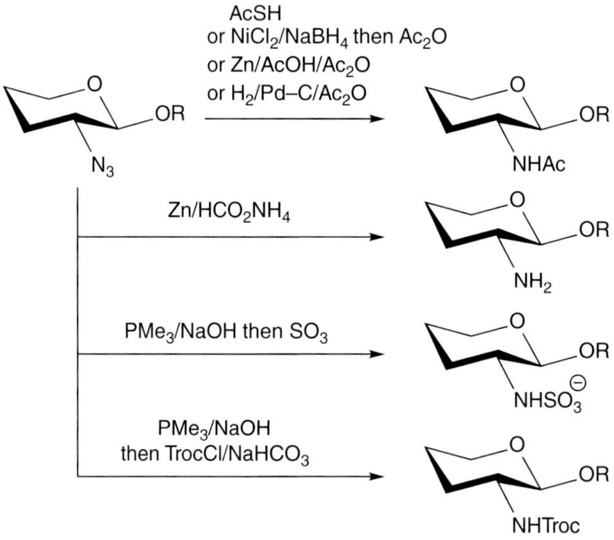

Scheme 6.28 Conversion of 2-azido-glycosides into 2-amino-glycosides.

6.7 From Glycals to 2-Azido Intermediates

An elegant approach to protect a reactive functionality in protecting group strategies is to remove the actual function and to *hide* it in another function that would be more stable or unreactive under the considered reaction conditions and would then be converted into the desired functionality in a minimum number of steps. The use of glycals as precursors of 2-amino-glycosides was first reported by Lemieux and Ratcliffe [182] for the synthesis of galactosamine derivatives (Scheme 6.29). The azido-nitration was selective for the 1,2-*trans*-diequatorial addition of the azido and nitrate to the double bond (55%), but a fair amount of the α-anomer was also isolated (22%) along with the 1,2-*trans*-diaxial product (8%). A series of later reports also used this pioneering synthetic strategy toward 2-azido-galactosides with several protecting groups such as acetyl [183–186], benzyl [187], or other protecting groups [188–190]. Conversion of these 2-azido-galactosides into the hemiacetal was achieved with thiophenol [186–188] or sodium nitrite [191, 192], into the corresponding 1-*O*-acetate with sodium acetate [185], or into the galactosyl bromide with lithium bromide [184, 193].

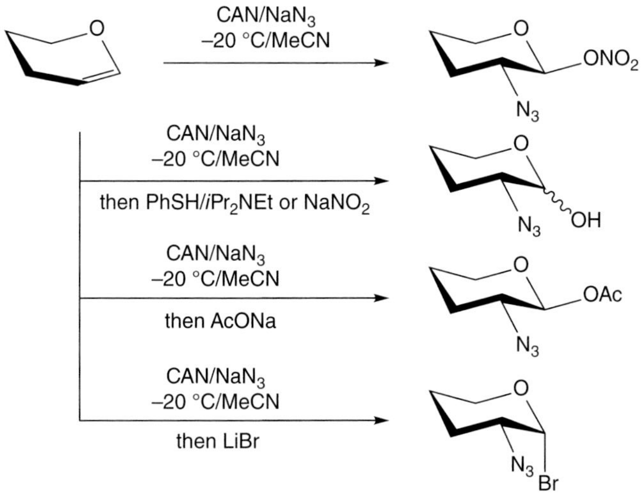

Scheme 6.29 Conversion of glycals into 2-azido-glycosides.

The same approach was also used in the 2-azido-glucoside series [191, 192, 194, 195] and was found less stereoselective than in the galacto-configured series [196] with the separation of the isomers sometimes difficult [197, 198].

Such azido-nitration appears as a robust access to gluco- and galacto-configured 2-amino-glycosides and more specifically as a reliable source of the more rare galactosamine derivatives (in comparison to glucosamine) from the less expensive and easily accessible galactal or galactose. The ceric ammonium nitrate and sodium azide used in this reaction both suffer from hazardous and/or explosive properties, and such azido-nitration reaction is better performed under small-scale synthesis (typically less than 5 g) and with appropriate protection and prevention tools (fire extinguisher, glass shield, or any other safety equipment). In addition, the disposal of the hazardous and toxic cerium and azide salts along with other by-products of the reaction must be considered with caution. Nevertheless, this reaction can be performed under safe conditions when cautiously considering these risks and has witnessed numerous applications in the literature.

6.8 From Glycals to 2-Sulfonamido Intermediates

Glycals also found applications in the access to 2-amino-glycosides through the approach initially reported by Lafont and Descotes [199, 200] for the addition of iodoazide (IN_3) to the double bond and further ring opening of the corresponding cyclic iodonium intermediate with the azide nucleophile to afford the 1,2-*trans*-2-deoxy-2-iodo-glycosides with an azido group at the anomeric position now referred as the Lafont intermediate [201]. Following these results, several applications of such synthetic strategy for the total synthesis of complex oligosaccharides have been reported by the Danishefsky group in which galactosamine [202] and glucosamine [203–206] were installed from their glycals, which were themselves used as precursors in the previous synthetic steps of the total synthesis. These syntheses were performed using benzenesulfonamide ($PhSO_2NH_2$) and iodonium di-*sym*-collidine perchlorate (IDCP) to afford a 1,2-*trans*-diaxial intermediate (Scheme 6.30). Further reaction with a base leads to the aziridine intermediate, and ring opening with a nucleophile provides the 1,2-*trans*-diequatorial 2-sulfonamido-glycosides. Conversion of the phenylsulfonamido moiety into the desired acetamido group can be performed using Birch conditions (Na/NH_3) and subsequent acetylation of the crude mixture with acetic anhydride. Even though such conditions can seem pretty harsh for carbohydrates, it was readily applied to tri- and tetrasaccharides in good yields [203–206].

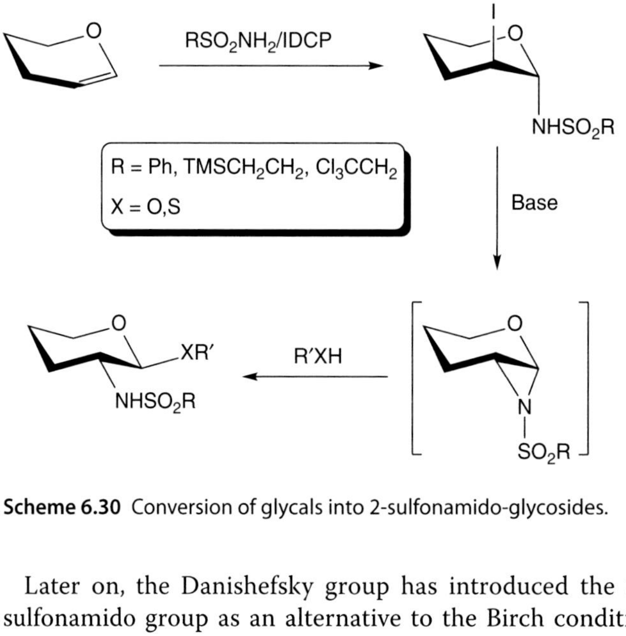

Scheme 6.30 Conversion of glycals into 2-sulfonamido-glycosides.

Later on, the Danishefsky group has introduced the 2-trimethylsilylethylsulfonamido group as an alternative to the Birch conditions required for the deprotection of the phenylsulfonamido group and now using smoother conditions such as cesium fluoride [31, 32, 207, 208]. Finally, the 2,2,2-trichloroethylsulfonamido group was also reported and can be removed with zinc dust and copper sulfate in acetic acid with or without acetic anhydride to provide the 2-amino or 2-acetamido derivatives, respectively [209]. The general access to 2-amino-glycosides from glycals has been reviewed by Goti and coworkers [210] and provides a more general and broader overview of such synthetic strategies.

6.9 Summary and Conclusions

Rapid and efficient access to orthogonally protected glycosamine derivatives is a major challenge because of the chemoselectivity and regioselectivity required for the differentiation of each alcohol and amine of the pyranose ring. A few recent reports have taken advantage of the numerous investigations presented above to design rapid access to such multifunctional glycosamines through 2-azido intermediates [162], *N*-phthalimido-glucosamine [211], or trichloroacetamido derivatives [212] with tin-acetals [213]. Such multistep syntheses of orthogonally protected glycosamines are the best illustration of the versatility of protecting groups developed for the 2-amino position of glycosides. Such developments have allowed the synthesis of numerous natural complex oligosaccharides to further decipher the implications of saccharides in biological processes. It is quite appealing that no general strategy can be drawn from the studies covered herein, but the reader will be able to have a broad overview of the possibilities offered to solve its specific problem and especially with a comparative aspect crucial for the assessment of the best strategy possible.

6.10 Experimental Section

6.10.1 1,3,4,6-Tetra-*O*-acetyl-2-phthalimido-2-deoxy-β-D-glucopyranose [211]

A 1 M NaOMe solution was prepared by adding Na metal (5.33 g, 0.232 mol) in small pieces to MeOH (232 ml) at −5 °C. This was slowly added at 0 °C to a 1 l round-bottomed flask containing glucosamine hydrochloride (50 g, 0.232 mol). The reaction mixture was mechanically agitated for 2 h at RT, then treated with finely ground phthalic anhydride (19 g, 0.128 mol), and mechanically stirred for another 45 min. A second portion of phthalic anhydride (19 g, 0.128 mol) was added with triethylamine (35.5 ml, 0.255 mol) and MeOH (230 ml) and vigorously stirred for another 24 h, during which it slowly changed from a milky white solution to a thick yellow paste. The intermediate phthalamate (and salts) was precipitated as a white solid by cooling the mixture to −20 °C for 4 h. These were filtered and thoroughly washed with cold MeOH and then dried overnight under reduced pressure. The solid was redispersed in pyridine (500 ml) with vigorous mechanical stirring and cooled to −5 °C, followed by treatment with acetic anhydride (330 ml). The mixture was mechanically stirred at RT for 48 h, during which it slowly changed from a translucent white to an opaque yellow solution. Cold EtOH (100 ml) was slowly added to the mixture to quench the excess acetic anhydride, which was then reduced by rotary evaporation. Azeotropic removal of pyridine was achieved by coevaporation with toluene (3 ml × 100 ml). The remaining slurry was redissolved in $CHCl_3$ (1 l) and washed with distilled H_2O (4 ml × 250 ml) and brine (250 ml), then dried (Na_2SO_4), and evaporated to dryness. The crude product was dissolved in minimal amount of hot EtOAc (100 ml), then diluted with hexanes (400 ml), and left to cool at −5 °C. The recrystallized product was collected by filtration, washed with cold hexanes, and dried to yield the desired tetra-acetate as a 8 : 1 mixture of anomers (85.3 g, 77%).

6.10.2 2-Acetamido-1,3-di-*O*-acetyl-4,6-*O*-benzylidene-2-deoxy-D-glucopyranose [79]

N-acetyl-D-glucosamine (5 g, 22.6 mmol), benzaldehyde (13.7 ml, 135.6 mmol), and $ZnCl_2$ (3.1 g, 22.6 mmol) were stirred at RT for 12 h under argon. The precipitate was filtered off, washed with petroleum ether (2 ml × 40 ml) and water (2 ml × 40 ml), and dried under reduced pressure. The crude 4,6-*O*-benzylidene-D-glucosamine (6.99 g) was used in the next step without further purification. This intermediate (500 mg, 1.62 mmol) and pyridine hydrochloride (243 mg, 2.10 mol) in pyridine (5 ml) were stirred at 100 °C for 1 h. To this solution, acetic anhydride (772 μl, 8.24 mmol) was added, and the reaction mixture was stirred at RT for 8 h. The volatiles were evaporated under reduced pressure, and the crude product was purified by chromatography on silica gel (heptane/EtOAc 6 : 4 to 0 : 1) to afford pure product (374 mg, 59%, α/β 1 : 1, white amorphous solid).

Abbreviations

AIBN	α,α′-azoisobutyronitrile
Alloc	allyloxycarbonyl
All	allyl
Bn	benzyl
Boc	*t*-butoxycarbonyl
Cbz or Z	benzyloxycarbonyl
CDI	1,1′-carbonyldiimidazole
DCP	dichlorophthalimide
Dde	4,4-dimethyl-2,6-dioxocyclohexylideneethyl
Ddm	4,4-dimethyl-2,6-dioxocyclohexylidenemethyl
DMA	*N*,*N*-dimethylacetamide
DMF	*N*,*N*-dimethylformamide
DMM	dimethylmaleoyl
DMSO	dimethylsulfoxide
DPM	diphenylmaleoyl

DTG	dithioglycolyl
DTPM	(1,3-dimethyl-2,4,6-trioxopyrimidine-5-ylidene)methyl
Dts	dithiasuccinyl
DTT	dithiothreitol
Fmoc	9-fluorenylmethoxycarbonyl
FucNAc	*N*-acetyl-fucosamine
GAG	glycosaminoglycan
GalNAc	*N*-acetyl-galactosamine
GlcNAc	*N*-acetyl-glucosamine
IDCP	iodonium di-*sym*-collidine perchlorate
Im	imidazole
OGT	*O*-GlcNAc-transferase
PMHS	poly(methylhydrosiloxane)
PMS	polymethylsiloxane
PNZ	*p*-nitrobenzyloxycarbonyl
RhaNAc	*N*-acetyl-rhamnosamine
RT	room temperature
TBAF	tetrabutylammonium fluoride
TBDPS	*t*-butyldiphenylsilyl
TCP	tetrachlorophthalimide
TFAA	trifluoroacetic anhydride
THF	tetrahydrofuran
TMSDMA	dimethylaminotrimethylsilane
Troc	2,2,2-trichloroethoxycarbonyl
Tf	trifluoromethanesulfonyl
TMS	trimethylsilyl

References

1 Hart, G.W., Slawson, C., Ramirez-Correa, G., and Lagerlof, O. (2011). *Ann. Rev. Biochem.* 80: 825–858.

2 Lefebvre, T., Guinez, C.L., Dehennaut, V. et al. (2005). *Exp. Rev. Proteomics* 2: 265–275.

3 Slawson, C. and Hart, G.W. (2011). *Nat. Rev. Cancer* 11: 678–684.

4 Wang, S. and Vidal, S. (2013). Recent design of glycosyltransferase inhibitors. In: *Carbohydrate Chemistry: Specialist Periodical Reports*, vol. 39 (ed. A.P. Rauter and T.K. Lindhorst), 78–101. Cambridge: RSC Publishing.

5 Zachara, N.E. and Hart, G.W. (2002). *Chem. Rev.* 102: 431–438.

6 Banoub, J., Boullanger, P., and Lafont, D. (1992). *Chem. Rev.* 92: 1167–1195.

7 Debenham, J., Rodebaugh, R., and Fraser-Reid, B. (1997). *Liebigs Ann.* 1997: 791–802.

8 Enugala, R., Carvalho, L.C.R., Dias Pires, M.J., and Marques, M.M.B. (2012). *Chem. Asian J.* 7: 2482–2501.

9 Aly, M.R.E.S. and El-Ashry, E.S.H. (2016). *Adv. Carbohydr. Chem. Biochem.* 73: 117–224.

10 Bartek, J., Müller, R., and Kosma, P. (1998). *Carbohydr. Res.* 308: 259–273.
11 Love, K.R. and Seeberger, P.H. (2005). *J. Org. Chem.* 70: 3168–3177.
12 Blatter, G., Beau, J.-M., and Jacquinet, J.-C. (1994). *Carbohydr. Res.* 260: 189–202.
13 Virlouvet, M., Gartner, M., Koroniak, K. et al. (2010). *Adv. Synth. Catal.* 352: 2657–2662.
14 El-Abadla, N., Lampilas, M., Hennig, L. et al. (1999). *Tetrahedron* 55: 699–722.
15 Blatter, G. and Jacquinet, J.-C. (1996). *Carbohydr. Res.* 288: 109–125.
16 Bélot, F. and Jacquinet, J.-C. (2000). *Carbohydr. Res.* 325: 93–106.
17 Barroca, N. and Jacquinet, J.-C. (2000). *Carbohydr. Res.* 329: 667–679.
18 Vibert, A., Lopin-Bon, C., and Jacquinet, J.-C. (2010). *Tetrahedron Lett.* 51: 1867–1869.
19 Pfister, H.B. and Mulard, L.A. (2014). *Org. Lett.* 16: 4892–4895.
20 Wolfrom, M.L. and Bhat, H.B. (1966). *Chem. Commun.* 146.
21 Wolfrom, M.L. and Bhat, H.B. (1967). *J. Org. Chem.* 32: 1821–1823.
22 Myszka, H., Bednarczyk, D., Najder, M., and Kaca, W. (2003). *Carbohydr. Res.* 338: 133–141.
23 Luo, X., Sugiura, T., Nakashima, R. et al. (2013). *Bioorg. Med. Chem. Lett.* 23: 4157–4161.
24 Silva, D.J., Wang, H., Allanson, N.M. et al. (1999). *J. Org. Chem.* 64: 5926–5929.
25 Wong, T.C., Townsend, R.R., and Lee, Y.C. (1987). *Carbohydr. Res.* 170: 27–46.
26 Svejgaard, L., Fuglsang, H., Jensen, P.B. et al. (2003). *J. Carbohydr. Chem.* 22: 179–184.
27 Kobayashi, Y., Ishida, N., Arai, M. et al. (1991). *Carbohydr. Res.* 222: 83–97.
28 Lin, H. and Walsh, C.T. (2004). *J. Am. Chem. Soc.* 126: 13998–14003.
29 Aly, M.R.E. and Schmidt, R.R. (2005). *Eur. J. Org. Chem.* 4382–4392.
30 Hirano, S. (1971). *Carbohydr. Res.* 16: 229–231.
31 Wang, Z.-G., Zhang, X., Visser, M. et al. (2001). *Angew. Chem. Int. Ed.* 40: 1728–1732.
32 Nagorny, P., Fasching, B., Li, X. et al. (2009). *J. Am. Chem. Soc.* 131: 5792–5799.
33 Madsen, R., Udodong, U.E., Roberts, C. et al. (1995). *J. Am. Chem. Soc.* 117: 1554–1565.
34 Bundle, D.R. and Josephson, S. (1979). *Can. J. Chem.* 57: 662–668.
35 Kanie, O., Crawley, S.C., Palcic, M.M., and Hindsgaul, O. (1993). *Carbohydr. Res.* 243: 139–164.
36 Durette, P.L., Meitzner, E.P., and Shen, T.Y. (1979). *Tetrahedron Lett.* 20: 4013–4016.
37 Shimizu, H., Ito, Y., Matsuzaki, Y. et al. (1996). *Biosci. Biotechnol. Biochem.* 60: 73–76.
38 Yamada, T., Kinjyo, S., Yoshida, J.-I., and Yamago, S. (2005). *Chem. Lett.* 34: 1556–1557.
39 Debenham, J.S., Madsen, R., Roberts, C., and Fraser-Reid, B. (1995). *J. Am. Chem. Soc.* 117: 3302–3303.
40 Debenham, J.S., Rodebaugh, R., and Fraser-Reid, B. (1997). *J. Org. Chem.* 62: 4591–4600.
41 Debenham, J.S., Debenham, S.D., and Fraser-Reid, B. (1996). *Bioorg. Med. Chem.* 4: 1909–1918.

42 Castro-Palomino, J.C. and Schmidt, R.R. (1995). *Tetrahedron Lett.* 36: 5343–5346.
43 Khiar, N., Fernández, I., Araújo, C.S. et al. (2003). *J. Org. Chem.* 68: 1433–1442.
44 Meinjohanns, E., Meldal, M., Paulsen, H., and Bock, K. (1995). *J. Chem. Soc., Perkin Trans. 1* 405–415.
45 Jensen, K.J., Hansen, P.R., Venugopal, D., and Barany, G. (1996). *J. Am. Chem. Soc.* 118: 3148–3155.
46 Barany, G. and Merrifield, R.B. (1980). *J. Am. Chem. Soc.* 102: 3084–3095.
47 Zehavi, U. (1977). *J. Org. Chem.* 42: 2819–2821.
48 Hesek, D., Suvorov, M., Morio, K.-I. et al. (2004). *J. Org. Chem.* 69: 778–784.
49 Hesek, D., Lee, M., Morio, K.-I., and Mobashery, S. (2004). *J. Org. Chem.* 69: 2137–2146.
50 Aly, M.R.E., Castro-Palomino, J.C., Ibrahim, E.-S.I. et al. (1998). *Eur. J. Org. Chem.* 2305–2316.
51 Bartolmäs, T., Heyn, T., Mickeleit, M. et al. (2005). *J. Med. Chem.* 48: 6750–6755.
52 Karst, N.A., Islam, T.F., and Linhardt, R. (2003). *J. Org. Lett.* 5: 4839–4842.
53 Aly, M.R.E., Ibrahim, E.-S.I., El-Ashry, E.S.H., and Schmidt, R.R. (2001). *Carbohydr. Res.* 331: 129–142.
54 Chiesa, M.V. and Schmidt, R.R. (2000). *Eur. J. Org. Chem.* 3541–3554.
55 Knuhr, P., Castro-Palomino, J., Grathwohl, M., and Schmidt, R.R. (2001). *Eur. J. Org. Chem.* 4239–4246.
56 Aly, M.R.E., Ibrahim, E.-S.I., El-Ashry, E.-S.H., and Schmidt, R.R. (1999). *Carbohydr. Res.* 316: 121–132.
57 Aly, M.R.E., Ibrahim, E.-S.I., El-Ashry, E.-S.H., and Schmidt, R.R. (2000). *Eur. J. Org. Chem.* 319–326.
58 Roussel, F., Takhi, M., and Schmidt, R.R. (2001). *J. Org. Chem.* 66: 8540–8548.
59 Boullanger, P., Jouineau, M., Bouammali, B. et al. (1990). *Carbohydr. Res.* 202: 151–164.
60 Ellervik, U. and Magnusson, G. (1996). *Carbohydr. Res.* 280: 251–260.
61 Ogawa, Y., Fujishima, Y., Ishida, H. et al. (1991). *Carbohydr. Res.* 220: 155–164.
62 Krishnamurthy, V.R., Dougherty, A., Kamat, M. et al. (2010). *Carbohydr. Res.* 345: 1541–1547.
63 Schwagerus, S., Reimann, O., Despres, C. et al. (2016). *J. Pept. Sci.* 22: 327–333.
64 Zhao, Q., Li, X., Li, W. et al. (2015). *Tetrahedron* 71: 1018–1022.
65 Wang, Z., Zhou, L., El-Boubbou, K. et al. (2007). *J. Org. Chem.* 72: 6409–6420.
66 Adinolfi, M., Iadonisi, A., Pezzella, A., and Ravidà, A. (2005). *Synlett* 1848–1852.
67 Dullenkopf, W., Castro-Palomino, J.C., Manzoni, L., and Schmidt, R.R. (1996). *Carbohydr. Res.* 296: 135–147.
68 Kumar, V., Locke, R.D., and Matta, K.L. (2009). *Synlett* 2633–2636.
69 Bednarczyk, D., Walczewska, A., Grzywacz, D. et al. (2013). *Carbohydr. Res.* 367: 10–17.
70 Ziegler, T. (1994). *Carbohydr. Res.* 262: 195–212.
71 Melean, L.G., Love, K.R., and Seeberger, P.H. (2002). *Carbohydr. Res.* 337: 1893–1916.
72 Paulsen, H. and Helpap, B. (1992). *Carbohydr. Res.* 216: 289–313.

73 Masahiro, I., Hiroyuki, Y., Tetsuo, S. et al. (1987). *Bull. Chem. Soc. Jpn.* 60: 2205–2214.
74 Huang, C.-Y., Wang, N., Fujiki, K. et al. (2010). *J. Carbohydr. Chem.* 29: 289–298.
75 Tokimoto, H. and Fukase, K. (2005). *Tetrahedron Lett.* 46: 6831–6832.
76 Ellervik, U. and Magnusson, G. (1997). *Tetrahedron Lett.* 38: 1627–1628.
77 Hayashi, M., Tanaka, M., Itoh, M., and Miyauchi, H. (1996). *J. Org. Chem.* 61: 2938–2945.
78 Yang, S., Wang, A.P., Zhang, G. et al. (2016). *Tetrahedron* 72: 5659–5670.
79 Xolin, A., Stevenin, A., Pucheault, M. et al. (2014). *Org. Chem. Front.* 1: 992–1000.
80 Kyas, A. and Feigel, M. (2005). *Helv. Chim. Acta* 88: 2375–2396.
81 Chiu, T.M.K., Sigillo, K., Gross, P.H., and Franz, A.H. (2007). *Synth. Commun.* 37: 2355–2381.
82 Xue, J., Shao, N., and Guo, Z. (2003). *J. Org. Chem.* 68: 4020–4029.
83 Hanessian, S., Szychowski, J., Adhikari, S.S. et al. (2007). *J. Med. Chem.* 50: 2352–2369.
84 Barroca-Aubry, N., Pernet-Poil-Chevrier, A., Domard, A., and Trombotto, S. (2010). *Carbohydr. Res.* 345: 1685–1697.
85 Hogendorf, W.F.J., Lameijer, L.N., Beenakker, T.J.M. et al. (2012). *Org. Lett.* 14: 848–851.
86 Hu, Y.-P., Lin, S.-Y., Huang, C.-Y. et al. (2011). *Nat. Chem.* 3: 557–563.
87 Kozioł, A., Lendzion-Paluch, A., and Manikowski, A. (2013). *Org. Process Res. Dev.* 17: 869–875.
88 Miller, D.C., Carbain, B., Beale, G.S. et al. (2015). *Org. Biomol. Chem.* 13: 5279–5284.
89 Qian, X. and Hindsgaul, O. (1997). *Chem. Commun.* 1059–1060.
90 Grann Hansen, S. and Skrydstrup, T. (2007). *Eur. J. Org. Chem.* 3392–3401.
91 Boullanger, P. and Descotes, O. (1986). *Tetrahedron Lett.* 27: 2599–2602.
92 Boullanger, P., Banoub, J., and Descotes, G. (1987). *Can. J. Chem.* 65: 1343–1348.
93 Becker, C., Hoben, C., and Kunz, H. (2007). *Adv. Synth. Catal.* 349: 417–424.
94 Lafont, D. and Boullanger, P. (2006). *Tetrahedron: Asymmetry* 17: 3368–3379.
95 Baisch, G. and Öhrlein, R. (1998). *Bioorg. Med. Chem.* 6: 1673–1682.
96 Bazin, H.G., Bess, L.S., Livesay, M.T. et al. (2006). *Tetrahedron Lett.* 47: 2087–2092.
97 Shimoyama, A., Saeki, A., Tanimura, N. et al. (2011). *Chem. Eur. J.* 17: 14464–14474.
98 Wang, Q., Matsuo, Y., Pradipta, A.R. et al. (2016). *Org. Biomol. Chem.* 14: 1013–1023.
99 Ramphal, J.Y., Hiroshige, M., Lou, B. et al. (1996). *J. Med. Chem.* 39: 1357–1360.
100 Bintein, F., Augé, C., and Lubineau, A. (2003). *Carbohydr. Res.* 338: 1163–1173.
101 Zeng, L., Xu, G., Gao, P. et al. (2015). *Eur. J. Med. Chem.* 93: 109–120.
102 Lin, Y., Li, Y., Wang, X. et al. (2013). *J. Controlled Release* 167: 148–156.
103 Billing, J.F. and Nilsson, U.J. (2005). *J. Org. Chem.* 70: 4847–4850.
104 Winterfeld, G.A., Khodair, A.I., and Schmidt, R.R. (2003). *Eur. J. Org. Chem.* 1009–1021.
105 Khodair, A.I. and Schmidt, R.R. (2011). *Eur. J. Org. Chem.* 7407–7413.

106 Lee, B.Y., Seeberger, P.H., and Varon Silva, D. (2016). *Chem. Commun.* 52: 1586–1589.
107 Capes, A.S., Crossman, A., Urbaniak, M.D. et al. (2014). *Org. Biomol. Chem.* 12: 1919–1934.
108 Yashunsky, D.V., Borodkin, V.S., Ferguson, M.A.J., and Nikolaev, A.V. (2006). *Angew. Chem. Int. Ed.* 45: 468–474.
109 Long, X., Deng, S., Mattner, J. et al. (2007). *Nat. Chem. Biol.* 3: 559–564.
110 Vishwakarma, R.A. and Menon, A.K. (2005). *Chem. Commun.* 453–455.
111 Pekari, K. and Schmidt, R.R. (2003). *J. Org. Chem.* 68: 1295–1308.
112 Meng, F.-C., Chen, K.-T., Huang, L.-Y. et al. (2011). *Org. Lett.* 13: 5306–5309.
113 Loft, K.J., Bojarová, P., Slámová, K. et al. (2009). *ChemBioChem* 10: 565–576.
114 Agarwal, J. and Peddinti, R.K. (2011). *J. Org. Chem.* 76: 3502–3505.
115 Agarwal, J. and Peddinti, R.K. (2012). *Eur. J. Org. Chem.* 6390–6406.
116 Adanitsch, F., Ittig, S., Stöckl, J. et al. (2014). *J. Med. Chem.* 57: 8056–8071.
117 Roy, S., El Hadri, A., Richard, S. et al. (2014). *J. Med. Chem.* 57: 4511–4520.
118 Zhang, Y., Gaekwad, J., Wolfert, M.A., and Boons, G.-J. (2007). *J. Am. Chem. Soc.* 129: 5200–5216.
119 Zhang, Y., Gaekwad, J., Wolfert, M.A., and Boons, G.-J. (2008). *Chem. Eur. J.* 14: 558–569.
120 Zhang, Y., Gaekwad, J., Wolfert, M.A., and Boons, G.-J. (2008). *Org. Biomol. Chem.* 6: 3371–3381.
121 Billing, J.F. and Nilsson, U.J. (2005). *Tetrahedron* 61: 863–874.
122 Di Fabio, G., De Capua, A., De Napoli, L. et al. (2001). *Synlett* 341–344.
123 Liu, L., Yi, L., Yang, X. et al. (2008). *Tetrahedron* 64: 5885–5890.
124 Ågren, J.K.M., Billing, J.F., Grundberg, H.E., and Nilsson, U.J. (2006). *Synthesis* 3141–3145.
125 Crich, D. and Vinod, A.U. (2005). *J. Org. Chem.* 70: 1291–1296.
126 Olsson, J.D.M., Eriksson, L., Lahmann, M., and Oscarson, S. (2008). *J. Org. Chem.* 73: 7181–7188.
127 Paulsen, H. (1982). *Angew. Chem. Int. Ed. Engl.* 21: 155–173.
128 Miyai, K. and Gross, P.H. (1969). *J. Org. Chem.* 34: 1638–1642.
129 Bohn, M.L., Colombo, M.I., Stortz, C.A., and Rúveda, E.A. (2006). *Carbohydr. Res.* 341: 1096–1104.
130 Boysen, M., Gemma, E., Lahmann, M., and Oscarson, S. (2005). *Chem. Commun.* 3044–3046.
131 Cirillo, L., Bedini, E., Molinaro, A., and Parrilli, M. (2010). *Tetrahedron Lett.* 51: 1117–1120.
132 Manabe, S., Satoh, H., Hutter, J. et al. (2014). *Chem. Eur. J.* 20: 124–132.
133 Yang, L. and Ye, X.-S. (2010). *Carbohydr. Res.* 345: 1713–1721.
134 Nagai, Y., Ito, N., Sultana, I., and Sugai, T. (2008). *Tetrahedron* 64: 9599–9606.
135 Benakli, K., Zha, C., and Kerns, R.J. (2001). *J. Am. Chem. Soc.* 123: 9461–9462.
136 Huang, L., Wang, Z., Li, X. et al. (2006). *Carbohydr. Res.* 341: 1669–1679.
137 Feng, J., Hevey, R., and Ling, C.-C. (2011). *Carbohydr. Res.* 346: 2650–2662.
138 Sugawara, T. and Narisada, M. (1989). *Carbohydr. Res.* 194: 125–138.
139 Yang, L., Zhu, J., Zheng, X.-J. et al. (2011). *Chem. Eur. J.* 17: 14518–14526.
140 Manabe, S., Ishii, K., and Ito, Y. (2011). *Eur. J. Org. Chem.* 2011: 497–516.

141 Lin, S.-C., Chao, C.-S., Chang, C.-C., and Mong, K.-K.T. (2010). *Tetrahedron Lett.* 51: 1910–1913.

142 Manabe, S., Ishii, K., and Ito, Y. (2006). *J. Am. Chem. Soc.* 128: 10666–10667.

143 Bodner, R., Marcellino, B.K., Severino, A. et al. (2005). *J. Org. Chem.* 70: 3988–3996.

144 Manabe, S. and Ito, Y. (2016). *Beilstein J. Org. Chem.* 12: 328–333.

145 Knapp, S., Kukkola, P.J., Sharma, S. et al. (1990). *J. Org. Chem.* 55: 5700–5710.

146 Hollinger, M., Abraha, F., and Oscarson, S. (2011). *Carbohydr. Res.* 346: 1454–1466.

147 Manabe, S., Ishii, K., and Ito, Y. (2007). *J. Org. Chem.* 72: 6107–6115.

148 Geng, Y., Zhang, L.-H., and Ye, X.-S. (2008). *Tetrahedron* 64: 4949–4958.

149 Sarkar, S. and Sucheck, S.J. (2011). *Carbohydr. Res.* 346: 393–400.

150 Irvine, J.C. and Earl, J.C. (1922). *J. Chem. Soc., Trans.* 121: 2376–2381.

151 Frontana-Uribe, B.A., Escárcega-Bobadilla, M.V., Juárez-Lagunas, J. et al. (2009). *Synthesis* 980–984.

152 Bergmann, M. and Zervas, L. (1931). *Ber. Dtsch. Chem. Ges.* 64: 975–980.

153 Henderson, A.S., Bower, J.F., and Galan, M.C. (2014). *Org. Biomol. Chem.* 12: 9180–9183.

154 Cunha, A.C., Pereira, L.O.R., de Souza, M.C.B.V., and Ferreira, V.F. (1999). *J. Chem. Educ.* 76: 79.

155 Shen, C., Xia, H., Zheng, H. et al. (2010). *Tetrahedron: Asymmetry* 21: 1936–1941.

156 Gómez-Sánchez, A., García-Martín, M.G., and Gasch, C. (1987). *Carbohydr. Res.* 164: 255–264.

157 Bornaghi, L., Dekany, G., Drinnan, N.B. et al. (2006). *J. Carbohydr. Chem.* 25: 1–18.

158 Dekany, G., Bornaghi, L., Papageorgiou, J., and Taylor, S. (2001). *Tetrahedron Lett.* 42: 3129–3132.

159 Singh, L. and Seifert, J. (2001). *Tetrahedron Lett.* 42: 3133–3136.

160 Alper, P.B., Hung, S.-C., and Wong, C.-H. (1996). *Tetrahedron Lett.* 37: 6029–6032.

161 Goddard-Borger, E.D. and Stick, R.V. (2007). *Org. Lett.* 9: 3797–3800.

162 Chang, K.-L., Zulueta, M.M.L., Lu, X.-A. et al. (2010). *J. Org. Chem.* 75: 7424–7427.

163 Ngoje, G. and Li, Z. (2013). *Org. Biomol. Chem.* 11: 1879–1886.

164 Xing, S. and Gleason, J.L. (2015). *Org. Biomol. Chem.* 13: 1515–1520.

165 Glibstrup, E. and Pedersen, C.M. (2016). *Org. Lett.* 18: 4424–4427.

166 Li, Y., Zhou, Y., Ma, Y., and Li, X. (2011). *Carbohydr. Res.* 346: 1714–1720.

167 Günther, K., Schips, C., and Ziegler, T. (2008). *J. Carbohydr. Chem.* 27: 446–463.

168 Laurent, N., Lafont, D., Boullanger, P., and Mallet, J.M. (2005). *Carbohydr. Res.* 340: 1885–1892.

169 Lindberg, J., Öhberg, L., Garegg, P.J., and Konradsson, P. (2002). *Tetrahedron* 58: 1387–1398.

170 Watanabe, Y., Miura, K., Shiozaki, M. et al. (2001). *Carbohydr. Res.* 332: 257–277.

171 Scaffidi, A., Stubbs, K.A., Dennis, R.J. et al. (2007). *Org. Biomol. Chem.* 5: 3013–3019.

172 Hansen, S.U., Miller, G.J., Baráth, M. et al. (2012). *J. Org. Chem.* 77: 7823–7843.

173 Hattie, M., Debowski, A.W., and Stubbs, K.A. (2012). *ChemBioChem* 13: 1128–1131.

174 Ciobanu, M., Huang, K.-T., Daguer, J.-P. et al. (2011). *Chem. Commun.* 47: 9321–9323.

175 Suárez, J.R., Trastoy, B., Pérez-Ojeda, M.E. et al. (2010). *Adv. Synth. Catal.* 352: 2515–2520.

176 Haller, M.F. and Boons, G.-J. (2002). *Eur. J. Org. Chem.* 2033–2038.

177 Berkin, A., Coxon, B., and Pozsgay, V. (2002). *Chem. Eur. J.* 8: 4424–4433.

178 Paulsen, H., Huffziger, A., and van Boeckel, C.A.A. (1988). *Liebigs Ann. Chem.* 419–426.

179 Tiruchinapally, G., Yin, Z., El-Dakdouki, M. et al. (2011). *Chem. Eur. J.* 17: 10106–10112.

180 Saito, A., Wakao, M., Deguchi, H. et al. (2010). *Tetrahedron* 66: 3951–3962.

181 Wang, Z., Chinoy, Z.S., Ambre, S.G. et al. (2013). *Science* 341: 379.

182 Lemieux, R.U. and Ratcliffe, R.M. (1979). *Can. J. Chem.* 57: 1244–1251.

183 Bukowski, R., Morris, L.M., Woods, R.J., and Weimar, T. (2001). *Eur. J. Org. Chem.* 2697–2705.

184 Liu, M., Young, V.G. Jr., Lohani, S. et al. (2005). *Carbohydr. Res.* 340: 1273–1285.

185 Lee, D.J., Harris, P.W.R., Kowalczyk, R. et al. (2010). *Synthesis* 2010: 763–769.

186 Parry, A.L., Clemson, N.A., Ellis, J. et al. (2013). *J. Am. Chem. Soc.* 135: 9362–9365.

187 Marra, A., Gauffeny, F., and Sinaÿ, P. (1991). *Tetrahedron* 47: 5149–5160.

188 Bosse, F., Marcaurelle, L.A., and Seeberger, P.H. (2002). *J. Org. Chem.* 67: 6659–6670.

189 Marra, A., Dong, X., Petitou, M., and Sinaÿ, P. (1989). *Carbohydr. Res.* 195: 39–50.

190 Kanie, O., Takeda, T., and Ogihara, Y. (1989). *Carbohydr. Res.* 190: 53–64.

191 D'Angelo, K.A. and Taylor, M.S. (2016). *J. Am. Chem. Soc.* 138: 11058–11066.

192 Kinzy, W. and Schmidt, R.R. (1985). *Liebigs Ann. Chem.* 1985: 1537–1545.

193 Mitchell, S.A., Pratt, M.R., Hruby, V.J., and Polt, R. (2001). *J. Org. Chem.* 66: 2327–2342.

194 Bertozzi, C.R. and Bednarski, M.D. (1992). *Tetrahedron Lett.* 33: 3109–3112.

195 Yeung, B.K.S., Hill, D.C., Janicka, M., and Petillo, P.A. (2000). *Org. Lett.* 2: 1279–1282.

196 Seeberger, P.H., Roehrig, S., Schell, P. et al. (2000). *Carbohydr. Res.* 328: 61–69.

197 Saxon, E., Luchansky, S.J., Hang, H.C. et al. (2002). *J. Am. Chem. Soc.* 124: 14893–14902.

198 Zhang, G.-S., Chen, J., Min, J.-M., and Zhang, L.-H. (2003). *Helv. Chim. Acta* 86: 2073–2081.

199 Lafont, D. and Descotes, G. (1987). *Carbohydr. Res.* 166: 195–209.

200 Lafont, D. and Descotes, G. (1988). *Carbohydr. Res.* 175: 35–48.

201 Peng, P., Liu, H., Gong, J. et al. (2014). *Chem. Sci.* 5: 3634–3639.

202 Griffith, D.A. and Danishefsky, S.J. (1990). *J. Am. Chem. Soc.* 112: 5811–5819.

203 Danishefsky, S.J., Gervay, J., Peterson, J.M. et al. (1995). *J. Am. Chem. Soc.* 117: 1940–1953.
204 Kim, H.M., Kim, I.J., and Danishefsky, S.J. (2001). *J. Am. Chem. Soc.* 123: 35–48.
205 Spassova, M.K., Bornmann, W.G., Ragupathi, G. et al. (2005). *J. Org. Chem.* 70: 3383–3395.
206 Dudkin, V.Y., Miller, J.S., Dudkina, A.S. et al. (2008). *J. Am. Chem. Soc.* 130: 13598–13607.
207 Wang, Z.-G., Warren, J.D., Dudkin, V.Y. et al. (2006). *Tetrahedron* 62: 4954–4978.
208 Wu, B., Hua, Z., Warren, J.D. et al. (2006). *Tetrahedron Lett.* 47: 5577–5579.
209 Murakami, T., Sato, Y., Yoshioka, K., and Tanaka, M. (2014). *RSC Adv.* 4: 21584–21587.
210 Mirabella, S., Cardona, F., and Goti, A. (2016). *Org. Biomol. Chem.* 14: 5186–5204.
211 Hernández-Torres, J.M., Liew, S.-T., Achkar, J., and Wei, A. (2002). *Synthesis* 2002: 487–490.
212 Mukherjee, C., Liu, L., and Pohl, N.L.B. (2014). *Adv. Synth. Catal.* 356: 2247–2256.
213 Ehret, M.S., Virlouvet, M., and Biskup, M.B. (2010). *Synthesis* 2010: 3481–3485.

7

One-pot Multistep Regioselective Protection of Carbohydrates Catalyzed by Acids

Jean-Marie Beau[1,2], Yann Bourdreux[1], Guillaume Despras[1], Alexandra Gouasmat[1], Géraldine San Jose[1], Dominique Urban[1], and Boris Vauzeilles[1,2]

[1] *Institut de Chimie Moléculaire et des Matériaux d'Orsay, Laboratoire de Synthèse de Biomolécules, Univ. Paris-Sud and CNRS, Université Paris-Saclay, 91405 Orsay, France*
[2] *Institut de Chimie des Substances Naturelles, CNRS UPR2301, Univ. Paris-Sud, Université Paris-Saclay, 91198 Gif-sur-Yvette, France*

7.1 Introduction

Ongoing developments in glycobiology, a field revealing the biological functions of carbohydrates, have stimulated the steady search for rational chemical transformations and tactics that provide oligosaccharides, glycoconjugates, or functional glycoprobes for investigation in biological research. Because of the multifunctional nature of carbohydrates (mostly hydroxyl and amino groups), any synthetic project using mono- or oligosaccharidic starting substrates necessitates the regioselective differentiation of these groups to yield the required building blocks for a selective synthetic construction. In carbohydrate chemistry, this regiodifferentiation benefits from a long-established knowledge using step-by-step selective protections relying on different hydroxyl reactivities influenced by chiral, steric, and electronic environments.

To improve the efficiency of the overall preparation of appropriate building blocks, these well-established individual procedures may also be combined in more atom-economical transformations, avoiding time-consuming protection/deprotection steps and the isolation of intermediates. In the search for such one-pot procedures, the selected experimental conditions should be conserved as much as possible throughout the transformations in tandem. Tandem (also called domino or cascade) chemical operations [1], in which a single product of a higher structural complexity is formed from very simple building blocks, are essential concerns in current organic chemistry. Further, the sequence of one-pot transformations may be preferably induced by catalytic reagents. Most of the procedures reported below are defined as auto-tandem catalysis [2], a process that refers to the use of a single catalyst promoting the sequential transformation

Protecting Groups: Strategies and Applications in Carbohydrate Chemistry, First Edition.
Edited by Sébastien Vidal.

of a substrate via two or more mechanistically distinct reactions in a single vessel.

In this chapter, it is not our intention to review all methods installing orthogonal sugar protecting groups using one-pot procedures. In addition, we will not consider one-pot multistep protocols including protection group manipulations combined with glycosylation toward the assembly of oligosaccharides. The discussion is illustrated by a short addition of synthetic procedures that cover a few transformations that have actually been carried out in our laboratories.

7.2 Examples of Early Developments of the One-pot Multistep Regioselective Hydroxyl Protection of Carbohydrates

One-pot protection of monosaccharides relying on the reactivity differences of the hydroxyl groups was, in fact, utilized long ago. One of the first illustrations of one-pot procedures of which we are aware of is the regioselective primary hydroxyl protection using basic catalysis conditions reported in the twenties (1923–1927) by Helferich et al. [3, 4]. When β-methyl-D-glucopyranoside **1** (XR = OMe) is successively treated with trityl chloride and acetic anhydride in pyridine, the selectively tritylated fully protected glucoside **2** is formed in a 65% yield of crystalline material (Scheme 7.1) [3b]. The regioselective protection also works well in acyclic systems [5], as shown with the D-galactose-dithioacetal **4** providing the selectively protected polyol **5** [5a]. This relies on a steric discrimination of the primary relative to the secondary hydroxyl groups, noticeable with hindered electrophiles such as trityl chloride.

This one-pot formation of differentially protected pyranosides was favorably updated by Kong and coworkers in 2000 to a more elaborated one-pot three-step sequence of tritylation, *tert*-butyldimethylsilylation, and acetylation of β-thiophenyl-D-glucopyranoside **1** (XR = SPh) [6]. This leads to the efficient preparation of fully protected glucoside **3**, the regioselectivity of the silylation step at O-3 being controlled by the β-orientation of the anomeric oxygen. Under the same conditions, the anomeric α-substrate **6** rather provides the silylated product at O-2 **9** in high yield. This method is also effective for other sugar derivatives, such as the manno- and galacto-pyranosides **7** and **8**, leading to building blocks **10** and **11** useful for oligosaccharide synthesis [6].

We do not record any "historical" examples of one-pot multistep methods induced by acids. More recently, the work of Field, using iodine as an electrophilic activator or reagent in the one-pot three-step synthesis of acetylated thioglycosides **12** from the unprotected reducing sugars is noteworthy [7]. Iodine first catalyzes the solvent-free per-*O*-acetylation of sugars that are further transformed, in the presence of hexamethyldisilane (hmds, an *in situ* equivalent of iodotrimethylsilane with iodine [8]), to the acetylated glycosyl iodides. In a one-pot manner, addition of a thiol (or dimethyl disulfide, via the probable formation of TMS-SCH_3) to the same reaction vessel stereoselectively provides the corresponding thioglycosides **12** in a good overall yield. The procedure is suitable for

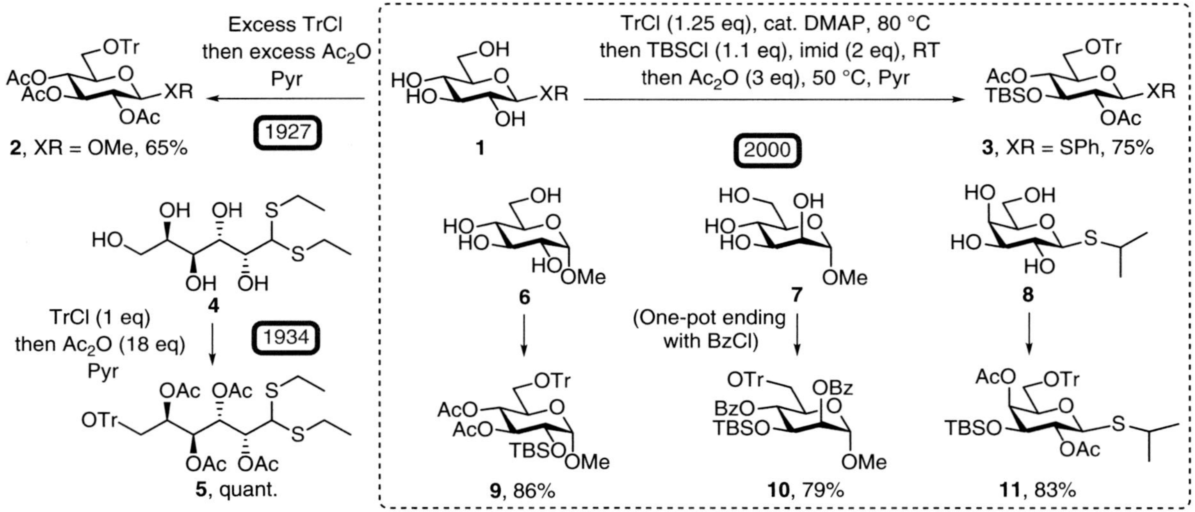

Scheme 7.1 One-pot preparation of fully protected glycopyranosides according to Helferich (1923–1927), Micheel or Wolfrom (1934), and Kong (2000).

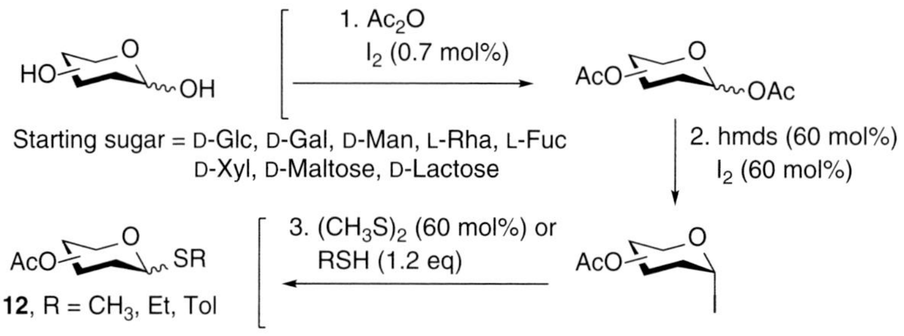

Scheme 7.2 One-pot preparation of per-*O*-acetylated thioglycosides from the unprotected reducing sugars.

larger scale preparations as with D-glucose, performed at a 100 mmol scale (Scheme 7.2).

The more recent advances in protic or Lewis acid catalyzed methodologies developed in organic synthesis stimulated the design and development of one-pot multistep methods for the regioselective installation of an orthogonal set of protecting groups on carbohydrates [9]. These approaches will be detailed in the following sections.

7.3 One-pot Multistep Methods from Silylated Substrates

Triflate salts and trimethylsilyltriflate are commonly used in carbohydrate chemistry to catalyze acetalation [10–13], reductive etherification [14, 15], regioselective reductive opening of arylidene acetals [16–19], and acylation [20–26] on polyol templates. They are also often utilized as catalysts or promoters in glycosylation reactions. By fine-tuning the experimental conditions to make all reactions well matched in the same vessel, a one-pot multistep tandem procedure can then be planned. The thermodynamic product stability and kinetic selectivity of the transformations on a given carbohydrate substrate would guide the regioselective functionalization.

7.3.1 Catalysis by Copper Triflate

The above conditions can be fulfilled using copper triflate as a catalyst [27]. For efficiency, the per-*O*-trimethylsilylated pyranoside derivatives are the starting substrates. Indeed, in most of the following procedures, the one-pot transformations are performed on the per-*O*-TMS starting carbohydrates. The use of the unprotected carbohydrates is not successful or efficient, partly as a consequence of their very low solubility. The per-*O*-TMS derivatives **14** are prepared in high yields from the corresponding glycopyranosides **13** using TMSCl and bases (usually pyridine or triethylamine) at room temperature and are routinely used without purification (Scheme 7.3). They can also be prepared by treatment with hexamethyldisilazane (HMDS, 2 eq.) and catalytic TMSOTf (10 mol%) in CH_2Cl_2

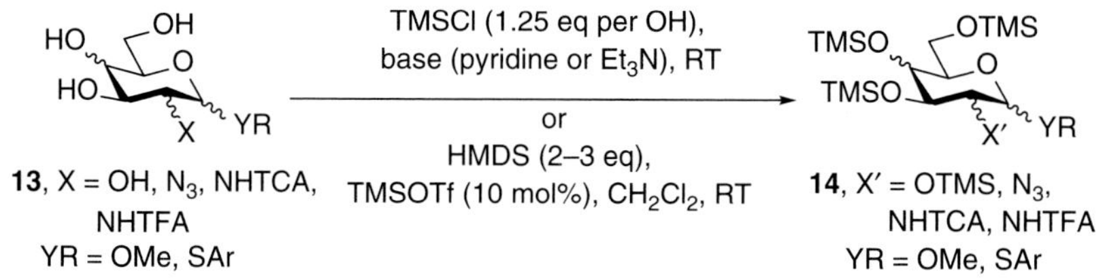

Scheme 7.3 Per-*O*-trimethylsilylation of the glycopyranosides for the one-pot regioselective protection.

at room temperature [28]. Interestingly, these last silylating conditions have been efficiently integrated as the first step in most of the one-pot protections catalyzed by TMSOTf that will be further discussed below [28].

The treatment of the per-*O*-trimethylsilylated α-methyl glucopyranoside **16** as the substrate sugar under the catalysis of $Cu(OTf)_2$ (1 mol%) at room temperature with an excess of benzaldehyde (3 eq.) and Et_3SiH (1.1 eq.) in a CH_2Cl_2/CH_3CN solvent mixture results in the clean formation of a 4,6-*O*-benzylidene-3-*O*-benzyl ether **18** in 78% yield (Scheme 7.4). The procedure may be used with thioaryl substrates, at 0 °C (as for the preparation of **17**), and carried out with further steps toward the required target products [27].

As will be discussed in the following sections, the "tandem" process can be extended further, as is done here, by sequential addition of reagents activated under copper triflate catalysis. Thus, further acylation is possible, providing the corresponding fully orthogonally protected carbohydrates **19** and **20** (Scheme 7.4), or additional regioselective ring opening of the 4,6-*O*-benzylidene ring in the presence of the $BH_3{\cdot}THF$ complex affords highly selectively the 3,4-di-*O*-benzyl derivatives **21** and **22**. Switching to the silane-promoted reductive cleavage in CH_3CN [18] furnishes regioisomeric 3,6-di-*O*-benzyl-D-glucopyranoside **23**.

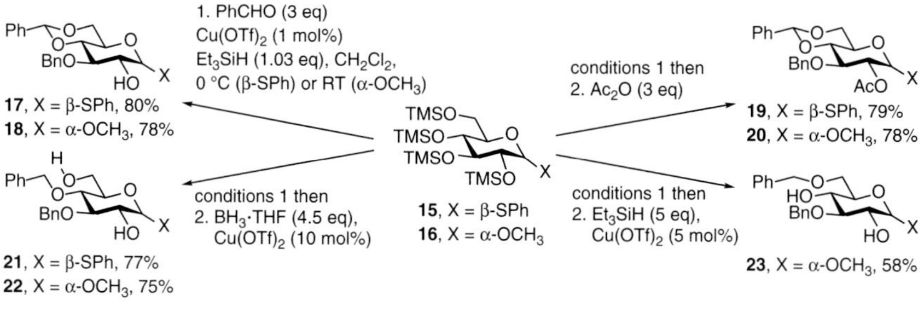

Scheme 7.4 Copper(II) triflate-catalyzed one-pot multistep regioselective protection of per-*O*-trimethylsilylated-D-glucopyranosides **15, 16.**

In the course of this tandem catalysis by copper triflate, the first reaction with benzaldehyde (aryl aldehyde) and the 4,6-*O*-TMS motif of the pyranoside readily provides the 4,6-*O*-benzylidene (arylmethylidene) **A** and TMSOTMS (Noyori acetalation [10a], Scheme 7.5). With a second equivalent of benzaldehyde

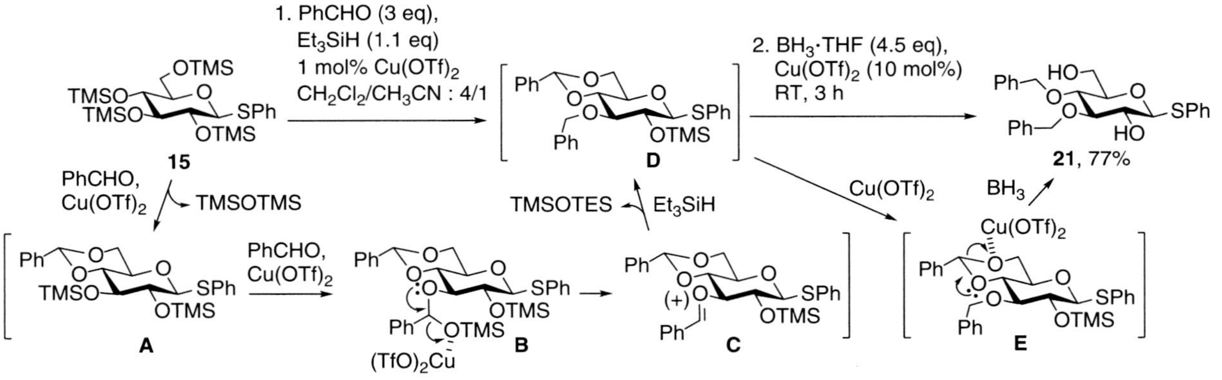

Scheme 7.5 Probable mechanism for the one-pot multistep preparation of the 3,4-di-*O*-benzyl-protected monosaccharide **21**.

(arylaldehyde), the catalyst then likely promotes the regioselective formation of the open mixed acetal at O-3 **B**, activated by the catalyst to oxonium **C**, which is reduced by Et_3SiH to benzyl (arylmethyl) ether **D** and TMSOTES (reductive etherification [14]). Copper triflate may assist further transformations in the same pot, as in the example chosen in Scheme 7.5, promoting the regioselective benzylidene (arylmethylidene) ring opening at O-6 by BH_3 (intermediate **E**, [18]) affording the 2,6-diol **21**. Three catalytic cycles are in operation for this specific example.

7.3.2 Catalysis by Trimethylsilyltriflate

7.3.2.1 Glucopyranosides

The above transformations are also efficiently performed by trimethylsilyltriflate catalysis [29]. Although $Cu(OTf)_2$ catalysis performs well at either 0 °C or room temperature depending on the substrate, the effectiveness of TMSOTf necessitates the catalysis to be achieved at −86 °C (obtained from liquid nitrogen in methyl ethyl ketone). In addition, larger amounts of catalyst are needed as in the preparation of the fully protected monosaccharides **25** and **26** which, at the end of the sequential one-pot transformation, needs 63 mol% of the promoter (Scheme 7.6).

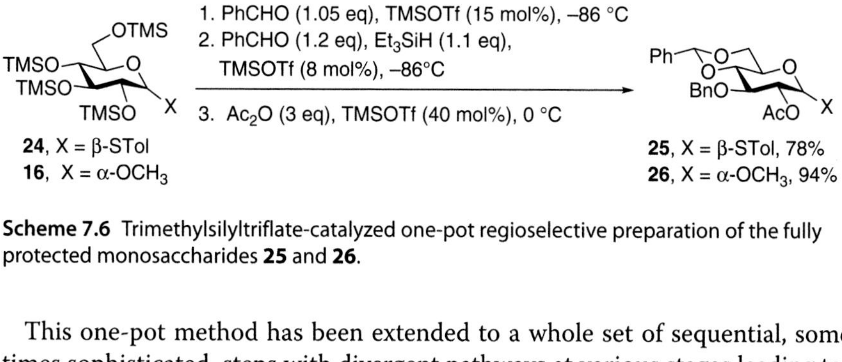

24, X = β-STol
16, X = α-OCH$_3$

25, X = β-STol, 78%
26, X = α-OCH$_3$, 94%

Scheme 7.6 Trimethylsilyltriflate-catalyzed one-pot regioselective preparation of the fully protected monosaccharides **25** and **26**.

This one-pot method has been extended to a whole set of sequential, sometimes sophisticated, steps with divergent pathways at various stages leading to a very large variety of protecting patterns [29]. A few examples of these transformations are depicted below (Scheme 7.7).

After the usual one-pot three steps (steps 1–3, Scheme 7.7), the 4,6-*O*-benzylidene motif can again be regioselectively opened at O-6 or O-4 (step 4 to alcohols **27** or **28**), and further addition of dichloro–dicyano–benzoquinone (DDQ, step 5) leads to the corresponding 3,6- or 3,4-diol **29** or **30**, respectively. In addition, greater diversity is gained when the one-pot acylation/alkylation step is performed after desilylation (TBAF) of the 2-OH and treatment under basic conditions with the acylating/alkylating agent (not shown, [29]).

7.3.2.2 Mannopyranosides

In the *glucopyrano* series, the advantage of the above one-pot multistep methods relies on the easy formation of the 4,6-*O*-benzylidene (arylmethylidene) alone because of the 2,3-*trans* configuration at C-2 and C-3 on the pyranose ring. With

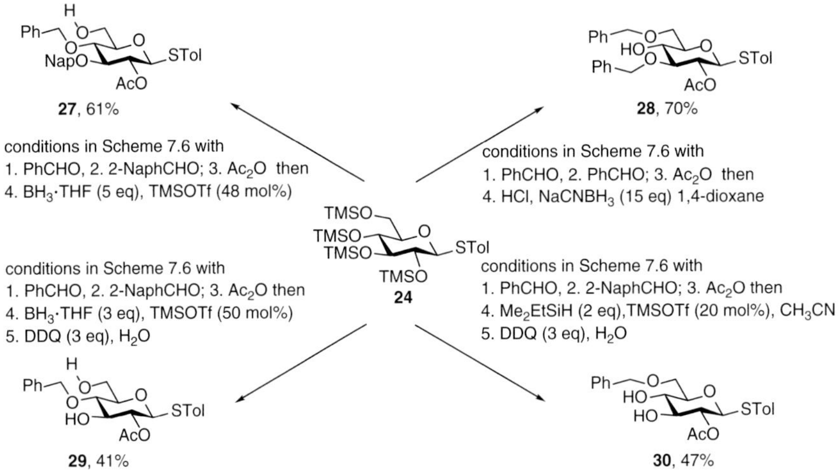

Scheme 7.7 Examples of trimethylsilyltriflate-catalyzed one-pot regioselective protection starting from per-*O*-trimethylsilylated 4-methylphenyl 1-thio-β-D-glucopyranoside **24**.

the 2,3-*cis* configuration in the *mannopyrano* series, the preferred 4,6-*O*-arylmethylidene formation often occurs with the formation of the endo- and exo-2,3-*O*-arylmethylidene cyclic acetals [30]. This makes a further one-pot modification more difficult to control. It is nevertheless possible to cleanly produce the 4,6-*O*-benzylidene only [31], and this is a requirement for further productive one-pot transformations.

Starting from the per-*O*-silylated 4-methylphenyl-1-thio-α-D-mannopyranoside **31**, acetalation conditions catalyzed by TMSOTf are available for cleanly preparing either the 4,6-*O*-mono-benzylidene derivative **32** regioselectively (1 eq. of benzaldehyde, 5 mol% of TMSOTf at −78 °C in dichloromethane, 92% yield) or, stereoselectively, the 2,3-exo-4,6-*O*-di-benzylidene **33** (2.1 eq. of benzaldehyde, 10 mol% of TMSOTf at 0 °C in acetonitrile, 91% yield) (Scheme 7.8) [32].

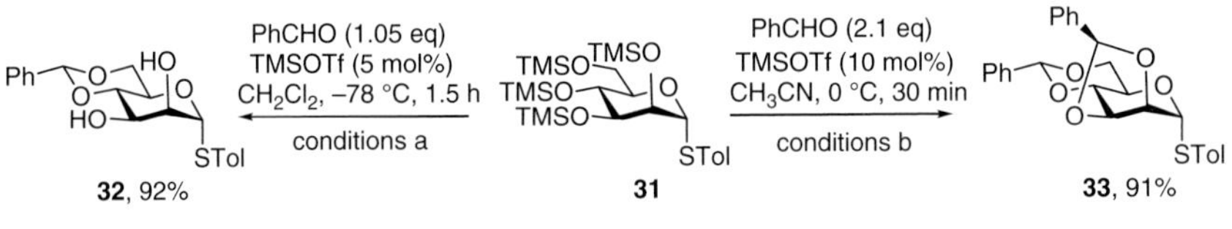

Scheme 7.8 Trimethylsilyltriflate-catalyzed benzylidenation of per-*O*-trimethylsilylated 4-methylphenyl 1-thio-α-D-mannopyranoside **31**.

Further applications of these conditions in a one-pot multistep protection procedure are then able to selectively provide the 2-, 3-, 4-, and 6-OH-protected D-mannopyranosides **34–37** (Scheme 7.9).

Thus, sequential TMSOTf-catalyzed monobenzylidenation and diacetylation followed by regioselective benzylidene ring opening with borane–tetrahydrofuran furnishes the 6-OH derivative **34**, while opening with dimethylethylsilane

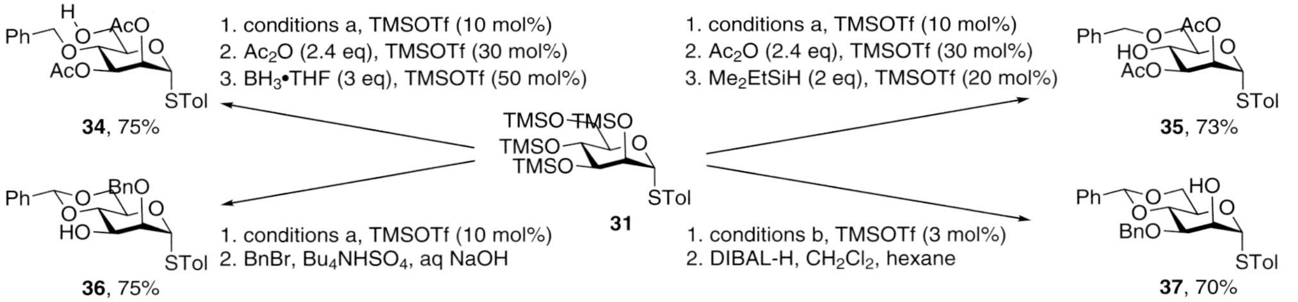

Scheme 7.9 Examples of trimethylsilyltriflate-catalyzed one-pot regioselective protection starting from per-*O*-trimethylsilylated 4-methylphenyl 1-thio-α-D-mannopyranoside **31**.

provides the 4-OH derivative **35**. The 3-OH derivative **36** is available by a subsequent regioselective benzylation using phase-transfer conditions [33] of the 4,6-*O*-benzylidene intermediate. The 2-OH derivative **37** is rather ingeniously derived from the intermediary 2,3-exo-4,6-*O*-dibenzylidene **33** by a regioselective opening of only the exo-5-membered acetal in **33** by diisobutylaluminum hydride (Scheme 7.9).

7.3.3 One-pot Catalysis with Other Glycopyranosides

The above one-pot processes may also be applied, using modified versions, to 2-azido (or 2-acetamido)-2-deoxy-D-glucose, D-xylose, and D-galactose. Similar successive additions of reagents on 2-azido-2-deoxy-1,3,4,6-tetra-*O*-trimethylsilyl-D-glucopyranose **38** regioselectively yield the partially or fully protected derivatives [34]. After the usual TMSOTf-catalyzed acetalation, the one-pot acetylation to the fully protected azido sugar **39** proceeds properly only with a large excess of acetic anhydride (10 eq.) in the presence 20 mol% of $Cu(OTf)_2$ (Scheme 7.10). A further one-pot reductive opening to the 4-OH derivative **40** by Me_2EtSiH requires a decrease in the amount of the preceding acylation reagent to 2.5 eq., in order to avoid acetylation at O-4 *in situ*.

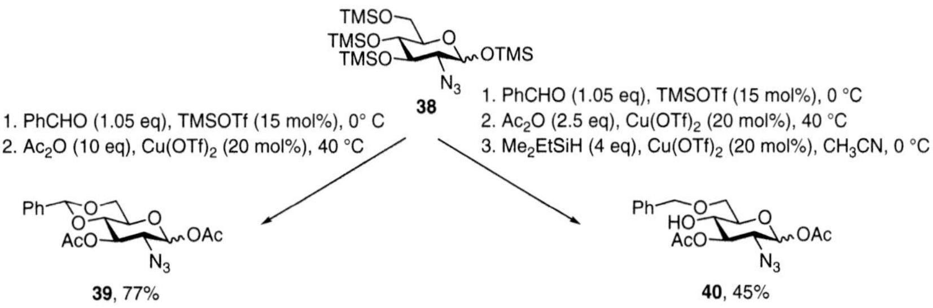

Scheme 7.10 One-pot regioselective protection starting from the per-*O*-silylated 2-azido-2-deoxy-D-glucopyranose **38**.

Instead of changing the promoter in the course of the one-pot protocol, a tandem catalysis with copper(II) triflate works well during the entire process. Starting from the phenyl 1-thio-β-D-glucosamine derivatives **41** and **42**, catalysis of the acetalation/acetylation steps proceeds properly with the azido substrate providing **43**, using only 3 mol% of $Cu(OTf)_2$ ([35], Scheme 7.11).

Using benzaldehyde for the acetalation step, a stoichiometric amount of the copper salt is required for completion of the reaction with the acetamido substrate **42**. Notwithstanding, using benzaldehyde dimethyl acetal enables a clean formation of **44** in a 85% yield under catalytic conditions (Scheme 7.11). In the azido series, extension of the copper catalysis to a three-step sequence ending with a selective reductive opening of the benzylidene ring provides the 4-OH derivative **45**.

In the recent assembly of the common β-D-Glc*p*A-(1→3)-β-D-Gal*p*-(1→3)-β-D-Gal*p*-(1→4)-β-D-Xyl*p* tetrasaccharide from the linkage region of proteoglycans,

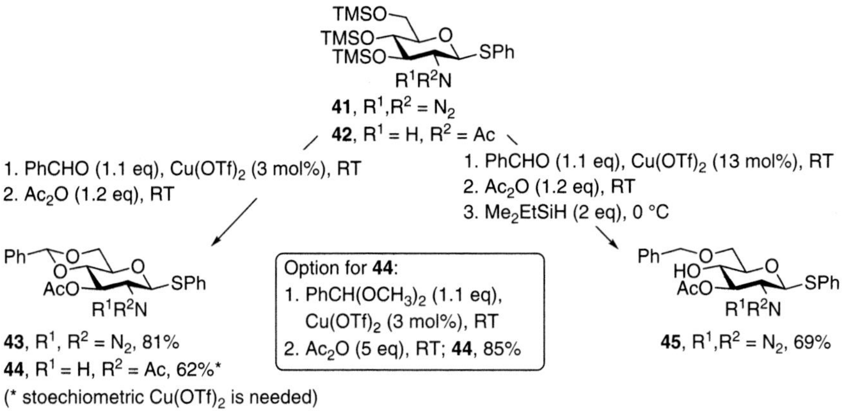

Scheme 7.11 One-pot regioselective protection starting from the per-*O*-silylated 2-azido (or 2-acetamido)-2-deoxy-β-D-glucopyranosides **41** and **42**.

some of the necessary monomer building blocks are also prepared applying similar regioselective one-pot protection approaches from the silylated substrates. Thus, the differentiation of the three hydroxyl groups of a D-xylopyranoside is possible starting from the persilylated thio-β-D-xylopyranoside **46** by a regioselective TMSOTf-catalyzed reductive benzylation at O-3 by Et_3SiH followed, in the same pot, by a regioselective $Yb(OTf)_3$-catalyzed benzoylation at O-2 providing the required 4-OH derivative **47** in a 45% yield (Scheme 7.12, [36]). The selective benzoylation may result from a coordination of the ytterbium ion to both the O-2 and the O-4 oxygen atoms, amplifying the reaction at the more nucleophilic O-2 position [36]. A lower than usual yield is explained by the formation of the minor regioisomers during both the steps.

The silylated thiogalactoside is first prepared by silylation of the primary hydroxyl in **48** with a TBDPS group and then trimethylsilylation of the remaining hydroxyl groups. The one-pot process that follows consists in a TMSOTf-catalyzed acetalation at the *cis*-diol, a benzoylation at O-2, and a deprotection at O-6 yielding the galactoside unit **50** (Scheme 7.12).

7.3.4 Catalysis by Iron(III) Chloride Hexahydrate

Iron salts and complexes have recently attracted much attention as environmentally friendly and inexpensive agents in a wide range of selective processes in organic synthesis [37]. Notably, iron(III) chloride hexahydrate represents an easy-to-handle, cheap, and convenient catalyst or promoter in many organic transformations. This simple hydrated salt, mostly applied for hydrolytic reactions [38–40], is also an effective substitute for programming the tandem transformations under the anhydrous conditions reported above. The anhydrous salt was previously reported for inducing acetal formation, reductive etherification, and acylation taken as single reactions [41–43]. The hexahydrate complex can also catalyze all these reactions in the same vessel, as well as the regioselective reductive opening of arylidenes [44].

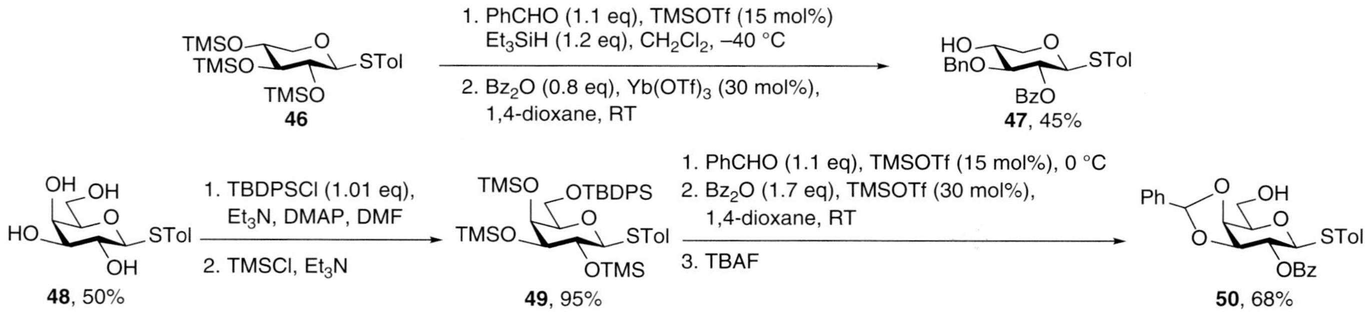

Scheme 7.12 Trimethylsilyltriflate-catalyzed one-pot regioselective protection starting from the per-*O*-silylated thio-β-D-xylo- and D-galacto- derivatives.

Thus, as mentioned above with $Cu(OTf)_2$ or TMSOTf, treatment of per-*O*-silylated β-thiophenyl or α-methyl-D-glucopyranoside **15** or **16** with benzaldehyde (3 eq.) and triethylsilane (1.1 eq.) in the presence of 5 mol% of $FeCl_3{\cdot}6H_2O$ in a CH_2Cl_2/CH_3CN solvent mixture led to 3-*O*-benzyl-4,6-*O*-benzylidene-D-glucopyranoside **17** or **18** (see Scheme 7.4) in a 77% or 75% yield, respectively [44a]. Anhydrous iron(III) chloride is as effective as catalyst, but other iron salts such as $Fe(acac)_3$, $(FeCl_3)_2(TMEDA)_3$, $Fe(NO_3)_3{\cdot}9H_2O$, or $FeCl_2{\cdot}4H_2O$ furnish either lower yields or are inefficient. Two examples of three-step one-pot transformations catalyzed by $FeCl_3{\cdot}6H_2O$ and leading to ester **51** or diol **52** are given in Scheme 7.13.

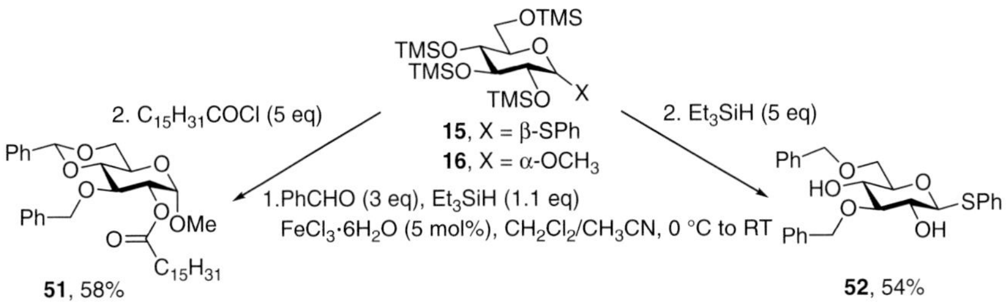

Scheme 7.13 Two examples of one-pot three-step transformations catalyzed by iron(III) chloride hexahydrate.

This catalytic option is not air sensitive, and the one-pot procedures can be easily performed on a multigram scale without an impact on the isolated yields. A real improvement over $Cu(OTf)_2$ is advantageous with oligosaccharidic substrates [44]. Hence, one-pot treatment of C_2-symmetric per-*O*-silylated-α,α-D-trehalose **53** (6 eq. of PhCHO, 2.2 eq. of Et_3SiH, and 5 mol% of $FeCl_3{\cdot}6H_2O$) provides the symmetric benzylated compound **54**, isolated in 61% yield (Scheme 7.14). This procedure can also be prolonged to a three-step process to obtain totally protected disaccharide **55** (Scheme 7.14). Thus, adding an excess of acetic anhydride (10 eq.) and 5 mol% of $FeCl_3{\cdot}6H_2O$ furnished the expected 2,2′-di-*O*-acetylated-α,α-D-trehalose **55** in an isolated 41% yield.

The shortcoming of this method is perceived with the one-pot bis-reductive benzylidene ring opening (10 eq. of Et_3SiH and 15 mol% of the catalyst are required) affording the expected compound **56** in only a moderate overall yield of 28%. The $FeCl_3{\cdot}6H_2O$-catalyzed tandem protocol is also successful with a nonsymmetrical disaccharide such as methyl α-maltoside **57**, converted to the monobenzylidene bis-benzyl derivative **58** (15 mol% of $FeCl_3{\cdot}6H_2O$, 51% yield) [44a].

Finally, a much-optimized one-pot regioselective protection of methyl β-maltotrioside **59** provides the expected 4,6-*O*-benzylidene-3-*O*-benzyl protecting pattern of the nonreducing unit. On the other two units, benzylation of the two primary hydroxyls and a selective 2-*O*-benzylation of the reducing unit are achieved, to selectively reach a tetraol structure **60** in a 45% yield (Scheme 7.15) [45].

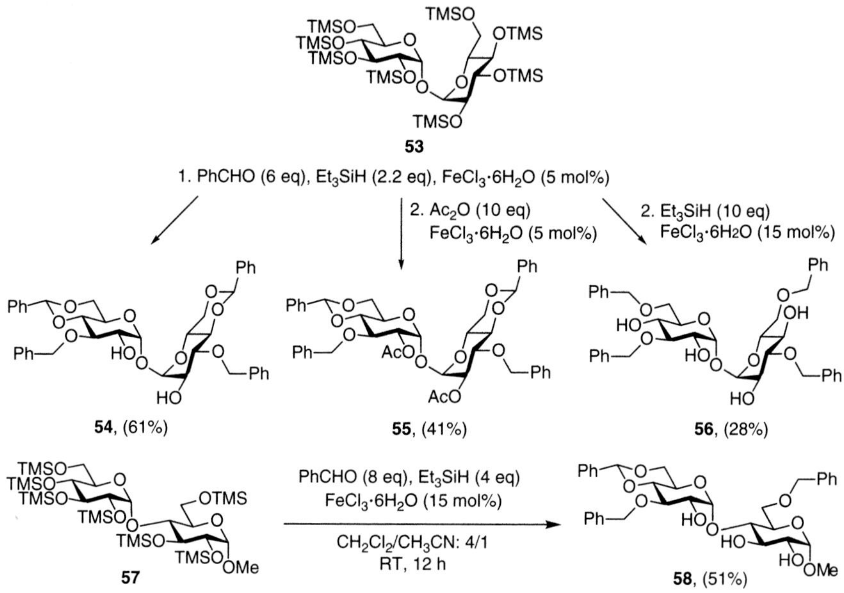

Scheme 7.14 Tandem iron(III) chloride hexahydrate-catalyzed regioselective protection of silylated α,α-D-trehalose **53** and methyl α-maltoside **57**.

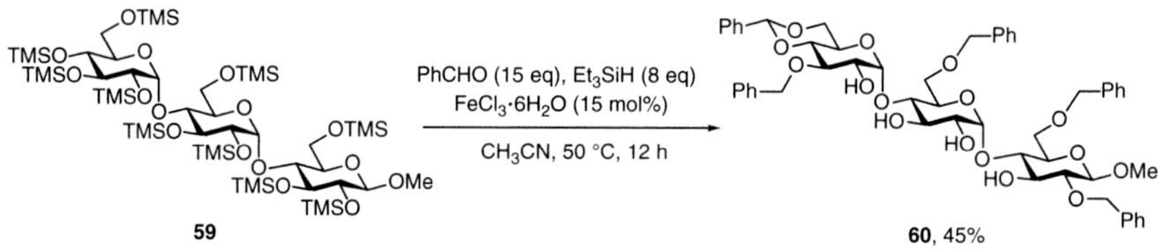

Scheme 7.15 Tandem iron(III) chloride hexahydrate-catalyzed regioselective protection of per-*O*-silylated methyl β-maltotrioside **59**.

7.3.5 Catalysis by Triflic Acid on Molecular Sieves

Glucosamine is a widely distributed and key sugar in many important oligosaccharides and glycoconjugates [46]. The implementation of one-pot processes to generate useful building blocks in a simple way is more challenging because of the amino function, and the above procedures, promoted by $Cu(OTf)_2$ or $FeCl_3·6H_2O$, present some limitations with these substrates [35, 47]. The major problem is that the metallic catalysts are likely to be captured by some of the *NHCO*-containing functionalities on carbohydrates, making the catalysis ineffective.

A solution is to use as a catalyst trifluoromethanesulfonic acid (triflic acid, TfOH) associated with molecular sieves. Activation of organic substrates by TfOH has been extensively studied [48]. With carbohydrate substrates, TfOH is particularly used in glycosylations [49] or, when associated with alkylsilanes, it promotes 4,6-*O*-benzylidene reductive opening to the C-4 alcohol [49b, 50].

A one-pot multistep transformation catalyzed by TfOH is a good system provided that powdered molecular sieves are added to the reaction mixture. For example, treatment of the 2-deoxy-2-trichloroacetamido-3,4,6-tri-*O*-trimethylsilyl-1-(2-methyl-5-*tert*-butylphenyl)-thio-β-D-glucopyranoside **61** (R = TCA, X = β-SMbp) under the catalysis of TfOH (5 mol%) and 3 Å molecular sieves (1 g g^{-1} of substrate) at room temperature with benzaldehyde (3 eq.) and Et_3SiH (1.2 eq.) in CH_2Cl_2 yields quickly (10 min) and cleanly the 4,6-*O*-benzylidene-3-*O*-benzyl derivative **62a** in 83% yield (Scheme 7.16) [51]. Various glucosamine units, displaying different N-protecting group and glycoside patterns, are obtained with efficiency in a single flask using the above reaction conditions.

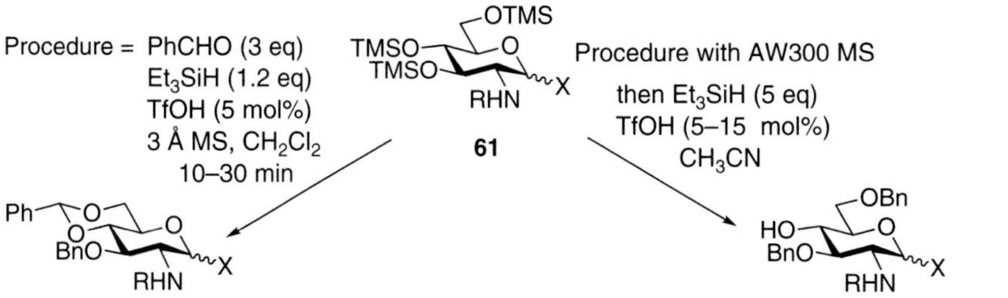

62a, R = TCA, X = β-SMbp, 83%
62b, R = TFA, X = β-SMbp, 85%
62c, R = TFA, X = β-OTBS, 92%
62d, R = Troc, X = β-SMbp, 90%
62e, R = Alloc, X = β-SMbp, 93%
62f, R = CO_2CH_3, X = β-SMbp, 79%

63a, R = TCA, X = β-SMbp, 72%
63b, R = TFA, X = α-OCH_3, 67%
63c, R = TFA, X = β-SMbp, 72%
63d, R = Troc, X = β-SMbp, 67%
63e, R = Alloc, X = β-SMbp, 60%
63f, R = CO_2Me, X = β-SMbp, 49%

Scheme 7.16 Triflic acid on molecular sieves-catalyzed one-pot regioselective protection of tri-*O*-trimethylsilylated glucosamine derivatives **61**.

The catalyst used is triflic acid associated with molecular sieves and may work as a heterogeneous acid catalysis. As shown by ^{1}H- and ^{19}F-NMR, triflic acid is not free in solution but is adsorbed on the surface of the sieves to provide an apparently better performing solid acid catalyst. Adsorbed TfOH on molecular sieves better represents this catalytic system rather than "supported" because there is no evidence that TfOH forms a stable solid acid catalyst with molecular sieves as previously shown with amorphous silica [52].

A further one-pot step of regioselective benzylidene ring opening gives good results after another fine-tuning of the reaction conditions (mol% of catalyst, solvent, reaction time, and temperature) optimized for every specific substrate (4-OH glucosamine derivatives **63**, Scheme 7.16) [51]. Notably, CH_3CN as a cosolvent and acid-washed molecular sieves AW300 are needed for the efficiency of this last step. Excellent results are also obtained using the TMSOTf-catalysis (total of 30–70 mol% of TMSOTf) to form the 4-OH compound **63** in a one-pot manner as well as the corresponding 6-OH derivative (87% and 88% yield, respectively), from the 2-deoxy-2-trichloroacetamido-1-thioethyl-β-D-glucopyranoside [28].

7.4 One-pot Multistep Methods Catalyzed by Copper Triflate on Unprotected Sugars

Rather than using the per-*O*-silylated starting substrates, unprotected glycopyranosides may be used as substrates with $Cu(OTf)_2$ catalysis [53]. This is applicable to the same sequential series of tandem transformations including arylmethylidene acetalation, esterification, and regioselective reductive acetal ring opening processes. Thus, phenyl 1-thio-β-glycopyranoside (glucosyl, galactosyl, and NHTroc-glucosaminyl) **64** subjected to the one-pot three-step procedure provided the expected products **65**, **66**, and **67** in the yields of 50–60% (Scheme 7.17). This sequential one-pot transformation is carried out at room temperature and demands a total of 10 mol% of $Cu(OTf)_2$.

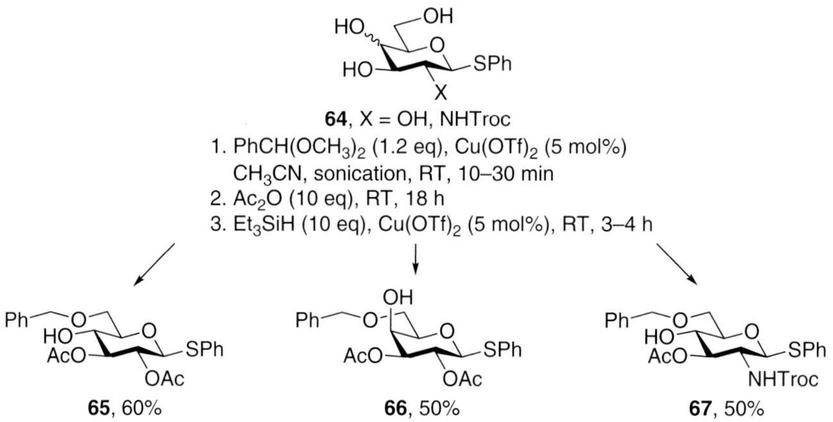

Scheme 7.17 Copper triflate-catalyzed one-pot multistep protection of unprotected D-glycopyranosides **64**.

7.5 Other One-pot Multistep Methods Catalyzed by Acids

The one-pot selective protections of carbohydrates reported above heavily rely on acid-catalyzed reductive etherification processes, from either acetalized diols (benzylidenes or arylmethylidenes) or from an alcohol or a TMS-ether, a much less employed but useful ether synthesis. Some other simpler catalyzed processes may be employed, promoting acetalation and esterification in the same vessel.

Thus, a solvent-free, iodine-catalyzed process with appropriate amounts of isopropenyl acetate on unprotected carbohydrates provides the isopropylidene ketals/acetates in one step [54]. This is a useful reaction with the reducing sugars

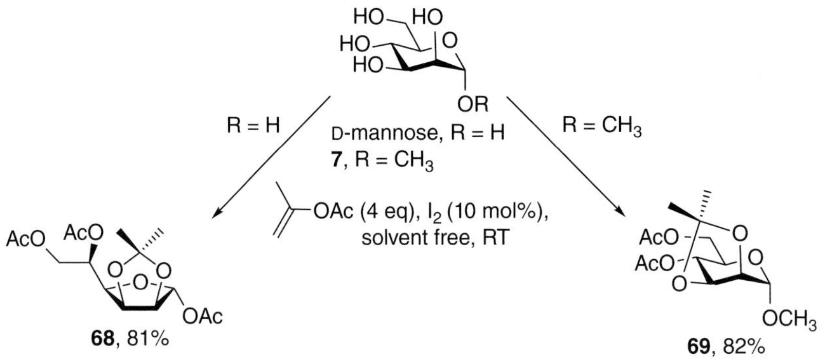

Scheme 7.18 Iodine-catalyzed tandem acetalation–acetylation of unprotected D-mannose.

(such as D-mannose leading to furanoside **68**, Scheme 7.18) or with the glycosides (such as methyl α-D-mannopyranoside **7** leading to pyranoside **69**). This interesting transformation is based on the known ability of molecular iodine to catalyze acetonide formation [55] and acetylation [56]. Molecular iodine may be seen as a mild alternative to many acid catalysts.

The reagent system reacts with the most favorable 1,2-*cis* diol available in the substrate to form the corresponding ketal, and the remaining hydroxyl groups are then acetylated.

Another option is a one-pot benzylidenation–acetylation combination, which can be utilized with sugar derivatives, promoted by solid acids such as $HClO_4$ [57] and H_2SO_4 [58] supported on silica gel (**70–71**, Scheme 7.19). With carbohydrates possessing a 1,2-*cis* diol motif (D-galactopyranoside, L-arabinopyranoside, and L-rhamnopyranoside), these conditions are also appropriate for the formation of isopropylidene ketals/acetates in sequence (**72–73**, Scheme 7.19).

The preparation of protected monosaccharide **75** from *p*-tolyl 1-thio-α-L-rhamnopyranoside **74** is given as an example (Scheme 7.19, [57]). These systems are viable on a larger scale and the H_2SO_4-silica catalyst can be recovered and reused after drying.

Alternatively, *p*-toluenesulfonic acid [59] or iodine [60] may also be used as a catalyst for a one-pot tandem acetalation–esterification reaction, extending the transformation to the thioaryl-glycopyranosides. It furnishes the fully protected glycosides **77** and **79** of D-glucose, D-galactose, D-mannose, or D-glucosamine from **76** to**78**, respectively (Scheme 7.20). With iodine, it is applicable to the formation of benzoate esters and the final addition of catalytic DMAP accelerates the esterification step.

A notable example is the one-pot preparation of the acid-sensitive 4,6-*O*-*p*-methoxybenzylidene acetal 2,3-di-acetate **81** from the unprotected thiogalactoside **80** in a 85% yield (Scheme 7.20, [60]). The TsOH catalysis (5–10 mol%) is also feasible with the isopropylidenation/acetylation protocol of carbohydrate substrates [59].

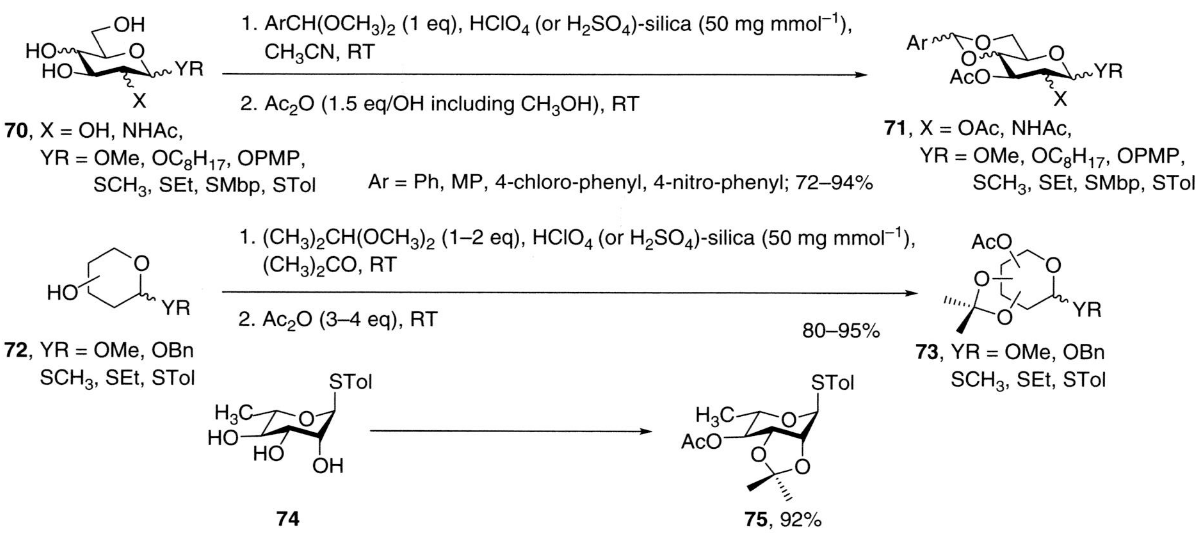

Scheme 7.19 $HClO_4$–silica or H_2SO_4–silica-catalyzed one-pot acetalation–esterification of *O*- and *S*-glycosides.

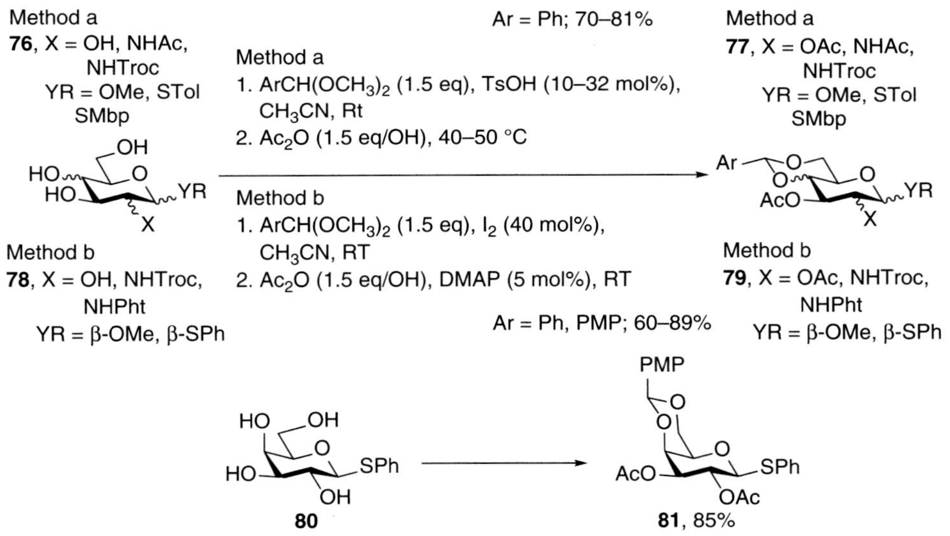

Scheme 7.20 *p*-Toluenesulfonic acid or iodine-catalyzed one-pot acetalation/esterification of *O*- and *S*-glycopyranosides.

7.6 Conclusions and Outlook

Diverse one-pot multistep procedures catalyzed by Lewis and protic acids have been exploited for the full differential protection of the numerous hydroxyl groups as well as the partial masking, with free hydroxyls at the C-2, C-3, C-4, and C-6 positions in the glycopyranosidic rings of carbohydrates. Most of the protocols are recent and result from an improved understanding of how different reagents can be compatible with each other in the same medium and activated by the same catalysts. It delivers straightforward preparations of sugar building blocks of interest for oligosaccharide assembly or for other synthetic projects starting from simple sugars.

An important consequence is an improved efficiency of the overall transformation, thus circumventing the traditional step-by-step protection–deprotection approaches associated with successive workups, as well as time- and solvent-consuming intermediate purifications. These catalyzed protocols also reduce wastes, notably by facilitating the use of stoichiometric reagents or amounts close to stoichiometry. An unavoidable step for success is, however, the fine-tuning of the operating conditions for each particular carbohydrate substrate.

The silylated substrates provide the most flexible routes for differentiating the hydroxyl groups because such routes have been developed using highly regioselective reductive benzylation (arylmethylation) of pyranosides from either benzylidenes (arylmethylidenes) or from isolated trimethylsilyl ethers. These approaches may also find useful applications on furanosides. Moreover, working with unprotected carbohydrate derivatives, the selective complexation by, for instance, transition metal or organic acid catalysts may be the origin for imaginative regioselective methods of activation.

7.7 Experimental Procedures

7.7.1 General Procedure for the Preparation of Compounds 3,9–11 [6]

To a solution of substrate **1,6–8** (1 mmol) in pyridine was added TrCl (1.25 eq.) and a catalytic amount of DMAP. The mixture was stirred at 80 °C for 16 h and then cooled to 0 °C. Imidazole (2 eq.) was added, and finally, TBSCl (1.1 eq.) in DMF was added portion by portion for 2 h. The mixture was stirred at RT overnight. Benzoyl chloride (2.5 eq.) or acetic anhydride (3 eq.) in pyridine was added. The reaction mixture was stirred at 50 °C overnight, poured into ice-cold water, and then extracted with EtOAc. The organic phase was concentrated to dryness with toluene. The residue was subjected to silica gel column chromatography (petroleum ether/EtOAc, 12 : 1) to give the fully protected glycosides **3,9–11**.

7.7.2 Procedure for the Preparation of Phenyl 2-*O*-acetyl-3-*O*-benzyl-4,6-*O*-benzylidene-1-thio-β-D-glucopyranoside 19 [27]

A 80 mM solution of freshly dried copper(II) trifluoromethanesulfonate in acetonitrile (0.50 ml) was added to an ice-cold solution of per-*O*-silylated thioglucoside **15**

(2.23 g, 3.98 mmol) and benzaldehyde (1.21 ml, 11.95 mmol) in dichloromethane (2 ml). Triethylsilane (0.70 ml, 4.38 mmol) was then added, and the solution was concentrated to dryness under argon flow for 30 min. The resulting yellow solid was diluted with dichloromethane (2 ml) and the acetic anhydride (1.13 ml, 11.95 mmol) was added. The mixture was stirred for 1 h at room temperature. A saturated aqueous $NaHCO_3$ solution was added, and the reaction mixture was extracted twice with dichloromethane. The combined organic layers were dried (Na_2SO_4), filtered, and concentrated *in vacuo*. Crystallization in EtOH (70 ml) gave **19** (1.56 g, 79%) as a white solid.

7.7.3 Procedure for the Preparation of 3,3′-di-*O*-benzyl-4,6;4′,6′-di-*O*-benzylidene-α,α-D-trehalose 54 [44a]

To a solution of per-*O*-silylated α,α-D-trehalose **53** (3 g, 3.26 mmol) in dichloromethane (5.8 ml) was added benzaldehyde (1.99 ml, 19.6 mmol, 6 eq.) at room temperature. A solution of $FeCl_3{\cdot}6H_2O$ (1.3 ml, 126 mM in acetonitrile, 5 mol%) and triethylsilane (1.15 ml, 7.2 mmol, 2.2 eq.) was then successively added at 0 °C. The reaction was stirred for 5 h at room temperature and tetra-*n*-butylammonium fluoride was added (~10 ml, 1 M in THF). The mixture was diluted with ethyl acetate and neutralized with a saturated aqueous $NaHCO_3$ solution. The aqueous layer was extracted twice with ethyl acetate. The combined organic layers were dried (Na_2SO_4), filtered, and concentrated under vacuum. The crude product was purified by silica gel column chromatography (cyclohexane/EtOAc, 7 : 3) to give the regioselectively protected derivative **54** in 61% yield (1.38 g). Similar results were provided starting from 52 g of per-*O*-silylated trehalose **53**.

7.7.4 General Procedure for the Tandem Acetalation/Benzylation; Preparation of *tert*-butyldimethylsilyl 3-*O*-benzyl-4,6-*O*-benzylidene-2-deoxy-2-trifluoroacetamido-β-D-glucopyranoside 62c [51]

To a 0.2 M solution of the per-*O*-silylated substrate in dry CH_2Cl_2 were added benzaldehyde (3 eq.) and 3 Å molecular sieves (1 g g^{-1} of substrate). The mixture was stirred at room temperature for 15 min before trifluoromethanesulfonic acid (5 mol%) and triethylsilane (1.2 eq.) were added. After stirring for 10–40 min, triethylamine was added, and the solution was then diluted with CH_2Cl_2, filtered through a celite pad, and washed with water. The aqueous layer was extracted with CH_2Cl_2, and the combined organic layers were dried (Na_2SO_4), filtered, and concentrated. The desired fully protected compound was purified by silica gel column chromatography.

Compound **61** (R = TFA, X = β-OTBS; 1 g, 1.65 mmol at 0 °C for 10 min) provided **62c** (0.866 g, 92%) as white foam after flash chromatography (cyclohexane/EtOAc, 9 : 1 to 4 : 1).

Acknowledgments

The authors thank Rosemary Green Beau for her comments on the manuscript. The authors are grateful to the French Agency for Research (Agence Nationale de la Recherche, ANR) and the Institut Universitaire de France (IUF) for the financial support of this study. The CHARM3AT program is also acknowledged for its support.

Abbreviations

acac	acetylacetonate
Alloc	allyloxycarbonyl
Bn	benzyl
Bz	benzoyl
DDQ	2,3-dichloro-5,6-dicyano-1,4-benzoquinone
DIBALH	diisobutylaluminum hydride
DMAP	4-(dimethylamino)pyridine
DMF	*N,N*-dimethylformamide
Gal*p*	galactopyranose
Glc*p*A	glucopyranuronic acid
hmds	hexamethyldisilane
HMDS	hexamethyldisilazane, (trimethylsilyl)amine
Mbp	2-methyl-5-*tert*-butylphenyl
MS	molecular sieves
Nap	(2-naphthyl)methyl
Naph	naphthyl
Pht	phthaloyl
PMB	4-methoxybenzyl
PMP	4-methoxyphenyl
Pyr	pyridine
RT	room temperature
SE	2-trimethylsilylethyl
TBAF	tetrabutylammonium fluoride
TBDPS	*t*-butyldiphenylsilyl
TBS	*t*-butyldimethylsilyl
TCA	trichloroacetyl
TES	triethylsilyl
Tf	trifluoromethylsulfonyl
TFA	trifluoroacetyl
TMEDA	*N,N,N',N'*,-tetramethylethylenediamine
TMS	trimethylsilyl
Tol	tolyl, toluenyl
Tr	triphenylmethyl
Troc	2,2,2-trichloroethyloxycarbonyl
Ts	*p*-toluenesulfonyl

References

1 Selected reviews: (a) Tietze, L.F. (1996). *Chem. Rev.* 96: 115–136. (b) Denmark, S.E. and Thorarensen, A. (1996). *Chem. Rev.* 96: 137–165. (c) Parsons, P.J.,

Penkett, C.S., and Shell, A.J. (1996). *Chem. Rev.* 96: 195–206. (d) Wasilke, J.-C., Obrey, S.J., Baker, R.T., and Bazan, G.C. (2005). *Chem. Rev.* 105: 1001–1020. (e) Nicolaou, K.C., Edmonds, D.J., and Bulger, P.G. (2006). *Angew. Chem. Int. Ed.* 45: 7134–7186. (f) Enders, D., Grondal, C., and Hüttl, M.R.M. (2007). *Angew. Chem. Int. Ed.* 46: 1570–1581. (g) Pellissier, H. (2013). *Chem. Rev.* 113: 442–524.

2 (a) Fogg, D.E. and dos Santos, E.N. (2004). *Coord. Chem. Rev.* 248: 2365–2379. (b) Shindoh, N., Takemoto, Y., and Takasu, K. (2009). *Chem. Eur. J.* 15: 12168–12179.

3 (a) Helferich, B., Speidel, P.E., and Toeldte, W. (1923). *Ber.* 56: 766–770. (b) Helferich, B. and Becker, J. (1924). *Ann.* 440: 1–18. (c) Helferich, B. and Schneidmuller, A. (1927). *Ber.* 60: 2002–2005.

4 For a review:Helferich, B. (1948). *Adv. Carbohydr. Chem.* 3: 79–111.

5 (a) Micheel, F. and Spruck, W. (1934). *Ber.* 67: 1665–1667.(b) Wolfrom, M.L., Quinn, J.L., and Christman, C.C. (1934). *J. Am. Chem. Soc.* 56: 2789.

6 Du, Y., Zhang, M., and Kong, F. (2000). *Org. Lett.* 2: 3797–3800.

7 Mukhopadhyay, B., Kartha, K.P.R., Russell, D.A., and Field, R.A. (2004). *J. Org. Chem.* 69: 7758–7760.

8 Olah, G.A. and Narang, S.C. (1982). *Tetrahedron* 38: 2225–2275.

9 For reviews see:(a)Hsu, C.-H., Hung, S.-C., Wu, C.-Y., and Wong, C.-H. (2011). *Angew. Chem. Int. Ed.* 50: 11872–11923.(b) Lee, D. and Taylor, M.S. (2012). *Synthesis* 44: 3421–3431.(c) Zulueta, M.M.L., Janreddy, D., and Hung, S.-C. (2015). *Isr. J. Chem.* 55: 347–359.

10 TMSOTf:(a)Tsunoda, T., Suzuki, M., and Noyori, R. (1980). *Tetrahedron Lett.* 21: 1357–1358.(b) Masaaki, K. and Wataru, H. (2003). *J. Org. Chem.* 68: 3413–3415.

11 $Sc(NTf_2)_3$:Ishihara, K., Karumi, Y., Kubota, M., and Yamamoto, H. (1996). *Synlett* 839–840.

12 $Bi(OTf)_3$:Leonard, N.M., Oswald, M.C., Freiberg, D.A. et al. (2002). *J. Org. Chem.* 67: 5202–5205.

13 $V(O)(OTf)_2$:Chen, C.-T., Weng, S.-S., Kao, J.-Q. et al. (2005). *Org. Lett.* 7: 3343–3346.

14 TMSOTf:(a)Hatakeyama, S., Mori, H., Kitano, K. et al. (1994). *Tetrahedron Lett.* 35: 4367–4370.(b) Wang, C.-C., Lee, J.-C., Luo, S.-Y. et al. (2002). *Angew. Chem. Int. Ed.* 41: 2360–2362.

15 $Cu(OTf)_2$:Yang, W.-C., Lu, X.-A., Kulkarni, S.S., and Hung, S.-C. (2003). *Tetrahedron Lett.* 44: 7837–7840.

16 Bu_2BOTf:(a)Jiang, L. and Chan, T.-K. (1998). *Tetrahedron Lett.* 39: 355–358. (b) Hernandez-Torres, J.M., Achkar, J., and Wei, A. (2004). *J. Org. Chem.* 69: 7206–7211.

17 $V(O)(OTf)_2$:Wang, C.-C., Luo, S.-Y., Shie, C.-R., and Hung, S.-C. (2002). *Org. Lett.* 4: 847–849.

18 $Cu(OTf)_2$:Shie, C.-R., Tzeng, Z.-H., Kulkarni, S.S. et al. (2005). *Angew. Chem. Int. Ed.* 44: 1665–1668.

19 TMSOTf:Daragics, K. and Fügedi, P. (2009). *Tetrahedron Lett.* 50: 2914–2916.

20 TMSOTf:Procopiou, P.A., Baugh, S.P.D., Flack, S.S., and Inglis, G.G.A. (1996). *Chem. Commun.* 2625–2626.

21 $Bi(OTf)_3$:Orita, A., Tanahashi, C., Kakuda, A., and Otera, J. (2000). *Angew. Chem. Int. Ed.* 39: 2877–2879.
22 $Ce(OTf)_3$:Bartoli, G., Dalpozzo, R., de Nino, A. et al. (2004). *Green Chem.* 6: 191–192.
23 $Sc(OTf)_3$:Lee, J.-C., Tai, C.-A., and Hung, S.-C. (2002). *Tetrahedron Lett.* 43: 851–865.
24 $Cu(OTf)_2$:Tai, C.-A., Kulkarni, S.S., and Hung, S.-C. (2003). *J. Org. Chem.* 68: 8719–8722.
25 $In(OTf)_3$:Chauhan, K.K., Frost, C.G., Love, I., and Waite, D. (1999). *Synlett* 1743–1745.
26 LiOTf:Karimi, B. and Maleki, J. (2003). *J. Org. Chem.* 68: 4951–4954.
27 Français, A., Urban, D., and Beau, J.-M. (2007). *Angew. Chem. Int. Ed.* 46: 8662–8665.
28 Joseph, A.A., Verma, V.P., Liu, X.-Y. et al. (2012). *Eur. J. Org. Chem.* 744–753.
29 (a) Wang, C.-C., Lee, J.-C., Luo, S.-Y. et al. (2007). *Nature* 446: 896–899. (b) Wang, C.-C., Kulkarni, S.S., Lee, J.-C. et al. (2008). *Nat. Protoc.* 3: 97–113. (c) Huang, T.-Y., Zulueta, M.M.L., and Hung, S.-C. (2014). *Org. Biomol. Chem.* 12: 376–382.
30 (a) Buchanan, S.G. and Schwarz, J.C.P. (1962). *J. Chem. Soc.* 4770–4777. (b) Shaban, M.A.E., Ary, I.E., Jeanloz, D.A., and Jeanloz, R.W. (1975). *Carbohydr. Res.* 45: 105–114.
31 (a) Albert, R., Dax, K., Pleschko, R., and Stuetz, A.E. (1985). *Carbohydr. Res.* 137: 282–290. (b) Chevalier, R., Esnault, J., Vandewalle, P. et al. (2005). *Tetrahedron* 61: 7669–7677. (c) Basu, N., Maity, S.K., Roy, S. et al. (2011). *Carbohydr. Res.* 346: 534–539. (d) Geng, Y., Faidallah, H.M., Albar, H.A. et al. (2013). *Eur. J. Org. Chem.* 7035–7040. (e) Liotta, L.J., Chalmers, J.F., Falco, Marshall, J.N. et al. (2014). *Carbohydr. Res.* 391: 31–36.
32 Patil, P.S., Lee, C.-C., Huang, Y.-W. et al. (2013). *Org. Biomol. Chem.* 11: 2605–2612.
33 (a) Garegg, P.J., Iversen, T., and Oscarson, S. (1976). *Carbohydr. Res.* 50: C12–C14. (b) Dan, A., Ito, Y., and Ogawa, T. (1995). *J. Org. Chem.* 60: 4680–4681. (c) Crich, D., Li, W., and Li, H. (2004). *J. Am. Chem. Soc.* 126: 15081–15086.
34 Chang, K.-L., Zulueta, M.M.L., Lu, X.-A. et al. (2010). *J. Org. Chem.* 75: 7424–7427.
35 San Jose, G. (2009). Molecular diversity from organic synthesis, associated with new mycorrhizal symbiosis tests. Research of active compounds as mycorrhizal factors. PhD thesis. University of Paris Sud, Orsay, France.
36 Huang, T.-Y., Zulueta, M.M.L., and Hung, S.-C. (2011). *Org. Lett.* 13: 1506–1509.
37 For selected reviews see:(a)Bolm, C., Legros, J., Le Paih, J., and Zani, L. (2004). *Chem. Rev.* 104: 6217–6254. (b) Diaz, D., Miranda, P., Padrón, J., and Martín, V. (2006). *Curr. Org. Chem.* 10: 457–476. (c) Sherry, D. and Fürstner, A. (2008). *Acc. Chem. Res.* 41: 1500–1511. (d) Bauer, E.B. (2008). *Curr. Org. Chem.* 12: 1341–1369. (e) Bauer, I. and Knölker, H.-J. (2015). *Chem. Rev.* 115: 3170–3387.
38 For cleavage bond reactions, see: (a) Wei, G., Zhang, L., Cai, C. et al. (2008). *Tetrahedron Lett.* 49: 5488–5491. (b) Sen, S.E., Roach, S.L., Boggs, J.K. et al.

(1997). *J. Org. Chem.* 62: 6684–6686. (c) Ding, X., Wang, W., and Kong, F. (1997). *Carbohydr. Res.* 303: 445–448.

39 For de-O-benzylation reactions using anhydrous $FeCl_3$, see: (a) Park, M.H., Takeda, R., and Nakanishi, K. (1987). *Tetrahedron Lett.* 28: 3823–3824. (b) Padrón, J.I. and Vázquez, J.T. (1995). *Tetrahedron: Asymmetry* 6: 857–858. (c) Rodebaugh, R., Debenham, J.S., and Fraser-Reid, B. (1996). *Tetrahedron Lett.* 37: 5477–5478.

40 Soulé, J.-F., Mathieu, A., Norsikian, S., and Beau, J.-M. (2010). *Org. Lett.* 12: 5322–5325.

41 Acetal formation promoted by anhydrous $FeCl_3$: (a) Singh, P.P., Gharia, Y.M., Dasgupta, F., and Srivastava, H.C. (1977). *Tetrahedron Lett.* 18: 439–440. (b) Danikiewicz, W., Olejnik, M., Wójcik, J. et al. (1997). *J. Mol. Catal. A* 123: 25–33. (c) Dasgupta, F., Singh, P.P., and Srivastava, H.C. (1980). *Ind. J. Chem.* 19B: 1056–1059. (d) Basu, N., Maity, S.K., Roy, S. et al. (2011). *Carbohydr. Res.* 346: 534–539.

42 Reductive etherification catalyzed by anhydrous $FeCl_3$: (a) Iwanami, K., Seo, H., Tobita, Y., and Oriyama, T. (2005). *Synthesis* 183–186. (b) Iwanami, K., Yano, K., and Oriyama, T. (2005). *Synthesis* 2669–2672. (c) Iwanami, K., Yano, K., and Oriyama, T. (2007). *Chem. Lett.* 36: 38–39.

43 Acylation: (a) Dasgupta, F., Singh, P.P., and Srivastava, H.C. (1980). *Carbohydr. Res.* 80: 346–349. (b) Sharma, G.V.M., Mahalingam, A.K., Nagarjan, M. et al. (1999). *Synlett* 1200–1202. with $FeCl_3{\cdot}6H_2O$ as a catalyst, see: (c) Komura, K., Ozaki, A., Ieda, N., and Sugi, Y. (2008). *Synthesis* 3407–3410.

44 (a) Bourdreux, Y., Lemétais, A., Urban, D., and Beau, J.-M. (2011). *Chem. Commun.* 47: 2146–2148. (b) Lemétais, A., Bourdreux, Y., Lesot, P. et al. (2013). *J. Org. Chem.* 78: 7648–7657. (c) Gau, B., Lemétais, A., Lepore, M. et al. (2013). *ChemBioChem* 14: 2413–2417.

45 Gouasmat, A., Lemétais, A., Solles, J. et al. (2017). *Eur. J. Org. Chem.* 3355–33361.

46 (a) Bongat, A.F.G. and Demchenko, A.V. (2007). *Carbohydr. Res.* 342: 374–406. (b) El Ashry, S.H. and Aly, M.R.E. (2007). *Pure Appl. Chem.* 79: 2229–2242. (c) Beau, J.-M. (2011). *Chimia* 65: 45–48.

47 Enugala, R., Carvalho, L.C.R., and Marques, M.M.B. (2010). *Synlett* 2711–2716.

48 (a) Olah, G.A., Prakash, G.K.S., and Sommer, J. (1985). *Superacids*, 2e. New York: Wiley. (b) Akiyama, T. (2007). *Chem. Rev.* 107: 5744–5758.

49 Trichloroacetimidates: (a) Schmidt, R.R. and Michel, J. (1985). *J. Carbohydr. Chem.* 4: 141–169. Glycosylations combined with reductive opening: (b) Vohra, Y., Vasan, M., Venot, A., and Boons, G.-J. (2008). *Org. Lett.* 10: 3247–3250. Acetates: (c) Sasaki, K., Matsumura, S., and Toshima, K. (2006). *Tetrahedron Lett.* 47: 9039–9043. Thioglycosides: (d) Veeneman, G.H., van Leeuwen, S.H., and van Boom, J.H. (1990). *Tetrahedron Lett.* 31: 1331–1334. (e) Konradsson, P., Udodong, U.E., and Fraser-Reid, B. (1990). *Tetrahedron Lett.* 31: 4313–4316.

50 (a) DeNinno, M.P., Etienne, J.B., and Duplantier, K.C. (1995). *Tetrahedron Lett.* 36: 669–672. (b) Sakagami, M. and Hamana, H. (2000). *Tetrahedron Lett.* 41: 5547–5551. (c) Mo, K.-F., Li, X., Li, H. et al. (2012). *J. Am. Chem. Soc.* 134: 15556–15562.

51 Despras, G., Urban, D., Vauzeilles, B., and Beau, J.-M. (2014). *Chem. Commun.* 50: 1067–1069.
52 (a) de Angelis, A., Flego, C., Ingallina, P. et al. (2001). *Catal. Today* 65: 363–371. (b) Liu, P.N., Xia, F., Wang, Q.W. et al. (2010). *Green Chem.* 12: 1049–1055. (c) Yan, S., Ding, N., Zhang, W. et al. (2012). *Carbohydr. Res.* 354: 6–20.
53 Tran, A.-T., Jones, R.A., Pastor, J. et al. (2011). *Adv. Synth. Catal.* 353: 2593–2598.
54 Mukherjee, D., Shah, B.A., Gupta, P., and Taneja, S.C. (2007). *J. Org. Chem.* 72: 8965–8968.
55 Kartha, K.P.R. (1986). *Tetrahedron Lett.* 27: 3415–3416.
56 Kartha, K.P.R. and Field, R.A. (1997). *Tetrahedron* 53: 11753–11766.
57 Mukhopadhyay, B., Russell, D.A., and Field, R.A. (2005). *Carbohydr. Res.* 340: 1075–1080.
58 Mukhopadhyay, B. (2006). *Tetrahedron Lett.* 47: 4337–4341.
59 Mong, K.-K.T., Chao, C.-S., Chen, M.-C., and Lin, C.-W. (2009). *Synlett* 603–606.
60 Jones, R.A., Davidson, R., Tran, A.T. et al. (2010). *Carbohydr. Res.* 345: 1842–1845.

8

Acyl Migrations in Carbohydrate Chemistry

Filip S. Ekholm[1] *and Reko Leino*[2]

[1] *University of Helsinki, Department of Chemistry, A. I. Virtasen aukio 1, 00014 Helsinki, Finland*
[2] *Åbo Akademi University, Johan Gadolin Process Chemistry Centre, Laboratory of Organic Chemistry, 20500 Åbo, Finland*

8.1 Introduction

Acyl groups are commonly used as protecting groups in organic synthesis, both because of their simple introduction and subsequent simple removal under either acidic or basic conditions [1]. Particularly in synthetic carbohydrate chemistry, acyl groups, typically acetyl (Ac) and benzoyl (Bz) or, in some cases, pivaloyl (Piv) or other designer groups, are frequently utilized for temporary protection of hydroxyl functionalities, in order to introduce other orthogonal protecting groups and for masking undesired hydroxyl sites in glycosylation reactions [2]. Acyl groups are also used as neighboring participating groups at 2-positions of sugars for controlling the stereochemistry at the anomeric center under glycosylation conditions, typically for the preparation of 1,2-*trans*-glycosides [3]. Stabilities of acyl protecting groups increase in the series Ac < Bz < Piv, with the last mentioned being removed, besides under suitably controlled acidic or basic conditions, also reductively.

It is well known in the literature, tracing back to the early reports of Fischer [4], Haworth [5], and others, that acyl groups have a facile tendency for intramolecular migration in polyhydroxyl compounds, reportedly under acidic or neutral, but especially under basic conditions. In addition, acyl group migration has been reported to take place under various other reaction conditions, besides in the presence of different reagents [6], also in the melt [7]. Consequently, the possibility of acyl group migration must be considered in all synthetic manipulations of partially acylated polyols, site-selective acylations of compounds containing multiple hydroxyl groups, as well as in structural characterization and isolation of naturally occurring oligosaccharides and polysaccharides [8]. Acyl migration may also occur in enzymatic reactions, both under acylating and deacylating conditions [9]. Besides *O*-acyl groups, *S*-acyl groups are also known to migrate in sulfur-containing compounds, including glycosides [10].

Protecting Groups: Strategies and Applications in Carbohydrate Chemistry, First Edition.
Edited by Sébastien Vidal.

8.2 Mechanism and Migration Kinetics

Mechanistically, acyl group migrations are often considered to proceed via cyclic five- or six-membered orthoester-type intermediates, which, after subsequent ring opening, produce the isomeric esters where the acyl group has migrated. Under basic conditions, the process is possibly catalyzed by deprotonation of the vicinal hydroxyl group (Scheme 8.1). Although the individual migration steps are probably reversible, reverse migrations are typically not observed, and the overall process is considered to be irreversible, proceeding clockwise in monosaccharides from the *exo*-anomeric oxygen toward position 6.

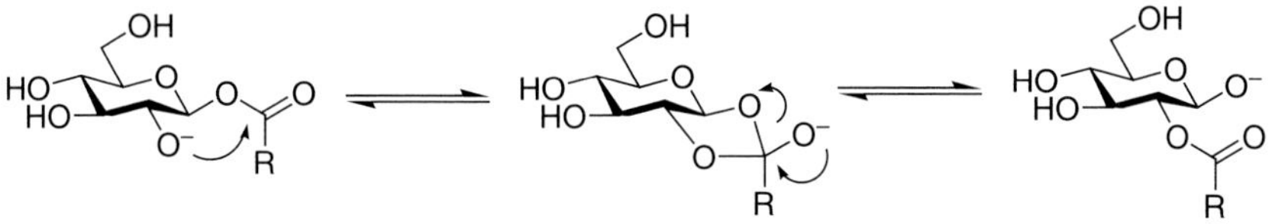

Scheme 8.1 Tentative mechanism of acyl group migration under basic conditions.

Although kinetic [11], mechanistic [12], and theoretical studies [13] or combinations thereof [14] have been performed, truly unifying theories do not exist, and much work could still be carried out in this area for a better understanding of the migration process and the underlying influencing factors and reaction conditions. In particular, nuclear magnetic resonance (NMR) spectroscopy has been proven as a useful tool for *in situ* monitoring of the migration processes [11, 12, 14].

In the literature, various model compounds have been utilized in more detailed migration studies. For example, intramolecular acyl migration from C2 to C6 in a series of 2-acylated galactose derivatives **1–3** (Figure 8.1) has been studied by NMR spectroscopy at variable pD [11]. Chemical shifts of the different migration products are distinguishable in their ^{1}H NMR spectra. Therefore, quantification of the products, here a typical example of consecutive reactions, becomes possible by

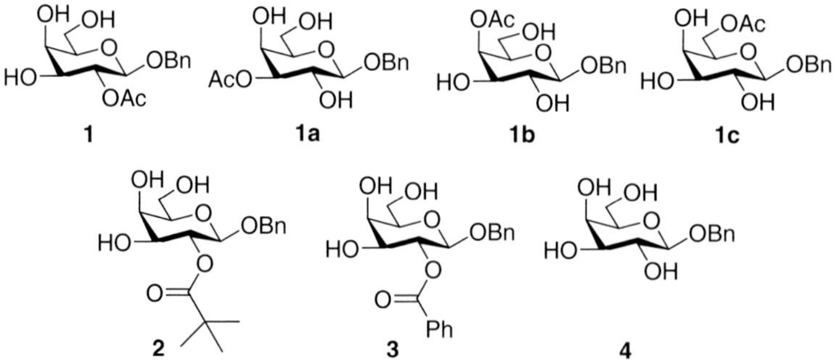

Figure 8.1 Acylated galactose model compounds utilized for acyl migration studies. *Source:* Roslund et al. 2008 [11]. Copyright 2008. Adapted with permission from American Chemical Society.

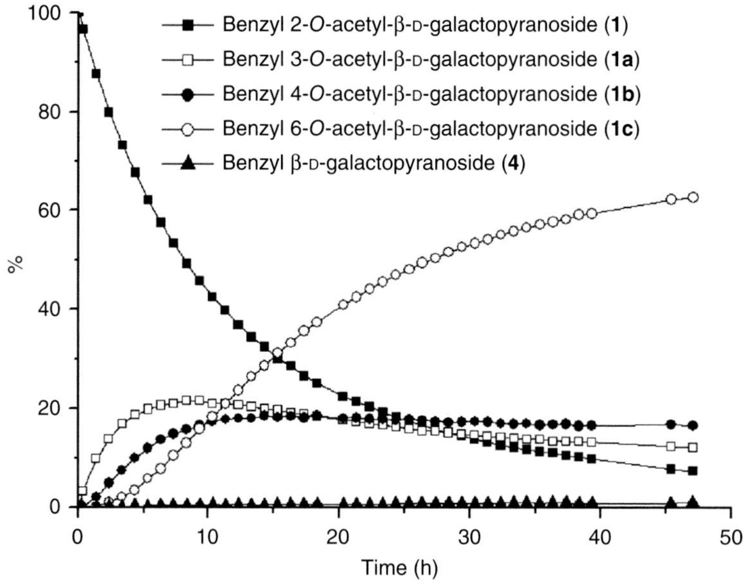

Figure 8.2 Migration of acetyl group (compound **1**) at pD = 8.0 and T = 25 °C in buffered D_2O. *Source:* Roslund et al. 2008 [11]. Copyright 2008. Reprinted with permission of American Chemical Society.

simple integration of the signal areas (Figure 8.2), providing a tool for more accurate determination of the migration phenomena and rates by kinetic calculations.

Some issues are worth noting: under strong acidic conditions (pD = 1.0–3.0), instead of migration, direct hydrolysis of the ester groups was observed. Although stable at pD = 5.0, migration of acyl groups from C2 → C3 → C4 → C6 took place at pD ≤ 6.8, followed by hydrolysis from the primary C6 position (Scheme 8.2) [11]. The sterically more congested pivaloyl group was shown to migrate considerably slower (approximately 20–80 fold at pD = 8.0) than the acetyl and benzoyl groups, of which the latter was observed to migrate faster than acetyl at high pD. Kinetic analysis under mildly basic conditions (pD = 8.0) indicated similar forward migration rates for both acetyl and benzoyl groups from C3 → C4 (k_2 = 17.6 ± 0.44 (Ac), 18.8 ± 2.1 (Bz), 0.61 ± 0.13 (Piv)), whereas the acetyl group migrated faster from C2 → C3 (k_1 = 4.91 ± 0.032 (Ac), 2.92 ± 0.07 (Bz), 0.06 ± 0.004 (Piv)), and C4 → C6 (k_3 = 10.4 ± 0.14 (Ac), 1.91 ± 0.06 (Bz)). Evidently, the migration is influenced, besides by pH/pD, also by steric and electronic factors induced by both the acyl group and sugar ring involved. In the β-D-galactopyranoside compound studied, an increase in the migration rates both to and from the axial C4 hydroxyl group over the equatorial positions was observed [11]. Thus, it becomes clear that in designing protecting group strategies for both synthesis and isolation of polyhydroxyl compounds, various parameters, including pH, steric bulk of the acyl groups, and stereochemical configurations of the target molecules, must be considered.

In related work, acyl migrations have also been studied and analyzed experimentally in, e.g. D-glucose derivatives [12, 15] and furanosides [6, 16].

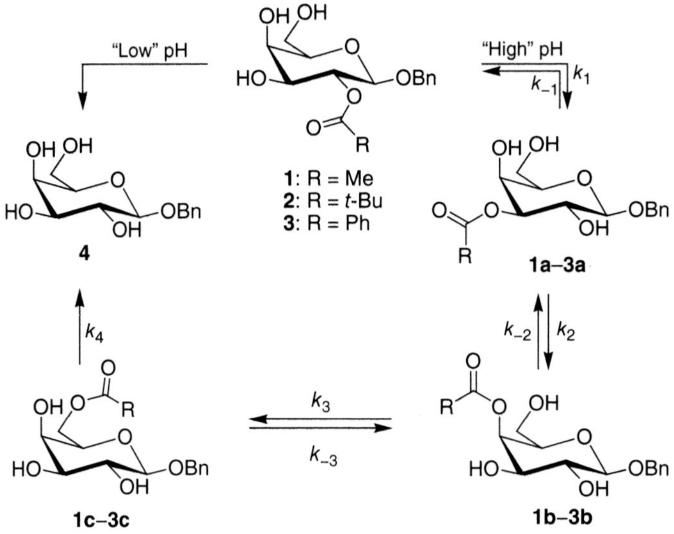

Scheme 8.2 Migration of acetyl groups in β-D-galactopyranoside model compounds. *Source:* Roslund et al. 2008 [11]. Copyright 2008. Reprinted with permission of American Chemical Society.

Notably, potential acyl migrations are also worth considering in glucuronides [15b, 17], having particular relevance in biology as metabolites, thereby also influencing safety assessments of drugs and drug development programs.

8.3 Acyl Group Migration – Synthetic Applications

The discussion on the synthetic applications of acyl group migration in carbohydrate chemistry is here limited to intramolecular migration reactions. In the literature, synthetic reports utilizing acyl group migration as a key transformation are scarce, and the field can be considered to be relatively unexplored. Therefore, this area offers several possibilities for the development of novel organic transformations and protecting group strategies that can compete with or even outperform the more traditional organic transformations applied frequently today. In order to provide a deeper understanding on the potential synthetic applications, the previous advances in the field are briefly addressed below.

8.3.1 Traditional Synthetic Applications

The title "traditional synthetic applications" refers to the straightforward nature of the acyl group migration studies discussed in this section. When approaching such a complex topic (and its potential applications in synthetic carbohydrate chemistry), it is preferable to start with appropriately protected substrates containing a limited amount of free hydroxyl groups. Therefore, an appropriate starting point is the work of Deng and Chang covering the synthesis of disaccharide building blocks of the blood group antigens [18]. The reported synthetic pathway featured acyl group migration reactions as key transformations. In the initial optimization of migration conditions, Chang and coworkers utilized phenyl 3-*O*-benzoyl-4,6-*O*-benzylidene-1-thio-β-D-galactopyranoside as a substrate. Silver oxide (Ag_2O) and cesium trifluoroacetate (CF_3CO_2Cs) were found to be the best Lewis base and acid initiators for the reaction, and tetrabutylammonium iodide (TBAI) was reported to enhance the rate of the reaction when used in catalytic amounts. The acyl migration was thermodynamically controlled and was found to proceed through a cyclic orthoester intermediate (as expected). Steric factors were considered to influence the reaction outcome. In general, moderate/high yields were observed in the acyl migration reactions. The migration of a number of acyl groups were investigated in different monosaccharides, for example, the O2 → O3 in glucopyranosides (Table 8.1, entries 1–4) and the O3 → O2 in galactopyranosides (Table 8.2, entries 5–9).

In addition to the acyl group migrations described above, Chang and coworkers investigated the migration patterns of pivaloyl groups in phenyl 3,6-di-*O*-pivaloyl-1-thio-α-D-mannopyranoside and methyl 2,6-di-*O*-pivaloyl-α-D-glucopyranoside (Scheme 8.3). In these reactions, the yields were moderate, and

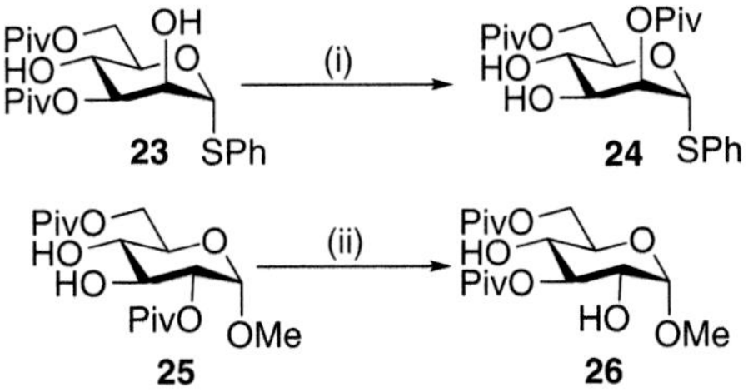

Scheme 8.3 Migration of pivaloyl groups in mannopyranosides and glucopyranosides. Reaction conditions: (i) Ag_2O, TBAI, DMF, 60 °C, 12 h, 58% or CF_3CO_2Cs, TBAI, DMF, 55 °C, 10 h, 67% or Ag_2O, TBAI, DMF, RT, sonication, 4 min, 81% and (ii) Ag_2O, TBAI, DMF, RT, 48 h, 49%.

Table 8.1 Summary of the O2 → O3 acyl group migration reactions in protected glucopyranosides.

Entry	Starting material	Product	Method[a]	Temperature (°C)	Time (h)	Yield (%)
1	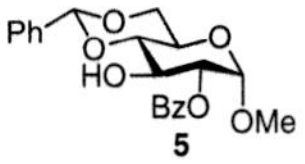 5	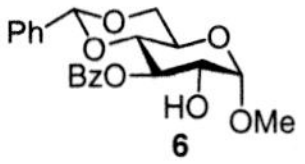 6	A B C	75 50 RT	7 17 5 min	74 78 73
2	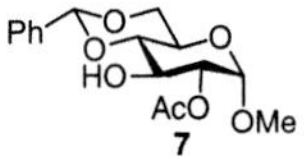 7	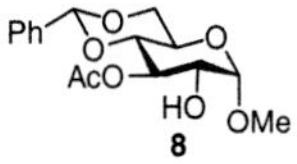 8	A B	RT RT	48 24	60 67
3	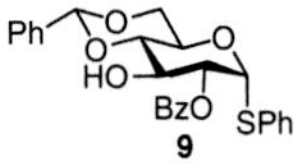 9	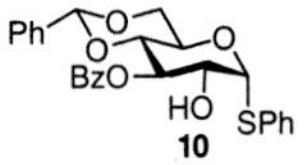 10	A	RT	48	70
4	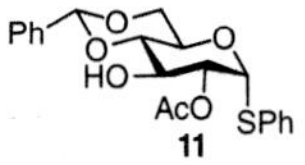 11	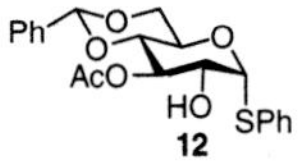 12	A B C	50 50 RT	24 12 3 min	71 72 71

a) Method A: Ag_2O (1 eq.), TBAI (0.1 eq.), DMF; method B: CF_3CO_2Cs (1 eq.), TBAI (0.1 eq.), DMF; and method C: Ag_2O (1 eq.), TBAI (0.1 eq.), DMF, sonication.

Table 8.2 Summary of the O3 → O2 acyl group migration reactions in protected galactopyranosides.

Entry	Starting material	Product	Method[a)]	Temperature (°C)	Time (h)	Yield (%)
5	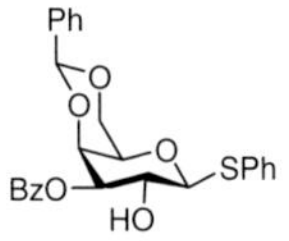 **13**	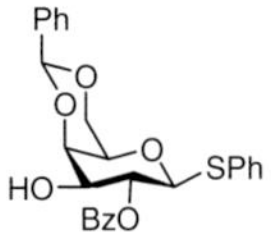 **14**	A B	RT RT	48 72	95 67
6	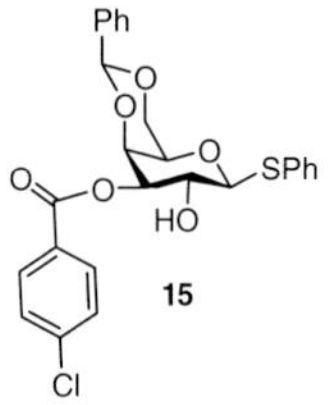 **15**	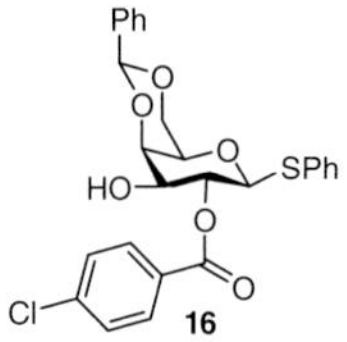 **16**	A B	75 75	7 8	80 83
7	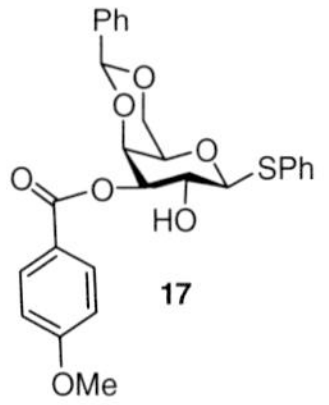 **17**	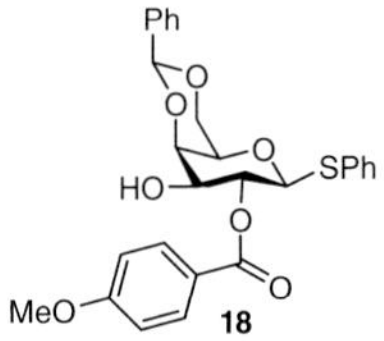 **18**	A B	40 50	10 12	77 79
8	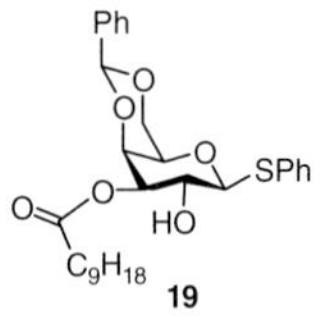 **19**	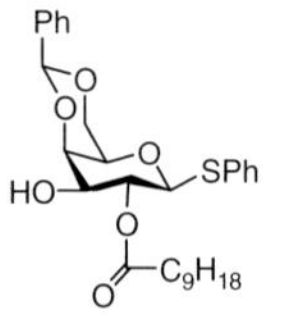 **20**	A B C	60 50 RT	7 12 13 min	76 76 80
9	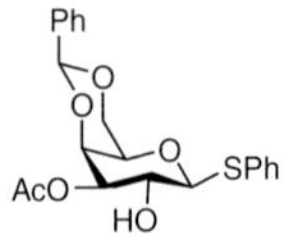 **21**	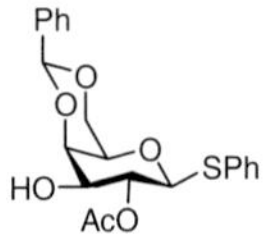 **22**	A B	RT 45	48 24	68 71

a) Method A: Ag_2O (1 eq.), TBAI (0.1 eq.), DMF; method B: CF_3CO_2Cs (1 eq.), TBAI (0.1 eq.), DMF; and method C: Ag_2O (1 eq.), TBAI (0.1 eq.), DMF, sonication.

the authors reported that further attempts to optimize the reaction conditions led to complex reaction mixtures.

Although the yields of the reactions were in the desired range, the reaction times were considered to be suboptimal. In a subsequent paper, it was reported that the reaction times could be significantly reduced by sonication [19]. With sonication, high yields were obtained in several minutes in contrast to the reaction times of several days needed previously.

On a general level, chemoenzymatic approaches to the synthesis of carbohydrates have become increasingly important in modern carbohydrate chemistry.

Enzymatic reactions can typically be performed on substrates containing a minimal amount of protecting groups [20], thereby eliminating the need for exhaustive protection/deprotection strategies [21]. Similar approaches can also be applied to acyl group migration reactions with carbohydrates. Palomo and coworkers have developed such an approach for the conversion of peracetylated monosaccharides into acceptors for glycosylation reactions (Table 8.3) [22]. The protocol featured a lipase catalyzed regioselective removal of the 6-*O*-acetyl group followed by acyl group migration from O4 → O6. In all cases, acyl group migrations resulted in the formation of two products. In addition to the desired product (yields ranging from 40% to 80%), a second product was obtained in yields ranging from 20% to 50% as a result of further migration of an acetyl group from O3 → O4. Furthermore, the amount of side product was found to be dependent on the monosaccharide structure. With glucose or *N*-acetyl-glucosamine, bearing an equatorial 4-OH group, the amount was significantly lower (20–30%, Table 8.3, entries 10–13). When galactose or *N*-acetyl-galactosamine, bearing an axial 4-OH group, was subjected to the reaction conditions, close to

Table 8.3 Summary of the investigated acyl group migrations from O4 → O6 (and O3 → O4) in different monosaccharides.

Entry	Starting material	Product 1	Product 2	Time (h)	Yield (%)
10	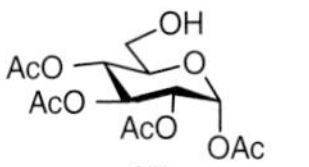 **27**	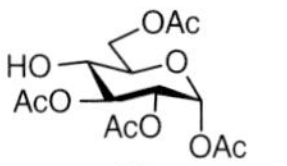 **28a**	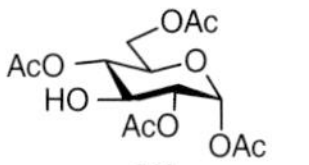 **28b**	5	**28a**: 80 **28b**: 20
11	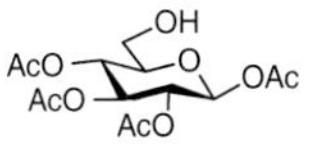 **29**	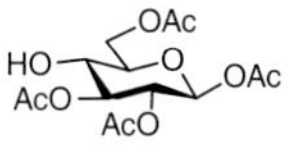 **30a**	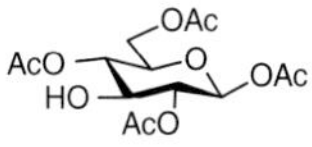 **30b**	8	**30a**: 70 **30b**: 18
12	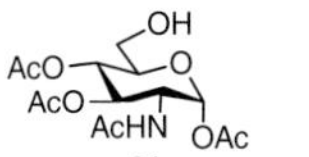 **31**	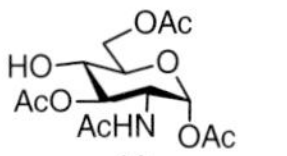 **32a**	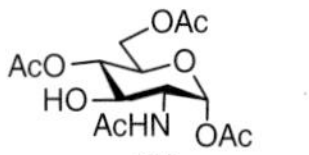 **32b**	3	**32a**: 79 **32b**: 20
13	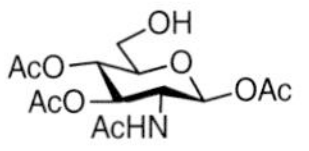 **33**	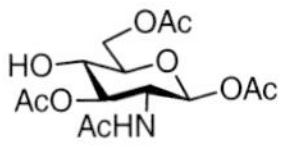 **34a**	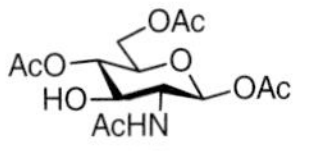 **34b**	10	**34a**: 78 **34b**: 20
14	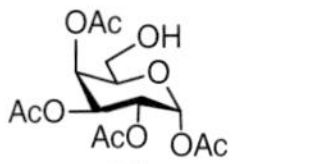 **35**	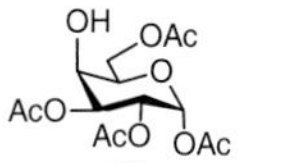 **36a**	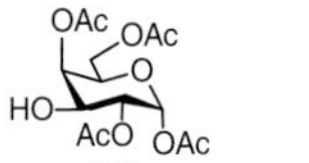 **36b**	3.5	**36a**: 42 **36b**: 50
15	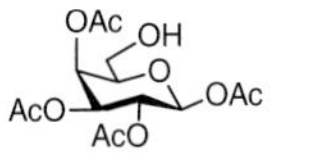 **37**	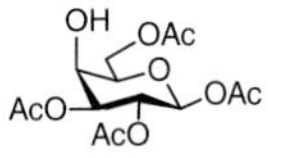 **38a**	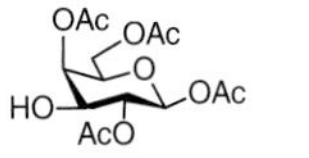 **38b**	1.5	**38a**: 45 **38b**: 50
16	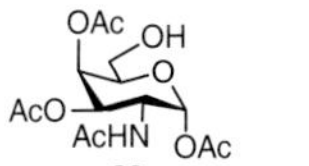 **39**	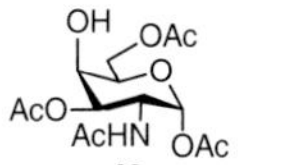 **40a**	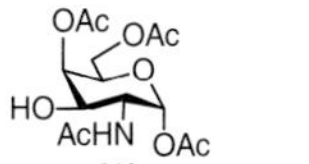 **40b**	2	**40a**: 42 **40b**: 48

a) Migration conditions: H_2O (50 mM KH_2PO_4 buffer): MeCN 5 : 1, pH = 8.5–9.5, 4 °C.

50% of the second product was observed (Table 8.3, entries 14–16). The anomeric composition did not have a considerable effect on the product distribution. The results indicate that stereochemical factors affect the migratory behavior of acyl groups and should be accounted for when planning such sequences. Although not reported in the paper, comparison of the product distribution using a suitably protected mannoside as a starting material (SM) would be a relevant topic for future studies.

8.3.2 Directed Evolution of Acyl Migrations

Although the development of traditional synthetic methods using acyl group migrations offers various possibilities, the challenges of uncontrollable acyl migrations with compounds containing multiple free hydroxyl groups remain. Ahn and Chang have suggested the use of external agents as selectors for amplifying the formation of a specific reaction species by modulation of the reaction equilibrium [23]. In their study, 1,4-di-*O*-benzoyl-*chiro*-inositol **41** was used as a model compound and phenylboronic acid as an external selection agent. It should be noted that the reaction conditions need to be strictly anhydrous as hydrolysis of the benzoyl groups is possible in the presence of water. In the absence of phenylboronic acid, a reaction mixture containing nine regioisomers in roughly equal amounts was obtained. When phenylboronic acid was utilized as a selector, 1,2-di-*O*-benzoyl-*chiro*-inositol **42** could be isolated in 72% yield (Scheme 8.4). The influence of phenylboronic acid is based on its ability to bind to 1,2-*cis*-diols (under basic conditions) [24], thus slowly blocking the other sites in *chiro*-inositol and ultimately resulting in a 10.3-fold enrichment on a time scale of 7 h. The authors have also demonstrated that similar reaction conditions can be applied to other substrates [25]. When 1,4,5-tri-*O*-benzoyl-*myo*-inositol **43** was used as the substrate, the enhanced formation of 2,4,6-tri-*O*-benzoyl-*myo*-inostiol **44** in yields of 83% under conditions favoring the thermodynamic product and 1,4,6-tri-*O*-benzoyl-*myo*-inositol **45** in isolated yields of 92% under conditions favoring the kinetic product was observed (Scheme 8.5). These exper-

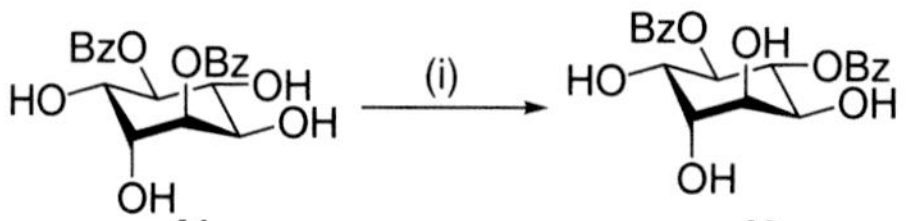

Scheme 8.4 Acyl group migration in *chiro*-inositol utilizing phenylboronic acid as a selector. Reaction conditions: (i) DBU, MeCN, mixture of compounds or DBU, $PhB(OH)_2$, MeCN, 72% of indicated structure.

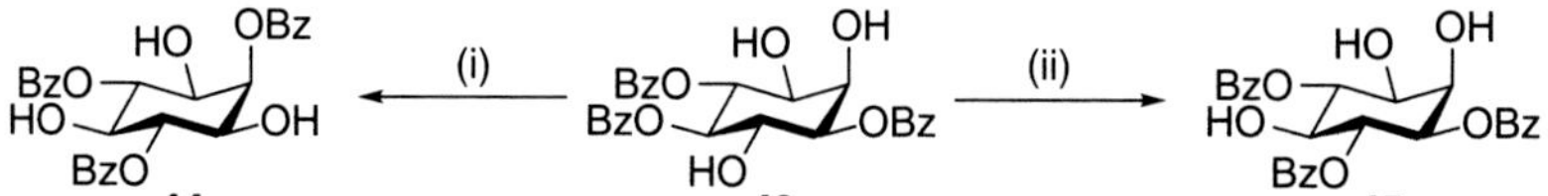

Scheme 8.5 Acyl group migration in *myo*-inositol utilizing phenylboronic acid or boric acid as a selector. Reaction conditions: (i) DBU, boric acid, MeCN, 83% of the thermodynamic product and (ii) DBU, $PhB(OH)_2$, MeCN, 92% of the kinetic product.

iments indicate that it is possible to obtain different migration products by careful control of the reaction conditions. Furthermore, development of improved selectors may turn out to be a fruitful future research topic.

8.3.3 Protecting Group Manipulation – Acyl Migration

This section focuses on protecting group manipulations followed by direct acyl group migrations. Such "one-pot" reaction sequences may prove useful in the preparation of building blocks as the amounts of reaction steps and purifications can be decreased. The majority of the reported procedures dealing with this subject have utilized substrates containing *O*-silyl-protected primary alcohols, ester protecting groups in a position close to the silyl protecting group, and tetrabutylammonium fluoride (TBAF) as a displacement agent [6, 16b, 26]. Passacantilli and coworkers described the displacement of a 6-*O*-triisopropylsilyl (TIPS) group with TBAF followed by an O4→O6 migration of a benzoyl group in several methyl pyranosides and glycals [26]. Yields of the reported 2-step procedures were excellent (Table 8.4).

Table 8.4 Summary of the displacement–migration reaction sequence of pyranosides and glycals.

Entry	Starting material	Product	Yield (%)
17	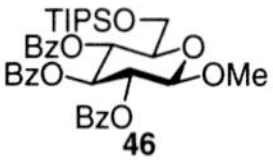 **46**	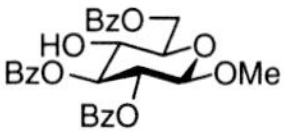 **47**	75–80
18	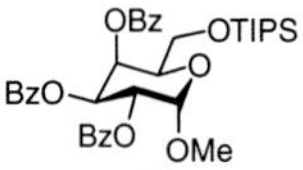 **48**	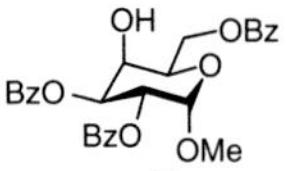 **49**	75–80
19	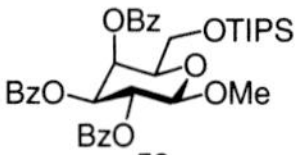 **50**	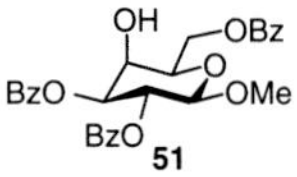 **51**	50–55
20	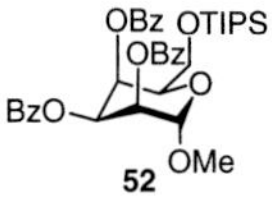 **52**	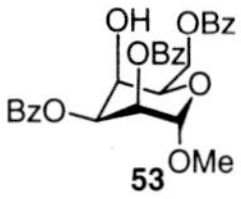 **53**	80–90
21	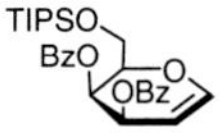 **54**	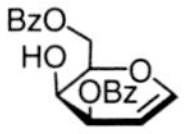 **55**	75–80
22	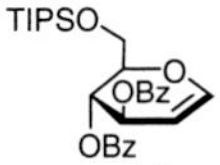 **56**	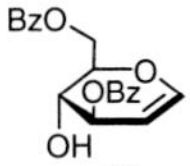 **57**	75–80

Reaction conditions: SM, 2 M TBAF (1 eq.), 0 °C.

Chevallier and Migaud investigated a similar displacement–migration sequence in ribo-, arabino-, and xylofuranosides [6]. The results reveal that migration occurs extensively in the ribofuranoside (Table 8.5, entry 23) and the xylofuranoside (Table 8.5, entry 24) while the rest of the molecules remained mostly unaffected by the migration events. When TBAF was replaced by ceric ammonium nitrate IV (CAN), no migration took place in the ribofuranoside (Table 8.5, entry 24). The same cannot, however, be said for the xylofuranoside (Table 8.5, entry 25) where migration occurred in a similar manner. It is therefore difficult to draw conclusions on the effects of the deprotection agent based on this limited set of substrates and more experiments are needed to confirm the substrate dependency of the migration reactions. Vanek and coworkers studied the displacement–migration behavior of 3-alkoxycarbonyl-substituted xylo- and ribofuranosides bearing 6-*O*-*tert*-butyldimethylsilyl (TBDMS) and 1,2-*O*-isopropylidene protecting groups [16b]. The displacement of the silyl group was investigated utilizing both CAN and TBAF (Table 8.5, entries 25–27). For the

Table 8.5 A summary of the displacement–migration reaction sequence in furanosides.

Entry	Starting material	Product	Yield (%)
23	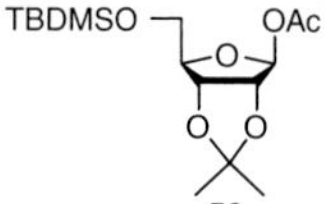 58	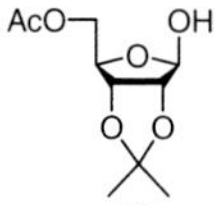 59	90[a)] 0[b)]
24	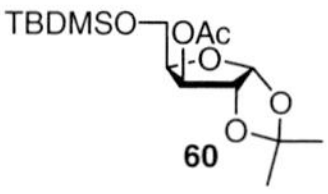 60	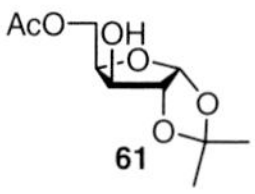 61	85[a)] 100[b)]
25	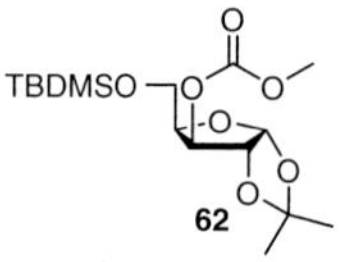 62	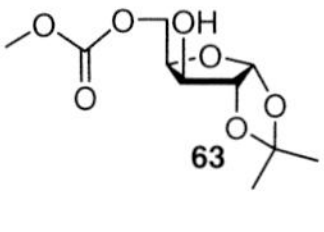 63	73[a)]
26	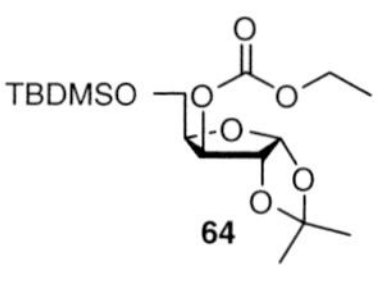 64	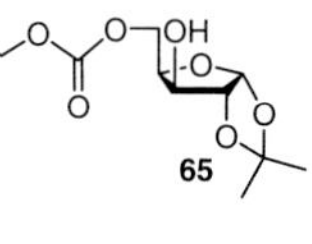 65	92[a)]
27	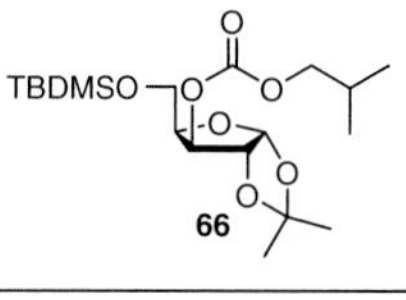 66	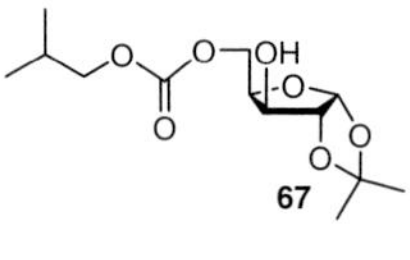 67	96[a)]

a) Reaction conditions: SM, TBAF, THF.
b) Reaction conditions: SM, CAN, THF.

xylofuranosides, the mechanism was suggested to be intramolecular while the mechanism for the ribofuranoside series was found to be intermolecular. On a general level, the results of these studies are encouraging as they indicate that acyl migration steps can be incorporated into multistep reaction sequences. However, a deeper understanding on the correlation between chemical structure and migratory behavior is necessary in order to effectively plan such reaction sequences.

Acyl protecting groups have been shown to act as nucleophiles when good leaving groups are present at neighboring positions. Therefore, acyl group migration reactions can be utilized to invert the stereochemistry on neighboring carbons with simultaneous protection of the newly inverted stereocenter. Several research groups have shown that a pivaloyl group can displace a triflate on a neighboring position through a S_N2 displacement reaction with inversion of stereochemistry (Scheme 8.6) [27]. Such an approach has been applied to the synthesis of a wide range of aminosugars present in naturally occurring compounds and analogs thereof.

Acyl group migration is not limited to *O*-linked acyl groups that have been discussed thus far. In fact, acyl group migration from other heteroatoms such as

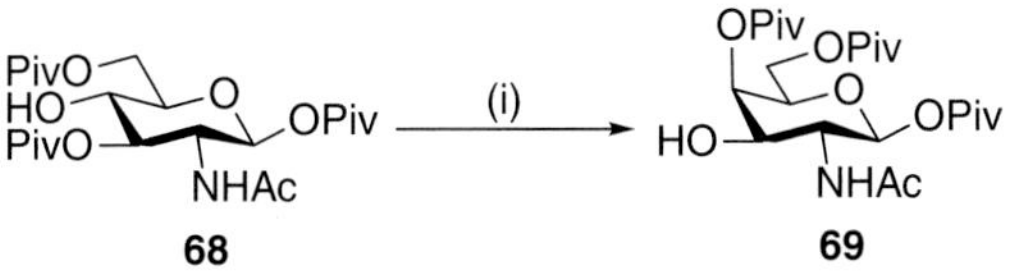

Scheme 8.6 Reaction conditions: (i) (1) Tf_2O, pyridine; (2) H_2O, Δ, 75%.

sulfur has been shown to occur easily in the synthesis of sulfur-containing glycosides. Pei and coworkers showed that unusual products are formed when monosaccharides containing *O*-tosyl protecting groups are treated with 5 eq. of potassium thioacetate (Table 8.6) [10]. In addition to containing the SAc-functional group, the product of the displacement reaction, the compounds were shown to contain an OAc group at a neighboring position (if the SM had a free hydroxyl group in such a position). The OAc group was suggested to form through a S6 → O4 acyl migration event. The free thiol was suggested to acquire a "new" acetyl group from the excess thioacetate reagent present in the solution. When 1.5 eq. of thioacetate was utilized, the acyl migration was inhibited, and high yields of the SAc compounds were obtained. These results are a further indication of the many possibilities to utilize acyl group migration reactions within synthetic carbohydrate chemistry.

8.4 Summary and Conclusions

To summarize, acyl group migration still remains as a relatively unexplored and mechanistically only partially understood area of (carbohydrate) chemistry with

Table 8.6 Summary of the S6 → O4 acyl migration reaction.

Entry	Starting material	Product	Yield (%)
28	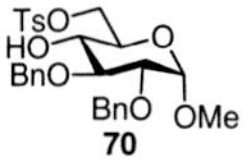 **70**	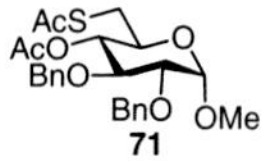 **71**	23[a)]
29	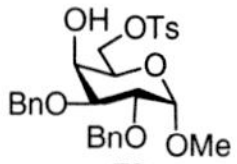 **72**	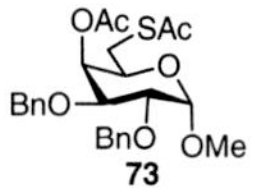 **73**	95[a)]
30	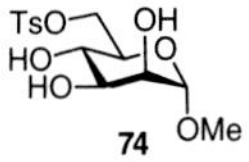 **74**	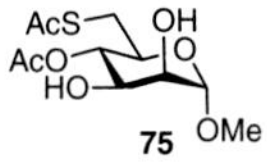 **75**	77[a)]
31	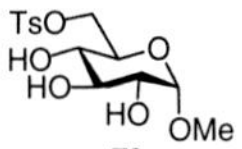 **76**	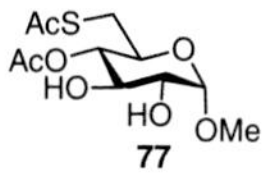 **77**	70[a)]
32	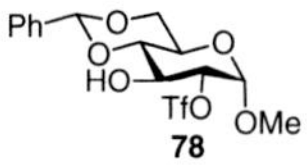 **78**	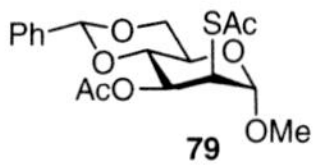 **79**	85[b)]
33	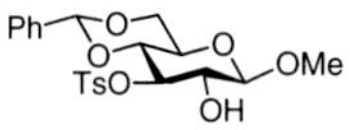 **80**	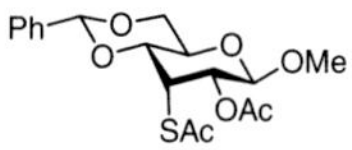 **81**	83[a)] 86[c)]
34	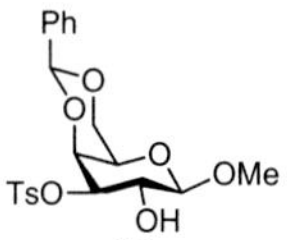 **82**	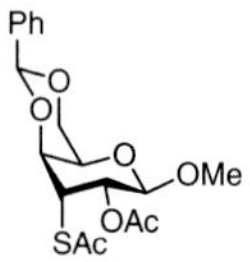 **83**	83[a)] 85[c)]

a) Reaction conditions: KSAc (5 eq.), DMF, 35–65 °C, 6 h.
b) Reaction conditions: KSAc (10 eq.), DMF, RT.
c) Reaction conditions: TBASAc (5 eq.), DMF, 35 °C, 8–12 h.

considerable room for new developments, particularly in the context of new synthetic applications. It has been shown that certain solid supports and purification conditions may influence acyl migration events [28]. It is therefore highly recommended that potential migration events are accounted for during isolation processes of synthetic compounds, natural products, and metabolites thereof. Furthermore, it is important to obtain a deeper understanding on the natural occurrence of the phenomenon and the related mechanisms in macromolecular systems. One example would be to study acyl migration processes occurring in wood/bacterial polysaccharides by the utilization of partially acylated oligosaccharide model compounds. Such studies would complement the previous literature studies on monosaccharides [11, 29]. On a more general level, the various synthetic strategies presented above (e.g. directed evolution by external selectors, displacement–migration sequences, and more traditional migration studies) highlight the many challenges associated with the development of synthetic applications based on acyl migration reactions. They also reveal the need for more unifying studies on the correlation between acyl groups, chemical structure, and migratory behavior. In the best case scenario, future development of reaction sequences featuring a combination of external selectors, enzymes, and other migration events could lead to one-pot cascades, which may incorporate other important reactions such as glycosylations.

8.5 Selected Experimental Procedures

General procedure for acyl migration with sonication [18]. Add silver oxide (1 eq.) and TBAI (0.1 eq.) to a solution containing dissolved SM (1 eq.) in anhydrous dimethylformamide (DMF) (0.12 mmol/5 ml). Sonicate the sample at ambient temperature until thin-layer chromatography (TLC) indicates that the reaction is complete. Filter the reaction mixture and purify the filtrate by silica gel column chromatography.

General procedure for directed evolution of acyl migration [23]. Add phenyl boronic acid (32 eq.) and 1,8-diazabicyclo[5.4.0]undec-7-ene (DBU = 79 eq.) to a solution containing SM (1 eq.) dissolved in anhydrous MeCN (1 ml mg^{-1} SM). Stir the reaction mixture until TLC indicates that the reaction is complete. Quench the reaction with AcOH, dilute with EtOAc, wash with 1 N HCl, satd. $NaHCO_3$, dry over $MgSO_4$, filter, concentrate, and purify by silica gel column chromatography.

General procedure for displacement of a silyl group migration sequence [26]. Add 2 M TBAF in tetrahydrofuran (THF = 1 eq.) to a solution containing SM (1 eq.) in THF (5 ml/0.5 mmol SM) at 0 °C under argon. Stir the solution for 15 min, then add another 2 M TBAF (1 eq.), and stir for 30 min. Dilute the reaction mixture with CH_2Cl_2 and wash the organic layer with satd. NH_4Cl solution. Extract the water phase with CH_2Cl_2. Dry the combined organic layer over $MgSO_4$, filter it, and concentrate the sample. Purify the crude product by silica gel column chromatography.

Abbreviations

Ac	acetyl
AcOH	acetic acid
Bn	benzyl
Bz	benzoyl
CAN	ceric ammonium nitrate
DBU	1,8-diazabicyclo[5.4.0]undec-7-ene
DMF	*N,N*-dimethylformamide
EtOAc	ethyl acetate
NMR	nuclear magnetic resonance

Piv	pivaloyl
SM	starting material
RT	room temperature
TBAF	tetrabutylammonium fluoride
TBAI	tetrabutylammonium iodide
TBDMS	*tert*-butyldimethylsilyl
THF	tetrahydrofuran
TIPS	triisopropylsilyl
TLC	thin layer chromatography
Ts	tosyl

References

1 (a) Wuts, P.G.M. and Greene, T.W. (2007). *Protection for the Hydroxyl Group, Including 1,2- and 1,3-Diols: Esters in Protective Groups in Organic Synthesis*, 4e (ed. P.G.M. Wuts and T.W. Greene), 222–271. New Jersey, US: Wiley.(b) Kocienski, P.J. (2005). *Hydroxyl Protecting Groups: Esters in Protecting Groups*, 3e (ed. P.J. Kocienski), 323–339. Stuttgart, Germany: Georg Thieme Verlag.

2 (a) Osborn, H.M.I. (ed.) (2003). *Carbohydrates*, 1e. Oxford, UK: Academic Press/ Elsevier Science Ltd.(b) Boons, G.-J. and Hale, K.J. (ed.) (2000). *Organic Synthesis with Carbohydrates*, 1e. Sheffield, UK: Sheffield Academic Press Ltd.(c) Levy, D.E. and Fügedi, P. (ed.) (2006). *The Organic Chemistry of Sugars*, 1e. Florida, US: Taylor & Francis Group.

3 (a) Whitfield, D.M. and Douglas, S.P. (1996). *Glycoconjugate J.* 13: 5–17.(b) Garegg, P.J. (2004). *Adv. Carbohydr. Chem. Biochem.* 59: 69–134.

4 Fischer, E. (1920). *Ber. Dtsch. Chem. Ges.* 53: 1621–1633.

5 Haworth, W.N., Hirst, E.L., and Teece, E.G. (1930). *J. Chem. Soc.* 1405–1409.

6 Chevallier, O.P. and Migaud, M.E. (2006). *Beilstein J. Org. Chem.* 2: doi: 10.1186/1860-5397-2-14.

7 Hann, R.M., Maclay, W.D., and Hudson, C.S. (1939). *J. Am. Chem. Soc.* 61: 2432–2442.

8 (a) Dutton, G.G.S. and Merrifield, E.H. (1982). *Carbohydr. Res.* 103: 107–128.(b) Gille, S. and Pauly, M. (2012). *Front. Plant. Sci.* 3: 1–7.(c) Ravenscroft, N., Cescutti, P., Gavini, M. et al. (2015). *Carbohydr. Res.* 404: 108–116.

9 (a) Filice, M., Guisan, J.M., Terreni, M., and Palomo, J.M. (2012). *Nat. Protoc.* 7: 1783–1796.(b) Filice, M., Ubiali, D., Pagani, G. et al. (2006). *Arkivoc* viii: 66–73.(c) Sweers, H.M. and Wong, C.-H. (1986). *J. Am. Chem. Soc.* 108: 6421–6422.

10 Zhou, Y., Zhang, X., Ren, B. et al. (2014). *Tetrahedron* 70: 5385–5390.

11 Roslund, M.U., Aitio, O., Wärnå, J. et al. (2008). *J. Am. Chem. Soc.* 130: 8769–8772.

12 Brecker, L., Mahut, M., Schwarz, A., and Nidetzky, B. (2009). *Magn. Reson. Chem.* 47: 328–332.

13 Rangelov, M.A., Vayssilov, G.N., and Petkov, D.D. (2006). *Int. J. Quantum Chem.* 106: 1346–1356.

14 Nicholls, A.W., Akira, K., Lindon, J.C. et al. (1996). *Chem. Res. Toxicol.* 9: 1414–1424.

15 (a) Yoshimoto, K. and Tsuda, Y. (1983). *Chem. Pharm. Bull.* 31: 4324–4334.(b) Iddon, L., Richards, S.E., Johnson, C.H. et al. (2011). *Org. Biomol. Chem.* 9: 926–934.
16 (a) Zanlungo, A.B., Deferrari, J.O., and Cadenas, R.A. (1969). *Carbohydr. Res.* 10: 403–416.(b) Dvorakova, M., Pribylova, M., Pohl, R. et al. (2012). *Tetrahedron* 68: 6701–6711.
17 (a) Stachulski, A.V., Harding, J.R., Lindon, J.C. et al. (2006). *J. Med. Chem.* 49: 6931–6945.(b) Stachulski, A.V. and Meng, X. (2013). *Nat. Prod. Rep.* 30: 806–848.
18 Deng, S. and Chang, C.-W.T. (2006). *Synlett* 756–760.
19 Deng, S., Gangadharmath, U., and Chang, C.-W.T. (2006). *J. Org. Chem.* 71: 5179–5185.
20 (a) Muthana, S., Cao, H., and Chen, X. (2009). *Curr. Opin. Chem. Biol.* 13: 573–581.(b) Guo, Z. and Shao, N. (2005). *Med. Res. Rev.* 25: 655–678.(c) Gillingham, D.G., Stallforth, P., Adibekian, A. et al. (2010). *Nat. Chem.* 2: 102–105.
21 (a) Codée, J.D.C., Ali, A., Overkleeft, H.S., and van der Marel, G.A. (2011). *C.R. Chim.* 14: 178–193.(b) Guo, J. and Ye, X.-S. (2010). *Molecules* 15: 7235–7265.(c) Nicolaou, K.C. and Mitchell, H.J. (2001). *Angew. Chem. Int. Ed.* 40: 1576–1624.
22 Filice, M., Bavaro, T., Fernandez-Lafuente, R. et al. (2009). *Catal. Today* 140: 11–18.
23 Ahn, Y.-H. and Chang, Y.-T. (2004). *J. Comb. Chem.* 6: 293–296.
24 (a) James, T.D., Sandanayake, S., and Shinkai, S. (1996). *Angew. Chem. Int. Ed. Engl.* 35: 1910–1922.(b) Yang, W., He, H., and Drueckhammer, D.G. (2001). *Angew. Chem. Int. Ed.* 40: 1714–1718.
25 Ahn, Y.-H. and Chang, Y.-T. (2004). *Chem. Eur. J.* 10: 3543–3547.
26 Graziani, A., Passacantilli, P., Piancatelli, G., and Tani, S. (2001). *Tetrahedron Lett.* 42: 3857–3860.
27 (a) Feng, J. and Ling, C.-C. (2010). *Carbohydr. Res.* 345: 2450–2457.(b) Cai, Y., Ling, C.-C., and Bundle, D.R. (2009). *J. Org. Chem.* 74: 580–589.(c) Laurent, N., Lafont, D., Boullanger, P., and Mallet, J.M. (2005). *Carbohydr. Res.* 340: 1885–1892.
28 (a) Compton, D.L., Laszlo, J.A., and Evans, K.O. (2013). *J. Am. Oil Chem. Soc.* 90: 1397–1403.(b) Kamaike, K., Uemura, F., Yamakage, S.-I. et al. (1987). *Nucleosides Nucleotides Nucleic Acids* 6: 699–736.
29 Ekholm, F.S., Ardá, A., Eklund, P. et al. (2012). *Chem. Eur. J.* 18: 14392–14405.

9

De Novo Asymmetric Synthesis of Oligosaccharides Using Atom-less Protecting Groups

Debarpita Ray and George A. O'Doherty

Northeastern University, Department of Chemistry and Chemical Biology, 360 Huntington Avenue, Boston, MA, 02115, USA

9.1 Introduction

From its outset, protecting groups play an integral role in the development and evolution of organic synthesis [1]. As the targets of these synthetic efforts grew in complexity, the use of protecting groups in these syntheses grew. Of particular importance is the use of alcohol protecting groups. This is particularly true in the field of carbohydrate synthesis. Despite the ubiquitous use of protecting groups by organic chemists, this has always occurred with a degree of trepidation. The use of protecting group incurs the use of additional protection and deprotection steps and the generation of unwanted by-products. In fact, synthetic efficiency is often defined by the minimization of protecting groups [2].

In this regard, the minimization of protection/deprotection steps is often a critical step in making synthetic processes greener. Minimization of protecting groups in synthesis appeals to atom economic concerns. However, there are many situations where the use of protecting group is considered unavoidable. This need for protecting groups is most easily seen in carbohydrate chemistry. Perhaps the clearest example of this is in the synthesis of oligosaccharides (Scheme 9.1). For example, the simple dimerization of glucose (**1**) to make the disaccharide **5** requires two separate multistep reaction sequences. The first requires the selective differentiation of the pyranose polyol of **1** to afford glucosyl-donor **2** with the anomeric alcohol selectively converted into a leaving group and the remaining tetraol suitably protected. The second sequence selectively leaves the C-6 alcohol free while protecting the remaining C-1 to C-4 tetraols. Once suitably protected, the two sugars are joined in a Lewis-acid-promoted glycosylation reaction to stereoselectively install the glycosidic bond in **4**, which after deprotection affords disaccharide **5**. Thus, it is not inconceivable that the construction of a single anomeric bond may require 10 steps. This issue can be particularly problematic when specific protecting groups are needed to control reactivity and/or direct anomeric diastereoselectivity in the glycosylation step.

Protecting Groups: Strategies and Applications in Carbohydrate Chemistry, First Edition.
Edited by Sébastien Vidal.

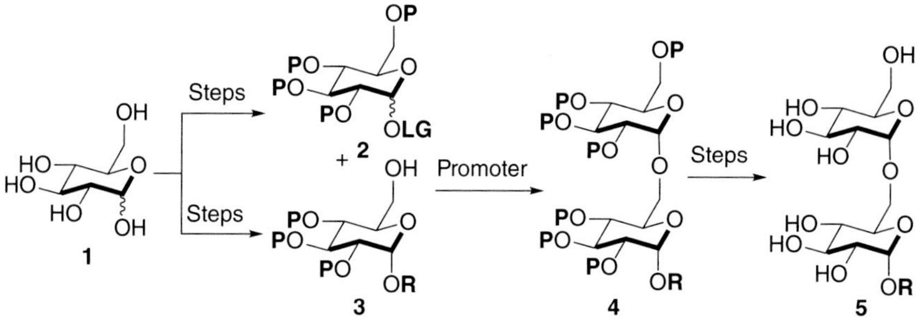

Scheme 9.1 The role of protecting groups in traditional oligosaccharide synthesis.

Herein, we review efforts to develop *de novo* asymmetric syntheses of oligosaccharides that try to minimize the use of protecting groups. Key to these approaches is the recognition that enone functionality can serve as atom-less protecting groups [3].

9.2 Atom-less Protecting Groups

Given the polyol structural realities of carbohydrates, it is reasonable to assume that protecting groups will play a critical role in oligosaccharide synthesis. Inspired by the fact that synthetic organic chemists want to minimize, if not completely remove, the use of protecting groups from oligosaccharide synthesis, the O'Doherty group has endeavored to devise an alternative approach to oligosaccharide assembly [4]. This effort has resulted in the successful development of a general approach to pyranose containing oligosaccharides [5]. Although the approach is flexible in terms of stereochemistry of the products, its greatest applicability lies in the synthesis of α-linked oligo-mannose and amicetose sugars [3, 6]. The success of this approach required the development of three novel concepts to traditional carbohydrate synthesis: (i) the use of *de novo* asymmetric synthesis of simplified glycosyl donors [7], (ii) the use of a Pd-π-allyl-catalyzed glycosylation reaction for installing the glycosidic bonds [8], and (iii) the use of a C-2 to C-4 enone functional group as a precursor to the desired sugar C-2 to C-4 polyol [9]. To help relate this latter approach to traditional carbohydrate chemistry, we coined the term "atom-less protecting groups." As we explore the potential of this approach, we often find that the use of atom-less protecting groups has the potential to significantly minimize the total number of steps used in a given synthetic sequence (*vide infra*).

9.3 *De Novo* Approach to Carbohydrates

The hexoses have long served as an inspiration for the development of new asymmetric methods. As a result, considerable effort has been made toward the synthesis of monosaccharides [10]. In addition to the development of new reactions, these efforts also provided material for the medicinal chemistry study of the structure–activity relationship (SAR) studies of carbohydrates. Of particular interest are the catalytic asymmetric routes to hexoses that start from achiral starting materials where asymmetric catalysis is used to install the stereochemistry. Traditionally, in the organic community, synthetic routes to the carbohydrates are described as "*de novo*" or "*de novo* asymmetric," whereas in the carbohydrate community, the term *de novo* often takes on a broader meaning. In this review, the term *de novo* asymmetric synthesis refers to the use of catalysis for the asymmetric synthesis of carbohydrates from achiral compounds [11]. This then precludes the inclusion of *de novo* process that produced sugars from molecules with pre-existing chiral centers (e.g. Seeberger, Reißig, and Dondoni) [12–14].

The challenge of a *de novo* synthetic approach to carbohydrates has been taken up by many groups (Scheme 9.2). This genre began with the seminal work by Masamune and Sharpless (**7** to **6**), which utilized iterative asymmetric epoxidation of allylic alcohols to prepare all eight possible hexoses [15]. Concurrent with this effort, Danishefsky demonstrated the power of asymmetric hetero-Diels–Alder reaction (**8** + **9**) for the synthesis of several sugars, which inspired further studies toward oligosaccharide synthesis [16]. In addition to chemical catalysis, enzyme catalysis has also been used for the *de novo* synthesis of monosaccharides. Examples can be seen from the use of enzymes that oxidize and/or desymmetrize achiral substrates (Johnson and Hudlicky) to achieve hexopyranoses (**10** to **6**) [17]. Alternatively, Wong and Sharpless used a combination of transition metal catalysis (asymmetric dihydroxylation) with an enzyme-catalyzed aldol reaction for the synthesis of several 2-keto-hexoses [18]. MacMillan joined this challenge by utilizing proline-catalyzed iterative aldol reaction to produce

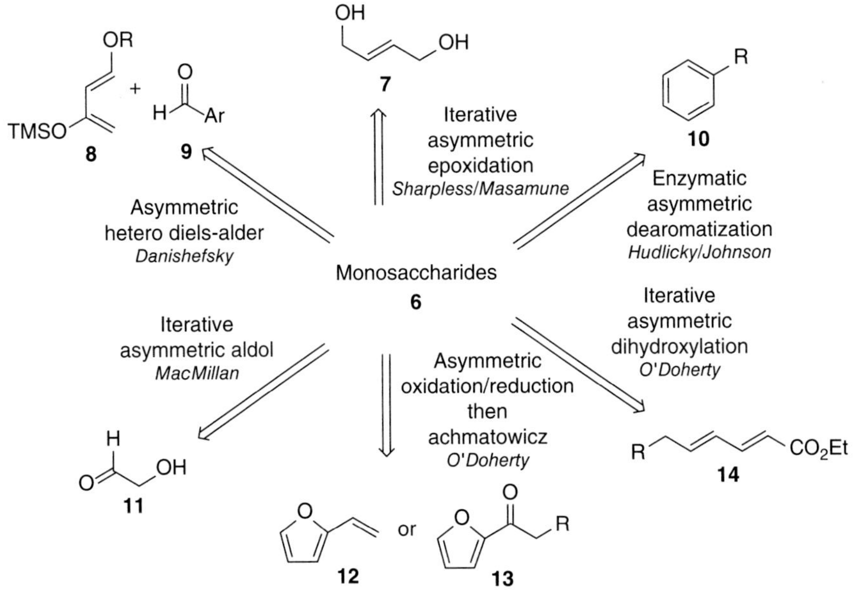

Scheme 9.2 The various *de novo* approaches to hexoses.

various hexoses (**11** to **6**) [15]. Of these approaches, only the iterative epoxidation strategy of Masamune and Sharpless provides access to all eight hexoses, but it is also noteworthy that their route required the most steps and protecting groups. It is this latter point, the reduction of steps, and avoidance of protecting groups that guided the development of these synthetic endeavors [19].

In addition to these efforts, the O'Doherty group has developed two *de novo* asymmetric syntheses of hexoses. These efforts have resulted in the discovery of two orthogonal approaches to hexopyranoses with variable C-6 substitutions. The first of these approaches involves an iterative asymmetric dihydroxylation of dienoates to produce sugar lactones (**14** to **6**) [20]. The second approach involves an Achmatowicz strategy [21], which is amenable to all eight hexose diastereomers (**12** or **13** to **6**) [22]. Of the two approaches, the dienoate approach is the most efficient in terms of steps and the minimal use of protecting groups. In contrast, the Achmatowicz approach is superior in terms of synthetic scope, as the route is particularly amenable to the synthesis of oligosaccharides. In fact, it was the application of the Achmatowicz approach to oligosaccharides that featured the use of atom-less protecting groups (*vide infra*).

9.4 O'Doherty Approach to Carbohydrates

9.4.1 The Achmatowicz Approach to the *De Novo* Synthesis of Pyranoses

A very practical aspect of the Achmatowicz approach is the easy access to enantiomerically pure furan alcohols from achiral furans (e.g. **16** and *ent*-**16**). The initial version of this approach involves the asymmetric Sharpless dihydroxylation of vinylfuran **7**, which worked well in many applications [23, 24]. However, the optimized approach synthesized the furan alcohol using a Noyori hydrogen transfer reaction for the asymmetric reduction of acylfurans (**15** to **16** or *ent*-**16**) [25]. Oxidation of the furan alcohols under aqueous conditions with many oxidants routinely provides pyranones **17** as a mixture of diastereomers (i.e. Achmatowicz rearrangements) in excellent yields. The Achmatowicz products can be diastereoselectively acylated to form α-pyranones **18**. This latter approach allowed us to readily prepare furan alcohols **16** and the corresponding pyranones **18** on 100 g scale (Scheme 9.3).

Although the Achmatowicz approach is applicable for many sugar diastereomers, it is most practically applied for the *de novo* synthesis of sugars with manno-/rhamno-stereochemistry (Scheme 9.4). Thus, in only 4 steps, furan alcohol *ent*-**16b** can be stereoselectively converted into manno-hexoses **20**, and in a few more steps to talo-sugar **21** and gulo-sugar **22** diastereomers. The route begins with the *N*-bromosuccinimide (NBS) oxidation of furan alcohol *ent*-**16b** to a 6-hydroxy-2H-pyran-3(6H)-one, which, after a α-selective benzoylation, affords benzoyl-protected **19**. A highly diastereoselective Luche reduction ($NaBH_4$/$CeCl_3$) converts **19** to allylic alcohol **20**. A subsequent Upjohn dihydroxylation ($OsO_{4(cat)}$/NMO (*N*-Methylmorpholine *N*-oxide)) gives manno-pyranose **22** as a single diastereomer. By incorporating a Mitsunobu/hydrolysis (**20** to **21**) into the

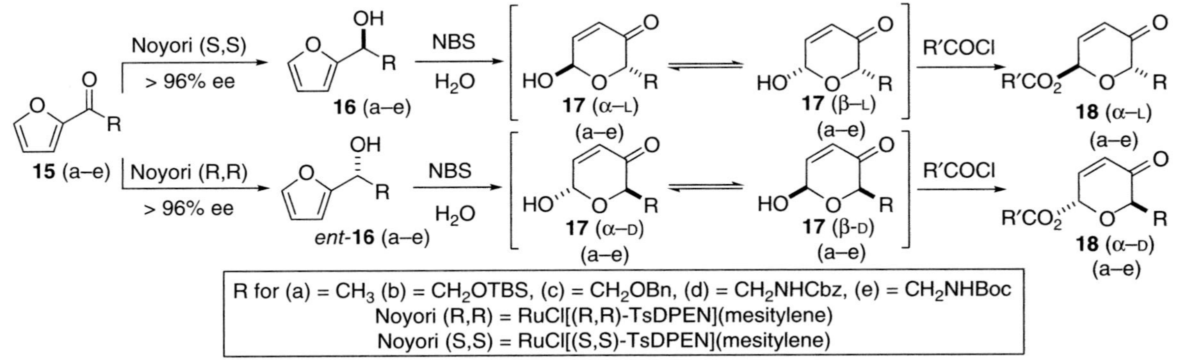

Scheme 9.3 Asymmetric synthesis of furan alcohols and pyranones.

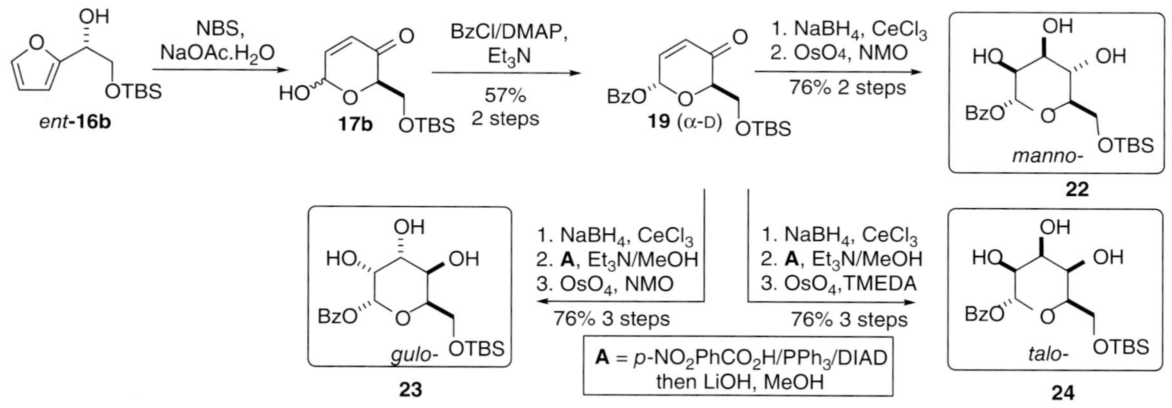

Scheme 9.4 *De novo* asymmetric approach to α-L-*manno*-, *talo*-, and *gulo*-pyranose.

sequence, the route can be expanded to the synthesis of talo- and gulo-pyranose. Exposure of the diastereomeric allylic alcohol **21** to the same Upjohn dihydroxylation ($OsO_{4(cat)}$/NMO) provides the gulo-pyranose **23**, whereas a hydroxy-directed dihydroxylation (OsO_4/TMEDA (tetramethylethylenediamine)) affords the diastereomeric talo-pyranose **24**. Other deoxysugars and deoxyamino sugars were also prepared by using this approach. An interesting feature of this approach is the use of C=C and C=O π-bonds functionality as triol precursor. This transformation, we will see, in the oligosaccharide settings allows the enone functionality to serve as an atom-less protecting group [9, 22].

In addition to serving as a triol precursor, the enone functionality in the pyranones can be converted into simple C-4 alcohol and C-3/4 diols (Scheme 9.5). This is most easily accomplished with the diimide reduction of allylic alcohol **26** to give the amecitose sugar **27**. By first including a Mitsunobu inversion of the C-4 alcohol before the diimide reduction, the enone can also be converted into rhodinose sugar **29**. By applying the Myers 1,3-allylic reduction on the allylic alcohol (**31** to **32**), pyrans with a C-3/4 double bond can be produced. A highly diastereoselective dihydroxylation of the β-pyrans such as **32** can be used to give sugars with digitoxose stereochemistry, as in **33**.

When the enone functionality is exposed to a nucleophilic epoxidation and Wharton reaction, rearranged C-2 allylic alcohols (**34** to **36** via **35**) can be readily produced (Scheme 9.6) [26]. The C-2 allylic alcohol **36** can serve as precursors to various rare sugar structural motifs. For instance, a diastereoselective dihydroxylation of **36** gives altro-sugar **37**, a diimide reduction gives 3,4-dideoxysugar **38**, and regioselective and diastereoselective hydration (*N*-iodosuccinimide (NIS)/H_2O then $LiAlH_4$) forms ascarylose sugar **39**.

The same C-2 allylic alcohol Wharton rearrangement products can also serve as precursors for C-2 enones after MnO_2 oxidation (**36** to **40**). Once again, a highly diastereoselective $NaBH_4$ reduction of the enone **40** affords the epimeric C-2 allylic alcohol **41**, which can serve as precursors to various rare sugar structural motifs (Scheme 9.7). For instance, a dihydroxylation of **41** gives a mixture of fuco- and allo-sugars (**42** and **43**), where the fuco-isomer is major. A 2-nitrobenzenesulfonylhydrazide (NBSH) or diimide reduction of **41** gives 3,4-dideoxysugar **44**, and regioselective and diastereoselective hydration (NIS/H_2O then $LiAlH_4$) forms 4-deoxy allose sugar **45** [27].

Finally, the epimeric C-2 allylic alcohol **36** can be diastereoselectively epoxidized to **46**, which can serve as precursors to various rare sugar structural motifs (Scheme 9.8). For instance, $LiAlH_4$ reduction of **46** affords the C-3 deoxy rare sugar 2-epi-colitose **47**. More interestingly, the epoxide can be regiodivergently opened with hydroxide/water nucleophile to give different sugar isomers. Thus, under acidic conditions, the epoxide in **46** can be hydrolyzed to selectively give sugars with rhamno-stereochemistry **48**. Alternatively, under basic conditions, the same epoxide **46** can be opened to give sugars with ido-stereochemistry **49**.

In addition to various monosaccharides, this Achmatowicz approach has been applied to various carbohydrate-containing natural products, for example, papulacandins [28], swainsonine [29], SL0101 [30], daumone [31], methymycin [32], gilvocarcin [33], and jadomycin [34]. A key feature to these approaches is the role of the enone group to serve as a precursor to a wide range of functionalities, as

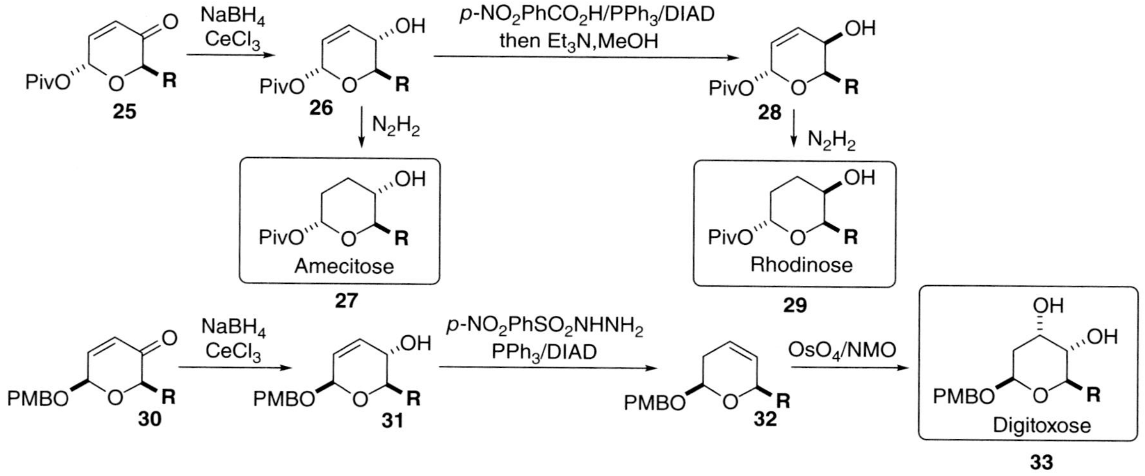

Scheme 9.5 *De novo* asymmetric approach to deoxy-pyranones.

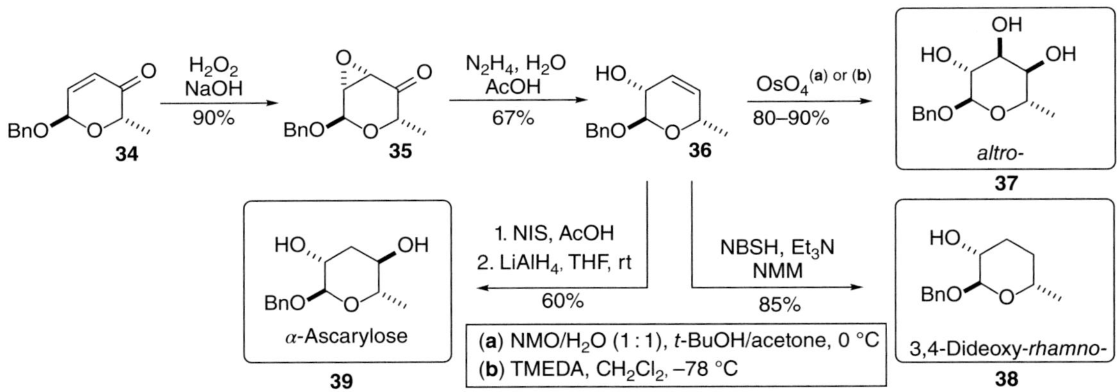

Scheme 9.6 Application of the Wharton rearrangement for the synthesis of rare sugars.

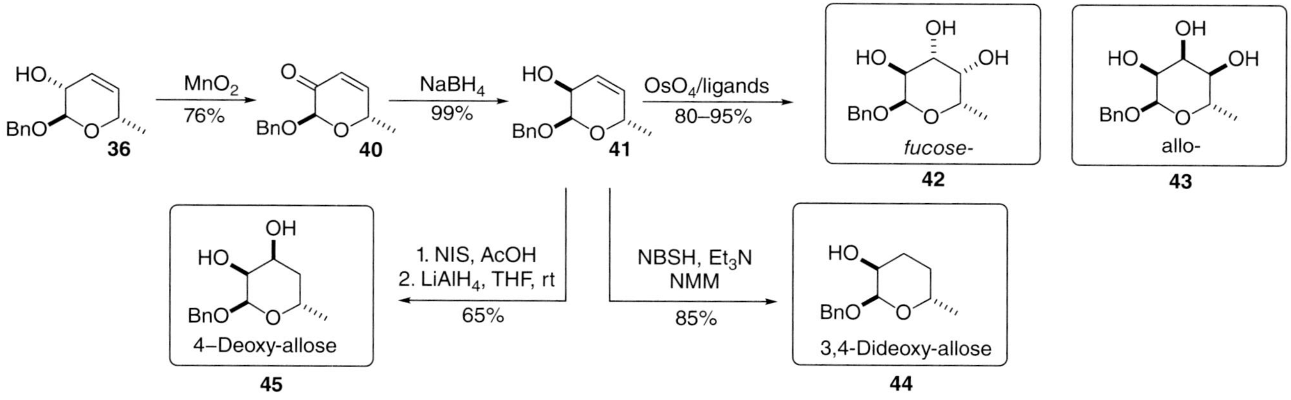

Scheme 9.7 Use of pyran-2-ones for the *de novo* synthesis of pyranoses.

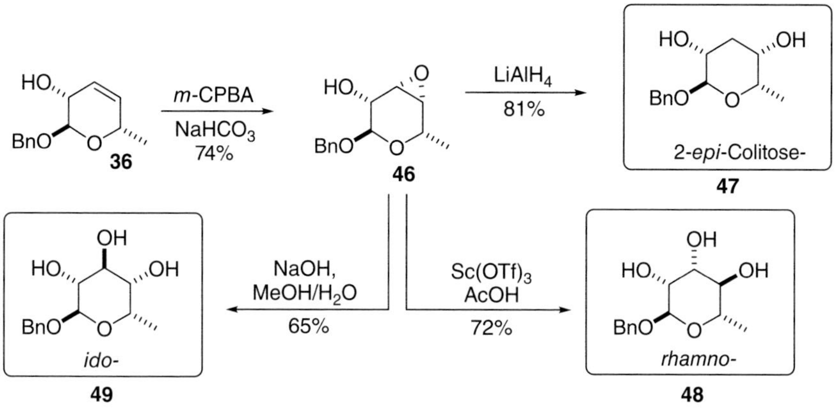

Scheme 9.8 *De novo* access to 2-epi-, idose-, rhamno-colitose.

opposed to serving as a polyol protecting group. In contrast, it is the synthetic application of this Achmatowicz approach to oligosaccharides where the atom-less protecting group can be seen. A key to applying this methodology to oligosaccharide-like targets is the ability for these pyranone precursors to serve as glycosyl donors in Pd-catalyzed glycosylation reactions.

9.4.2 Palladium-catalyzed Glycosylation of Pyranoses

Of all the various *de novo* routes to carbohydrates, the Achmatowicz approach uniquely distinguishes itself from the other approaches when it comes to access to oligosaccharides. These features result from its compatibility with the Pd-π-allyl-catalyzed glycosylation for the stereospecific formation of the glycosidic bond (Scheme 9.9) [8, 35]. In the Pd(0)-glycosylation reaction, the anomeric ester/carbonate group serves as the leaving group, which can be replaced with a wide range of alcohol nucleophiles in a general and stereospecific manner [36, 37]. The reaction occurs rapidly and in high yields for both the **50(α-L)** to **51(α-L)** and **50(β-L)** to **51(β-L)**. Although the reaction (**50** to **51**) does not, at first glance, look like a typical glycosylation reaction, as previously discussed, the resulting enone in these glycosylation products can be converted into mono-, di-, and triol products. These post-glycosylation reactions transform two diastereomeric pyranone transfer reactions into a wide range of traditional glycosylation products (*vide infra*). Critical to the success of these approaches is the chemoselective use of C=C and C=O π-bond functionality, as atom-less protecting groups (i.e. enone as a protected triol) as well as an anomeric directing group (via a Pd-π-allyl) [8, 35–37]. The power of this approach becomes apparent when it is applied in oligosaccharide synthesis.

9.4.3 Unnatural Oligosaccharides

The first applications of this approach to oligosaccharide assembly were in the synthesis of α-linked unnatural oligosaccharides in either D-, L-, or mixed

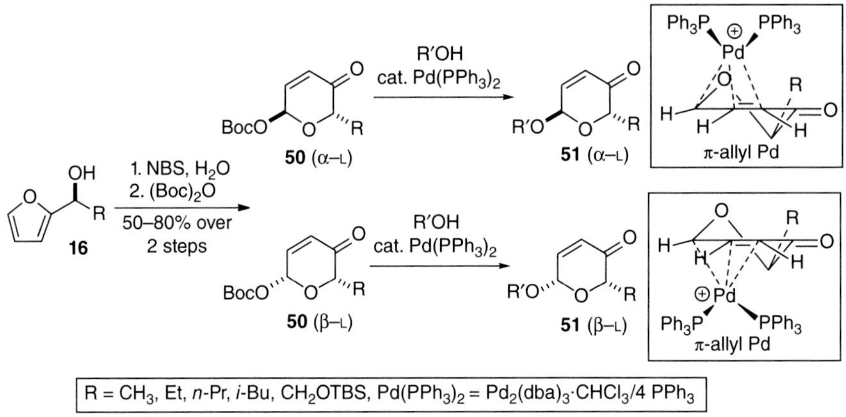

R'OH	Yield of α	Yield of β	R'OH	Yield of α	Yield of β
MeOH	87	85	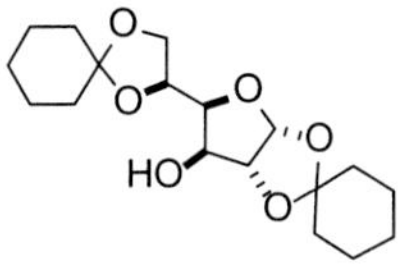	82	79
BnOH	89	85			
t-BuOH	78	75			
Cyclohexanol	88	80			
Menthol	82	72			

Scheme 9.9 Palladium-catalyzed glycosylation.

D-/L-stereochemistry (Schemes 9.10–9.14) [38]. In addition to providing access to difficult-to-access rare oligosaccharide motifs, this approach was noteworthy in its overall synthetic efficiency in terms of number of steps and minimization of protecting groups. Specifically, the step savings occur from the bidirectional use of post-glycosylation transformations (e.g. one-pot iterative Luche reductions and Upjohn dihydroxylation).

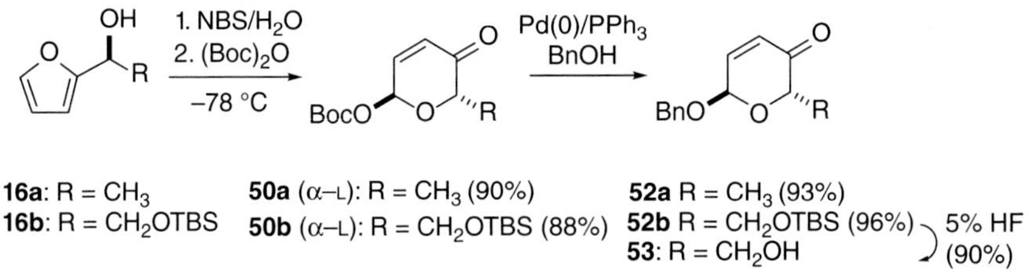

Scheme 9.10 *De novo* synthesis and use of pyranone-based glycosyl donor.

The synthesis of the L-sugar glycosyl donors **50a** and **50b** is outlined in Scheme 9.10. Thus, in only 2 steps, glycosyl donors **50a** and **50b** could be prepared from furan alcohols **16a** and **16b**. Using the Pd(0)-catalyzed glycosylation of benzyl alcohol (**50a**/**50b** to **52a**/**52b**), an anomeric benzyl protecting group could be installed. The pyranone **52b** could be used for the synthesis of 1,6-linked oligosaccharides (Scheme 9.11) [6], which began with the removal of the *tert*-butyldimethylsilyl (TBS) group with HF (**52b** to **53**). A Pd-glycosylation of

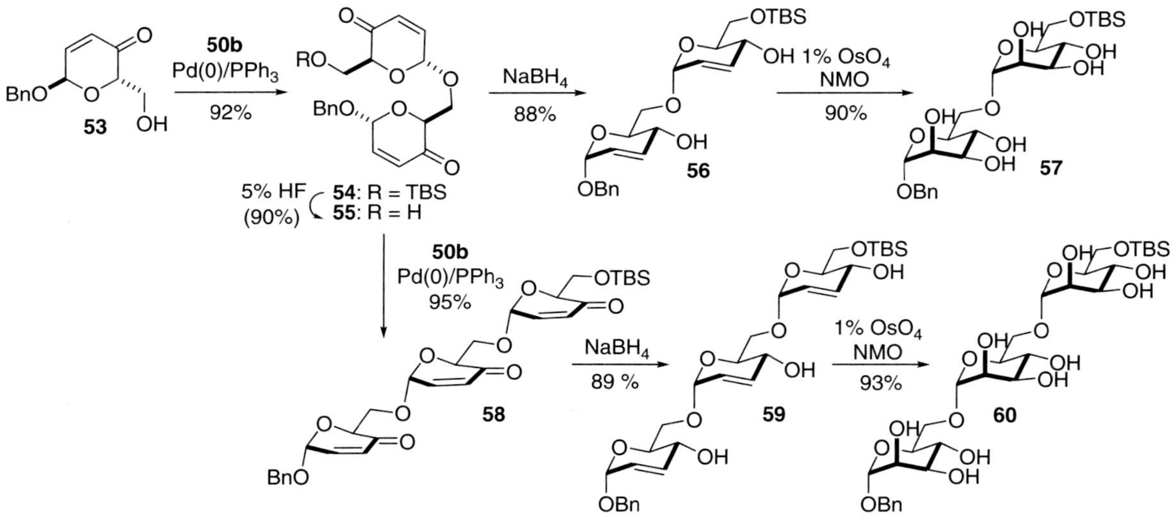

Scheme 9.11 *De novo* synthesis of 1,6-linked mannose disaccharide and trisaccharides.

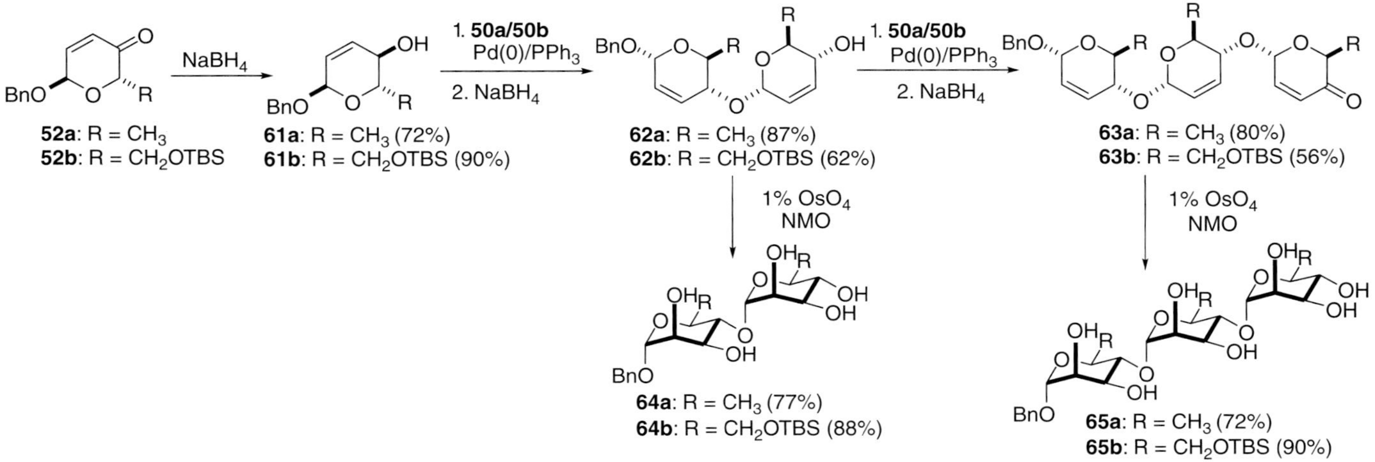

Scheme 9.12 *De novo* synthesis of 1,4-linked mannose disaccharide and trisaccharides.

53 with **50b** gave bis-pyranone **54**, which, upon a subsequent TBS-deprotection and Pd-glycosylation, gave tris-pyranone **58**. Exposing both bis-pyranone **54** and tris-pyranone **58** to the $NaBH_4$ reduction and OsO_4 dihydroxylation cleanly gave the desired di- and trisaccharides **57** and **60** in excellent yields and diastereoselectivities. Although TBS-protecting groups are still required, its use was kept to a minimum (i.e. one TBS group/sugar). This highly atom economical approach avoids the extensive use of protection/deprotection steps. For example, the 1,6-manno-trisaccharide **60** was prepared from enone **52b** in 6 steps (nine from furan **15b**). A key to the step savings is the use of bidirectional post-glycosylation Luche and Upjohn reactions.

The same step saving and protecting group minimization advantages can be seen in the *de novo* asymmetric synthesis of 1,4-linked oligosaccharides (Scheme 9.12). The application of this approach to 1,4-oliogsacchirdes begins, by simply switching the order of the reduction and glycosylation steps. For example, in only 6 steps from pyranones **52a/52b,** 1,4-manno-trisaccharides **65a/65b** can be selectively prepared. A key to the success of this sequence was the highly stereoselective reduction and dihydroxylation reaction, which installed six stereocenters in one transformation (**63** to **65**).

To demonstrate the synthetic power of this approach, we applied this approach to the synthesis of a library of 1,4-linked oligosaccharides (Scheme 9.13).

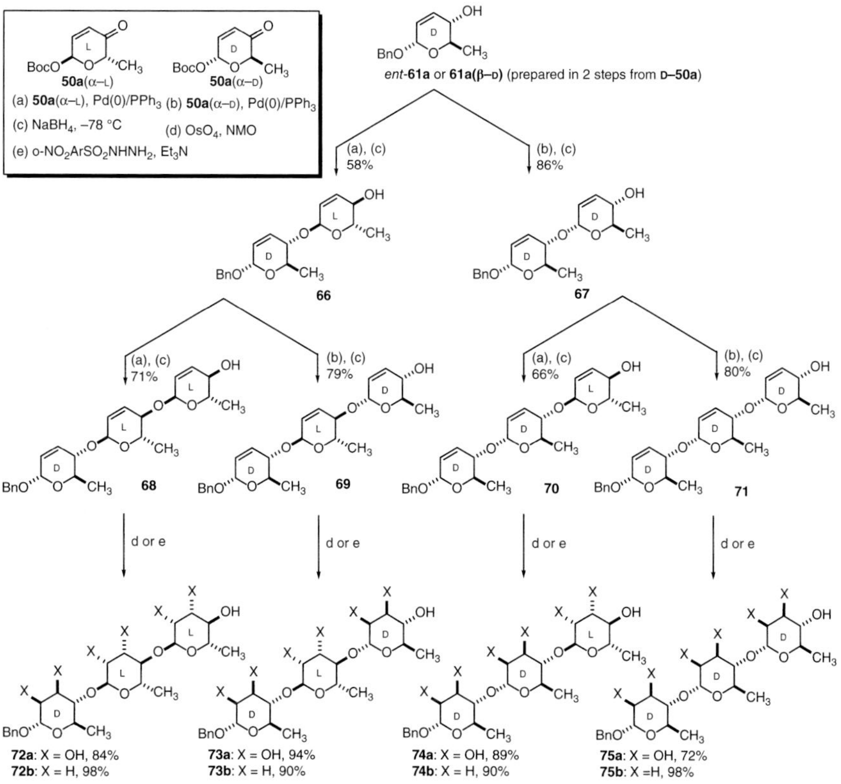

Scheme 9.13 *De novo* synthesis of a library of D-/L-1,4-linked α-*rhamno*-trisaccharides.

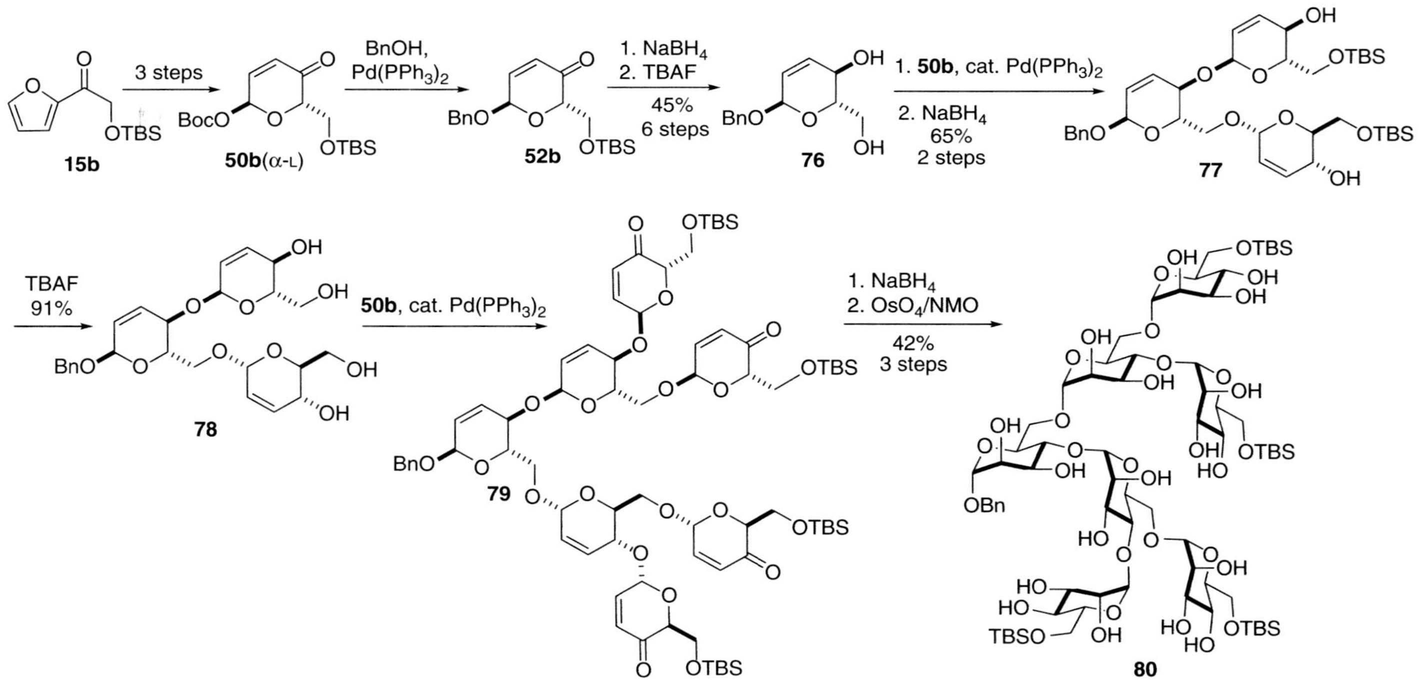

Scheme 9.14 Synthesis of highly branched 1,4-/1,6-heptasacharides.

Specifically, we prepared all the possible D-/L-diastereomers of **65a** in the same 6 steps. All that is required is having access to both enantiomers of **50a**. More recently, this approach was successfully used in the medicinal chemistry SAR study of digitoxin oligosaccharides, an anticancer agent [39].

The synthetic advantage of bidirectional synthesis can also be gained by applying it to the glycosylation step to the C-4 and C-6 positions (Scheme 9.14). This began with enone reduction and removal of the TBS group of pyran **52b** to form diol **76**. A subsequent bis-glycosylation of diol **76** with 2 eq. of **50b** gave *tris*-pyran **77** after enone reduction. Once again, a bidirectional TBS deprotection gave tetraol **78**. The tetraol **78** can be per-glycosylated with excess pyranone **50b** to give heptasaccharide **79**. Finally, a ketone per-reduction and double-bond per-dihydroxylation gave heptasaccharide **80**. It is worth noting that, although there is a similar local symmetry around each alkene, they exist in different stereochemical environments. The synthesis of **80** is accomplished from the α-L-donor **50b** in 9 steps by simply applying the Pd-glycosylation/post-glycosylation strategy three times in a tandem 1,4- and 1,6-bidirectional synthesis. Thus, the heptasaccharide **8** is produced with 35 stereocenters (14 stereocenters in 1 step, **79** to **80**) in an excellent overall yield [3].

Clearly, these *de novo* approaches can provide routes to complex carbohydrate structures in routes that are commensurate to traditional carbohydrate approaches. These *de novo* synthetic approaches also assemble oligosaccharides without the generation of by-products associated with traditional protecting group strategy.

9.4.4 Natural Oligosaccharides

In addition to unnatural oligosaccharide targets, the *de novo* Achmatowicz methodology is also amenable for the synthesis of the oligosaccharide portion of natural products (e.g. anthrax tetrasaccharide [40] and merremoside [41]). These approaches can often allow for the synthesis of these oligosaccharides with the minimal use of protecting groups as it features the use of enones as atom-less protecting groups.

An early example of this approach can be seen in the synthesis of the disaccharide fragment of the mannopeptimycins (Scheme 9.15). Mannopeptimycin-ε, is a cyclic hexapeptide that was isolated from *Streptomyces hygroscopicus* LL-AC98 [42]. The key structural feature of the natural product is the *O*-glycosylated tyrosine with an α-(1,4-linked)-bis-manno-pyranose disaccharide attached (e.g. **86**). Our *de novo* approach to disaccharide **86** began with the Pd-glycosylation of protected tyrosine **81** with the C-6 OTBS-protected glycosyl donor **50b**. Although the route required the use of two TBS groups, the bidirectional nature of post-glycosylation keeps the protecting groups to a minimum as well as the total steps. As such, the linear route involved the application of the iterative glycosylation and reduction (**81** to **83** and **83** to **84**), and C-4 acylation with isovaleryl acid gave disaccharide **85**. Finally, a bis-dihydroxylation of the two alkene protecting groups and a tetrabutylammonium fluoride (TBAF) bis-deprotection of the two C-6 TBS-protecting groups gave the desired tyrosine disaccharide target **86** in only 7 steps (Scheme 9.15) [43, 44]. As part of further

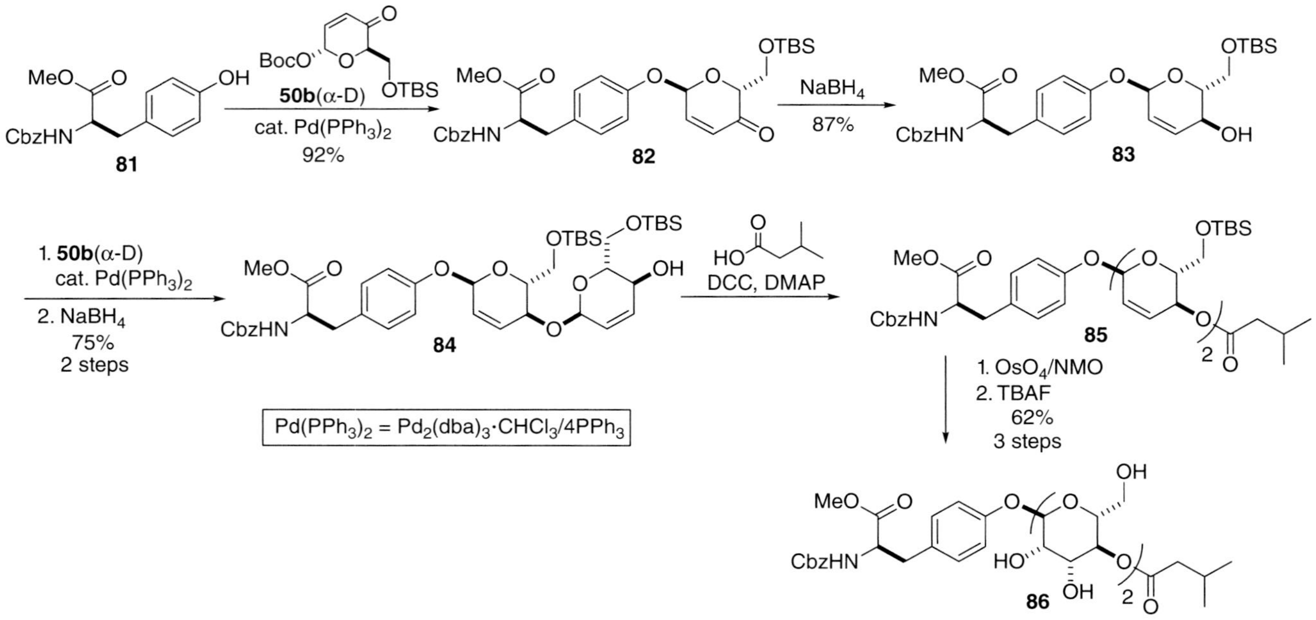

Scheme 9.15 Synthesis of glycosylated tyrosine portion of mannopeptimycin-ε.

SAR studies of the antibiotic mannopeptimycin, access to the C-4 amide analog was also prepared [43].

This *de novo* approach can also be used for the synthesis of the trisaccharide portion of the glycosylated angucycline antibiotic natural products PI-080 (Schemes 9.16). This approach was built on our post-glycosylation strategy that installed the β-D-digitoxose sugar stereochemistry in digitoxin [45], which is modified so as to address the first β-D-olivose sugar of the trisaccharide **94**. Thus, in 4 steps, the β-D-pyranone glycosyl-donor **50a(β-D)** was used to glycosylate *p*-methoxybenzyl alcohol to form **87**. An unselective 1,2-Luche reduction followed by Myers reductive rearrangement converted **87** into the C-2 deoxy-β-D-pyran **88**. A highly diastereoselective dihydroxylation of **88** was used to install the digitoxose stereochemistry in **89**. A highly regioselective and stereoselective Mitsunobu-like inversion of the C-3 axial alcohol of the *cis*-diol in **89** converted it into **90** with its olivose stereochemistry. It is important to note that this transformation in combination with a highly diastereoselective dihydroxylation reaction becomes a nice solution to the problem of 1,2-*trans*-diequatorial addition to a cyclohexene. In addition to a regioselective inversion, the Mitsunobu chemistry also differentially protects the diol for further glycosylation with α-L-glycosyl donor **50a(α-L)** to give **91**. A 1,2-reduction and Mitsunobu inversion (**91** to **92**) was used to install the rhodinose stereochemistry of the second sugar in disaccharide **92**. A regioselective ester hydrolysis and diimide reduction were used to give disaccharide **93**. Finally, a Pd-glycosylation with α-L-glycosyl donor **50a(α-L)** gives the PI-080 trisaccharide **94**, where only the *p*-methoxybenzyl (PMB) group functioned exclusively as a protecting group [46]. Although the *p*-NO_2-benzoate group in olivose sugar **90** functions as a protecting group, it played a primary synthetic function of inverting the C-3 alcohol stereochemistry. We have found this approach to be a particularly powerful approach to deoxy sugars. In fact, a related approach was also used to prepare the trisaccharide carbohydrate portion of landomycin A [47].

This *de novo* approach can also be used for the synthesis of more complex oligosaccharide structural motifs. We believe that this was nicely displayed in the synthesis of cleistriosides (**95** and **96**) and cleistetrosides (**97** to **102**), which are reported to possess interesting antibacterial properties [48, 49, 50]. In addition to clarifying the structural issues, the *de novo* synthesis of these oligosaccharides can be used to supply sufficient material for SAR-type studies as well as serve as a test for this synthetic methodology. The approach is outlined in Schemes 9.17–9.20.

In contrast to the other 1,4- and 1,6-linked oligosaccharide motifs that have been addressed with *de novo* approach, the cleistriosides and cleistetrosides have a 1,3-linked di- and trisaccharide portion that traditionally need protecting groups to address this regiochemistry problem.

The *de novo* approach to the trisaccharide portion for these two classes of natural products is outlined in Scheme 9.19. Although there is a minimal amount of protecting groups, the route did require the use of a C-2/3 acetonide protecting group for the first sugar and a C-2 chloroacetate group for the second sugar. The route begins with pyranone α-L-**50**, which, in 4 steps (glycosylation, reduction,

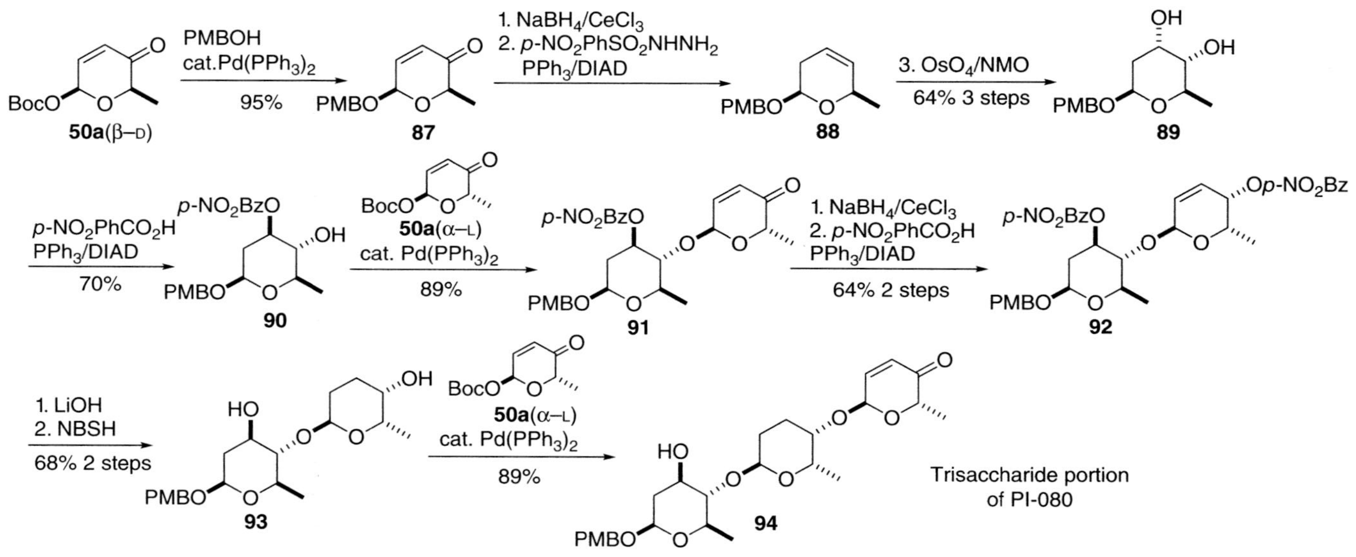

Scheme 9.16 Synthesis of the trisaccharide portion of PI-080.

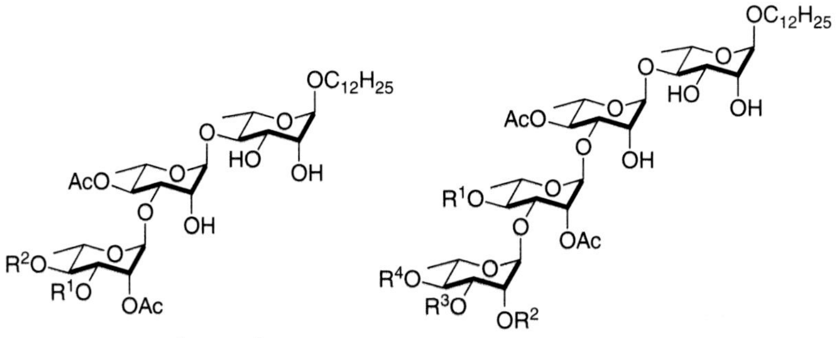

Cleistrioside-5 (**95**): R^1 = H, R^2 = Ac
Cleistrioside-6 (**96**): R^1 = Ac, R^2 = H

Cleistetroside-2 (**97**): R^1 = R^4 = Ac, R^2 = R^3 = H
Cleistetroside-3 (**98**): R^1 = R^2 = Ac, R^3 = R^4 = H
Cleistetroside-4 (**99**): R^1 = Ac, R^2 = R^3 = R^4 = H
Cleistetroside-5 (**100**): R^1 = R^2 = R^3 = R^4 = H
Cleistetroside-6 (**101**): R^1 = R^2 = R^4 = Ac, R^3 =H
Cleistetroside-7 (**102**): R^1 = R^2 = R^3 = R^4 =Ac

Scheme 9.17 Cleistriosides and cleistetrosides family of natural products.

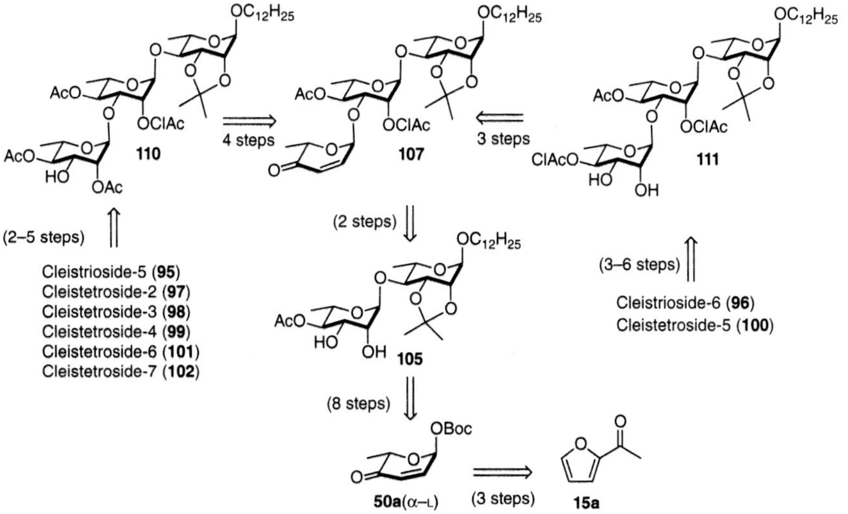

Scheme 9.18 Retrosynthetic route to cleistetrosides and cleistriosides.

dihydroxylation, and acetonide protection), can be converted into protected rhamnose **104**. A subsequent glycosylation (with α-L-**50**), reduction, acylation, and dihydroxylation affords disaccharide **105**. A tin-directed regioselective (7 : 1) glycosylation of the diol in **105** was used to install a pyrano-sugar at the C-3 position to give **106**. Finally, a C-2 chloroacylation was used to give trisaccharide **107**, which was the linchpin molecule for the synthesis of the two classes of natural products.

Subsequent elaborations of the pyranone sugar in **107** into the required rhamnose stereochemistry were pursued for the synthesis of the cleistriosides

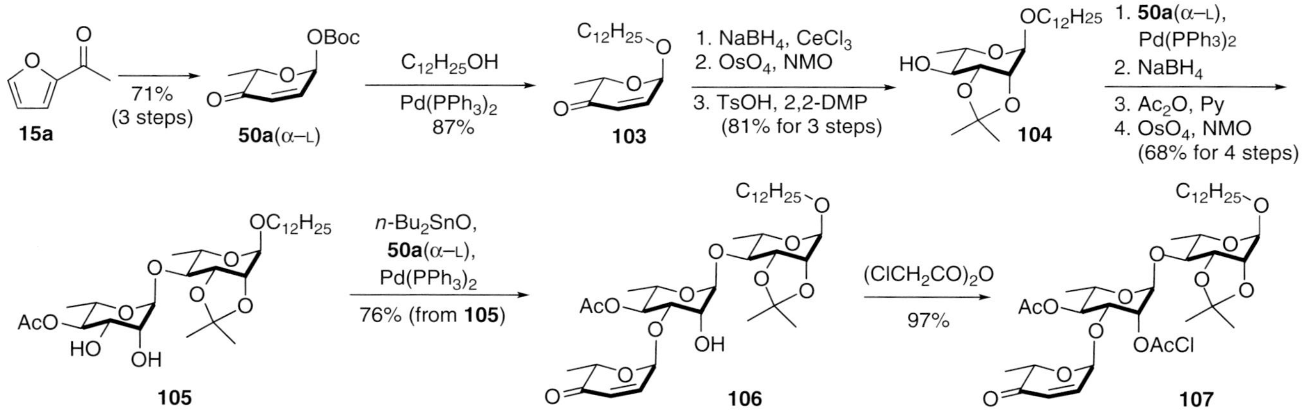

Scheme 9.19 Preparation of trisaccharide intermediate **107** for the syntheses of cleistriosides and cleistetrosides.

and cleistetrosides without the further use of protecting groups. This was accomplished for cleistrioside 5 and cleistetrosides 2 and 6, by taking advantage of highly regioselective acylation chemistry (Scheme 9.20). For example, in only 4 steps, **107** can be converted into **110** (e.g. reduction, acylation, dihydroxylation, and C-2 acylation), which served as the linchpin molecule for the synthesis of cleistrioside and cleistetrosides. A 2-step deprotection of **110** gave cleistrioside 5 (**95**). Another Pd-catalyzed glycosylation, reduction, acylation, and dihydroxylation gave **112**, which, in 2 steps, could be converted into cleistetroside 2 (**97**). Finally, a C-2 acylation of **112** gave **113**, which, in 2 steps, could be converted into cleistetroside 6 (**101**). With the addition of one more chloroacylation step, the same route was also amenable for the synthesis of the remaining cleistrioside and cleistetroside. It is worth noting that the route to any of the cleistetrosides is quite comparable to the two previously reported routes to cleistetroside 2, in terms of total number of steps. What distinguished it from these more traditional routes is its flexibility to diverge to any of the possible natural product isomers [51, 52].

With the successful synthesis and biological investigation of the cleistrioside and cleistetrosides, we next pursued the syntheses of the related oligo-rhamnose natural products, the mezzettiasides (Scheme 9.21). The mezzettiaside family of natural products represents a large class of ten 1,3-linked oligo-rhamnose-containing natural products. The class consists of three disaccharide, four trisaccharide, and three tetrasaccharide members, where the specific acylation pattern distinguishes the members of each set of di-, tri-, and tetrasaccharides. Once again, our *de novo* synthesis of all the members of this class of natural products featured our minimal use of protecting groups, which was enabled by a combination of highly regioselective catalytic reactions [53].

Our retrosynthetic approach to the three sets of mezzettiasides is outlined in Scheme 9.22. A key to the success of the approach was the recognition that both the trisaccharides (**117**–**120**) and the tetrasaccharides (**121**–**123**) could come from common intermediate pyranone **132**. In turn, **132** could come from disaccharide allylic alcohol **130**, which could also be used for the synthesis of the disaccharide mezzettiasides (**114**–**116**). By using the *de novo* asymmetric method, the allylic alcohol **130** could be readily prepared from acetylfuran **15a**.

The synthesis of the key disaccharide **130** began with the 3-step asymmetric synthesis of ubiquitous Pd(0) α-L-glycosyl donor **50a** from **15a** (Scheme 9.23). The first rhamnose sugar was introduced onto *n*-octanol by a Pd-glycosylation, Luche reduction, acylation, and dihydroxylation sequence (**50a** to **126**). The second sugar was regioselectively introduced onto the diol **126** by one of the two methods. The first being the stoichiometric tin acetal procedure, similar to what was described in the cleistetroside synthesis. The second method utilized a catalytic boron and palladium protocol, which provided the desired disaccharide **128** in greater yield (74%) and regioselectivity (~8 : 1). A subsequent chloroacylation and Luche reduction gave the first linchpin molecules for the synthesis of disaccharide **130**. Our preferred precursor for the boron catalyst was the Taylor catalyst ($Ph_2BOCH_2CH_2NH_2$) [54], which was

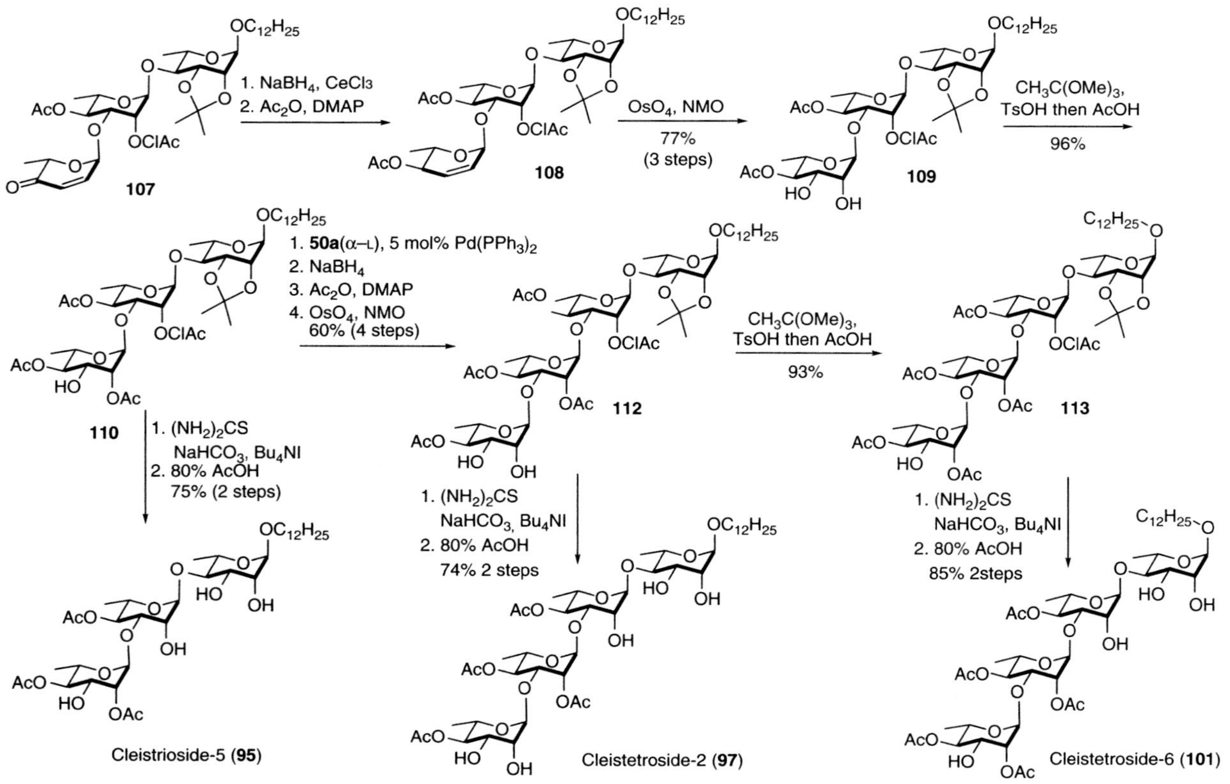

Scheme 9.20 Example of divergent synthesis of cleistrioside and cleistetrosides.

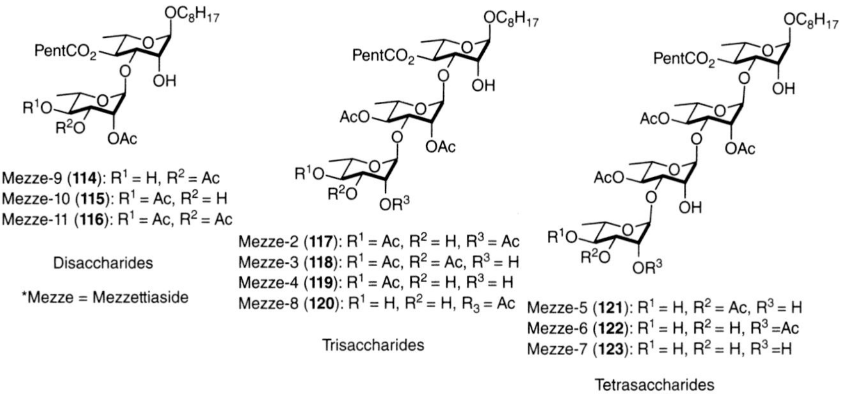

Scheme 9.21 Mezzettiaside family of natural products.

also used for subsequent regioselective acylation and glycosylation steps (Schemes 9.24–9.26).

The synthesis of the four disaccharide mezzettiasides began with disaccharide **130** (Scheme 9.24). Thus, in only 4 steps (chloroacylation, dihydroxylation, acylation, and deprotection), **130** was converted into mezzettiaside-9 (**114**). In 3 steps (dihydroxylation, per-acylation, and deprotection), **130** was converted into mezzettiaside-111 (**116**). Finally, in only 4 steps (acylation, dihydroxylation, C-2 acylation, and deprotection), **130** was converted into mezzettiaside-10 (**115**). The penultimate intermediate **131** in this latter synthesis (i.e. before the chloroacetate deprotection was used in the synthesis of the larger mezzettiasides) (Schemes 9.24 and 9.25).

The synthesis of the three trisaccharide mezzettiasides began with the Pd-glycosylation of disaccharide **131** to give trisaccharide **132** (Scheme 9.25), which in 3 steps (reduction, acylation, and dihydroxylation) was converted to the trisaccharide **133** with a C-4 acyl-group on the terminal rhamnose ring. From these two trisaccharides **132** and **133**, all the trisaccharide mezzettiasides can be prepared. Thus, from **132**, a 4-step reduction, dihydroxylation, C-2 acylation, and deprotection, gave mezzettiaside-8 (**120**). Simply deprotecting the chloroacetate group of **133** gave mezzettiaside-4 (**119**). Addition of a C-2 acylation before the deprotection of chloroacetate was used to convert **133** to mezzettiaside-2 (**117**), whereas the addition of a C-3 acylation (via borinate catalysis) before the deprotection of the chloroacetate was used to convert **133** to mezzettiaside-3 (**118**).

The synthesis of the remaining three tetrasaccharide mezzettiasides began with disaccharide **133** (Scheme 9.26). A regioselective boron-mediated C-3 glycosylation **133** gave tetrasaccharide **134**. A Luche reduction, dihydroxylation (**134** to **135**), C-3 acylation, and chloroacetate deprotection was used to obtain mezzettiaside-5 (**121**). The other two tetrasaccharides came from the common intermediate **135**. A simple thiourea-promoted removal of the chloroacetate from **135** gave mezzettiaside-7 (**123**), whereas a C-3 acylation and chloroacetate deprotection were used to obtain mezzettiaside-6 (**122**).

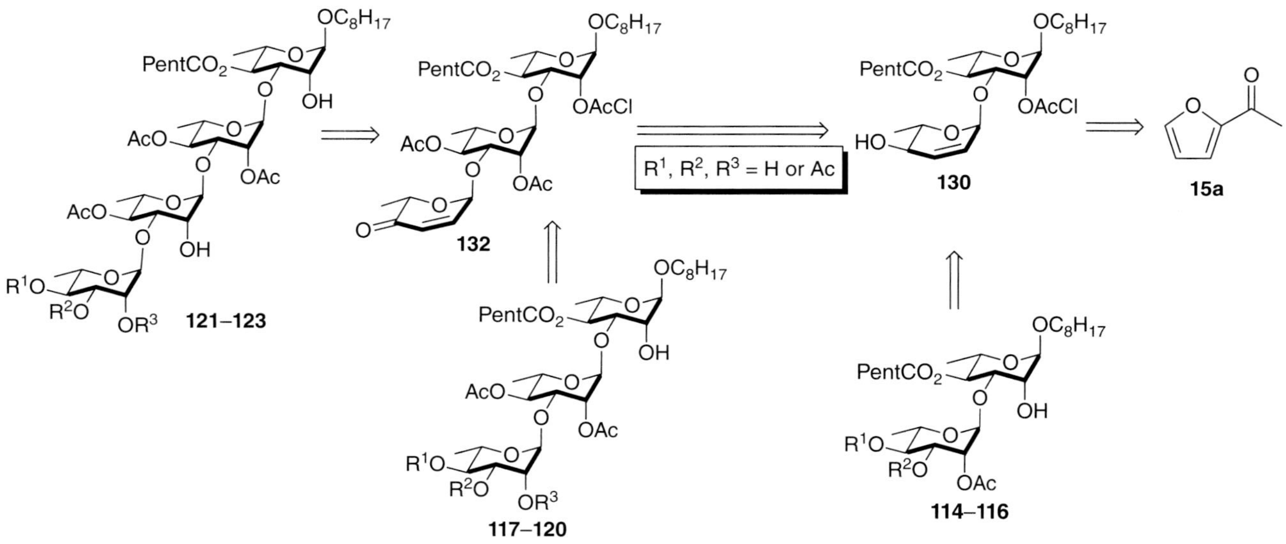

Scheme 9.22 Retrosynthetic scheme for the divergent synthetic route to mezzettiasides.

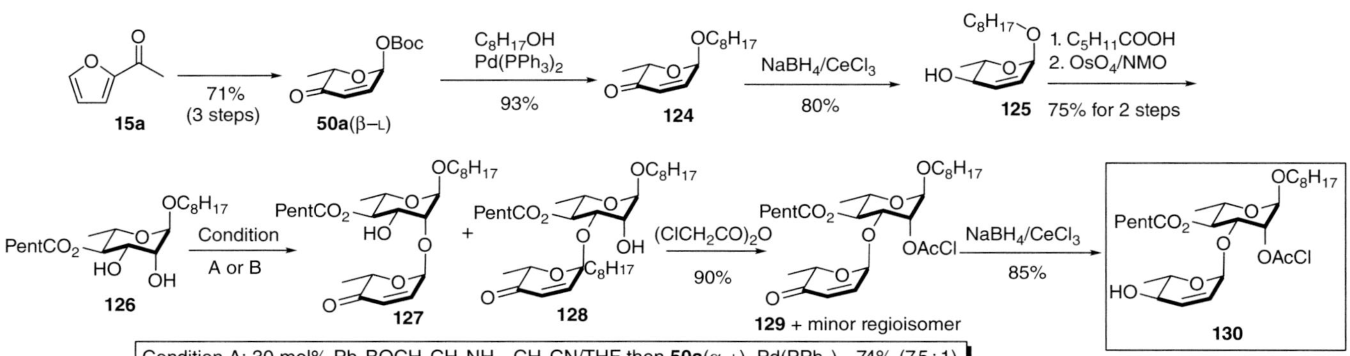

Condition A: 30 mol% $Ph_2BOCH_2CH_2NH_2$, CH_3CN/THF then **50a**(α–L), $Pd(PPh_3)_2$, 74% (7.5 : 1)
Condition B: *n*-Bu_2SnO, toluene/reflux then **50a**(α–L) $Pd(PPh_3)_2$, CH_2Cl_2, 68% (5 : 1)
$Pd(PPh_3)_2$ = 2.5 mol% $Pd_2(dba)_3 \cdot CHCl_3/4PPh_3$

Scheme 9.23 Synthesis of key intermediate toward divergent synthesis of mezzettiasides.

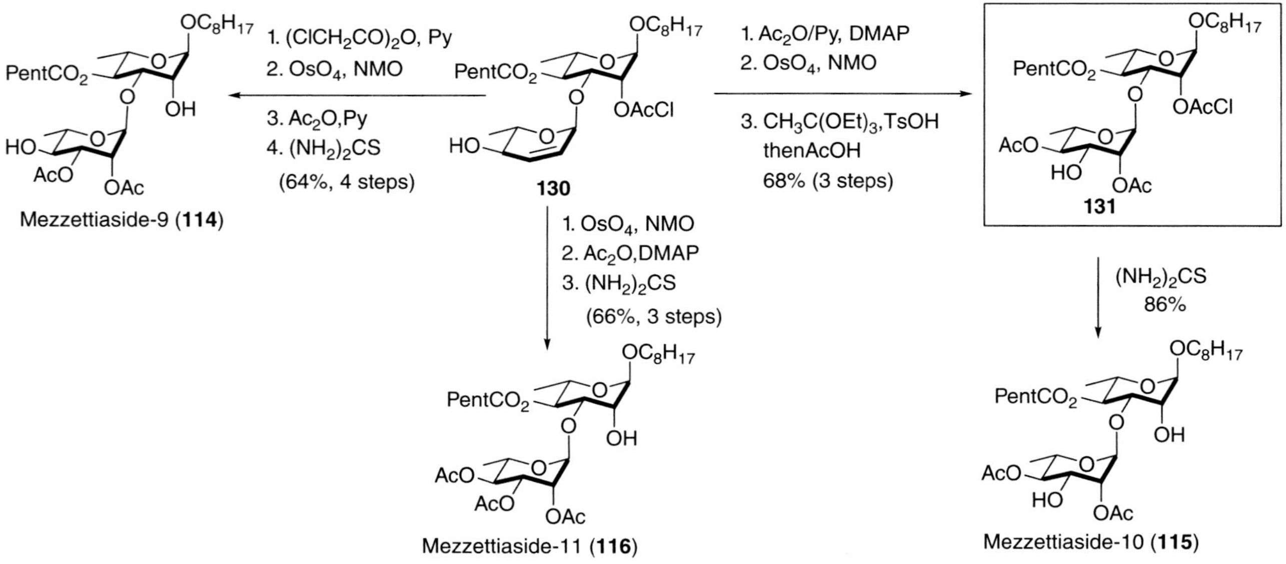

Scheme 9.24 Divergent synthesis of mezzettiaside disaccharides.

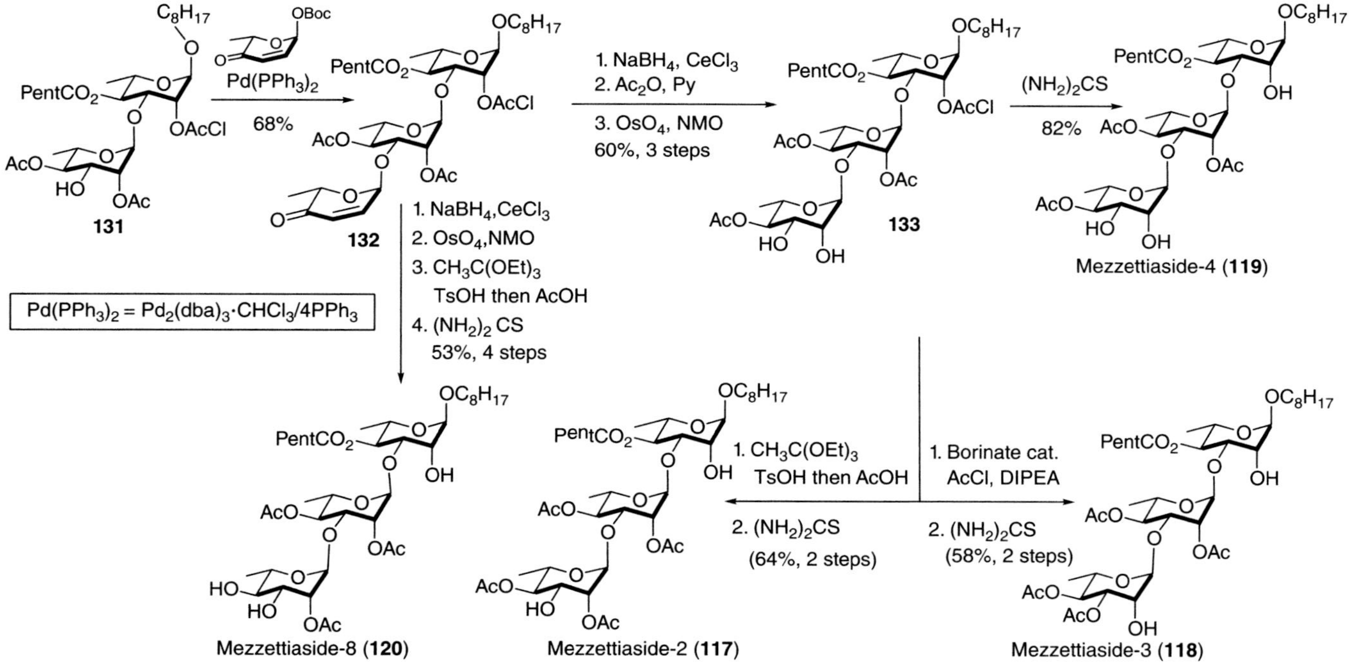

Scheme 9.25 Divergent route to mezzettiaside trisaccharides.

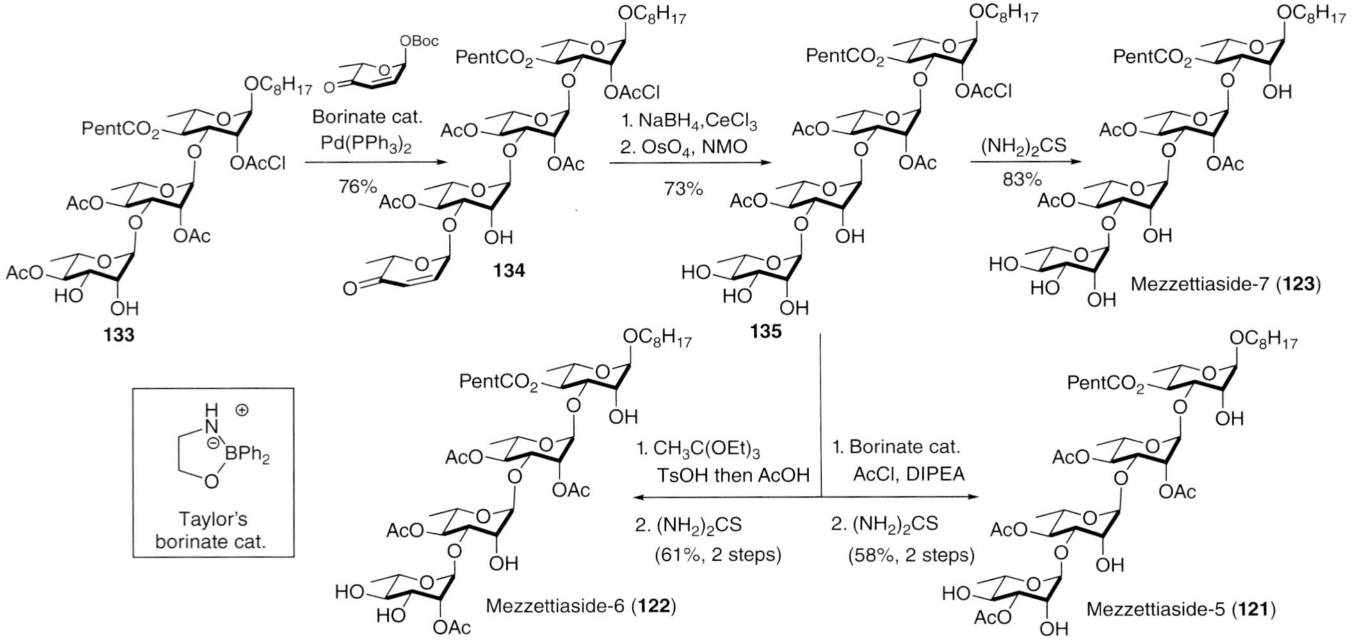

Scheme 9.26 Synthesis of mezzettiaside tetrasaccharide.

9.5 Conclusion

In this chapter, we reviewed the use of an alternative approach to carbohydrates and oligosaccharides. This alternative route avoids the pre-existing stereochemistry of traditional carbohydrate approaches, which gives these routes great compatibility to address targets with rare stereochemistry and functional groups. The approach relies on the use of asymmetric synthesis to generate pyranone-based glycosyl donors, a Pd-catalyzed glycosylation reaction, and post-glycosylation transformations to install the required sugar functionality. The strategic use of these three reaction types can enable the discovery of highly efficient synthetic approaches. When applied toward oligosaccharide targets, this approach often can minimize the use of protecting groups. In this context, the enone functionality can serve as atom-less protecting groups, where the post-glycosylation transformations function as enone deprotection reactions that provide the carbohydrate functionality. The power and uniqueness of this approach is probably best exemplified in our synthesis and study of various sugar stereoisomers and, in turn, the medicinal chemistry SAR study of these structures, which until now were not obtainable by traditional carbohydrate routes.

9.6 Experimentals [3]

9.6.1 Benzyl 2,3-Dideoxy-4,6-bis-*O*-[(2*R*,6*S*)-6-[[[(1,1-dimethylethyl) dimethylsilyl]oxy]methyl]-5,6-dihydro-5-oxo-2*H*-pyran-2-yl]-L-*erythro*-hex-2-enopyranoside (136)

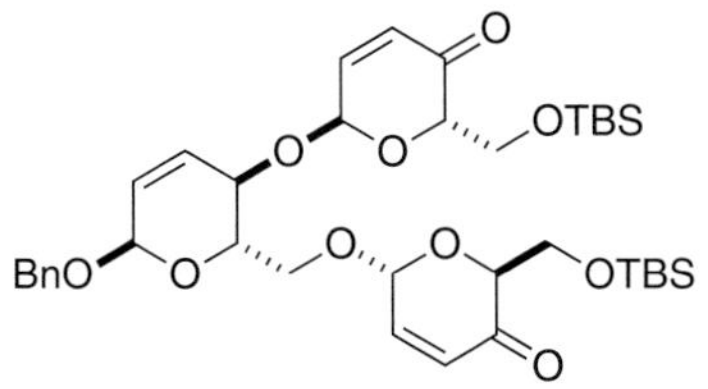

A CH_2Cl_2 (0.6 ml) solution of **50b(α-L)** (954 mg, 2.66 mmol) and **76** (262 mg, 1.11 mmol) was cooled to 0 °C. A CH_2Cl_2 (0.6 ml) solution of $Pd_2(dba)_3 \cdot CHCl_3$ (28 mg, 2.5 mol%) and PPh_3 (28 mg, 10 mol%) was added to the reaction mixture at 0 °C. The reaction mixture was stirred at 0 °C for 6 h, then quenched with satd. aq. $NaHCO_3$ (2 ml), extracted with Et_2O (3 × 10 ml), dried (Na_2SO_4), and concentrated under reduced pressure. The crude product was purified by silica gel flash chromatography eluting with 20% EtOAc/hexanes to give **136** (576 mg, 0.803 mmol, 72%) as a clear yellowish liquid: R_f = 0.52 (30% EtOAc/hexanes); $[\alpha]_D^{21}$ = +21.8 (c = 1.0, CH_2Cl_2); IR (thin film, cm^{-1}) 2951, 2941, 1701, 1695, 1473,

1462, 1254, 1043, 1034, and 836; ^{1}H NMR (600 MHz, $CDCl_3$) δ 7.32 (m, 5H), 6.87 (dd, *J* = 10.2, 3.6 Hz, 1H), 6.78 (dd, *J* = 10.2, 3.6 Hz, 1H), 6.13–6.08 (m, 3H), 5.86 (ddd, *J* = 10.2, 2.4, 2.4 Hz, 1H), 5.60 (d, *J* = 3.0 Hz, 1H), 5.46 (d, *J* = 3.6 Hz, 1H), 5.08 (dd, *J* = 1.2, 1.2 Hz, 1H), 4.76 (d, *J* = 11.4 Hz, 1H), 4.55 (d, *J* = 11.4 Hz, 1H), 4.50 (dd, *J* = 10.2, 1.2 Hz, 1H), 4.48 (dd, *J* = 4.2, 2.4 Hz, 1H), 4.41 (dd, *J* = 4.2, 2.4 Hz, 1H), 4.07–3.98 (m, 6H), 3.93 (dd, *J* = 11.4, 2.4 Hz, 1H), 3.82 (dd, *J* = 10.2, 3.6 Hz, 1H), 0.86 (s, 9H), 0.85 (s, 9H), 0.06 (s, 3H), 0.04 (s, 6H), and 0.03 (s, 3H); ^{13}C NMR (150 MHz, $CDCl_3$) δ 194.8, 194.3, 144.1, 143.8, 137.7, 128.8, 128.5, 128.4 (2C), 128.2, 127.9 (2C), 127.7, 127.6, 93.4, 93.1, 89.9, 76.8, 76.4, 70.0, 69.2, 67.5, 67.2, 62.91, 62.88, 25.8 (6C), 18.24, 18.23, −5.38, −5.43, −5.45, and −5.5; and CI-HRMS calcd for [$C_{37}H_{56}O_{10}Si_2Na^+$]: 739.3304, found 739.3313.

9.6.2 Benzyl *O*-2,3-Dideoxy-6-*O*-[(1,1-dimethylethyl)dimethylsilyl]-α-L-*erythro*-hex-2-enopyranosyl-(1→4)-*O*-[2,3-dideoxy-6-*O*-[(1,1-dimethylethyl)dimethylsilyl]-α-L-*erythro*-hex-2-enopyranosyl-(1→6)]-2,3-dideoxy-L-*erythro*-hex-2-enopyranoside (77)

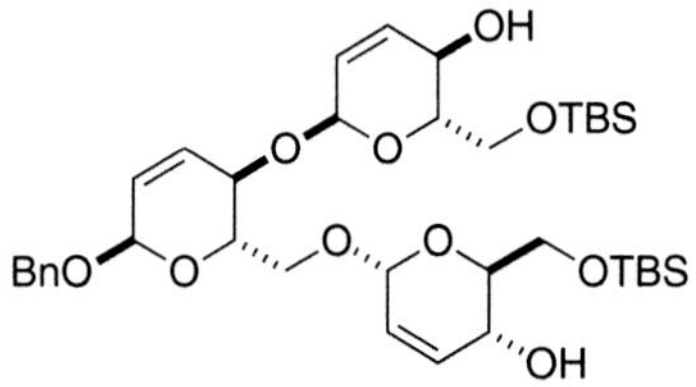

A CH_2Cl_2 (0.2 ml) solution of **136** (120 mg, 0.167 mmol) and MeOH (0.2 ml) was cooled to −78 °C. $NaBH_4$ (7.0 mg, 0.16 mmol) was added, and the reaction mixture was stirred at −78 °C for 8 h, then diluted with Et_2O (10 ml) and quenched with satd. aq. $NaHCO_3$ (5 ml), extracted with Et_2O (3 × 5 ml), dried (Na_2SO_4), and concentrated under reduced pressure. The crude product was purified by silica gel flash chromatography eluting with 40% EtOAc/hexanes to give **77** (108 mg, 0.150 mmol, 90%) as a clear liquid: R_f = 0.34 (40% EtOAc/hexanes); $[\alpha]_D^{21}$ = +32.5 (*c* = 1.0, CH_2Cl_2); IR (thin film, cm^{-1}) 3445, 2927, 1462, 1456, 1369, 1255, 1053, and 837; ^{1}H NMR (600 MHz, $CDCl_3$) δ 7.34 (m, 5H), 6.09 (d, *J* = 10.2 Hz, 1H), 5.93 (d, *J* = 10.2 Hz, 1H), 5.90 (d, *J* = 10.2 Hz, 1H), 5.82 (ddd, *J* = 10.2, 2.4, 2.4 Hz, 1H), 5.75 (ddd, *J* = 10.2, 1.8, 1.8 Hz, 1H), 5.64 (ddd, *J* = 10.2, 2.4, 2.4 Hz, 1H), 5.19 (d, *J* = 1.2 Hz, 1H), 5.07 (m, 2H), 4.80 (d, *J* = 11.4 Hz, 1H), 4.56 (d, *J* = 11.4 Hz, 1H), 4.38 (dd, *J* = 9.6, 1.2 Hz, 1H), 4.16 (m, 2H), 4.02 (ddd, *J* = 9.0, 2.4, 2.4 Hz, 1H), 3.92 (m, 3H), 3.85 (m, 2H), 3.79 (m, 2H), 3.71 (dd, *J* = 12.0, 2.4 Hz, 1H), 3.16 (bs, 1H), 3.11 (bs, 1H), 0.91 (s, 18H), 0.11 (s, 6H), and 0.11 (s, 6H); ^{13}C NMR (150 MHz, $CDCl_3$) δ 138.0, 133.1, 132.5, 129.3, 128.4(2C), 128.0(2C), 127.6, 127.0, 126.0, 125.7, 94.3, 93.6, 90.3, 70.6, 70.0, 69.9, 69.4, 67.4, 67.0, 66.9, 66.1, 65.4, 64.8, 25.9(3C), 25.8(3C), 18.3, 18.2, −5.3, −5.4, −5.5, and −5.6; and CI-HRMS calcd for [$C_{37}H_{60}O_{10}Si_2Na]^+$: 743.3617, found 743.3637.

9.6.3 Benzyl *O*-2,3-Dideoxy-α-L-*erythro*-hex-2-enopyranosyl-(1→4)-*O*-[2,3-dideoxy-α-L-*erythro*-hex-2-enopyranosyl-(1→6)]-2,3-dideoxy-L-*erythro*-hex-2-enopyranoside (**78**)

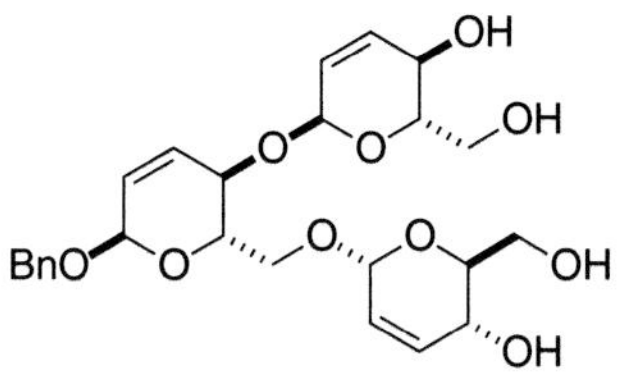

A tetrahydrofuran (THF, 3.2 ml) solution of **77** (1.2 g, 1.666 mmol) was cooled to 0 °C. TBAF (3.5 ml, 3.5 mmol) was added, and the reaction mixture was stirred at 0 °C for 2 h. The reaction mixture was diluted with EtOAc (15 ml) and quenched with satd. aq. $NaHCO_3$ (10 ml), extracted with EtOAc (3 × 10 ml), dried (Na_2SO_4), and concentrated under reduced pressure. The crude product was purified by silica gel flash chromatography eluting with 60% EtOAc/hexanes to give **78** (590 mg, 1.64 mmol, 91%) as a colorless oil: R_f = 0.28 (60% EtOAc/hexanes); $[\alpha]_D^{21}$ = −32.1 (*c* = 1.0, MeOH); IR (thin film, cm^{-1}) 3445, 2927, 1462, 1456, 1369, 1255, 1053, and 837; ^1H NMR (600 MHz, $CDCl_3$) δ 7.33 (m, 5H), 6.04 (d, *J* = 10.2 Hz, 1H), 5.91 (d, *J* = 10.2 Hz, 2H), 5.78 (d, *J* = 10.2 Hz, 1H), 5.74 (d, *J* = 10.2 Hz, 1H), 5.61 (d, *J* = 10.2 Hz, 1H), 5.15 (d, *J* = 1.2 Hz, 1H), 5.11 (d, *J* = 1.2 Hz, 1H), 5.05 (d, *J* = 1.2 Hz, 1H), 4.74 (d, *J* = 12.0 Hz, 1H), 4.55 (d, *J* = 12.0 Hz, 1H), 4.47–4.30 (m, 3H), and 4.04–3.70 (m, 13H); ^{13}C NMR (150 MHz, $CDCl_3$) δ 137.7, 133.7, 133.5, 129.4, 128.3 (2C), 127.8 (2C), 127.6, 126.7, 125.8, 125.5, 94.2, 93.7, 90.5, 71.9, 71.8, 70.0, 69.2, 66.4, 66.3, 64.2, 63.8, 62.7, and 62.3; and CI-HRMS calcd for $[C_{25}H_{32}O_{10}Na]^+$: 515.1888, found 515.1892.

9.6.4 Benzyl *O*-2,3-Dideoxy-4,6-bis-*O*-[(2*R*,6*S*)-6-[[[(1,1-dimethylethyl)dimethylsilyl]oxy] methyl]-5,6-dihydro-5-oxo-2*H*-pyran-2-yl]-α-L-*erythro*-hex-2-enopyranosyl-(1→4)-*O*-[2,3-dideoxy-4,6-bis-*O*-[(2*R*,6*S*)-6-[[[(1,1-dimethylethyl)dimethylsilyl]oxy] methyl]-5,6-dihydro-5-oxo-2*H*-pyran-2-yl]-α-L-*erythro*-hex-2-enopyranosyl-(1→6)]-2,3-dideoxy-L-*erythro*-hex-2-enopyranoside (**79**)

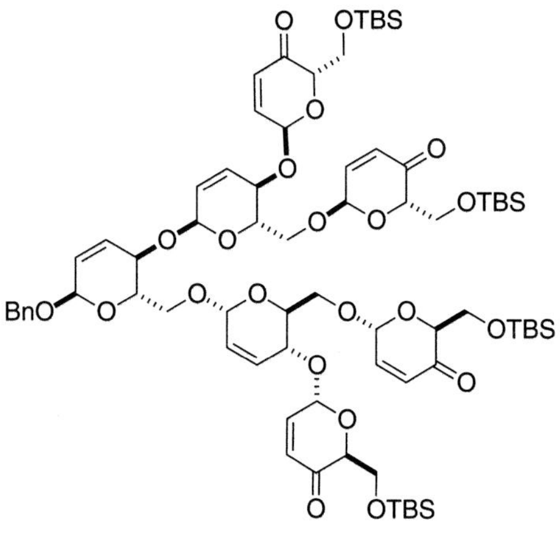

A CH_2Cl_2 (0.6 ml) solution of **50b(α-L)** (1.47 g, 4.10 mmol) and **78** (504 mg, 1.02 mmol) was cooled to 0 °C. A CH_2Cl_2 (0.2 ml) solution of $Pd_2(dba)_3{\cdot}CHCl_3$ (105 mg, 2.5 mol%), and PPh_3 (107 mg, 10 mol%) was added to the reaction mixture at 0 °C. The reaction mixture was stirred at 0 °C for 6 h, then quenched with satd. aq. $NaHCO_3$ (2 ml), extracted with Et_2O (3 × 10 ml), dried (Na_2SO_4), and concentrated under reduced pressure. The crude product was purified by silica gel flash chromatography eluting with 40% EtOAc/hexanes to give **79** (1.04 g, 0.71 mmol, 70%) as a white solid: R_f = 0.34 (50% EtOAc/hexanes); mp: 56–57 °C; $[\alpha]_D^{21}$ = +10.2 (*c* = 1.0, CH_2Cl_2); IR (thin film, cm^{-1}) 3445, 2927, 1462, 1456, 1369, 1255, 1053, and 837; ^1H NMR (600 MHz, $CDCl_3$) δ 7.29 (m, 5H), 6.90 (dd, *J* = 10.2, 3.0 Hz, 1H), 6.89 (dd, *J* = 10.2, 3.0 Hz, 1H), 6.81 (dd, *J* = 10.2, 3.0 Hz, 1H), 6.71 (dd, *J* = 10.2, 3.0 Hz, 1H), 6.14–6.04 (m, 7H), 5.87 (dq, *J* = 10.2, 2.4 Hz, 1H), 5.82 (dq, *J* = 10.2, 2.4 Hz, 1H), 5.76 (dq, *J* = 10.2, 2.4 Hz, 1H), 5.62 (d, *J* = 3.6 Hz, 1H), 5.60 (d, *J* = 3.6 Hz, 1H), 5.48 (d, *J* = 3.6 Hz, 1H), 5.42 (d, *J* = 3.6 Hz, 1H), 5.22 (d, *J* = 1.2 Hz, 1H), 5.08 (d, *J* = 2.4 Hz, 1H), 5.02 (d, *J* = 2.4 Hz, 1H), 4.79 (d, *J* = 11.4 Hz, 1H), 4.52 (d, *J* = 11.4 Hz, 1H), 4.50–4.38 (m, 5H), 4.26 (dd, *J* = 9.6, 1.8 Hz, 1H), 4.07–3.76 (m, 15H), 0.852 (s, 18H), 0.847 (s, 9H), 0.845 (s, 9H), 0.05 (s, 6H), 0.04 (s, 6H), 0.03 (s, 6H), and 0.02 (s, 6H); ^{13}C NMR (150 MHz, $CDCl_3$) δ 194.8, 194.7, 194.4, 194.3, 144.2, 144.0 (2C), 143.8, 137.8, 129.1, 129.0, 128.8, 128.7, 128.6, 128.4 (2C), 128.2 (2C), 128.1, 127.8, 127.7 (2C), 127.5, 127.4, 94.2, 93.2, 93.1, 92.9, 90.7, 90.0, 89.8, 76.9 (2C), 76.5, 76.4, 69.8, 69.5, 69.3, 68.7, 67.7, 67.6, 67.4, 67.32, 67.28, 67.2, 63.1, 63.05, 63.00, 62.9, 25.8 (12C), 18.3 (4C), −5.4 (4C), and −5.5 (4C); CI-HRMS calcd for $[C_{73}H_{112}O_{22}Si_4Na]^+$: 1475.6615, found 1475.6686.

9.6.5 (2*S*,3*R*,6*R*)-6-(((2*S*,3*R*,6*R*)-6-(((2*S*,3*R*,6*R*)-6-(Benzyloxy)-2-((((2*R*,5*R*,6*S*)-5-(((2*R*,5*R*,6*S*)-6-(((*tert*-butyldimethylsilyl)oxy)methyl)-5-hydroxy-5,6-dihydro-2*H*-pyran-2-yl)oxy)-6-((((2*R*,5*R*,6*S*)-6-(((*tert*-butyldimethylsilyl)oxy)methyl)-5-hydroxy-5,6-dihydro-2*H*-pyran-2-yl)oxy)methyl)-5,6-dihydro-2*H*-pyran-2-yl)oxy)methyl)-3,6-dihydro-2*H*-pyran-3-yl)oxy)-2-((((2*R*,5*R*,6*S*)-6-(((*tert*-butyldimethylsilyl)oxy)methyl)-5-hydroxy-5,6-dihydro-2*H*-pyran-2-yl)oxy)methyl)-3,6-dihydro-2*H*-pyran-3-yl)oxy)-2-(((*tert*-butyldimethylsilyl)oxy)methyl)-3,6-dihydro-2*H*-pyran-3-ol (137)

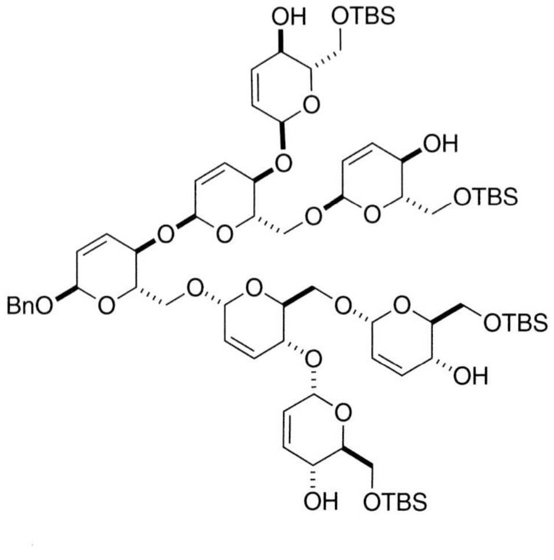

A CH_2Cl_2 (0.6 ml) solution of **79** (450 mg, 0.31 mmol) and MeOH (0.6 ml) was cooled to −78 °C. $NaBH_4$ (47 mg, 1.24 mmol) was added, and the reaction mixture was stirred at −78 °C for 8 h, then diluted with Et_2O (10 ml) and quenched with satd. aq. $NaHCO_3$ (5 ml), extracted with Et_2O (3 × 5 ml), dried (Na_2SO_4), and concentrated under reduced pressure. The crude product was purified by silica gel flash chromatography eluting with 60% EtOAc/hexanes to give **137** (315 mg, 0.22 mmol, 70%) as a colorless oil: $R_f = 0.30$ (60% EtOAc/hexanes); $[\alpha]_D^{21} = -28.6$ ($c = 1.0$, CH_2Cl_2); IR (thin film, cm^{-1}) 3406, 2928, 1462, 1253, 1035, and 990; ^1H NMR (600 MHz, $CDCl_3$) δ 7.35 (m, 5H), 6.12–5.56 (m, 14H), 5.17 (d, J = 1.2 Hz, 1H), 5.17 (d, J = 1.8 Hz, 1H), 5.16 (d, J = 1.8 Hz, 1H), 5.07 (d, J = 1.8 Hz, 1H), 5.07 (d, J = 1.8 Hz, 1H), 5.04 (d, J = 1.2 Hz, 1H), 5.02 (d, J = 1.2 Hz, 1H), 4.80 (d, J = 12.0 Hz, 1H), 4.52 (d, J = 12.0 Hz, 1H), 4.35 (d, J = 9.6 Hz, 1H), 4.26 (d, J = 9.6 Hz, 1H), 4.20–3.60 (m, 26H), 3.30 (bs, 1H), 3.22 (bs, 1H), 3.10 (bs, 2H), 0.91 (s, 18H), 0.90 (s, 18H), 0.11 (s, 6H), 0.11 (s, 6H), 0.10 (s, 6H), and 0.09 (s, 6H); ^{13}C NMR (150 MHz, $CDCl_3$) δ 137.8, 133.4, 133.1, 132.9, 132.6, 129.7, 129.5, 128.6, 128.33 (2C), 128.28 (2C), 127.6, 127.4, 127.2, 126.8, 126.2, 126.0, 125.8, 125.5, 94.4, 94.0, 93.7, 92.8, 91.0, 90.5, 90.2, 70.90, 70.87, 70.85, 70.66, 69.89, 69.64, 69.13, 68.92, 67.73, 67.68, 67.66, 67.16, 67.11, 66.72, 66.24, 66.12, 65.94, 65.87, 65.00, 64.87, 64.72, 64.66, 25.95 (3C), 25.94 (3C), 25.92 (3C), 25.91 (3C), 18.37, 18.35, 18.32, 18.31, −5.20, −5.26, −5.29, −5.33, and −5.45 (4C); and CI-HRMS calcd for $[C_{73}H_{120}O_{22}Si_4Na]^+$: 1483.7241, found 1483.7330.

9.6.6 Benzyl *O*-6-*O*-[(1,1-dimethylethyl)dimethylsilyl]-α-L-mannopyranosyl-(1→4)-*O*-[6-*O*-[(1,1-dimethylethyl)dimethylsilyl]-α-L-mannopyranosyl-(1→6)]-*O*-α-L-mannopyranosyl-(1→4)-*O*-[*O*-6-*O*-[(1,1-dimethylethyl)dimethylsilyl]-α-L-mannopyranosyl-(1→4)-*O*-[6-*O*-[(1,1-dimethylethyl)dimethylsilyl]-α-L-mannopyranosyl-(1→6)]-α-L-mannopyranosyl-(1→6)]-α-L-mannopyranoside (80)

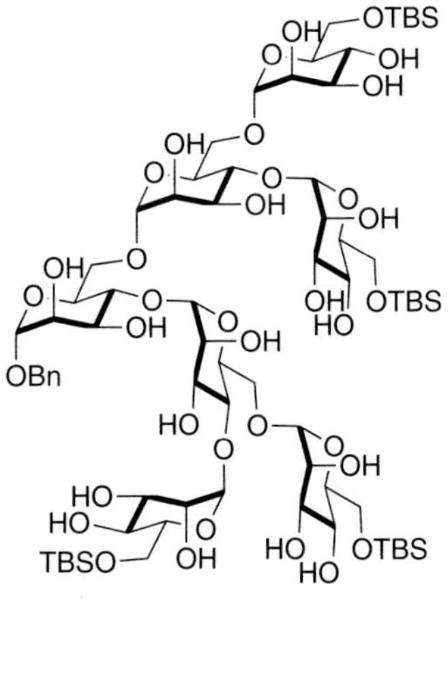

To a solution of **137** (50 mg, 0.0342 mmol) in *t*-butanol/acetone (0.4 ml, 1 : 1) was added a solution of (50% w/v) of *N*-methyl morpholine *N*-oxide/water (0.17 ml, 0.822 mmol) at 0 °C. Crystalline OsO_4 (0.4 mg, 5 mol%) was added, and the reaction was stirred at room temperature for 12 h. The reaction mixture was concentrated and pipetted directly on to a silica gel column using a small amount of CH_2Cl_2 in three portions. Impurities were eluted with ether, and the product was eluted with 30% MeOH/ether to give **23** (50 mg, 0.0294 mmol, 86%): R_f = 0.32 (20% methanol/ether); $[\alpha]_D^{21}$ = −35.8 (*c* = 1.0, CH_2Cl_2); IR (thin film, cm^{-1}) 3382, 2927, 1457, 1254, 1042, 836, and 780; ^1H NMR (400 MHz, CD_3OD) δ 7.35 (m, 5H), 5.27 (d, *J* = 1.2 Hz, 1H), 5.25 (m, 2H), 4.83 (m, 4H), 4.75 (d, *J* = 11.6 Hz, 1H), 4.52 (d, *J* = 12.0 Hz, 1H), 4.00–3.56 (m, 42H), 0.93 (s, 12H), 0.93 (s, 18H), 0.91 (s, 6H), 0.12 (s, 6H), 0.12 (9H), 0.10 (s, 3H), 0.10 (s, 3H), and 0.09 (s, 3H); ^{13}C NMR (100 MHz, $(CD_3)_2CO$) δ 138.59, 129.28 (2C), 128.83 (2C), 128.49, 101.57 (2C), 100.77 (3C), 100.03, 99.57, 75.05 (3C), 73.96 (2C), 73.42, 73.03, 72.47 (3C), 72.30 (6C), 71.83 (3C), 71.61 (3C), 71.40 (3C), 69.48, 68.54 (3C), 68.41 (2C), 67.21, 64.58, 64.15 (3C), 26.60 (3C), 26.58 (3C), 26.56 (3C), 26.52 (3C), 19.14, 19.12, 19.04, 19.03, −4.72 (2C), −4.75 (2C), and −4.79 (4C); and CI-HRMS calcd for $[C_{73}H_{135}O_{36}Si_4]^+$: 1699.7805, found 1699.7802.

Abbreviations

Ac	acetyl
Ar	aryl
Bn	benzyl
Boc	*tert*-butyloxycarbonyl
Bu	butyl
Bz	benzoyl
Cat	catalytic
ClAc	chloroacetyl
D-	dextrorotatory
dba	dibenzylideneacetone
DCC	*N,N'*-dicyclohexylcarbodiimide
DIAD	diisopropyl azodicarboxylate
DMAP	4-(dimethylamino)pyridine
DMP	2,2-dimethoxypropane
ent-	enantiomeric
L-	levorotatory
mCPBA	*meta*-chloroperoxybenzoic acid
Me	methyl
NBS	*N*-bromosuccinimide
NBSH	2-nitrobenzenesulfonylhydrazide
NIS	*N*-iodosuccinimide
NMM	*N*-methylmorpholine
NMO	*N*-methylmorpholine *N*-oxide
o-	*ortho-*

p-	*para-*
Pent	pentyl
Ph	phenyl
Piv	2,2-dimethylpropanoyl
PMB	*p*-methoxybenzyl
Py	pyridine
SAR	structure–activity relationship
t-	*tertiary*
TBAF	tetrabutylammonium fluoride
TBS	*tert*-butyldimethylsilyl
Tf	trifluoromethanesulfonyl
THF	tetrahydrofuran
TMEDA	tetramethylethylenediamine
Ts	4-toluenesulfonyl
TsDPEN	*N-p*-tosyl-1,2-diphenylethylenediamine

References

1 Greene, T.W. and Wuts, P.G.M. (1999). *Protective Groups in Organic Synthesis*, 3e. New York: Wiley.
2 Baran, P.S., Maimone, T.J., and Richter, J.M. (2007). *Nature* 446: 404–408.
3 Babu, R.S., Chen, Q., Kang, S.-W. et al. (2012). *J. Am. Chem. Soc.* 134: 11952–11955.
4 Aljahdali, A.Z., Shi, P., Zhong, Y., and O'Doherty, G.A. (2013). *Adv. Carbohydr. Chem. Biochem.* 69: 55–123.
5 Cuccarese, M.F. and O'Doherty, G.A. (2012). *Application of the Achmatowicz Rearrangement for the Synthesis of Oligosaccharides in Asymmetric Synthesis II: More Methods and Applications* (ed. M. Christmann and S. Braese), 249–259. Weinheim: Wiley-VCH.
6 Babu, R.S., Zhou, M., and O'Doherty, G.A. (2004). *J. Am. Chem. Soc.* 126: 3428–3429.
7 Bajaj, S.O., Farnsworth, J.R., and O'Doherty, G.A. (2014). *Org. Synth.* 91: 338–355.
8 Babu, R.S. and O'Doherty, G.A. (2003). *J. Am. Chem. Soc.* 125: 12406–12407.
9 Harris, J.M., Keranen, M.D., and O'Doherty, G.A. (1999). *J. Org. Chem.* 64: 2982–2983.
10 (a) Bertozzi, C.R. and Kiessling, L.L. (2001). *Science* 291: 2357–2364. (b) Danishefsky, S.J., McClure, K.F., Randolph, J.T., and Ruggeri, R.B. (1993). *Science* 260: 1307–1309. (c) Plante, O.J., Palmacci, E.R., and Seeberger, P.H. (2001). *Science* 291: 1523–1527. (d) Sears, P. and Wong, C.-H. (2001). *Science* 291: 2344–2350. (e) Yamada, H., Harada, T., Miyazaki, H., and Takahashi, T. (1994). *Tetrahedron Lett.* 35: 3979–3982. (f) Wang, L.X. (2006). *Curr. Opin. Drug Discovery Dev.* 9: 194–202. (g) Ni, J., Song, H., Wang, Y. et al. (2006). *Bioconjugate Chem.* 17: 493–500. (h) Zeng, Y., Wang, J., Li, B. et al. (2006). *Chem. Eur. J.* 12: 3355–3364. (i) Wang, J., Le, N., Heredia, A. et al. (2005). *Org. Biomol. Chem.* 3: 1781–1786.

11 For reviews of other approaches to hexoses, see: (a) Gijsen, H.J.M., Qiao, L., Fitz, W., and Wong, C.-H. (1996). *Chem. Rev.* 96: 443–473. (b) Hudlicky, T., Entwistle, D.A., Pitzer, K.K., and Thorpe, A.J. (1996). *Chem. Rev.* 96: 1195–1220. (c) Yu, X. and O'Doherty, G.A. (2008). *ACS Symp. Ser.* 990: 3–28.
12 (a) Timmer, M.S.M., Adibekian, A., and Seeberger, P.H. (2005). *Angew. Chem. Int. Ed.* 44: 7605–7607. (b) Adibekian, A., Bindschädler, P., Timmer, M.S.M. et al. (2007). *Chem. Eur. J.* 13: 4510–4527. (c) Stallforth, P., Adibekian, A., and Seeberger, P.H. (2008). *Org. Lett.* 10: 1573–1576.
13 (a) Bouché, L. and Reißig, H.-U. (2012). *Pure Appl. Chem.* 84: 23–36. (b) Pfrengle, F. and Reißig, H.-U. (2010). *Chem. Soc. Rev.* 39: 549–557.
14 Dondoni, A. and Perrone, D. (1997). *Aldrichimica Acta* 30: 35–46.
15 (a) Ko, S.Y., Lee, A.W.M., Masamune, S. et al. (1983). *Science* 220: 949–951. (b) Ko, S.Y., Lee, A.W.M., Masamune, S. et al. (1990). *Tetrahedron* 46: 245–264.
16 (a) Danishefsky, S.J. (1989). Chemtracts. *Org. Chem.* 273–297. For improved catalysis see: (b) Schaus, S.E., Branalt, J., and Jacobsen, E.N. (1998). *J. Org. Chem.* 63: 403–405.
17 (a) Johnson, C.R., Golebiowski, A., Steensma, D.H., and Scialdone, M.A. (1993). *J. Org. Chem.* 58: 7185–7194. (b) Hudlicky, T., Pitzer, K.K., Stabile, M.R. et al. (1996). *J. Org. Chem.* 61: 4151–4153.
18 Henderson, I., Sharpless, K.B., and Wong, C.-H. (1994). *J. Am. Chem. Soc.* 116: 558–561.
19 For examples of protecting group free synthesis, see: (a) Crabtree, R.H. (2007). *Science* 318: 756. (b) Chen, M.S. and White, M.C. (2007). *Science* 318: 783.
20 (a) Ahmed, M.M. and O'Doherty, G.A. (2005). *J. Org. Chem.* 67: 10576–10578. (b) Ahmed, M.M. and O'Doherty, G.A. (2005). *Tetrahedron Lett.* 46: 4151–4155. (c) Gao, D. and O'Doherty, G.A. (2005). *Org. Lett.* 7: 1069–1072. (d) Zhang, Y. and O'Doherty, G.A. (2005). *Tetrahedron* 61: 6337–6351. (e) Ahmed, M.M. and O'Doherty, G.A. (2005). *Tetrahedron Lett.* 46: 3015–3019. (f) Ahmed, M.M., Berry, B.P., Hunter, T.J. et al. (2005). *Org. Lett.* 7: 745–748.
21 Harris, J.M., Li, M., Scott, J.G., and O'Doherty, G.A. (2004). Achmatowicz approach to 5,6-dihydro-2H-pyran-2-one Containing Natural Products. In: *Strategies and Tactics in Organic Synthesis*, vol. 5 (ed. M. Harmata), 221–253. Amsterdam: Academic Press (Elsevier).
22 Harris, J.M., Keranen, M.D., Nguyen, H. et al. (2000). *Carbohydr. Res.* 328: 17–36.
23 Bushey, M.L., Haukaas, M.H., and O'Doherty, G.A. (1999). *J. Org. Chem.* 64: 2984–2985.
24 Haukaas, M.H., Li, M., Starosotnikov, A.M., and O'Doherty, G.A. (2008). *Heterocycles* 76: 1549–1559.
25 (a) Li, M. and O'Doherty, G.A. (2004). *Tetrahedron Lett.* 45: 6407–6411. (b) Li, M., Scott, J.G., and O'Doherty, G.A. (2004). *Tetrahedron Lett.* 45: 1005–1009. (c) Haukaas, M.H. and O'Doherty, G.A. (2001). *Org. Lett.* 3: 3899–3992.
26 (a) Wang, H.-Y.L. and O'Doherty, G.A. (2011). *Chem. Commun.* 47: 10251–10253. (b) Cuccarese, M.F., Wang, H.-Y.L., and O'Doherty, G.A. (2013). *Eur. J. Org. Chem.* 3067–3075. (c) Hinds, J.W., McKenna, S.B., Sharif, E.U. et al. (2013). *ChemMedChem* 8: 63–69.
27 Shan, M., Xing, Y., and O'Doherty, G.A. (2009). *J. Org. Chem.* 74: 5961–5966.

28 (a) Balachari, D. and O'Doherty, G.A. (2000). *Org. Lett.* 2: 863–866. (b) Balachari, D. and O'Doherty, G.A. (2000). *Org. Lett.* 2: 4033–4036. (c) Balachari, D., Quinn, L., and O'Doherty, G.A. (1999). *Tetrahedron Lett.* 40: 4769–4773.
29 Haukaas, M.H. and O'Doherty, G.A. (2011). *Org. Lett.* 3: 401–404.
30 (a) Shan, M. and O'Doherty, G.A. (2006). *Org. Lett.* 8: 5149–5152. (b) Shan, M. and O'Doherty, G.A. (2010). *Org. Lett.* 12: 2986–2989. (c) Mrozowski, R.M., Vemula, R., Wu, B. et al. (2013). *ACS Med. Chem. Lett.* 4: 175–179. (d) Mrozowski, R.M., Sandusky, Z.M., Vemula, R. et al. (2014). *Org. Lett.* 16: 5996–5999. (e) Li, M., Li, Y., Mrozowski, R.M. et al. (2015). *ACS Med. Chem. Lett.* 16: 95–99.
31 (a) Guo, H. and O'Doherty, G.A. (2005). *Org. Lett.* 7: 3921–3924. (b) Baiga, T.J., Guo, H., Xing, Y. et al. (2008). *ACS Chem. Biol.* 3: 294–304. (c) Guo, H., LaClair, J.J., Masler, E.P. et al. (2016). *Tetrahedron* 72: 2280–2286.
32 Borisova, S.A., Guppi, S.R., Kim, H.J. et al. (2010). *Org. Lett.* 12: 5150–5153.
33 (a) Tibrewal, N., Downey, T.E., Van Lanen, S.G. et al. (2012). *J. Am. Chem. Soc.* 134: 12402–12405. (b) Sharif, E.U. and O'Doherty, G.A. (2014). *Heterocycles* 88: 1275–1285.
34 (a) Shan, M., Sharif, E.U., and O'Doherty, G.A. (2010). *Angew. Chem. Int. Ed.* 49: 9492–9495. (b) Sharif, E.U. and O'Doherty, G.A. (2012). *Eur. J. Org. Chem.* 2095–2108.
35 Concurrent with these studies was the similar discovery by Feringa, Lee: (a) Comely, A.C., Eelkema, R., Minnaard, A.J., and Feringa, B.L. (2003). *J. Am. Chem. Soc.* 125: 8714–8715. (b) Kim, H., Men, H., and Lee, C. (2004). *J. Am. Chem. Soc.* 126: 1336–1337.
36 The poor reactivity in Pd-catalyzed allylation reaction of alcohols as well as a nice solution to this problem was reported, see: Ki, H. and Lee, C. (2002). *Org. Lett.* 4: 4369–4372.
37 For a related Rh system, see: Evans, P.A. and Kennedy, L. (2000). *J. Org. Lett.* 2: 2213–2215.
38 Babu, R.S. and O'Doherty, G.A. (2005). *J. Carbohydr. Chem.* 24: 169–177.
39 (a) Wang, H.-Y.L., Rojanasakul, Y., and O'Doherty, G.A. (2011). *ACS Med. Chem. Lett.* 2: 264–269. (b) Wang, H.-Y.L., Xin, W., Zhou, M. et al. (2011). *ACS Med. Chem. Lett.* 2: 73–78.
40 (a) Guo, H. and O'Doherty, G.A. (2007). *Angew. Chem. Int. Ed.* 46: 5206–5208. (b) Guo, H. and O'Doherty, G.A. (2008). *J. Org. Chem.* 73: 5211–5220. (c) Wang, H.-Y.L., Guo, H., and O'Doherty, G.A. (2013). *Tetrahedron* 69: 3432–3436.
41 Sharif, E.U., Wang, H.-Y.L., Akhmedov, N.G., and O'Doherty, G.A. (2014). *Org. Lett.* 16: 492–495.
42 He, H., Williamson, R.T., Shen, B. et al. (2002). *J. Am. Chem. Soc.* 124: 9729–9736.
43 Guppi, S. and O'Doherty, G.A. (2007). *J. Org. Chem.* 72: 4966–4969.
44 Babu, R.S., Guppi, S.R., and O'Doherty, G.A. (2006). *Org. Lett.* 8: 1605–1608.
45 (a) Zhou, M. and O'Doherty, G.A. (2006). *Org. Lett.* 8: 4339–4342. (b) Zhou, M. and O'Doherty, G.A. (2007). *J. Org. Chem.* 72: 2485–2493. (c) Iyer, A., Zhou, M., Azad, N. et al. (2010). *ACS Med. Chem. Lett.* 1: 326–330.

46 (a) Yu, X. and O'Doherty, G.A. (2008). *Org. Lett.* 10: 4529–4532. (b) Yu, X., Li, M., and O'Doherty, G.A. (2011). *Heterocycles* 82: 1577–1584.

47 Zhou, M. and O'Doherty, G.A. (2008). *Org. Lett.* 10: 2283–2286.

48 Tané, P., Ayafor, J.P., Sondengam, B.L. et al. (1988). *Tetrahedron Lett.* 29: 1837–1840.

49 Seidel, V., Baileul, F., and Waterman, P.G.J. (1999). *Phytochemistry* 52: 465–472.

50 Hu, J.-F., Garo, E., Hough, G.W. et al. (2006). *J. Nat. Prod.* 69: 585–590.

51 Shi, P., Silva, M., Wu, B. et al. (2012). *ACS Med. Chem. Lett.* 3: 1086–1090.

52 Wu, B., Li, M., and O'Doherty, G.A. (2010). *Org. Lett.* 12: 5466–5469.

53 (a) Bajaj, S.O., Sharif, E.U., Akhmedov, N.G., and O'Doherty, G.A. (2014). *Chem. Sci.* 5: 2230–2234. (b) Bajaj, S.O., Shi, P., Beuning, P.J., and O'Doherty, G.A. (2014). *Med. Chem. Commun.* 5: 1138–1142.

54 (a) Taylor, M.S. (2015). *Acc. Chem. Res.* 48: 295–305. (b) Mancini, R.S., McClary, C.A., Anthonipillai, S., and Taylor, M.S. (2015). *J. Org. Chem.* 80: 8501–8510. (c) Tanveer, K., Jarrah, K., and Taylor, M.S. (2015). *Org. Lett.* 17: 3482–3485.

10

Protecting Group Strategies for Sialic Acid Derivatives

Harsha Amarasekara, Szymon Buda, Appi R. Mandhapati, and David Crich

Wayne State University, Department of Chemistry, 5101 Cass Avenue, Detroit, MI 48202, USA

10.1 Introduction

The sialic acids with four different functional groups in the form of multiple alcohols, an amide, the anomeric carbon in its various forms, and a carboxylic acid provide ample opportunities for the demonstration of selective protection strategies, many of which have been covered in previous reviews [1–16]. This chapter focuses systematically on selectivity in the introduction and removal of alcohol, amide, and carboxylate protecting groups in *N*-acetyl neuraminic acid and its derivatives, as a typical and the most common example of the sialic acids. The field was previously reviewed in 2004 [17]. The chemistry of the anomeric carbon, sialidation reactions, and the influence of protecting groups on selectivity in such processes are not discussed in view of the numerous and extensive reviews of this general topic [1–16]. Coverage is not comprehensive but is intended to be illustrative of the range of existing strategies for the selective protection of sialic acids and so to serve as a guide to workers in the field. The specific examples used to illustrate the chapter have been selected with an eye to priority and the inclusion of experimental details. No attempt is made to discuss the underlying chemistry of the protecting groups employed, nor the conditions used for their introduction and removal, as these are the focus of the standard compendia of protecting groups and their chemistry [18, 19].

10.2 Protection of the Carboxylate Group

10.2.1 Methyl Ester Formation

The carboxylate group of the sialic acids is most commonly protected in the form of a methyl ester. Two methods for its installation are in common use. In the first, *N*-acetyl neuraminic acid is stirred in methanol in the presence of a

Protecting Groups: Strategies and Applications in Carbohydrate Chemistry, First Edition.
Edited by Sébastien Vidal.

sulfonic acid-type resin to install the ester followed by peracetylation with acetic anhydride and pyridine giving an overall excellent yield of the peracetylated methyl ester (Scheme 10.1) [20–23]. Alternatively, peracetylation may be conducted on the methyl ester using acetic anhydride and perchloric acid [20, 23]. The second common approach involves peracetylation followed by installation of methyl ester with diazomethane (Scheme 10.1) [24–26].

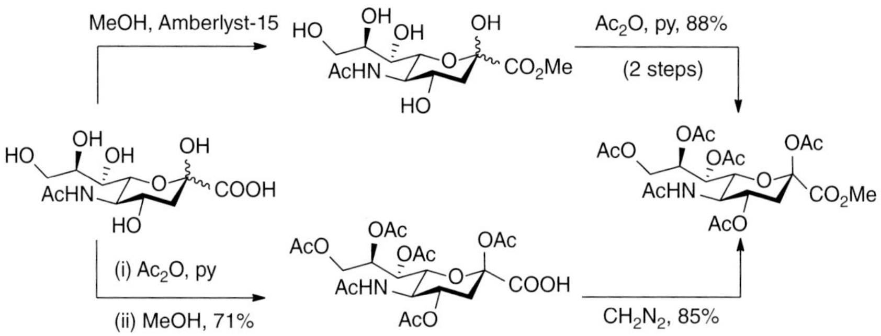

Scheme 10.1 Formation of the peracetyl methyl ester.

The methyl ester may be selectively cleaved in the presence of the acetyl groups by treatment with lithium chloride [27, 28] or lithium iodide [29–31] in pyridine at reflux (Scheme 10.2). It is noteworthy that while the lithium iodide protocol proceeds cleanly in the presence of sialyl phenylthio glycosides, yields are lower with the methylthio analogs [31], presumably because of nucleophilic demethylation. Alternatively, triphenylsilyl mercaptan and cesium carbonate in the presence of catalytic 2,6-di-*tert*-butylcresol provides an effective means of demethylation in *N*,*N*-dimethylformamide (DMF) at 80 °C (Scheme 10.2) [32].

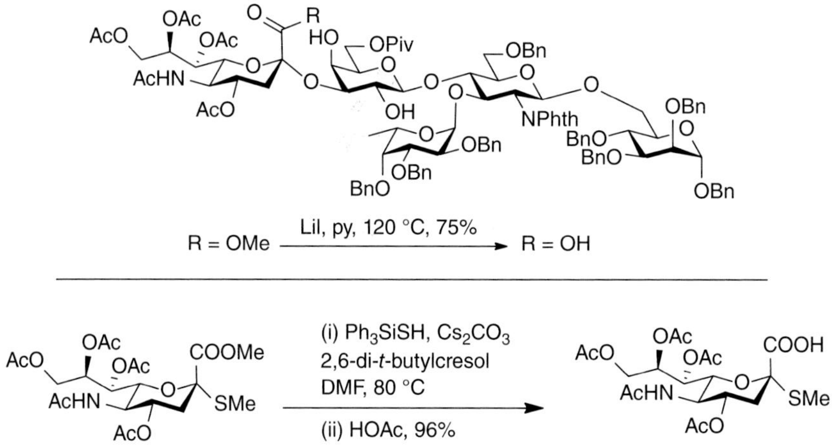

Scheme 10.2 Selective cleavage of methyl ester.

10.2.2 Functionalized Ester Formation

Starting from peracetylated *N*-acetyl neuraminic acid, a variety of different ester types have been introduced to the carboxylate group. Thus, alkylation of the carboxylate with allyl bromide in the presence of cesium carbonate affords the allyl ester, whose principal advantage stems from the ease and mildness of its removal under palladium-catalyzed conditions in the presence of a suitable nucleophile such as morpholine (Scheme 10.3) [33, 34]. The benzyl ester is similarly formed in high yield by treatment of the potassium carboxylate with benzyl bromide; it is removed by hydrogenolysis over palladium on charcoal [35].

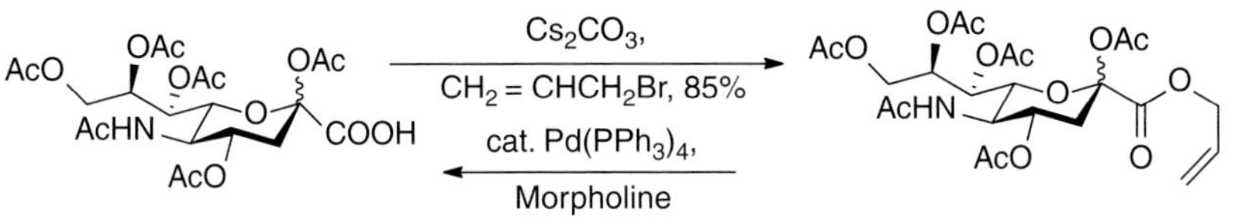

Scheme 10.3 Formation and cleavage of allyl esters.

More highly functionalized esters have been introduced with a view to controlling anomeric stereoselectivity through participation by the functionality present in the ester. Such esters include the 2-methylthioethyl ester [36] and the *N,N*-dimethylcarboxamidomethyl ester [37, 38], which were introduced by alkylation of the carboxylic acid with 2-methylthioethyl chloride in the presence of silver oxide and with the combination of the methanesulfonate of *N,N*-dimethylglyoxamide and cesium carbonate. Further examples of esters derived by carboxylate alkylation include cyanomethyl and phenacyl esters [32].

The phenacyl esters are especially noteworthy as they are compatible with Boc solid-phase peptide synthesis and are removed under very mild conditions with piperidine and 2-mercaptoethanol or with aqueous sodium hydroxide (Scheme 10.4) [39, 40].

Alternatively, functionalized esters may be introduced to the carboxylate group, again with the goal of participation in subsequent glycosylation reactions,

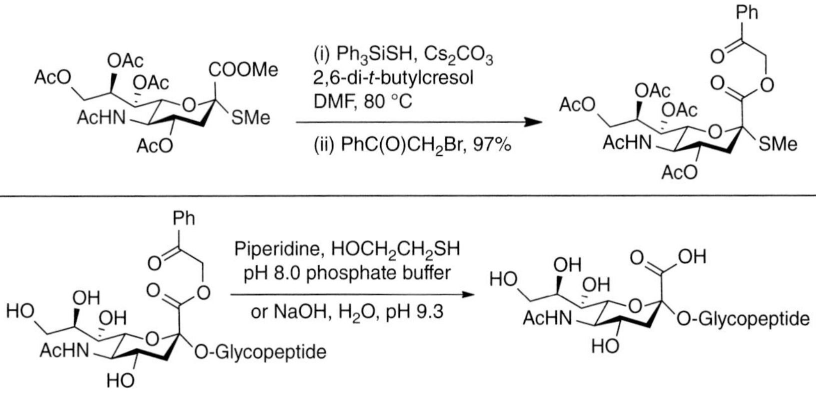

Scheme 10.4 Formation and cleavage of phenacyl esters.

by carbodiimide-mediated coupling of the carboxylic acid and the appropriate alcohol. Esters assembled in this manner include the 2-cyanoethyl, 2-cyanobenzyl, and 2-nitrobenzyl esters [32].

10.2.3 Thioesterification

N-Acetylneuraminic acid thioesters have been prepared from the carboxylic acid by the formation of a mixed anhydride using ethyl chloroformate and subsequent displacement with thiols in the presence of trimethylamine. A range of phenyl thioesters substituted by electron donating and withdrawing groups were prepared in this manner (Scheme 10.5) [27] and investigated for their ability to enhance glycosylation selectivity in the presence of acetonitrile through participative stabilization of the glycosyl nitrilium ion.

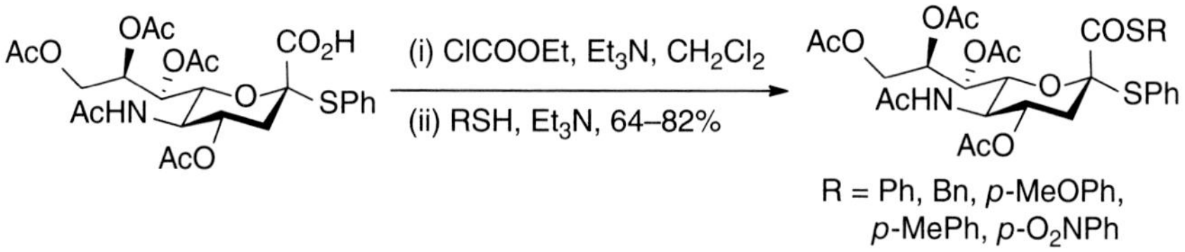

Scheme 10.5 Preparation of thioesters.

10.2.4 Amide Formation

A series of per-*O*-acetyl-*N*-acetyl amidosialyl acetates was prepared by standard methods involving either reaction of methyl esters with ammonium hydroxide or methylamine followed by peracetylation or by carbodiimide-mediated coupling of the carboxylic acid with dimethylamine. Subsequent conversion of anomeric acetate to the corresponding diethylphosphites enabled the influence of amide nitrogen substitution on glycosylation stereochemistry to be examined. Subsequent deprotection of the amide function was achieved with 2 N sodium hydroxide at 100 °C followed by reinstallation of the *N*-acetyl group at N5 (Scheme 10.6) [41].

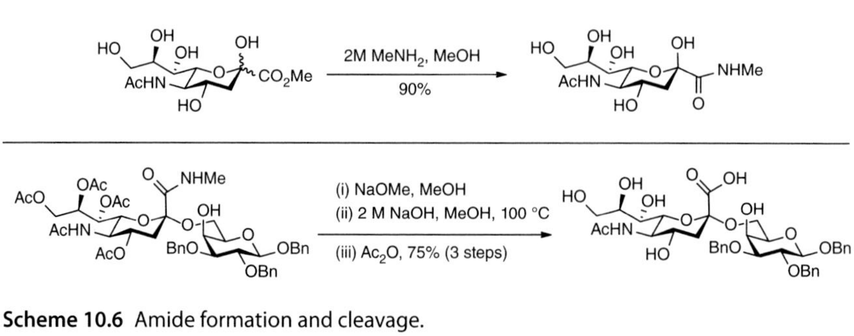

Scheme 10.6 Amide formation and cleavage.

10.2.5 Lactonization

Esterification of *N*-acetylneuraminic acid by means of acyl chlorides or anhydrides is frequently accompanied and even dominated by lactone formation arising from the formation of an initial mixed anhydride with the C1 carboxylate [31]. The 1,7-lactone predominates under most conditions but is often accompanied by 1,4-lactone (Scheme 10.7) [42, 43]. The potential for the exploitation of such intra-residue lactones in selective protection of sialic acids exists [44]. For example, ring opening of 2,7,8,9-tetra-*O*-ethoxycarbonyl-1,4-lactone with methanol affords the methyl ester of 2,7,8,9-tetra-*O*-ethoxycarbonyl-*N*-acetyl neuraminic acid with a single free hydroxyl group at the 4-position [43].

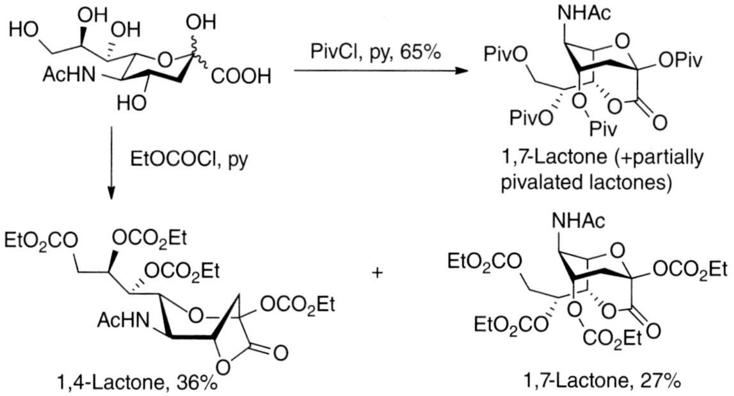

Scheme 10.7 Intra-residue lactone formation.

Controlled acidic hydrolysis of the bacterial capsular polysaccharide colominic acid, the α-2,8-linked homopolymer of *N*-acetylneuraminic acid, can be optimized to give mono- and disaccharides as the major products [45]. Stirring of the dimer with the acidic form of an ion exchange resin in methanol followed by acetylation and finally methanolysis of the reducing end mixed anhydride affords a derivative in which the internal carboxylate group is protected in the form of an inter-residue δ-lactone in high yield (Scheme 10.8) [46]. As the anomeric acetate can be readily converted into a thioglycoside, this protocol affords a useful building block for the synthesis of higher oligosaccharides [46]. Comparable procedures applied to the trimer of *N*-acetylneuraminic acid give the analogous trisaccharides in which the two internal carboxylates are both protected in the form of inter-residue δ-lactones [47].

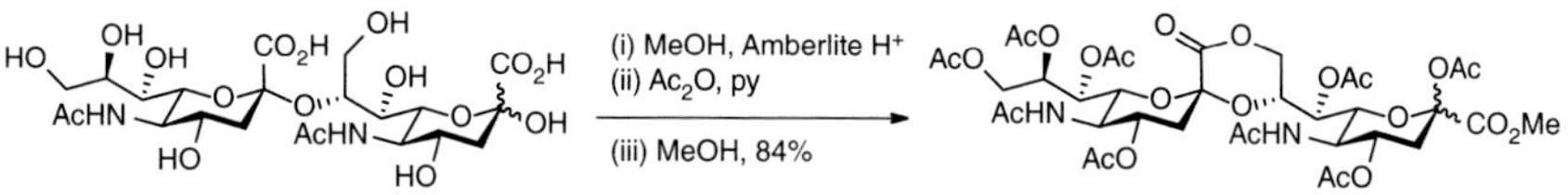

Scheme 10.8 Inter-residue lactone formation.

10.2.6 Lactam Formation

Concomitant protection of the carboxylic acid and amino functions can be achieved by lactam formation. This system has the advantage of imposing a boat conformation on the pyranose ring and enhancing the reactivity of the 4- and 8-hydroxyl groups in subsequent reactions. The lactam is formed from the methyl ester by treatment of a free N5 amine or of the trifluoroacetamide with sodium methoxide in methanol at reflux in the presence of the desiccant Drierite® (Scheme 10.9) [48–50]. If needed, the lactam can be *N*-acylated with acetic anhydride and pyridine in the presence of 4-dimethylaminopyridine (DMAP) [48], or by the action of benzyl chloroformate and DMAP in pyridine [49]. When desired, the *N*-acyl lactams are ring-opened by the action of trimethylamine and water in acetonitrile followed by reinstallation of the methyl ester with methyl iodide and potassium carbonate (Scheme 10.9) [49]. Alternatively, the *N*-Boc lactam may be cleaved and the methyl ester reinstalled directly with sodium methoxide in methanol [50]. The lactam moiety is compatible with the reduction of an 8,9-*O*-benzylidene acetal with borane-trimethylamine and aluminum trichloride [49].

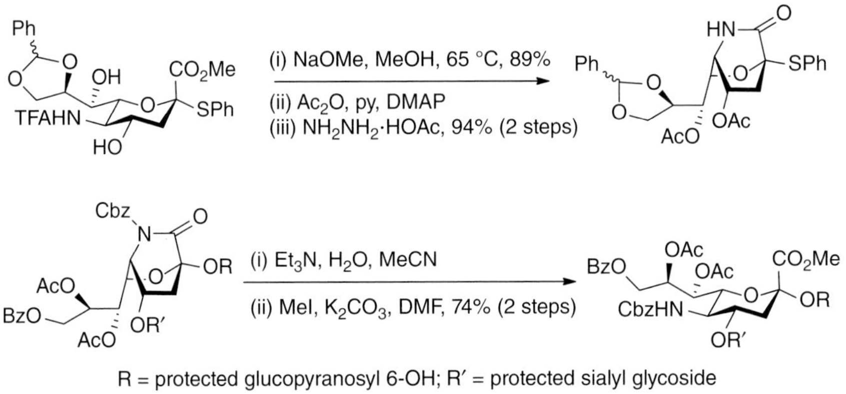

Scheme 10.9 Lactam formation and cleavage.

10.2.7 Other Cyclic Derivatives

Reaction of per-*O*-acetyl *N*-acetylneuraminic acid with carbodiimides in the absence of an external nucleophile gives *N*-acylureas, which cyclize on treatment with boron trifluoride-etherate to give spirocyclic hydantoins (Scheme 10.10) [51].

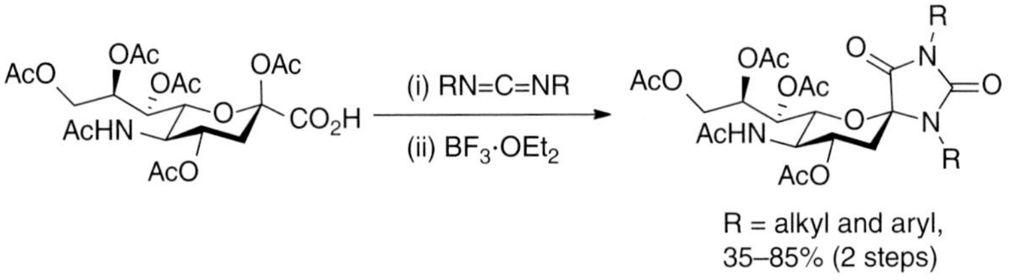

Scheme 10.10 Formation of spirocyclic hydantoins.

10.3 Protection of Amine Function

10.3.1 Amide Cleavage

Modification of the N5 amino group has played a dominant role in the search for more highly selective sialyl donors. As the most available sialic acid is *N*-acetyl neuraminic acid, this necessitates prior cleavage of the N5 acetamide for which a variety of conditions have been developed. Thus, heating a per-*O*-acetyl methyl ester to 60 °C in methanol in the presence of methanesulfonic acid results in the removal of the acetate esters and the acetamido group (Scheme 10.11) [52]. After neutralization and concentration, the so-obtained free amine is typically protected *in situ* and the acetate esters are reinstalled as in the example. It is noteworthy that the thioglycoside moiety survives the rather vigorous conditions of the amide hydrolysis.

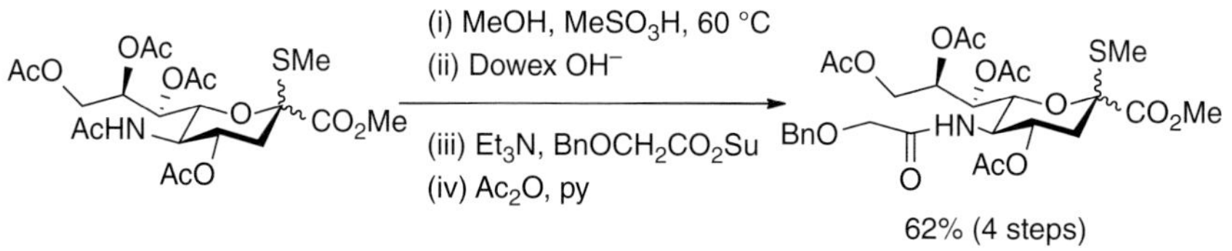

Scheme 10.11 Acidic cleavage of the acetamide and reinstallation of an amide group.

Under basic conditions, the acetamide may be cleaved with barium hydroxide in water at 90 °C followed by neutralization and lyophilization enabling isolation of the zwitterionic amino acid (Scheme 10.12) [26]. Cleavage of the acetamide has also been reported on heating of the free acid with 100% hydrazine hydrate at 85 °C [53]. Alternatively, the acetamide may be cleaved by treatment with sodium methoxide in hot methanol [41], or on heating with aqueous sodium hydroxide [45, 54].

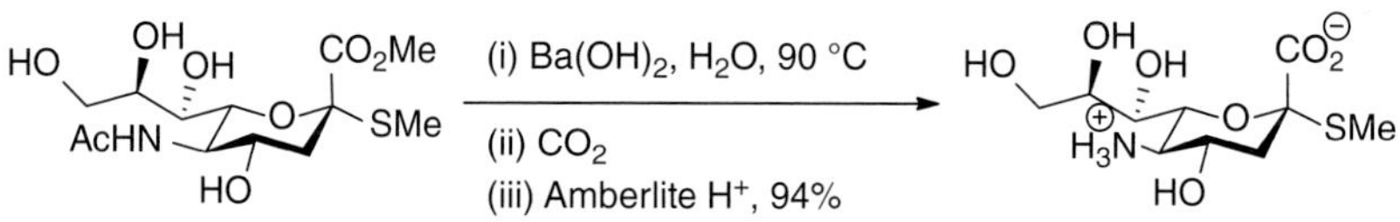

Scheme 10.12 Amide cleavage under basic conditions.

Milder conditions for cleavage of the acetamide group consist of adaptation of the Grieco method [55] for amide cleavage according to which a *tert*-butoxycarbonyl group is installed on the nitrogen to give an imide-type system, from which the acetyl group is cleaved selectively to give the *N*-Boc amine. The Boc group may then be removed with trifluoroacetic acid (Scheme 10.13) [54, 56]. The Boc group cleavage at the end of this sequence may also be advantageously achieved with HCl in ether when the amine is isolated in the form of its hydrochloride salt [57].

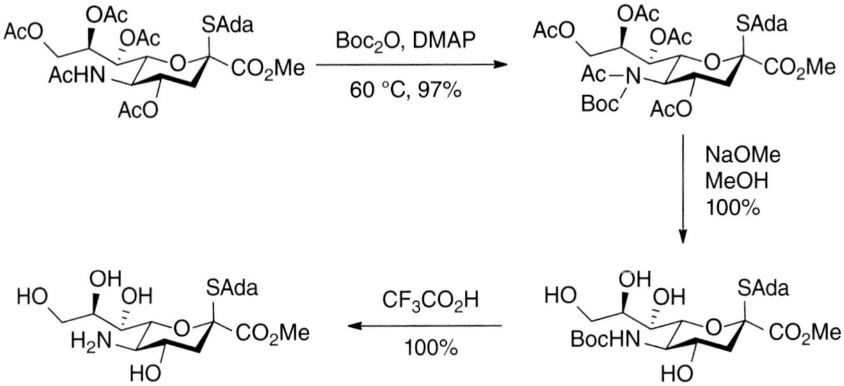

Scheme 10.13 Cleavage of the acetamide by prior installation of a Boc group.

10.3.2 Amine Protection with Acyclic Groups

With free N5 amines in hand, many different protecting groups have been installed on the nitrogen atom and explored for their ability to influence the stereoselectivity of sialidation reactions. Such protecting groups include the trifluoroacetamide function [48, 58, 59], the *N*-(9-fluorenylmethyl) [60], *N*-allyl [60, 61], and *N*-trichloroethyl carbamates [48, 60, 62, 63], the *N*-phthalimido group [29, 64, 65], the *N*-*tert*-butoxycarbonylacetamide [66, 67], and the azido function [68, 69]. As the installation, stereodirecting influence, and removal of such groups have been reviewed recently [11, 12], they will not be covered further here, with the exception of the acetimide and the more recent isothiocyanate moieties.

The acetimide function is unique as a protecting group for the N5 acetamide as it is installed directly onto the amide, concomitantly with acetylation of any free alcohols and without cleavage of a benzylidene acetal, by treatment with isopropenyl acetate in the presence of toluenesulfonic acid (Scheme 10.14) [48, 70]. The acetimide function is compatible with the reductive cleavage of a benzylidene acetal by means of the borane-trimethylamine and aluminum trichloride system (*vide infra*) and is considered to afford a more reactive and α-selective sialyl donor than the acetamide itself [70]. The acetimide possesses the second distinct advantage of cleavage directly to the acetamide function, concomitantly with the removal of any acetate esters, on simple treatment with sodium methoxide in methanol [70].

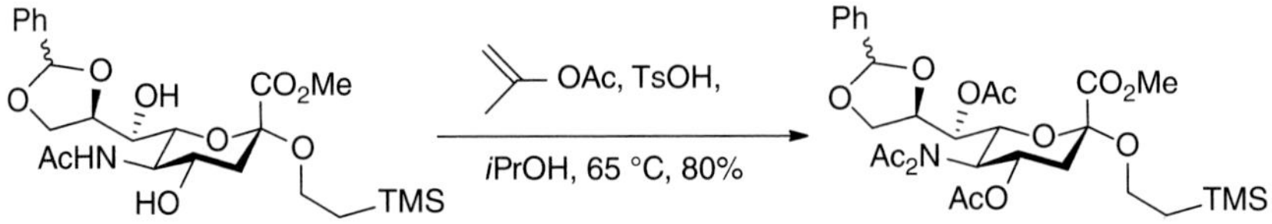

Scheme 10.14 Installation of the acetamide function.

More recently, the isothiocyanato group has been revealed to be highly α-directing in sialidation reactions [71]. Installation of the isothiocyanate moiety with phenyl chlorothionoformate is accompanied by the formation of an oxazolidinthione

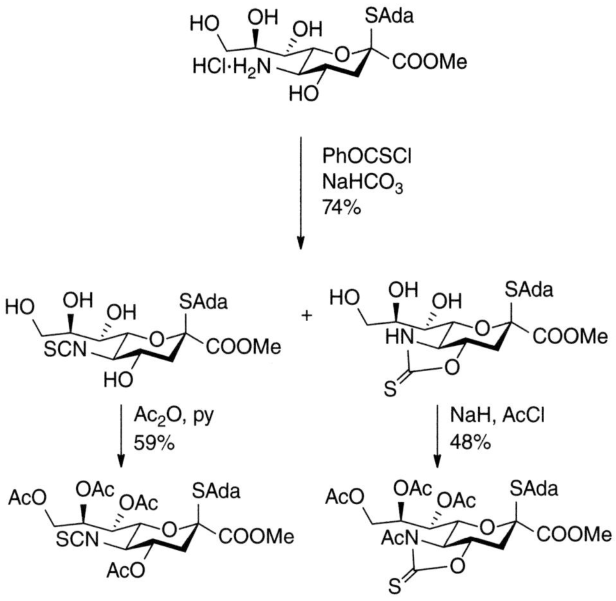

Scheme 10.15 Installation of the isothiocyanate moiety.

(Scheme 10.15) and is followed by peracetylation under usual conditions (Scheme 10.15) [57]. Acetylation of the oxazolidinthione with sodium hydride and acetyl chloride affords the corresponding *N*-acetyl system (Scheme 10.15) [57]. After glycosylation, the isothiocyanate is readily converted into a number of other functional groups, which renders it a very versatile protecting group

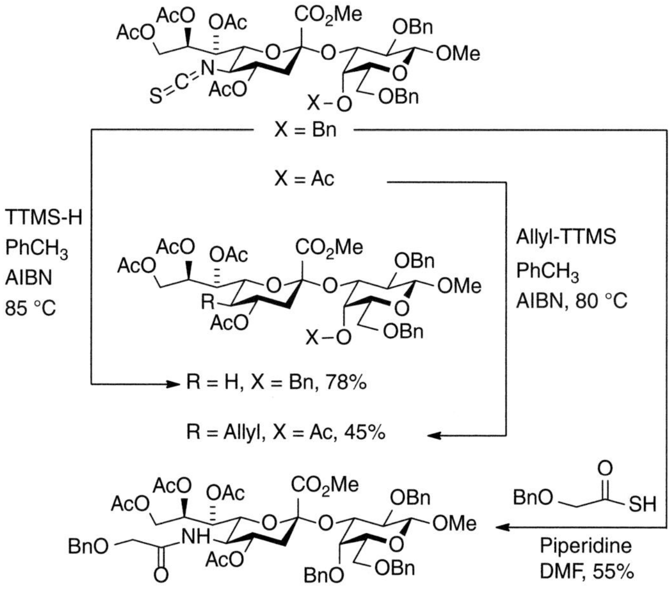

Scheme 10.16 Isothiocyanate manipulation.

(Scheme 10.16) [71]. It is noteworthy that the isothiocyanate is compatible with the *N*-iodosuccinimide/triflic acid conditions frequently employed in sialidation reactions based on the use of thiosialosides as donors.

10.3.3 Dual Protection of O4 and N5 as the Oxazolidinone and *N*-Acetyloxazolidinone

The *O*4,*N*5-oxazolidinone moiety is readily installed in high yield by treatment of sialyl thioglycosides with 4-nitrophenyl chloroformate under mild basic conditions [72–74]. Subsequent reaction with acetic anhydride and pyridine acetylates the remaining hydroxyl groups, and treatment with acetyl chloride and Hünig's base then gives the *N*-acetyl oxazolidinone (Scheme 10.17) [74]. The oxazolidinone moiety, although not the more reactive *N*-acetyl variant, is compatible with the regioselective reductive cleavage of benzylidene acetals by means of the borane-trimethylamine and aluminum trichloride combination [72]. The advantage of the oxazolidinone and *N*-acetyl oxazolidinone moieties derives from their ability to afford highly α-selective sialidation reactions either by direct employment of the thioglycosides [72–75] or after conversion to the anomeric dibutylphosphates and other related groups [76].

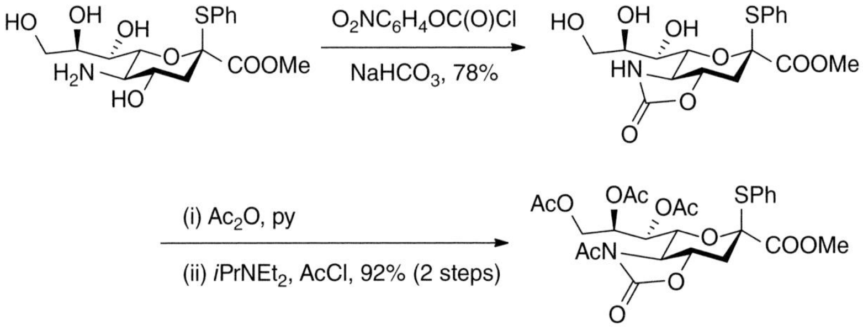

Scheme 10.17 Installation of *O*4,*N*5-oxazolidinone and *N*-acetyloxazolidinone.

After glycosylation, the oxazolidinone group is cleaved, along with any esters present, by heating to either 80 °C with aqueous ethanolic lithium hydroxide or 65 °C with barium hydroxide in ethanol [72, 73]. The native acetamide is then reinstalled by treatment with acetic anhydride in methanol, followed, if necessary, by Zemplén reaction to remove any reintroduced acetates (Scheme 10.18) [72]. Direct treatment of the oxazolidinone with sodium methoxide in methanol results in ring opening, to give the 5-*N*-methoxycarbamate, with concomitant cleavage of any esters present [73]. With the *N*-acetyl oxazolidinone, on the other hand, simple treatment with sodium methoxide in methanol at room temperature directly leads to the native acetamide in parallel with cleavage of all acetyl esters (Scheme 10.18) [74]. The direct cleavage of the *N*-acetyl oxazolidinone to acetamide in this manner offsets the extra step required for the installation of the *N*-acetyl group on the oxazolidinone before glycosylation, albeit the methyl ester is retained and requires an eventual saponification step. Note, however, that the

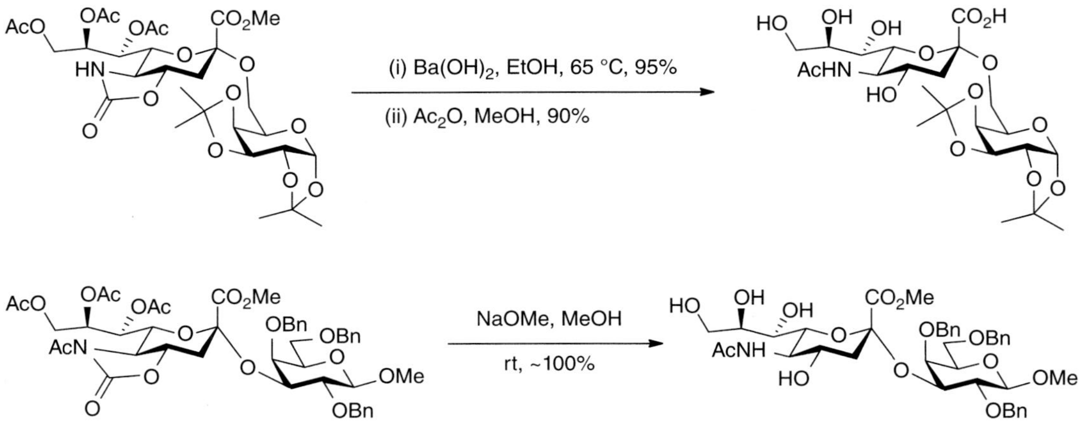

Scheme 10.18 Cleavage of the *O*4,*N*5-oxazolidinone and *N*-acetyloxazolidinone.

selective cleavage of the *N*-acetyl oxazolidinone ring does not extend to *N*-acyl oxazolidinones in which the acyl group is more reactive than the acetyl group [77]. In the *N*-acetyl oxazolidinthiones (Scheme 10.15), the *N*-acetyl group is removed selectively under Zemplén conditions leaving the oxazolidinthione in place [57].

10.3.4 Dual Protection of O7 and N5 as an Oxazinone

The protection of C5 amine as an oxazinone may be achieved by treatment of a selectively protected 7-hydroxy derivative with carbonyldiimidazole in tetrahydrofuran (THF) at reflux (Scheme 10.19) [78]. 4-Nitrophenylchloroformate is also effective for this transformation [79]. Alternatively, the oxazinone may be accessed by treatment of a 7-hydroxy 5-*N*-tert-butyloxycarbamate with sodium hydride in THF (Scheme 10.19) [80]. It is noteworthy that this latter process functions in the presence of a free hydroxyl group at the 4-position, thereby providing a rare example of derivatization of the 7-OH in preference to the normally more reactive 4-OH (*vide infra*). The oxazinone group proved to be strongly disarming in glycosylation and to afford little, if any, useful selectivity under a variety of conditions [78–80]. The oxazinone group is removed with lithium hydroxide in hot aqueous ethanol [78].

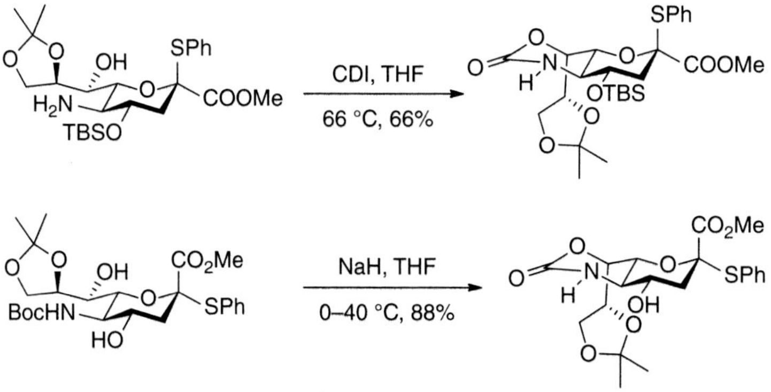

Scheme 10.19 Installation of the *O*7,*N*5-oxazinone.

10.4 Selective Protection of Alcohols

10.4.1 Acylation and Alkylation

Selective protection of the primary 9-OH group in sialic acids can be achieved by treatment with the appropriate trimethyl orthoester with catalytic *p*-toluenesulfonic acid (TsOH) in dimethyl sulfoxide (DMSO). This reaction can be conducted in the presence of a free carboxylate group (Scheme 10.20) [81]. As an alternative to TsOH, the reaction can be conducted in methanol with catalysis by silica gel-supported trivalent lanthanide salts (Scheme 10.20) [82]. Controlled acetylation with acetyl chloride and pyridine in dichloromethane at −40 °C also affords the 9-*O*-monoacetate in excellent yield [35].

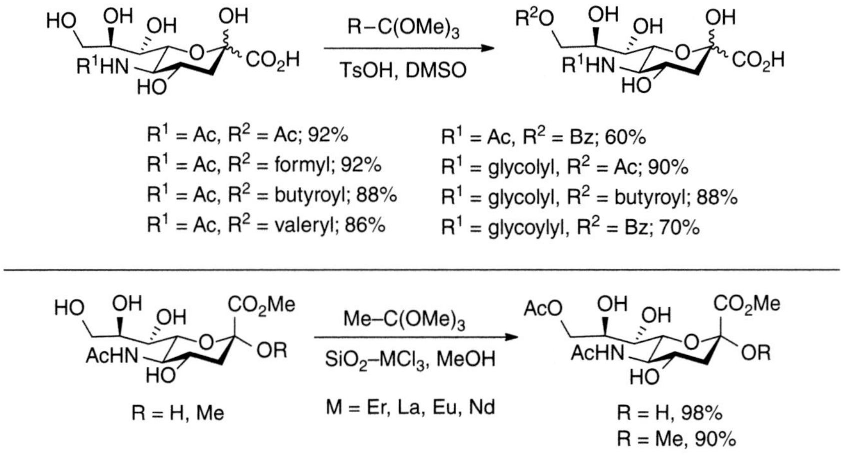

Scheme 10.20 Selective acylation of the primary alcohol.

10.4.2 Alkylation

Trimethyloxonium tetrafluoroborate enables the selective methylation of the 9-OH group (Scheme 10.21) [83].

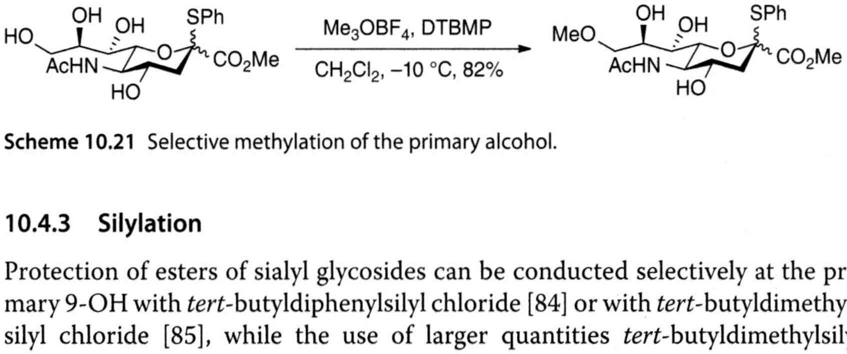

Scheme 10.21 Selective methylation of the primary alcohol.

10.4.3 Silylation

Protection of esters of sialyl glycosides can be conducted selectively at the primary 9-OH with *tert*-butyldiphenylsilyl chloride [84] or with *tert*-butyldimethylsilyl chloride [85], while the use of larger quantities *tert*-butyldimethylsilyl chloride enables the formation of a 4,9-bis-*O*-silyl ether (Scheme 10.22) [84]. The use of *tert*-butyldimethylsilyl trifluoromethanesulfonate permits persilylation at the 4, 7, 8-, and 9-positions [84].

10.4.4 Selective Acetal Formation and Manipulation

Reaction of methyl *N*-acetylneuraminate with 2,2-dimethoxypropane (DMP) in acetone with catalysis by TsOH selectively affords the 8,9-*O*-acetonide, with minor amounts of the 7,9-*O*-acetonide reported in some instances [35]. After saponification of the ester, treatment with acetic anhydride and pyridine followed by aqueous work-up selectively afforded the 4-*O*-acetyl derivative leaving the less-reactive 7-OH unchanged (Scheme 10.23). Subsequent treatment with aqueous acetic acid removed the acetonide and yielded 4-*O*-acetyl-*N*-acetyl neuraminic acid [35, 81, 86]. Selective silylation of the initial 8,9-*O*-acetonide with TBDMSCl and imidazole

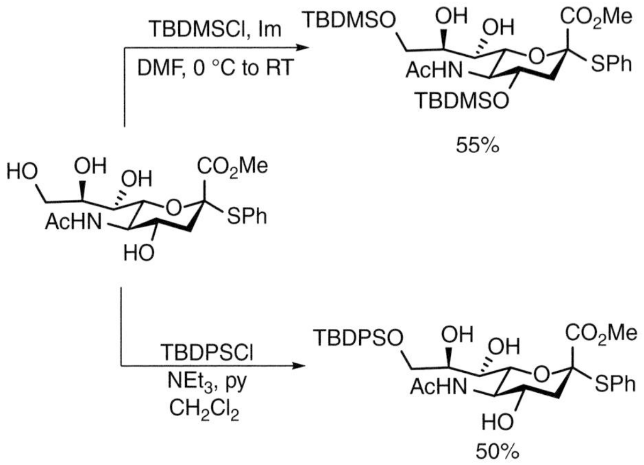

Scheme 10.22 Selective silylation reactions.

[44] followed by de-esterification with trimethyltin hydroxide then afforded the 4-*O*-silyl 8,9-*O*-acetonide. Selective protection of the hemiacetal with triethylsilyl triflate was followed by acetylation of the remaining free hydroxyl group at the 7-position. Subsequent hydrolysis of the acetonide and desilylation afforded 7-*O*-acetyl-*N*-acetylneuraminic acid [86]. Alternatively, treatment of the 4-*O*-silyl 8,9-*O*-acetonide with acetic anhydride and DMAP enabled selective installation of the acetate group at the anomeric position. Cleavage of the acetonide and desilylation finally gave 2-*O*-acetyl-*N*-acetylneuraminic acid (Scheme 10.23) [86]. In related work, it has been demonstrated that sialyl glycosides and thioglycosides carrying a 8,9-*O*-acetonide can be selectively silylated at the 4-position with TBDMSCl and imidazole [87, 88]; the use of the more potent *tert*-butyldimethylsilyl triflate enables the installation of a second silyl group at the 7-position [88]. The greater reactivity of the 4-OH as compared to the 7- and 8-hydroxy groups is further documented by the ability to conduct selective alkylations at that position by means of a dialkyl sulfate and sodium hydride when the primary 9-OH and anomeric positions are protected beforehand (Scheme 10.23) [89]. 4,9-Di-*O*-acetates may be obtained in high yield from the 4-*O*-acetyl-7,8,9-triols by acetylation at −40 °C with acetyl chloride in pyridine and dichloromethane [35].

Other manipulations of 8,9-*O*-isopropylidene derivatives of *N*-acetyl neuraminic acid glycosides exploiting the differential reactivity of the 4- and 7-hydroxyl groups in combination with acetonide removal and subsequent esterification permit the formation of 7-*O*-, 7,9-di-*O*-, and 7,8,9-tri-*O*-acetyl derivatives (Scheme 10.24) [90].

Under kinetic conditions, benzylidene and *p*-methoxybenzylidene acetals can be introduced selectively onto the 8,9-diols of sialic acid derivatives in the presence of the free 7-OH group (Scheme 10.25) [70, 84]. This protecting group strategy has the advantage that subsequent reduction with trimethylamine-borane in the presence of aluminum trichloride, either with or without prior acetylation of the remaining hydroxyl groups, selectively yields the 9-*O*-benzyl ether or the corresponding 4-methoxybenzyl derivative (Scheme 10.25) [70, 84].

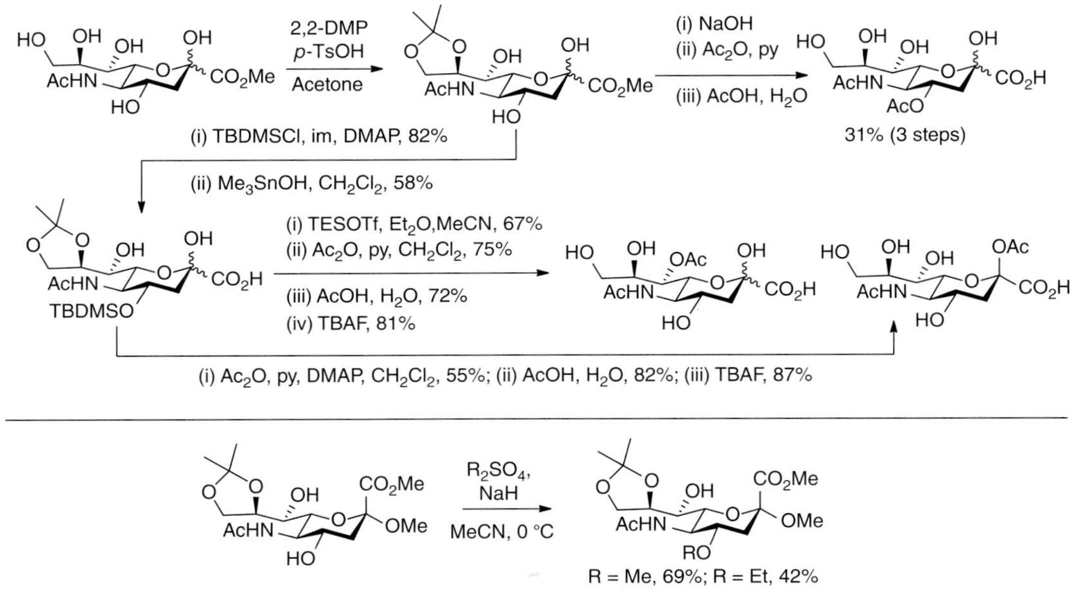

Scheme 10.23 Selective acetonide formation and subsequent selective acetylation, silylation, and alkylation.

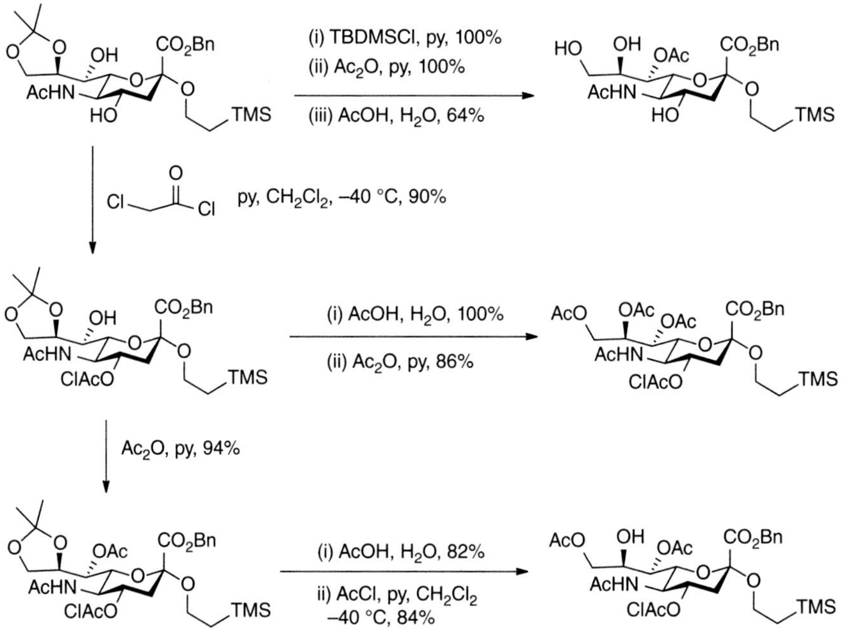

Scheme 10.24 Selective formation of 7-*O*-, 7,9-di-*O*-, and 7,8,9-tri-*O*-acetyl derivatives.

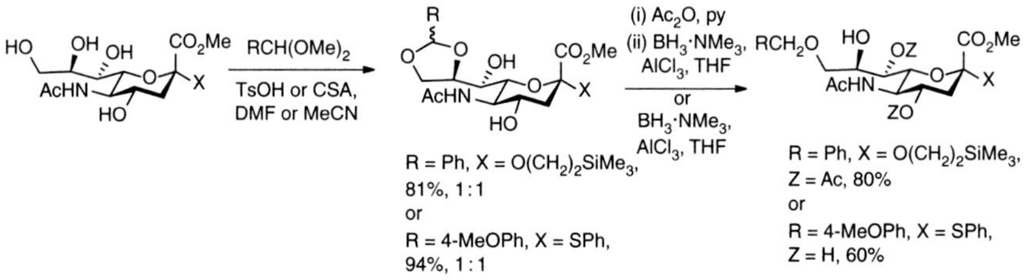

Scheme 10.25 Installation of arylidene acetals under kinetic conditions with subsequent reductive ring opening.

In some instances, the installation of an 8,9-*O*-arylidene acetal by the transacetalization method has been reported to give mixtures with the corresponding 7,9-*O*-acetal [91]. Certainly, under the more forcing conditions of treatment of 7,8,9-triol with benzaldehyde and TsOH in DMF at room temperature, the more stable 7,9-*O*-benzylidene acetal is formed as the major product (Scheme 10.26) [86, 92]. Such 7,9-*O*-benzylidene acetals are formed as single isomers that exist in the chair conformation with all substituents equatorial [86, 91, 92]; they are not reduced by the combination of $BH_3 \cdot NMe_3$ and aluminum trichloride that is so effective for the 8,9-isomers [91]. However, selective acetylation of the 4-OH in a 7,9-*O*-benzylidene-4,8-diol may be conducted with acetic anhydride and pyridine in the presence of DMAP to give, after removal of the benzylidene group, the 4-*O*-acetyl-7,8,9-triol [86]. Alternatively, silylation followed by acetylation

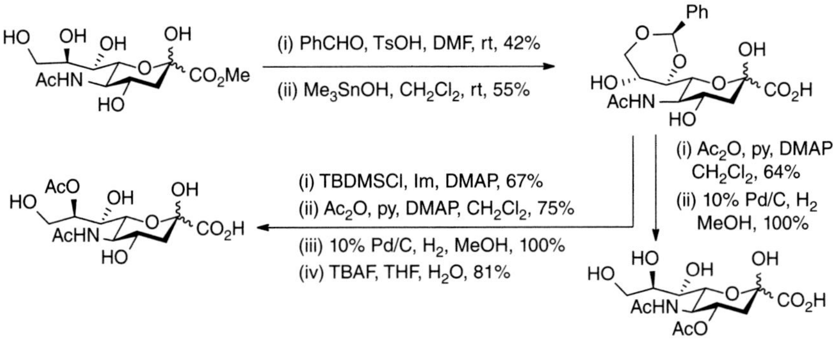

Scheme 10.26 Installation and manipulation of a 7,9-*O*-benzylidene acetal.

and removal of the benzylidene and silyl groups affords the 8-*O*-acetyl-4,7,9-triol (Scheme 10.26) [86].

10.4.5 Silylene Acetal Formation and Manipulation

As illustrated in the context of oxazolidinones (Scheme 10.27), both 8,9-*O*-(di-*tert*-butylsilylene) and (tetraisopropyl disiloxyl) acetals can be installed selectively in the presence of the 7-OH and provide an alternative to the use of the more common isopropylidene and benzylidene adducts. Further silylation with ditrifloxy-di-*tert*-butylsilane affords 5*N*,7*O*-silyleneaminals [93]. Both the silylene and disiloxane groups were compatible with typical glycosylation reactions, which afforded moderate to excellent α-selectivity depending on the acceptor [93].

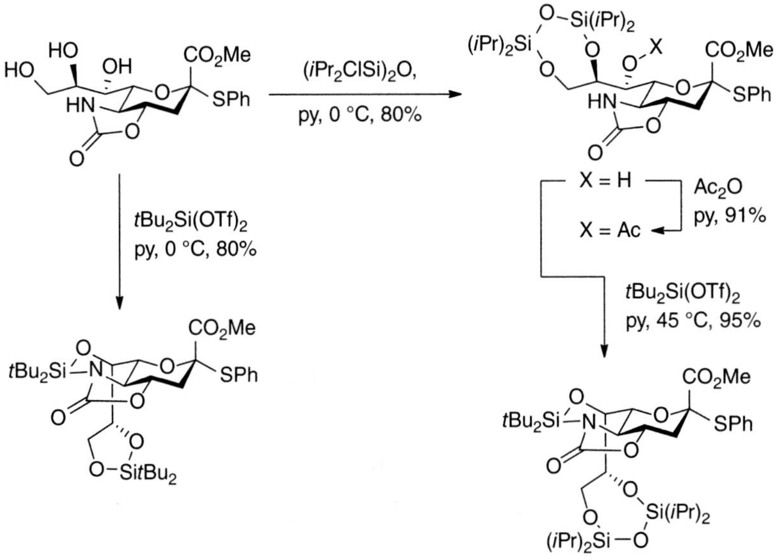

Scheme 10.27 Silylene and disiloxane installation.

10.4.6 Acetate Migration

Removal of the trichloroethoxycarbonyl group from 4,7,8,9-tetra-*O*-acetyl neuraminic acid derivatives with zinc and acetic acid affords the corresponding amine, presumably in the form of ammonium acetate salt. However, when the reaction is conducted in a solution of acetic acid in DMF, dioxane or acetonitrile acetyl migration from the 8-position occurs to give the 4,7,9-tri-*O*-acetyl-*N*-acetyl derivative in yield, thereby providing a facile means of access to the 8-OH, which was exploited in follow-up *O*-methylation, sulfonylation, and acylation reactions (Scheme 10.28) [48, 94]. It is noteworthy that the reverse migration of a single acetyl group from an *N*,*N*-diacetylimide to a free 8-OH giving the 8-*O*-acetyl-*N*-acetyl derivative is also facile [94].

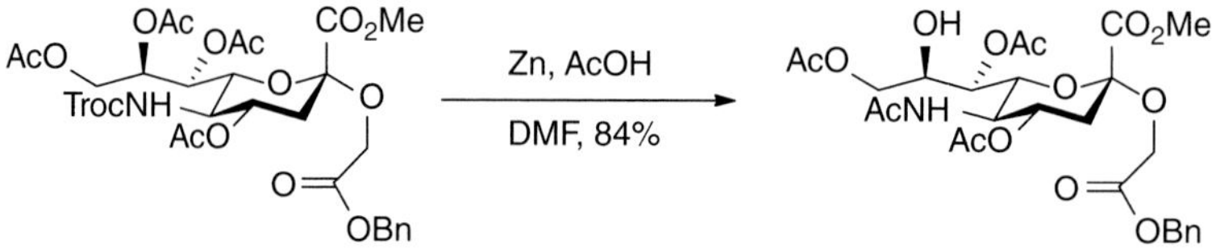

Scheme 10.28 Selective *O*→*N*-acetyl migration.

10.4.7 Acetylative Desilylation of Per-*O*-silyl Ethers

Reaction of the benzyl ester of *N*-acetyl neuraminic acid with trimethylsilyl chloride and hexamethyldisilazane in pyridine affords a per-*O*-trimethylsilyl ether in high yield [95]. Subsequent treatment with mixtures of acetic acid and acetic anhydride in pyridine under microwave irradiation leads to the progressive replacement of silyl ethers by acetyl groups. The order of reactivity is such that the 4-*O*-silyl group is first replaced by an acetyl group, followed by the 9-, 8-, and anomeric *O*-silyl groups, with the more hindered 7-*O*-trimethylsilyl group being the most resistant (Scheme 10.29). After desilylation and hydrogenolysis of the benzyl ester, the method affords ready access to the 4-*O*-acetyl, 4,9-di-*O*-acetyl, 4,8,9-tri-*O*-acetyl, and 2,4,8,9-tetra-*O*-acetyl derivatives of *N*-acetyl neuraminic acid [95].

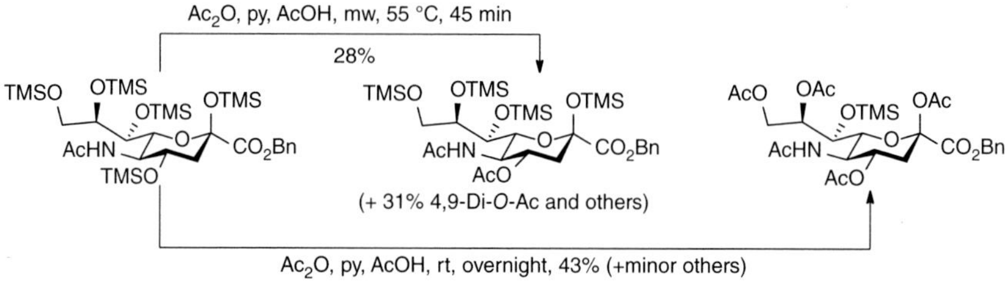

Scheme 10.29 Selective replacement of trimethylsilyl ethers by acetyl groups.

10.5 Access to Protected Sialic Acid Derivatives by Total Synthesis

The sialic acids and their derivatives and analogs have been a popular target for total and hemisynthesis over a period of several decades, with much of the work covered in the review literature [96–99]. Such total syntheses frequently provide access to selectively protected derivatives. However, with rare exceptions, such *de novo* approaches cannot compete in efficiency with derivatization of sialic acids themselves and accordingly are not covered here.

10.6 Access to Protected Sialic Acid Derivatives by Chemoenzymatic Synthesis

In both eukaryotes and prokaryotes, the biosynthesis of *N*-acetyl neuraminic acid proceeds via the aldolase-mediated coupling of *N*-acetylmannosamine and pyruvic acid (Scheme 10.30). Although not covered here, this biosynthetic machinery is sufficiently accommodating that a variety of mannose derivatives at the 2, 4, and 6-positions are processed, thereby affording direct access to the neuraminic acids modified at the 5, 7, and 9-positions and an effective alternative to chemical modification of *N*-acetylneuraminic acid [100–103].

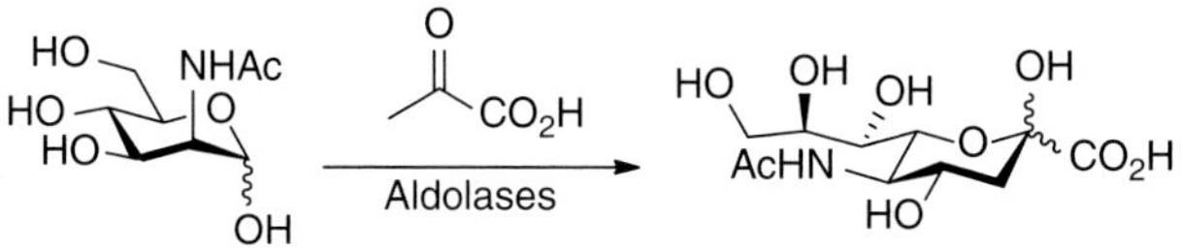

Scheme 10.30 Biosynthesis of *N*-acetylneuraminic acid.

10.7 Preparation of Methyl (methyl 5-acetamido-3,5-dideoxy-D-glycero-β-D-galacto-non-2-ulopyranosid)onate [20, 22, 104]

N-Acetyl neuraminic acid (5 g, 16.17 mmol) was suspended in dry methanol (250 ml), and Dowex® 50 × 8 H$^+$ ion exchange resin (12 g) was added. The mixture was heated to reflux for 3 days, then cooled to room temperature, and filtered through Celite®. The solvent was evaporated and the residue was purified by silica gel column chromatography eluting with AcOEt:MeOH:H_2O to give the title compound (3.87 g, 71%) as colorless oil; $[\alpha]_D^{21}$ −45.2 (c = 1.0, CH_3OH); ^{1}H NMR (400 MHz, CD_3OD) δ 4.03–3.95 (m, 1H), 3.84 (m, 4H), 3.79 (s, 3H), 3.64 (dd, J = 12.1, 6.1 Hz, 1H), 3.49 (d, J = 9.0 Hz, 1H), 3.32–3.29 (m, 1H), 3.26 (s, 3H), 2.34 (dd, J = 12.9, 4.9 Hz, 1H), 2.00 (s, 3H), and 1.63 (dd, J = 12.9, 11.2 Hz, 1H); ^{13}C NMR (101 MHz, CD_3OD) δ 173.4, 169.4, 98.9, 70.86, 69.9, 68.7, 66.3, 63.9, 52.3, 40.2, and 21.3. ESI-HRMS ($C_{13}H_{23}NNaO_9$): $[M + Na]^+$ m/z: 360.1271; found: 360.1280.

Abbreviations

Ac	acetyl
Ada	1-adamantanyl
AIBN	azobis(isobutyronitrile)
Alloc	allyloxycarbonyl
Bn	benzyl
Boc	*tert*-butyloxycarbonyl
Bz	benzoyl
Cat	catalytic
CDI	1,1′-carbonyldiimidiazole
ClAc	chloroacetyl
DMAP	4-(dimethylamino)pyridine
DMF	*N,N*-dimethylformamide
DMP	2,2-dimethoxypropane
DMSO	dimethyl sulfoxide
DTBMP	2,6-di-*tert*-butyl-4-methylpyridine
Et	ethyl
Fmoc	9-fluorenylmethoxycarbonyl
Im	imidazole
Me	methyl
μW	microwave irradiation
p-	*para*-
Ph	phenyl
Phth	phthaloyl
*i*Pr	2-propyl
Piv	2,2-dimethylpropanoyl
Py	pyridine
Su	succinimidyl
t-	*tertiary*
tert-	*tertiary*
TBAF	tetrabutylammonium fluoride
TBDMS	*tert*-butyldimethylsilyl
TBDPS	*tert*-butyldiphenylsilyl
TES	triethylsilyl
Tf	trifluoromethanesulfonyl
THF	tetrahydrofuran
Troc	2,2,2-trichloroethoxycarbonyl
Ts	4-toluenesulfonyl
TMS	trimethylsilyl
TTMS	tris(trimethylsilyl)silyl

References

1 Schauer, R. (1982). *Adv. Carbohydr. Chem. Biochem.* 40: 131–234.

2 Hasegawa, A. and Kiso, M. (1997). Synthesis of sialyl glycosides. In: *Preparative Carbohydrate Chemistry* (ed. S. Hanessian), 357–379. New York: Marcel Dekker.
3 Hasegawa, A. (1996). Synthesis of sialyl glycoconjugates. In: *Modern Methods in Carbohydrate Synthesis* (ed. S.H. Khan and R.A. O'Niell), 277–300. Amsterdam: Harwood Academic Publishers.
4 Kiso, M., Ishida, H., and Ito, H. (2000). Special problems in glycosylation reactions: sialidations. In: *Carbohydrates in Chemistry and Biology: Chemistry of Saccharides*, vol. 1 (ed. B. Ernst, G.W. Hart and P. Sinaÿ), 345–365. Weinheim: Wiley-VCH.
5 Boons, G.-J. and Demchenko, A.V. (2000). *Chem. Rev.* 100: 4539–4565.
6 Magnusson, G. and Nilsson, U.J. (2001). Regio- and stereoselective methods of glycosylation. In: *Glycoscience: Chemistry and Chemical Biology*, vol. 2 (ed. B. Fraser-Reid, T. Kuniaki and J. Thiem), 1543–1587. Berlin: Springer-Verlag.
7 Halcomb, R.L. and Chappell, M.D. (2001). Recent developments in technology for glycosylation with sialic acid. In: *Glycochemistry: Principles, Synthesis, and Applications* (ed. P.G. Wang and C.R. Bertozzi), 177–220. New York: Dekker.
8 Kiso, M., Ishida, H., and Ando, H. (2003). Complex carbohydrate synthesis. In: *Carbohydrate-Based Drug Discovery*, vol. 1 (ed. C.-H. Wong), 37–54. Weinheim: Wiley-VCH.
9 Boons, G.-J. and Demchenko, A.V. (2003). The chemistry of sialic acid. In: *Carbohydrate-Based Drug Discovery*, vol. 1 (ed. C.-H. Wong), 55–102. Weinheim: Wiley-VCH.
10 De Meo, C., Boons, G.J., and Demchenko, A.V. (2007). *Comprehensive Glycoscience*, vol. 1 (ed. J. Kamerling), 583–604. Amsterdam: Elsevier.
11 De Meo, C. (2007). Synthesis of N-5-derivatives of neuraminic acid and their application as sialosyl donors. In: *Frontiers in Modern Carbohydrate Chemistry*, vol. 960 (ed. A.V. Demchenko), 118–131. Washington, DC: ACS.
12 De Meo, C. and Priyadarshani, U. (2008). *Carbohydr. Res.* 343: 1540–1552.
13 Hanashima, S. (2011). *Trends Glycosci. Glycobiol.* 23: 111–121.
14 Kononov, L.D. (2013). *Adv. Chem. Res.* 18: 143–177.
15 Ando, H., Pal, R., Ishida, H., and Kiso, M. (2014). Novel approaches to complex glycosphingolipids. In: *Modern Synthetic Methods in Carbohydrate Chemistry: From monosaccharides to Complex Glycoconjugates* (ed. D.B. Werz and S. Vidal), 313–333. Weinheim: Wiley-VCH.
16 Navuluri, C. and Crich, D. (2016). Recent advances in the stereocontrolled synthesis of sialic acid glycosides. In: *Glycochemical Synthesis: Strategies and Applications* (ed. S.-C. Hung), 131–154. New York: Wiley.
17 Furuhata, K. (2004). *Trends Glycosci. Glycobiol.* 16: 143–169.
18 Greene, T.W. and Wuts, P.G.M. (1999). *Protective Groups in Organic Synthesis*, 3rde. New York: Wiley.
19 Kocienski, P.J. (2005). *Protecting Groups*, 3rd ed. Stuttgart: Thieme.
20 Kuhn, R., Lutz, P., and Macdonald, D.L. (1966). *Chem. Ber.* 99: 611–617.
21 Yu, P.K. and Ledeen, R. (1969). *J. Biol. Chem.* 244: 1306–1313.
22 Chopra, P., Thomson, R.J., Grice, D., and von Itzstein, M. (2012). *Tetrahedron Lett.* 53: 6254–6256.
23 Baggett, N. and Marsden, B.J. (1982). *Carbohydr. Res.* 110: 11–18.
24 Paulsen, H. and Tietz, H. (1984). *Carbohydr. Res.* 125: 47–64.

25 Myers, R.W., Lee, R.T., Lee, Y.C. et al. (1980). *Anal. Biochem.* 101: 166–174.
26 Fujita, S., Numata, M., Sugimoto, M. et al. (1992). *Carbohydr. Res.* 228: 347–370.
27 Hanashima, S., Akai, S., and Sato, K.-I. (2008). *Tetrahedron Lett.* 49: 5111–5114.
28 Ren, C.-T., Chen, C.-S., Yu, Y.-P. et al. (2003). *Chem. Eur. J.* 9: 1085–1095.
29 Fujita, S., Numata, M., Sugimoto, M. et al. (1994). *Carbohydr. Res.* 263: 181–196.
30 Huang, B.-G., Locke, R.D., Jain, R.K., and Matta, K.L. (1997). *Bioorg. Med. Chem. Lett.* 7: 1157–1160.
31 Kononov, L.O., Shpirt, A.M., Ito, Y., and Ogawa, T. (2003). *Russ. Chem. Bull. Int. Ed.* 52: 1442–1446.
32 Ishiwata, A. and Ito, Y. (2003). *Synlett* 1339–1343.
33 Kunz, H., Waldmann, H., and Klinkhammer, U. (1988). *Helv. Chim. Acta* 71: 1868–1874.
34 Pezzotta, J., Urban, D., Guillot, R. et al. (2014). *Synlett* 25: 375–380.
35 Hasegawa, A., Murase, T., Ogawa, M. et al. (1990). *J. Carbohydr. Chem.* 9: 415–428.
36 Takahashi, T., Tsukamoto, H., and Yamada, H. (1997). *Tetrahedron Lett.* 38: 8223–8226.
37 Haberman, J.M. and Gin, D.Y. (2003). *Org. Lett.* 5: 2539–2541.
38 Haberman, J.M. and Gin, D.Y. (2001). *Org. Lett.* 3: 1665–1668.
39 Murakami, M., Okamoto, R., Izumi, M., and Kajihara, Y. (2012). *Angew. Chem. Int. Ed.* 51: 3567–3572.
40 Murakami, M., Kiuchi, T., Nishihara, M. et al. (2016). *Sci. Adv.* 2: e1500678.
41 Okamoto, R., Souma, S., and Kajihara, Y. (2008). *J. Org. Chem.* 73: 3460–3466.
42 Furuhata, K., Sato, S., Anazawa, K. et al. (1987). *Chem. Pharm. Bull.* 35: 3609–3614.
43 Sato, S., Furuhata, K., and Ogura, H. (1988). *Chem. Pharm. Bull.* 36: 4678–4688.
44 Zbiral, E., Phadtare, S., and Schmid, W. (1987). *Liebigs Ann. Chem.* 39–43.
45 Roy, R. and Pon, R.A. (1990). *J. Carbohydr. Chem.* 7: 3–12.
46 Ishida, H., Ohta, Y., Tsukada, Y. et al. (1993). *Carbohydr. Res.* 246: 75–88.
47 Ando, H., Ishida, H., Kiso, M., and Hasegawa, A. (1997). *Carbohydr. Res.* 300: 207–217.
48 Ando, H., Koike, Y., Ishida, H., and Kiso, M. (2003). *Tetrahedron Lett.* 44: 6883–6886.
49 Ando, H., Koike, Y., Koizumi, S. et al. (2005). *Angew. Chem. Int. Ed.* 44: 6759–6763.
50 Iwayama, Y., Ando, H., Tanaka, H.-N. et al. (2011). *Chem. Commun.* 9726–9728.
51 Zhang, D., Ye, D., Feng, E. et al. (2010). *J. Org. Chem.* 75: 3552–3557.
52 Sugata, T., Kan, Y., Nagaregawa, Y. et al. (1997). *J. Carbohydr. Chem.* 16: 917–925.
53 Schreiner, E., Zbiral, E., Kleineidam, R.G., and Schauer, R. (1992). *Carbohydr. Res.* 216: 61–66.
54 Gervay, J., Flaherty, T.M., and Nguyen, C. (1997). *Tetrahedron Lett.* 38: 1493–1496.
55 Flynn, D.L., Zelle, R.E., and Grieco, P.A. (1983). *J. Org. Chem.* 48: 2424–2426.
56 Crich, D. and Li, W. (2007). *J. Org. Chem.* 72: 7794–7797.
57 Rajender, S. and Crich, D. (2013). *J. Carbohydr. Chem.* 32: 324–335.

58 Byramova, N.E., Tuzikov, A.B., and Bovin, N.V. (1992). *Carbohydr. Res.* 237: 161–175.

59 De Meo, C., Demchenko, A.V., and Boons, G.J. (2001). *J. Org. Chem.* 66: 5490–5497.

60 Adachi, M., Tanaka, H., and Takahashi, T. (2004). *Synlett* 609–614.

61 Tanaka, H., Adachi, M., and Takahashi, T. (2005). *Chem. Eur. J.* 11: 849–862.

62 Ren, C.-T., Chen, C.-S., and Wu, S.-H. (2002). *J. Org. Chem.* 67: 1376–1379.

63 Tanaka, H., Nishiura, Y., Adachi, M., and Takahashi, T. (2006). *Heterocycles* 67: 107–112.

64 Tanaka, K., Goi, T., and Fukase, K. (2005). *Synlett* 2958–2962.

65 Tanaka, S.-I., Goi, T., Tanaka, K., and Fukase, K. (2007). *J. Carbohydr. Chem.* 26: 369–394.

66 Sherman, A.A., Yudina, O.N., Shashkov, A.S. et al. (2001). *Carbohydr. Res.* 330: 445–458.

67 Ikeda, K., Miyamoto, K., and Sato, M. (2007). *Tetrahedron Lett.* 48: 7431–7435.

68 Schreiner, E. and Zbiral, E. (1990). *Liebigs Ann. Chem.* 581–586.

69 Schneider, R., Freyhardt, C.C., and Schmidt, R.R. (2001). *Eur. J. Org. Chem.* 1655–1661.

70 Demchenko, A.V. and Boons, G.-J. (1999). *Chem. Eur. J.* 5: 1278–1283.

71 Mandhapati, A.R., Rajender, S., Shaw, J., and Crich, D. (2015). *Angew. Chem. Int. Ed.* 54: 1275–1278.

72 Tanaka, H., Nishiura, Y., and Takahashi, T. (2006). *J. Am. Chem. Soc.* 128: 7124–7125.

73 Farris, M.D. and De Meo, C. (2007). *Tetrahedron Lett.* 48: 1225–1227.

74 Crich, D. and Li, W. (2007). *J. Org. Chem.* 72: 2387–2391.

75 Harris, B.N., Patel, P.P., Gobble, C.P. et al. (2011). *Eur. J. Org. Chem.* 4023–4027.

76 Hsu, C.-H., Chu, K.-C., Lin, Y.-S. et al. (2010). *Chem. Eur. J.* 16: 1754–1760.

77 Crich, D. and Wu, B. (2008). *Org. Lett.* 10: 4033–4035.

78 Crich, D. and Wu, B. (2008). *Tetrahedron* 64: 2042–2047.

79 Kancharla, P.K. and Crich, D. (2013). *J. Am. Chem. Soc.* 135: 18999–19007.

80 Tanaka, H., Ando, H., Ishihara, H., and Koketsu, M. (2008). *Carbohydr. Res.* 343: 1585–1593.

81 Ogura, H., Furuhata, K., Sato, S., and Anazawa, K. (1987). *Carbohydr. Res.* 167: 77–86.

82 Bianco, A., Brufani, M., Melchioni, C., and Romagnoli, P. (1997). *Tetrahedron Lett.* 38: 651–652.

83 Wu, Y.-F., Tsai, Y.-F., Guo, J.-R. et al. (2014). *Org. Biomol. Chem.* 12: 9345–9349.

84 Yuan, X., Ress, D.K., and Linhardt, R.J. (2007). *J. Org. Chem.* 72: 3085–3088.

85 Hartmann, M. and Zbiral, E. (1989). *Monatsh. Chem.* 120: 899–906.

86 Clarke, P.A., Mistry, N., and Thomas, G.H. (2012). *Org. Biomol. Chem.* 10: 529–535.

87 Brandstetter, H.H., Zbiral, E., and Schulz, G. (1982). *Liebigs Ann. Chem.* 1–13.

88 Premathilake, H.D., Gobble, C.P., Pornsuriyasak, P. et al. (2012). *Org. Lett.* 14: 1126–1129.

89 Liav, A. (1995). *Carbohydr. Res.* 271: 241–245.

90 Hasegwa, A., Murase, T., Ogawa, M. et al. (1990). *J. Carbohydr. Chem.* 9: 429–439.

91 Meinke, S., Schroven, A., and Thiem, J. (2011). *Org. Biomol. Chem.* 9: 4487–4497.

92 Castro-Palomino, J.C., Tsvetkov, Y.E., Schneider, R., and Schmidt, R.R. (1997). *Tetrahedron Lett.* 38: 6837–6840.

93 Hanashima, S., Sato, K.-I., Ito, Y., and Yamaguchi, Y. (2009). *Eur. J. Org. Chem.* 4215–4220.

94 Tamai, H., Ando, H., Tanaka, H.-N. et al. (2011). *Angew. Chem. Int. Ed.* 50: 2330–2333.

95 Park, S.S. and Gervay-Hague, J. (2014). *Org. Lett.* 16: 5044–5047.

96 Danishefsky, S.J. and DeNinno, M.P. (1987). *Angew. Chem. Int. Ed.* 26: 15–23.

97 DeNinno, M.P. (1991). *Synthesis* 583–593.

98 Hemeon, I. and Bennet, A.J. (2007). *Synthesis* 1899–1926.

99 Kiefel, M.J. and von Itzstein, M. (2002). *Chem. Rev.* 102: 471–490.

100 Blixt, O. and Paulson, J.C. (2006). *Adv. Synth. Catal.* 345: 687–690.

101 Yu, H. and Chen, X. (2009). Enzymatic synthesis of carbohydrate-containing biomolecules. In: *Wiley Encyclopedia of Chemical Biology*, vol. 1 (ed. T.P. Begley), 631–652.

102 Chen, X. and Varki, A.P. (2010). *ACS Chem. Biol.* 5: 163–176.

103 Rillahan, C.D., Macauley, M.S., Schwartz, E. et al. (2014). *Chem. Sci.* 5: 2398–2406.

104 Buda, S. and Crich, D. (2016). *J. Am. Chem. Soc.* 138: 1084–1092.

11

Strategies Toward Protection of 1,2- and 1,3-Diols in Carbohydrate Chemistry

Marie Schuler and Arnaud Tatibouët

Université d'Orléans and CNRS, Institut de Chimie Organique et Analytique (ICOA), UMR 7311, 45067 Orléans, France

11.1 Introduction

The choice of protecting groups is a crucial step in the synthesis of complex natural products [1]. This is particularly true when coming to carbohydrate synthons and oligosaccharides, which can possess multiple hydroxyl groups to be orthogonally protected. The use of cyclic protecting groups is an efficient approach to isolate at least two hydroxyl groups from the others. This chapter focuses on the protection of 1,2- and 1,3-diols in carbohydrate chemistry, by giving the most useful methods for protection and deprotection recently developed. For practical reasons, this chapter will focus mainly on the cyclic acetals that have been used for the protection of aldo-/keto-pentoses and hexoses over the last years from 2005.

11.2 Protection as Cyclic Acetals

11.2.1 Aliphatic Acetals

11.2.1.1 Methylene

Methylene acetals are quite robust under a wide range of both acidic and basic conditions but have been scarcely used for diol protection in carbohydrate chemistry, probably because of the strong acidic conditions needed for their deprotection. They are classically formed by reaction with formaldehyde in an acidic medium or by condensation with dihalomethanes in basic conditions. However, some efforts were recently made toward the development of mild conditions of formation and/or deprotection.

Formation:

1) $(HCHO)_n$ 1.1 eq., $p\text{TsOH}_{\text{cat.}}$, toluene, 110 °C, 67% [2].
2) $PhSCH_2OMe$ 2 eq., 1,3-dibromo-5,5-dimethylhydantoïn (DBDMH) 2 eq., dibutylhydroxytoluene (BHT), 2 eq., MeCN 85% (Scheme 11.1) [2].

Protecting Groups: Strategies and Applications in Carbohydrate Chemistry, First Edition.
Edited by Sébastien Vidal.

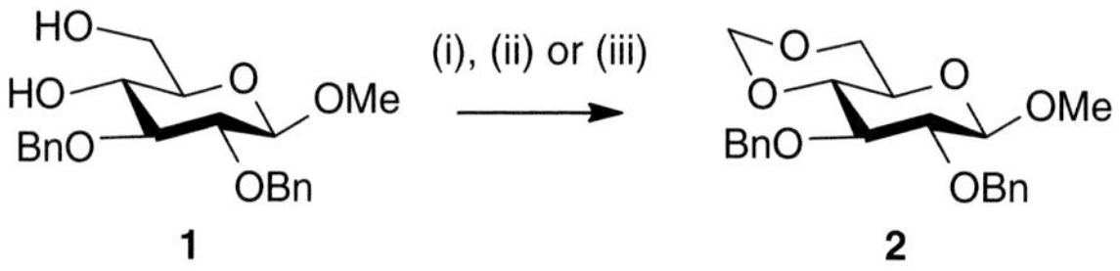

Scheme 11.1 Methylene acetal formation from methyl tri-*O*-benzyl-β-D-glucopyranoside (i) $(HCHO)_n$, *p*TsOH, toluene, 110 °C, 67%; (ii) CH_2Br_2 2 eq., KOH 8 eq., DMSO, RT, 39%; (iii) $PhSCH_2OMe$, DBDMH, BHT, MeCN, RT, 85%. *Source:* Tomohiro et al. 2013 [2]. Reproduced with permission of American Chemical Society.

3) $CH_2(OMe)_2$, $CSA_{cat.}$, 40 °C, 41% [3, 4].
4) CH_2Br_2, NaOH $50\%_{aq.}$, $Et_4N^+Br^-{}_{cat.}$, 60 °C, 60% [5].

Deprotection:

1) Dowex 50WX8, 77% deprotection of a 1,2:3,4-di-*O*-methyleneglucofuranose derivative [6].
2) AcOH/TFAA (trifluoroacetic anhydride) hydrolyzes 4,6-*O*-methylene **3** into 6-OAc derivative **5** and 4-OCH_2OAc acetal **4**, which can undergo further basic hydrolysis (NaOH 1 N, MeOH/H_2O) to give the corresponding diol **6** (Scheme 11.2) [7].

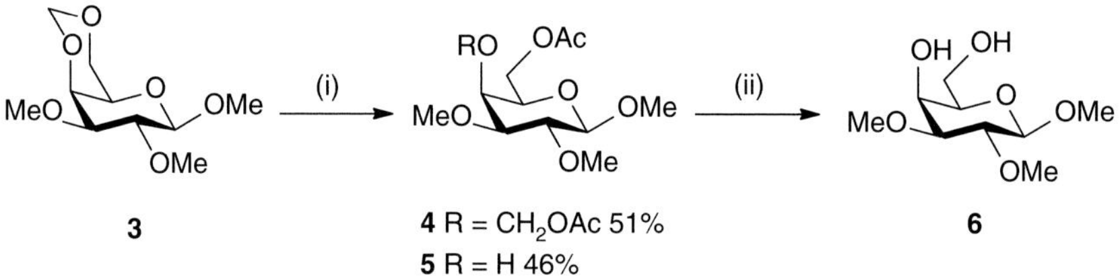

Scheme 11.2 Hydrolysis of methylenedioxy acetals (i) AcOH, TFAA, 26 °C, 3 h and (ii) NaOH 1 N, MeOH/H_2O 9/1, 1 h (90% from **4**, 89% from **5**). *Source:* Boto et al. 2006 [7]. Reproduced with permission of American Chemical Society.

11.2.1.2 Phenylsulfonylethylidene (PSE): $PhSO_2CH_2CH_2CH(OR)_2$

Recently, phenylsulfonylethylidene (PSE) acetal, was introduced by Rollin et al. as a new protecting group for 1,2 and 1,3-diols. On carbohydrate substrates, 1,2-diols form dioxolane-type acetals with good yields but poor diastereoselectivity. 1,3-Diols on the other hand lead exclusively to equatorially configured dioxanes, similarly to what was observed for 4,6-*O*-benzylidene derivatives. The reaction proceeds through a double Michael addition pathway with extrusion of a sulfinate ion. PSE acetals are resistant to both standard acidic hydrolysis (AcOH/H_2O, trifluoroacetyl (TFA)/H_2O, $BF_3{\cdot}OEt_2$/MeOH) and alcoholysis conditions, whereas they can be removed under reductive or basic conditions (Scheme 11.3). In addition, they also display a remarkable stability under oxidative conditions (2,3-dichloro-5,6-dicyano-1,4-benzoquinone, DDQ) or in the presence of strong Lewis acids.

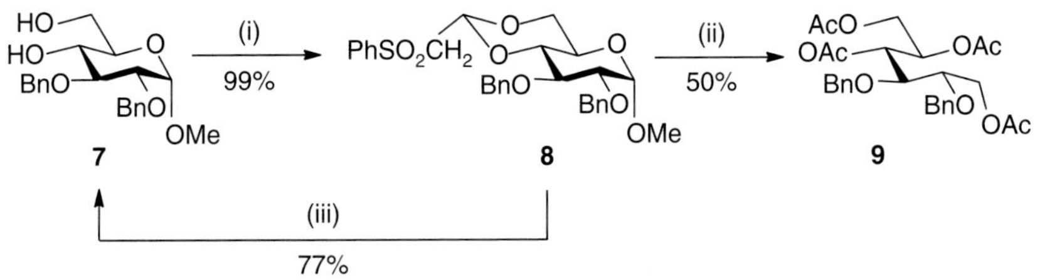

Scheme 11.3 Formation, deprotection, and reduction of a PSE-glucopyranose derivative (i) NaH, 1,2-bis(phenylsulfonyl)ethylene, Bu_4NBr; (ii) DIBAL-H, toluene then acetylation; and (iii) $LiAlH_4$ or $LiAlH_4$, $AlCl_3$. *Source:* From Chéry 2000 and Chéry 2001 [8].

Formation:

Z- or *E*-1,2-bis(phenylsulfonyl)ethylene 1 eq., NaH 2 eq., $Bu_4NBr_{cat.}$, tetrahydrofuran (THF), 12 h, RT, or *t*BuOK 2 eq., dimethylformamide (DMF), RT, 12 h: 70–99% for various furano- and pyranosides [8].

Deprotection:

1) $LiAlH_4$ 6 eq., Et_2O, 2 h, RT: 77% of **7** from **8** or $LiAlH_4$ 5 eq., $AlCl_3$ 4 eq., 1 h, RT, 89% [8b].
2) Diisobutyl aluminum hydride (DIBAL-H) 9 eq., toluene, reflux, 3 h: deprotection, over-reduction: 50% of **9** after acetylation [8b].
3) KOH/EtOH 0.6 M, reflux, 8 h, 75%: adenine xylonucleoside derivative [9].
4) Cs_2CO_3, EtOH, reflux, 5 h, 70%: methyl 2,3-di-*O*-benzyl-α-D-glucopyranoside **7** [10].
5) SmI_2 6 eq., THF/HMPA (hexamethylphosphoramide) 20/1, RT, 1 h: single electron transfer (SET)-type reductive desulfonylation providing a 1/1 mixture of 4-*O*-vinyl ether and ethylidene derivatives with yields ranging from 60% to 96% [11].

11.2.1.3 Cycloalkylidene (Cyclopentyl- and Cyclohexyl-)

Formation:

1) Cyclopentanone, H_2SO_4, DMF, RT, 4 h, 83%: 2,3-*O*-cyclopentylidene ribofuranose from D-ribose [12a]. Similar conditions in methanol as the solvent allow simultaneous protection of the anomeric hydroxyl [12b].
2) Cyclopentanone, $(MeO)_3CH$, *p*TsOH, dioxane, 1,2 : 3,5-di-*O*-cyclopentylidenexylofuranose [13].
3) Cyclohexanone, *p*TsOH, RT, 12 h, 97%: 2,3-*O*-cyclohexylideneribofuranose [14].
4) Cyclohexanone, $H_2SO_{4cat.}$, dioxane, RT, 61% [15]: 1,2 : 5,6-di-*O*-cyclohexylidene-glucofuranose.
5) 1,1-Dimethoxycyclohexanone (DMC) 12 eq., *p*TsOH, 60 °C, 140 mbar (rotary evaporator), 69%: 2,3:4,6:4′,6′-tri-*O*-cyclohexylidene-D-trehalose (Scheme 11.4) [16].

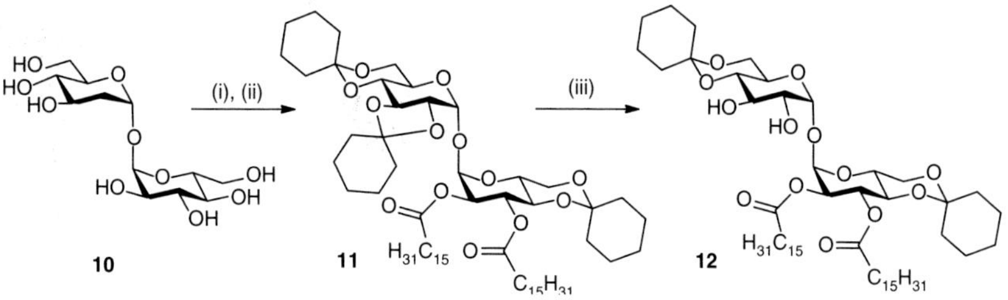

Scheme 11.4 Synthesis of diacyltrehalose **12** (i) DMC, *p*TsOH, 60 °C, 140 mbar, 69%; (ii) palmitic acid, DCC, DMAP, CH_2Cl_2, 92%; and (iii) AcOH, MeOH; 60 °C, 50%. *Source:* Sarpe and Kulkarni 2014 [16]. Reproduced with permission of American Chemical Society.

Deprotection:

1) AcOH/H_2O, 3 h, RT: selective hydrolysis of 3,5-*O*-cyclopentylidene [13].
2) HCl 3 M, MeOH, RT, 58–92% [13, 17].
3) AcOH/H_2O 2/1, RT, 65%: chemoselective hydrolysis of 5,6-acetal from 1,2:5,6-di-*O*-cyclohexylidene-glucofuranose [18].
4) AcOH 80%, 80 °C, 2 h, 79% [16].
5) TFA/H_2O 1/1, THF, 0 °C to RT, 15 min, 81% [16].
6) AcOH, MeOH, 60 °C, 50%: selective hydrolysis of 2,3-ketal in 2,3:4,6:4′,6′-tri-*O*-cyclohexylidene-D-trehalose [16].
7) HF-pyridine, $CHCl_3$, 81% (over two steps after benzylation) [14].
8) Imidazolium-TFA based protic ionic liquid 1 eq., H_2O, 70 °C, 30 min, 98%, the catalytic reaction (10 mol%) is also feasible within 2 h, 90%: selective hydrolysis of 3,5-ketal from 1,2 : 3,5-di-*O*-cyclohexylidene-xylofuranose [19].

11.2.1.4 Isopropylidene (Acetonide)

Isopropylidene acetals are among the most commonly used protecting groups for 1,2- and 1,3-diols. They are usually formed by reaction with acetone, 2,2-dimethoxypropane (DMP), or 2-methoxypropene (MP) under acidic conditions such as catalytic amounts of *p*TsOH or CSA. Under these conditions, the thermodynamic product is most often formed: 1,2-derivatives (1,3-dioxolane) are often preferred over 1,3-derivatives (1,3-dioxane), 1,2-*cis*-diols are usually favored over 1,2-*trans*-diols. Hence, lyxo- and mannopyranosides give 2,3-cyclic products, whereas arabino- and galactopyranosides tend to give 3,4-cyclic acetals. However, using a catalyst other than H^+ or under forcing conditions such as acetal exchange, thermodynamically unfavored products can be obtained such as 4,6-*O*-isopropylidenegalacto- and glucopyranoside. The reaction of free sugars is more complex, but some acetals can be prepared in good yields: for example, D-glucose, D-mannose, and D-galactose usually lead to the 1,2:5,6-, 2,3:5,6-di-*O*-isopropylidenefuranose and 1,2:3,4-di-*O*-isopropylidenepyranose, respectively. D-Xylose and D-lyxose will preferentially lead to the 1,2 : 3,5-diacetal and D-ribose to the 3,4-monoacetal in furanose forms.

Isopropylidene removal can be achieved with a large variety of catalysts, such as protic acids (HCl, HBr, AcOH, H_2SO_4, and TFA), Lewis acid reagents, and silica-supported reagents ($FeCl_3{\cdot}6H_2O$, H_2SO_4, $HClO_4$, and $NaHSO_4$) ([20] and references cited therein). Regioselective hydrolysis of bis-ketals is possible, 1,3-isopropylidene are hydrolyzed before the 1,2-derivatives, and usually the least hindered or the more electron-rich acetonide will react first (5,6- before 1,2-). Here are some recent examples of isopropylidene formation and hydrolysis.

Formation:

1) Acetone, $H_2SO_{4conc.}$, RT, 97%: 2,3-*O*-isopropylidene-D-ribofuranose [21].
2) Acetone, TsOH, 93%: 2,3:5,6-di-*O*-isopropylidene-L-gulonolactone [22].
3) Acetone, $I_{2cat.}$, 2 h, RT: 85% of 2,3:5,6-di-*O*-isopropylidene-α-D-mannofuranose; PPh_3-polymer-bound/iodine complex 1 eq./acetonide, acetone, RT, 30 min, 90–97% [23].
4) Acetone, tetrabutylammonium tribromide (TBATB) 2 mol%, RT, 1–10 h, 72–96% (no aqueous work-up required) [24].
5) $HCl_{conc.\ cat.}$, acetone, methanol, 75 °C, 4 h: 84% of methyl 2,3-*O*-isopropylidene-β-D-ribofuranoside [25].
6) MP, pTsOH$_{cat.}$, DMF, 70–92%: *p*-methoxyphenyl 4,6-*O*-isopropylidene-α/β-D-glucopyranoside [26].
7) MP, $CSA_{cat.}$, DMF, RT, 3 h, 67% of 2,3- and 3,4-*O*-isopropylidene-D-xylopyranosides (77 : 23) [27].
8) DMP, $CSA_{cat.}$, 48 h, RT then MeOH/-H_2O, reflux: 93% (Scheme 11.5) [27].
9) DMP, pTsOH$_{cat.}$, acetone or DMF, RT, o/n: 38% of allyl 3,4-*O*-isopropylideneribopyranoside (two-steps from D-ribose) [28a]; 48% of benzyl 2,3-*O*-isopropylidene-α-D-lyxopyranoside [28b]; 2,3:5,6-di-*O*-isopropylidene-D-mannofuranose (no purification) [28c].

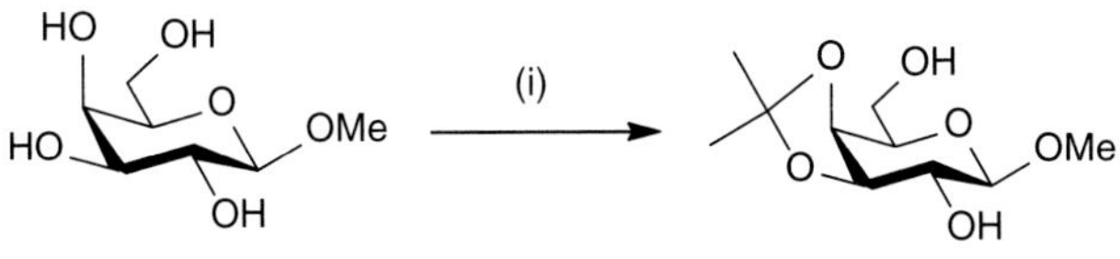

13 14

Scheme 11.5 Regioselective protection of 1,2-*cis*-diol: synthesis of 3,4-*O*-isopropylidene-β-D-galactopyranoside **14**. *Source:* Zong et al. 2015 [26]. Reproduced with permission of American Chemical Society.

Deprotection:

1) $TFA_{aq.}$ 60%, 0 °C to RT, 80%: ribofuranoside derivative [21].
2) TFA, RT, 2 h, 76%: hydrolysis of 2,3:5,6-di-*O*-isopropylidenemannose derivative; with TFA/MeOH 1/1, RT, 2 h, 90%: regioselective deprotection of a manno-derived pyrrolidine (Scheme 11.6) [29].
3) AcOH/H_2O 8/2, 110 °C, 85 min, 97% [28b]; hydrolysis of 2,3-*O*-isopropylidene iminosugars $AcOH_{aq.}$ 85%, 80 °C, 2–4 h, 31–77% [30].
4) Dowex 50WX8, H^+, H_2O, 12–24 h, 74% of L-fucose [22].

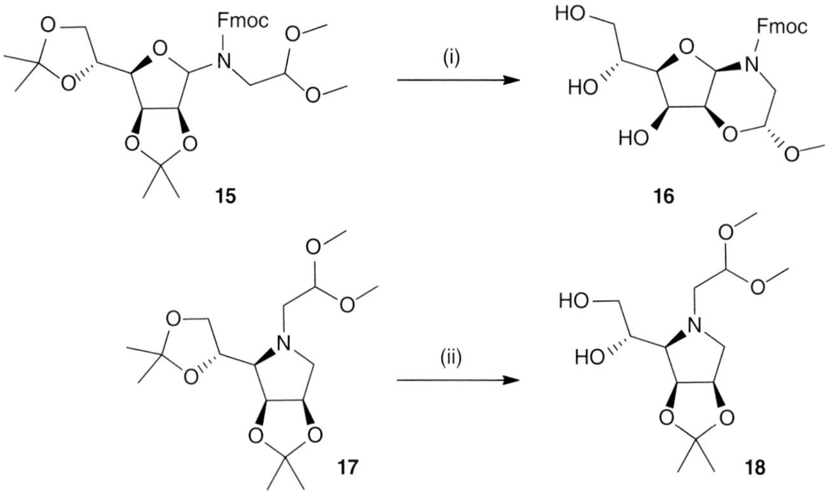

Scheme 11.6 D-Mannose-based scaffolds (i) TFA, 76% and (ii) TFA/MeOH 1/1, 90% [29]. *Source:* Lenci et al. 2015 [29]. Reproduced with permission of American Chemical Society.

5) H_5IO_6 1.3 eq., EtOAc, 93%: oxidative cleavage of 3-*O*-protected 1,2:5,6-di-*O*-isopropylidene-α-D-glucofuranose into the 4-formyl derivative [31].
6) $HCO_2H/AcOH/H_2O$, RT, 30 min, 94%: selective removal of the 5,6-acetonide over the 1,2- in a D-gulofuranose derivative [32].
7) $H_2SO_{4conc.}$, MeOH, 100 °C, 5 h, 99%, $\alpha/\beta = 1/3$; 2,3-*O*-isopropylidene hydrolysis from a ribofuranoside [25].
8) $Zn(NO_3)_2{\cdot}6H_2O$ 3.2 eq., MeCN, RT, 12 h, 80%: 3-deoxy-D-manno-2-octulosonic acid (KDO) glycal (selective hydrolysis of 7,8-isopropylidene vs 4,5-) [33].
9) *p*TsOH 3 eq., MeOH, RT, 12 h, 90%: removal of both 4,5 and 7,8-acetonides in KDO disaccharide [33].
10) $BF_3{\cdot}SiO_2$ 5 mol%, MeOH, ultrasounds, RT, 30–40 min, 85–95%: selective removal of terminal acetonide (5,6 or 3,5) *vs* 1,2 and 2,3 [20].
11) H_2O, 90 °C, 4–6 h, 85–95%: chemoselective deprotection of terminal acetonides [34].
12) $H_2SO_4{\cdot}SiO_2$, MeOH, 30–45 min, 91–95% [35].

11.2.2 1,2-Acetals: Dispiroketals (Spoke), Cyclohexane-1,2-diacetal (CDA), and Butane-1,2-diacetal (BDA)

The pioneering work of Ley et al. led to the discovery of cyclic diacetal as selective protecting groups for *trans* diequatorial-1,2-diols [36]. Dispiroketals were first introduced in 1992 [37], but their utility as protecting groups in carbohydrate chemistry is hampered by the high cost of bis-DHP (3,4-dihydropyran) and also by the low solubility of polyols in the required solvents (toluene or chloroform). In 1994, 1,1,2,2-tetramethoxycyclohexane (TMC) was introduced as a cheaper alternative to bis-DHP. Finally, butane-1,2-diacetals [38] is the most widely used (Scheme 11.7). The protection proceeds with high *trans*

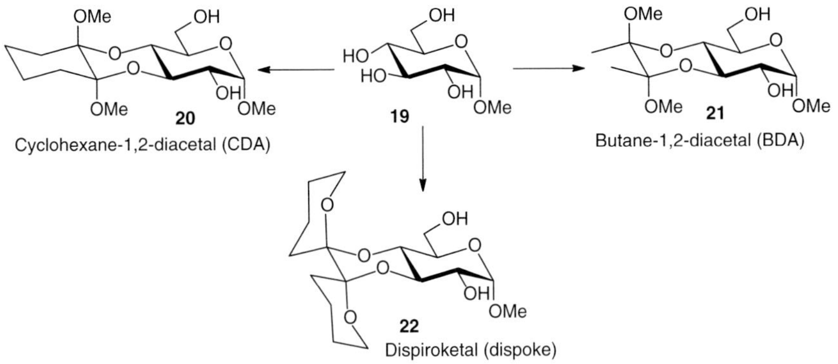

Scheme 11.7 Protection of 1,2-*trans* diols with diacetals

regioselectivity because of the formation of the less sterically demanding *trans* ring junction between the dioxane and the pyranoside. In addition, the configuration of the acetal centers is controlled by the anomeric effect, so only one diastereoisomer is formed. Butane-2,3-diacetal (BDA) thus confers a strong conformational rigidity to the carbohydrate ring. On a practical point of view, it is worth mentioning that diacetal protected products are usually crystalline materials with simple NMR spectra.

Formation:

1) bis-3,4-DHP, $CSA_{cat.}$, chloroform, reflux, 76%: methyl 2,3-dispoke-α-D-galactopyranoside [39].
2) TMC, $CH(OMe)_3$, $CSA_{cat.}$, methanol, reflux: 45% of methyl 3,4-CDA-α-D-galactopyranoside (CDA, cyclohexane-1,2-diacetal) along with 11% of 2,3-*cis*-derivative [40a], 51% of benzyl 2,3-CDA-L-arabinopyranoside [40b].
3) 2,3-Butanedione, $CH(OMe)_3$, $CSA_{cat.}$, methanol, reflux: 82% of methyl 3,4-BDA-α-D-mannopyranoside, 80% of methyl 2,3-CDA-β-D-galactopyranoside [41a]; 70% of benzyl 2,3-CDA-L-arabinopyranoside [41b].
4) 2,3-Butanedione, $CH(OCH_3)_3$, $BF_3{\cdot}OEt_2$, RT, 12 h: 75% of phenylthio-3,4-BDA-xyloside [42].
5) 2,2,3,3-Tetramethoxybutane (TMB) then $CSA_{cat.}$, MeOH, reflux: 70% of phenylthio-3,4-BDA-glucopyranoside (the 2,3-diacetal is also initially formed but is recycled by refluxing in methanol with CSA) [43a]; 77% of benzyl-2,3-BDA-L-arabinoside [43b].
6) TMB, $BF_3{\cdot}OEt_{2cat.}$, MeCN, Nap-3,4-BDA- (61%) and 2,3-BDA-xylopyranoside (39%) [44].

Deprotection:

1) *diSpoke*: TFA 95%, RT, 58% or ethylene glycol, $CSA_{cat.}$, reflux [37a].
2) *CDA*: TFA/H_2O 20/1, 5 min, 76% or $AcOH/THF/H_2O$ 4/1/1, 4 days, 90% (mannose derivative) [45].
3) *BDA*: $FeCl_3$ 5 mol%, H_2O, AcOH, RT, 10 h, 60% [41b, 46].
4) $TFA_{aq.}$ 95%, 5 min, RT, 96% [47].

5) HCl, 60 °C, 24 h, 81% [48].
6) TFA 95%/CH_2Cl_2 1/1 or 1/7, 15 min: 21–100% yield of variously substituted pentopyranosides [44].

11.2.3 Aromatic Acetals

Among protecting groups, aromatic acetals such as benzylidene acetals are well known key protecting groups for their flexibility and selectivity in protection and deprotection of 1,2- and mostly 1,3-diols and for their impact on glycosylation steps. The efficiency of introduction of such acetals and their selective deprotection over primary or secondary position allowed their use to selectively introduce aromatics pharmacophores for biological studies [49].

It appears over recent years that the *p*-methoxybenzylidene protection is by far the most abundant in literature. However, still the benzylidene group proved useful in innovative approaches of protection and selective deprotection. As an example, a one pot process has been reported recently, in which a tandem selective deprotection and oxidative functionalization is disclosed [50]. This approach is closely related to the known method of fragmentation, the Hanessian-Hullar reaction, leading to a selective opening of this acetal with the formation of a benzoyl group on the 4-position and a bromine atom on the 6-position of the carbohydrate backbone (Scheme 11.8).

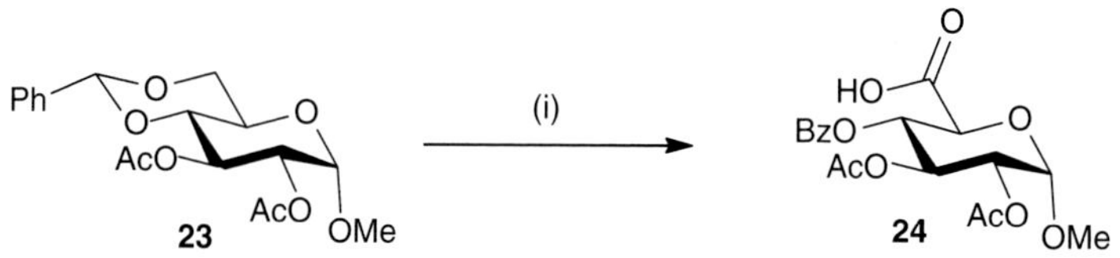

Scheme 11.8 Selective benzylidene deprotection and *in situ* oxidation of the primary position of the D-glucopyranoside **23** to a glucuronic derivative **24**: (i) $RuCl_3$ (20 mol%), $NaIO_4$ (10 eq.) $CH_3CN/CCl_4/H_2O$, 3.5 h, 61%. Banerjee et al. 2016 [50]. Reproduced with permission of John Wiley and Sons.

11.2.3.1 Benzylidene Acetals

A compilation of the most recent conditions to obtain selective reactivities is described here.

Formation: incorporation of the benzylidene acetal used anhydrous conditions with acid catalysis. Instead of using benzaldehyde, dimethoxytoluene is often used.

1,2-Diol: benzaldehyde dimethylacetal, *p*-TsOH, DMF, 95 °C, reduced pressure. 12 h [51a].

1,3-Diol: benzaldehyde dimethylacetal, *p*-TsOH, DMF, RT, 12 h [51b] or benzaldehyde dimethylacetal, CSA, DMF, RT, 8 h [51c].

Deprotection: removal of this acetal can be performed under protic acidic conditions. Selective opening of the acetal can be performed toward the primary or secondary alcohol depending on the conditions. The standard procedure aims at reducing the acetal into a benzyl derivative.

The reagents used for the regioselective benzylidene acetal opening either at the fourth or sixth position are numerous [52a]. Slight changes could induce a shift in the selective opening, for example, with DIBAL-H with a change of solvent from toluene to CH_2Cl_2 leading to a complete change in selectivity. Oxidative conditions could also be used to selectively open this acetal.

Full deprotection:

CH_2Cl_2, AcOH (aq. 80%) (1/1), 60 °C, 48 h [51c]; MeOH, *p*-TsOH (0.1 eq.) 9 h, RT [3]; 70% aq TFA/CH_2Cl_2 (1/8), 0 °C to RT, 5 h, 87% [53a]; AcOH 80%, 55 °C, 7 h [52b].

Selective deprotection:

1) *1,3-Diol*: Et_3SiH, I_2, CH_3CN, 5 °C, 30 min regioselective and reductive opening to a 6-*O*-benzyl derivative [52c].
2) *1,3-Diol*: $BH_3{\cdot}NMe_3$ (6 eq.) then $AlCl_3$ (6 eq.), THF 45 min regioselective and reductive opening to a 6-*O*-benzyl derivative [43a].
3) *1,3-Diol*: CH_2Cl_2, $BH_3{\cdot}THF$ (5 eq.) then $Sc(OTf)_3$ (0.1 eq.) regioselective and reductive opening to a 4-*O*-benzyl derivative [3].
4) *1,3-Diol*: $LiAlH_4$ (4.5 eq.)–$AlCl_3$(1.5 eq.), CH_2Cl_2/Et_2O (3/2), 0 °C, 30 min, 83% regioselective and reductive opening to a 4-*O*-benzyl derivative [52b].
5) *1,3-Diol*: (D-glucopyranoside) DIBAL-H in toluene (5 eq.), toluene, −18 °C to RT, regioselective, and reductive opening to a 4-*O*-benzyl derivative 89% [52e].
6) *1,3-Diol*: (D-glucopyranoside) DIBAL-H in CH_2Cl_2 (5 eq.), neat, 0 °C, reverse selectivity to a 4-*O*-benzyl derivative 60%, and 6-*O*-benzyl in 25% [52e].
7) *1,3-Diol*: a $RuCl_3$–$NaIO_4$ reagent system leads to a regioselective oxidation approach to uronic acid derivatives [50].
8) *1,2-Diol*: DIBAL-H in toluene (3 eq.), toluene, 0 °C, regioselective, and reductive opening to 2-*O*-benzyl mannopyranoside [52d].

11.2.3.2 *p*-Methoxybenzylidene Acetals

Formation: incorporation of *p*-methoxybenzylidene acetal used anhydrous conditions with acid catalysis. In most cases, the dimethylacetal (*para*-anisaldehyde dimethyl acetal, *p*-ADA) is used. An interesting approach used very weak acid conditions to incorporate the acetal under organocatalysis using thiourea and squaramide as a catalyst [54a] (Scheme 11.9).

1) *1,3-Diol*: (general conditions, D-glucopyranose, D-mannopyranose, D-galactopyranose, and D-glucal) *p*-ADA (1.5 eq.), squaramide or thiourea catalysts (0.05 eq.), CH_3CN, RT, 5 h to 7 days [54a].
2) *1,3-Diol*: (D-glucopyranoside) *p*-ADA (2 eq.), CSA (0.04 eq.), CH_3CN/DMF (3.6/1), RT, 2.5 h [54b].
3) *1,3-Diol*: *p*-ADA (1.05 eq.), *p*-TsOH (0.015 eq.), DMF, 50 °C, reduced pressure, 4 h (D-glucosaminopyranoside), [54c] (D-glucopyranoside) [53a].
4) *1,3-Diol*: (D-galactopyranoside) *p*-ADA (1.5 eq.), CSA (0.02 eq.), CH_3CN, RT, 5 h [54d].

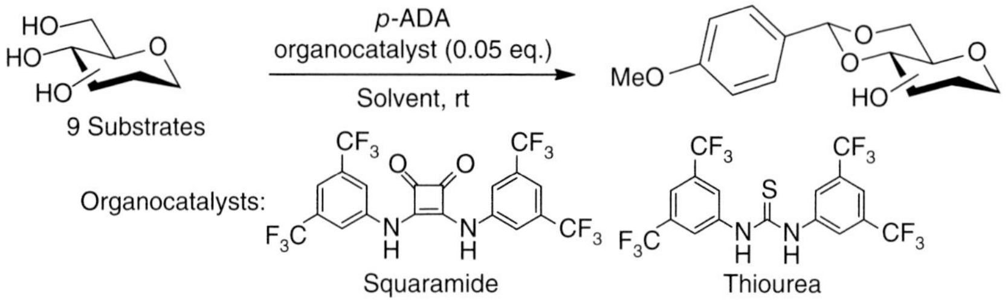

Scheme 11.9 Thiourea and squaramide as organocatalysts in benzylidene acetal formation. *Source:* Geng et al. 2013 [54a]. Reproduced with permission of John Wiley and Sons.

5) *1,3-Diol*: (D-mannosaminopyranoside) *p*-ADA (2.4 eq.), *p*-TsOH (0.1 eq.), CH_3CN, 0 °C to RT, 5 h [54e].

Deprotection: removal of this acetal could be performed under various types of conditions from protic acidic to oxidative conditions. Selective opening of this benzylidene follow the same line of reactivities observed with the benzylidene protecting group.

Full deprotection:

1) *1,3-Diol*: (D-galactopyranoside) AcOH 80%, 40 °C, 1 h [53b, 55a].
2) *1,3-Diol*: (D-glucopyranoside), AcOH 80%, 50 °C, 1.5 h, quantitative, [54b].
3) *1,3-Diol*: (D-mannopyranoside) DDQ (4 eq.), CH_2Cl_2, H_2O, RT, 1 h (89%) [55b].
4) *1,3-Diol*: (D-glucopyranoside) $AlCl_3$ (6.2 eq.), CH_2Cl_2/Et_2O (1/3), reflux, 4 h, (87%) [55c].
5) *1,3-Diol*: (D-mannosaminopyranoside) $EtOH/H_2O/AcOH$ (8/4/3), 50 °C, 27 h [54e].
6) *1,3-Diol*: (D-galactopyranoside), *p*-TsOH (0.02 eq.), MeOH, 0 °C, 89% [54d].

Selective deprotection:

1) *1,3-Diol*: regioselective and reductive opening to a 6-*O*-*p*-methoxybenzyl derivative: (D-galactopyranoside, $NaBH_3CN$ (5 eq.) then TFA (10 eq.), DMF, 0 °C to RT 91% [54d]. $NaBH_3CN$ (10 eq.), HCl (2 M in Et_2O, 7.6 eq.), THF, molecular sieves (MS) (3 Å), 0 °C [55d].
2) *1,3-Diol*: regioselective and reductive opening to a 4-*O*-*p*-methoxybenzyl derivative (D-glucopyranoside) 1 M $LiAlH_4$ in THF (5 eq.), $AlCl_3$ (4 eq.), CH_2Cl_2/Et_2O (1/5), reflux, 4 h, 83% [55c]. (D-Glucopyranoside), $LiAlH_4$ (4.6 eq.)–$AlCl_3$ (1.5 eq.), CH_2Cl_2/Et_2O (3/2.3), 0 °C, 30 min, 83% [53a, 43a]. (D-glucopyranoside), DIBAL-H in toluene (3 eq.), toluene, −18 °C to RT, 89% [52d].

11.2.3.3 Halobenzylidene Acetals

Halobenzylidene acetals are quite uncommon acetals used in carbohydrate chemistry

Formation: incorporation of the halobenzylidene acetal used anhydrous conditions with acid catalysis in most cases but it also could be introduced under basic conditions.

1) *1,3-Diol*: *p*-iodobenzylideneacetal: α,α-dibromo-4-iodotoluene (2.5 eq.) pyridine, reflux, 24 h, 65% [56a].
2) *1,3-Diol*: *p*-fluorobenzylideneacetal: α,α-dimethoxy-4-fluorotoluene (1.3 eq.), *p*-TsOH (0.13 eq.) CH_3CN, RT, 20 h, 90% [56b, c].
3) *1,3-Diol*: *m*-fluorobenzylideneacetal: α,α-dimethoxy-4-fluorotoluene (1.3 eq.), *p*-TsOH (0.13 eq.) CH_3CN, RT, 20 h, 86% [56c]. *o*-Fluorobenzylideneacetal: α,α-dimethoxy-2-fluorotoluene (1.3 eq.), *p*-TsOH (0.13 eq.), CH_3CN, RT, 17 h, 94% [56c]. *p*-Fluorobenzylideneacetal, *o*-, *m*-, *p*-chlorobenzylideneacetal: benzaldehyde, CSA, $CuSO_4$ anh., CH_3CN, reflux, 33–90% [56d].
4) *1,3-Diol*: *p*-chlorobenzylideneacetal, α,α-dimethoxy-4-chlorotoluene (1.14 eq.), *p*-TsOH (0.01 eq.), DMF, 15 mmHg, 5 h, 60 °C, 52% [56e].
5) *1,3-Diol*: *p*-chlorobenzylideneacetal, 4-chlorobenzaldehyde (3 eq.), *p*-TsOH (catalytic), DMF, 30 mmHg, 3.5 h, 55 °C, 90% [56f].
6) *1,3-Diol*: *p*-bromo, *p*-chlorobenzylideneacetal, 4-bromo- or 4-chlorobenzaldehyde (0.9 eq.), *p*-TsOH (0.025 eq.) , triethylorthoformate (0.9 eq.) DMF, 30 mmHg, 3.5 h, 55 °C, 90% [56f].

Deprotection:

Full deprotection: removal of this acetal could be performed under protic acidic conditions, but also using photolysis reaction.

1,3-Diol: *o*- and *m*-fluorobenzylideneacetal; H_2 (1 atm), Pd/C (10%, 1.3 and 4.1 eq.), AcOH (91% and 97%) [56c].

1,3-Diol: *p*-chlorobenzylideneacetal; H_2, Pd/C (catalytic), THF/H_2O/tBuOH (2/2/1), 2 h, 91% [56f].

Selective deprotection:

1,3-Diol: *p*-fluorobenzylideneacetal, resin supported acetal, $NaBH_3CN$ (50 eq.), satd. Et_2O HClg, α,α-dimethoxy-4-fluorotoluene (1.3 eq.), *p*-TsOH (0.13 eq.) CH_3CN, RT, 20 h, 90%. Regioselective and reductive opening to a 4-*O*-*o*-nitrobenzyl derivative in 89–96% yield depending on the carbohydrate series [56b, c].

11.2.3.4 Nitrobenzylidene Acetals

Nitrobenzylidene acetals are quite uncommon acetals in carbohydrate chemistry.

***o*-Nitrobenzylidene Acetal** The orthonitrobenzylidene acetal has been used to develop and use the photochemistry abilities of such groups to orthogonal and bioorthogonal applications in protecting groups. Selective deprotections were achieved, but no "full classical deprotection" under acidic and protic conditions have been reported. Even a failure was noticed.

Formation conditions: incorporation of 2-nitrobenzylidene acetal used anhydrous conditions with acid catalysis in most cases described in 1974 by Collins. The benzaldehyde is used in these cases:

1) *1,3-Diol*: *o*-nitrobenzaldehyde 0.75 eq., $H_2SO_{4conc.}$ (0.75 eq.) dioxane, RT, 8 h, 70% to 90% [57a].
2) *1,3-Diol*: 6-nitropiperonal (3 eq.), $H_2SO_{4conc.}$ (2 eq.), dimethyl sulfoxide (DMSO), 0 °C to RT, 78% [57b].

Deprotection:

Full deprotection: removal of this acetal could be performed under protic acidic conditions. But also using photolysis reaction:

1) *1,3-Diol*: NaOH in MeOH, irradiation Hg Lamp, 95% [58a].

Selective deprotection:

1) *1,3-Diol*: Et_3SiH (12 eq.) then $BF_3{\cdot}OEt_2$, CH_2Cl_2, 0 °C to RT, 3 h. Regioselective and reductive opening to a 6-*O*-*o*-nitrobenzyl derivative in 73–88% depending on the carbohydrate series [58b].
2) *1,3-Diol*: Et_3SiH (12 eq.) then $BF_3{\cdot}OEt_2$ (6 eq.), CH_2Cl_2, 0 °C to RT, 16 h. Regioselective and reductive opening to a 6-*O*-*o*-nitrobenzyl derivative in 52–78% depending on the carbohydrate series [58c].
3) *1,3-Diol*: Et_3SiH (3 eq.) then $PhBCl_2$ (3 eq.), MS 4 Å, CH_2Cl_2, −78 °C, 10 min. Regioselective and reductive opening to a 4-*O*-*o*-nitrobenzyl derivative in 89–96% yield depending on the carbohydrate series [58c].
4) *1,3-Diol*: irradiation Hg Lamp, 95% [58a].

m-Nitrobenzylidene Acetal

Formation: incorporation of a 3-nitrobenzylidene acetal used anhydrous conditions with acid catalysis.

1,3-Diol: *m*-nitrobenzylidenedimethylacetal (2.5 eq.), CSA (0.05 eq.), drierite, CH_3CN, 55 °C, 2.5 h (95%) [59].

Full deprotection: removal of this acetal could be performed under protic acidic conditions.

Selective deprotection is also obtained under classical benzylidene opening conditions:

1,3-Diol: $NaBH_3CN$ (15 eq.) then HCl (4 M in dioxane, until pH ~ 1), THF, 0 °C to RT. Regioselective and reductive opening to a 6-*O*-*m*-nitrobenzyl derivative in quantitative yield [59].

p-Nitrobenzylidene Acetal

Formation: incorporation of the 4-nitrobenzylidene acetal used anhydrous conditions with acid catalysis. In most cases, the dimethoxybenzylidene acetal is used.

1) *1,3-Diol*: *p*-nitrobenzylidenedimethylacetal (1 eq.), H_2SO_4-silica (50 mg), CH_3CN, RT, (75%) [52, 60].

2) *1,3-Diol*: *p*-nitrobenzylidenedimethylacetal (2.1 eq.), CSA (pH 3), DMF, 84% [4].

11.2.3.5 Polyaromatic Methylene Acetals

Polyaromatic methylene acetals are less common in carbohydrate chemistry. The most well known is the 2-naphthylmethylene acetal. An important increase in applications has been witnessed recently in the literature for the 2-naphthylmethyl (NAP) derivatives.

1- or 2-Naphthyl Methylene Acetal

Formation: incorporation of 2-naphthyl methylene acetal used anhydrous conditions with acid catalysis starting with 2-naphtaldehyde dimethylacetal.

1) *1,3-Diol*: 2-naphtaldehyde dimethylacetal (1.5 eq.), DMF, TsOH (0.2 eq.), 50 °C, 150 mbar, 2 h [43a].
2) *1,3-Diol*: 2-naphthaldehyde (1.5 eq.), CH_3CN, CSA (0.2 eq.), RT, overnight [61a].
3) *1,3-Diol*: L-idopyranoside, 1-naphthaldehyde dimethylacetal (1.5 eq.), CSA (0.18 eq.), DMF, 98% [61b].
4) *1,3-Diol*: 2-naphthaldehyde (2 eq.), CH_3CN, *p*-TsOH (0.1 eq.), RT, 2 h [61c].

Deprotection:

Full deprotection: removal of this acetal could be performed under protic acidic conditions

1) *1,3-Diol*: $CH_2Cl_2/H_2O/TFA$ (9/0.75/1), RT, 20 min, 62% over two steps after peracetylation [61a].
2) *1,3-Diol*: *p*-TsOH, MeOH, CH_2Cl_2, RT, 86% [61d].

Selective deprotection:

1) *1,3-Diol*: Et_3SiH (12 eq.) then $BF_3{\cdot}OEt_2$ (1 eq.), CH_2Cl_2, 0 °C to RT, 45 min. Regioselective and reductive opening to a 6-*O*-NAP derivative in 69% on a glucopyranoside [43a].
2) *1,3-Diol*: $LiAlH_4$ (4 eq.), $AlCl_3$ (1.5 eq.), CH_2Cl_2, Et_2O, RT, 4 h. Regioselective and reductive opening to a 4-*O*-NAP derivative in 69% on a D-glucopyranoside [43a].
3) *1,3-Diol*: $BH_3{\cdot}THF$ (2.45 eq.), TMSOTf (0.15 eq.), CH_2Cl_2, RT, 3 h (76%). Regioselective and reductive opening to a 4-*O*-NAP derivative in 76% on a D-galactopyranoside [61e].
4) *1,3-Diol*: $BH_3{\cdot}NMe_3$ (2.45 eq.), $AlCl_3$ (0.15 eq.), THF, $H_2O_{cat.}$, RT, 3 h. Regioselective and reductive opening to a 4-*O*-NAP derivative in 92% on a D-galactopyranoside [61c].
5) *1,3-Diol*: L-idopyranoside, $BH_3{\cdot}THF$, TMSOTf, CH_2Cl_2, Regioselective and reductive opening to a 4-*O*-NAP derivative in 92% [61e].

6) *1,3-Diol*: D-glucopyranoside, DIBAL-H in toluene (5 eq.), toluene, −18 °C to RT, 7 h. Regioselective and reductive opening to a 4-*O*-NAP derivative 81% [52d].
7) *1,3-Diol*: D-glucopyranoside, DIBAL-H in CH_2Cl_2 (5 eq.), neat, −30 °C, reverse selectivity to a 6-*O*-NAP derivative 58% and 4-*O*-NAP in 8% [52d].
8) *1,3-Diol*: D-glucopyranoside, $LiAlH_4$, $AlCl_3$, CH_2Cl_2/Et_2O, RT, regioselective and reductive opening to a 4-*O*-NAP derivative 75% [61d].
9) *1,3-Diol*: D-glucopyranoside, Et_3SiH, TFA, CH_2Cl_2, RT, regioselectivity to a 6-*O*-NAP derivative 73% [61d].
10) *1,3-Diol*: (D-galactopyranoside), $BH_3{\cdot}THF$, TMSOTf, CH_2Cl_2, RT, 3 h regioselectivity to a 4-*O*-NAP derivative 76% [61e].

9-Anthryl Methylene Acetal

Formation: incorporation of the 9-anthryl methylene acetal used anhydrous conditions with acid catalysis starting with 9-anthraldehyde dimethylacetal

1) *1,3-Diol*: D-glucopyranoside and D-galactopyranoside, 9-anthraldehyde dimethylacetal (1.5 eq.), CH_3CN, TsOH (0.02 eq.), RT, 3 h, 96 and 94% [62a].
2) *1,3-Diol*: 9-anthraldehyde dimethylacetal (1.5 eq.), CH_3CN, CSA (0.02 eq.), RT, 3 h, 79% [62b].
3) *1,2-Diol*: L-rhamnopyranoside, 9-anthraldehyde dimethylacetal (1.5 eq.), CH_3CN, CSA (0.02 eq.), RT, 3 h: two stereoisomers (*exo*: 46%, *endo*: 9%) [62b].

Selective deprotection:

1) *1,3-Diol*: $NaBH_3CN$ (10 eq.), HCl (2 M in Et_2O, 7.6 eq.), THF, 2 h, D-glucopyranoside 91%, D-galactopyranoside, 57%. Regioselective and reductive opening to a 6-*O*-anthrylmethyl derivative [62a].
2) *1,3-Diol*: $NaBH_3CN$ (9 eq.), HCl (in Et_2O), THF, 5 min. Regioselective and reductive opening to a 6-*O*-anthrylmethyl derivative in 48% together with 4-*O*-derivative in 12% yield [62b].
3) *1,2-Diol*: $LiAlH_4$, $AlCl_3$, Et_2O, CH_2Cl_2, 0 °C, 2 h. Regioselective and reductive opening to a 2-*O*-anthrylmethyl derivative in 58% from the *endo* isomer, no selectivity from the *exo*-isomer [62b].

11.3 Protection as Orthoesters

Cyclic orthoesters [63] were developed to protect 1,2-*cis*-diols. They are resistant to basic conditions, which allows, for example, the selective deprotection of esters or some etherification reactions. However, because of their acetal-like character, they are highly labile in acidic media, where acyloxonium intermediates are formed, which can hydrolyze and be trapped by a nucleophile such as alcohol (reaction with internal hydroxyls gives triorthobenzoates). Careful hydrolysis leads to their rearrangement into the corresponding monobenzoate or monoacetate derivatives, which usually lead to axial ester-equatorial alcohol

[64]. 1,2-Orthoesters are probably the best known and most used class of orthoesters; they are often formed from 2-*O*-acetyl glycosyl halides in the presence of a base such as 2,6-lutidine and water or any simple alcohols (e.g. methanol, ethanol, and pentenol). They are mainly used in glycoside synthesis. Over the last decade, orthoesters have been mostly used to protect 1,2-diols on pyranose derivatives and more rarely 2,3- and 5,6-positions of furanose derivatives.

11.3.1 Methoxy-, Ethoxymethylene Orthoesters

Formation: methylorthoformate is sometimes preferred over ethylorthoformate because of the easier removal of the by-products and solvents upon evaporation.

1) Trimethylorthoformate, CH_2Cl_2, ceric ammonium nitrate (CAN), RT, 2 h (Scheme 11.10) [65].
2) Trimethylorthoformate, p-TsOH·$H_2O_{cat.}$, RT, 24 h [66a].

Deprotection and/or further functionalization:

1) DIBAL-H, −78 °C, 92%: introduction of a methoxymethylether (MOM) protecting group by reduction of the orthoester intermediate 23 (not regioselective on D-ribose scaffold – Scheme 11.10) [65].

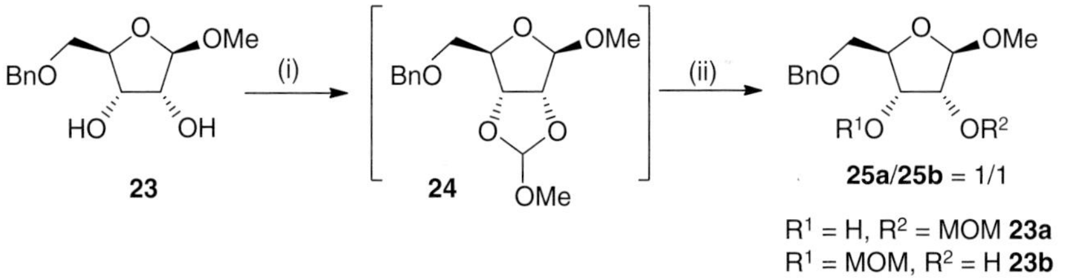

Scheme 11.10 Introduction of a MOM protecting group via an orthoester intermediate **24**: (i) $(MeO)_3CH$, CH_2Cl_2, CAN then (ii) DIBAL-H, −78 °C, 92% of **25a/25b** as a 1/1 mixture. *Source:* Comin et al. 2004 [65]. Reproduced with permission of Elsevier.

2) Ac_2O, 140–150 °C, 4 h, 98%: thermal rearrangement of methyl orthoester to form 5-*exo*-methylene (reported both for gluco- and mannofuranose derivatives) (Scheme 11.11) [66].

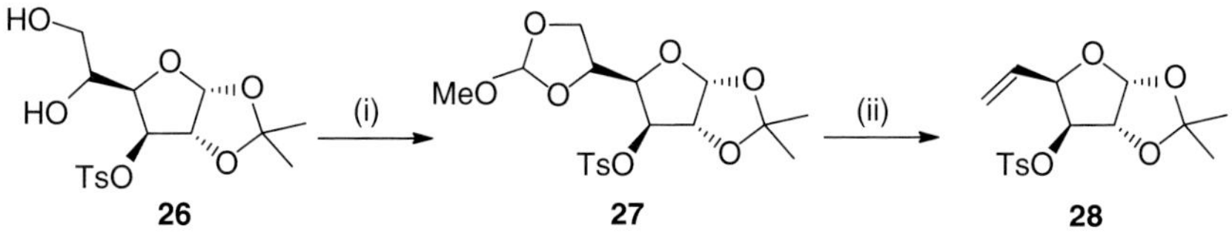

Scheme 11.11 Synthesis of 5-*exo*-methylene mannofuranose derivative (i) $(MeO)_3CH$, p-TsOH·$H_2O_{cat.}$ then (ii) Ac_2O, reflux, 88%. *Source:* Furukawa et al. 2013 [3]. Reproduced with permission of American Chemical Society.

11.3.2 1-Methoxy, 1-Ethoxyethylidene and α-Methoxy, α-Ethoxybenzylidene Ortho Esters

Formation:

1) *1-Methoxyethylidene*: 1,1-dimethoxyethene, *p*-TsOH$_{cat.}$, DMF. Kinetically controlled conditions, reaction occurs preferentially on primary hydroxyls and nonanomeric position. From both free pyranoses and pyranosides, the 4,6-*O*-methoxyethylidene is formed. In furanose forms, the formation of a quadricyclic orthoester can be observed because of an intramolecular reaction of another free hydroxyl (Scheme 11.12) [67].

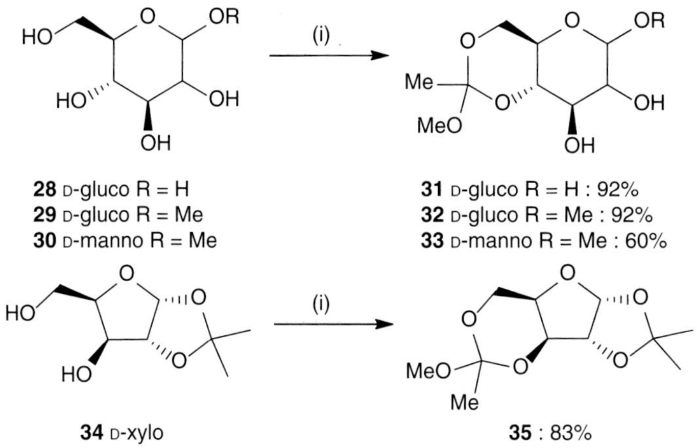

Scheme 11.12 Kinetically controlled orthoesterification: (i) 1,1-dimethoxyethene, *p*-TsOH$_{cat.}$, DMF, 0 °C. *Source:* From Bouchra,1995 [67].

2) *1-Methoxyethylidene*: $CH_3C(OMe)_3$, *p*-TsOH$_{cat.}$, MeCN, RT: 4,6-*O*-methoxyethylidene of GlcNAc [68].
3) *1-Ethoxyethylidene*: $CH_3C(OEt)_3$, *p*-TsOH·H_2O, toluene, RT, 1 h: 84% of 3,4-orthoester from *n*-pentenyl 6-*O*-TBDPS-β-D-galactopyranoside [69a]; $CH_3C(OEt)_3$, CSA$_{cat.}$, MeCN, RT [69b].
4) *α-Methoxy/ethoxybenzylidene*: $PhC(OEt)_3$, *p*-TsOH, TFA, MeCN, RT, 15 min [70a, b] or $PhC(OEt)_3$, CSA$_{cat.}$, CH_2Cl_2 [70c] or $PhC(OMe)_3$, *p*-TsOH, toluene, RT, 1 h [71].

Deprotection:

1) *p*-TsOH, $CHCl_3$–H_2O, RT, 20 min: regioselective hydrolysis of 5,6-*O*- or 3,5-*O*-methoxyethylidene into the 6-*O*-acetyl (D-glucofuranoside, 83%) or 5-*O*-acetyl product (D-xylofuranoside, 82%) [67a].
2) H_2O, 25 °C, 82%: regioselective hydrolysis of 2,3-ethoxybenzylidene mannoheptopyranoside into the 2-*O*-benzoyl product [70c].
3) AcOH/H_2O (4/1), 25 °C, 68%: 6-*O*-acetyl product from regioselective hydrolysis of 2,3-di-*O*-acetyl-4,6-orthoacetate glucopyranoside (4-*O*-acetyl: 19%) [68], 4-*O*-acetyl galactopyranoside from 3,4-orthoacetate [69b, 72].

4) TFA/H_2O, MeCN, RT, 10 min: hydrolysis of bis-orthoester **37** into 2,6-dibenzoate **38** (32%) and 2,4-di-*O*-benzoyl-mannopyranoside **39** (65%) (Scheme 11.13) [70a].

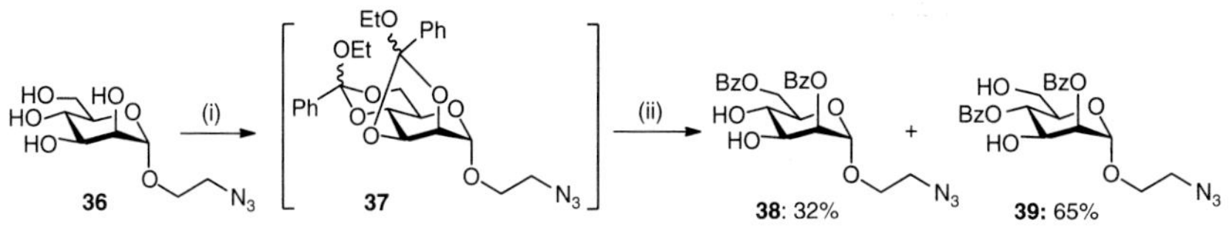

Scheme 11.13 Hydrolysis of bis-orthoester **37**: (i) $PhC(OEt)_3$, *p*-TsOH, TFA, MeCN, RT, 2 h; (ii) TFA/H_2O, MeCN, RT, 10 min [70a]. *Source:* Lindhorst et al. 2010 [70a]. https://www.ncbi.nlm.nih.gov/pmc/articles/PMC2956480/.Licensed Under CC BY 2.0.

11.3.3 Methylidene Ortho Ester

Methylidene ortho esters are protecting groups in which three hydroxyl groups are simultaneously protected. They can also be found as intermediates during the formation of 1,2-orthoesters [52a]. Methylidene orthoesters seem to be mainly used in furanose series to protect 1,2,5 and 3,5,6 positions. By simultaneously blocking three positions of the furanose ring, the manipulation of the remaining free alcohol is then possible. These rigid tricyclic orthoesters are easily cleaved in acidic conditions. However, as little information has been reported recently about the use of these protecting groups, we will briefly summarize the last results reported in the literature about the use of tri-orthobenzoate and tri-orthodichloroacetate orthoesters.

Formation: tricyclic dichloroacetyl orthoester can be obtained using trichloroethylidene acetals: the reaction probably proceeds through the formation of dichloroethylidene ketene acetal by dehydrochlorination of the trichloroethylidene acetal group, followed by the nucleophilic attack of the stereochemically suitable hydroxyl group.

1) *t*BuOK, *t*BuOH, reflux, 77%: 1,2,3-*O*-orthodichloroacetyl-α-D-ribofuranose from β-ribochloralose [73]; 1,2,5-*O*-orthodichloroacetyl-α-D-galactofuranose (67%) and 1,2,5-*O*-orthodichloroacetyl-α-D-arabinofuranose (62%) [74]; 1,2,3-*O*-orthodichloroacetyl-β-D-mannofuranose **42** (48%) and 1,2,5-*O*-orthodichloroacetyl-β-D-mannofuranose **41** (65%) (Scheme 11.14) [75].

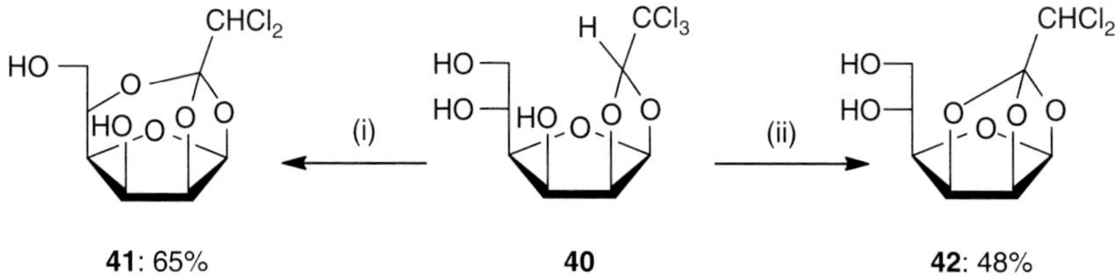

Scheme 11.14 Tricyclic dichloroacetyl orthoesters: (i) *t*BuOK (3 eq.), *t*BuOH, reflux, 40 min; (ii) *t*BuOK (1.5 eq.), reflux, 6 h. *Source:* Salman et al. 2004 [75]. Reproduced with permission of Elsevier.

2) *$CSA_{cat.}$*, CH_2Cl_2, MS 4 Å, RT, 30 min, 86%: β-D-arabinofuranose 1,2,5-orthobenzoate from the corresponding 1,2-orthoester [76]; $MgBr_2$, THF, 78% [77].
3) *Bu_3SnH* (5 eq.), α,α′-azoisobutyronitrile (AIBN) (0.7 eq.), toluene, reflux, 7 h, 84%: 3-*O*-benzyl-β-D-arabinose 1,2,5-orthoacetate by radical reduction of the corresponding orthodichloroacetate [76].
4) 1,2-Dimethoxyethene (1.5 eq.), DMF, *p*-$TsOH_{cat.}$, 0 °C, 1 h, 88%: 1,2-*O*-isopropylidene-3,5,6-orthoacetyl-α-D-glucofuranose [67a].

Deprotection:

1) IR-120 H^+, H_2O/MeOH, RT, 2 days, 56%: hydrolysis of 1,2,5-*O*-orthodichloroacetyl in 6-deoxy-galactofuranoside derivative **43** (Scheme 11.15) [78].

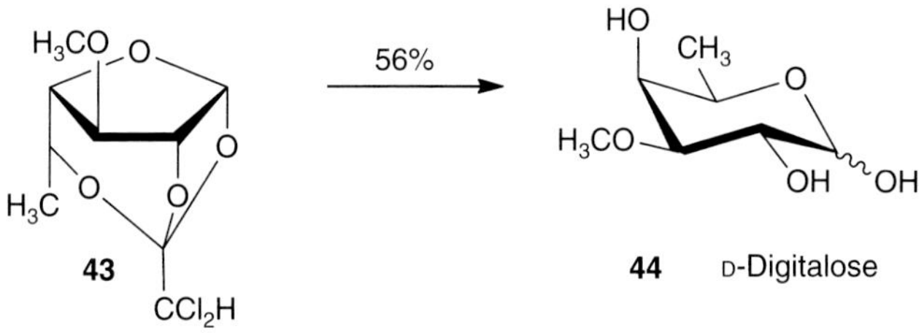

Scheme 11.15 Synthesis of D-digitalose **44**: IR-120 H^+, H_2O/MeOH. *Source:* Kök and Salman 2012 [78]. Reproduced with permission of Taylor & Francis.

2) $AcOH/H_2O$ (1/3), 0 °C, 30 min, 95%: regioselective hydrolysis of 1,2-*O*-isopropylidene-3,5,6-orthoacetyl-α-D-glucofuranose into the 6-*O*-acetyl product [67a].
3) ROH or RSH, $SnCl_4$ or $BF_3{\cdot}OEt_2$ or TMSOTf: acid-catalyzed ring opening of 1,2,5-arabinose derivatives with oxy-, thio-, and seleno-nucleophiles to give the corresponding glycosides in good yields (60–95%; R = octyl, Me, Ph, Et) [76, 79].

11.4 Silylene Acetals as Protecting Groups

1,2- and 1,3-Diols protected via silylene derivatives are quite common in glycochemistry. These protecting groups were firstly applied on furanose templates and dedicated to nucleoside chemistry. On a more strict level, applications to glycochemistry of these types of compounds started in the early 1990s and are nowadays fully integrated in glycochemistry. Three main protecting groups have been explored and studied. Here is a glimpse of the different recent applications dedicated to glycochemistry found in the literature

11.4.1 Di-*t*-butylsilylene DTBS $(OR)_2$

The di-*t*-butylsilylene (DTBS) group is the most prominent protecting group in this short family of compounds. Numerous recent publications could be found in which this protecting group has been used for its armed effect during glycosylation. It has, for example, been applied in glycosylation and exhibited strong stereocontrol, greater than benzylidene or isopropylidene, during the condensations and complete α-anomeric selectivity was obtained for the products (Scheme 11.16) [80]. No β-stereoisomer was found even after chromatographic purification in these reactions.

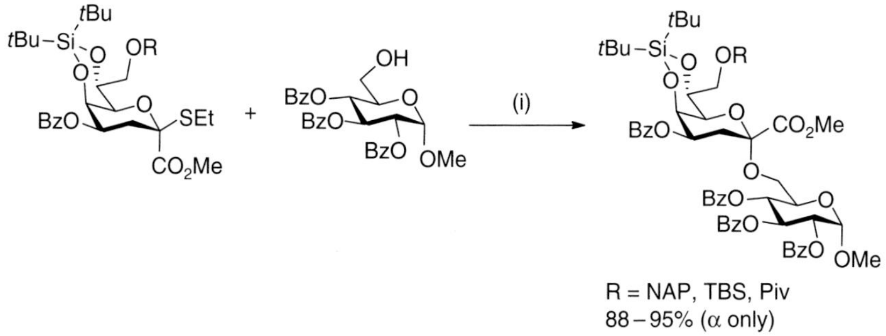

Scheme 11.16 Application of a di-*t*-butylsilene in glycosylation methodology: (i) NIS, TfOH, MS 4 Å, −78 to −50 °C, 2–3 h, CH_2Cl_2.

Formation: incorporation of silylene acetals used anhydrous conditions with the bis-triflated silyl reagent mostly at low temperature. The conditions used did not change much over the time.

1,3-Diol: $(t\text{-Bu})_2Si(OTf)_2$ (2.0 eq.), 2,6-lutidine (5.1 eq.), CH_2Cl_2, 0 °C, 30 min, 85% [81a].

1,3-Diol: $(t\text{-Bu})_2Si(OTf)_2$ (1.1 eq.), pyridine (1.2 eq.), DMF, −40 °C, 50 min, 96% [81b].

1,3-Diol: $(t\text{-Bu})_2Si(OTf)_2$ (1.0 eq.), 2,6-lutidine (2.5 eq.), DMF, −20 °C, 30 min, 84% [81c].

1,3-Diol (furanoside backbone): $(t\text{-Bu})_2Si(OTf)_2$ (1.1 eq.), 4-(dimethylamino) pyridine (DMAP) (0.05 eq.) pyridine, 0 °C, 2 h, 84% [81d].

1,3-Diol D-*mannopyranoside*: $(t\text{-Bu})_2Si(OTf)_2$ (1.1 eq.), pyridine (3 eq.), DMF, −35 °C, 60 min, 89% for two steps [81e].

Deprotection:

Full deprotection: removal of this protection is usually performed using fluoride ion.

1) *1,3-Diol*: tris(dimethylamino)sulfonium difluorotrimethylsilicate (TASF, 6 eq.), THF, 0 °C, 64% [81a].
2) *1,3-Diol*: TBAF (2 eq.), THF, 0 °C, 3 h, 98% [81b]
3) *1,3-Diol* D-*mannopyranoside*: HF·Py (70%, 50 μl) at 0 °C, THF, 30 min [81e].
4) *1,3-Diol* D-*glucopyranoside*: 48% aq HF (11.6 eq.), CH_3CN, RT, overnight [81f].
5) *1,3-Diol* D-*glucopyranoside*: TBAF·$3H_2O$ in THF (1.0 M, 1.6 ml, 3 eq.), THF, RT, 24 h [81g].

6) *1,3-Diol seven membered ring*: triethylamine trihydrofluoride ($Et_3N{\cdot}HF$) (1.1 eq.), THF, RT, 3 h, 90% [81h].
7) *1,3-Diol* L-*gulopyranoside*: HF·Py (5 eq.), THF/pyridine (1/1), 0 °C to RT, overnight, 85% [81i].

No selective deprotection has been disclosed.

11.4.2 Methylene 1,3-bis-Diisopropylsilanyl Group

No application in carbohydrate chemistry, mostly applied in 3,5-diol protection of pentoses in nucleoside chemistry.

Formation: incorporation of the silyl group used anhydrous conditions with the bis-chlorinated silyl reagent (Scheme 11.17). The conditions used are very similar to those applied with the classical TBDMS group: $TIPDSCl_2$ (1.2 eq.), imidazole (4.9 eq.), DMF, 0 °C to RT, 6 h, 63% [82].

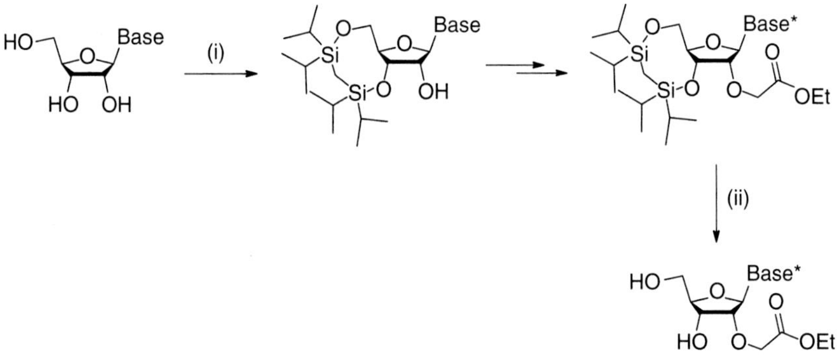

Scheme 11.17 Diisopropylsilanyl group application in nucleoside chemistry [81]: formation (i) $TIPDSCl_2$, imidazole, DMF; removal (ii) TBAF, THF.

Deprotection: TBAF (1.5 eq.), THF anhydrous, RT, 27 h (89%) [82].

11.4.3 1,1,4,4-Tetraphenyl-1,4-disilabuta-1,4-diyl group (SIBA)

Formation: to our knowledge, only one example has been developed for a selective glycosylation on a L-rhamnopyranoside donor using the Kahne protocol (Scheme 11.18). Silylation of the *trans* diol was performed using $SIBACl_2$ (1.4 eq.), imidazole (5.5 eq.), DMF, 0 °C to RT, 6 h, 92% [83].

Deprotection: TBAF (2.3 eq.), THF anhydrous, 0 °C to RT, 1 h, 99% [83].

11.4.4 1,1,3,3-Tetraisopropyldisiloxane-1,3-diyl (TIPDS) Group

This 1,2- and 1,3-diol protecting group is the most used bis-silylated group described and used in the literature (Scheme 11.19).

Formation:

1) $TIPDSCl_2$ (1.15 eq.), imidazole (2.6 eq.), DMF, −40 °C, 68% [84a].
2) $TIPDSCl_2$ (1.2 eq.), pyridine, RT, 48 h [84b].

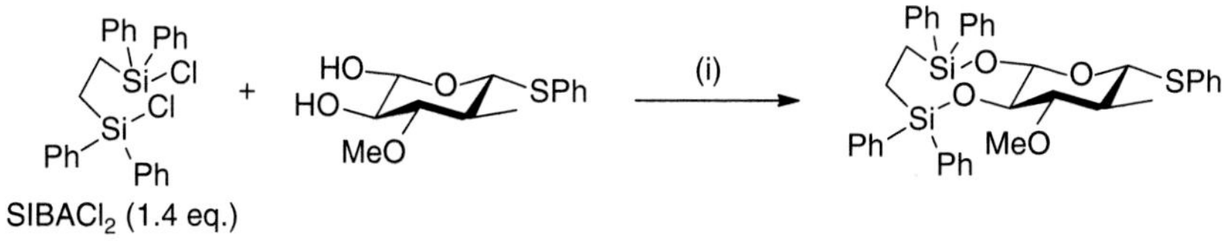

Scheme 11.18 Incorporation of the bis-silyl (SIBA) protecting group: (i) imidazole, DMF, 0 °C to RT. *Source:* Wehlan et al. 2006 [83]. Reproduced with permission of John Wiley and Sons.

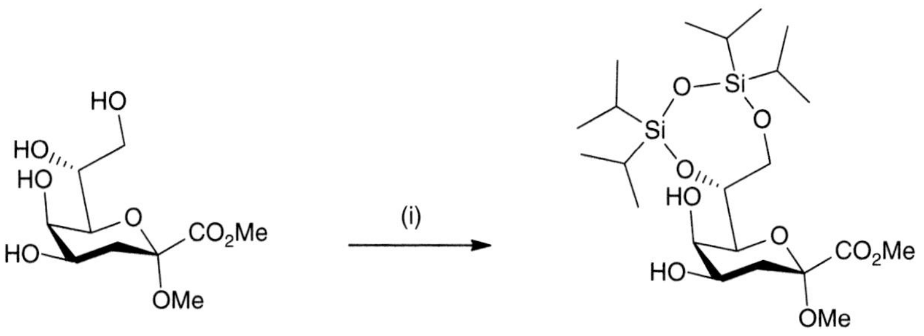

Scheme 11.19 1,2-Diol protection with a TIPDS group: (i) $TIPDSCl_2$, imidazole, DMF, −40 °C. *Source:* Pokorny et al. 2014 [84a]. Reproduced with permission of Elsevier.

3) $TIPDSCl_2$ (1.2 eq.), pyridine, 0 °C to RT, 5 h. (61%) [84c].
4) *6-O-Benzyl*: imidazole (8.2 eq.), $TIPDSCl_2$ (1.067 eq.), DMF, 0 °C to RT, 5 h (83%) [84d] or 6-*O*-TIPS: imidazole (1.9 eq.), $TIPDSCl_2$ (1.48 eq.), pyridine, 0 °C to RT, 18 h (83%) [84d].

11.4.5 1,1,3,3-Tetramethyldisiloxane-1,3-diyl Group

This rare protecting group in glycochemistry has been applied for specific reactivity comparisons in glycosylation.

Formation:

6-O-TIPS: imidazole (9.9 eq.), 1,3-dichloro-1,1,3,3-tetramethyldisiloxane (1.02 eq.), DMF, 0 °C to RT, 18 h, 52% [84d].

11.5 Cyclic Carbonate

Cyclic carbonate attract a lot of research interest, not only as protecting groups of 1,2-diols or 1,3-diols but also as a mean to trap CO_2 or as precursors in ROP (ring opening polymerization).

Formation: various methods have been described starting with phosgene as the primary reagent but also catalytic version using carbon monoxide has been disclosed.

1,3-Diol: (D-gluco, L-idofuranose) Im_2CO (2.5 eq.), toluene, reflux, 1.5 h, 76% and 95% [85a].

1,3-Diol: (D-gluco and L-idofuranose), triphosgene (0.6 eq.), pyridine (7.3 eq.) CH_2Cl_2, 0 °C to RT, 2 h, 81% and 83% [85a].

1,3-Diol: (D-xylofuranose) a two-step sequence a- CO_2 (1 atm), 1,8-diazabicycloundec-7-ene (DBU) (1 eq.), $CHCl_3$, 2 h, RT then Et_3N (1 eq.), TsCl (1 eq.) 16 h, RT, one example 11% [85b].

1,3-Diol: (D-glucopyranoside) $(F_5C_6O)_2CO$, CsF, THF, 60 °C, 25 h, 36% [85c].

1,2-Diol (*cis*): (noviosyl synthesis) Im_2CO (1.1 eq.), 1,2-dichloroethane (DCE), reflux, 3 h, 67% for a two-step sequence [86a].

1,2-Diol (*cis*): (from D-glucopyranoside to D-mannopyranoside via intramolecular epimerization using a carbamate) (Scheme 11.20) [86b].

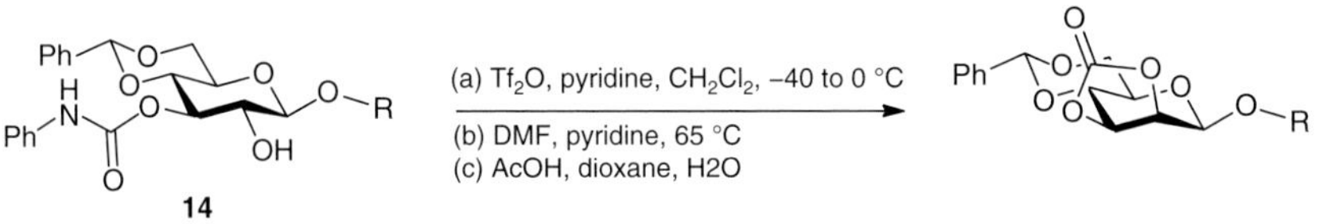

R = protected disaccharide

Scheme 11.20 Cyclic carbonate formation via intramolecular substitution of transient triflate derivative: (i) Tf_2O, pyridine, CH_2Cl_2, −40 to 0 °C; (ii) DMF, pyridine, 65 °C; and (iii) AcOH, dioxane, H_2O. *Source:* Ott et al. 2012 [86b]. Reproduced with permission of John Wiley and Sons.

1,2-Diol (*cis*): (D-lyxose) diphosgene ($Cl_3COCOCl$) (1.2 eq.), pyridine, 0 °C to RT, 0.5 h, 75% [86c].

1,2-Diol (*trans*): (D-glucopyranoside) EtOCOCl (105 eq.), Et_3N (34 eq.), 1,4-dioxane, 0 °C 89% [87a, b].

1,2-Diol (*trans*): (D-glucopyranoside, D-galactopyranoside) triphosgene $(Cl_3COCO)_2O$ (0.65–1.5 eq.), Et_3N (3–7.5 eq.), CH_2Cl_2, 4 °C, 1 h, 88–96% yield (influence on epimerization) [87c–e].

1,2-Diol (*trans*): phosgene (20% in toluene, 2 eq.), Et_3N (3 eq.), 0 °C, 2 h. D-Glucopyranoside, (glycopyranosylation reactions) [87f].

1,2-Diol: CH_3OCOCl (4.74 eq.), Et_3N (4.74 eq.), CH_2Cl_2, 0 °C, 3 h. D-Glucofuranose, D-allofuranose (97%) [88].

1,2-Diol: PdI_2 (2 mol%), KI (2 mol%), $MeC(OMe)_3$ (3 eq.), CO (16 atm), *N,N*-dimethylacetamide (DMA), 40% (catalytic method to incorporate carbon monoxide and form cyclic carbonate (Scheme 11.21) [89].

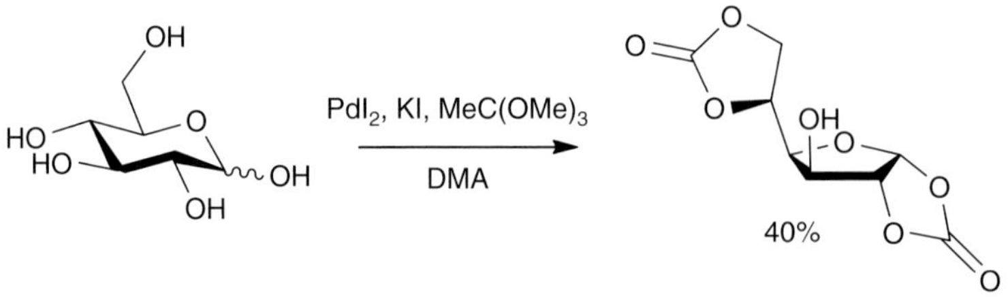

Scheme 11.21 Catalytic formation of a bis-cyclic carbonate on D-glucose: (i) PdI_2, KI, $MeC(OMe)_3$, DMA. *Source:* Gabriele et al. 2011 [89]. Reproduced with permission of John Wiley and Sons.

Deprotection:

Full deprotection: removal of the cyclic carbonate is usually performed under basic conditions assuming a transcarbonatation process, for example, using NaOMe as a catalyst in MeOH as the solvent [86b, 88], or more recently using Barton's base with wet dichloromethane [86a].

11.6 Summary and Conclusions

Protecting groups for 1,2 and 1,3-diols consist of a large panel reagents and solutions to help in the design of complex carbohydrate synthesis. Nonetheless, there is still room to develop original and specifically design entities to not only protect but also add selective deprotection and/or functions.

11.7 Experimental Part: Procedure for Regioselective and Reductive Benzylidene Opening Synthesis of Methyl 2,3,4-Tri-*O*-benzyl-α-D-glucopyranoside

DIBAL-H 1.2 M in cyclohexane (2.7 ml, 3.24 mmol, 3 eq.) was added dropwise at −10 °C to a solution of the benzylidene derivative (500 mg, 1.08 mmol, 1 eq.) in toluene (5.4 ml). The reaction mixture was stirred at −10 °C for 48 h. The reaction was quenched by addition of MeOH (5 ml) at −10 °C and then allowed to warm to RT before the addition of an aqueous solution of 10% KOH (5 ml). The aqueous layer was extracted with Et_2O (3 × 10 ml), and the combined organic layers were washed with water (3 × 10 ml) and with brine (10 ml), dried ($MgSO_4$), filtered, and the solvent removed *in vacuo* to afford the crude product (99%) with a fairly good purity and could directly be used without purification in the next step (91%) (Ref. [90]).

Abbreviations

Ac	acetyl
AIBN	α,α′-azoisobutyronitrile
BDA	butane-2,3-diacetal
BHT	dibutylhydroxytoluene
Bn	benzyl
Bu	butyl
Bz	benzoyl
CAN	ceric ammonium nitrate
CDA	cyclohexane-1,2-diacetal
CSA	camphor sulfonic acid
DBDMH	1,3-dibromo-5,5-dimethylhydantoïn
DCC	*N,N′*-dicyclohexylcarbodiimide
DDQ	2,3-dichloro-5,6-dicyano-1,4-benzoquinone
DHP	3,4-dihydropyran

DIBAL-H	diisobutyl aluminum hydride
DMA	*N,N*-dimethylacetamide
DMAP	4-(dimethylamino)pyridine
DMC	1,1-dimethoxycyclohexanone
DMP	dimethoxypropane
DMF	dimethylformamide
DMSO	dimethyl sulfoxide
DTBS	di-*t*-butylsilylene
Gal	galactose
GalN	galactosamine
Glc	glucose
GlcNAc	*N*-acetyl-glucosamine
GlcA	glucuronic Acid
HMPA	hexamethylphosphoramide
KDO	3-deoxy-D-manno-2-octulosonic acid
MOM	methoxymethylether
MP	2-methoxypropene
MS	molecular sieves
NAP	2-naphthylmethyl
O/N	overnight
p-ADA	*para*-anisaldehyde dimethyl acetal
PMB	*p*-methoxybenzyl
Pn	pent-4-enyl
Ph	phenyl
RT	room temperature
SET	single electron transfer
Spoke	spiroketal
TBAF	tetrabutylammonium fluoride
TBATB	tetrabutylammoniun tribromide
TBDMS	*t*-butyldimethylsilyl
TBDPS	*t*-butyldiphenylsilyl
TCA	trichloroacetyl
Tf	trifluoromethanesulfonyl
TFA	trifluoroacetyl
TFAA	trifluoroacetic anhydride
THF	tetrahydrofuran
TMB	2,2,3,3-tetramethoxybutane
TMC	1,2,2,2-tetramethoxycyclohexane
Ts	toluenesulfonyl

References

1 (a) Greene, T.W. and Wuts, P.G.M. *Protective Groups in Organic Synthesis*, 4th, 2007. New York: Wiley-VCH. (b) Kociensky, P.J. *Protecting Groups*, 2004. Stuttgart: Georg Thieme.

2 Tomohiro, M., Yasuyuki, K., Kazuki, O., and Hiromishi, F. (2013). *J. Org. Chem.* 78: 3384–3390.

3 Furukawa, T., Mondal, S., and Crich, D. (2013). *J. Am. Chem. Soc.* 135: 14249–14255.
4 Hevey, R., Chen, X., and Ling, C.-C. (2013). *Carbohydr. Res.* 376: 37–48.
5 Maryanoff, B.E., Costanzo, M.J., Nortey, S.O. et al. (1998). *J. Med. Chem.* 95: 1315–1343.
6 Dumas, C., Schibli, R., and Schubiger, P.A. (2003). *J. Org. Chem.* 68: 512–518.
7 Boto, A., Hernandez, D., Hernandez, R., and Suarez, E. (2006). *J. Org. Chem.* 71: 1938–1948.
8 (a) Chéry, F., Rollin, P., De Lucchi, O., and Cossu, S. (2000). *Tetrahedron Lett.* 41: 2357–2360. (b) Chéry, F., Rollin, P., De Lucchi, O., and Cossu, S. (2001). *Synthesis* 2: 286–292.
9 Pellizzaro, L., Cabianca, E., Tatibouët, A. et al. (2007). *Tetrahedron Lett.* 48: 3699–3703.
10 Chéry, F., Cabianca, E., Tatibouët, A. et al. (2012). *Tetrahedron* 68: 544–551.
11 Chéry, F., Cabianca, E., Tatibouët, A. et al. (2015). *Carbohydr. Res.* 417: 117–124.
12 (a) Hirano, S., Ichikawa, S., and Matsuda, A. (2007). *J. Org. Chem.* 9936–9946. (b) Chen, C., Ye, W., Liu, C., and Schneller, S.W. (2012). *Tetrahedron* 68: 3908–3914.
13 Jin, Y., Biancotto, G., and Just, G. (1996). *Tetrahedron Lett.* 37: 973–976.
14 Azad, C.S. and Saxena, A.K. (2015). *Org. Chem. Front.* 2: 665–669.
15 Jamois, F., Le Goffic, F., Yvin, J.C. et al. (2008). *Open Glycoscience* 19–24.
16 Sarpe, V.A. and Kulkarni, S.S. (2014). *Org. Lett.* 16: 5732–5735.
17 Liu, P. and Chu, C.K. (2006). *Can. J. Chem.* 754: 748–754.
18 (a) Tomooka, K., Sakamaki, J., Harada, M., and Wada, R. (2008). *Synlett* 683–686. (b) Avanzo, R.E., Anesini, C., Fascio, M.L. et al. (2012). *Eur. J. Med. Chem.* 47: 104–110.
19 Majumdar, S., Chakraborty, M., Maiti, D.K. et al. (2014). *RSC Adv.* 4: 16497–16502.
20 Xiong, J., Yan, S., Ding, N. et al. (2013). *J. Carbohydr. Chem.* 32: 184–192.
21 Kim, E. and Krishnamurthy, R. (2015). *Chem. Commun.* 51: 5618–5621.
22 Liu, Z., Yoshihara, A., Wormald, M.R. et al. (2014). *Org. Lett.* 16: 5663–5665.
23 (a) Clarion, L., Jacquard, C., Sainte-Catherine, O. et al. (2014). *J. Med. Chem.* 57: 8293–8306. (b) Pedatella, S., Guaragna, A., D'Alonzo, D. et al. (2006). *Synthesis* 305–308.
24 Khan, A.T., Khan, M.M., and Adhikary, A. (2011). *Carbohydr. Res.* 346: 673–677.
25 Lambrecht, M.J., Brichacek, M., Barkauskaite, E. et al. (2015). *J. Am. Chem. Soc.* 137: 3558–3564.
26 Zong, G., Barber, E., Aljewari, H. et al. (2015). *J. Org. Chem.* 80: 9279–9291.
27 Manner, S. and Ellervik, U. (2014). *Synlett* 25: 1271–1274.
28 (a) Das, S.N., Chowdhury, A., Tripathi, N. et al. (2015). *J. Org. Chem.* 80: 1136–1148. (b) Biela-Banaś, A., Oulaïdi, F., Front, S. et al. (2014). *ChemMedChem* 9: 2647–2652. (c) Feng, Y., Dong, J., Xu, F. et al. (2015). *Org. Lett.* 17: 2388–2391.
29 Lenci, E., Menchi, G., Guarna, A., and Trabocchi, A. (2015). *J. Org. Chem.* 80: 2182–2191.
30 Yang, X., Xiong, D.-C., Song, C. et al. (2015). *Org. Biomol. Chem.* 13: 9364–9368.
31 Benedeković, G., Kovačević, I., Popsavin, M. et al. (2015). *Tetrahedron* 71: 4581–4589.
32 Blade, H., Bradley, D., Diorazio, L. et al. (2015). *J. Org. Chem.* 80: 5337–5343.

33 Pradhan, T.K., Lin, C.C., and Mong, K.T. (2014). *Org. Lett.* 2: 6–9.
34 Mukherjee, S., Sengupta, A., and Roy, S.C.J. (2013). *Chem. Sci.* 125: 1493–1496.
35 Verma, P.R. and Mukhopadhyay, B.J. (2015). *J. Carbohydr. Chem.* 34: 319–337.
36 (a) Ziegler, T. (1994). *Angew. Chem. Int. Ed.* 33: 2272–2275. (b) Lence, E., Castedo, L., and González-Bello, C. (2008). *Chem. Soc. Rev.* 37: 1689–1708.
37 (a) Ley, S.V., Leslie, R., Tiffin, P.D., and Woods, M. (1992). *Tetrahedron Lett.* 33: 4767–4770. (b) Ley, S.V., Downham, R., Edwards, P.J., and Innes, J.E. (1995). *Contemp. Org. Synth.* 2: 365–392.
38 Ley, S.V., Baeschlin, D.K., Dixon, D.J. et al. (2001). *Chem. Rev.* 101: 53–80.
39 Ley, S.V., Boons, G.-J., Leslie, R. et al. (1993). *Synthesis* 689–692.
40 (a) Ley, S.V., Osborn, H.M.I., Priepke, H.W.M., and Warriner, S.L. (1998). *Org. Synth.* 75: 170–173. (b) Shing, T.K.M. and Luk, T. (2009). *Tetrahedron: Asymmetry* 20: 883–886.
41 (a) Douglas, N.L., Ley, S.V., Osborn, H.M.I. et al. (1996). *Synlett* 793–795. (b) Wang, C.-C., Tang, Y., Yang, K. et al. (2013). *Tetrahedron* 69: 8284–8290.
42 Kongkathip, B., Kongkathip, N., and Rujirawanich, J. (2014). *Synth. Commun.* 44: 2248–2255.
43 (a) Herczeg, M., Demeter, F., Mező, E. et al. (2015). *Eur. J. Org. Chem.* 5730–5741. (b) Shing, T.K.M., So, K.H., and Kwok, W.S. (2009). *Org. Lett.* 11: 5070–5073.
44 Siegbahn, A., Aili, U., Ochocinska, A. et al. (2011). *Bioorg. Med. Chem.* 19: 4114–4126.
45 Grice, P., Ley, S.V., Pietruszka, J. et al. (1997). *J. Chem. Soc., Perkin Trans. 1* 351–363.
46 Tzschucke, C., Pradidphol, N., Diéguez-Vázquez, A. et al. (2008). *Synlett* 1293–1296.
47 Shing, T., Chen, Y., and Wu, H. (2012). *Synlett* 23: 1793–1796.
48 Lindbäck, E., Lopez, O., Fernandez-Bolanos, J.G. et al. (2011). *Org. Lett.* 13: 2908–2911.
49 Sum, P.-E., How, D., Torres, N. et al. (2003). *Bioorg. Med. Chem. Lett.* 13: 2805–2808.
50 Banerjee, A., Senthilkumar, S., and Baskaran, S. (2016). *Chem. Eur. J.* 22: 902–906.
51 (a) Cañeque, T., Gomes, F., Mai, T.T. et al. (2015). *Nat. Chem.* 7: 744–751. (b) Dhara, D., Kumar Kar, R., and Bhunia, A. (2014). Misra A. *K. Eur. J. Org. Chem.* 4577–4584. (c) Cox, D.J., Singh, G.P., Watson, A.J.A., and Fairbanks, A.J. (2014). *Eur. J. Org. Chem.* 4624–4644.
52 (a) Ohlin, M., Johnsson, R., and Ellervik, U. (2011). *Carbohydr. Res.* 346: 1358–1370. (b) Panchadhayee, R. and Misra, A.K. (2010). *Synlett* 1193–1196. (c) Herczeg, M., Demeter, F., Mezo, E. et al. (2015). *Eur. J. Org. Chem.* 5730–5741. (d) Tanaka, N., Ogawa, I., Yoshigase, S., and Nokami, J. (2008). *Carbohydr. Res.* 343: 2675–2679. (e) Sarpe, V.A. and Kulkarni, S.S. (2013). *Org. Biomol. Chem.* 11: 6460–6465.
53 (a) Mezo, E., Herczeg, M., Eszenyi, D., and Borbás, A. (2014). *Carbohydr. Res.* 388: 19–29. (b) Pett, C. and Westerlind, U. (2014). *Chem. Eur. J.* 20: 7287–7299.
54 (a) Geng, Y., Faidallah, H.M., Albar, H.A. et al. (2013). *Eur. J. Org. Chem.* 7035–7040. (b) Konishi, M., Imamura, A., Fujikawa, K. et al. (2013). *Molecules*

18: 15153–15181. (c) Goyal, N., Mangunuru, H.P.R., Parikh, B. et al. (2014). *Beilstein J. Org. Chem.* 10: 3111–3121. (d) Budhadev, D. and Mukhopadhyay, B. (2015). *Tetrahedron* 71: 6155–6163. (e) Iwata, R., Doi, A., Maedaa, Y., and Wada, T. (2015). *Org. Biomol. Chem.* 13: 9504–9515.

55 (a) Pett, C., Schorlemer, M., and Westerlind, U. (2013). *Chem. Eur. J.* 19: 17001–17010. (b) Tsutsui, N., Tanabe, G., Gotoh, G. et al. (2013). *Tetrahedron* 69: 9917–9930. (c) Rodriguez Lavado, J., Sestito, S.E., Cighetti, R. et al. (2014). *J. Med. Chem.* 57: 9105–9123. (d) Tamai, H., Imamura, A., Ogawa, J. et al. (2015). *Eur. J. Org. Chem.* 5199–5211.

56 (a) Desruet, M.-D., Morin, C., and Ogier, L. (1997). *Carbohydr. Res.* 301: 225–229. (b) Mogemark, M., Elofsson, M., and Kihlberg, J. (2001). *Org. Lett.* 3: 1463–1466. (c) Mogemark, M., Elofsson, M., and Kihlberg, J. (2003). *J. Org. Chem.* 68: 7281–7288. (d) Van Hoof, S., Ruttens, B., Hubrecht, I. et al. (2006). *Bioorg. Med. Chem. Lett.* 16: 1495–1498. (e) Nokami, T., Shibuya, A., Tsuyama, H. et al. (2007). *J. Am. Chem. Soc.* 129: 10922–10928. (f) de Jong, A.R., Volbeda, A.G., Hagen, B. et al. (2013). *Eur. J. Org. Chem.* 6644–6655.

57 (a) Collins, P.M. and Oparaeche, N.N. (1974). *Carbohydr. Res.* 33: 35–46. (b) Young, D.D. and Deiters, A. (2007). *Angew. Chem. Int. Ed.* 49: 4290–4292.

58 (a) Sebej, P., Solomek, T., Hroudna, L. et al. (2009). *J. Org. Chem.* 74: 8647–8658. (b) Watanabe, S., Sueyoshi, T., Ichihara, M. et al. (2001). *Org. Lett.* 3: 255–257. (c) Zhu, C.-J., Yi, H., Chen, G.-R., and Xie, J. (2008). *Tetrahedron* 64: 10687–10693.

59 Ando, H., Manabe, S., Nakahara, Y., and Ito, Y. (2001). *J. Am. Chem. Soc.* 123: 3848–3849.

60 Mukhopadhyay, B. (2006). *Tetrahedron Lett.* 47: 4337–4341.

61 (a) Chinoy, Z.S., Schafer, C.M., West, C.M., and Boons, G.-J. (2015). *Chem. Eur. J.* 21: 11779–11787. (b) Tatai, J., Osztrovszky, G., Kajtar-Peredy, M., and Fügedi, P. (2008). *Carbohydr. Res.* 343: 596–606. (c) Takeda, N. and Tamura, J.-I. (2014). *Biosci. Biotechnol., Biochem.* 78: 29–37. (d) Mulani, S.K., Guh, J.-H., and Tony Mong, K.-K. (2014). *Org. Biomol. Chem.* 12: 2926–2937. (e) Agoston, K., Watt, G.M., and Fügedi, P. (2015). *Tetrahedron Lett.* 56: 5010–5012.

62 (a) Ellervik, U. (2003). *Tetrahedron Lett.* 44: 2279–2281. (b) Jakab, Z., Mándi, A., Borbás, A. et al. (2009). *Carbohydr. Res.* 344: 2444–2453.

63 Lebel, H. and Grenon, M. (2005). *Science of Synthesis*, vol. 22 (ed. A.B. Charette and E.J. Thomas), 669–747. Stuttgart and New York: Georg Thieme Verlag.

64 Lemieux, R.U. and Driguez, H. (1975). *J. Am. Chem. Soc.* 97: 4069–4074.

65 Comin, J., Elhalem, E., Rodriguez, J.B., and Comin, M.J. (2004). *Tetrahedron* 60: 11851–11860.

66 (a) Tian, X.B., Min, J.M., and Zhang, L.H. (2000). *Tetrahedron: Asymmetry* 11: 1877–1889. (b) Vonlanthen, D. and Leumann, C.J. (2003). *Synthesis* 7: 1087–1090.

67 (a) Bouchra, M., Calinaud, P., and Gelas, J. (1995). *Synthesis* 561–565. (b) Bouchra, M., Calinaud, P., and Gelas, J. (1995). *Carbohydr. Res.* 267: 227–237.

68 Baba, A. and Yoshioka, T. (2012). *J. Org. Chem.* 77: 1675–1684.

69 (a) Vol, T. and Britain, G. (1992). *Tetrahedron* 48: 4713–4724. (b) Kuir, D. and Auzanneau, F.-I. (2015). *J. Org. Chem.* 80: 5004–5013.

70 (a) Lindhorst, T.K., Bruegge, K., Fuchs, A., and Sperling, O. (2010). *Beilstein J. Org. Chem.* 6: 801–809. (b) Oscarson, S. (1993). *Carbohydr. Res.* 241: 323–328. (c) Stanetty, C., Walter, M., and Kosma, P. (2014). *J. Org. Chem.* 79: 582–598.
71 Belot, F., Otter, A., Fukuda, M., and Hindsgaul, O. (2003). *Synlett* 1315–1318.
72 Mandal, P.K., Dhara, D., and Misra, A.K. (2014). *Beilstein J. Org. Chem.* 10: 293–299.
73 Ay, K. and Halay, E. (2013). *J. Chem.* 1–7. doi: 10.1155/2013/748161.
74 Salman, Y.G., Makinabakan, Ö., and Yüceer, L. (1994). *Tetrahedron Lett.* 35: 9233–9236.
75 Salman, Y.G., Kök, G., and Yüceer, L. (2004). *Carbohydr. Res.* 339: 1739–1745.
76 Sanchez, S., Bamhaoud, T., and Prandi, J. (2002). *Eur. J. Org. Chem.* 3864–3873.
77 Podvalnyy, N.M., Sedinkin, S.L., Abronina, P.I. et al. (2011). *Carbohydr. Res.* 346: 7–15.
78 Kök, G. and Salman, Y.G. (2012). *J. Carbohydr. Chem.* 31: 1–9.
79 (a) Bamhaoud, T., Sanchez, S., and Prandi, J. (2000). *Chem. Commun.* 659–660. (b) Sanchez, S., Bamhaoud, T., and Prandi, J. (2000). *Tetrahedron Lett.* 41: 7447–7452.
80 Huang, J.-S., Huang, W., Meng, X. et al. (2015). *Angew. Chem. Int. Ed.* 54: 7670–7673.
81 (a) Barett, A.G.M., Pena, M., and Willardsen, J.A. (1996). *J. Org. Chem.* 61: 1082–1100. (b) Hoberg, J.O. (1997). *Carbohydr. Res.* 300: 365–367. (c) Mori, Y. and Hayashi, H. (2001). *J. Org. Chem.* 66: 8666–8668. (d) Mezo, E., Herczeg, M., Eszenyi, D., and Borbás, A. (2014). *Carbohydr. Res.* 396: 25–36. (e) Adanitsch, F., Ittig, S., Stöckl, J. et al. (2014). *J. Med. Chem.* 57: 8056–8071. (f) Schroeder, B.R., Ghare, M.I., Bhattacharya, C. et al. (2014). *J. Am. Chem. Soc.* 136: 13641–13656. (g) Parkan, K., Pohl, R., and Kotora, M. (2014). *Chem. Eur. J.* 20: 4414–4419. (h) Habibian, M., Martínez-Montero, S., Portella, G. et al. (2015). *Org. Lett.* 17: 5416–5419. (i) Zhang, Q., van Rijssel, E.R., Walvoort, M.T.C. et al. (2015). *Angew. Chem. Int. Ed.* 54: 7670–7673.
82 Casu, F., Harston, R.K., Chiacchio, M.A., and Gumina, G. (2012). *Nucleosides Nucleotides Nucleic Acids* 31: 224–235.
83 Wehlan, H., Dauber, M., Mujica Fernaud, M.T. et al. (2006). *Chem. Eur. J.* 12: 7378–7397.
84 (a) Pokorny, B., Müller-Loennies, S., and Kosma, P. (2014). *Carbohydr. Res.* 391: 66–81. (b) Guiard, J., Collmann, A., Gilleron, M. et al. (2008). *Angew. Chem. Int. Ed.* 47: 9734–9738. (c) Farrell, M., Zhou, J., and Murphy, P.V. (2013). *Chem. Eur. J.* 19: 14836–14851. (d) Balmond, E.I., Benito-Alifonso, D., Coe, D.M. et al. (2014). *Angew. Chem. Int. Ed.* 53: 8190–8194.
85 (a) Kovačević, I., Popsavin, M., Benedeković, G. et al. (2016). *J. Med. Chem.* 108: 594–604. (b) Gregory, G.L., Ulmann, M., and Buchard, A. (2015). *RSC Adv.* 5: 39404–39408. (c) Mikami, K., Lonnecker, A.T., Gustafson, T.P. et al. (2013). *J. Am. Chem. Soc.* 135: 6826–6829.
86 (a) Kaufmann, E., Hattori, H., Miyatake-Ondozbal, H., and Gademann, K. (2015). *Org. Lett.* 17: 3514–3517. (b) Ott, D., Seifert, J., Prahl, I. et al. (2012). *Eur. J. Org. Chem.* 5054–5068. (c) Abbot, J.R. and Roush, W.R. (2013). *Org. Lett.* 15: 62–64.

87 (a) Azechi, M., Matsumoto, K., and Endo, T. (2013). *J. Polym. Sci., Part A: Polym. Chem.* 51: 1651–1655. (b) Doane, W.M., Shasha, B.S., Stout, E.I. et al. (1967). *Carbohydr. Res.* 4: 445–451. (c) Manabe, S., Ishii, K., Hashizume, D. et al. (2009). *Chem. Eur. J.* 15: 6894–6901. (d) Manabe, S., Ishii, K., Satoh, H., and Ito, Y. (2011). *Tetrahedron* 67: 9966–9974. (e) Geng, Y., Qin, Q., and Ye, X.-S. (2012). *J. Org. Chem.* 77: 5255–5270. (f) Crich, D. and Jayalath, P. (2005). *J. Org. Chem.* 70: 7252–7259.

88 Yu, H., Cao, H., Tiwari, V.K. et al. (2011). *Bioorg. Med. Chem. Lett.* 21: 5037–5040.

89 Gabriele, B., Mancuso, R., Salerno, G. et al. (2011). *ChemSusChem* 4: 1778–1786.

90 Lipták, A., Jodál, I., and Nánási, P. (1975). *Carbohydr. Res.* 44: 1–8.

12

Protecting Group Strategies Toward Glycofuranoses

Vincent Ferrières, Laurent Legentil, and Loïc Lemiègre

Ecole Nationale Supérieure de Chimie de Rennes, 11 Allée de Bealieu, CS 50837 35708 Rennes Cedex 7, France

12.1 Introduction

One of the peculiarities of carbohydrates is their ability to cyclize in different forms, thus leading to six-membered rings (pyranose forms, pyranosides) or five-membered derivatives (furanose forms, furanosides). This behavior, common to all reducing carbohydrates, is well known since the pioneering works of Emil Fischer who was awarded the Nobel Prize in 1902 [1]. Indeed, he demonstrated that acid-catalyzed glycosylation of methanol afforded methyl furanosides and methyl pyranosides, both in diastereoisomeric mixtures. Few years later, direct acylation of unprotected monosaccharide also gave possible mixture of peracylated derivatives. In both reactions, the results are highly carbohydrate dependent.

Based on this, and considering further developments in organic chemistry, glycochemistry could evolve and was mainly dedicated to the assembly of pyranosidic building blocks [2]. In the meantime, developments of analytical methods helped to revisit and/or to corroborate the structures of the glycosyl part of natural compounds. Moreover, glycobiology has become a science of the highest order, and many studies on natural and synthetic glycoconjugates emphasized the important roles played by these molecules in the life kingdom. The main advances related to carbohydrates in a furanose form are certainly the structural elucidation of nucleic acids. Nevertheless, pentofuranosyl- and hexofuranosyl-containing conjugates are also found in bacteria, mycobacteria, archaea, fungi, protozoa, plants, marine sponges, or starfishes [3]. One exception still persists today, that of finding hexofuranosides in mammals [3, 4]. Because such rare derivatives are exclusively found in microorganisms, and even in pathogenic ones, the last three decades have been the opportunity for a renewed interest for these compounds and in their corresponding biosynthetic and metabolic pathways [3–5]. Among numerous examples, *Mycobacterium tuberculosis*, the agent that affects millions of patients in the world, concentrates in its arabinogalactan

Protecting Groups: Strategies and Applications in Carbohydrate Chemistry, First Edition.
Edited by Sébastien Vidal.

cell wall both D-arabinose and D-galactose in a furanose form (D-Ara*f* and D-Gal*f*, respectively) [6].

Therefore, from the standpoint of the glycochemist, the main challenges to address are to lock and ensure the control of the ring size and also to find specific and differentiating analytical data for these highly functionalized cycles. This chapter will describe reactions starting from unprotected and nonactivated monosaccharides to give furanose building blocks for further activation and protecting group manipulations. It will also present contribution of enzymatic approaches and more recent strategies based on ring contractions.

12.2 What About Chemistry Without Protecting Groups?

As implied above, the first step required in carbohydrate chemistry necessarily involves an ose neither protected on its hydroxyl functions nor activated on its anomeric center. Several approaches were developed on the basis that the acetalic hydroxyl is the more reactive under either acidic or basic conditions. Once this anomeric position is fixed, the chemistry of furanosides can go on.

12.2.1 Special Reactivity at the Anomeric Center

To avoid ring size modulation, hydroxyl at position 1, 4, or 5 must not remain free. Consequently, many efforts were focused on the possibility to first block the anomeric position with simple groups in order to lock the carbohydrate in the desired furanose form. The aglycon thus introduced can be seen as a temporary protecting group or can activate the sugar.

12.2.1.1 Fischer Glycosylation

The main advantages of the Fischer glycosylation are its simplicity, its low cost, and ease reaction workup when starting from volatile alcohols. It proceeds through an oxycarbenium ion intermediate that results from the interaction of a catalytic acid with the acetal hydroxyl. Subsequent reaction of the alcohol gave α- and β-furanosides that isomerized under these conditions to yield the more stable α- and β-pyranosides (Scheme 12.1) [7].

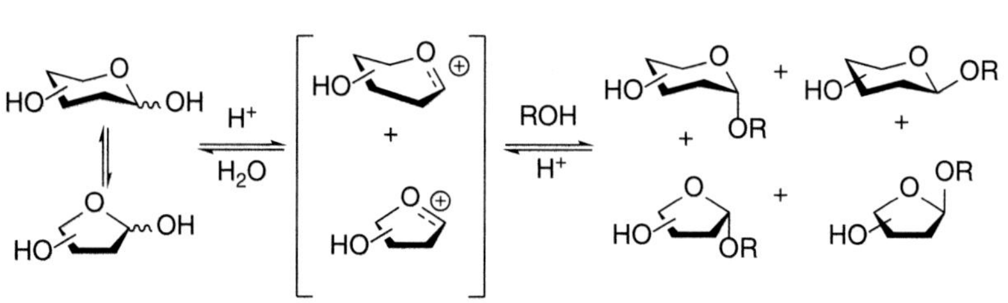

Scheme 12.1 Fischer glycosylation.

The reaction outcome depends on (i) the reactants (carbohydrate and alcohol), (ii) the catalyst, and (iii) the conditions (temperature, concentration, and reaction time). In 1966, Green has written that the *availability* [of furanosides] *depends to a great extent on the diligence of the worker* [8], thus underlying the requirements for challenging preparative aspects in glycochemistry but also for increasing basic knowledge in this family of biomolecules. It was thus established that this *a priori* simple reaction could be kinetically controlled to afford furanosides, while the corresponding pyranosides are obtained under thermodynamic conditions. Nevertheless, under acidic catalysis, pentoses gave the desired furanosides much more efficiently than hexoses. Instead of standard acids, zeolites [9] or mesoporous materials [10] could also be used. Amongst all hexoses, galactose was probably the most studied considering its biological importance, and fortunately, it was the most prone to afford the desired furanosides [11]. When starting from free uronic acids, esterification occurred along with glycosidation so that bilipidic furanosides [12] or the corresponding amphiphilic lactone [13] were obtained in good yields (Figure 12.1).

Moreover, it was shown that complexation of some cations by monosaccharides or polyols positively helped to the chromatographic separation of carbohydrate mixtures [14]. This finding contributed to the use of inorganic cation-containing Lewis acids, and alkyl furanosides were indeed synthesized with really high selectivity starting from galactose, glucose, mannose, glucuronic, and galacturonic acids [15]. The reaction was best promoted by ferric chloride and, when it was performed in tetrahydrofuran as a solvent, the diastereoselectivity was improved by adding calcium or barium chloride (Figure 12.1). The iron(III) cation was a key component as the use of boron trifluoride–diethyl etherate complex mainly yielded the pyranosides. Under these heterogeneous conditions, the reaction temperature should not exceed 25 °C to avoid the formation of pyranosides and coloration of final furanosides [16]. Glycosylation of longer aliphatic alcohols (from C5 to C11) was efficiently

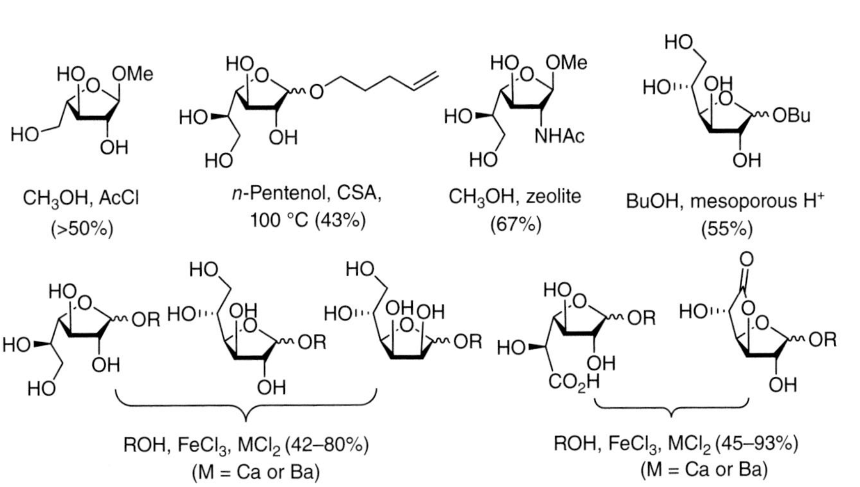

Figure 12.1 Examples of glycosylation of simple alcohols.

performed, was compatible with ethylenic functions, and microwave assistance significantly decreased reaction time from 72 h. to 30 min.

The aglycon chain thus introduced can be seen as a temporary protecting group. After suitable protection of remaining free hydroxyl, or at least protection of OH-5, the alkoxy chain can be easily removed by acetolysis (see Chapter 5) using acetic anhydride and an acid catalyst [17] (Scheme 12.2).

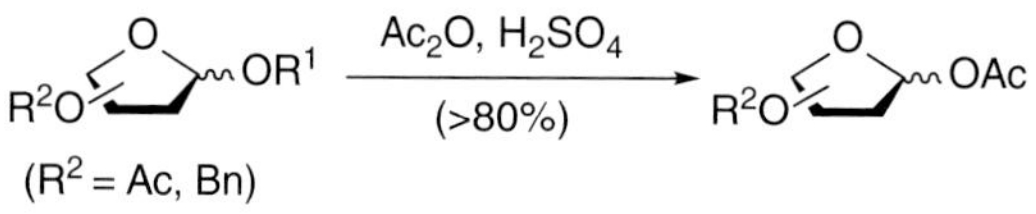

Scheme 12.2 Acetolysis of furanosides.

Nevertheless, the Fischer approach suffers from some limitations: it is not suitable for synthesizing either aryl glycosides or thioglycosides, starting from phenolic derivatives or thiols, respectively. The higher liability of aryl glycosides compared to alkyl glycosides under acidic conditions can explain the first result [7a], while thiols markedly react according to different pathways providing stable acyclic dithioacetals [18]. Nevertheless, kinetically controlled ring closure was performed in aqueous hydrochloric acid thanks to partial hydrolysis of the starting materials. This cyclization was improved by using mercury salts. The thiofuranosides thus obtained are suitable for further protection steps and glycosylation reactions as other thioglycosides [2c]. More recently, 1,3-dibromo-5,5-dimethylhydantoin (DBDH) was proposed for a two-step cyclization–glycosylation reaction (Scheme 12.3) [19].

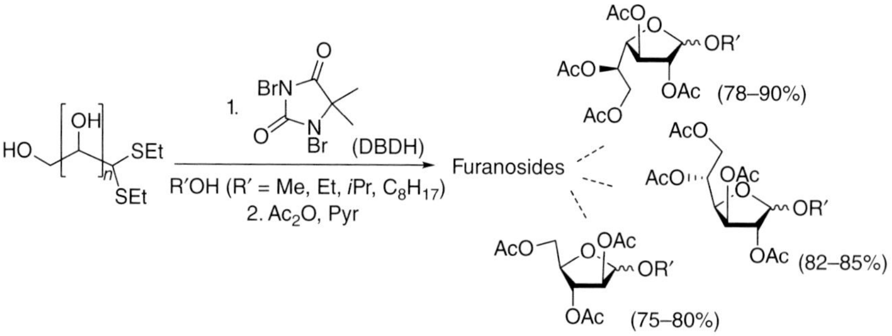

Scheme 12.3 Synthesis of 1-thiofuranosides.

12.2.1.2 1-*O*-alkylation of Galactose

According to a reverse reactivity, the reducing carbohydrate, which becomes the nucleophilic species, can react with electrophiles [20]. In the presence of sodium hydride in *N,N′*-dimethylpropylurea (DMPU), all pentoses and hexoses studied yielded the desired glycosides. All gave pyranosides, except arabinose, which gave a mixture of pyranosides and furanosides and, very surprisingly, galactose

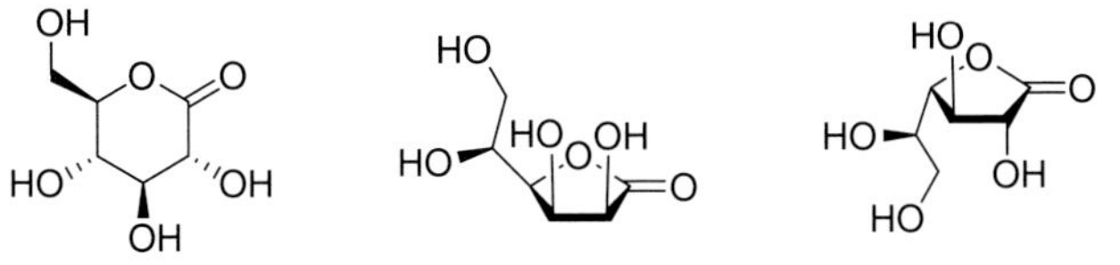

D-Glucono-1,5-lactone L-Gulono-1,4-lactone D-Galactono-1,4-lactone

Figure 12.2 Structure of some lactones.

that afforded pure alkyl α-D-galactofuranosides (Scheme 12.4). Thus, the 1-*O*-alkylation approach complements well the standard Fischer reaction for which the β-furanoside was generally favored.

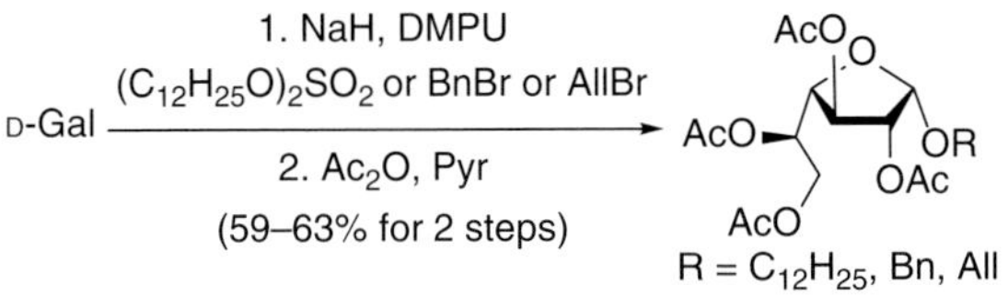

Scheme 12.4 Preparation of alkyl α-D-galactofuranosides.

12.2.1.3 Furanolactones

As oxidation of the anomeric position is efficiently performed by bromine, some furanolactones are commercially available [21]. This lactonic function offers many opportunities for synthesizing derivatives thereof. For instance, glucono-1,5-lactone (pyranose form, Figure 12.2) was used as a precursor of sugar-based nanotubes presenting hydrogel properties [22] or could also be involved in the preparation of polymers [23]. *C*-nucleophiles were used to react with the diacetonide derivative of the L-gulono-1,4-lactone [24]. Nevertheless, the more exploited lactone was the D-galactono-1,4-lactone. This chemistry, explored by de Lederkremer's team, will be presented in Section 12.4.1.

12.2.2 Direct Protection of Hexoses by Acetalization

Among the numerous methodologies described for the protection of diols [25], several are applicable in carbohydrate chemistry. Indeed, ketalization is one of the simpler strategies that permit to protect 1,2- or 1,3-diols (see Chapter 11). Since the work of Schmidt [26] or Morgenlie [27] for instance, several reaction conditions have been found to protect the sugar in its furanose or pyranose form (Scheme 12.5). Depending mainly on the starting sugar and also on the reaction conditions, protected furanosides can be reached following this strategy. Most of the ketalizations involve acetone or cyclohexanone in the presence of a suitable catalyst. In addition, transketalization from 2,2-dimethoxypropane (DMP) or reaction of 2-methoxypropylene (MP) also provides convenient transformations.

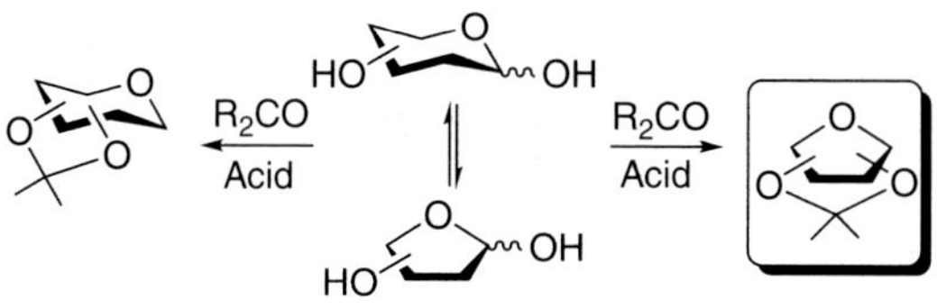

Scheme 12.5 Representative ketalization of sugars toward furanoside or pyranoside forms.

12.2.2.1 Preparation of Glucofuranoside Ketals

The protection of D-glucose by ketalization provides the thermodynamic product 1,2:5,6-diketalglucofuranose leaving the hydroxyl in position 3 free (Scheme 12.6). The research in the aim of finding the best catalyst led to numerous works on the subject (Table 12.1), including the use of various catalysts such as Brönsted acids, Lewis acids, supported catalysts, or inorganic materials, each condition giving rise to the same diketal compound.

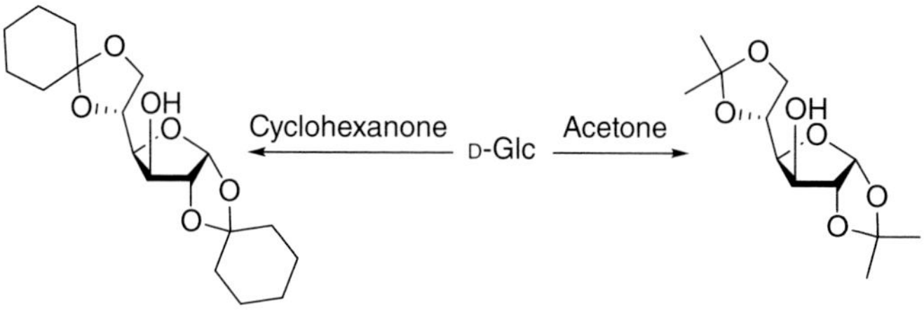

Scheme 12.6 Ketalization of D-glucose toward 1,2:5,6-diketalglucofuranose.

The original Schmidt's conditions [26] has been applied to the preparation of both isopropylidenes and cyclohexylidenes [28] (Table 12.1, entry 1), which were enhanced by combining Brönsted acid with ultrasound reducing the reaction time to less than 1 h in 62% (R = Me) and 75% (R = C_6H_{10}) yields (Table 12.1, entries 2 and 3) [29]. Sulfuric acid can also be turned into a reusable catalyst facilitating the reaction workup and providing diketalglucofuranose (Table 12.1, entry 4) [30]. Ferric chloride was found to be a suitable catalyst since 1977 [31], and again, ultrasound conditions reduced the reaction time with similar yields [32] (Table 12.1 entries 5 and 6). Iodine [33] as such or combined as a supported phosphonium salt [34] induced the acetonation in good yields (80% and 95%, respectively) (Table 12.1, entries 7 and 8). The *in situ* production of HBr from bromodimethylsulfonium bromide (BDMS) [35] or TBATB [36] leads to a sufficient acidic media that promote the acetonation of glucose in 84% and 88% yields (Table 12.1, entries 9 and 11). Under such reaction conditions, the use of DMP instead of acetone increased the yield to 90% (Table 12.1, entry 10). As $FeCl_3$, various other Lewis acids catalyze the ketalization of glucose: strong $BF_3{\cdot}OEt_2$ [37] (Table 12.1, entry 12), air- and water-stable $In(OTf)_3$ [38] (Table 12.1, entry 13), transition metal complex $[Cp^*IrCl_2]_2$ [39] (Table 12.1, entries 14 and 15), or mild vanadyl triflate $VO(OTf)_2$ [40] (Table 12.1, entries 16–18). All provided smooth conversions of D-glucose in acetone and yields as good as 90%. $[Cp^*IrCl_2]_2$

Table 12.1 Reaction conditions for the preparation of 1,2:5,6-diketalglucofuranose from D-glucose and 2,3:5,6-diketalmannofuranose from D-mannose.

Entry	Reaction conditions[a)]	Yield (%)	References
1	Cyclohexanone, H_2SO_4, RT	Glc: 50	[28]
2	Acetone, H_2SO_4 cat.,))), 50 min	Glc: 62 Man: 86	[29]
3	Cyclohexanone (4 eq.), H_2SO_4 cat.,))), MeCN, 45 min	Glc: 75 Man: 43	[29]
4	Acetone, H_2SO_4 on silica, Δ, 3 h	Glc: 89 Man: 30	[30]
5	Acetone, $FeCl_3$ cat., Δ, 5 h Acetone, $FeCl_3$ cat., Δ, 30 min	Glc: 76 Man: 70	[31]
6	Acetone, $FeCl_3$ cat., Δ,)))	Glc: 76	[32]
7	Acetone, I_2, Δ, 2 h or RT, 4 h Acetone, I_2, Δ, 20 min or RT, 25 min	Glc: 80 Man: 85	[33]
8	Acetone, Resin-PPh_2, I_2, RT, 30 min	Glc: 95 Man: 95	[34]
9	Acetone, BDMS 5 mol%, RT, 10 h Acetone, BDMS 5 mol%, RT, 2 h	Glc: 84 Man: 90	[35]
10	DMP (4 eq.), BDMS 5 mol%, MeCN, 4 h DMP (4 eq.), BDMS 5 mol%, MeCN, 1 h	Glc: 90 Man: 95	[35]
11	Acetone, TBATB 2 mol%, RT, 8 h Acetone, TBATB 2 mol%, RT, 2 h	Glc: 88 Man: 96	[36]
12	Acetone, $BF_3{\cdot}OEt_2$, Δ, MS 4 Å, 30 min	Glc: 85	[37]
13	Acetone, $In(OTf)_3$ 3 mol%, Δ, 8 h Acetone, $In(OTf)_3$ 3 mol%, Δ, 10 min	Glc: 90 Man: 100	[38]
14	Acetone, $[Cp^*IrCl_2]_2$ 3 mol%, 2 h	Glc: 89 Man: 81	[39]
15	DMP, $[Cp^*IrCl_2]_2$ 3 mol%, MeCN, 40 °C, 45 min	Glc: 90	[39]
16	Acetone, $VO(OTf)_2$ 5 mol%, MeCN, RT, 10 h Acetone, $VO(OTf)_2$ 5 mol%, MeCN, RT, 3 h	Glc: 89 Man: 92	[40]
17	DMP (10 eq.), $VO(OTf)_2$ 5 mol%, MeCN, RT, 2 h	Glc: 90 Man: 93	[40]
18	MP (10 eq.), $VO(OTf)_2$ 5 mol%, MeCN, RT, 2 h	Glc: 90 Man: 92	[40]
19	Acetone, Zeolite HY, 50 °C, 48 h	Glc: 50	[41]
20	Acetone, $H_3PW_{12}O_{40}$ nH_2O, 5 mol%, RT, 10 h Acetone, $H_3PW_{12}O_{40}$ 5 mol%, RT, 5 h	Glc: 48 Man: 62	[42]
21	DMP (2 eq.), $H_3PW_{12}O_{40}$ nH_2O, 5 mol%, acetone, RT, 7 h DMP (2 eq.), $H_3PW_{12}O_{40}$ nH_2O, 5 mol%, acetone, RT, 5 h	Glc: 86 Man: 89	[42]
22	Acetone, propylsulfonic nano Fe_2O_3, Δ, 2 h	Glc: 97 Man: 96	[43]

a) Δ, reflux;))), ultrasound; DMP, 1,2-dimethoxypropane; MP, 2-methoxypropene.

and $VO(OTf)_2$ also promoted the ketalization of D-glucose in the presence of DMP or MP in acetonitrile with similar yields but in shorter reaction times (Table 12.1, entries 15, 17–18). The ability of inorganic material to catalyze the formation of diketal from D-glucose has been evaluated in three cases. Zeolite HY was able to promote the transformation in a moderate yield of 50% (Table 12.1, entry 19) [41]. Phosphotungstic acid ($H_3PW_{12}O_{40}$, nH_2O) induced smooth diketalization of D-glucose in several hours. It is worth noting that the reaction in the presence of DMP gave a much better yield (86%) compared to the reaction in acetone only (48%) (Table 12.1, entries 20 and 21). This result could be explained by the presence of water within the catalyst. The last protocol used γ-Fe_2O_3 magnetic nanoparticles decorated with propylsulfonic acid residues. After removing the reusable catalyst by the action of an external magnet, the diketal was isolated in 97% yield (Table 12.1, entry 22).

12.2.2.2 Preparation of Mannofuranose Ketals

As for D-glucose, ketalization with D-mannose provides a diketalfuranose. It takes place on positions 2,3 and 5,6 leading to the 2,3:5,6-diketalmannofuranose (Scheme 12.7). Indeed, the latter is the thermodynamic product of the reaction leaving the anomeric position free. In general, reactions from D-mannose are faster than from D-glucose and provide high yields whatever the catalyst used (Brönsted acids, Lewis acids, supported catalyst, or inorganic materials) (Table 12.1).

Scheme 12.7 Ketalization of D-mannose toward 2,3:5,6-diketalmannofuranose.

Brönsted acids such as H_2SO_4 [29], immobilized H_2SO_4 [30], or *in situ* generated HBr [35, 36] gave a readily access to the diisopropylidene product from acetone, or DMP (Table 12.1, entries 2, 4, and 9–11). The preparation of the dicyclohexylidene counterpart has not been extensively studied, and it seems to be hardly prepared; for example, under ultrasound acetone/H_2SO_4 conditions, only 43% yield was obtained [29]. Lewis acid such as common $FeCl_3$ [31] provided a clean transformation in acetone with a moderate yield (70%). Indium triflate [38] allowed a full conversion of D-mannose in acetone solution to the diketal. Iridium complex [39] gave an 81% yield in 2 h and vanadyl triflate [40] provided good yields (92–93%) in 2–3 h, regardless of the ketalization reagent (acetone, DMP, or MP) (Table 12.1, entries 5, 13–14, and 16–18). Iodine [33] was also able to catalyze the reaction in acetone with 85% yield, and when a supported phosphine [34] was used as a combination with I_2, the yield increased to 95% in a more or less same reaction time (Table 12.1, entries 7 and 8). Hydrated phosphotungstic acid [42] ($H_3PW_{12}O_{40}$, nH_2O) required 5 h reaction in acetone; if the presence of water molecules was circumvented by addition of 2 eq. of DMP, a good yield

(89%) was obtained (Table 12.1, entries 20 and 21). Propylsulfonic acid and γ-Fe_2O_3 magnetic nanoparticles [43] were also applied for this transformation at reflux of acetone in 2 h (96% yield) (Table 12.1, entry 22).

12.2.2.3 Preparation of Galactofuranose Ketals

In such ketalization toward furanosides, D-galactose is harder to handle compared to D-glucose and D-mannose. Indeed, the relative orientation of the hydroxyl groups does not favor the furanose form and, without paying attention to the reaction conditions, 1,2:3,4-diketalpyranose is most of the time obtained. In some few cases, a furanose/pyranose mixture was reached in favor of the five-membered ring (Scheme 12.8). These examples are presented in Table 12.2.

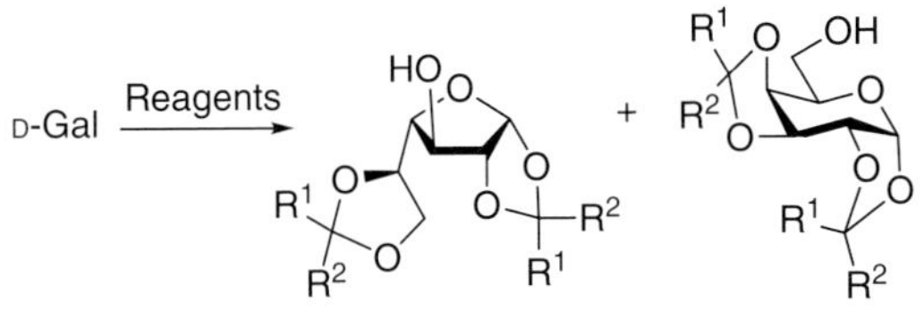

Scheme 12.8 Ketalization of D-galactose toward 1,2:5,6-diketalgalactofuranose and 1,2:3,4-diketalgalactopyranose.

One of the first examples that considerably increased the furanose/pyranose ratio involved the use of Zeolite HY [41] that allowed to reach a 2 : 1 ratio in acetone with a global yield of 60% (Table 12.3, entry 23). Addition of DMF, in order to increase the reflux temperature, in the presence a dry Dowex H^+-form, enhanced the ratio to 4 : 1 in 50% yield [44] (Table 12.2, entry 24). Another factor that can favor the formation of the furanose form was the use of cyclohexanone instead of acetone. In that case, several reaction conditions and catalysts have been investigated, and the best conditions involved PPTS (pyridinium *p*-toluenesulfonate) catalysis at reflux of DMF [45]. Dicyclohexylidenefuranose was isolated in 66% yield in addition to a trace amount of the pyranose counterpart (Table 12.2, entry 25).

Table 12.2 Reaction conditions for the preparation of galactofuranose ketals from D-galactose.

Entry	Reaction conditions[a)]	Yield (%)	Furanose/ pyranose	References
23	Acetone, Zeolite HY, 50 °C, 48 h	60	4 : 2	[41]
24	Acetone, Dowex 50 H^+, Δ, MS 4 Å, DMF/ acetone (4 : 1), 15 h	50	>4 : 1	[44]
25	Cyclohexanone, PPTS, Δ, DMF, 8 h	66	99 : 1	[45]

a) Δ, reflux.

Table 12.3 Reaction conditions for the preparation of other furanose ketals from L-rhamnose, D- or L-ribose, and D-xylose.

Entry	Sugar	Reaction conditions[a)]	Yield (%)	References
26	L-rhamnose	Acetone, H_2SO_4 on silica, Δ, 2 h	87	[30]
27	D-xylose	Acetone, H_2SO_4 on silica, Δ, 1.5 h	83	[30]
28	L-rhamnose	MP (2 eq.), PTSA, 0 °C, 2 h	85	[46]
29	L-rhamnose	Acetone, BDMS 5 mol%, RT, 2 h	88	[35]
30	D-ribose	Acetone, BDMS 5 mol%, RT, 2 h	86	[35]
31	D-rylose	Acetone, BDMS 5 mol%, RT, 2 h	91	[35]
32	L-rhamnose	Acetone, TBATB 2 mol%, RT, 1 h	88	[36]
33	D-ribose	Acetone, TBATB 2 mol%, RT, 1.5 h	86	[36]
34	D-xylose	Acetone, TBATB 2 mol%, RT, 3 h	89	[36]
35	D-xylose	Acetone, $FeCl_3$ cat., Δ, 10 min	80	[31]
36	D-xylose	Acetone, $In(OTf)_3$ 3 mol%, Δ, 10 min	100	[38]
37	L-rhamnose	Acetone, $[Cp^*IrCl_2]_2$ 3 mol%, 1.5 h	87	[39]
38	D-ribose	Acetone, $[Cp^*IrCl_2]_2$ 3 mol%, 1.5 h	85	[39]
39	D-xylose	Acetone, $[Cp^*IrCl_2]_2$ 3 mol%, 1.5 h	82	[39]
40	D-xylose	Acetone, I_2, Δ, 20 min or RT, 3 h	75	[33]
41	L-rhamnose	DMP, Smopex-101, DMF, RT, 23 h	82	[47]
42	L-ribose	DMP, Smopex-101, DMF, RT, 19 h	Furanose (45%)/ pyranose (47%)	[47]
43	L-rhamnose	Acetone, zeolite HY, 50 °C, 48 h	47	[41]
44	D-xylose	Acetone, zeolite HY, 50 °C, 48 h	55	[41]
45	D-ribose	Acetone, propylsulfonic nano-γ-Fe_2O_3, Δ, 1.5 h	98	[43]
46	L-rhamnose	DMP (10 eq.), CAN 10 mol%, RT, 3 h	88	[48]

a) Δ, reflux; DMP, 1,2-dimethoxypropane; MP, 2-methoxypropene.

12.2.2.4 Preparation of Other Furanoside Ketals

Other sugars such as pentoses (D-xylose, L-arabinose, and D-ribose), L-rhamnose, and D-sorbose can also be locked under their furanose forms when ketal or diketal moieties are present as protecting groups. Because of the relative orientation of their hydroxyl groups, L-rhamnose and D-ribose have in common that their protection as ketal leads to 2,3-monoketalfuranoses (Scheme 12.9).

D-Xylose, however, could accept an extra ketal that afforded the 1,2 : 3,5-diisopropylidenexylofuranose. The main reaction conditions for these transformations are presented in Table 12.3. Immobilized H_2SO_4 on silica has been investigated for the protection of L-rhamnose and D-xylose with acetone. In both cases, the corresponding products were obtained in good 87% and 83% yields, respectively [30] (Table 12.3, entries 26 and 27). 2-Methoxypropene also reacted with L-rhamnose in the presence of pTSA (*p*-toluenesulfonic acid) [46]

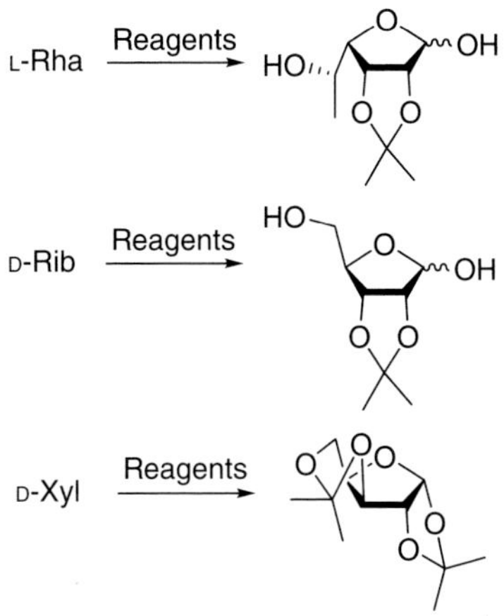

Scheme 12.9 Ketalization of D-rhamnose, D-ribose, and D-xylose.

at 0 °C with a similar yield (85%) (Table 12.3, entry 28). *In situ* HBr generation in acetone (BDMS [35] or TBATB [36]) leads to high conversion of the three sugars in a few hours without changing the selectivity toward the furanose compounds (Table 12.3, entries 29–34). Until now, only few examples described the use of Lewis acids for L-rhamnose, D-ribose, or D-xylose. $FeCl_3$ [31] and $In(OTf)_3$ [38] have been envisaged for the protection of D-xylose in acetone, indium(III) triflate giving a comparative better yield up to 100% (Table 12.3, entries 35 and 36). Iridium-based catalyst [39] provided a clean conversion (82–87%) for all of these sugars (Table 12.3, entries 37–39). Iodine [33] has been used for the preparation of D-xylose diketal in 75% in short reaction time (20 min) at reflux of acetone (Table 12.3, entry 40). The polyethylene-based fiber catalyst grafted with sulfonic acid residues (Smopex-101) [47] allowed an efficient ketalization from DMP in DMF in 19–23 h at room temperature (Table 12.3, entry 41), whereas other materials such as zeolite HY [41] gave moderates yields from acetone (Table 12.3, entries 43 and 44). Interestingly, Smopex-101 yielded poor furanose/pyranose selectivity for L-ribose (Table 12.3, entry 42). Conversely, γ-Fe_2O_3 magnetic nanoparticles [43] bearing sulfonic acid moieties selectively converted D-ribose in acetone to the corresponding 2,3-isopropylidene furanose (Table 12.3, entry 45). Surprisingly, CAN (cerium ammonium nitrate) [48] also catalyzed the ketalization of sugar, however only efficiently toward L-rhamnofuranose (Table 12.3, entry 46).

Diacetone L-sorbose (Scheme 12.10) is also another important sugar regarding its usefulness in the preparation of ascorbic acid for instance. It was first synthesized in 1934 [49] through a sulfuric-acid-catalyzed reaction as a 1,2 : 3,5-diketalfuranose in 84% yield [49b] (Table 12.4, entry 47). Since 1934, other catalysts such as the immobilized H_2SO_4 on silica have been considered that offered a similar yield (86%) in 2 h reaction (Table 12.4, entry 48) [30], or oleum that proceeded at low temperature (Table 12.4, entry 49) [50]. Supported phosphine/iodine [34] was also able to provide diacetone L-sorbose in high yield (95%) and in only 30 min at room temperature (Table 12.4, entry 50). Phosphotungstic acid

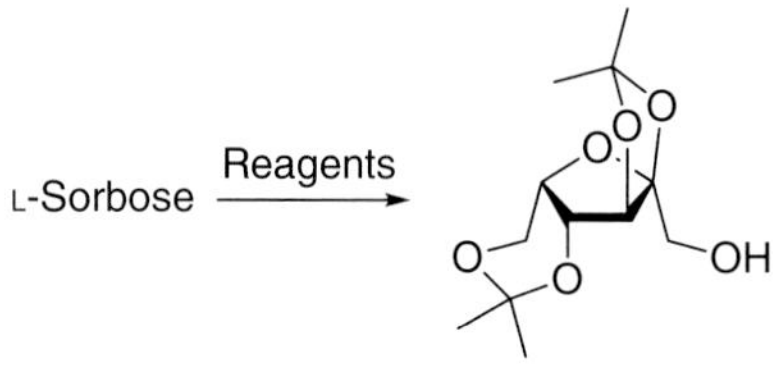

Scheme 12.10 Ketalization of L-sorbose toward 1,2 : 3,5-disopropylidenesorbofuranose.

Table 12.4 Reaction conditions for the preparation of diacetone sorbose from L-sorbose.

Entry	Reaction conditions[a)]	Yield (%)	References
47	Acetone, H_2SO_4, RT, 22 h	84	[49b]
48	Acetone, H_2SO_4 on silica, Δ, 2 h	86	[30]
49	Acetone, oleum, −15 to 25 °C, 2 h	80	[50]
50	Acetone, Resin-PPh_2, I_2, RT, 30 min	95	[34]
51	Acetone, $H_3PW_{12}O_{40}$ 2 mol%, RT, 4 h	82	[51]

a) Δ, reflux.

in acetone was used for this transformation, providing the corresponding diketal in 82% yield in acetone (Table 12.4, entry 51).

12.2.2.5 Selective Deprotection of Furanose Ketals

The protection of sugars in their furanose forms is usually useful in the aim of both freezing the sugars in this particular five-membered cycle and differentiating the reactivity of a few hydroxyl groups from the others. In the case of diketalfuranoses, the synthetic strategies developed so far often involved a selective cleavage of the most labile ketal giving new selectively protected derivatives and enlarging the scope of functionalization. This is particularly relevant in the case of D-xylose, for which the ketalization leads to the protection of all hydroxyls. Whatever the derivatives, the most labile isopropylidene involves a primary alcohol. This ketal is thus easily removed by hydrolysis in the presence of various acid catalysts. Interestingly, isopropylidene is enough labile to permit the milder catalyst to perform the deprotection in the presence of other functions, such as esters, silyloxy, or tetrahydropyranyloxy. Among the selective catalysts, $BF_3{\cdot}OEt_2O$ [37], $BiCl_3$ [52], VCl_3 [53], $In(OTF)_3$ [38], Zeolite [54], and phosphotungstic acids [42] provided the best results and desired orthogonality. When ketalization did not offer a total selectivity toward the furanose, this kind of selective deprotection could simplify the purification and isolation of the furanose product as it has been described, for example, in the D-altrose series [55].

12.2.3 Direct Acylation

Direct acylation of hexoses and pentoses occurs mostly by action of anhydride in pyridine. It gives rise mainly to per-*O*-acylated pyranose with contamination by

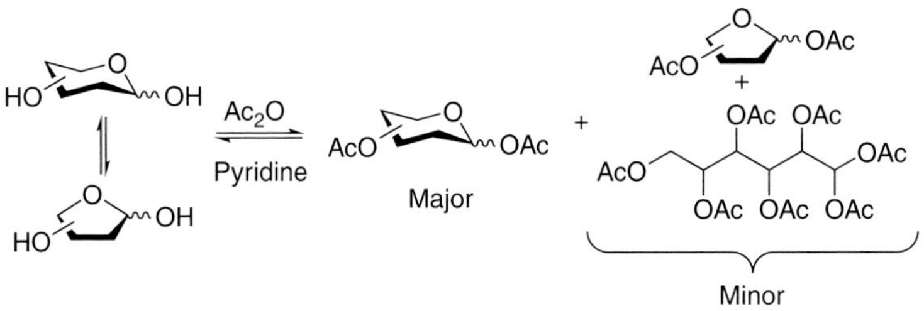

Scheme 12.11 Products of acetylation of pentoses and hexoses.

the furanose form and the acyclic acetal (Scheme 12.11). The proportion of furanoses depends on the considered carbohydrate and is directly linked to the composition of the free sugar in the solution [56]. The amount of furanose form is often higher with the pentoses, and subsequently, their acetylation gives a higher amount of per-*O*-acyl pentofuranose. In particular, acetylation of D-ribose [57] or the hexose analog D-talose [58] at high temperature gave the corresponding tetra-acetate or penta-acetate furanoses as the major compounds. Interestingly penta-acetate D-talofuranose could also be obtained under low-temperature acetylation conditions [59].

The operating conditions used for acylation greatly influence the outcome of the reaction. The nature of the acylating agent, the temperature, and the solvent are of prime importance. This effect is particularly striking for galactose. Indeed, per-*O*-acetylation of D-galactose in standard conditions gives a mixture of the five isomers of per-*O*-acetyl galactose with the pyranose form as the major products. Nevertheless, de Lederkremer and coworkers described the synthesis of per-*O*-benzoylated D-galactofuranose from D-Gal by the action of benzoyl chloride in pyridine at 60 °C (Scheme 12) [60]. The compound slowly crystallized from EtOH and the overall yield reached 35%. These conditions were somehow difficult to reproduce, but an improved procedure was recently reported and allowed the isolation of the pure per-*O*-benzoyl-galactofuranose with a 75% yield ($\alpha/\beta = 55:45$) [61]. This result with galactose can be explained by the fact that in pyridine at 65 °C, the tautomeric equilibrium of D-Gal favors the kinetic furanoid forms. Benzoylation then occurs first on the more reactive hydroxyl groups, 1-OH and 6-OH, thus leading rapidly to the 1,6-dibenzoate intermediate in which the five-membered ring is fixed. This intermediate presents a 5-OH outside the ring more reactive than the axial 4-OH of the corresponding pyranose. Further benzoylation on this position was thus favored. The prompt reactivity of the benzoylating agent compared to acetic anhydride enhances this kinetic effect. Interestingly, when the bulky *N*-pivaloyl imidazole was used, 1,2,3,6-tetra-*O*-pivaloyl-α-D-galactofuranose, as the major product (55%), and a mixture of α,β-anomers of per-*O*-pivaloyl-D-galactofuranose, as the minor product (19%), were isolated (Scheme 12) [62]. The reaction proceeds best at 60 °C for 24 h in DMF with 5 eq. of the acylating agent. Incomplete acylation is probably due to steric hindrance generated by the pivaloyl group on 6-OH, thus demonstrating that the acylation is also reactive dependent.

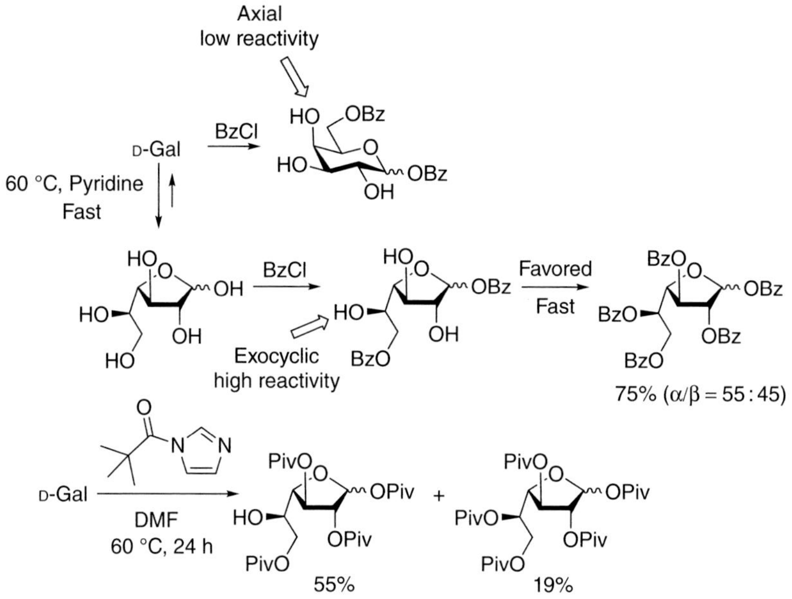

Scheme 12.12 Obtention of per-*O*-acyl galactofuranoses.

The same trend was observed with D-fructose. It is particularly difficult to prepare one particular form of the peracetyl-D-fructose in a single step. A complex mixture of the five isomers is generally obtained. However, if D-fructose is previously dissolved in pyridine at 60 °C before adding benzoyl chloride, the resulting products turn out to be the tetrabenzoyl fructofuranose (60%) and the pentabenzoyl one with minor contamination by the acyclic acetal (Scheme 12.13) [63]. In addition, using the bulky pivaloyl chloride at −10 °C afforded only the tetrapivaloyl fructofuranose in 70% yield and a β-selectivity. At such low temperature, almost all the D-fructose is in β-pyranoid form and the first hydroxyl to react would be the anomeric hydroxyl. Nevertheless, there will be a competition for pivaloylation between the 5-OH of the 1-*O*-pivaloyl fructopyranose (axial) and the 6-OH of the furanose one (primary). The second one would be the more reactive, thus driving slowly the equilibrium toward the tetra-*O*-pivaloyl β-fructofuranose. The β-anomer is greatly favored because of obvious steric reasons.

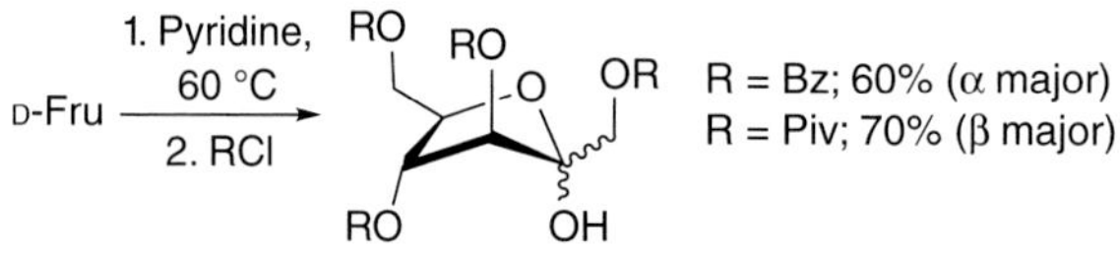

Scheme 12.13 Conditions of acylation of fructose that favor the furanose forms.

Noteworthy, the extension of the methodology to other sugars, pentoses or hexoses, failed. For example, when D-arabinose was acylated with benzoyl chloride in pyridine, a mixture of isomers was formed with arabinofuranoses as the major products (ratio furanose/pyranose 57 : 43) [64]. Interestingly, Furneaux et al. managed to selectively obtain the per-*O*-acetyl furanose form of D-glucose as a mixture of isomers ($\alpha/\beta = 1:1.8$) by first reaction of glucose with boric acid followed by treatment with acetic anhydride and a catalytic amount of sulfuric acid (Scheme 12.14) [65]. The trace amount of the pyranose counterpart (6%) and the acyclic derivative (1%) were formed following this protocol. To explain this selectivity, the authors hypothesized that complexation between D-glucose and boric acid in acetic acid leads mostly to the 1,2 : 3,5 diborate intermediate (see Chapter 7) that could be smoothly interconverted into the per-*O*-acetyl derivatives without ring extension. As the borate bridge on C-1/C-2 is certainly the most reactive toward transesterification, it must first interconvert, thus locking the glucose into the desired five-membered ring. If the reaction proceeds in two steps, first borate complex formation and then acetylation by acetic anhydride in pyridine, the anomeric ratio reached 1 : 52 for the β-anomer. The procedure was successfully applied to D- and L-xylose and L-idose to give the corresponding per-*O*-acetyl furanoses in 90% and 87% yields, respectively. However, partial epimerization at C-2 occurs with D-mannose, D-allose, D-ribose, and D-lyxose.

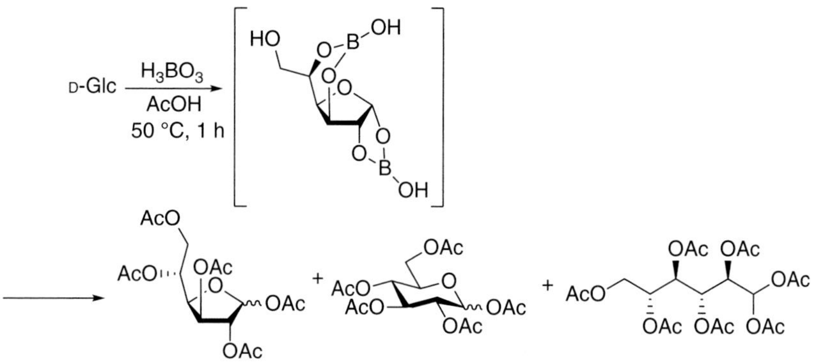

Ac_2O/H_2SO_4	93% (α:β 1 : 1.8)	6%	1%
Ac_2O/Pyridine	92% (α:β 1 : 52)	6%	1%

Scheme 12.14 Acetylation of glucose in the presence of boric acid leads to per-*O*-acetyl glucofuranose.

12.2.4 Direct Silylation

There are only a few examples in the literature of direct silylation of free oses. Nevertheless, it is possible to selectively protect the primary alcohol of D-arabinose by the action of *t*-butyldiphenylsilylchloride (TBDPSCl) in pyridine. Similarly, 3,5-*O*-tetraisopropyldisiloxane D-arabinofuranose could be obtained from arabinose by the action of dichlorotetraisopropyldisiloxane (Scheme 12.15) [66]. Both reactions proceeded with moderate yield.

Per-*O*-silylation of D-arabinose was performed by Besra and coworkers in the mid-1990s by action of *t*-butyldimethylsilylchloride (TBDMSCl) in DMF in the

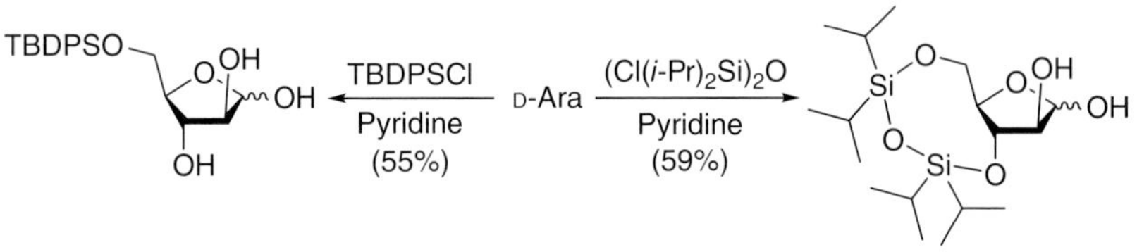

Scheme 12.15 Selective silylation of D-arabinofuranose.

presence of imidazole at 70 °C [6a]. The corresponding per-*O*-TBDMS-α-D-arabinofuranose dropped out of the solution after cooling overnight. Interestingly, selective cleavage of the TBDMS group at the anomeric position was performed by the action of TFA followed by quick quenching with ammonium hydroxide to limit the hydrolysis at position 5. Later on, Marino and Baldoni extended the protocol to D-galactose [67]. They showed that the sterically hindered TBDMSCl reacts slowly with D-Gal in DMF at RT but favors the selective formation of per-*O*-TBDMS β-D-galactofuranose. This result was recently rationalized by Ferrières and coworkers who demonstrated that small silylating agent such as TMSCl or TIPSCl led to mostly per-*O*-silyl galactopyranose unlike TBDMSCl (Table 12.5) [68]. In addition, the presence of imidazole accelerated the tautomeric equilibrium of the carbohydrate in DMF and up to 30% of the furanose was generated within a minute. Mechanism of the reaction is certainly similar to the one of the per-*O*-benzoylation of galactose. Rapid silylation of O-1 and O-6 locks the sugar in the five-membered ring and then the other positions follow. 1,2-*trans* Anomer is also favored as the steric constraints are minimized. Unfortunately, this result is not transferable to other ose series but only to carbohydrates with structural mimicry with galactose. Again, vicinal 1,2-*trans* orientation is compulsory to obtain selectively the furanose form. Therefore, arabinose, fucose, methyl galacturonate, and *N*-acetylgalactosamine were efficiently transformed into per-*O*-TBDMS furanose in yields ranging from 30% to 80%.

Table 12.5 Persilylation of D-galactose by TBDMSCl leads to the sole per-*O*-TBS β-D-galactofuranose.

			Yield (%)	
Entry	**SiR$_3$**	**Time (h)**	**Furanose**	**Pyranose**
52	TMS	3	—	100
53	TIPS	72	72	10
54	TBDMS	48	48	73

12.3 Protecting Group Interconversion

12.3.1 Selective Acylation and Deacylation

12.3.1.1 Chemical Methods

Different strategies are now available for the selective protection of alcohol and subsequent cleavage of the acyl or silyl groups. Carbohydrates as polyhydroxylated scaffolds have been the object of numerous studies to draw rules regarding differences in reactivity of the different hydroxyl groups. Such studies confirmed the enhanced nucleophilicity of equatorial hydroxyl vs the axial one and the particular reactivity of the anomeric hemiacetal, often considered as the more reactive one (see Chapter 1). Although such rules are well exemplified concerning hexopyranoside, they could, however, not be always applied when the sugar is constrained as five-membered ring. In this case, steric and polar factors have to be taken into account to explain the enhanced reactivity of a particular alcohol.

The primary hydroxyl group of aryl 1-thio-pentofuranoside (5-OH) or hexofuranoside (6-OH) can be selectively acylated at low temperature with 1 eq. of the acylating agent and catalytic amount of DMAP [69]. The process can be improved using dibutyltinoxide as inductor of regioselectivity [70]. First, a tin complex was formed between 5-OH and 3-OH of the aryl-1-thio-D-arabinofuranoside. The nucleophilicity of the primary hydroxyl group is, therefore, exacerbated, and only this position will react with acetic anhydride (Scheme 12.16).

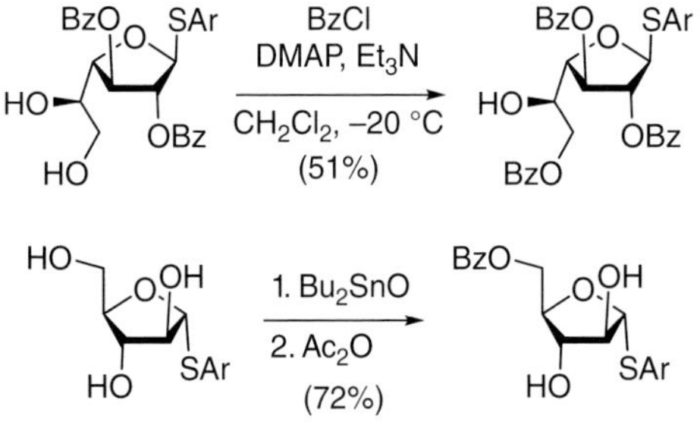

Scheme 12.16 Examples of selective acylation of primary alcohol of furanoside.

de Lederkremer and coworkers described the selective pivaloylation of D-gulono-1,4-lactone with acyl chloride and pyridine. It resulted in partial acylation that mainly afforded the 3-*O*-unsubstituted derivatives in 65% yield and with low contamination by the per-*O*-pivaloyl lactone [71]. Later on, benzyl 5,6-*O*-isopropylidene-α-D-galactofuranoside, thus protected at positions 5 and 6, was subjected to acylation or silylation. This predominantly gave the 3-*O*-unsubstituted product [72]. The higher nucleophilicity of position 2 was further confirmed using the β-galactofuranoside and the galactonolactone [73]. The anomeric configuration thus did not influence the regioselectivity of the protection. Nevertheless, one can hypothesize that a α-benzyl aglycon will enhance the nucleophilicity of the hydroxyl via hydrogen bonding. Also, increasing steric

Table 12.6 Regioselective acylation of C-2 position of galactofuranoside and lactone analogs.

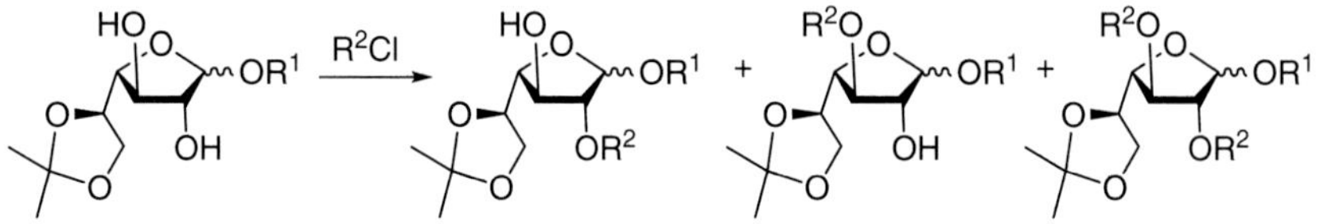

Entry	OR[1]	R^2	Yields (%)	Ratio
55	=O	Piv[a)]	79	100 : 0 : 0
56	=O	Bz[b)]	45	60 : 0 : 40
57	=O	TBDPS[c)]	83	100 : 0 : 0
58	α-OBn	Lev[d)]	38	86 : 0 : 14
59	α-OBn	TBDPS[c)]	84	100 : 0 : 0
60	β-OBCB[a)]	TBS[c)]	90	90 : 10 : 0

a) 2′-(Benzyloxycarbonyl)benzyl (BCB)
b) R^2Cl
c) R^2Cl, imidazole, DMF
d) R^2OH, DCC, DMAP.

hindrance of the acylating agent reinforced the regioselectivity of the reaction. Best results were obtained with pivaloyl chloride or *tert*-butyldiphenylsilyl chloride (Table 12.6).

Interestingly, chloroacetylation of allyl 3,6-di-*O*-benzyl-β-D-galactofuranoside produced either the 5-*O*-acetylated compound when pyridine was used as a nucleophilic base, or a majority of 2-*O*-acetylated product in the presence of triethylamine (Scheme 12.17) [74]. As the base is involved in the activation of the acylating agent and the deprotonation of the oxonium intermediate, its basicity and steric hindrance could greatly influence the regioselectivity. Regioselective C-5 esterification was also observed following the reaction of 3-(trifluoromethyl) benzoyl chloride on methyl D-galactofuranuronate in the presence of pyridine [75]. Here, the vicinal carboxyl group enhanced the reactivity of 5-OH through a hydrogen bond.

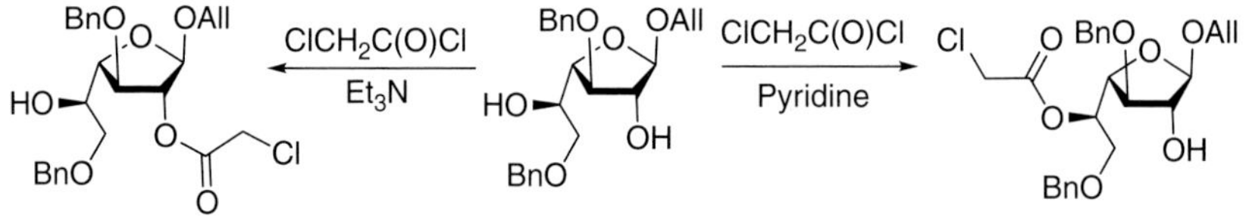

Scheme 12.17 Regioselective acylation at positions 2 and 5 according to the base.

12.3.1.2 Chemoenzymatic Methods

Alternative methodologies to perform selective acylation of free oses or alkyl and aryl osides have arisen in the 1990s based on the use of lipases or esterases. Biocatalytic processes in addition to their high performance require less toxic solvents, are compatible with the presence of water, and need less energy to provide the product(s). This is why lipase- and esterase-catalyzed acylations or deacylations are now commonly used in the industry. Selective acylation of glycopyranosides by lipases were obviously extensively described in the literature (see Chapter 2). Nevertheless, their furanose analogs have been the object of fewer reports. No literature precedent has yet been published on selective acylation of hexofuranosides catalyzed by enzymes. However, starting from free D-ribose or 2-deoxy-D-ribose, regioselective propionylation catalyzed by lipase from *Candida antarctica* Novozym 435 in the presence of propionic anhydride in THF provided 5-*O*-propionyl-ribose and deoxyribose with 95% yield [76]. This regioselective acylation was also performed with *Candida antarctica* lipase B immobilized on polyacrylate (CAL-B) and with butyric anhydride as a acylating agent [77]. No acylation of the anomeric hemiacetal occurred. The regioselectivity of the acylation for the primary 5-OH was also reported with methyl ribo-, arabino-, and xylofuranosides as well as with *p*-nitrophenyl α-L-arabinofuranoside using porcine pancreatic lipase (PPL) and the labile 2,2,2-trifluoroethyl or vinyl acetates (Scheme 12.18) [78]. The nature of the anomeric position does not influence the yield and regioselectivity of the reaction. The same methodology was further applied on a mixture of 1-(D-xylofuranosyl)thymine and 1-(D-xylopyranosyl)thymine [79]. The sole furanosyl entity was acylated on 5′-OH position upon action of CAL-B with butanoic anhydride as the best acylating agent. As no primary hydroxyl group exists on the pyranosyl nucleoside, no biocatalyzed acylation occurs. This is a convenient strategy for the separation of both furanosyl and pyranosyl nucleosides. Surprisingly, no lipase or esterase was able to selectively acylate positions 2 and 3 of pentosides.

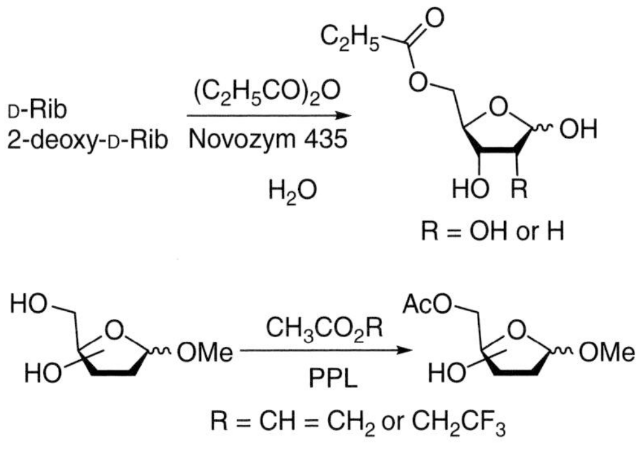

Scheme 12.18 Regioselective acylation of position 5 of pentoses.

The reverse reaction, meaning the selective hydrolysis of one acetate upon the other, could be conducted thanks to lipases. Starting from per-*O*-acetyl ribofuranose and xylofuranose, the action of lipase from *Aspergillus niger* selectively removed the ester at the anomeric position in 63% and 50% yields, respectively (Table 12.7, entries 61 and 62) [78]. When using *Candida rugosa* lipase, selective removal of the acetate at the 5-OH position occurred starting from

Table 12.7 Regioselective deacetylation of different pentoses and methyl pentosides.

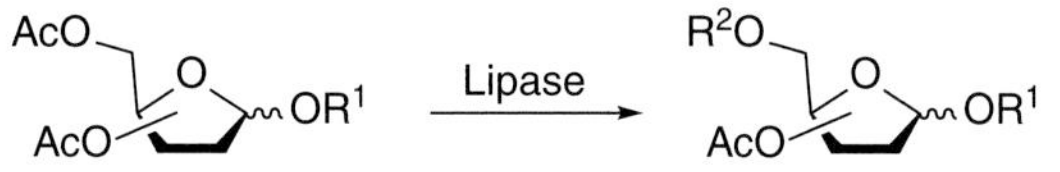

				Final product		
Entry	**Substrate**	**R^1**	**Lipase[a)]**	**R^1**	**R^2**	**Yield (%)**
61	Per-*O*-Ac-β-D-Rib*f*	Ac	ANL	H	Ac	63
62	Per-*O*-Ac-D-Xyl*f*	Ac	ANL	H	Ac	50
63	Per-*O*-Ac-α,β-D-Rib*f*	Ac	CRL	Ac	H	80
64	Per-*O*-Ac-α-D-Rib*f*	Ac	Lipozyme	Ac	H	92
65	Per-*O*-Ac-β-D-Rib*f*	Ac	Lipozyme	Ac	Ac	—
66	Methyl per-*O*-Ac-α-D-Rib*f*	CH_3	CAL-B	CH_3	Ac	65
67	Methyl per-*O*-Ac-α-D-Rib*f*	CH_3	CAL-B	Methyl β-D-ribofuranoside		100

a) ANL, *Aspergillus niger* lipase; CRL, *Candida rugosa* lipase; Lipozyme, *Thermomyces lanuginosus* lipase; CAL-B, *Candida antarctica* lipase.

1,2,3,5-tetra-*O*-acetyl-α-D-ribofuranose (Table 12.7, entry 63) [80]. Interestingly, *Thermomyces lanuginosus* lipase immobilized on silica (Lipozyme® TL IM) catalyzed the hydrolysis of per-*O*-acyl α-D-ribofuranoside selectively on position C-5 but not of the β-anomer analogs (Table 12.7, entries 64 and 65) [77]. Such behavior allowed the easy separation of both anomers. This finding was extended to the methyl per-*O*-acetyl D-ribofuranoside using CAL-B as a lipase [81]. Again, CAL-B catalyzed the selective alcoholysis at position 5-OH of the α-anomer with 81% yield (Table 12.7, entry 66). As for the β-anomer, the deacetylaton was slow and lacked regioselectivity. Upon time, the sole methyl β-D-ribofuranoside was formed quantitatively (Table 12.7, entry 67). Among all pentoses, only anomers of ribose showed such differential behavior. The alkyl and aryl per-*O*-acetyl α,β-pentofuranosides from D-ribose and D/L-arabinose were also subjected to selective removal of one acetyl in the presence of *Caulerpa cylindracea*, *C. rugosa*, *Aspergillus niger*, or porcine pancreatic lipases [78]. For all carbohydrates, only the 5-OH position was hydrolyzed in excellent yields. As for xylose, the hydrolysis of methyl per-*O*-acetyl α,β-xylofuranoside by these lipases provided two products, the expected 2,3-di-*O*-acetyl-α-D-xylofuranoside and the 2,5-di-*O*-acetyl-β-D-xylofuranoside.

In the case of the ketose fructose, two primary hydroxyls exist, the 1-OH and the 6-OH. They can be selectively acetylated using CAL-B to give quantitatively 1,6-di-*O*-acetyl-D-fructofuranose starting from D-fructose (Scheme 12.19) [82]. Interestingly, one of the remaining hydroxyl, namely the 4-OH, was regioselectively acetylated in the presence of lipase of *C. rugosa*. Deacetylation of the corresponding per-*O*-acetyl α,β-fructose catalyzed by *C. rugosa* lipase gave only the 2,3,4,6-tetra-*O*-acetyl-β-D-fructofuranose and the starting per-*O*-acetyl α-fructose. Again, specific recognition of the C-1β position is outlined. When

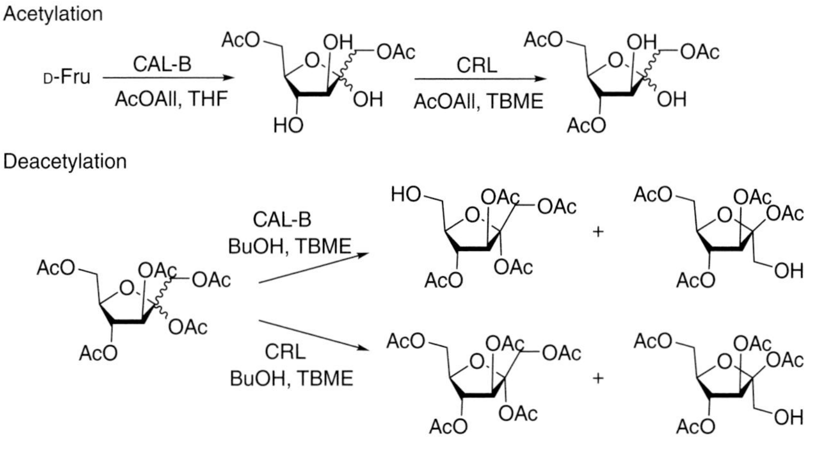

Scheme 12.19 Regioselective acetylation and deacetylation of D-fructose.

CAL-B was used, differential recognition of the anomers also occurred. Primary C-6 position of the α-anomer was cleaved on one hand and primary C-1 position of the β-anomer was hydrolyzed on the other hand. Such results highlight the critical role of the configuration of the anomeric center in the regioselectivity of alcoholysis catalyzed by lipases.

12.3.2 The Silyl Groups

The silane group is another popular protecting group in carbohydrate chemistry. It can be selectively introduced on the primary position of alkyl and aryl furanosides by action of the corresponding silyl chloride in the presence of imidazole. More importantly, simultaneous protection of positions 3 and 5 of aryl or alkyl pentoside was achieved thanks to the tetraisopropyldisiloxane-1,3-diyl (TIPDS) group. This group was developed by Markiewic and Wiewiorowski [83] and is considered as the most popular and useful group in the field of nucleoside chemistry. It is admitted that a first silylation on the primary 5-hydroxyl occurred followed by the formation of a 8-membered ring with the 3-OH position. Such group is removed under strong acidic conditions or by action of fluorine anion (Scheme 12.20). In the case of nucleoside, migration of the TIPDS from position 3′,5′ to position 2′,3′ can be catalyzed by trimethylsilyl trifluoromethanesulfonate [84].

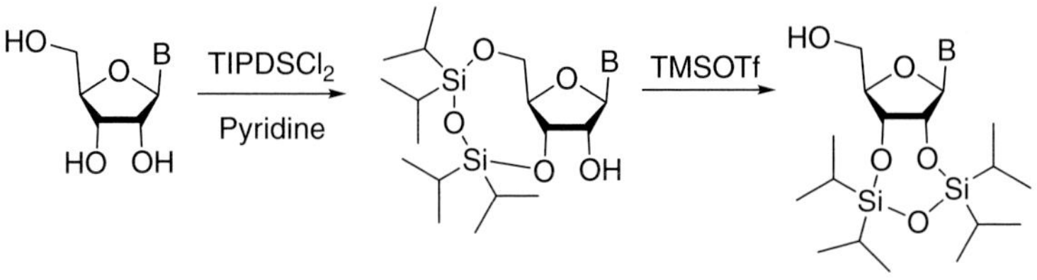

Scheme 12.20 The TIPS protecting group for nucleoside synthesis.

More constrained six-membered ring 3,5-*O*-di-*tert*-butylsilane protecting group is introduced by the action of di-*tert*-butylsilane bis-(trifluoromethanesulfonate) in the presence of 2,6-lutidine. Interestingly, starting from phenyl 1-thioarabinofuranoside, the resulting 3,5-*O*-di-*tert*-butylsilylene (DTBS) arabinofuranoside adopt a E_3 conformation where the α-face is hindered (steric hindrance with H-2) and the β-face rather exposed [85]. It allowed the synthesis of 1,2-*cis* glycosidic bond between the arabinofuranoside and various acceptors (Scheme 12.21). Access to such 1,2-*cis* configuration is highly sought as such configuration is found in the exopolysaccharide of mycobacteria, for example.

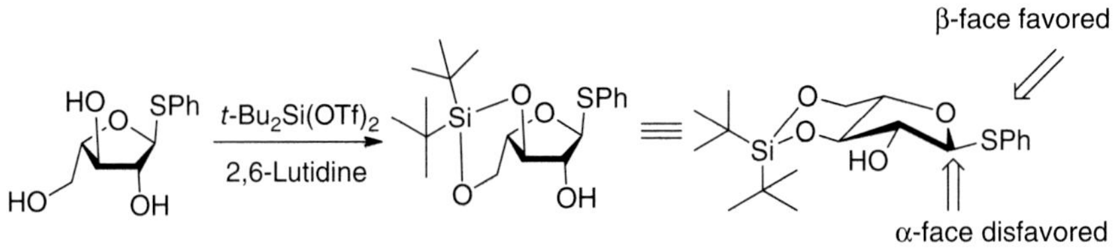

Scheme 12.21 Influence of the DTBS group on the anomeric position.

Liang and collaborators have studied the difference of reactivity between the three silyl protecting groups, the TBS group, the 3,5-TIPDS one, and the 3,5-DTBS one [86]. Although per-*O*-silylation of arabinothiofuranoside greatly increased the reactivity of the resulting donor, introduction of DTBS group generated a donor with low reactivity. The constraint induced by the five-membered silylene group increased the energy gap between the starting donor and the oxocarbenium transition state. The 3,5-TIPDS arabinofuranoside present a reactivity between these two extreme cases (Scheme 12.22).

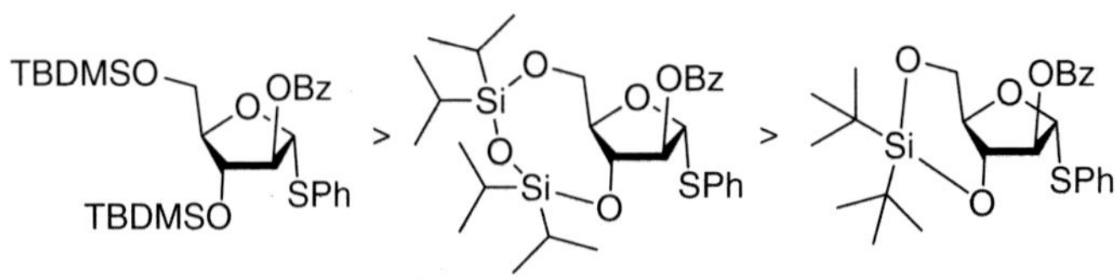

Scheme 12.22 Influence of the silyl protecting group on the reactivity of the anomeric position.

12.4 Multistep Synthesis of Some Furanosyl-containing Glycosides and Conjugates

As for standard glycosidic synthesis, the elaboration of a strategy generally relies on an appropriate use of protecting groups. Because many efforts have been done for the multistep preparation of furanosides, this chapter will describe only some relevant examples.

12.4.1 The Galactono-1,4-lactone as a Key Substrate

The highly rich chemistry in this domain comes from Marino's team [87] because galactono-1,4-lactone was used as a key building block for both acceptor and donor derivatives for the synthesis of rare galactofuranosyl-containing glycosides. For instance, it was established that, after 5,6-protection, the free OH-2 could be more nucleophilic than OH-3, which opens the way for further selective protections or glycosylations (Scheme 12.23). The second most important point of this chemistry is the highly efficient reduction of the lactone by diisoamylborane (DSB) [88]. The hemiacetal was then reactive for subsequent activation, so that the lactonic derivatives were converted into furanosyl donors.

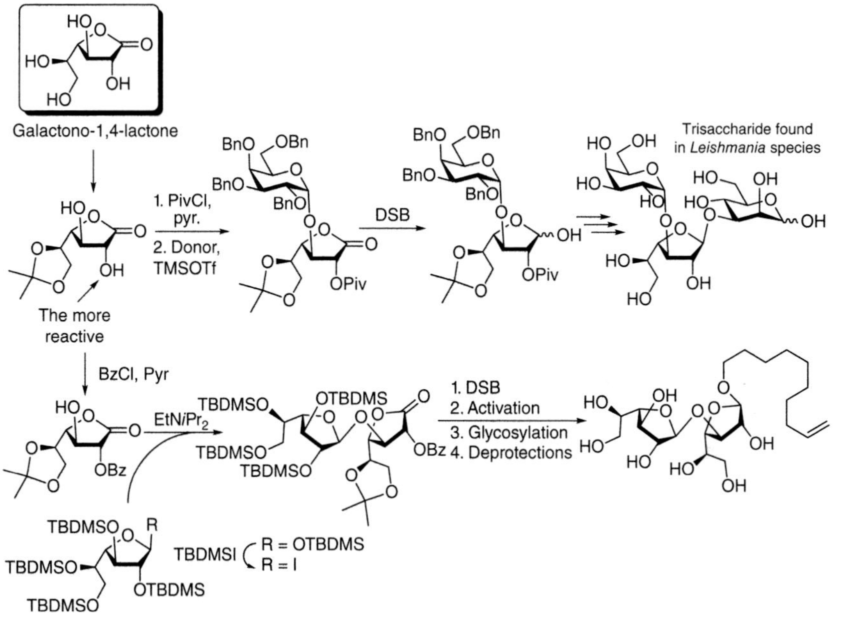

Scheme 12.23 The galactono-1,4-lactone as a key intermediate in the synthesis of galactofuranosyl-containing saccharides and conjugates.

Sinay and coworkers have synthesized silylated lactones with a free OH-5 [89]. Condensation with *C*-nucleophilic species afforded lactol intermediates that were converted into *exo*-glycals, which could be cyclized under acidic conditions before yielding nonhydrolyzable analog of UDP-Gal*f* with a constrained conformation (Scheme 12.24).

12.4.2 Synthesis of 1,2-*cis*-Furanosides

12.4.2.1 Intramolecular Aglycon Delivery (IAD)

The synthesis of α-D-fucosyl-containing disaccharide found in *Eubacterium saburreum* started from 1-*O*-alkylation of galactose as described above (Scheme 12.25) [90]. After tritylation of the primary position, regioselective

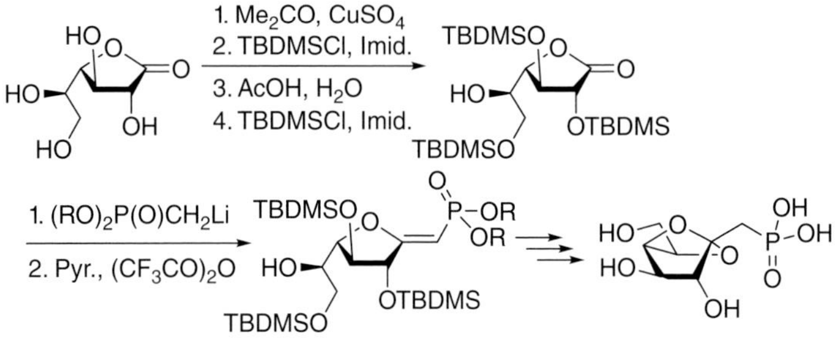

Scheme 12.24 Synthesis of analogs of UDP-Gal*f*.

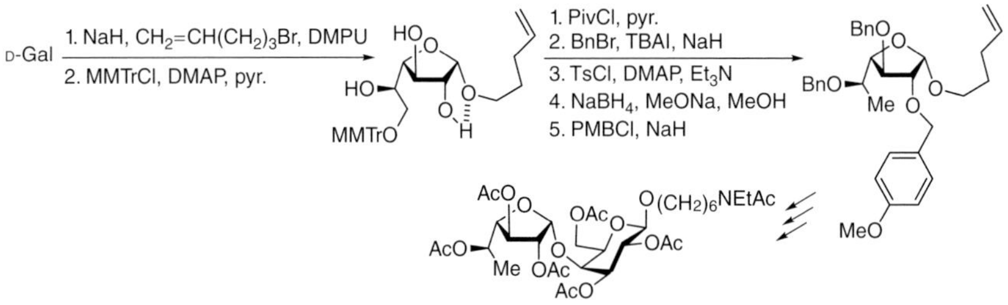

Scheme 12.25 Synthesis of α-D-fucofuranosides.

pivaloylation was efficiently performed thanks to intramolecular OH-2–O-1 hydrogen bonding. Benzylation was followed by conversion of pivaloyl group to *p*-methoxybenzyl one. The methyl group at C-6 was obtained according to the sequence detritylation–tosylation–reduction. This afforded the intermediate suitable for tethering of a galactopyranoside and subsequent IAD [91] promoted by *N*-iodosuccinimide (NIS).

12.4.2.2 Synthesis of Agelagalastatin

Agelagalastatin is a rare glycosphingolipid isolated from the marine sponge *Agelas* species. The glycosyl part is characterized by the presence of two galactofuranosyl residues, one with a α-configuration and the other with a β-anomeric center. The main features of the synthesis published by Kim and coworkers relied on (i) protection–deprotection sequences on a furanosidic scaffold using standard groups (acetonide, silyl, benzyl, and pivaloyl) and (ii) nonactivated 2′-(benzyloxycarbonyl)benzyl (BCB) aglycon, which can be advantageously converted into 2′-carboxybenzyl (CB) glycoside as a glycosyl donor (Scheme 12.26) [73b]. For this synthesis, it is noteworthy that silylation of the di-isopropylidene intermediate regioselectively occurred on position 3, contrary to that observed in lactonic series.

12.4.2.3 Epoxide Opening

At the beginning of 2000s, Lowary's team has proposed to use epoxides as a potential protection of hydroxyl at C-2 and C-3 in several furanosides (lyxose, arabinose, galactose, etc.) [92]. The corresponding 2,3-anhydro-thioglycosides

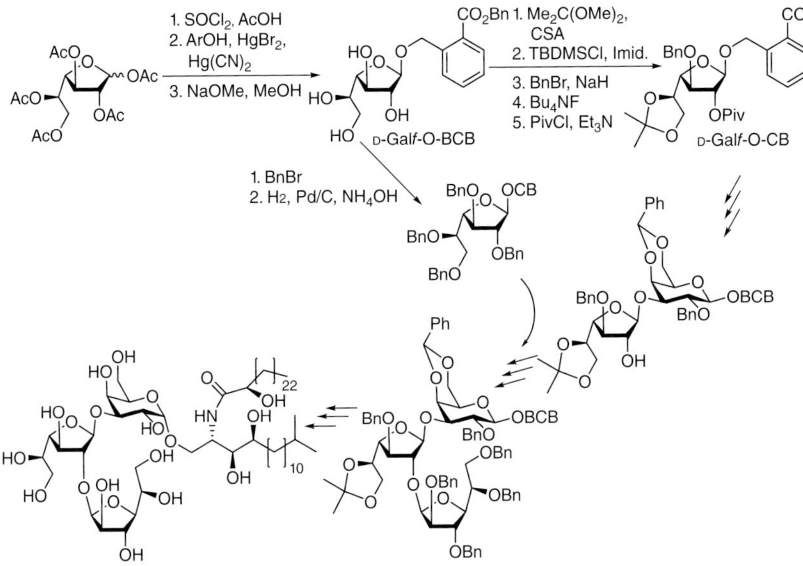

Scheme 12.26 Synthesis of agelagalastatin.

were easily obtained from free or partially protected glycosides under Mitsunobu conditions and were oxidized into sulfoxides (Scheme 12.27). Activation of the sulfoxides by triflic anhydride afforded the more stable anomeric triflate, *trans* to the epoxide [92b]. A S_N2-type attack by an acceptor (ROH) thus yielded 1,2-*cis*

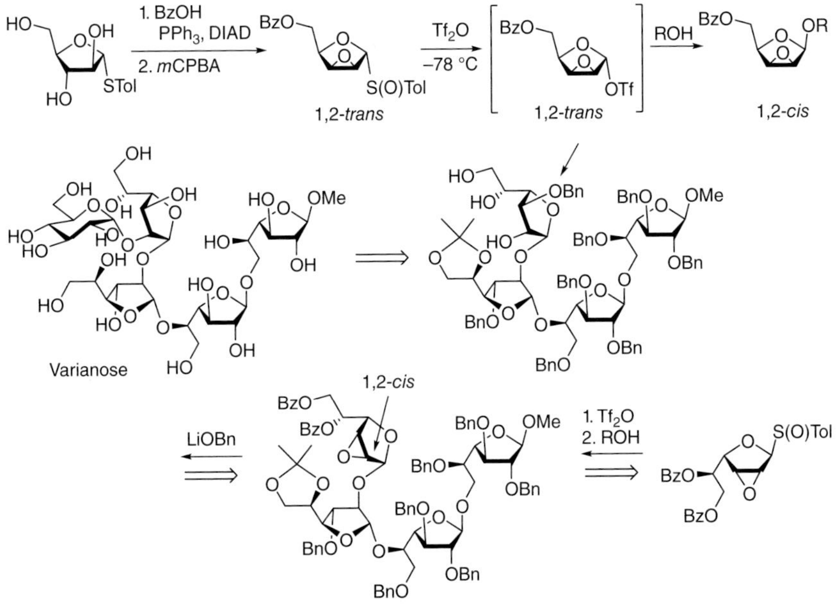

Scheme 12.27 2,3-Epoxides as protecting group.

furanosides with excellent diastereoselectivity. Final opening of the epoxide by benzyloxy anion generally took place on the less hindered face, so that 1,2-*cis* furanosides were obtained in very good yields and excellent selectivities. The later was interestingly improved by using (−)-spartein. This two-step approach allowed the authors to synthesize from simple furanosides up to complex oligofuranosides [92c].

12.5 The Striking Ring Contraction Strategy

As furanosidic derivatives are generally considered as less stable than their pyranosidic counterparts, it is therefore not very likely to conceive an easy way to prepare the formers starting from larger six-membered cycles. Nevertheless, we found a few examples of such a strategy through the scientific literature.

12.5.1 An Enzymatic Pathway: The Unprecedented Mechanism from the Mutase

During the last two decades, numerous efforts have aimed at elucidating the biological mechanisms involved in the formation of the membrane cell wall of *M. tuberculosis*. There is no need to underline the requirement for fighting this parasite responsible for millions of infections each year worldwide. This wall is highly complex and notably relies on a hydrophobic barrier consisting in mycolic acids, and a noteworthy saccharidic polymer based on a arabinogalactan in which D-arabinosyl (and not L-arabinose as in plants) and D-galactosyl residues are present in their furanose form [5a]. The glycosidic elongation therefore involves glycofuranosyl transferases and thus activated donors. Although decaprenylphosphoryl arabinofuranose is the result of a two-step epimerization of the ribose isomer [93], uridine diphosphogalactofuranose (UDP-Galf) comes from the ring contraction of the native UDP-Gal*p* catalyzed by the enzyme UDP galactopyranose mutase (Scheme 12.28) [94], One of the obstacles for understanding the currently accepted mechanism has been the identification of the role of the cofactor NADPH. The latter plays an important role in this interconversion because a

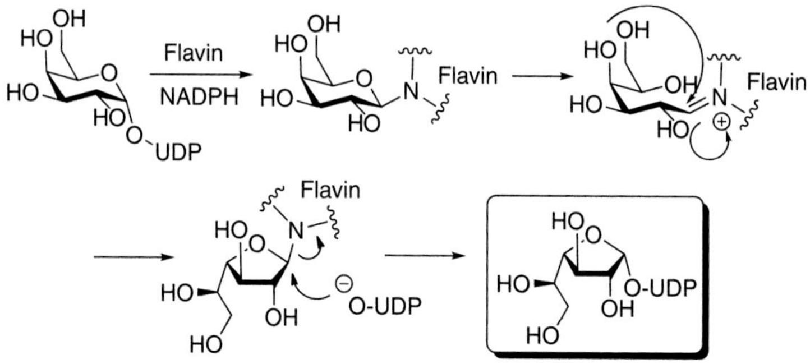

Scheme 12.28 Isomerization of UPD-Gal*p* into UDP-Gal*f* assisted by a mutase in the presence of NADPH.

covalent flavin–galactose conjugate is formed before an iminium intermediate. These discoveries related to the mutase of *M. tuberculosis* allowed further elucidations of other mutases, for instance, in *Trypanosoma cruzi* [95], the causative agent of Chagas disease, or in *Aspergillus fumigatus* [96]. It is interesting to note that a similar bifunctional mutase was also characterized from *Campylobacter jejuni*, which is able to perform the transformation of UDP-Gal*p*NAc into UDP-Gal*f*NAc [97]. Considering the high reactivity of the resulting UDP-Gal*f*, mutases were not used on a preparative scale.

12.5.2 Chemical Approaches

12.5.2.1 From 1,4-Anhydrogalactopyranose

1,4-Anhydrogalactopyranose derivatives were regioselectively opened by O- or C-nucleophiles under acidic conditions. On the assumption that one of the lone-pair electrons of O-4 is antiperiplanar to the C-1–O-4 σ bond, the C-1–O-5 linkage is weaker than the C-1–O-4 one. This resulted in desired ring contraction giving either *O*-galactofuranoside with the assistance of the Brönsted camphorsulfonic acid or *C*-galactofuranoside using allyltrimethylsilylane and TMSOTf as a Lewis catalyst (Scheme 12.29) [89, 98]. A bicyclic intermediate structurally close to 1,4-anhydrogalactopyranose was proposed to rationalize the pyranose–furanose rearrangement observed during glycosylation of 3-deoxy-D-*glycero*-D-*galacto*-2-nonulosonic acid (KDN) thanks to the participation of O-5 (Scheme 12.30) [99].

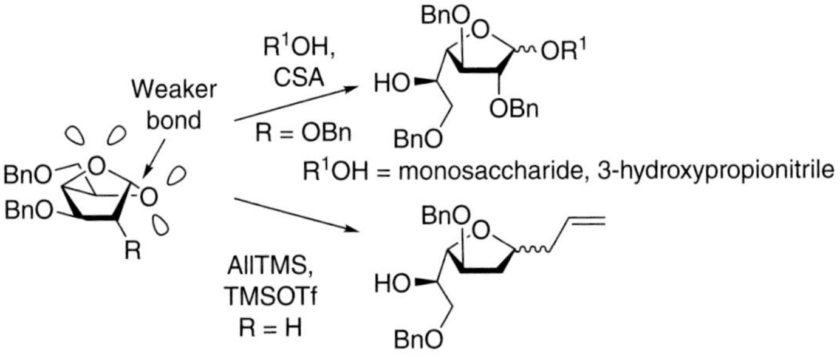

Scheme 12.29 Selective opening of 1,4-anhydrogalactopyranose.

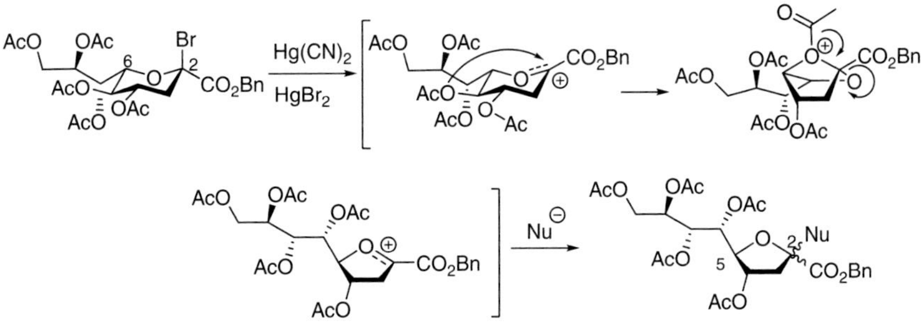

Scheme 12.30 Pyranose–furanose rearrangement from KDN donor.

12.5.2.2 The Role of Sulfate Groups

Significant progresses were performed thanks to the method proposed by Nifantiev and coworkers [100]. They indeed observed that sulfated galactopyranosides smoothly rearranged at room temperature. Contrary to standard procedures of sulfation, the authors have added chlorosulfonic acid to the SO_3·Pyridine complex (Scheme 12.31). The resulting acid was thus supposed to interact with the endocyclic oxygen atom and the axial O-4 sulfate group to initiate ring opening also assisted by the anchimeric assistance of the 2-*O*-sulfate group. Subsequent O-4–C-1 cyclization occurred leading to sulfate group migration from O-4 to O-5, opening of the 1,2-cyclic sulfate, and so release of the targeted furanoside with retention of configuration. This finding allowed the preparation of key building block with a free OH-3 suitable for the synthesis of disaccharides.

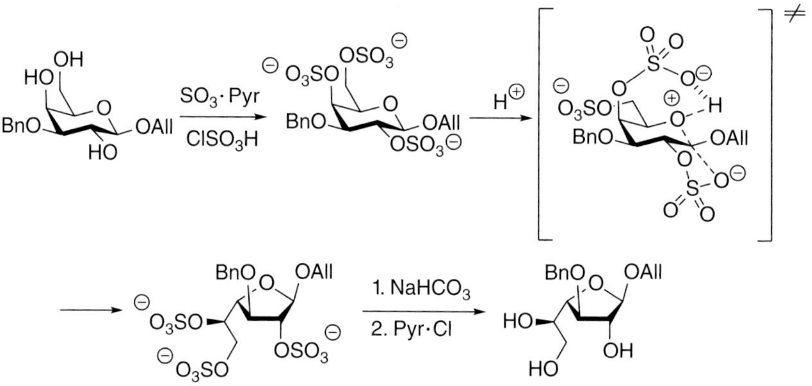

Scheme 12.31 The pyranose-to-furanose rearrangement.

12.6 Conclusion Strategy for Synthesizing 4-Amino-4-deoxy and 4-Deoxy-4-thio-aldose Derivatives

Many efforts have been done during the last three decades to improve the chemical synthesis of furanosyl-containing glycoconjugates and oligosaccharides. This chapter does not claim to reference all data in this area. Thus, 4-amino-4-deoxy-, 4-deoxy-4-thio-, and 4-deoxy-4-seleno-furanose derivatives were also prepared to be further evaluated as inhibitors for instance. Some bibliographic references are therefore available [101] and underline the richness of the chemistry of furanosides.

Abbreviations

Ac acetyl
All allyl

Ar	aryl
BCB	2′-(benzyloxycarbonyl)benzyl
BDMS	bromodimethylsulfonium bromide
Bn	benzyl
Bu	butyl
Bz	benzoyl
CAL-B	*Candida antarctica* lipase-B
CAN	cerium ammonium nitrate
CB	2′-carboxybenzyl
CRL	*Candida rugosa* lipase
CSA	camphorsulfonic acid
Cp	cyclopentenyl
DBDH	1,3-dibromo-5,5-dimethylhydantoin
DCC	dicyclohexylcarbodiimide
DIAD	diisopropyl azodicarboxylate
DMAP	*N,N*-dimethyl-aminopyridine
DMF	*N,N*-dimethylformamide
DMP	2,2-dimethoxypropane
DMPU	dimethylpropylurea
DTBS	di-*tert*-butylsilylene
Et	ethyl
h	hour
IAD	intramolecular aglycon delivery
*i*Pr	*i*-propyl
*m*CPBA	*m*-chloroperbenzoic acid
Me	methyl
min	minute
MMTr	monomethoxytrityl
MP	2-methoxypropylene
NADP	nicotinamide adenine dinucleotide phosphate
Piv	pivaloyl
PMB	*p*-methoxybenzyl
PPL	pig pancreatic lipase
PPTS	pyridinium *p*-toluenesulfonate
Pyr	pyridine
TFA	trifluoroacetic acid
*p*TSA	*p*-toluenesulfonic acid
RT	room temperature
TBATB	tetrabutylammonium tribromide
TBDMS	*t*-butyldimethylsilyl
TBDPS	*t*-butyldiphenylsilyl
TIPDS	tetraisopropyldisiloxane-1,3-diyl
TIPS	triisopropylsilyl
Tf	trifluoromethanesulfonyl
THF	tetrahydrofuran
Tol	toluidyl
UDP	uridine diphosphate

References

1 (a) Kunz, H. (2002). *Angew. Chem. Int. Ed.* 41: 4439–4451.(b) Lichtenthaler, F.W. (2002). *Eur. J. Org. Chem.* 1095–4122.

2 (a) Jung, K.-H., Mueller, M., and Schmidt, R.R. (2000). *Chem. Rev.* 100: 4423–4442. (b) Pachamuthu, K. and Schmidt, R.R. (2006). *Chem. Rev.* 106: 160–187. (c) Zhu, X. and Schmidt, R.R. (2009). *Angew. Chem. Int. Ed.* 48: 1900–1934. (d) Bennett, C.S. (2014). *Org. Biomol. Chem.* 12: 1686–1698. (e) Seeberger, P.H. (2015). *Acc. Chem. Res.* 48: 1450–1463.

3 (a) Peltier, P., Euzen, R., Daniellou, R. et al. (2008). *Carbohydr. Res.* 343: 1897–1923. (b) Richards, M.R. and Lowary, T.L. (2009). *ChemBioChem* 10: 1920–1938.

4 de Lederkremer, R.M. and Colli, W. (1995). *Glycobiology* 5: 547–552.

5 (a) Besra, G.S., Khoo, K.-H., McNeil, M.R. et al. (1995). *Biochemistry* 34: 4257–4266. (b) Imamura, A. and Lowary, T. (2011). *Trends Glycosci. Glycotechnol.* 23: 134–152. (c) Chlubnova, I., Legentil, L., Dureau, R. et al. (2012). *Carbohydr. Res.* 356: 44–61.

6 (a) Lee, R.E., Mikusova, K., Brennan, P.J., and Besra, G.S. (1995). *J. Am. Chem. Soc.* 117: 11829–11832. (b) Chatterjee, D. (1997). *Curr. Opin. Chem. Biol.* 1: 579–588. (c) Barlow, J.N., Girvin, M.E., and Blanchard, J.S. (1999). *J. Am. Chem. Soc.* 121: 6968–6969. (d) Lu, J. and Fraser-Reid, B. (2005). *Chem. Commun.* 862–864. (e) Rose, N.L., Completo, G.C., Lin, S.-J. et al. (2006). *J. Am. Chem. Soc.* 128: 6721–6729. (f) Berg, S., Kaur, D., Jackson, M., and Brennan, P.J. (2007). *Glycobiology* 17: 35R–56R. (g) Yuan, Y., Bleile, D.W., Wen, X. et al. (2008). *J. Am. Chem. Soc.* 130: 3157–3168. (h) Tam, P.-H. and Lowary, T.L. (2009). *Curr. Opin. Chem. Biol.* 13: 618–625. (i) Levengood, M.R., Splain, R.A., and Kiessling, L.L. (2011). *J. Am. Chem. Soc.* 133: 12758–12766. (j) Xia, L. and Lowary, T.L. (2013). *Biopolymers* 99: 697–712.

7 (a) Capon, B. (1969). *Chem. Rev.* 69: 407–498. (b) Pater, R.H., Coelho, R.A., and Mowery, D.F. Jr. (1973). *J. Org. Chem.* 38: 3272–3277. (c) Mowery, D.F. Jr. and Coelho, R.A. (1986). *Carbohydr. Res.* 155: 33–44.

8 Green, J.W. (1966). *Adv. Carbohydr. Chem.* 21: 95–142.

9 Rauter, A.P., Almeida, T., Xavier, N.M. et al. (2007). *J. Mol. Catal. A: Chem.* 275: 206–213.

10 Climent, M.J., Corma, A., Iborra, S. et al. (1999). *J. Catal.* 183: 76–82.

11 Arasappan, A. and Fraser-Reid, B. (1995). *Tetrahedron Lett.* 36: 7967–7970.

12 Allam, A., Behr, J.-B., Dupont, L. et al. (2010). *Carbohydr. Res.* 345: 731–739.

13 Richel, A., Nicks, F., Laurent, P. et al. (2012). *Green Chem. Lett. Rev.* 5: 179–186.

14 (a) Evans, M.E. and Angyal, S.J. (1972). *Carbohydr. Res.* 25: 43–48. (b) Angyal, S.J., Bethell, G.S., and Beveridge, R.J. (1979). *Carbohydr. Res.* 73: 9–18.

15 (a) Lubinueau, A. and Fischer, J.-C. (1991). *Synth. Commun.* 21: 815–818. (b) Bertho, J.-N., Ferrières, V., and Plusquellec, D. (1995). *Chem. Commun.* 1391–1393. (c) Ferrières, V., Bertho, J.-N., and Plusquellec, D. (1995). *Tetrahedron Lett.* 36: 2749–2752. (d) Velty, R., Benvegnu, T., Gelin, M. et al. (1997). *Carbohydr. Res.* 299: 7–14.

16 Legentil, L., Audic, J.-L., Daniellou, R. et al. (2011). *Carbohydr. Res.* 346: 1541–1545.

17 (a) Ferrières, V., Gelin, M., Boulch, R. et al. (1998). *Carbohydr. Res.* 314: 79–83. (b) Gelin, M., Ferrières, V., and Plusquellec, D. (2000). *Eur. J. Org. Chem.* 1423–1431.

18 Wolfrom, M.L. and Thompson, A. (1963). *Methods Carbohydr. Chem.* 2: 427–430.

19 Madhusudan, S.K. and Misra, A.K. (2005). *Carbohydr. Res.* 340: 497–502.

20 Klotz, W. and Schmidt, R.R. (1993). *Liebigs Ann. Chem.* 6: 683–690.

21 Xavier, N.M., Rauter, A.P., and Queneau, Y. (2010). *Top. Curr. Chem.* 295: 19–62.

22 Ma, Z., Zhang, P., Yu, X. et al. (2015). *J. Mater. Chem. B* 3: 7366–7371.

23 Cerrada, M.L., Bordegé, V., Muñoz-Bonilla, A. et al. (2013). *Carbohydr. Polym.* 94: 755–764.

24 (a) Srivastava, V.K. and Lerner, L.M. (1979). *J. Org. Chem.* 44: 3368–3373. (b) Chapleur, Y. (1984). *J. Chem. Soc., Chem. Commun.* 449–450.

25 Greene, T.W. and Wuts, P.G.M. (2002). *Protective Groups in Organic Synthesis*, 17–245. Wiley.

26 Schmidt, O.T. (1963). *Methods Carbohydr. Chem.* 318–325.

27 Morgenlie, S. (1975). *Carbohydr. Res.* 41: 77–83.

28 Sakamoto, I., Ichimura, K., and Ohrui, H. (2000). *Biosci. Biotechnol., Biochem.* 64: 1915–1922.

29 Einhorn, C. and Luche, J.L. (1986). *Carbohydr. Res.* 155: 258–261.

30 Rajput, V.K. and Mukhopadhyay, B. (2006). *Tetrahedron Lett.* 47: 5939–5941.

31 Singh, P.P., Gharia, M.M., Dasgupta, F., and Srivastava, H.C. (1977). *Tetrahedron Lett.* 18: 439–440.

32 Tang, Y., Zhu, X., Zhu, J. et al. (2012). *Adv. Mater. Res.* 441: 408–412.

33 Kartha, K.P.R. (1986). *Tetrahedron Lett.* 27: 3415–3416.

34 Pedatella, S., Guaragna, A., D'Alonzo, D. et al. (2006). *Synthesis* 305–308.

35 Khan, A.T. and Khan, M.M. (2010). *Carbohydr. Res.* 345: 154–159.

36 Khan, A.T., Khan, M.M., and Adhikary, A. (2011). *Carbohydr. Res.* 346: 673–677.

37 Albert, R., Dax, K., Pleschko, R., and Stutz, A.E. (1985). *Carbohydr. Res.* 137: 282–290.

38 Giri, S.K., Verma, M., and Kartha, K.P.R. (2008). *J. Carbohydr. Chem.* 27: 464–478.

39 Mandal, S., Verma, P.R., Mukhopadhyay, B., and Gupta, P. (2011). *Carbohydr. Res.* 346: 2007–2010.

40 Lin, C.-C., Jan, M.-D., Weng, S.-S. et al. (2006). *Carbohydr. Res.* 341: 1948–1953.

41 Rauter, A.P., Ramoa-Ribeiro, F., Fernandes, A.C., and Figueiredo, J.A. (1995). *Tetrahedron* 51: 6529–6540.

42 Vanlaldinpuia, K. and Bez, G. (2011). *Tetrahedron Lett.* 52: 3759–3764.

43 Zhang, X. and Zhang, C. (2013). *J. Carbohydr. Chem.* 32: 249–258.

44 (a) Wang, H., Zhang, G., and Ning, J. (2003). *Carbohydr. Res.* 338: 1033–1037. (b) Wang, H. and Ning, J. (2003). *J. Org. Chem.* 68: 2521–2524. (c) Ning, J., Wang, H., and Yi, Y. (2002). *Tetrahedron Lett.* 43: 7349–7352.

45 Sakamoto, I. and Ohrui, H. (2000). *Biosci. Biotechnol. Biochem.* 64: 1974–1977.

46 Barbat, J., Gelas, J., and Horton, D. (1983). *Carbohydr. Res.* 116: 312–316.

47 Forsman, J.J. and Leino, R. (2010). *Carbohydr. Res.* 345: 1548–1554.

48 Manzo, E., Barone, G., and Parrilli, M. (2000). *Synlett* 887–889.

49 (a) Reichstein, T. and Grüssner, A. (1934). *Helv. Chim. Acta* 17: 311–328. (b) Gurjar, M.K., Yellol, G.S., and Mohapatra, D.K. (2012). *Eur. J. Org. Chem.* 1753–1758.
50 Nadtochii, M.A., Burova, L.E., Vasil'eva, I.B., and Melent'eva, T.A. (2001). *Pharm. Chem. J.* 35: 229–230.
51 Nadtochii, M.A., Burova, L.E., Vasil'eva, I.B., and Melent'eva, T.A. (2001). *Pharm. Chem. J.* 35: 282–283.
52 Swamy, N.R. and Venkateswarlu, Y. (2002). *Tetrahedron Lett.* 43: 7549–7552.
53 Sabitha, G., Reddy, G.S.K.K., Reddy, K.B. et al. (2005). *J. Mol. Catal. A: Chem.* 238: 229–232.
54 Xavier, N.M., Lucas, S.D., and Rauter, A.P. (2009). *J. Mol. Catal. A: Chem.* 305: 84–89.
55 Araújo, N., Jenkinson, S.F., Martínez, R.F. et al. (2012). *Org. Lett.* 14: 4174–4177.
56 Davis, B.G. and Fairbanks, A.J. (2002). *Carbohydr. Chem.* .
57 Brown, G.B., Davoll, J., and Lowy, B.A. (1955). *Biochem. Prep.* 4: 70–76.
58 Stevens, J.D. (2012). *Carbohydr. Res.* 347: 9–15.
59 Banwell, M.G., Edwards, A.J., Lambert, J.N. et al. (2002). *Aust. J. Chem.* 55: 95–103.
60 Gallo-Rodriguez, C., Varela, O., and de Lederkremer, R.M. (1997). *Carbohydr. Res.* 305: 163–170.
61 Cattiaux, L., Sendid, B., Collot, M. et al. (2011). *Bioorg. Med. Chem.* 19: 547–555.
62 Santoyo-González, F., Uriel, C., and Calvo-Asín, J. (1998). *Synthesis* 1787–1792.
63 Lichtenthaler, F.W., Klotz, J., and Flath, F.-J. (1995). *Liebigs Ann.* 2069–2080.
64 Gandolfi-Donadio, L., Gallo-Rodriguez, C., and de Lederkremer, R.M. (2006). *Can. J. Chem.* 84: 486–491.
65 Furneaux, R.H., Rendle, P.M., and Sims, I.M. (2000). *J. Chem. Soc., Perkin Trans. 1* 2011–2014.
66 Zhang, J., Khoo, K.-H., Wu, S.-W., and Chatterjee, D. (2007). *J. Am. Chem. Soc.* 129: 9650–9662.
67 Baldoni, L. and Marino, C. (2009). *J. Org. Chem.* 74: 1994–2003.
68 Dureau, R., Legentil, L., Daniellou, R., and Ferrieres, V. (2012). *J. Org. Chem.* 77: 1301–1307.
69 Cattiaux, L., Sendid, B., Collot, M. et al. (2011). *Bioorg. Med. Chem.* 19: 547–555.
70 Naresh, K., Bharati, B.K., Avaji, P.G. et al. (2011). *Glycobiology* 21: 1237–1254.
71 Gallo, C., Jeroncic, L.O., Varela, O., and de Lederkremer, R.M. (1993). *J. Carbohydr. Chem.* 12: 841–851.
72 Kashiwagi, G.A., Mendoza, V.M., de Lederkremer, R.M., and Gallo-Rodriguez, C. (2012). *Org. Biomol. Chem.* 10: 6322–6332.
73 (a) Mendoza, V.M., Kashiwagi, G.A., de Lederkremer, R.M., and Gallo-Rodriguez, C. (2010). *Carbohydr. Res.* 345: 385–396. b) Lee, Y.J., Lee, B.-Y., Jeon, H.B., and Kim, K.S. (2006). *Org. Lett.* 8: 3971–3974.
74 Argunov, D.A., Krylov, V.B., and Nifantiev, N.E. (2015). *Org. Biomol. Chem.* 13: 3255–3267.
75 Bordoni, A., de Lederkremer, R.M., and Marino, C. (2008). *Tetrahedron* 64: 1703–1710.
76 Prasad, A.K., Sørensen, M.D., Parmar, V.S., and Wengel, J. (1995). *Tetrahedron Lett.* 36: 6163–6166.

77 Sharma, R.K., Aggarwal, N., Arya, A. et al. (2009). *Indian J. Chem., Sect. B* 48B: 1727–1731.
78 Hennen, W.J., Sweers, H.M., Wang, Y.F., and Wong, C.H. (1988). *J. Org. Chem.* 53: 4939–4945.
79 Maity, J., Shakya, G., Singh, S.K. et al. (2008). *J. Org. Chem.* 73: 5629–5632.
80 Fernandez-Lorente, G., Palomo, J.M., Cocca, J. et al. (2003). *Tetrahedron* 59: 5705–5711.
81 Iñigo, S., Porro, M.T., Montserrat, J.M. et al. (2005). *J. Mol. Catal. B: Enzym.* 35: 70–73.
82 D'Antona, N., El-Idrissi, M., Ittobane, N., and Nicolosi, G. (2005). *Carbohydr. Res.* 340: 319–323.
83 Markiewicz, W.T. and Wiewiorowski, M. (1986). *Nucleic Acid Chemistry. Improved and New Synthetic Procedures and Techniques*. Vol. Part 3 (ed. L.B. Towsend and R.S. Tipson), 229–231. New York: Wiley.
84 Kulikova, I.V., Muradova, D.A., and Mikhailov, S.N. (2009). *Arkivoc* 158–170.
85 Zhu, X., Kawatkar, S., Rao, Y., and Boons, G.-J. (2006). *J. Am. Chem. Soc.* 128: 11948–11957.
86 Liang, X.-Y., Bin, H.-C., and Yang, J.-S. (2013). *Org. Lett.* 15: 2834–2837.
87 Marino, C. and Baldoni, L. (2014). *ChemBioChem* 15: 188–204.
88 Gandolfi-Donadio, L., Gallo-Rodriguez, C., and de Lederkremer, R.M. (2002). *J. Org. Chem.* 67: 4430–4435.
89 Caravano, A., Mengin-Lecreulx, D., Brondello, J.-M. et al. (2003). *Chem. Eur. J.* 9: 5888–5898.
90 Gelin, M., Ferrières, V., Lefeuvre, M., and Plusquellec, D. (2003). *Eur. J. Org. Chem.* 1285–1293.
91 (a) Nigudkar, S.S. and Demchenko, A.V. (2015). *Chem. Sci.* 6: 2687–2704. (b) Ishiwata, A., Lee, Y.J., and Ito, Y. (2010). *Org. Biomol. Chem.* 8: 3596–3608.
92 (a) Gadikota, R.R., Callam, C.S., and Lowary, T.L. (2001). *Org. Lett.* 3: 607–610. (b) Callam, C.S., Gadikota, R.R., Krein, D.M., and Lowary, T.L. (2003). *J. Am. Chem. Soc.* 125: 13112–13119. (c) Gadikota, R.R., Callam, C.S., Wagner, T. et al. (2003). *J. Am. Chem. Soc.* 125: 4155–4165. (d) Bai, Y. and Lowary, T.L. (2006). *J. Org. Chem.* 71: 9658–9671. (e) Bai, Y. and Lowary, T.L. (2006). *J. Org. Chem.* 71: 9672–9680. (f) Hou, D., Taha, H.A., and Lowary, T.L. (2009). *J. Am. Chem. Soc.* 131: 12937–12948.
93 Mikusová, K., Huang, H., Yagi, T. et al. (2005). *J. Bacteriol.* 18: 8020–8025.
94 Tanner, J.J., Boechi, L., Andrew McCammon, J., and Sobrado, P. (2014). *Arch. Biochem. Biophys.* 544: 128–141.
95 Oppenheimer, M., Valenciano, A.L., Kizjakina, K. et al. (2012). *PLoS One* 7: e32918.
96 Oppenheimer, M., Poulin, M.B., Lowary, T.L. et al. (2010). *Arch. Biochem. Biophys.* 502: 31–38.
97 Poulin, M.B., Nothaft, H., Hug, I. et al. (2010). *J. Biol. Chem.* 285: 493–501.
98 Kovensky, J. and Sinaÿ, P. (2000). *Eur. J. Org. Chem.* 3523–3525.
99 Sun, X.-L., Kai, T., Fujita, S. et al. (1997). *Chem. Pharm. Bull.* 45: 795–798.
100 Krylov, V.B., Argunov, D.A., Vinnitskiy, D.Z. et al. (2014). *Chem. Eur. J.* 20: 16516–16522.

101 (a) Varela, O., Cicero, D., and de Lederkremer, R.M. (1989). *J. Org. Chem.* 54: 1884–1890. (b) Desvergnes, S., Desvergnes, V., Martin, O.R. et al. (2007). *Bioorg. Med. Chem.* 15: 6443–6449. (c) Guo, L., Liu, Y., Wan, Y. et al. (2013). *Tetrahedron Lett.* 54: 2579–2582. (d) Compain, P., Chagnault, V., and Martin, O.R. (2009). *Tetrahedron: Asymmetry* 20: 672–711. (e) Randell, K.D., Johnston, B.D., Lee, E.E., and Pinto, B.M. (2000). *Tetrahedron: Asymmetry* 11: 207–222. (f) Choubdar, N., Bhat, R.G., Stubbs, K.A. et al. (2008). *Carbohydr. Res.* 343: 1766–1777. (g) Bhat, R.G., Kumar, N.S., and Pinto, B.M. (2007). *Carbohydr. Res.* 342: 1934–1942. (h) Veerapen, N., Yuan, Y., Sanders, D.A.R., and Pinto, B.M. (2004). *Carbohydr. Res.* 339: 2205–2217. (i) Nasi, R., Patrick, B.O., Sim, L. et al. (2008). *J. Org. Chem.* 73: 6172–6181. (j) Liu, H., Nasi, R., Jayakanthan, K. et al. (2007). *J. Org. Chem.* 72: 6562–6572.

13

Cyclodextrin Chemistry via Selective Protecting Group Manipulations

Juan M. Benito and José M. García Fernández

CSIC – Universidad de Sevilla, Institute for Chemical Research (IIQ), Avenida Américo Vespucio 49, E-41092 Sevilla, Spain

13.1 Introduction

First identified from bacterial digests of starch by Villiers at the end of nineteenth century and later isolated by Schardinger during the first decades of the past century, cyclodextrins (CDs) largely remained as chemical curiosities for nearly 50 years [1]. Their structure, a cyclic arrangement of α(1→4)-linked glucopyranosyl units, defines a conical frustum with a nanometric hydrophobic cavity flanked at both rims by a dense array of hydroxyl groups (Figure 13.1). The unique abilities of these *in–out* amphiphiles to host nonpolar species in their cavity, first noticed by Cramer [2], have not yet been imitated by any other natural or artificial species [3]. In fact, this feature has historically dominated the academic and industrial interest on CDs. Being water soluble and essentially biocompatible, native CDs have gained a prominent position in pharmaceutical technology labs as *off-the-shelf* tools to manipulate the pharmacokinetics and dynamics of a broad range of drugs [4], and nowadays, they are widely employed in food [5] and cosmetic industry [6] as well as for analytical purposes [7]. This is quite remarkable considering that only three CD representatives, namely α-, β-, and γ-CD, are industrially available. The large number of hydroxyl groups, the relatively constrained space toward which they are oriented by the rigid cyclo-oligosaccharide framework, and the minute differences in their reactivity exacerbate the difficulties to selectively manipulate their topology, which has probably frustrated further development of the field for over 30 years.

In spite of the recalcitrant limitations of classical carbohydrate chemistry protecting group strategies when facing CDs, the last two decades have witnessed an intense effort to revert this scene. Chemists have answered to the demand for selectively functionalized CDs by mastering molecular inclusion properties and steric constraints to control and direct reactivity [8]. As a result, the most diverse array of selectively functionalized CD scaffolds has been produced. This chapter

Protecting Groups: Strategies and Applications in Carbohydrate Chemistry, First Edition.
Edited by Sébastien Vidal.

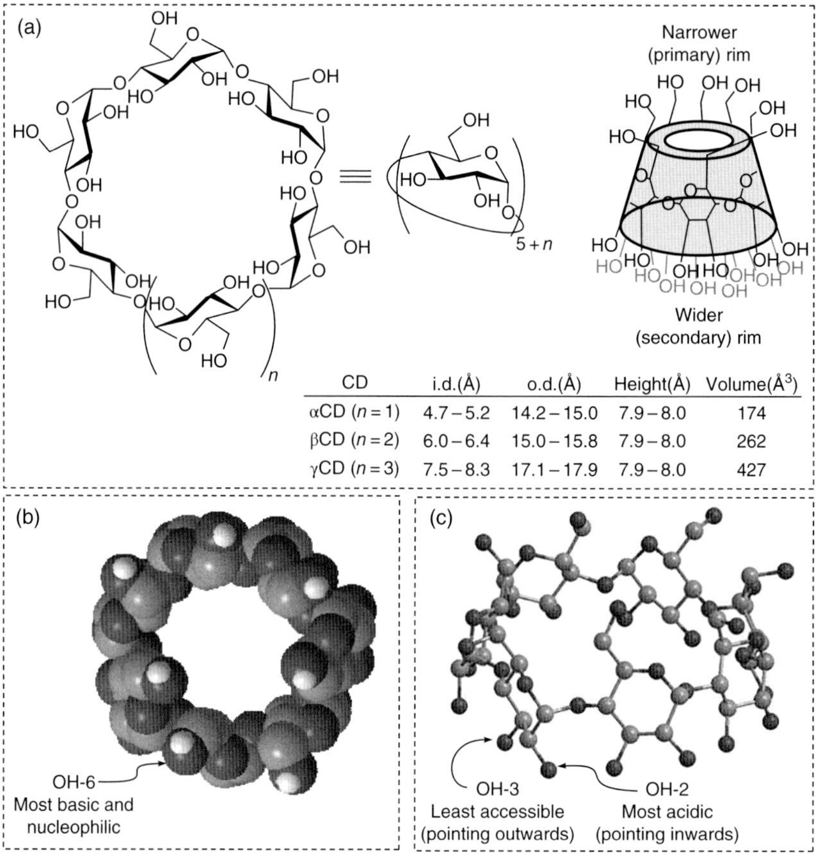

CD	i.d.(Å)	o.d.(Å)	Height(Å)	Volume(Å^3)
αCD ($n = 1$)	4.7–5.2	14.2–15.0	7.9–8.0	174
βCD ($n = 2$)	6.0–6.4	15.0–15.8	7.9–8.0	262
γCD ($n = 3$)	7.5–8.3	17.1–17.9	7.9–8.0	427

Figure 13.1 (a) General structure and pictographic representation of native CDs, with an indication of averaged dimensions (i.d. and o.d. are the inner and outer diameter, respectively), (b) axial, and (c) side views of βCD highlighting hydroxyl location orientation and main features.

focuses at dissecting the contribution of protecting group chemistry to the field and aims at tutoring their implementation toward the preparation of CD derivatives with unambiguous structures and precise modification patterns. Some selected cases are highlighted at the light of their conceptual novelty, the properties of the resulting derivatives, or their applications. Strategies not directly related with protecting group manipulation schemes are not covered herein, the reader being referred to excellent accounts [8a, b].

13.1.1 Structure–Reactivity Interplay in Native CDs

CDs represent a paradigmatic example of how molecular architecture defines chemical reactivity, quite contrasting with that of *conventional* carbohydrates (i.e. monosaccharides). Native α-, β-, and γ-CDs are rigid frameworks in which glucose units cannot flip. The α- and β-faces of the constitutive D-glucopyranosyl units are always pointing inward and outward the cavity defined by the macrocyclic structure, respectively, and primary (those at glucose C-6 position) and secondary (those at C-2 and C-3) hydroxyls flank the narrower and wider rim of the torus, respectively (Figure 13.1). Each hydroxyl set is confined into a relatively small space. The relative orientation of the OH-2 and OH-3 groups, inside- and outside-directed with respect to the cavity, respectively, facilitates the formation of an intramolecular H-bonding network that further rigidifies the structure and, in the case of βCD, diminishes water solubility. The height of the CD torus is invariably dictated by the glucopyranoside motif width, but the cavity diameter and volume depends on the number of glucose units. Narrower rim primary hydroxyls are comparatively more nucleophilic (and basic) because of the lesser steric congestion. On the other hand, as a consequence of the intramolecular H-bonding network, OH's-2 are the most acidic (easily deprotonated, $pK_a = 12.2$) [9]. Accessibility to OH-3 is annoyingly difficult as compared to conventional glucose derivatives. These conceptually simple rules, however, tend to mess up when certain OHs are modified and can turn the most trivial protecting group manipulation into a synthetic challenge.

13.2 Per-*O*-protection of Cyclodextrins

Practicing functional group protection in synthetic chemistry usually aims at preventing the reaction of a certain type of functional group when manipulating others. In this context, full CD hydroxyl protection may sound worthless, considering that no additional functional groups would remain left to work out. Nevertheless, homogeneous per-O-protection of CDs has served well to understand the principles governing hydroxyl reactivity in CD platforms. Complete hydroxyl protection is also a necessary first step in the case of regioselective deprotection tactics.

Acyl halides in the presence of pyridine or a tertiary amine can be used to indiscriminately acylate all CD hydroxyls [10]. Yet, long reaction times are often required during which overacylation side products may form [11]. As a consequence of their hard electrophile character, with very few exceptions (see Section 13.3.1 [12]), steric constraints cannot be easily exploited to control the reactivity of acyl halides toward the different hydroxyl types in CDs, limiting the synthetic utility of this reaction.

Reaction of CD alkoxides (preformed *in situ* by the treatment of CD with a strong base, e.g. NaH, in aprotic anhydrous media) with alkyl halides easily furnishes the homogeneously alkylated derivatives [13]. Although apparently naive, such protecting scheme was profusely exploited at the early stage of CD chemistry to modulate their solubility (e.g. methylated derivatives feature enhanced water solubility, while benzylated ones prefer organic solvents) minimally affecting their molecular inclusion capabilities [13b]. Moreover, the development of (regio)selective deprotection schemes (see Section 13.6 [14]) has renewed the interest of per-O-alkylation reactions.

13.3 Face-selective Differentiation: Primary vs Secondary Hydroxyl Protection

Facial differentiation represents the kick-off of genuine protecting group chemistry with CDs in the late 1980s. Quite obviously, most strategies aim at taking advantage of the larger nucleophilicity (and accessibility) of primary hydroxyls as compared to the sterically congested secondary ones. The protecting group reagents have to be carefully chosen in terms of size and electrophilicity to ensure (i) that their preference for primary hydroxyls is sufficiently distinct and (ii) that progressive steric crowding because of successive OH-6 protection does not switch this preference toward secondary positions. Indeed, currently available options are not that many, and experience on protecting group manipulation on *more conventional* carbohydrate scaffolds does not always help [15].

13.3.1 Strategies Toward Regioselective Protection of All Hydroxyls at the Primary Rim

The first and still probably the most reliable strategy for selective protection of primary vs secondary CD hydroxyls consists in the use of medium-sized silyl ethers. Following earlier work on the use of silicon-based protecting groups in CD chemistry [16], Takeo et al. conducted a systematic study on the effect of variations of several key parameters in the reaction outcome [17]. They found that native CDs can be silylated at all OH-6 with *tert*-butyldimethylsilyl chloride (TBDMSCl) in dimethylformamide (DMF) in the presence of imidazole by modulating the number of the silylating reagent equivalents and the reaction temperature, minimally affecting the secondary hydroxyls (Scheme 13.1a). The choice of the weak base critically determines regioselectivity. Indeed, soon later, Fügedi noticed that primary vs secondary hydroxyl selectivity could be further improved by using pyridine as a solvent and base [21]. Alternatively, Coleman introduced the use of the more hindered thexyldimethylsilyl (TDS) ethers to minimize concomitant OH-2 silylation [18]. This method is claimed to simplify workup (no column chromatography is required for isolation of the target product) and better suited for large-scale preparations (>100 g of βCD). Yet, per-(*O*-6)-TBDMS derivatives continue to be the most popular primary face-silylated CD products. Remarkably, less bulky silylating agents (e.g. TMSCl) are unable to differentiate at all between primary or secondary hydroxyls [22, 23].

Differentiating primary vs secondary hydroxyls by acylation is not that easy because of the high electrophilicity of acyl donors. Indeed, only the use of sterically demanding acyl donors, such as the pivaloylation procedure reported by Santoyo-González et al. [12], has proven efficient enough (Scheme 13.1b). Alternatively, the indirect, 2-step strategy consisting in (i) regioselective primary rim halogenation [24] and (ii) nucleophilic halogen substitution by the appropriate cesium carboxylate [19] is a feasible option.

As for acylation reactions, playing with the steric congestion to selectively alkylate primary vs secondary hydroxyls is also delicate. When alkylating reagents are sufficiently bulky to show a preference for the 6-position (e.g. trityl ethers), the growing steric congestion at the primary rim may impair complete

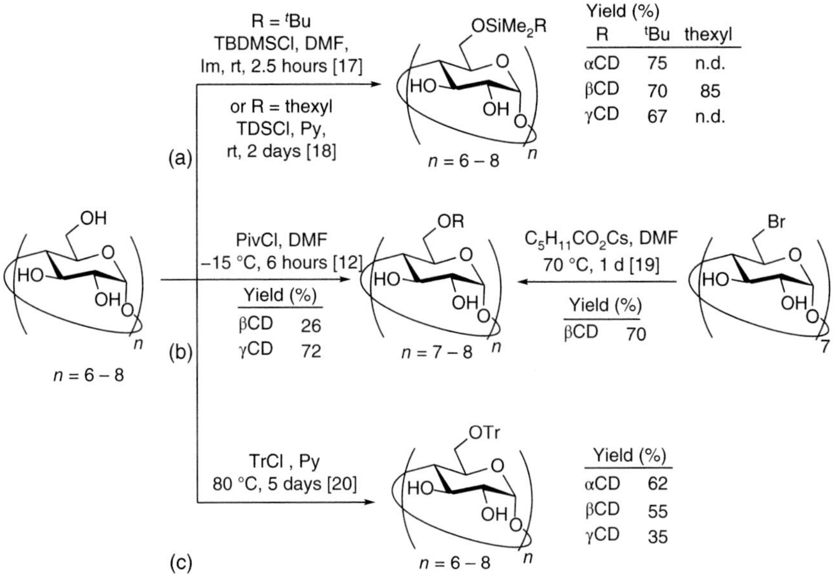

Scheme 13.1 Strategies toward regioselective protection of all hydroxyls at primary rim with bulky (a) silylating [17, 18], (b) acylating [12, 19], and (c) alkylating agents [20].

primary hydroxyl protection. Only very recently, conditions to synthesize per-(O-6)-tritylated CDs in satisfactory yields (35–62%) were reported by Ling and coworkers using a large excess of reagent at high temperature for prolonged times [20] (Scheme 13.1c). The tritylated CD torus in the final product is largely distorted to accommodate all trityl groups on the narrower face. The steric congestion imposed by the presence of trityls is so large that, indeed, trityl and trityl-like protecting groups have been intensively investigated in the concerted protection of alternate hydroxyl patterns in CD scaffolds (see Section 13.5.1) [25].

13.3.2 Strategies Toward Full Protection of Hydroxyls at the Secondary Rim

Face-selective differentiation of CDs by regioselective protection of the secondary hydroxyls, keeping the primary hydroxyls free, is more challenging than the previously discussed reverse strategy. Selectively reacting all OH's-2 is affordable by taking advantage of their higher acidity. However, the OH's-3 are far less reactive, and their accessibility is gradually reduced over reaction course; the harsh conditions required to complete protection of all of them are generally not compatible with the presence of primary OH's-6.

13.3.2.1 Strategies Toward Full Protection of Secondary Hydroxyls

To circumvent difficulties associated with secondary face-selective hydroxyl protection, selective deprotection schemes have been implemented. For instance, Angibeaud and Utille observed that TMSOTf-catalyzed (trimethylsilyl triflate) acetolysis of benzyl ethers at low temperatures (c. −35 °C) preferentially takes place at primary positions and demonstrated the utility of this reaction to produce per-(2,3-di-O-benzylated)-αCD in remarkable yield (>90% from perbenzylated αCD) [26] (Scheme 13.2a). In addition, Lehn [10a] and later Stoddart and coworkers [27] showed that a bulky alkoxide (^{i}PrOK) may selectively transesterify benzoate esters at the primary rim of CDs (Scheme 13.2b). Both strategies efficiently furnish secondary face-protected CD derivatives, albeit their substrate scope is rather limited and reaction conditions must be carefully adjusted.

The classical "long-way" sequence, involving protection/deprotection steps of the more reactive primary rim hydroxyls, has turned definitely more versatile for the installation of functional groups at the secondary OH's on CD frameworks. The most successful strategy, pioneered by Takeo et al., relies on TBDMS protection of the OH's-6 before homogeneous alkylation (methyl or benzyl) or acylation (acetyl) [17b] (Scheme 13.2c) followed by orthogonal cleavage of the silyl ether groups. In fact, most common commercial CD derivatives are prepared on the basis of this chemistry.

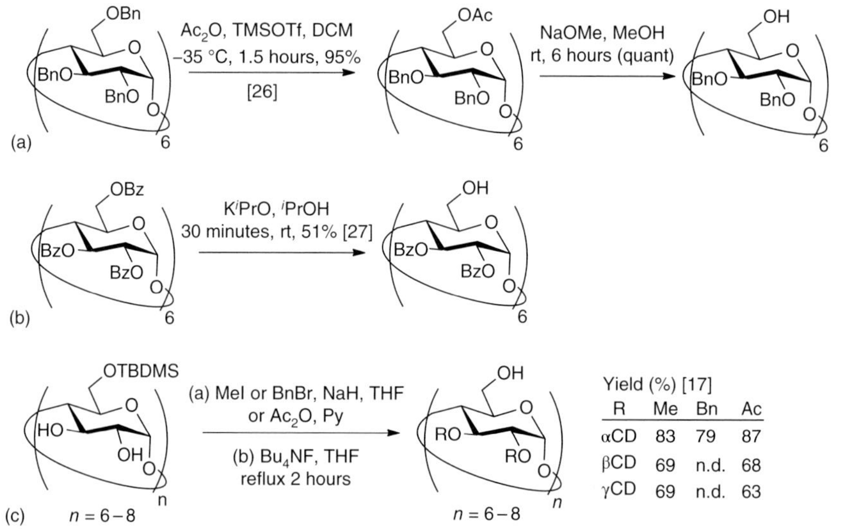

Scheme 13.2 Strategies toward protection of secondary rim hydroxyls implementing (a) TMSOTf-catalyzed acetolysis of benzyl ethers at the primary rim [26]. *Source:* Angibeaud and Utille 1992 [26]. Reproduced with permission of Thieme. (b) Regioselective transesterification of benzoates from the primary to the secondary rim [27]. *Source:* Ellwood et al. 1992. Reproduced with permission of Springer Nature. (c) Indirect orthogonal protection/deprotection of primary rim hydroxyls [17]. *Source:* From Takeo 1988, 1989 [17].

Even in the case of primary rim-blocked CD derivatives, secondary rim protecting group manipulation is far from evident [10b, 28]. Complete protection of the least accessible OH's-3 is not always achievable, and attempts to use harsher conditions may lead to concomitant-side reactions. For instance, acylation of primary position-modified CD derivatives with highly reactive acyl halides as acyl donors has been shown to produce overacylation at the α position of the acyl chain [11]. Rewardingly, this drawback has been recently overcome by a combination of milder acid donors such as anhydrides in the presence of 4-dimethylaminopyridine (DMAP) as a base catalyst [29]. This methodology has proven highly efficient at producing CD derivatives with homogeneous acylation patterns and is compatible with a range of acyl donor chain lengths and the presence of functional groups bearing acidic protons (e.g. carbamates).

Full homogeneous silylation at the secondary rim of CDs is even more difficult. Only trimethylsilyl (TMS) groups can be installed in all OH's-2 and OH's-3 positions [23], but they are reported to readily hydrolyze even in neutral conditions [30]. Secondary rim alkylation is comparatively less tedious. Williamson conditions (e.g. NaH and alkyl halide in anhydrous aprotic media) quite efficiently furnish homogeneously per-(2,3-di-O-alkylated) CDs [17b, 31], and the methodology is equally efficient for the synthesis of allyl and propargyl ethers, which can be further exploited as cores for further elaboration through "click chemistry" approaches [32]. Moreover, provided sufficiently large excesses of the deprotonating base and the alkyl halide are used, homogeneous CD derivatives bearing long hydrocarbon chains at the secondary face can be accessed [33].

13.3.2.2 Strategies Toward Differentiation Between Secondary Hydroxyls

As a consequence of the larger accessibility of OH-2 as compared to OH-3 CD hydroxyls, differentiation between them is relatively easy. Thus, native CDs can be regioselectively methylated at positions O-2 and O-6 simultaneously in the presence of $BaO/Ba(OH)_2$ and dimethyl sulfate [13b] (Scheme 13.3a). Implementation of similar reactions using allyl bromide as an alkyl donor furnishes a complex mixture that hampers isolation of the desired product [31, 39]. An interesting alternative to achieve O-2- and O-6- vs O-3-differentiated patterns is the method recently reported by Sollogoub and coworkers, taking advantage of I_2/Et_3SiH-mediated debenzylation to selectively cleave benzyl ethers at position O-3 on a perbenzylated CD precursor [35]. This methodology allows the straightforward access to per-(2,6-di-*O*-benzyl)-αCD in 58% yield.

Interestingly, alkylation conditions can be tuned to promote exclusive etherification of the more acidic OH-2 hydroxyls in CDs. In a seminal contribution, Rong and D'Souza reported that stepwise treatment of βCD with stoichiometric amounts of NaH and MeI furnished the per-(O-2)-methylated derivative in over 80% yield [34] (Scheme 13.3a).

Silyl ethers have also been used to differentiate among the secondary OH's-2 and 3. For instance, treatment of αCD with an excess of TBDMSCl in anhydrous DMF/pyridine mixture at high temperature affords exclusively the per-(2,6-di-*O*-TBDMS) derivative in virtually quantitative yield [16, 21] (Scheme 13.3b). Similar results were described by Coleman et al. using TDSCl as an silylating agent [18]. As commented above, a similar strategy using smaller silyl halides is unpractical for the purpose of regioselective differentiation of CD hydroxyls. Interestingly, Bukowska et al. described that milder versions of TMSCl (e.g. *N*-TMS acetamide) selectively furnish the corresponding per-(2,6-di-O-silylated) derivatives in over 90% yield [40].

Attempts to achieve regioselective silylation of the OH's-2 in CDs, leaving unprotected OH's-3 and 6, have proven unsuccessful. Tian and D'Souza observed

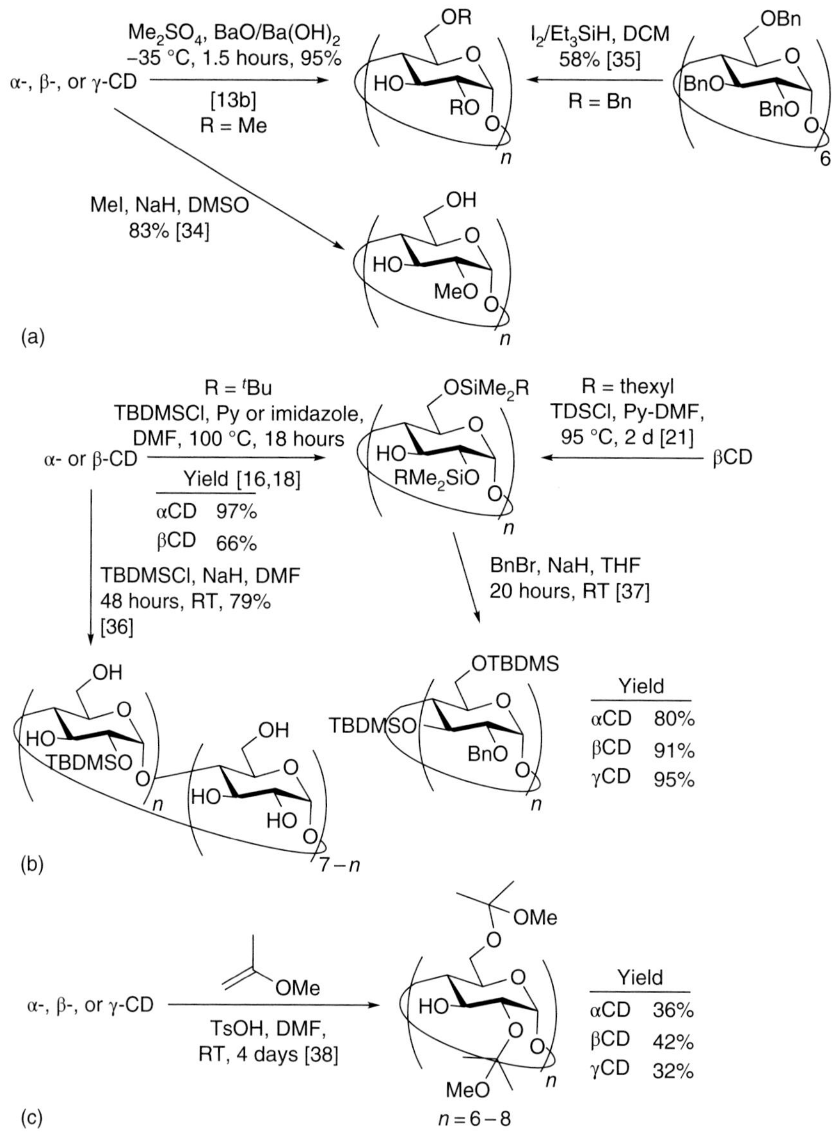

Scheme 13.3 Strategies toward differentiation of secondary OH-2 vs OH-3 hydroxyls by (a) regioselective alkylation [13b, 34] and dealkylation [35], (b) silylether group manipulation [16, 18, 21, 36, 37], and (c) asymmetric ketal formation [38].

that selective deprotonation of the OH-2 hydroxyls in βCD, using stoichiometric amounts of NaH, followed by treatment with TBDMSCl afforded a mixture of partially (O-2)-silylated products [36] (Scheme 13.3b). The authors demonstrated that steric constrains prevent the remaining OH-2 groups from reacting under conditions that prevent silylation of the primary OH's-6. Discrimination between the OH-2 and OH-6 hydroxyls performed worse in the case of αCD, which the authors attributed to the less likely formation of the hexa-(*O*-2)-alkoxide while preventing deprotonation of any of the OH's-6 because of the smaller size of αCD. Later on, Stoddart and coworkers showed that under these strong basic conditions, O2 → O3-silyl migration readily takes place. Indeed, this observation led the authors to describe a synthetic procedure toward per-(3,6-di-*O*-TBDMS)-βCD [37] (Scheme 13.3b). The procedure has been generalized to other CDs and mechanistically interpreted [15, 41], turning into a highly valuable synthetic tool in CD chemistry [31, 42].

Liptak et al. have also reported regioselective protection at positions O-2 and O-6 in α- and βCD by reaction with 2-methoxypropene using *p*-toluenesulfonic acid (TsOH) as a catalyst, affording the corresponding per-(*O*-2,O-6)-acetonide [38] (Scheme 13.3c). The authors claim that this protection scheme leaves OH's-3 readily accessible for further manipulation, whereas the linear ketal protecting groups can be easily cleaved under acid hydrolysis. Although attractive, this strategy has been scarcely exploited.

13.4 Single Hydroxyl Protection Strategies

Protection of a single CD hydroxyl has typically being achieved using minored amounts of protecting group reagent. This practice, thus, inherently implies more or less tedious purifications and often modest yields.

13.4.1 (Regio)selective Protection of a Primary Hydroxyl Group

Initial sophistication involved the use of bulky reagents, thereby hindering the incorporation of a second molecule and increasing the differences in the chromatographic or solubility properties as compared with the unreacted CD. This strategy preferentially targets at modifying a single primary hydroxyl. The method developed by Fügedi for the synthesis of a mono-(*O*-6)-silyl ether βCD derivative, as a precursor of mono-glycosylated conjugates, is a representative example. Slightly over-stoichiometric amounts of TBDMSCl in pyridine and *in situ* acetylation furnished the mono-(*O*-6)-TBDMS βCD peracetate in 40% yield (Scheme 13.4a) [43]. Teranishi and Ueno later observed that this selectivity is temperature-dependent and can be switched to position O-2 upon heating [44]. Although intuitively similar, trityl ethers were not that handy [48]. Although tritylation preferentially takes place at primary hydroxyls, often multiple tritylation is observed, which indeed has been exploited to produce particular CD protection patterns [25] (see Section 13.5.1). Direct monoprotection at the primary rim with less bulky alkylating agents is unpractical. For instance, mono-(O-6)-benzylation can only be efficiently achieved after temporary blocking of the OH's-2, for instance as the corresponding TBDMS ethers, as described by Tian and D'Souza [36] (Scheme 13.4b), because of the strong basic conditions required. Indeed, exploiting preferential deprotonation of OH-2 hydroxyls is a second level of sophistication driving the monoprotection reaction to OH-2.

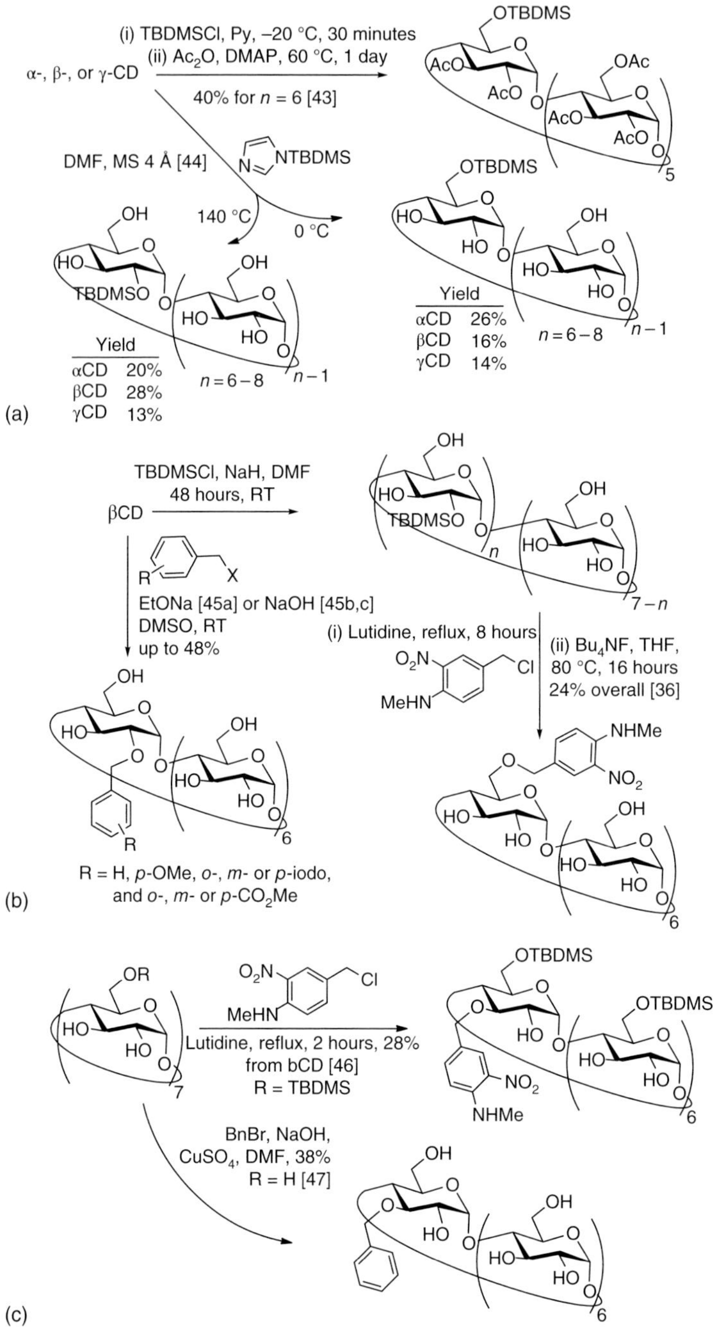

Scheme 13.4 Strategies toward selective hydroxyl monoprotection (a) at OH-6 [43, 44], (b) OH-2 [36, 45], and (c) OH-3 hydroxyls [46, 47].

13.4.2 (Regio)selective Protection of a Secondary Hydroxyl Group

Rong and D'Souza initially set up benzyl ether formation conditions to selectively install a 4-methylamino-3-nitrobenzyl group at a single βCD 2-position [34]. Later, this methodology has been implemented for other benzyl ether groups, with yields reaching 50% [45] (Scheme 13.4b). Interestingly, the inherent molecular inclusion capabilities of the target CD can be exploited to promote a precise orientation of the reactive benzylation reagent and force the counterintuitive OH-3 functionalization pattern [49]. D'Souza and coworkers realized the concept by selectively installing the 4-methylamino-3-nitrobenzyl group at a single OH-3 on per-(*O*-6)-TBDMS βCD by conducting the reaction in lutidine at 80 °C [46] (Scheme 13.4c). More recently, Estour and coworkers have implemented a strategy for regioselective O-3 monobenzylation of βCD (40% yield) that combines (i) molecular inclusion of the alkylating agent (benzyl bromide) and (ii) Cu(II) chelation of the secondary hydroxyls [47]. Confluence of both reagent inclusion and Cu(II) chelation is essential to control the reaction outcome, as the same reaction with allyl halide or in EtOH (where no inclusion complex is formed) or in the absence of $CuSO_4$ preferentially rendered the mono-(O-2)-protected regioisomer.

13.5 Concerted Protection of Hydroxyl Sets (Pairs or Triads)

The straitjacketed structure of native CDs imposes a rigid hydroxyl display, with minute variation in their relative disposition and distances. This is a favorable scenario to implement strategies that simultaneously target hydroxyl sets with particular arrangements. Pioneering work in the field is due to Tabushi et al. using conformationally restricted sulfonylation reagents [50]. Initial efforts were directed to the manipulation of inclusion capabilities by capping CD rims [25b]. Later, this knowledge has been revealed as an extremely useful synthetic tool.

13.5.1 Concerted Protection of Primary Hydroxyl Sets

Trityls were the first groups used in concerted protection of several CD hydroxyls. As commented above, their steric volume prevents from reacting with secondary hydroxyls. Despite trityls cannot block the reactivity of distal hydroxyls at the primary rim, they can impair that of adjacent OHs, as recently demonstrated in a detailed study on the regioselectivity of the reaction [51]. In the nascence of regioselective CD chemical manipulation, Knowles and coworkers implemented this notion to synthesize an αCD derivative featuring three trityl groups on alternating primary OHs ($6^{I,III,V}$-tri-*O*-trityl-αCD) in a remarkable 23% yield considering all statistically possible arrangements (13 in total; 1-mono-, 3-di-, 4-tri-, 3-tetra-, 1-penta-, and 1-hexa-tritylated patterns) [25a]

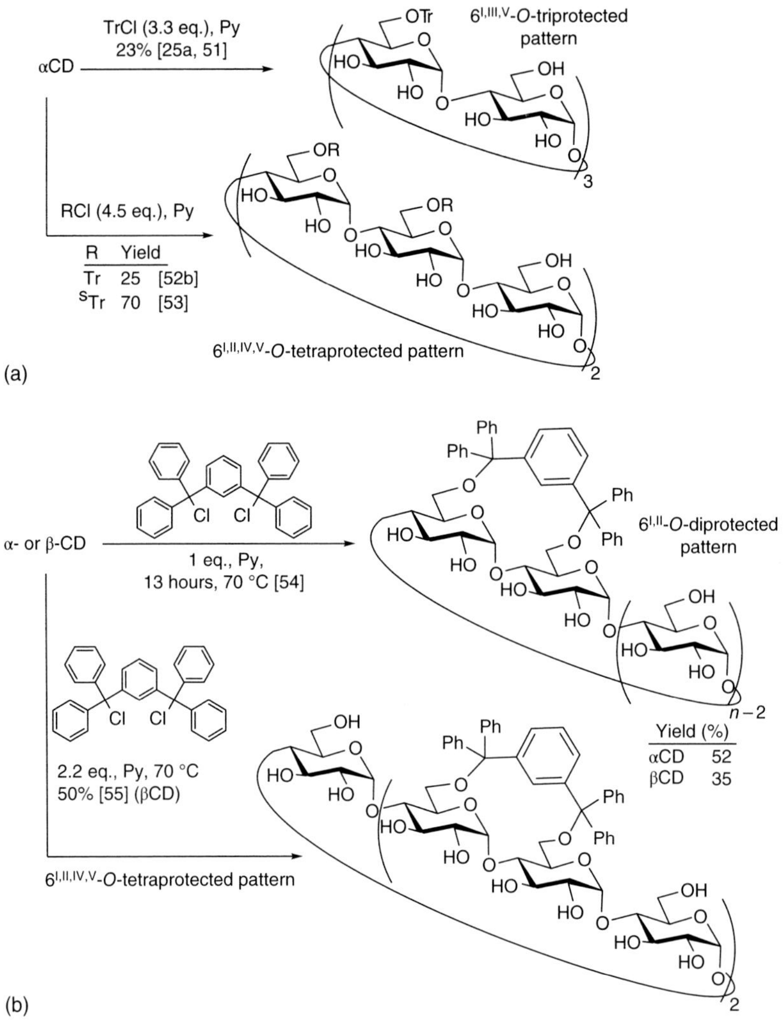

Scheme 13.5 Strategies for the concerted protection of hydroxyl sets using (a) tritylating [25a, 51, 52b, 53] and (b) bis-tritylating [54, 55] reagents.

(Scheme 13.5a). Coleman and Ling further observed that the resulting tritylation patterns are intimately dependent on reaction temperature and trityl chloride amount [52a]. Thus, by increasing the TrCl:αCD molar ratio from 3.3 to 4.5, the $6^{I,II,IV,V}$-tetra-O-tritylated derivative could be isolated in 25% yield [52b]. Unfortunately, experimental implementation of the concept is tricky and purifications cumbersome [52, 56], limiting its utility to the αCD platform.

Armspach and Matt developed tris(*tert*-butylphenyl)methyl (*supertrityl*, ^{S}Tr) as a bulkier group aimed in principle at promoting the regioselective protection of distal hydroxyls (e.g. $6^{I,IV}$ pattern in αCD). Although it fails for that purpose, affording a mixture of the target $6^{I,IV}$- and the $6^{I,III}$-di-*O*-^{S}Tr isomers in equal amounts (c. 30% each) [57], they found that increasing the amount of reagent nicely produced the corresponding $6^{I,II,IV,V}$-tetra-*O*-^{S}Tr derivative in 70% yield (Scheme 13.5a) [53].

Aimed at obtaining vicinal di-OH-6 protection patterns, the same authors have designed *bis*-tritylating agents (1,3-*bis*[*bis*(aryl)chloromethyl]benzenes, Scheme 13.5b) fitting the average distance between adjacent primary hydroxyls [54]. Their strategy conveniently afforded $6^{I,II}$-di-O-tritylated αCDs in c. 50% yield and, interestingly, fairly performed in the more demanding βCD scaffold. X-Ray diffraction revealed that c. 60% of the αCD primary rim surface is covered by the *bis*-trityl group, hindering access of the reagent to the rest of the primary hydroxyls. In the case of βCD, however, the space is enough to host a second *bis*-trityl unit. Indeed, reacting βCD with 2 eq. of the tritylating agent selectively produced the $6^{I,II,IV,V}$-tetra-O-tritylated derivative in a remarkable 50% yield (Scheme 13.5b) [55].

13.5.2 Concerted Protection of Secondary Hydroxyl Sets

Mobility restrictions on the CD secondary rim are even larger compared to the primary rim. Distances and spatial arrangement of secondary hydroxyl groups are tightly fixed, which should, in principle, facilitate the development of concerted hydroxyl protection patterns. Unfortunately, their number – double the primary OH's – and the difficulties associated with the precise control of reactivity differences remain serious limitations. Insights on the utility of α,α′-dibromo *o*-xylene for the protection of *vic*-diols in carbohydrates [58] rendered one of the most appealing strategies for concerted protection of two secondary CD hydroxyls located on the same glucopyranosyl subunit. Alkylation of α-, β-, or γCD with this reagent in the presence of a strong base in aprotic solvent selectively furnished the corresponding $2^{I},3^{I}$-di-O-(*o*-xylylenated) derivative, out of the 27, 33, or 36 possible diprotected regioisomers, in 24–30% yields [59] (Scheme 13.6a). Cyclic diether formation is hypothesized to take place by initial reaction at the more acidic OH-2 hydroxyl, which is selectively deprotonated in the presence of lithium diisopropylamide (LDA) or NaH. Cyclization involving the contiguous OH-3 is then kinetically favored. In addition to the synthetic interest of this unprecedented *trans*-diequatorial *vic*-diol protection pattern, xylylenated CDs feature remarkable molecular inclusion and self-assembling capabilities [61].

In a previous work, Sakairi and Kuzuhara reported that oligomaltose benzylidenation produced cyclic ketals involving hydroxyls OH-2 and OH-3 from contiguous glucose units [62]. Rewardingly, they later observed that the same eight-membered cyclic arrangement is selectively formed when treating per-(O-6)-protected CDs with dimethoxytoluene in the presence of camphorsulfonic acid (CSA) [60] (Scheme 13.6b). The benzylidene cap prevents from further acetalation even in the presence of an excess of reagent, the isolated yields (37–54%) depending on the CD size. Moreover, regioselective benzylidene opening allows subsequent manipulation of the O-2 and O-3 positions initially involved in cyclic acetal formation individually [63].

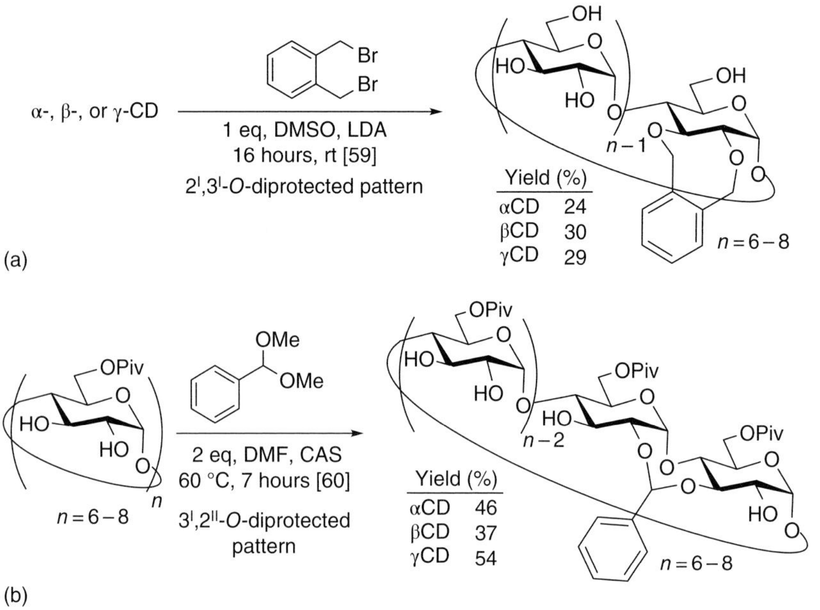

Scheme 13.6 Strategies for the concerted protection of hydroxyl sets using (a) benzylating reagents. *Source:* Balbuena et al. 2007 [59]. Reproduced with permission of Royal Society of Chemistry. (b) Benzylidene ketals. *Source:* Sakairi and Kuzuhara 1993 [60a] and Sakairi et al. 1996 [60b]. Reproduced with permission of Royal Society of Chemistry.

13.6 Regioselective Deprotection of Symmetric Cyclodextrins

With few exceptions [10a, 26], selective chemical manipulation of CD topology remained for years a one-way route focused at implementing the subtle disparities in hydroxyl reactivity of the native CDs. However, it was conceivable that the other way around, i.e. differentiating among the position and relative arrangement of protecting groups on the CD scaffold, could also help to fulfill the challenge.

13.6.1 The DIBALH Scalpel: Concerted DIBALH-Promoted Primary Rim Debenzylations

Approximately 15 years ago, a totally new concept for accessing CD derivatives with precise sets of hydroxyl groups selectively differentiated from the rest was disclosed by the group of Sinaÿ and Sollogoub. When developing an original strategy for the synthesis of carbasugars by *exo*-glycal rearrangement [64], they serendipitously observed an unprecedented trialkylaluminum-promoted de-O-benzylation reaction operating with intriguing selectivity on their substrates. A series of structure–reactivity studies led the authors to rationalize this methodology [65] (Scheme 13.7a). De-O-benzylation is only observed at positions where a 1,2-*cis* di-*O*-aluminum chelate can be formed and a second aluminum species could approach. This condition, by virtue of the steric volume of alkylaluminum species (e.g. diisobutylaluminum hydride, DIBALH), turns into a nanometric scalpel to selectively remove benzyl ethers at diametrically opposed glucose units of the primary CD rim with surgical precision (82% and 83% for α and βCD, respectively [66]) (Scheme 13.7b). In γCD, the larger primary rim diameter may allow different positional isomers to be formed as well as a third debenzylation to occur [66]. Debenzylation at OH-2 is also possible, but it requires much harsher reaction conditions given the higher steric constraint provoked by the large volume of two DIBALH species at the secondary rim. Interestingly, under modified conditions, the mono-(O-6)-debenzylated products can also be selectively isolated [66].

Further refinement has encumbered this methodology as the most elegant approach to selectively modify CD topology. Obtaining αCD mono-, tri-, and tetraols in reasonable yields is possible by acting on the reaction conditions [65, 68]. However, more importantly, Sollogoub's group has demonstrated that DIBALH reactivity can be selectively guided to selected positions by controlling sterics or frustrating the normal sequential debenzylation preference through chemical modification. The synthesis of heterotri- [69], tetra- [70], and up to the Holy Grail hexa-functionalized αCD derivatives [67] has been achieved by implementing stepwise the same regioselective DIBALH-mediated debenzylation reaction (Scheme 13.7c).

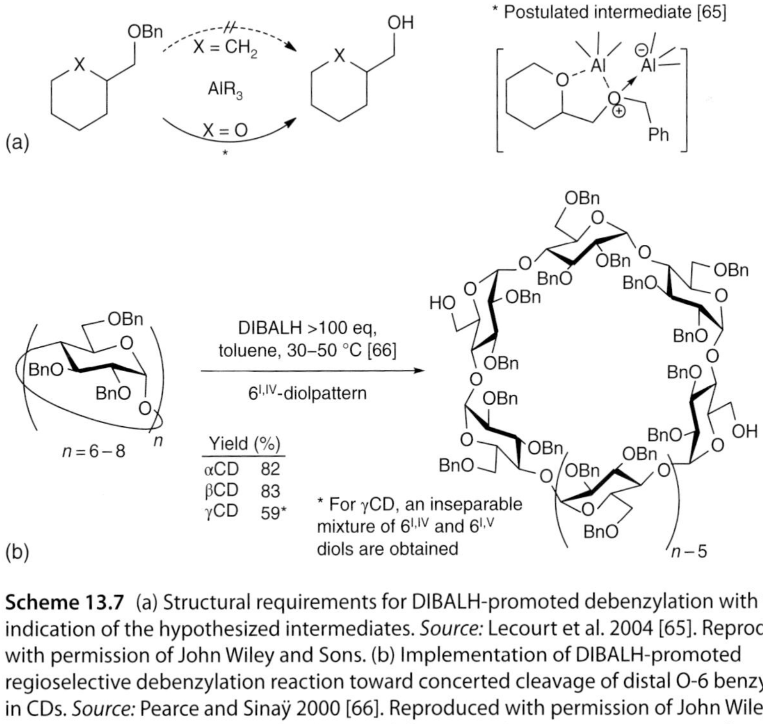

Scheme 13.7 (a) Structural requirements for DIBALH-promoted debenzylation with the indication of the hypothesized intermediates. *Source:* Lecourt et al. 2004 [65]. Reproduced with permission of John Wiley and Sons. (b) Implementation of DIBALH-promoted regioselective debenzylation reaction toward concerted cleavage of distal O-6 benzyl groups in CDs. *Source:* Pearce and Sinaÿ 2000 [66]. Reproduced with permission of John Wiley and Sons. (c) Tri-, tetra-, and orthogonally hexadifferentiated αCDs. *Source:* Wang et al. 2014 [67]. Reproduced with permission of Springer Nature.

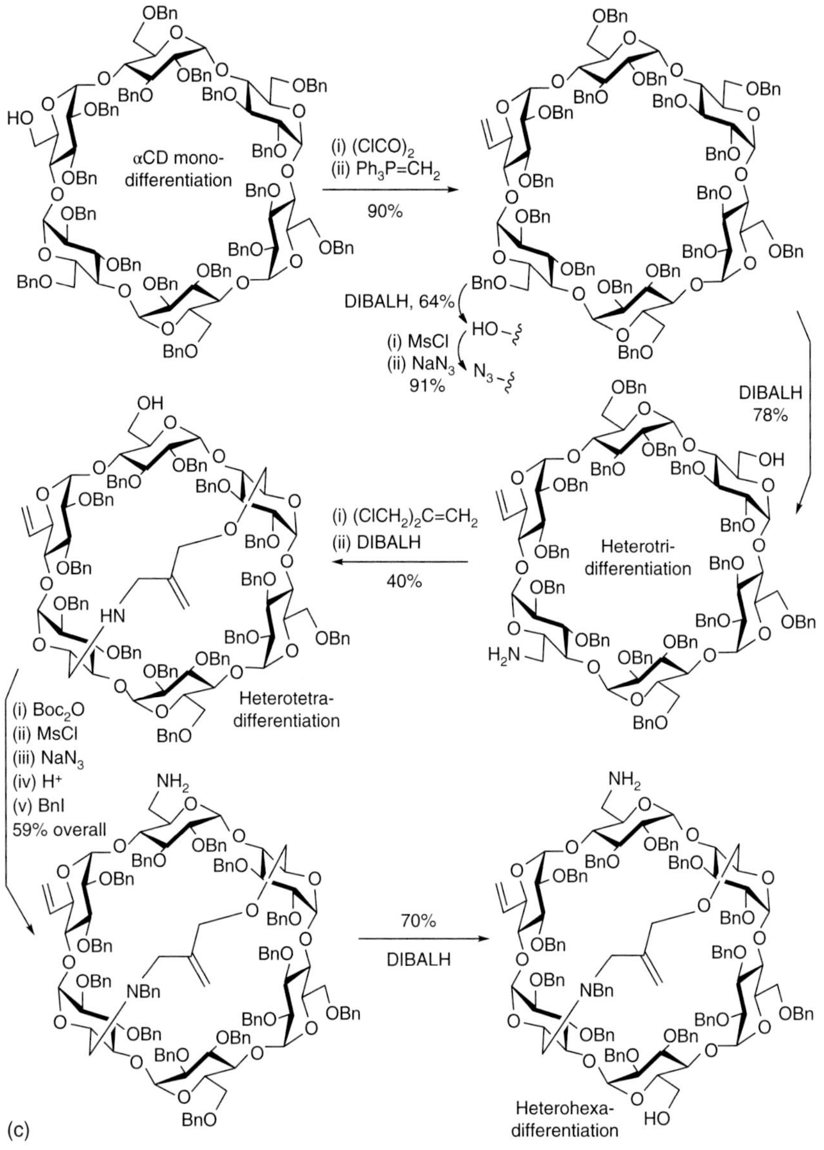

Scheme 13.7 (*Continued*)

The DIBALH methodology is not only very efficient in terms of selectivity but also predictable. Although the higher degree of development has been reached for αCD, in most cases, it can be applied to perbenzylated βCD [69b]. It represents with no doubt a major breakthrough in CD chemistry that is widening the scope of applications of this host molecules in fields such as asymmetric catalysis [71], metallocavitands [72], supramolecular self-assembling systems [73], or artificial enzymes [74] among others.

13.6.2 Regioselective DIBALH-Promoted Primary Rim Desilylation

The methodology introduced by Sinaÿ and Sollogoub can also be implemented for selective cleavage of silyl ethers. Ling and coworkers have reported that deprotection of per-(*O*-6)-silyl CDs (TBDMS or TDS) proceeds with the same regioselectivity than for benzylated CDs [75]. Depending on reaction conditions, mono- or di-(O-6)-desilylated α- and βCD derivatives can be favored, while γCD substrates lead to regioisomeric mixtures (Scheme 13.8). Noteworthy, the consumption of DIBALH is much lower, the strategy now being compatible with the presence of more *vulnerable* alkyl protecting groups at the secondary rim (e.g. methyl or allyl) in addition to robust benzyls. Unfortunately, the strategy does not work for secondary TBDMS group cleavage, but recently, Ling and coworkers have developed a simple procedure for regioselective monodesilylation at O3 in the presence of HCl (pH 1~2) [76].

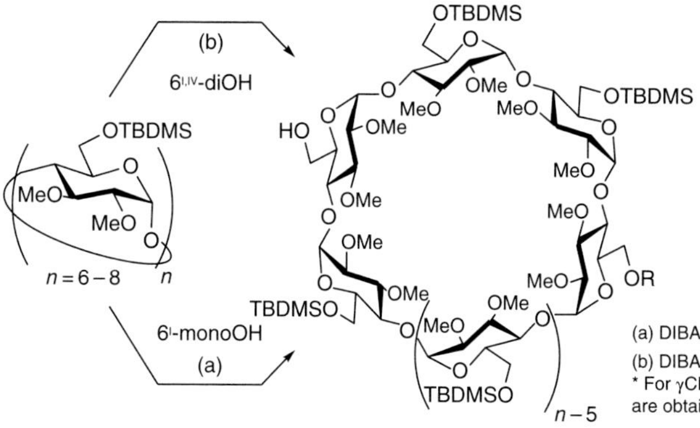

Yield (%)		monoOH (R = TBDMS)	diOH (R = H)
αCD	(a)	71	16
	(b)	10	68
βCD	(a)	62	22
	(b)	71	12
γCD	(a)	—	26
	(b)	—	53

(a) DIBALH 3 eq, toluene, 3 hours, 0 °C
(b) DIBALH 6 eq, toluene, 4 hours , 0 °C
* For γCD, inseparable mixtures of diols are obtained

Scheme 13.8 DIBALH-promoted regioselective desilylation of primary hydroxyls.
Source: From Ghosh 2012, 2013 [75].

13.6.3 DIBALH-Promoted Regioselective Demethylation of Secondary Hydroxyl Pairs

Efforts on DIBALH-promoted regioselective deprotection of alkylated CD derivatives at the secondary rim have also rendered interesting strategies. Methyl ethers are hydrolyzed by DIBALH at higher rates as compared to benzyl ethers [77]. Quite unexpectedly, in the case of fully methylated CD substrates, reactivity is directed to the wider secondary rim, once more with excellent regioselectivity. DIBALH is chelated between O-2 and O-3 of contiguous glucose units (the closest methoxyl pair on the CD scaffold) and the second DIBALH species destabilizes and cleaves the less hindered methyl group at O-3. The steric decompression caused allows the immediate action of a third DIBALH species over O-2 to selectively produce the di-(O-3^{I},O-2^{II})-demethylated pattern [78] (Scheme 13.9), which has been recently exploited by Estour and coworkers for the design of artificial enzymes capable of degrading organophosphorus nerve agents [79]. Further demethylations at contiguous positions are hindered by

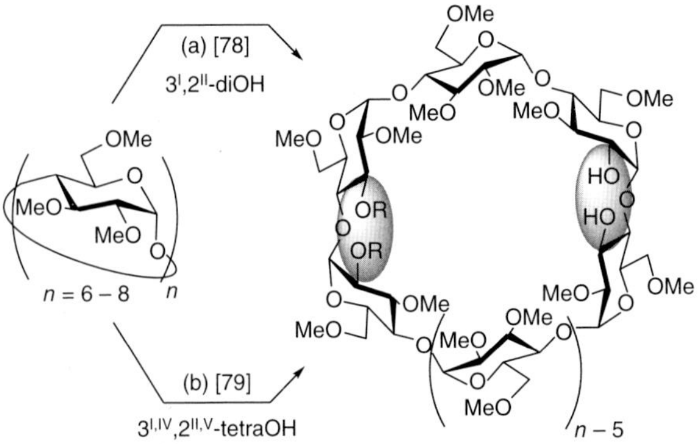

Yield (%)		diOH (R = Me)	tetraOH (R = H)
αCD	(a)	55	—
	(b)	—	45
βCD	(a)	56	—
	(b)	—	51

(a) DIBALH 9 eq, toluene
(b) DIBALH 40–50 eq, toluene

Scheme 13.9 DIBALH-promoted regioselective demethylation of secondary hydroxyls [78, 79].

residual OH-DIBAL chelates. Interestingly, controlling the reaction conditions di-(O-3^{I}, O-2^{II})- or tetra-(O-$3^{I,IV}$, O-$2^{II,V}$)-hydroxylated αCD patterns can be regioselectively accessed [80]. In the case of βCD, the second di-demethylation is not selective, whereas the strategy was impractical for γCD [81].

13.7 Summary and Conclusions

The last two decades have witnessed an unprecedented development of CD chemical functionalization methodologies. A significant part of these advances relies on specifically tailored protecting group manipulation schemes. Either serendipity or purposely conceived, the synthetic toolbox is now capable of complying with the demands of an array of applications ranging from the pharmatech arena (drug delivery or controlled release), analytical sciences, or material technologies. Moreover, such a prompt reward is serving as the most efficient feedback, inspiring new applications based on previously inaccessible molecular entities and new solutions based on this synthetic toolbox.

13.8 Experimental Procedures

13.8.1 *O*-TBDMS Protection at the Primary Rim of α-, β-, and γCD

To a stirred mixture of corresponding CD (vacuum-dried overnight at 80 °C) and imidazole (1.5 eq.) in anhydrous DMF (30 ml per mmol of CD), a solution of TBDMSCl (1.1 eq.) in DMF was dropwise added for 30 min at RT. The mixture was stirred for c. 2.5 h at RT until a major spot evolved in thin-layer chromatography (TLC) (R_f 0.3–0.4 in $CHCl_3$–MeOH–water 50 : 10 : 1, depending on the parent CD). Then, the reaction mixture was further stirred for 30 min and poured into ice water (1.5 l); the solid was filtered-off and taken into $CHCl_3$; washed with 1 N HCl, aq. $NaHCO_3$, and water; dried over Na_2SO_4; filtrated; and concentrated to dryness. The residue was triturated with petroleum ether, filtered off, and further chromatographed in silica gel using a 8 : 1 : 0→ 4 : 1 : 0→ 4 : 1 : 0.1 $CHCl_3$–MeOH–water eluent gradient. Per-(*O*-6)-TBDMS CD derivatives can thus be obtained in 67–75% yields [17].

13.8.2 Regioselective Per-(*O*-2,*O*-6)-allylation of α-, β-, and γCD

Allyl bromide was added to a stirred mixture of the corresponding CD, BaO (2.5 eq. per OH group), and $Ba(OH)_2{\cdot}8H_2O$ (1.2 eq. per OH group) in a 1 : 1 mixture of anhydrous DMF and dimethylsulfoxide (DMSO) (40 ml per mmol of CD). The mixture was stirred at RT for 5 days under N_2 and then quenched by adding aq. NH_4OH and further stirred for 30 min. The mixture was poured into $CHCl_3$, the salts were precipitated by the addition of pentane, and filtered off. The organic layer was washed with water (×3), dried over $MgSO_4$, and evaporated. The resulting residue was purified by column chromatography (1 : 4 EtOAc—$CHCl_3$) to afford the per(2,6-di-*O*-allyl)-CD derivatives in 22–40% yield [39].

13.8.3 Regioselective Per-(*O*-2)-methylation of α- and βCD

A solution of βCD (dried overnight at 117 °C) in anhydrous DMSO was reacted with NaH (1 eq. per OH-2), and the mixture was stirred until a clear solution was obtained (eventually overnight). Then, MeI (1 eq. per OH-2) was added, and the reaction mixture was stirred at RT for 30 min. The reaction was quenched by addition of acetone, and the target product was collected by filtration in 83% yield [34].

13.8.4 Regioselective *bis*(Debenzylation) at the Primary Rim of Per-*O*-benzylated α and βCD

A solution of the corresponding per-benzylated CD in anhydrous toluene was treated at RT with a solution of DIBALH (30 eq.) in toluene, so that the final DIBALH concentration is 1 M. The mixture was then warmed up to 50 °C and stirred for 2 h, after which the reaction was carefully quenched by dropwise addition of water at 0 °C and further stirred for 15 min. Extraction with EtOAc (×3) followed by evaporation of the solvent under reduced pressure afforded the $6^{I,IV}$-di-O-debenzylated derivatives that were finally purified by silica gel column chromatography (1 : 5 EtOAc-cyclohexane). Yields were in the range of 85% [65, 70].

Abbreviations

Ac	acetyl
Bn	benzyl
CSA	camphorsulfonic acid
CD	cyclodextrin
DIBALH	diisobutylaluminum hydride

DMAP	4-dimethylaminopyridine
DMF	*N,N*-dimethylformamide
DMSO	dimethylsulfoxide
Im	imidazole
LDA	lithium diisopropylamide
MS	molecular sieves
n.d.	not described
n.i.	not isolated
Piv	pivaloyl
Py	pyridine
TBDMS	*tert*-butyldimethylsilyl
TDS	thexyldimethylsilyl, (2,3-dimethyl-2-butyl)dimethylsilyl
THF	tetrahydrofuran
TLC	thin layer chromatography
TMS	trimethylsilyl
TMSOTf	trimethylsilyl triflate
TsOH	*p*-toluenesulfonic acid
Tr	triphenylmethyl
^{S}Tr	tris(*tert*-butylphenyl)methyl, *supertrityl*

References

1 (a) Loftsson, T. and Duchêne, D. (2007). *Int. J. Pharm.* 329: 1–11. (b) Crini, G. (2014). *Chem. Rev.* 114: 10940–10975.
2 (a) Cramer, F. (1952). *Angew. Chem. Int. Ed. Engl.* 64: 437–447. (b) Cramer, F. and Kampe, W. (1965). *J. Am. Chem. Soc.* 87: 1115–1120.
3 Dodziuk, H. (ed.) (2006). *Cyclodextrins and Their Complexes*. Weinheim, Germany: Wiley-VCH.
4 (a) Loftsson, T. and Brewster, M.E. (2010). *J. Pharm. Pharmacol.* 62: 1607–1621. (b) Ortiz Mellet, C., García Fernández, J.M., and Benito, J.M. (2013). *Supramolecular Systems in Biomedical Fields* (ed. J.-H. Schneider), 94–139. Cambridge, UK: RSC.
5 Astray, G., Gonzalez-Barreiro, C., Mejuto, J.C. et al. (2009). *Food Hydrocolloids* 23: 1631–1640.
6 Tarimci, N. (2011). *Cyclodextrins in Pharmaceutics, Cosmetics, and Biomedicine: Current and Future Industrial Applications* (ed. E. Bilensoy), 131–144. Weinheim, Germany: Wiley-VCH.
7 Szente, L. and Szemán, J. (2013). *Anal. Chem.* 85: 8024–8030.
8 (a) Guieu, S. and Sollogoub, M. (2014). *Synthetic Methods in Carbohydrate Chemistry: From Monosaccharides to Complex Glycans* (ed. D.B. Werz and S. Vidal), 241–283. Weinheim, Germany: Wiley-VCH. (b) Khan, A.R., Forgo, P., Stine, K.J., and D'Souza, V.T. (1998). *Chem. Rev.* 98: 1977–1996. (c) Martínez, A., Ortiz Mellet, C., and García Fernández, J.M. (2013). *Chem. Soc. Rev.* 42: 4746–4773. (d) Ortiz Mellet, C., García Fernández, J.M., and Benito, J.M. (2011). *Chem. Soc. Rev.* 40: 1586–1608.

9 Gelb, R.I., Schwartz, L.M., Bradshaw, J.J., and Laufer, D.A. (1980). *Bioorg. Chem.* 9: 299–304.
10 (a) Boger, R.J., Corcoran, R.J., and Lehn, J.M. (1978). *Helv. Chim. Acta* 61: 2190–2218. (b) Hirayama, F., Yamanaka, M., Horikawa, T., and Uekama, K. (1995). *Chem. Pharm. Bull.* 43: 130–136.
11 Lesieur, S., Charon, D., Lesieur, P. et al. (2000). *Chem. Phys. Lipids* 106: 127–144.
12 Santoyo-González, F., Isac-García, J., Vargas-Berenguel, A. et al. (1994). *Carbohydr. Res.* 262: 271–282.
13 (a) Sato, T., Nakamura, H., Ohno, Y., and Endo, T. (1990). *Carbohydr. Res.* 199: 31–35. (b) Szejtli, J., Lipták, A., Jodál, I. et al. (1980). *Starch* 32: 165–169. (c) Badi, N. and Guégan, P. (2007). *Carbohydr. Res.* 342: 1989–1991.
14 Sollogoub, M. (2013). *Synlett* 2629–2640.
15 Icheln, D., Gehrcke, B., Piprek, I. et al. (1996). *Carbohydr. Res.* 280: 237–250.
16 (a) Michalski, T.J., Kendler, A., and Bender, M.L. (1983). *J. Inclusion Phenom.* 1: 125–128. (b) Spencer, C.M., Stoddart, J.F., and Zarzycki, R. (1988). *J. Chem. Soc., Perkin Trans. 2* 319–324.
17 (a) Takeo, K., Ueraura, K., and Mitoh, H. (1988). *J. Carbohydr. Chem.* 7: 293–308. (b) Takeo, K., Mitoh, H., and Uemura, K. (1989). *Carbohydr. Res.* 187: 203–221.
18 Coleman, A.W., Zhang, P., and Ling, C.-C. (1992). *Carbohydr. Res.* 224: 307–309.
19 Ringard-Lefebvre, C., Bochot, A., Memişoğlu, E. et al. (2002). *Colloids Surf., B* 25: 109–117.
20 Zhang, P., Wang, A., Cui, L., and Ling, C.-C. (2012). *Org. Lett.* 14: 1612–1615.
21 Fügedi, P. (1989). *Carbohydr. Res.* 192: 366–369.
22 Harabagiu, V., Simionescu, B.C., Pinteala, M. et al. (2004). *Carbohydr. Polym.* 56: 301–311.
23 Cramer, F., Mackensen, G., and Sensse, K. (1969). *Chem. Ber.* 102: 494–508.
24 Gadelle, A. and Defaye, J. (1991). *Angew. Chem. Int. Ed. Engl.* 30: 78–80.
25 (a) Boger, J., Brenner, D.G., and Knowles, J.R. (1979). *J. Am. Chem. Soc.* 101: 7930–7931. (b) Engeldinger, E., Armspach, D., and Matt, D. (2003). *Chem. Rev.* 103: 4147–4173.
26 Angibeaud, P. and Utille, J.-P. (1992). *Synthesis* 737–738.
27 Ellwood, P., Spencer, C.M., Spencer, N. et al. (1992). *J. Inclusion Phenom. Mol. Recognit. Chem.* 12: 121–150.
28 (a) Zhang, P., Ling, C.-C., Coleman, A.W. et al. (1991). *Tetrahedron Lett.* 32: 2769–2770. (b) Grachev, M.K., Edunov, A.V., Kurochkina, G.I. et al. (2011). *Russ. J. Gen. Chem.* 81: 322–329.
29 (a) Díaz-Moscoso, A., Balbuena, P., Gómez-García, M. et al. (2008). *Chem. Commun.* 2001–2003. (b) Bienvenu, C., Martínez, Á., Jiménez Blanco, J.L. et al. (2012). *Org. Biomol. Chem.* 10: 5570–5581. (c) Méndez-Ardoy, A., Gómez-García, M., Gèze, A. et al. (2012). *Med. Chem.* 8: 524–532.
30 Fragoso, A., Cao, R., and Villalonga, R. (1995). *J. Carbohydr. Chem.* 14: 1379–1386.
31 Fulton, D. and Stoddart, J.F. (2001). *J. Org. Chem.* 66: 8309–8319.
32 (a) Ward, S. and Ling, C.-C. (2011). *Eur. J. Org. Chem.* 4853–4861. (b) Ortega-Caballero, F., Ortiz Mellet, C., Le Gourriérec, L. et al. (2008). *Org. Lett.* 10: 5143–5146.

33 (a) Parrot-Lopez, H., Ling, C.-C., Zhang, P. et al. (1992). *J. Am. Chem. Soc.* 114: 5479–5480. (b) Wazynska, M., Temeriusz, A., Chmurski, K. et al. (2000). *Tetrahedron Lett.* 41: 9119–9123. (c) Badi, N., Jarroux, N., and Guégan, P. (2006). *Tetrahedron Lett.* 47: 8925–8927.

34 Rong, D. and D'Souza, V.T. (1990). *Tetrahedron Lett.* 31: 4275–4278.

35 Guitet, M., de Beaumais, S.A., Blériot, Y. et al. (2012). *Carbohydr. Res.* 356: 278–281.

36 Tian, S. and D'Souza, V.T. (1994). *Tetrahedron Lett.* 35: 9339–9342.

37 Ashton, P.R., Boyd, S.E., Gattuso, G. et al. (1995). *J. Org. Chem.* 60: 3898–3903.

38 (a) Liptak, A., Jánossy, L., Batta, G. et al. (2000). *Carbohydr. Lett.* 4: 111–116. (b) Liptak, A., Jánossy, L., Borbas, A., and Szejtli, J. (2002). *Carbohydr. Res.* 337: 93–96.

39 (a) Fulton, D. and Stoddart, J.F. (2000). *Org. Lett.* 2: 1113–1116. (b) Eskandani, Z., Huin, C., and Guégan, P. (2011). *Carbohydr. Res.* 346: 2414–2420.

40 Bukowska, M., Maciejewski, M., and Prejzner, J. (1998). *Carbohydr. Res.* 308: 275–279.

41 Teranishi, K. and Ueno, F. (2003). *Tetrahedron Lett.* 44: 4843–4848.

42 (a) Ortega-Caballero, F., Giménez-Martínez, J.J., and Vargas-Berenguel, A. (2003). *Org. Lett.* 5: 2389–2392. (b) Gu, J., Chen, T., Zhang, P., and Ling, C.-C. (2014). *Eur. J. Org. Chem.* 5793–5805.

43 Fügedi, P., Nánásy, P., and Szejtli, J. (1988). *Carbohydr. Res.* 175: 173–181.

44 Teranishi, K. and Ueno, F. (2002). *J. Inclusion Phenom.Macrocyclic Chem.* 44: 307–311.

45 (a) Masurier, N., Estour, F., Lefèvre, B. et al. (2006). *Carbohydr. Res.* 341: 935–940. (b) Cousin, H., Cardinael, P., Oulyadi, H. et al. (2001). *Tetrahedron: Asymmetry* 12: 81–88.(c) Tisse, S., Peulon-Agasse, V., Oulyadi, H. et al. (2003). *Tetrahedron: Asymmetry* 14: 2259–2266.

46 Tian, S., Forgo, P., and D'Souza, V.T. (1996). *Tetrahedron Lett.* 37: 8309–8312.

47 Masurier, N., Lafont, O., Le Provost, R. et al. (2009). *Chem. Commun.* 589–591.

48 Melton, L.D. and Slessor, K.N. (1971). *Carbohydr. Res.* 18: 29–37.

49 Ueno, A. and Breslow, R. (1982). *Tetrahedron Lett.* 23: 3451–3454.

50 (a) Tabushi, I., Shimokawa, K., Shimizu, N. et al. (1976). *J. Am. Chem. Soc.* 98: 7855. (b) Tabushi, I. and Tatsuya, N. (1985). *J. Org. Chem.* 50: 2638–2643.

51 Yoshikiyo, K., Matsui, Y., and Yamamoto, T. (2015). *Beilstein J. Org. Chem.* 11: 1530–1540.

52 (a) Ling, C.-C., Coleman, A.E., and Miocque, M. (1992). *Carbohydr. Res.* 223: 287–291. (b) Ward, S. and Ling, C.-C. (2009). *Carbohydr. Res.* 344: 808–814.

53 Poorters, L., Armspach, D., and Matt, D. (2003). *Eur. J. Org. Chem.* 1377–1381.

54 Armspach, D., Poorters, L., Matt, D. et al. (2005). *Org. Biomol. Chem.* 3: 2588–2592.

55 Gramage-Doria, R., Rodriguez-Lucena, D., Armspach, D. et al. (2011). *Chem. Eur. J.* 17: 3911–3921.

56 Spencer, C., Stoddart, J.F., and Zarzycki, R. (1987). *J. Chem. Soc., Perkin Trans. 2* 1323–1336.

57 Armspach, D. and Matt, D. (1998). *Carbohydr. Res.* 310: 129–133.

58 (a) Rubio, E.M., García-Moreno, M.I., Balbuena, P. et al. (2005). *Org. Lett.* 7: 729–731. (b) Balbuena, P., Gonçalves-Pereira, R., Jiménez Blanco, J.L. et al. (2013). *J. Org. Chem.* 78: 1390–1403.

59 Balbuena, P., Lesur, D., González Álvarez, M.J. et al. (2007). *Chem. Commun.* 3270–3272.
60 (a) Sakairi, N. and Kuzuhara, H. (1993). *Chem. Lett.* 2077–2080. (b) Sakairi, N., Nishi, N., Tokura, S., and Kuzuhara, H. (1996). *Carbohydr. Res.* 291: 53–62.
61 (a) González-Álvarez, M.J., Balbuena, P., Ortiz Mellet, C. et al. (2008). *J. Phys. Chem. B* 114: 13717–13729. (b) Gallego-Yerga, L., González-Álvarez, M.J., Mayordomo, N. et al. (2014). *Chem. Eur. J.* 20: 6622–6627.
62 Sakairi, N. and Kuzuhara, H. (1993). *Carbohydr. Res.* 246: 61–73.
63 Matsuoka, K., Shiraishi, Y., Terunuma, D., and Kuzuhara, H. (2001). *Tetrahedron Lett.* 42: 1531–1533.
64 (a) Sollogoub, M., Mallet, J.M., and Sinaÿ, P. (1998). *Tetrahedron Lett.* 39: 3471–3472. (b) Pearce, A.J., Sollogoub, M., Mallet, J.-M., and Sinaÿ, P. (1999). *Eur. J. Org. Chem.* 2103–2117.
65 Lecourt, T., Herault, A., Pearce, A.J. et al. (2004). *Chem. Eur. J.* 10: 2960–2971.
66 Pearce, A.J. and Sinaÿ, P. (2000). *Angew. Chem. Int. Ed.* 39: 3610–3612.
67 Wang, B., Zaborova, E., Guieu, S. et al. (2014). *Nat. Commun.* 5: 5354.
68 (a) Rawal, G.K., Rani, S., and Ling, C.-C. (2009). *Tetrahedron Lett.* 50: 4633–4636. (b) Rawal, G.K., Rani, S., Ward, S., and Ling, C.-C. (2010). *Org. Biomol. Chem.* 8: 171–180.
69 (a) Sollogoub, M. (2009). *Eur. J. Org. Chem.* 1295–1303. (b) Guieu, S. and Sollogoub, M. (2008). *Angew. Chem. Int. Ed.* 47: 7060–7063. (c) Bistri, O., Sinaÿ, P., Jiménez-Barbero, J., and Sollogoub, M. (2007). *Chem. Eur. J.* 13: 9757–9774. (d) Zaborova, E., Guitet, M., Prencipe, G. et al. (2013). *Angew. Chem. Int. Ed.* 52: 639–644.
70 Guieu, S. and Sollogoub, M. (2008). *J. Org. Chem.* 73: 2819–2828.
71 Guieu, S., Zaborova, E., Blériot, Y. et al. (2010). *Angew. Chem. Int. Ed.* 49: 2314–2318.
72 Guitet, M., Zhang, P., Marcelo, F. et al. (2013). *Angew. Chem. Int. Ed.* 52: 7213–7218.
73 Kretschmann, O., Choi, S.W., Miyauchi, M. et al. (2006). *Angew. Chem. Int. Ed.* 45: 4361–4365.
74 (a) Rousseau, C., Ortega-Caballero, F., Nordstrøm, L.U. et al. (2005). *Chem. Eur. J.* 11: 5094–5101. (b) Bjerre, J., Rousseau, C., Marinescu, L., and Bols, M. (2008). *Appl. Microbiol. Biotechnol.* 81: 1–11.
75 (a) Ghosh, R., Zhang, P., Wang, A., and Ling, C.-C. (2012). *Angew. Chem. Int. Ed.* 51: 1548–1552. (b) Ghosh, R., Hennigan, C., and Ling, C.-C. (2013). *Tetrahedron* 69: 5227–5233.
76 Gu, J., Chen, T., Wang, Q. et al. (2015). *Carbohydr. Res.* 410: 36–46.
77 Wang, W., Pearce, A.J., Zhang, Y., and Sinaÿ, P. (2001). *Tetrahedron: Asymmetry* 12: 517–523.
78 (a) Xiao, S., Yang, M., Sinaÿ, P. et al. (2010). *Eur. J. Org. Chem.* 1510–1516. (b) du Roizel, B., Baltaze, J.-P., and Sinaÿ, P. (2002). *Tetrahedron Lett.* 43: 2371–2373.
79 Letort, S., Mathiron, D., Grel, T. et al. (2015). *Chem. Commun.* 2601–2604.
80 Xiao, S., Yang, M., Yu, F. et al. (2011). *Tetrahedron Lett.* 52: 5273–5276.
81 (a) Luo, X., Chen, Y., Huber, J.G. et al. (2004). *C.R. Chim.* 7: 25–28. (b) Chen, Y., Huber, J.G., Zhang, Y., and Sinaÿ, P. (2005). *C.R. Chim.* 8: 27–30.

14

Protecting Group Strategies Toward Sulfated Glycosaminoglycans

Hélène Ledru[1], *Pascal Matton*[2], *Jean-Maurice Mallet*[2], *and Chrystel Lopin-Bon*[1]

[1] *Université d'Orléans and CNRS, Institut de Chimie Organique et Analytique (ICOA), UMR 7311, rue de Chartres, 45067 Orléans, France*
[2] *Laboratoire des biomolécules, LBM, Département de chimie, École Normale Supérieure, PSL University, Sorbonne Université, CNRS, 24 rue Lhomond, 75005 Paris, France*

14.1 Introduction

Glycosaminoglycans (GAGs) are a family of complex polyanionic macromolecules linked to a core protein through a tetrasaccharidic linkage region to form what is called proteoglycans (PGs). GAGs play important roles in a plethora of biological processes such as cell growth and proliferation, embryonic development, and coagulation cascade. They are also involved in several diseases including arthropathies, Alzheimer's disease, and cancer. GAGs are linear polysaccharides composed of repeating disaccharide units containing a hexuronic acid and a hexosamine sulfated at various positions of the sugar residues. GAGs are classified into two types depending on the nature of the hexosamine (gluco- or galactosamine): heparin/heparin sulfate (Hep/HS) (*gluco*saminoglycans) and chondroitin sulfate/dermatan sulfate (CS/DS) (*galacto*saminoglycans). Biosynthesis of PG-GAG chains starts with the formation of the tetrasaccharidic linkage region (GlcA-β-1,3-Gal-β-1,3-Gal-β-1,4-Xyl) attached to a L-serine residue of a core protein. From this linkage region, the divergence between the two types of GAGs occurs, and addition of β-D-GalNAc initiates the formation of CS/DS, whereas addition of α-D-GlcNAc initiates the Hep/HS formation. It has been reported that the linkage region may be sulfated at position 4 or 6 of both D-Gal units, but the exact role of these substitutions is not fully understood (Figure 14.1).

Because biosynthetic mechanisms involved in the PG-GAG biosynthesis remain obscure, syntheses of various sulfoforms of linkage region or of GAGs have been of great interest for many years. This chapter will focus on the synthesis of sulfated chondroitin (CS), heparin (HS), and linkage region. The importance of protecting group strategy to introduce regio- and /or chemoselectively sulfate will be pointed out.

Protecting Groups: Strategies and Applications in Carbohydrate Chemistry, First Edition.
Edited by Sébastien Vidal.

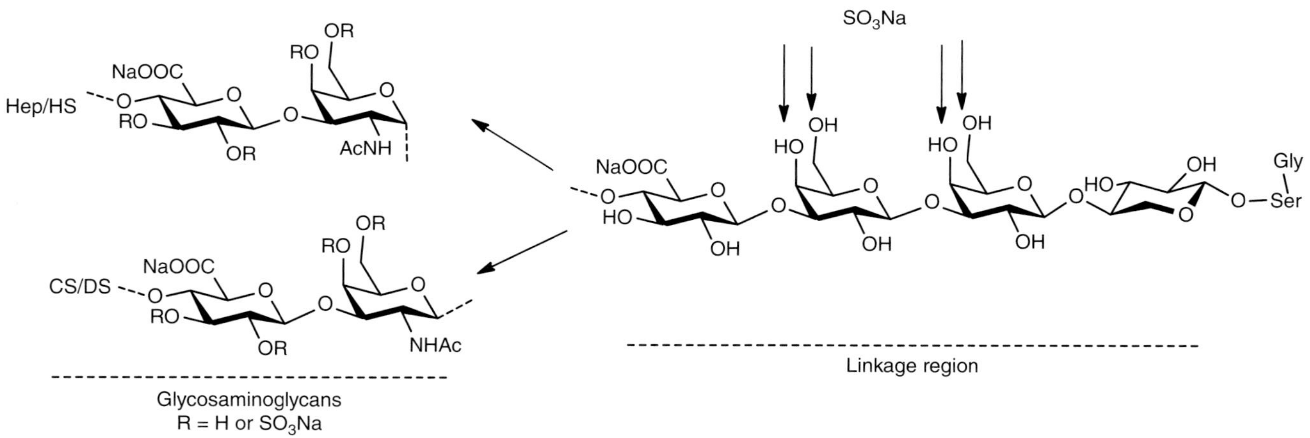

Figure 14.1 Schematic representation of principal steps of PG-GAG chain synthesis.

14.2 *O*- and *N*-sulfation in Glycosaminoglycan Glycosaminoglycans Synthesis

O- and *N*-sulfonation – also called sulfation – is specific reactions of GAG syntheses (see Chapter 17) and is generally not an easy task. Sulfates are fragile groups, much more than the related phosphate groups: sulfates are sensitive to aqueous acid conditions leading to sulfate hydrolysis, and in basic conditions, they are potential leaving groups prone to side reaction such as epoxide formation and elimination. That is the reason why they are introduced in the very last steps of the synthesis. Sulfation reagents are activated sulfuric acid derivatives [1]. Classical *O*-sulfation reagents are $SO_3{\cdot}Pyr$, $SO_3{\cdot}NEt_3$, and $SO_3{\cdot}NMe_3$ in dry DMF (at 50 °C). They are solid, easy to handle, and generally used in large excess. Selective *N*-sulfation is generally performed in water (pH 9.5 Na_2CO_3) and $SO_3{\cdot}Pyr$ or $SO_3{\cdot}NMe_3$ [2]. *O*- and *N*-sulfation are generally achieved in two distinct steps, mainly because of the choice of the protecting groups. Chlorosulfonic acid has been used in early works in polysaccharides but was not found suitable for GAG synthesis [3] (extensive degradation because of acidity of the reagent). Attempts to prepare and use protected sulfate (such as phenyl ester using phenylchlorosulfate) did not find any application, probably because of their instability [4].

14.3 Protecting Group Strategies for the Synthesis of Sulfated Oligosaccharides of the Proteoglycans Linkage Region

The linkage region of PGs is a tetrasaccharide sequence, β-D-GlcA-(1 → 3)-β-D-Gal-(1→3)-β-D-Gal-(1→4)-β-D-Xyl. During the biosynthesis, sulfations may occur on D-Gal units. Synthetic oligosaccharides of the linkage region are sulfated at positions 4 or 6 of one or both D-Gal unit(s). This involves distinguishing not only the two D-Gal units but also positions 4 and 6. It is therefore necessary to develop synthetic strategies based on orthogonal protecting groups.

14.3.1 Sulfations via a 4,6-Diol Precursor

In this first approach, the D-Gal units are distinguished by a particular system of protections P_1 and P_2. These protective groups, which have to be selectively removable to make regioselective sulfation, are usually introduced during the synthesis of the monosaccharide building block (Figure 14.2).

Once the two D-Gal units are correctly distinguished, the synthesis of a 4,6-diol intermediate, precursor of the 4- or 6-sulfated products, occurs. Based on the different reactivities of these two hydroxyl groups, regioselective 6-sulfation affords the 6-sulfated product. 4-Sulfated compound can be obtained by regioselective 6-benzoylation followed by a 4-sulfation (Scheme 14.1).

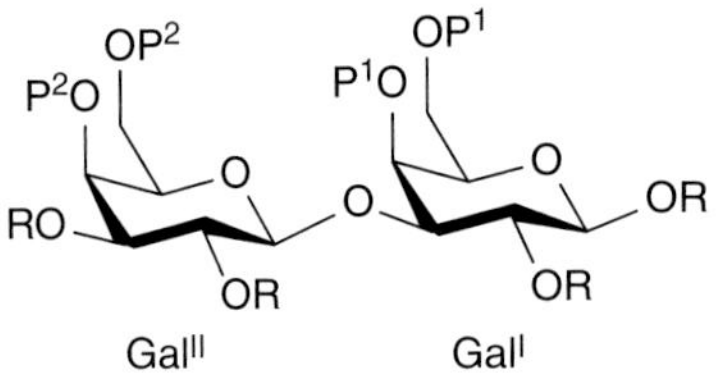

Figure 14.2 Structure of the GalII–GalI sequence.

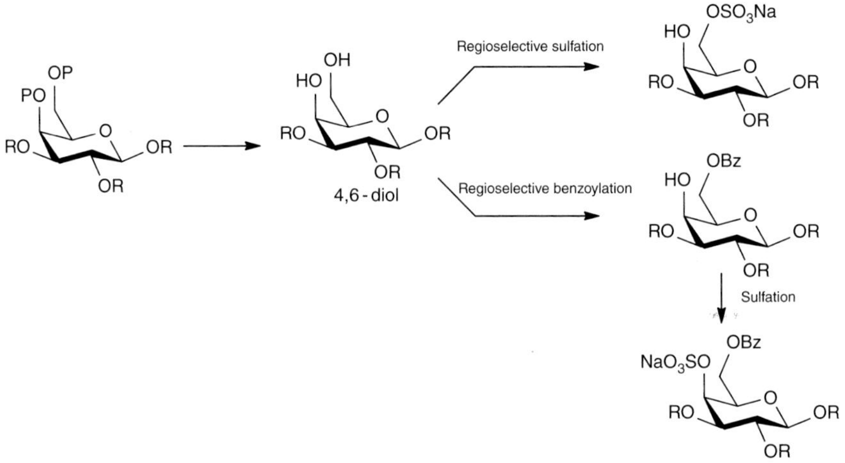

Scheme 14.1 Synthesis of 4- and 6-sulfated products via a 4,6-diol.

Several examples following this strategy were reported by Jacquinet and coworker in 2004 with the syntheses of various sulfoforms of β-D-Gal-(1→3)-β-D-Gal-OMP [5] and β-D-GlcA-(1→3)-β-D-Gal-(1→3)-β-D-Gal-OMP [6] (Figure 14.3).

The disaccharide **1**, bearing a benzylidene acetal on GalI and levulinoyl groups (Lev) on GalII, was selected as the starting material. Selective acid hydrolysis of the benzylidene acetal followed by benzoylation (Bz) gave the

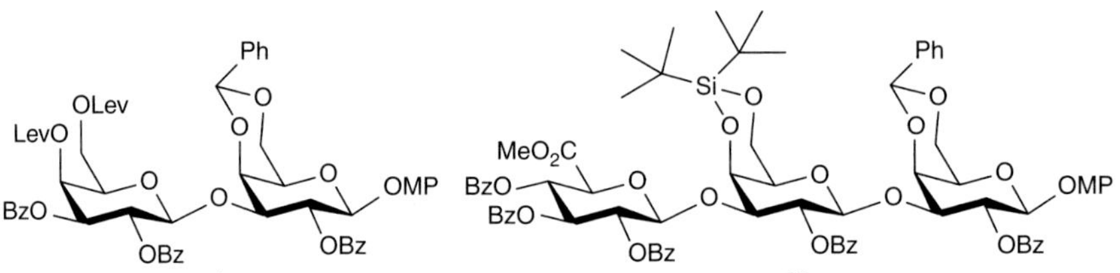

Figure 14.3 Structure of disaccharide **1** and trisaccharide **10** synthesized by Jacquinet's group.

disaccharide **2**. Deprotection of levulinoyl groups with hydrazinium acetate and regioselective 6-sulfation afforded the sulfoform **3**, 6-sulfated on Gal^{II}, whereas a regioselective 6-benzoylation with benzoyl cyanide followed by a 4-sulfation gave the sulfoform **4**, 4-sulfated on Gal^{II}. In order to have access to the compound sulfated on Gal^{I}, disaccharide **5** was obtained after selective delevulinoylation and benzoylation. After acid hydrolysis of the benzylidene acetal, the same sulfation strategy was applied, and 4- and 6-sulfated Gal^{I} disaccharides **6** and **7** were obtained. Full deprotection of the two temporary protecting groups followed by regioselective sulfation gave the sulfoform **9**, 6-sulfated on both Gal^{I} and Gal^{II} (Scheme 14.2).

The same method was applied to obtain various sulfoforms of β-D-GlcA-(1→3)-β-D-Gal-(1→3)-β-D-Gal-OMP sequence [6] starting from trisaccharide **10** (see Figure 14.3), using selective deprotection of the benzylidene acetal (with acid hydrolysis) or silylene acetal, which can be selectively removed with $Et_3N{\cdot}3HF$.

In 2012, the same group reported the syntheses of various sulfoforms of the biotinylated trisaccharide β-D-GlcA-(1→3)-β-D-Gal-(1→3)-β-D-Gal [7]. Different protecting groups were used to differentiate the two D-Gal moieties. Starting from trisaccharide **11** bearing a silylene acetal on Gal^{I} and benzoates on Gal^{II} and trisaccharide **13** bearing Bz groups on Gal^{I} and chloroacetyl groups (ClAc) on Gal^{II}, the various sulfoforms could be obtained, respectively, sulfated on Gal^{I} or Gal^{II} (Figure 14.4). Selective deprotection of the silylene acetal of the compound **11** gave the 4,6-diol **12**, which can be directly 6-sulfated or 6-benzoylated followed by the 4-sulfation. Deprotection of the chloroacetate groups of **13** with thiourea gave the 4,6-diol **14**, which can be similarly regioselectively sulfated.

14.3.2 Sulfations via Temporary Protection on Specific Positions

The second strategy is based on the use of orthogonal protecting groups P_4 and P_6, with a temporary protection that can be selectively removed to introduce sulfates. This approach allows the access to 4- or 6-sulfoforms through selective deprotections (Scheme 14.3).

Tamura and Nishihara, in 1999, prepared the disaccharide β-D-Gal(6S)-(1→4)-β-D-Xyl-Ser **22** using *tert*-butylchlorodiphenylsilane (TBDPS) as temporary protection on the hydroxyl to be sulfated [8] (Table 14.1, entry 1). Later, they applied this strategy to the synthesis of two trisaccharides (β-D-Gal-(1→3)-β-D-Gal-(1→4)-β-D-Xyl-Ser) **16** and **17**, precursors of compounds mono-6-sulfated on each Gal unit [9]. After deprotection of TBDPS with tetrabutylammonium fluoride (TBAF) and acetic acid and 6-sulfation, the molecules **23** and **24** were obtained (Table 14.1, entries 2 and 3).

In 2006, Jacquinet reported the synthesis of 6-sulfated compounds, β-D-Gal(6S)-(1→4)-β-D-Xyl-MN **25** and β-D-Gal-(1→3)-β-D-Gal(6S)-(1→4)-β-D-Xyl-MN **26** [10]. For the preparation of **25**, an imidate of D-galactose with a trityl as temporary protection on the primary alcohol (see Chapter 2) was prepared

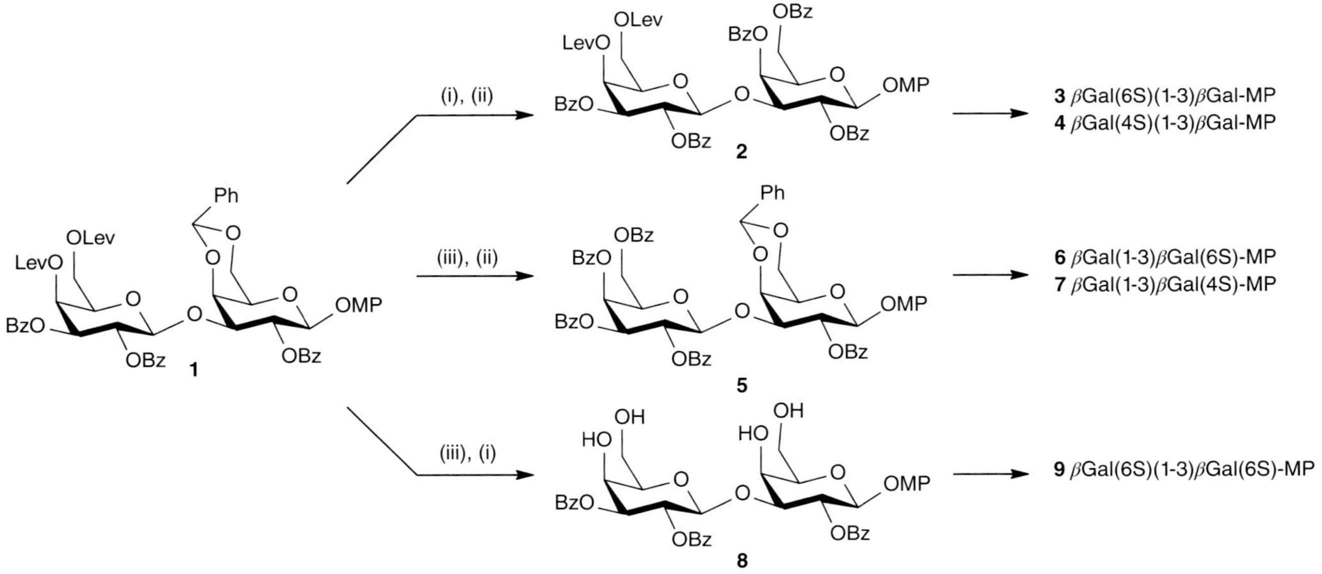

Scheme 14.2 Synthesis of various sulfoforms following the first strategy. (i) 90% TFA, RT, 15 min; (ii) PhCOCl, pyridine, RT, 16 h; and (iii) hydrazinium acetate, pyridine, RT, 8 min.

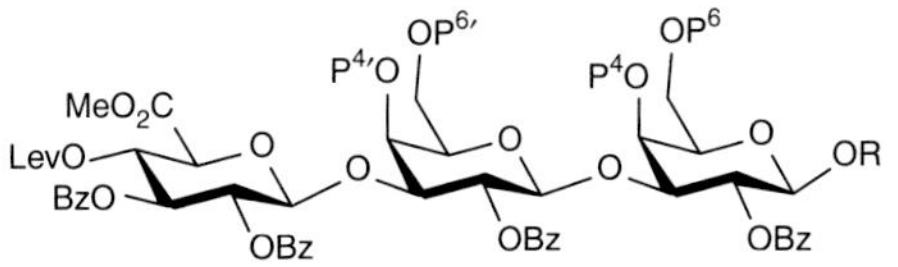

Figure 14.4 Structure of the GlcA-Gal^{II}–Gal^{I} compounds synthesized by Jacquinet's group.

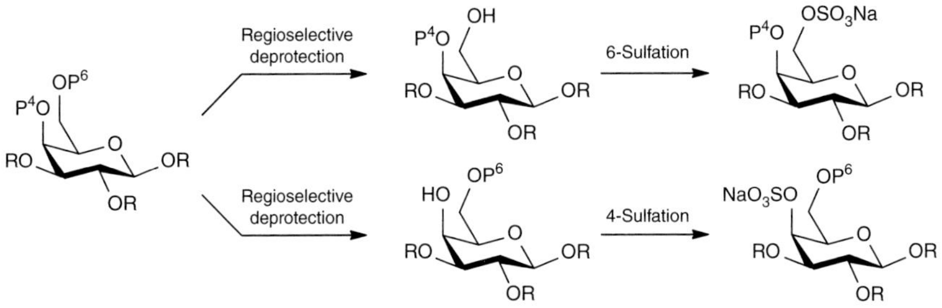

Scheme 14.3 Synthesis of 4- and 6-sulfated products based on temporary protection.

for subsequent sulfation of this alcohol. This trityl was replaced by a levulinoyl group, (resistant enough to the next steps of the synthesis), *via* a treatment with boron trifluoride diethyletherate and methanol followed by levulinoylation in classical conditions providing compound **18**. The 6-sulfation could be performed after delevulinoylation (Table 14.1, entry 4).

The target trisaccharide **26** was obtained by a similar strategy. A disaccharide was synthesized with a silylene on Gal^{I} and benzoyl groups on Gal^{II}. After deprotection of the silylene, the primary alcohol was levulinoylated to give compound **19**, a precursor of **26** (Table 14.1, entry 5).

The same protection allows access to 4-sulfated compound, as described by Goto and Ogawa [11] who synthesized the 4-sulfated tetrasaccharide β-D-GlcA-(1→3)-β-D-Gal(4S)-(1→3)-β-D-Gal-(1→4)-β-D-Xyl-Ser **27**, after delevulinoylation and 4-sulfation of **20**. A disaccharide with benzyl groups (Bn) on Gal^{I} and a benzylidene acetal on Gal^{II} was prepared. The synthesis of **20** was achieved after deprotection of benzylidene acetal, selective acetylation on the primary alcohol followed by the levulinoylation of the remaining alcohol, and exchange of benzoyl groups into acetyl groups (Table 14.1, entry 6).

To prepare various sulfoforms of β-D-GlcA-(1→3)-β-D-Gal(4S)-(1→3)-β-D-Gal-(1→4)-β-D-Xyl-protein with a peptide aglycon [12], Jacquinet's group started from the disaccharide GlcA-Gal-OBn, with benzylidene as P_4 and P_6 protections of Gal^{II}. This protection was removed, and a regioselective benzoylation followed by the chloroacetylation of the remaining alcohol gave a precursor of the disaccharide **21**. At the end of their synthesis, the ClAc group was removed with thiourea in ethanol–pyridine, enabling sulfation at this position to obtain compound **28** (Table 14.1, entry 7).

Table 14.1 Structures of Gal^{II}–Gal^{I} sequences reported in this part.

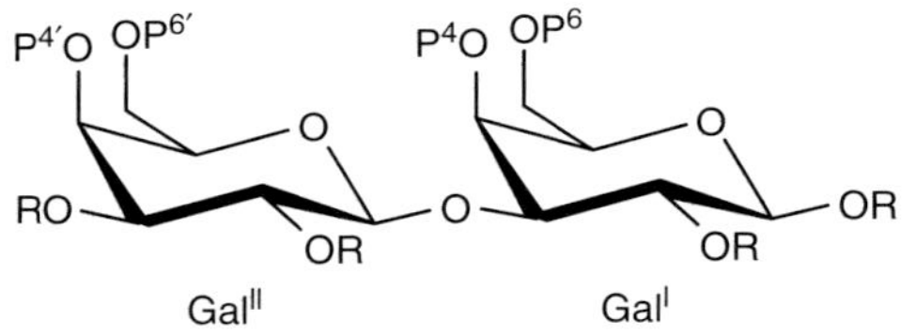

			Gal^{I}		Gal^{II}			Sulfated products			
								Gal^{I}		Gal^{II}	
Entry	**References**	**No.**	**P^4**	**P^6**	**P$^{4'}$**	**P$^{6'}$**	**No.**	**P^4**	**P^6**	**P$^{4'}$**	**P$^{6'}$**
1	[8]	**15**	MBz	TBDPS	—	—	**22**	H	SO_3Na	—	—
2	[9]	**16**	PhCH		MBz	TBDPS	**23**	H	H	H	SO_3Na
3	[9]	**17**	MBz	TBDPS	Ac	Ac	**24**	H	SO_3Na	H	H
4	[10]	**18**	Bz	Lev	—	—	**25**	H	SO_3Na	—	—
5	[10]	**19**	Bz	Lev	Bz	Bz	**26**	H	SO_3Na	H	H
6	[11]	**20**	Ac	Ac	Lev	Ac	**27**	H	H	SO_3Na	H
7	[12]	**21**	Bz	Bz	ClAc	Bz	**28**	H	H	SO_3Na	H

14.4 Protecting Group Strategy for the Synthesis of Chondroitin Sulfate

Chondroitin sulfate (CS) consists of a disaccharide repeating unit composed of a D-glucuronic acid (D-GlcA) and a 2-acetamido-2-deoxy-D-galactose (D-GalNAc) arranged in the sequence [→4)-β-D-Glc*p*A-(1→3)-β-D-Gal*p*NAc-(1→]$_n$ and contains, on average, one sulfate per disaccharide unit [13]. Chondroitin sulfates are sulfated only at the hydroxyl group, and the major variants contain the 4- and/or 6-sulfo group on the D-GalN residue: CS-A (4-*O*-sulfo group), CS-C (6-*O*-sulfo group), and CS-E (4,6-di-*O*-sulfo group). Other oversulfated variants including sulfation on the D-GlcA are known such as CS-D (6,2′-di-*O*-sulfo group), CS-K (4,3′-di-*O*-sulfo group), CS-L (6,3′-di-*O*-sulfo group), and CS-M (4,6,3′-tri-*O*-sulfo group) (Figure 14.5) [14]. These differences in sulfation patterns, also called "sulfo code" [15], result in significantly different biological properties such as growth factor, signaling, osteoarthritis, or morphogenesis [16].

For many years, the chemical syntheses of CS oligosaccharides have been of great interest. The choice of suitable orthogonal protecting groups is crucial to introduce the sulfate groups on the desired carbohydrate residue and also on the right position. It has to be noted that sulfation patterns are installed at the last stage of the syntheses under thermal conditions. Very recently, synthesis of CS-C and -E disaccharides has been reported using sulfation under microwaves [17].

14.4.1 Sulfation at Position 4 or/and 6 of the GalN: Synthesis of CS-A, -C, and -E

14.4.1.1 4- or 6-Sulfation from a Single 4,6-Diol

One of the reported approaches starts with the synthesis of a 4,6-diol intermediate, which is a single precursor for the of CS-A, -C, and -E variants. Regioselective 6-benzoylation (Bz) of the primary alcohol using benzoyl cyanide followed by 4-sulfation of the free hydroxyl group affords precursor of the CS-A sequence. 6-Sulfated variant (CS-C) can be obtained by regioselective 6-sulfation of the 4,6-diol under mild conditions, whereas sulfation under hard conditions generates disulfated CS-E precursors (Scheme 14.4).

Several protecting groups on the GalN unit (P) are reported as a precursor of the 4,6-diol, and one of the most commonly used is the benzylidene acetal, which can be hydrolyzed under acid conditions. This protecting group has been used for the synthesis of CS-A and -C oligosaccharides. One of the first examples was

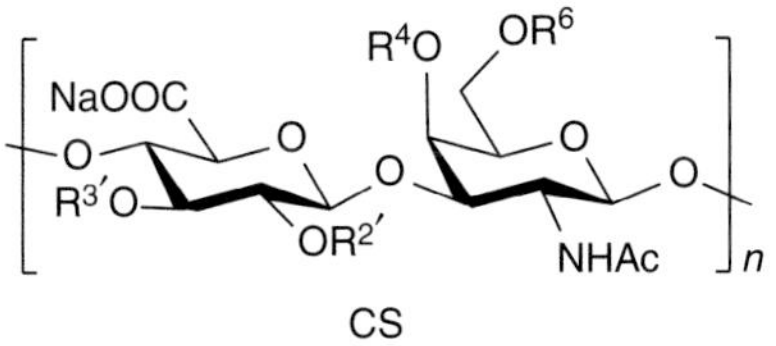

CS

Figure 14.5 Structure of the CS disaccharide units.

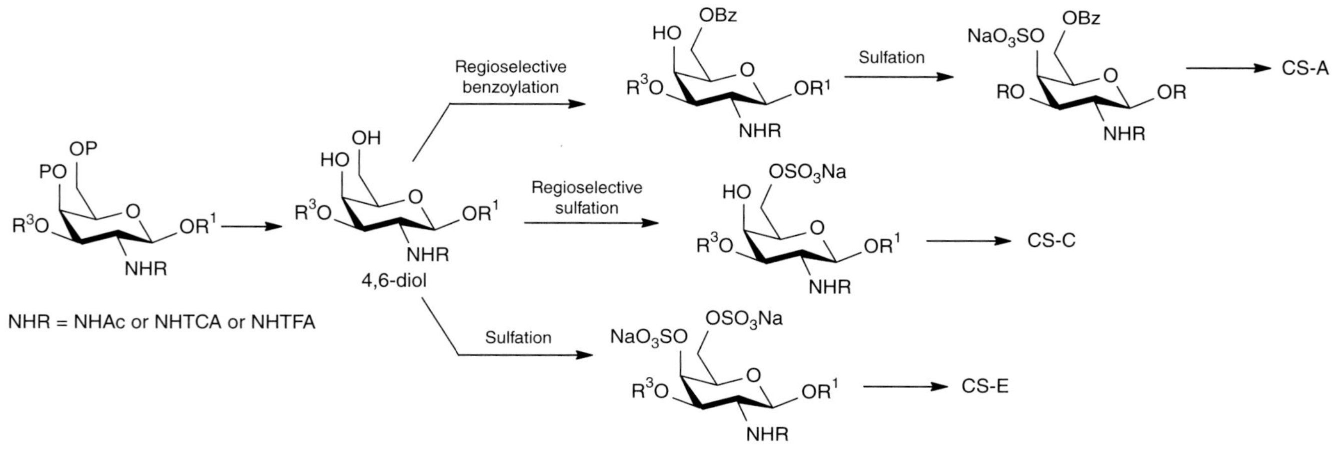

Scheme 14.4 Synthesis of CS-A, -C, and -E starting from a 4,6-diol.

reported in 1998 by Jacquinet et al. in their synthesis of CS-A and -C disaccharides having a GlcA-GalN motif starting from the disaccharide **29** bearing benzylidene acetal on the GalN unit and benzyl (Bn) aglycon [18] (Scheme 14.5). Acid hydrolysis of the acetal afforded the corresponding 4,6-diol **30** which, after treatment with BzCN and 4-sulfation, gave the disaccharide **31** in good yields. Regioselective 6-sulfation under mild conditions afforded the disaccharide **33**. Complete deprotection yielded the final CS-A and -C disaccharides **32** and **34**. Following the same strategy, CS-A and -C pentasaccharides were isolated starting from a protected pentasaccharide bearing two benzylidene acetals [19].

Hydrolysis of benzylidene acetal was also reported for synthesis of CS-E oligosaccharides. For example, Tamura et al. described the synthesis of di-, tri-, tetra-, hexa-, and octasaccharide CS-E having a GalN-GlcA motif from azide or acetamide-type building blocks **35** [20]. The presence of orthogonal levulinoyl and methoxyphenyl (MP) groups allowed the divergent synthesis of acceptor and donor, which were engaged in elongation strategy to obtain the oligosaccharides **36–38** with defined size. Hydrolysis of benzylidene acetal followed by exhaustive sulfation and deprotection afforded CS-E oligosaccharides **39–41** (Scheme 14.6). Very recently, similar approach was applied on a disaccharide **74** for the synthesis of CS-E disaccharide (see Scheme 14.13, Section 14.4.2) [21].

Benzylidene acetal is a protecting group of choice to have access to a 4,6-diol intermediate, but a few other protecting groups are described. The *para*-methoxybenzylidene (PMB) has been reported by Hsieh-Wilson and coworkers in their synthesis of the tetrasaccharide **43** having GlcA-GalN motif [15, 22]. Oxidative cleavage using DDQ afforded the target 4,6-diol **44** ($n = 1$), which was then submitted to the above sulfation strategy to give tetrasaccharide with a distinct sulfation sequence (CS-A, -C, and -E) (Scheme 14.7). More recently, an innovative strategy based on hydrolysis of commercially available CS polymer was described, allowing the preparation on a large scale of a key disaccharide **45** possessing a benzylidene acetal and a 2-naphthylmethyl (NAP) aglycon [23]. However, as glycosylation reaction in the presence of benzylidene acetal leads to undesired 1,2-*cis*-glycosides, benzylidene acetal was replaced by chloroacetyl groups as temporary protection of the hydroxyl to be sulfated and after elongation. A set of protected oligosaccharides **46** with different aglycons was obtained. Selective deprotection of chloroacetyl groups with thiourea and then reduction of trichloroacetamides (TCA) afforded the 4,6-diol **44** ($n = 1, 2$) as a precursor of a library of CS-A, -C, and -E oligosaccharides bearing biotin or fluorescent aglycons (Scheme 14.7) [24]. Very recently, Nieto and coworkers described the synthesis of the tetrasaccharide **48** having a GalN-GlcA motif and bearing a silylene acetal on the GalN unit. Selective deprotection of the silylene group followed by exhaustive sulfation under microwave irradiation, deprotection, and selective N-acetylation gave the tetrasaccharide CS-E **39** (Scheme 14.8) [25].

14.4.1.2 4- or 6-Sulfation After Reductive Ring Opening of 4,6-Benzylidene Acetal

In order to differentiate the positions 4 and 6 of the GalN unit, another approach has been described based on the regioselective ring opening of benzylidene acetal (see Chapter 11). Selective cleavage of a benzylidene group on a

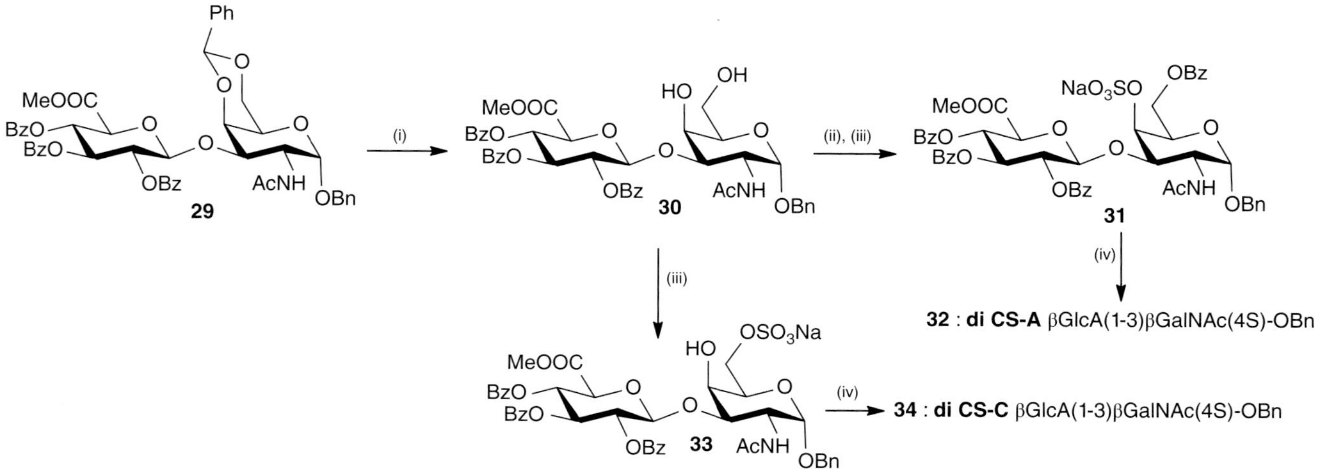

Scheme 14.5 Synthesis of CS-A and -C disaccharides. (i) AcOH-H_2O, 100 °C, 87%; (ii) BzCN, pyridine, 93%; (iii) $SO_3 \cdot NMe_3$, DMF, 50 °C, then [Na^+] ion exchange, 90%; and (iv) LiOH/H_2O_2, THF/H_2O, then NaOH, 80–82%.

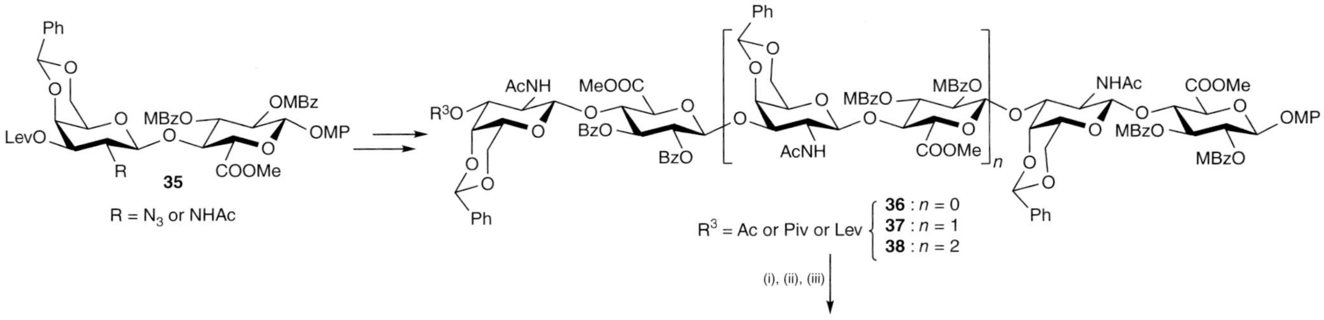

39 : **tetra CS-E** βGalNAc(4,6S)(1-4)βGlcA(1-3)βGalNAc(4,6S)(1-4)βGlcA-OMP
40 : **hexa CS-E** [βGalNAc(4,6S)(1-4)βGlcA(1-3)]$_2$βGalNAc(4,6S)(1-4)βGlcA-OMP
41 : **octa CS-E** [βGalNAc(4,6S)(1-4)βGlcA(1-3)]$_3$βGalNAc(4,6S)(1-4)βGlcA-OMP

Scheme 14.6 Synthesis of CS-E oligosaccharides. (i) CSA, CH_2Cl_2-MeOH, 80%; (ii) $SO_3{\cdot}NMe_3$, DMF, 50 °C, then [Na^+] ion exchange, 91–95%; and (iii) LiOH, THF/H_2O, then NaOH, 70–75%.

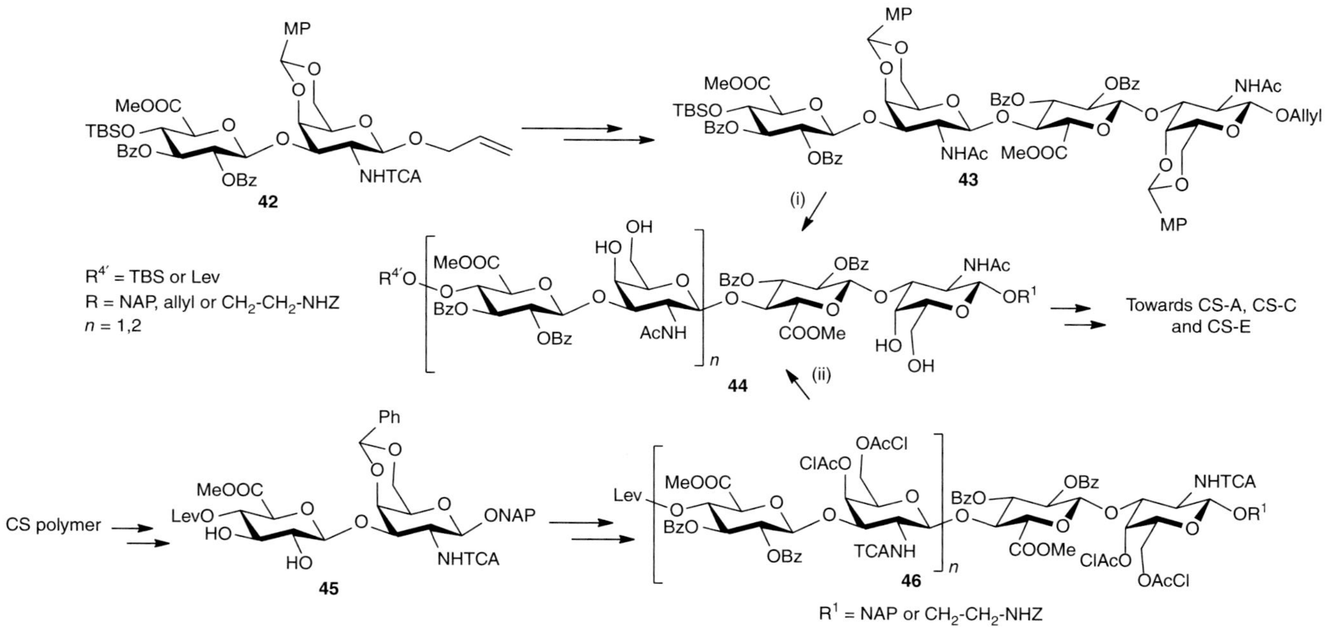

Scheme 14.7 Synthesis of CS-A, -C, or -E using different protecting groups on the GalN unit. (i) DDQ, H_2O-CH_3CN, 93%. (ii) Thiourea, pyridine-EtOH, then Bu_3SnH, AIBN, benzene, 71–91%.

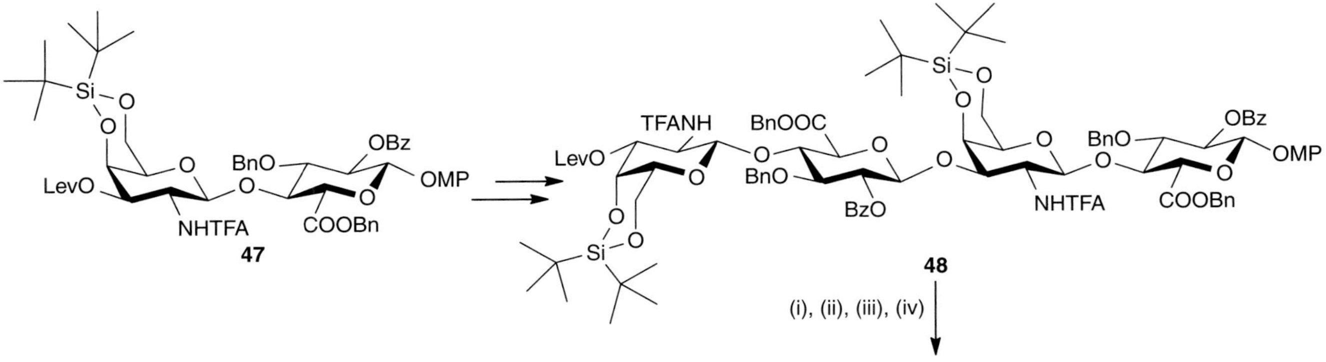

39 : **tetra CS-E** βGalNAc(4,6S)(1-4)βGlcA(1-3)βGalNAc(4,6S)(1-4)βGlcA-OMP

Scheme 14.8 Synthesis of CS-E tetrasaccharide. (i) $(HF)_n$·Pyr, THF, 0 °C, 100%; (ii) SO_3·NMe_3, DMF, 100 °C, microwave (MW) irradiation, then [Na^+] ion exchange, 95%; (iii) LiOH, THF/H_2O, then NaOH, then Ac_2O, MeOH, Et_3N, 62%; and (iv) H_2, $Pd(OH)_2/C$, H_2O/MeOH, 92%.

GalN residue is limited, and few examples are reported for the synthesis of CS-A or -C. Tamura and coworkers described the synthesis of CS-A tetrasaccharide starting from the protected compound **36** (Scheme 14.9) [16a, 26]. Reductive opening of the benzylidene acetal allowed the formation of corresponding 4-OH/6-OBn intermediate **49**. Sulfation and deprotection afforded the CS-A tetrasaccharide. As regioselective 6-*O*-sulfation failed on these structures, compound **49** served as a starting material for the synthesis of 6-sulfated compound **51** after acetylation of the free 4-hydroxy group followed by hydrogenolysis of the 6-benzyl group and subsequent 6-sulfation. This reductive opening procedure has also been applied on di- and trisaccharides [26]. Very recently, regioselective reductive opening of benzylidene acetal **74** has been explored for the synthesis of CS-A and -C disaccharides, and two regioisomers were obtained depending on the choice of the promoters (see Scheme 14.13, Section 14.4.2) [21].

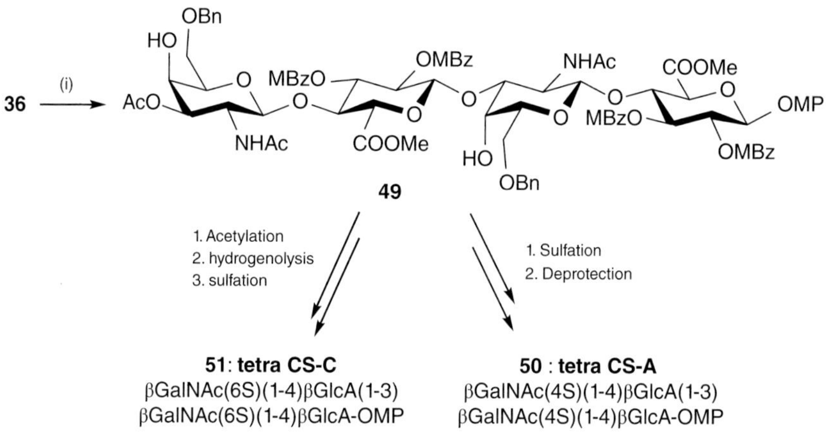

Scheme 14.9 Reductive opening of benzylidene acetal. (i) HCl, $NaBH_3CN$, THF, 51%.

14.4.1.3 4- or 6-Sulfation from Temporary Protecting Groups

The last approach described in the literature to have access to CS-A or -C motif is based on the design of oligosaccharides bearing a temporary protection at position 4 or 6, which could be selectively removed to introduce sulfates. This strategy was recently reported starting from the disaccharide **45** (obtained by hemisynthesis from CS polymer). Benzoylation followed by hydrolysis of benzylidene gave the corresponding 4,6-diol **52**. A one-pot process using acid hydrolysis of 4,6-orthoester followed by chloroacetylation produced both the 6-benzoate **53** and the 4-benzoate **54**. After elongation and selective deprotection of chloroacetyl, tri- and tetrasaccharides **55–58** were obtained as precursors of the biotinylated CS-A and -C tri- and tetrasaccharides **59–62** (Scheme 14.10) [27].

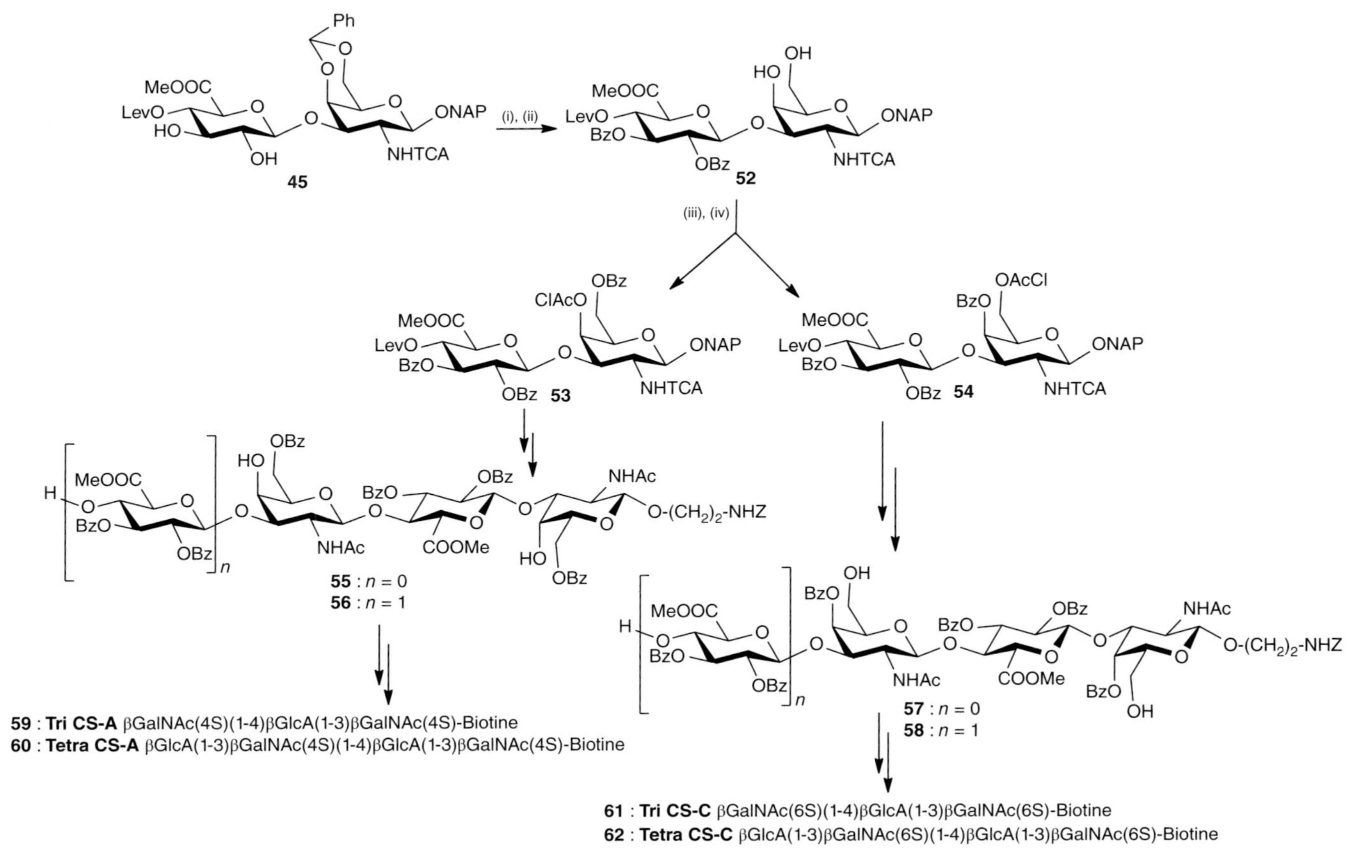

Scheme 14.10 Synthesis of CS-A and -C using orthoester strategy. (i) BzCl, CH_2Cl_2-pyridine, 91%; (ii) TFA-H_2O, CH_2Cl_2, 84%; (iii) $PhC(OMe)_3$, CSA then 80% AcOH, 35% for 6-Bz, 37% for 4-Bz; and (iv) $(ClAc)_2O$, CH_2Cl_2-pyridine, 90% for **53**, 87% for **54**.

14.4.2 Sulfation at Position 2 or 3 of the GlcA and 4 and/or 6 of the GalN: Synthesis of CS-D, -K, -L, and -M

Although CS-A, -C, and -E are only sulfated on the GalN unit, CS-D, -K, -L, and -M also implies 2′- or 3′-sulfation. The first example of CS-D synthesis was focused on GlcA-GalN-OMe and GlcA-GalN-OH disaccharides [28a]. Later, CS-D tetra- and hexasaccharides were prepared from the elaborated disaccharide **63** bearing benzoate as temporary protection for the future sulfated position, benzyl as permanent protection, and 4′-ClAc for elongation of the chain. Selective deprotection of the benzoate followed by sulfation and hydrogenation afforded the CS-D 6,2′-sulfated oligosaccharides **66** and **67** (Scheme 14.11) [28b].

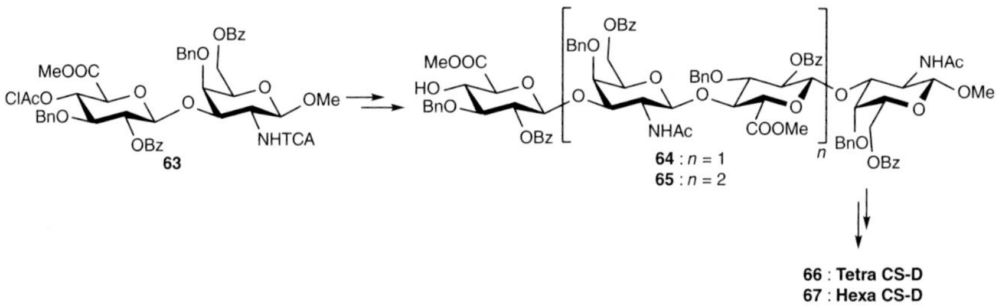

Scheme 14.11 Synthesis of CS-D oligosaccharides.

As discussed above, the disaccharide **45**, bearing a benzylidene acetal, was a starting material of choice for the synthesis of CS-A, -C, and -E oligosaccharides (Scheme 14.7). This disaccharide was also used to access CS-K, -L, -M disaccharides. Regioselective benzoylation of **45** using benzoyl cyanide afforded the 3′-benzoate **68**, which was converted into the disaccharide **69** (precursor of CS-D) having 2′,6-ClAc as temporary protecting groups and 3′,4-Bz as permanent groups. Benzoylation of **45** using tin-methodology gave the 2′-benzoate which, after hydrolysis of benzylidene and reduction of the TCA, afforded the triol **70**. Exhaustive sulfation allowed the access to 4,6,3′-trisulfated compound **73** (precursor of CS-M), whereas carefully controlled selective sulfation afforded the 6,3′-disulfated disaccharide **72** (precursor of CS-L). Selective 6-benzoylation followed by sulfation gave the 4,3′-disulfated **71** (precursor of CS-K) (Scheme 14.12) [24a].

Very recently, Wakao et al. reported the synthesis of all types of CS disaccharides with a glucose aglycon [21]. This library was obtained from the key building block **74** having a benzylidene acetal on the GalN unit and different orthogonal groups on the GlcA: a 4-acetaminobenzyl (AAB) group on position 3′, benzoate ester on position 2′, and benzyl ether as permanent protection on the other positions. Hydrolysis of benzylidene gave the 4,6-diol **75** as a precursor of CS-E disaccharide. Selective deprotection of the AAB group using DDQ afforded 4,6,3′-triol from which CS-M disaccharide could be obtained. Regioselective reductive opening of the benzylidene acetal provided either the 4-OH disaccharide **76** (a precursor of CS-A) or the 6-OH disaccharide **77** (precursor of CS-C)

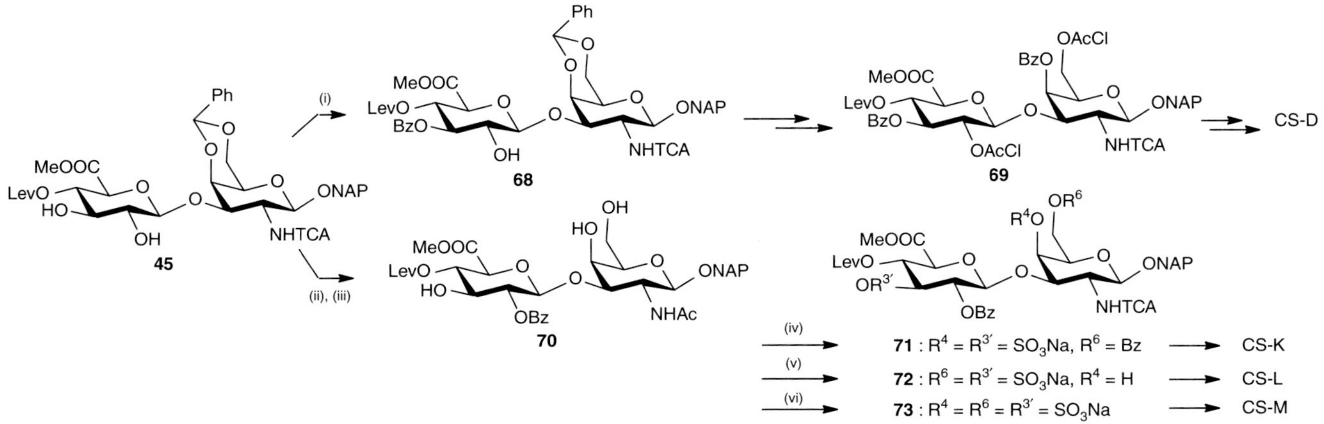

Scheme 14.12 Synthesis of CS-D, -K, -M, -L disaccharides. (i) BzCN, pyridine, 80%; (ii) Bu_2SnO, dioxane–benzene, then BzCl, 68%; (iii) Bu_3SnH, AIBN, benzene–DMAC, then 80% AcOH, 80%; (iv) BzCN, pyridine, then $SO_3{\cdot}NMe_3$, DMF, 60 °C then [Na^+] ion exchange, 80%; (v) $SO_3{\cdot}NMe_3$, DMF, 40 °C, then [Na^+] ion exchange, 75%; and (vi) $SO_3{\cdot}NMe_3$, DMF, 60 °C, then [Na^+] ion exchange, 90%.

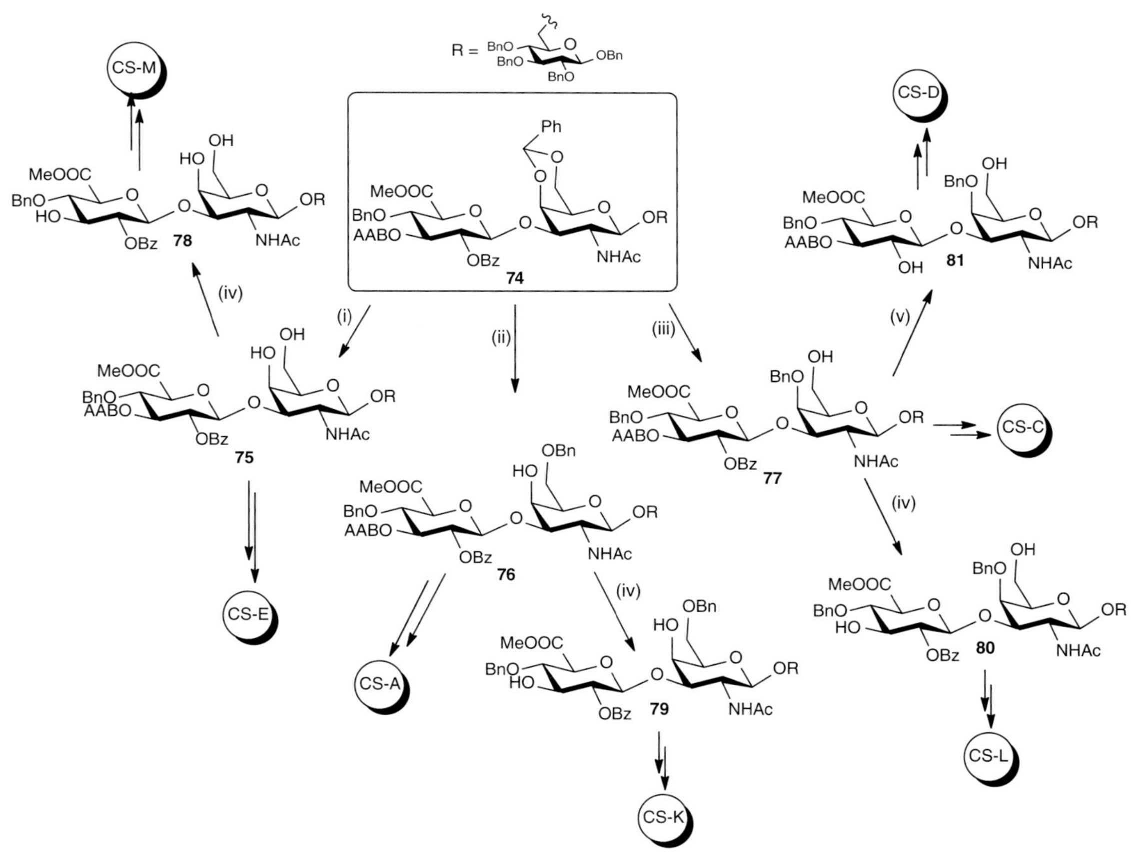

Scheme 14.13 Synthesis of all types of CS disaccharides. (i) TFA/H_2O, CH_2Cl_2, 95%; (ii) Et_3SiH, TfOH, CH_2Cl_2, 88%; (iii) Et_3SiH, $PhBCl_2$, CH_2Cl_2, 97%; (iv) DDQ, H_2O, CH_2Cl_2; and (v) NaOH, MeOH/THF.

depending on the promoter used. Selective deprotection of the AAB group on **76** and 77 gave 4,3′-diol **79** and 6,3′-diol **80** (precursors of CS-K and -L, respectively), whereas selective deprotection of benzoate ester on **77** afforded the 6,2′-diol **81** as a starting material for the synthesis of CS-D disaccharide (Scheme 14.13). Combinatorial approaches (not detailed in this chapter) have also been reported for the synthesis of various CS disaccharides using different orthogonal protecting groups [29].

14.5 Protecting Groups in Heparin and HS Synthesis

In comparison to chondroitin sulfate, heparin possesses three main differences. It contains (i) a 1,2-*cis* linkage on an α-D-glucosamine, (ii) *N*-sulfamido groups, and (iii) iduronic groups (with a 1,2-*trans* linkage compatible with a participating group at position 2). The selection of protecting groups includes commonly a *nonparticipating protecting group* for the amine of the glucosamine (1,2-*cis* glycoside desired) and azido is generally selected. The disaccharide at the reducing end may differ, and an NHZ group may be used as no glycosylation will take place at this unit. Esters (acetate and benzoate) for position to be sulfated, benzyl for free OH in the final product will be used.

Thus, the main difference with CS chemistry is nitrogen chemistry. Scheme 14.14 shows an example of synthesis of regular heparin [30] with this set of protecting groups but also two auxiliary protecting groups: a levulinoyl for the position to be glycosylated and a *p*-methoxybenzyl (PMB) for the protection of position 4 of the nonreducing end. PMB was preferred to a benzyl based on an easy introduction on an iduronic building block via the PMB imidate. The synthesis relies on three building blocks assembled through imidate glycosylations (Scheme 14.14).

The present discussion will focus on the four final deprotection steps:

a) Saponification of esters (LiOH, H_2O_2, then NaOH THF/MeOH).
b) *O*-sulfation of the free hydroxyl groups ($SO_3{\cdot}NEt_3$, DMF, 50 °C 16 h).
c) Hydrogenolysis of benzyl ethers and reduction of azides (H_2, Pd/C, *t*BuOH, H_2O, 60 atm). As the generated amines poison the catalyst, high pressure is used in this step to complete the reaction.
d) Selective N-sulfation ($SO_3{\cdot}$Pyr, H_2O, pH 9.5) of the oligosaccharide in water.

Martin Lomas and coworkers [31], in the synthesis of hexa- and octasaccharides, used a similar strategy with two particular points (Scheme 14.15). A pivaloate was selected in position 2 of uronate as a participating group. This group may avoid the orthoester side product easily formed in uronate series with acetates and benzoates. At the nonreducing end of the protected oligosaccharide, a 4,6-benzylidene was used to protect position 4 to be glycosylated. The building block was then transformed into a glycosyl acceptor in two steps (benzylidene hydrolysis followed by selective benzoylation).

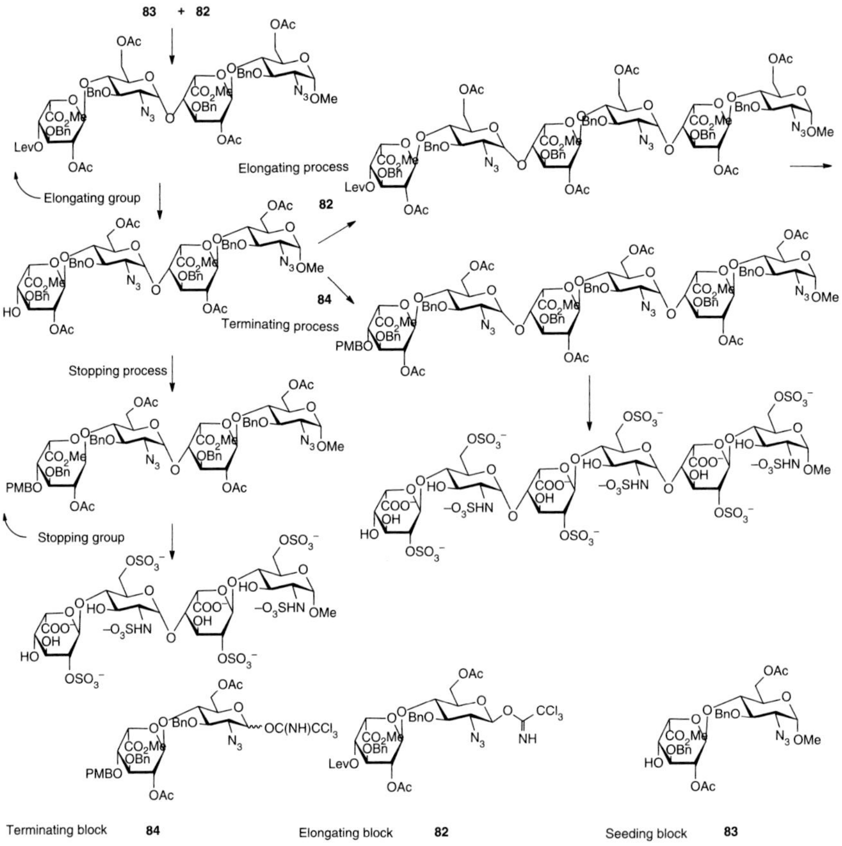

Scheme 14.14 Regular heparin synthesis. *Source:* From Tabeur et al. 1999 [30].

Among the heparin fragments, the pentasaccharide fondaparinux plays a prominent role as the first synthetic GAG fragments to reach the drug market. Several syntheses have been reported in the literature and may serve as an example of synthesis optimization. Manikowski et al. [3] described a strategy to combine N- and O-sulfation, usually achieved in two separated steps. The challenge was to reduce azido groups, without reducing the benzyl groups. First, the azide and NHZ groups are reduced by Pd/C in MeOH/THF in the presence of triethylamine and dihydrogen. Triethylamine poisons the catalyst and avoids benzyl hydrogenolysis. The next step is the critical N- and O-sulfation, and various set of conditions and reagents were tried: chlorosulfuric acid (9 eq.) and SO_3·Pyr (40 eq.) led, respectively, to degradation or complex mixtures of under-sulfated saccharides. The best conditions (SO_3·NEt_3, 40 eq., 15 h, 55 °C in DMF) gave a good yield (89%). The purity of SO_3·NEt_3 is critical in this reaction, and the presence of hydrolysis product was found detrimental. The last step in their synthesis was the easy removal of *O*-benzyl groups. Sulfated amines are no more interfering with the catalyst. This study presents an interesting alternative to the common sequence of reactions and provides a focus on the problems encountered in catalytic hydrogenation reactions (see Chapter 17).

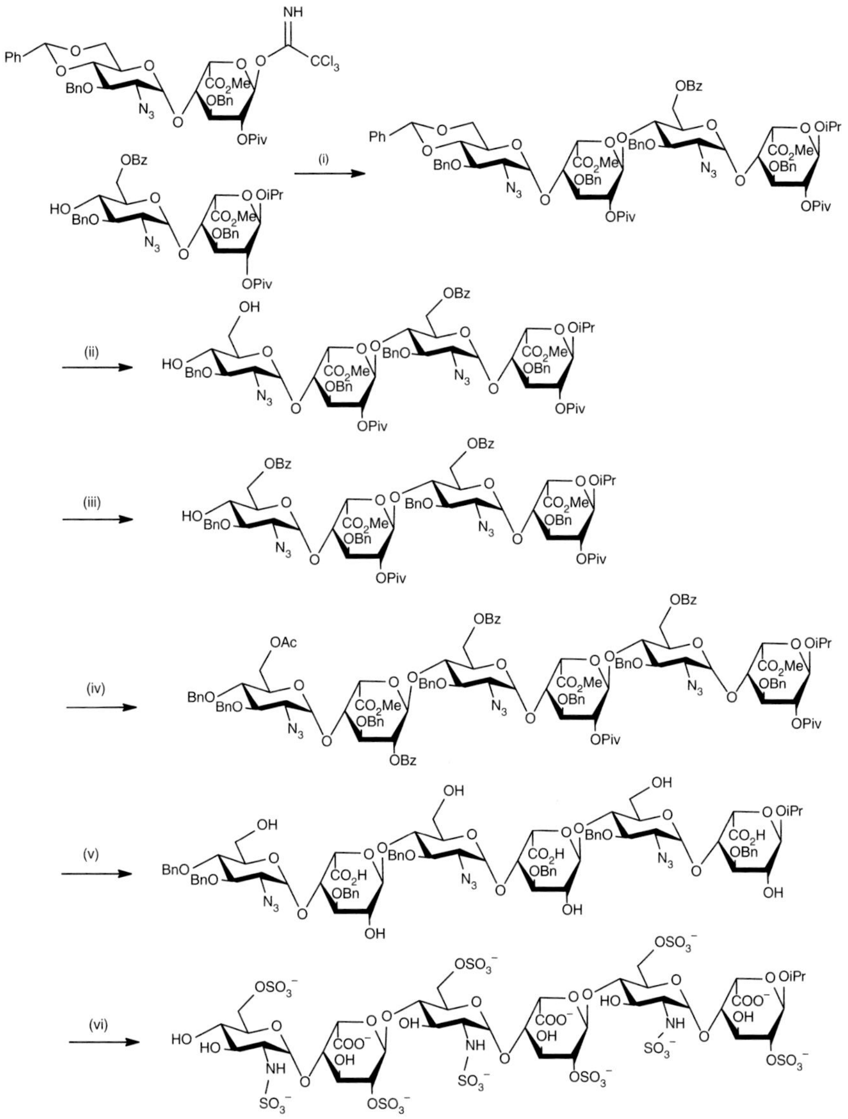

Scheme 14.15 Reagents and conditions: (i) 3% TMSOTf, CH_2Cl_2, 79%; (ii) EtSH, PTSA (cat.), 75%; (iii) BzCN, Et_3N (cat.), MeCN, −40 °C, 93%; (iv) 3% TMSOTf, 58%; (v) KOH, 74%; and (vi) SO_3·NMe_3; Dowex 50WX4 (Na^+), 71%, then H_2 10% Pd/C then SO_3·Pyr, 87%.

The main objective in HS synthesis is to prepare selectively sulfated heparin. The use of benzyl ethers is reliable, but it requires harsh conditions (NaH, DMF) for their introduction, and a more convenient protecting group is needed. A recent example was designed to selectively introduce a sulfate at position 6 of glucosamine (Scheme 14.16) [32]. An extended set of protecting groups was used, completed by Lev and Fmoc. Lev was used for positions to be sulfated, acetates and benzyl ethers for free hydroxyls, and Fmoc as a protecting group

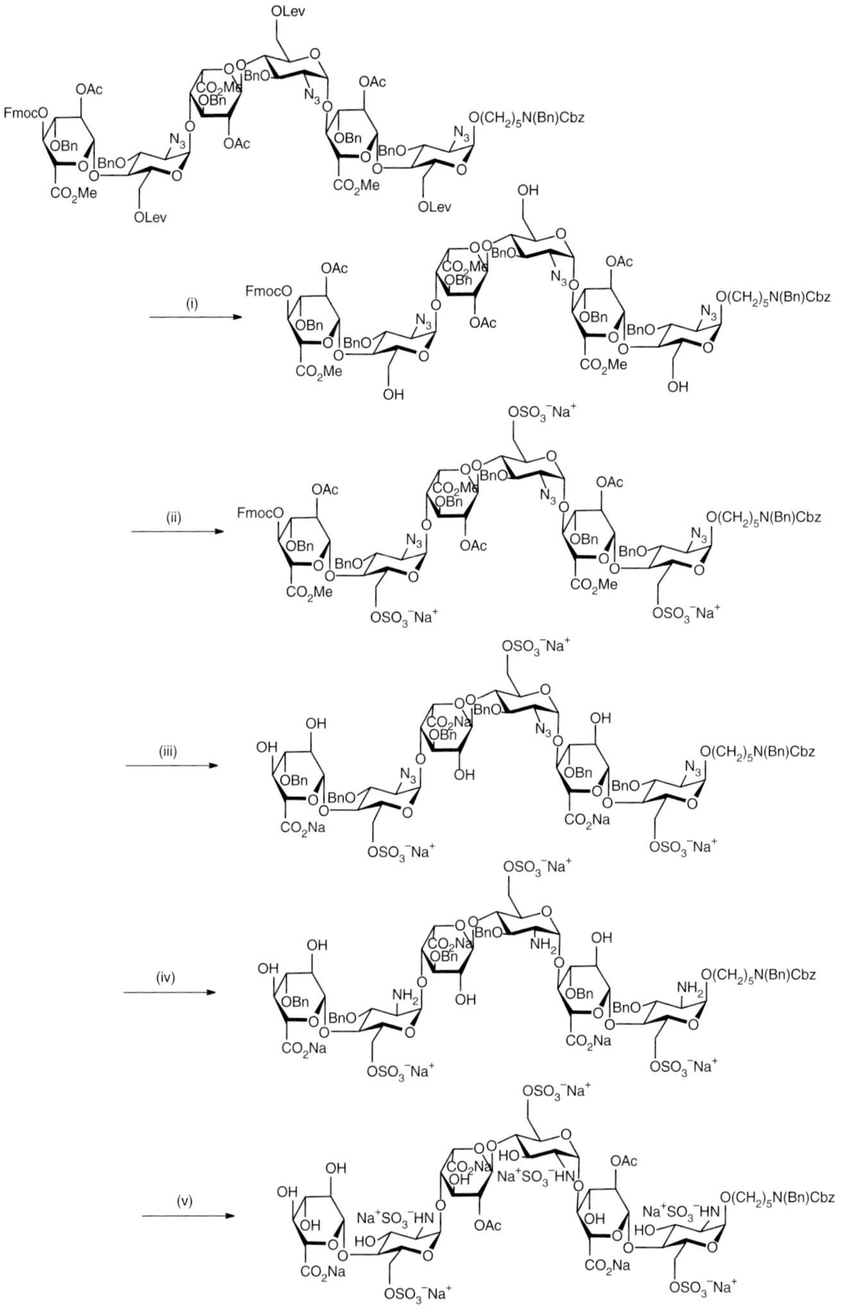

Scheme 14.16 Reagents and conditions (i) N_2H_4·HOAc, toluene/EtOH, 90%; (ii) SO_3·Pyr, DMF; (iii) (a) LiOH, H_2O_2, THF, then 4 M NaOH, MeOH, 58%; (iv) PMe_3, THF, NaOH, 65%; and (v) SO_3·Pyr, MeOH, Et_3N, 0.1 M NaOH, 50%, then H_2, Pd/C, MeOH/H_2O.

Table 14.2 Examples of protection groups used in HS synthesis.

Nitrogen protection	Position to be glycosylated	Position to be sulfated	Other hydroxyls	References
N_3	Lev (Ido)	Ac	Bn	[30]
N_3	Benzylidene (GlcN)	Ac, Bz, Piv	Bn	[31]
N_3	Fmoc (Ido)	Lev	Bn, Ac	[32]

for the positions to be glycosylated (Scheme 14.16). In this synthesis, to avoid difficulties in hydrogenolysis, azido groups were first reduced using trimethylphosphine (Staudinger reaction) and the resulting amines were chemoselectively sulfated in water. This set of protecting groups (Fmoc, Lev, Ac, Bn, N_3) allows successful synthesis of regioselectively sulfated heparin fragments (Table 14.2).

14.6 Summary and Conclusions

Chemical synthesis of glycosaminoglycans fragments combines many difficulties in oligosaccharide synthesis: troublesome glycosyl donors (2-amino sugars and uronic acids) and rare carbohydrates (galactosamine and iduronic acid) with controlled sulfoforms. The control of the sulfation pattern imposes an extended set of protecting groups to distinguish between positions to be sulfated or not, resulting in an important increase in the number of synthetic steps required in comparison to neutral oligosaccharides. Fortunately, well-established and reliable methods exist and are to be considered when designing a new synthesis to limit the risks.

14.7 Experimental Part: Procedure for Regioselective 6-*O*-benzoylation Followed by 4-Sulfation, an Example of the Synthesis of Disaccharide 31 [18]

A solution of diol **30** (8.14 g, 10 mmol) and benzoyl cyanide (2.62 g, 20 mmol) in dry pyridine (100 ml) was stirred for 16 h at RT under argon. MeOH (10 ml) was added, and the mixture was concentrated and then evaporated with toluene (3 × 30 ml). The residue was crystallized from MeOH-CH_2Cl_2 to give the 6-*O*-benzoylated intermediate (8.58 g, 93%). A mixture of this 6-*O*-benzoylated intermediate (4.59 g, 5 mmol) and $SO_3{\cdot}NMe_3$ (5.57 g, 50 mmol) in dry DMF (70 ml) was stirred for 48 h at 50 °C under argon and then cooled to RT. MeOH (10 ml) was added, and the mixture was concentrated. The residue was eluted from a silica gel column with CH_2Cl_2-MeOH (11 : 1), then from a Sephadex SP C25

(Na^+) column with CH_2Cl_2-MeOH-water (5 : 9 : 1), and crystallized from aq. EtOH to give compound **32** (4.59 g, 90%).

Abbreviations

AAB	4-acetaminobenzyl
Ac	acetyl
AIBN	azobisisobutyronitrile
Bn	benzyl
Bu	butyl
Bz	benzoyl
CS	chondroitin sulfate
CSA	camphorsulfonic acid
DDQ	2,3-dichloro-5,6-dicyano-1,4-benzoquinone
DMAc	dimethylacetamide
DMF	dimethylformamide
DS	dermatan sulfate
Fmoc	fluorenylmethyloxycarbonyl
GAG	glycosaminoglycan
Gal	galactose
GalN	galactosamine
Glc	glucose
GlcN	glucosamine
GlcA	glucuronic acid
Hep	heparin
HS	heparan sulfate
Ido	iduronic acid
Lev	levulinyl
MBz	*para*-methoxybenzoyl
Me	methyl
MN	7-methoxy-2-naphthyl
MP	*para*-methoxyphenyl
NAP	2-naphthylmethyl
PG	proteoglycan
Ph	phenyl
PMB	*para*-methoxybenzylidene
Pyr	pyridine
RT	room temperature
Ser	serine
TBAF	tetrabutylammonium fluoride
TBDPS	*tert*-butyldimethylsilyl
TCA	trichloroacetyl
Tf	trifluoromethanesulfonyl
TFA	trifluoroacetyl
THF	tetrahydrofuran
Xyl	xylose

References

1 Al-Horani, R.A. and Desai, U.R. (2010). *Tetrahedron* 66: 2907–2918.
2 Holme, K.R. and Perlin, A.S. (1997). *Carbohydr. Res.* 302: 7–12.
3 Manikowski, A., Kozioł, A., and Czajkowska-Wojciechowska, E. (2012). *Carbohydr. Res.* 361: 155–161.
4 (a) Takiura, K. and Honda, S. (1967). *Yakugaku Sasshi* 87: 1248–1255. (b) Penney, C.L. and Perlin, A.S. (1981). *Carbohydr. Res.* 93: 241–246.
5 Jacquinet, J.-C. (2004). *Carbohydr. Res.* 339: 349–359.
6 Thollas, B. and Jacquinet, J.-C. (2004). *Org. Biomol. Chem.* 2: 434–442.
7 Aït-Mohand, K., Lopin-Bon, C., and Jacquinet, J.-C. (2012). *Carbohydr. Res.* 353: 33–48.
8 Tamura, J.-I. and Nishihara, J. (1999). *Bioorg. Med. Chem. Lett.* 9: 1911–1914.
9 Tamura, J.-I. and Nishihara, J. (2001). *J. Org. Chem.* 66: 3074–3083.
10 Jacquinet, J.-C. (2006). *Carbohydr. Res.* 341: 1630–1644.
11 Goto, F. and Ogawa, T. (1992). *Tetrahedron Lett.* 33: 5099–5102.
12 Rio, S., Beau, J.-M., and Jacquinet, J.-C. (1994). *Carbohydr. Res.* 255: 103–124.
13 Kjellen, L. and Lindahl, U. (1991). *Annu. Rev. Biochem.* 60: 443–475.
14 Sugahara, K. and Yamada, S. (2000). *Trends Glycosci. Glycotechnol.* 12: 321–349.
15 Gama, C.I., Tully, S.E., Sotogaku, N. et al. (2006). *Nat. Chem. Biol.* 9: 467–473.
16 (a) Sugahara, K., Mikami, T., Uyama, T. et al. (2003). *Curr. Opin. Struct. Biol.* 13: 612–620. (b) Murrey, H.E. and Hsieh-Wilson, L.C. (2008). *Chem. Rev.* 108: 1708–1731. (c) Yamada, S. and Sugahara, K. (2008). *Curr. Drug Discov. Technol.* 5: 289–301. (d) Uebelhart, D., Malaises, M., Marcolongo, R. et al. (2004). *Osteoarthritis Cart.* 12: 269–276.
17 Macchione, G., Maza, S., Kayser, M. et al. (2014). *Eur. J. Org. Chem.* 3868–3884.
18 Jacquinet, J.-C., Rochepeau-Jobron, L., and Combal, J.-P. (1998). *Carbohydr. Res.* 314: 283–288.
19 Bélot, F. and Jacquinet, J.-C. (2000). *Carbohydr. Res.* 326: 88–97.
20 (a) Tamura, J.-I., Neumann, K.W., Kurono, S., and Ogawa, T. (1998). *Carbohydr. Res.* 305: 43–63. (b) Tamura, J.-I. and Tokuyoshi, M. (2004). *Biosci. Biotechnol. Biochem.* 12: 2436–2443. (c) Tamura, J.-I., Nakada, Y., Taniguchi, K., and Yamane, M. (2008). *Carbohydr. Res.* 343: 39–47.
21 Wakao, M., Obata, R., Miyachi, K. et al. (2015). *Bioorg. Med. Chem. Lett.* 25: 1407–1411.
22 (a) Tully, S.E., Mabon, R., Gama, C.I. et al. (2004). *J. Am. Chem. Soc.* 126: 7736–7737. (b) Tully, S.E., Rawat, M., and Hsieh-Wilson, L.C. (2006). *J. Am. Chem. Soc.* 128: 7740–7741.
23 Lopin, C. and Jacquinet, J.-C. (2006). *Angew. Chem. Int. Ed.* 45: 2574–2578.
24 (a) Jacquinet, J.-C., Vibert, A., and Lopin-Bon, C. (2009). *Chem. Eur. J.* 15: 9579–9595. (b) Jacquinet, J.-C. and Lopin-Bon, C. (2015). *Carbohydr. Res.* 402: 35–43.
25 Solera, C., Macchione, G., Maza, S. et al. (2016). *Chem. Eur. J.* 22: 2356–2369.
26 Tamura, J.-I., Neumann, K.W., and Ogawa, T. (1995). *Bioorg. Med. Chem. Lett.* 5: 1351–1354.

27 Vibert, A., Lopin-Bon, C., and Jacquinet, J.-C. (2011). *Eur. J. Org. Chem.* 4183–4204.
28 (a) Karst, N. and Jacquinet, J.-C. (2000). *J. Chem. Soc., Perkin Trans. 1* 2709–2717. (b) Karst, N. and Jacquinet, J.-C. (2002). *Eur. J. Org. Chem.* 815–825.
29 Lubineau, A. and Bonnaffé, D. (1999). *Eur. J. Org. Chem.* 2523–2532.
30 Tabeur, C., Mallet, J.-M., Bono, F. et al. (1999). *Bioorg. Med. Chem.* 7: 2003–2012.
31 de Paz, J.L., Angulo, J., Lassaletta, J.M. et al. (2001). *ChemBioChem.* 2: 673–685.
32 Arungundram, S., Al-Mafraji, K., Asong, J. et al. (2009). *J. Am. Chem. Soc.* 131: 17394–17405.

15

Applications of Fluorous and Ionic Liquid Tags in Oligosaccharide Synthesis

Imke Sittel and M. Carmen Galan

School of Chemistry, University of Bristol, Cantock's Close, Bristol, BS8 1TS, UK

15.1 Introduction

Carbohydrates are one of the most diverse and important classes of biomolecules in nature. The multitude of biological roles that carbohydrates and their glycoconjugates play has stimulated researchers to devote efforts to the elucidation of their role and mechanisms of interaction in living processes. To accomplish that, it is essential to have access to structurally defined carbohydrate sequences to use as probes for biological exploration. However, approaches to prepare diverse libraries of complex carbohydrates in a rapid manner are lacking, and glycobiology research has had to rely on isolated materials, target-oriented lengthy chemical syntheses, or enzymatic approaches. Chemical synthesis offers the advantage of producing pure and in-scale oligosaccharides. Thus, many efforts have been devoted to the development of improved and automated synthetic methodologies to access these complex molecules [1] (see Chapter 16).

The issues associated with oligosaccharide synthesis, e.g. the need for chromatography-based purification after each reaction step, laborious protecting group manipulations needed to prepare orthogonally protected building blocks, and the requirement for efficient and regioselective and stereoselective glycosylation reactions, make the chemical synthesis of each target a very laborious and time-consuming enterprise. To circumvent the above issues, chemists have devoted efforts to the development of supported oligosaccharide synthesis as viable alternatives to traditional methodologies, where purification is simplified by the use of covalently attached purification labels (Tags) to either the glycosyl donor or acceptor, which allows for chromatography-free isolation of the product after each reaction step. In addition, the purification labels offer a dual function as temporary protecting groups, and as such, consideration must be paid to the type of functional group used as a linker, the position around the ring where the Tag will be attached, and the compatibility with other functional groups in the molecule.

Protecting Groups: Strategies and Applications in Carbohydrate Chemistry, First Edition.
Edited by Sébastien Vidal.

This chapter highlights the most relevant examples of the past two decades in the development, attachment, and application of fluorous tags (FTags) [2] and ionic tags (ITags) [3], in their dual role as soluble supports and protecting groups. The bifunctional labels incorporate all the advantages of traditional solution-phase oligosaccharide chemistry, such as atom economy (e.g. low amounts of glycosyl donor and reagents are needed) and easy reaction monitoring, without the need for lengthy and laborious chromatography-based purification, while at the same time temporarily masking key hydroxy groups.

15.2 Fluorous Supports

Fluorous support (FTag)-based strategies have been devised as supported methodologies that benefit from the facile purification features of solid-phase synthesis, but with the advantages of traditional solution-phase chemistry. The strategy relies on the selective affinity of fluorinated species for fluorinated solvents or fluorine-derivatized surfaces. Upon installation of the FTag at key positions in the desired glycan molecule, purification is accomplished by biphasic or triphasic extraction with a fluorous layer and an appropriate organic and aqueous layer (Scheme 15.1a). Alternatively, fluorous solid-phase extraction (FSPE) can be employed, in which a solid phase is modified with a fluorous substrate that selectively retains fluorinated species, which can later on be eluted with an appropriate solvent (Scheme 15.1b). The viability of FTags has also been shown in automated oligosaccharide synthesis [4], as well as in fluorine-modified microarrays for the noncovalent attachment of oligosaccharides [5].

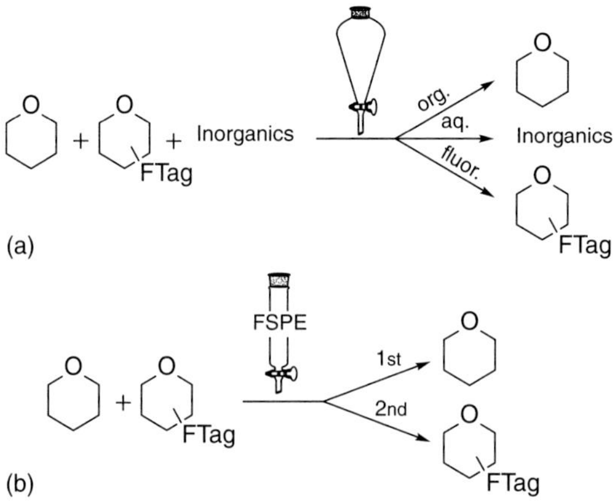

Scheme 15.1 Two different simple purification strategies: (a) liquid–liquid extraction; (b) fluorinated solid-phase extraction (FSPE).

Two main considerations for the FTags have to be made. Firstly, the fluorine content in the FTag has to be chosen carefully in order to act as an efficient purification handle. In that context, FTags can be divided into heavy and light FTags, depending on the fluorine content (i.e. the FTag holds greater or lesser than 60% of the total molecular weight) [6]. Both light [7] and heavy [8] FTags have been successfully employed in oligosaccharide chemistry. Secondly, the FTag also acts as a temporary masking functionality that needs to be introduced chemoselectively and regioselectively in the monosaccharide building block of choice. It needs to be stable to glycosylation chemistry and protecting group manipulations and must also be selectively cleaved at the end of the synthesis to release the final product.

A variety of linkers have been designed to be used as FTags, the most common functional groups being ether and ester linkages at different positions around the pyranose ring. By far, the most prevalent linkage used entails the incorporation of an ether linkage at the C-1 position of the reducing end monosaccharide. For instance, the FTag can be coupled to the monosaccharide unit via a standard glycosylation of a glycosyl donor with the corresponding FTag aglycon acceptor under Lewis acid (LA) catalysis (Scheme 15.2a). Alternatively, the ether-linked FTag can be incorporated to a different hydroxy group on the glycan ring (typically C-4 or C-6) via base-catalyzed S_N2-reaction with the desired halogen-bearing FTag (Scheme 15.2b). Another common method to install the FTag relies on employing an ester linkage at either C-1 or C-4 of the glycoside. In this instance, the FTag contains a terminal acid group, which is coupled to the corresponding free hydroxy group using a suitable coupling reagent and base to form the ester-linked product (Scheme 15.2c). Ester or amide linkages can also

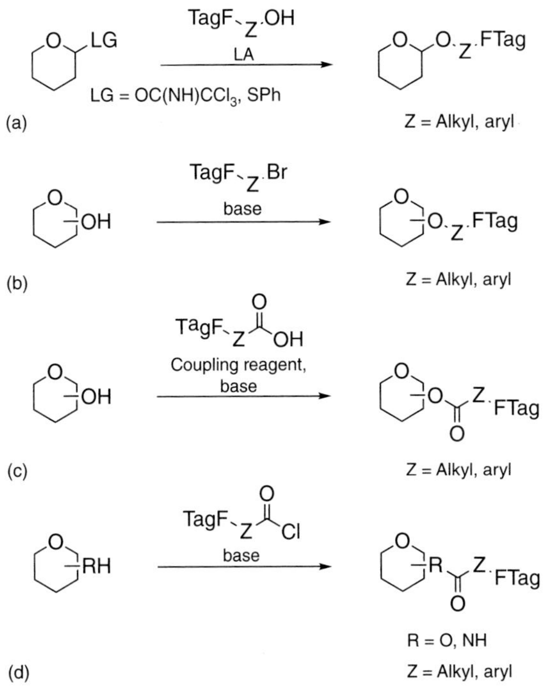

Scheme 15.2 Most common reactions to install an FTag to a monosaccharide unit.

be prepared by reacting either a free hydroxy or amino group with an acid chloride-bearing FTag under base catalysis (Scheme 15.2d).

The choice of linker in the FTag is often determined by the other protecting groups present in the carbohydrate molecule and the glycosylation strategy used. However, most approaches agree that installation of the FTag at the anomeric position (C-1) of the glycoside acceptor is the most advantageous alternative, as the FTag can help mask the reactive anomeric position and minimize the formation of unwanted FTagged side products during the glycosylation process (e.g. hydrolyzed materials), which can complicate the final purification.

15.2.1 Ether-linked FTags

In 1998, Curran et al. were the first to use a modified protecting group as a purification handle in oligosaccharide synthesis [9]. The group used fluorous benzyl ether as a protecting group (Bnf) on glucal derivative **2** to ensure the selective solubility of the product, upon *p*-toluenesulfonic acid (TsOH)-catalyzed glycosylation with di-*O*-isopropylidene-protected galactose acceptor **3** in perfluorohexane (FC-72) (Scheme 15.3). The modified benzyl protecting group contained a heavily fluorinated silyl ether in the *para*-position and was installed via S_N2 displacement upon reaction of D-glucal **1** with BnfBr in the presence of sodium hydride in dimethylformamide (DMF). Isolation of FTagged disaccharide **4** from the reaction mixture was accomplished by triphasic extraction with water/CH_2Cl_2/FC-72, which separated all nonfluorinated material from FTagged disaccharide **4**.

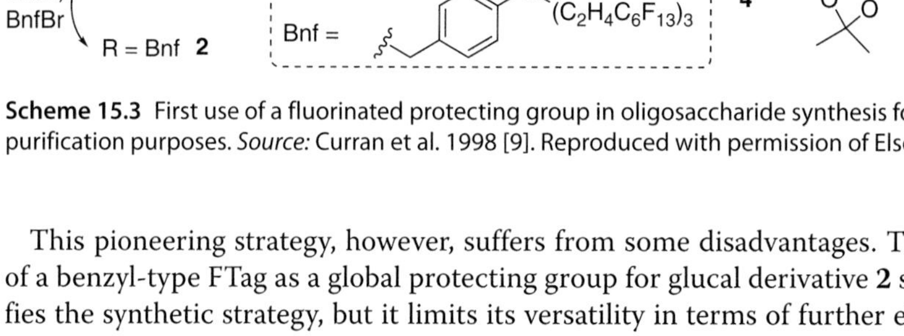

Scheme 15.3 First use of a fluorinated protecting group in oligosaccharide synthesis for purification purposes. *Source:* Curran et al. 1998 [9]. Reproduced with permission of Elsevier.

This pioneering strategy, however, suffers from some disadvantages. The use of a benzyl-type FTag as a global protecting group for glucal derivative **2** simplifies the synthetic strategy, but it limits its versatility in terms of further elongation and lacks chemical orthogonality with respect to other protecting groups. It also significantly increases the molecular weight (MW ~ 3700 g mol^{-1}) of the compound, which, in turn, broadens NMR signals and might hinder the visualization of smaller impurities in the ^{19}F spectrum. In addition, the presence of the FTags in the glycosyl donor allows for the formation of side products, which will also bear the FTag and that will have to be separated by time-consuming column chromatography.

To address some of the issues associated with Curran's original work on the application of FTags in oligosaccharide synthesis, other groups developed different fluorous protecting groups, most of them designed for the anomeric position of the glycoside.

In 2007, Mizuno and coworker reported the FTag-supported synthesis of an orthogonally protected monosaccharide building block ready to be used in glycosylation reactions, employing an ether-linked FTag installed at the anomeric position of the desired target glycoside [10]. Upon installation of the FTag by LA catalyzed glycosylation of per-*O*-acetyl-β-D-galactopyranose with FTag-OH, the group performed a number of protecting group manipulations, which included deacetylation, acetate removal, followed by selective benzylation at C-3, and acetylation of the remaining hydroxy groups; selective debenzylation and subsequent chloroacetylation of the free hydroxy group formed the orthogonally protected glycoside, which was purified by simple triphasic extractions in between steps. Once the final protected monosaccharide was obtained, the FTag was cleaved using ZnBr and AcBr, and the hemiacetal was converted into a glycosyl bromide in 58% overall yield (Scheme 15.4).

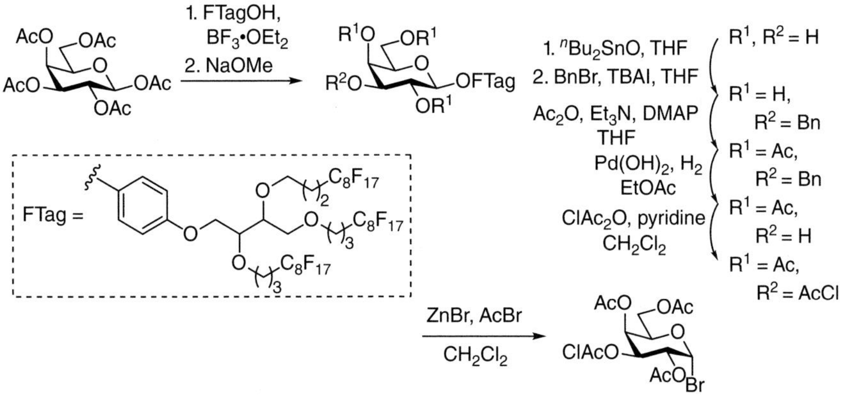

Scheme 15.4 FTag-assisted monosaccharide building block synthesis. *Source:* Goto and Mizuno 2007 [10]. Reproduced with permission of Elsevier.

Another contribution to this area comes from the group of Seeberger who also developed an ether-linked FTag for the anomeric position. In this example, the group demonstrated the application of fluorous-phase chemistry in combination with the use of a microreactor for the synthesis of β-1,6-linked tetraglucan **5** [11]. The key feature of the FTag is the presence of a *trans*-alkene-type linker, which can be converted into different functional groups once the desired oligosaccharide is synthesized (Scheme 15.5). For instance, olefin cross-metathesis reaction of the double bond in the presence of ethylene, under Grubb's catalysis, leads to cleavage of the FTag and affords *n*-pentenyl glucoside **6**. Alternatively, ozonolysis of FTagged glycoside **7** forms aldehyde **8**, whereas acetal hydrolysis,

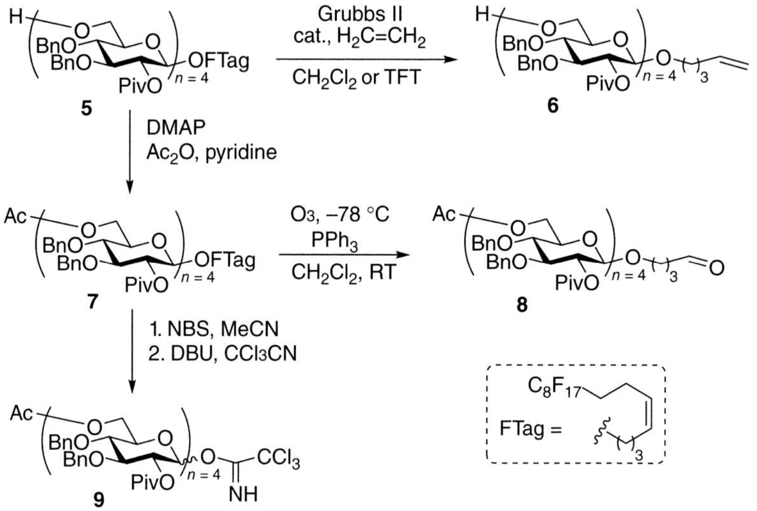

Scheme 15.5 Different cleavage strategies with a pentenyl-based FTag. *Source:* Carrel et al. 2007 [11]. Reproduced with permission of American Chemical Society.

using *N*-bromosuccinimide (NBS) in acetonitrile [12], followed by reaction with 1,8-diazabicyclo[5.4.0]undec-7-ene (DBU) and CCl_3CN yields trichloroacetimidate **9**. Moreover, the FTag used in this example has a significantly lower molecular weight (only 17 fluorines) when compared to the previously described ones, which simplifies the analysis by NMR.

In 2010, Huang and coworkers reported a catch-and-release one-pot approach to synthesize the biologically relevant Lewis^X trisaccharide [13]. In this strategy, the FTag is only installed after the desired trisaccharide is assembled to simplify purification of the final product via FSPE. The group used a reactivity-based one-pot approach [14] that relies on the difference in reactivity between the hydroxy group at C-4 vs the one at C-3 in the glucosamine acceptor (Scheme 15.6a). In order to obtain an efficient catch-and-release FTag, it was necessary to develop a suitable handle on the glycoside that would allow efficient tagging with the FTag, but also facile and selective cleavage of the final product.

Therefore, the team introduced a 2-butanonyl linker at the anomeric position of the glucosamine moiety, which could then be trapped by an FTag-bearing hydrazine to form the corresponding hydrazide. After purification, the FTag could be released with trifluoroacetic acid (TFA) to reveal the original ketone (Scheme 15.6b). The team further exemplified their approach in the synthesis of a linear β-1,6-linked galactose tetrasaccharide.

FTags have also been used to synthesize more complex oligosaccharides [15]. For example, in 2015, Boons and coworkers described the FTag-assisted synthesis of the GPI anchor hexasaccharide of *Trypanosoma brucei* [15c]. The biggest challenge in the preparation of hexasaccharide **17** was to obtain 1,2-*cis*-selectivity, which was accomplished by using glycoside donors with a C-2 chiral auxiliary that, via neighboring group participation during the glycosylation process, led to the formation of a *trans*-decalin anomeric sulfonium ion

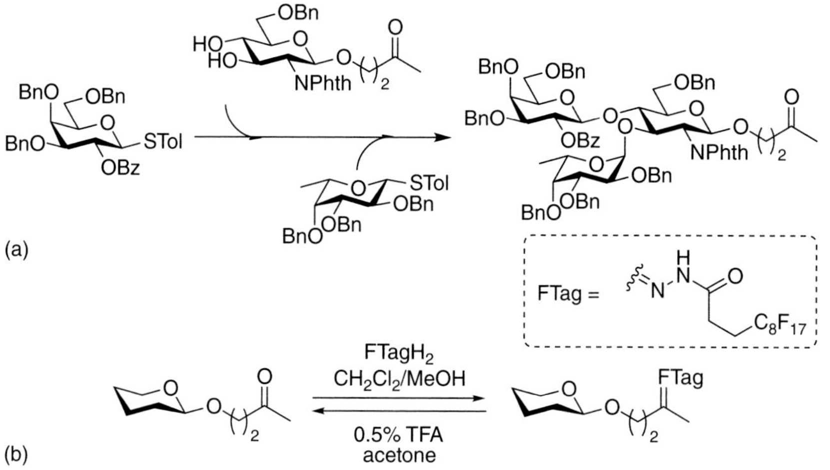

Scheme 15.6 One-pot synthesis of Lewis^X (a) and general catch-and-release mechanism (b). *Source:* Yang et al. 2010 [13]. Reproduced with permission of John Wiley and Sons.

intermediate that favors the formation of α-glycosides [12]. Their approached started by the incorporation of a benzyl ether-based FTag at the anomeric position of the first mannoside building block by glycosylation of the FTag-modified benzyl alcohol with orthogonally protected thiomannoside **10** using *N*-iodosuccinimide (NIS) and trifluoromethanesulfonic acid (TfOH), which upon selective deprotection of the naphthyl (Nap) group at C-3 using 2,3-dichloro-5,6-dicyano-1,4-benzoquinone (DDQ) in a mixture of CH_2Cl_2 and water, yielded α-linked FTagged glycoside acceptor **11** ready for subsequent glycosylations (Scheme 15.7). Coupling between **11** and preformed glycosyl donor **12**, which bears a C-2′ auxiliary to ensure 1,2-*cis*-selective glycosylation (I), followed by removal of the chiral auxiliary under acidic conditions, afforded the FTagged disaccharide in 63% overall yield. The disaccharide was then glycosylated with **13** (II) in the presence of catalytic TfOH in 71% yield. After selective removal of the naphthyl protecting group at C-6″, an α-linked galactoside was incorporated using the C-2‴ chiral auxiliary strategy with donor **14** (III) in 67% and with complete stereocontrol. Upon unmasking of the hydroxy group at C-2‴ by removal of the auxiliary, followed by trimethylsilyl trifluoromethanesulfonate (TMSOTf)-mediated glycosylation with **15** (IV) and subsequent cleavage of the auxiliary, the FTagged pentasaccharide was isolated in 76% over the 2 steps. Finally, after selective removal of the levulinoyl (Lev) protecting group at C-6, the pentasaccharide was glycosylated with mannoside **16** (V), to give hexasaccharide **17**. The FTag allowed for fast, chromatography-free purification after each reaction step using FSPE. Untagged compounds were removed with 20% water in methanol, while the desired FTag-materials were eluted with acetone. The ether-linked FTag remained intact during the whole synthetic quest, until its final cleavage using catalytic hydrogenolysis. The only drawback, according to the authors, is that during the fluorous-assisted glycan assembly of this complex hexasaccharide, each glycosylation step had to be performed twice to

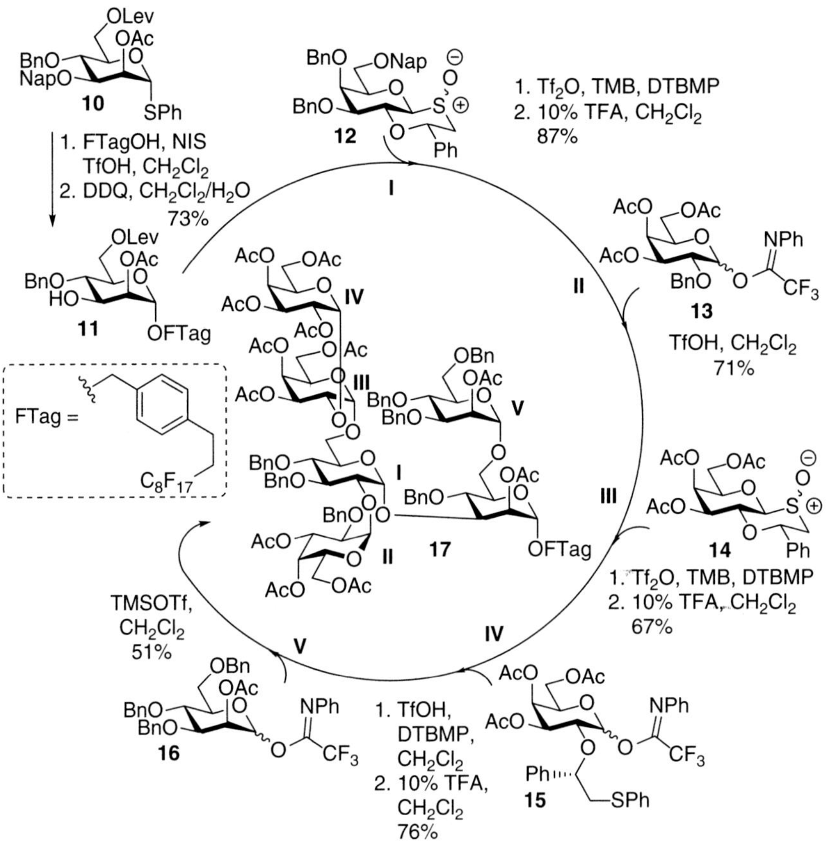

Scheme 15.7 Step-wise assembly of branched hexasaccharide with full stereocontrol on a fluorous support. *Source:* Huang et al. 2015 [15c]. Reproduced with permission of John Wiley and Sons.

ensure completion of these critical steps. The fully unprotected glycan target was isolated in an overall yield of 17% with minimal chromatography-based purification, which highlights the great potential of these supported methodologies to quickly prepare a given oligosaccharide target, once the synthetic strategy is optimized.

15.2.2 Ester-linked FTags

Ester-based FTags are not as widely used as ether-based FTags because of their incompatibility with other temporary ester-protecting groups that might be present in the glycoside building block. However, if no ester protecting groups, e.g. acetates or benzoates, are used or if those protecting groups are not removed during the synthesis, ester-linked FTags are a viable alternative to ether-linked FTags that benefit from facile installation and cleavage.

Inazu et al. reported in 2003 the application of a highly fluorinated FTag in oligosaccharide synthesis. The new FTag contained several fluorous chains (six), with the expectation that a larger fluorine content will provide a higher efficiency of extraction into the fluorous solvent. To showcase the approach, the hexakisfluorous chain-type butanoic acid-based FTag was attached to the anomeric hydroxy group of orthogonally protected glucose derivative **18** via an ester linkage using benzotriazol-1-yl-oxytripyrrolidinophosphonium hexafluorophosphate (PyBOP) and 4-dimethylaminopyridine (DMAP) to yield FTagged glucosyl acceptor **19** [16]. Selective removal of the silyl ether group at C-6 using HF in pyridine revealed the glucosyl acceptor **20**. Glycosylations with trichloroacetimidate donor **21** in the presence of catalytic TMSOTf afforded the corresponding disaccharide **22** in good yield (42% from the hemiacetal) (Scheme 15.8). Finally, the FTag was cleaved at the end of the synthesis concomitantly with the other ester groups in the molecule using sodium methoxide in a mixture of methanol and $EtOC_4F_9$ to give non-FTagged disaccharide **23**. Although no stereocontrol was achieved in the glycosylation reaction as no directing groups was used at the C-2 position of the glycosyl donor, the results here further emphasize that upon reaction optimization, supported methodologies should be applicable to the large-scale synthesis of oligosaccharides as they are performed in solution phase.

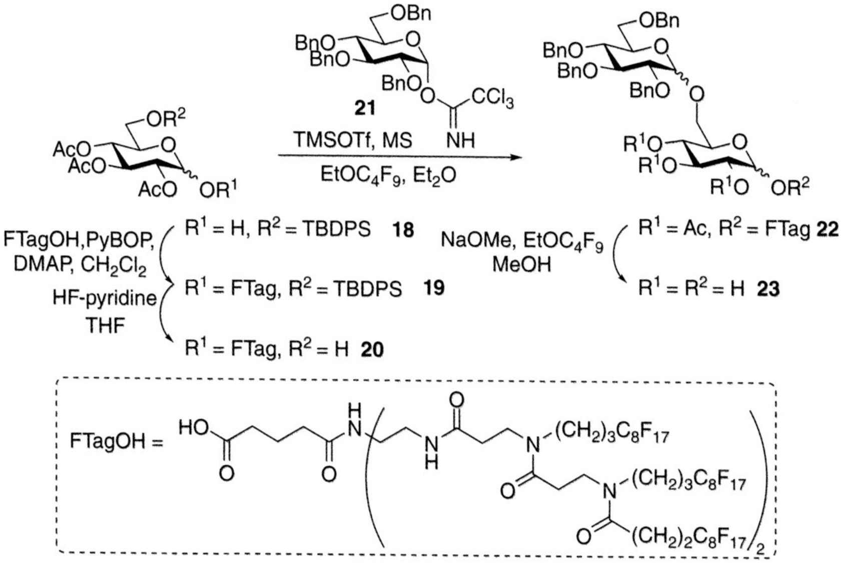

Scheme 15.8 Ester-based, heavy FTag-assisted synthesis of disaccharides. *Source:* Miura et al. 2003 [16]. Reproduced with permission of John Wiley and Sons.

15.2.3 Difluorous Tagged Strategies

In order to facilitate the purification of FTagged products from FTagged starting materials and/or side products, the group of Pohl developed in 2008 a strategy in which the product of the reaction would become the only species bearing two FTags, and thus, it is expected that their separation would be facilitated [17].

The group combined the use of their previously reported fluorous allyl group for light FTag-based oligosaccharide synthesis [18], which is attached to the anomeric position, with the use of fluorous acetate that acts as an ester protecting group. To test their hypothesis, a C-3 FTag-ester bearing glycosyl donor was synthesized, which, upon glycosylation with a fluorous allyl group, formed the di-FTagged product (Scheme 15.9). The group was able to show that any mono-FTagged material was separable by FSPE from the di-FTagged product, which only eluted at 100% methanol. These findings showed that a di-FTag strategy in which both the acceptor and the donor bear an FTag could be a useful method to facilitate purification of the product when undesired mono-FTagged materials are present in the reaction mixture.

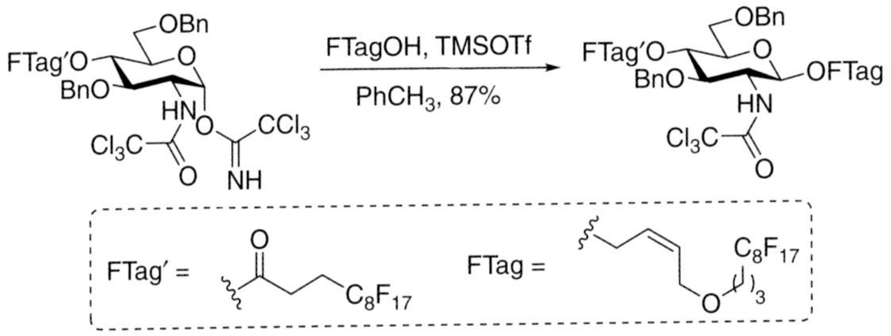

Scheme 15.9 Di-FTagged strategy: donor and acceptor bound. *Source:* Park et al. 2008 [17]. Reproduced with permission of Elsevier.

Subsequently, in 2013, Liu and coworkers demonstrated the utility of the approach in oligosaccharide synthesis with the preparation of a branched tetrasaccharide, in which the di-FTagged strategy was used to purify the product from the reaction mixture [19]. Starting from central non-FTagged disaccharide **24**, which bears free hydroxy groups at C-3′ and C-6′ of the mannoside moiety, double glycosylation with C-6 FTagged mannosyl trichloroacetimidate **25** afforded doubly FTagged tetrasaccharide **26** in 65% yield, which could be easily isolated from the unreacted starting materials and any mono-FTagged glycosylation product by FSPE (Scheme 15.10).

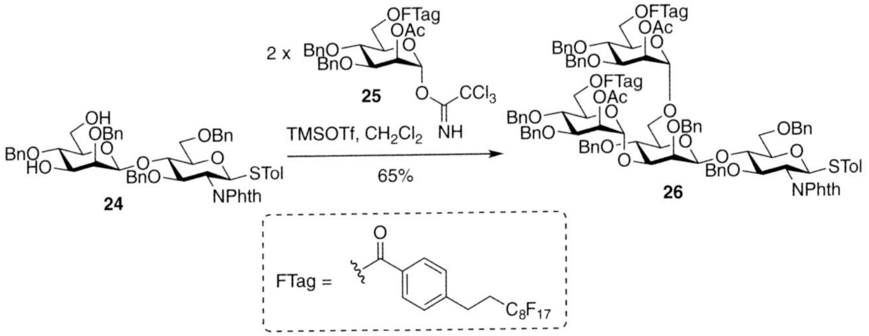

Scheme 15.10 Di-FTagged strategy: donor bound. *Source:* Zhang et al. 2013 [19]. Reproduced with permission of Springer.

Another elegant example of the doubly FTagged strategy originally reported by the Pohl's group [17] comes from Mizuno and coworker, who further extended the methodology in 2008 to the synthesis of oligosaccharide libraries [20]. The group employed *para*-alkoxyphenyl FTags with different degrees of fluorination as a means to differentiate between the different glycosides and thus aid in the deconvolution of the final mixture. Glycosylation of three different mono-hydroxylated glycoside acceptors, bearing different fluorous alkyl labels at the anomeric position, with two different glycosyl donors, one of which also contained an ester-linked FTag, produced a mixture of six different species (Scheme 15.11). The group was able to separate the different disaccharides on a reverse-phase Fluophase HPLC column, taking advantage of the different fluorine content in each compound, demonstrating the feasibility and versatility of the approach.

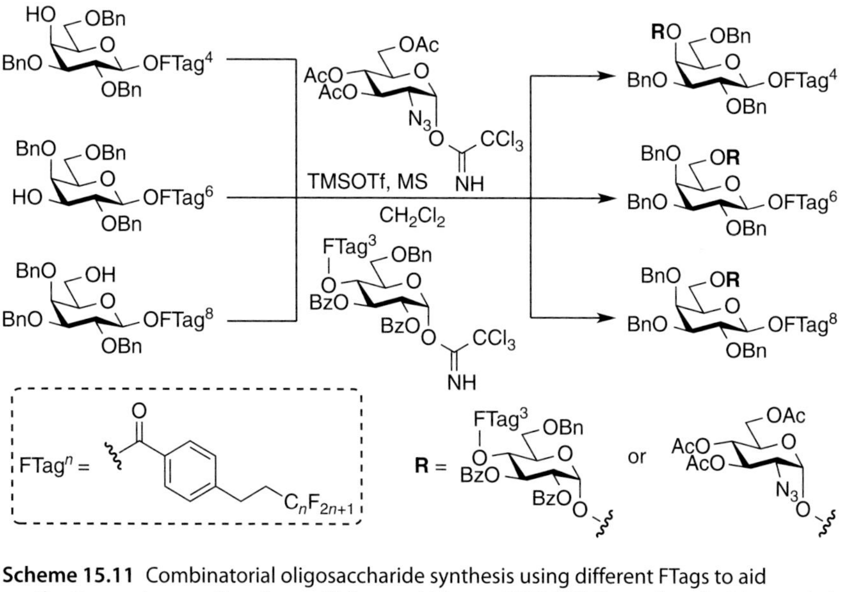

Scheme 15.11 Combinatorial oligosaccharide synthesis using different FTags to aid purification and separation. *Source:* Tojino and Mizuno 2008 [20]. Reproduced with permission of Elsevier.

15.2.4 Carbamate FTags

Although ethers or ester linkages are the most common functional groups used to attach the FTags to any glycoside, in 2006, the Manzoni group developed a modified fluorous-based carbamate protecting group for amino functionalities. The "Froc" FTag is based on the structure of the trichloroethoxycarbonyl (Troc) protecting group, which is commonly used in glycosamine chemistry (see Chapter 6), because of its stability under various reaction conditions and facile cleavage under conditions that are orthogonal to most common protecting groups (e.g. benzyl ethers, acetals, and silyl ethers) [21]. FrocCl can be synthesized in 3 steps and in excellent yield from 1*H*,1*H*,2*H*-perfluoro-1-decene, which, under mild basic conditions such as sodium bicarbonate in water, can be reacted with an amine (Scheme 15.12).

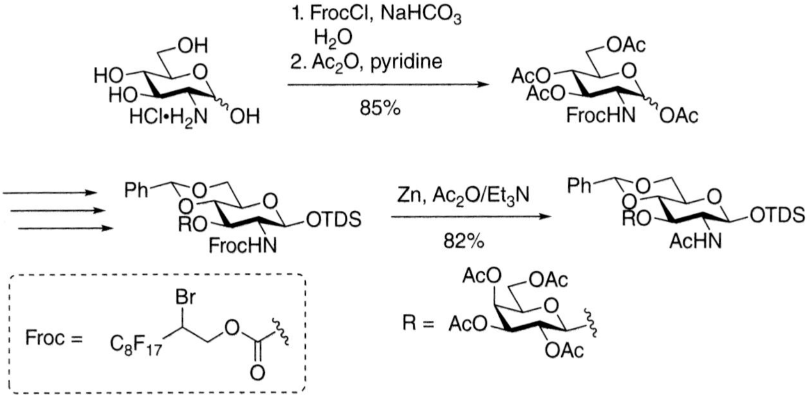

Scheme 15.12 Froc as fluorinated amino protecting group in oligosaccharide synthesis. *Source:* Manzoni and Castelli 2006 [21]. Reproduced with permission of American Chemical Society.

Once the Froc group was installed, the team was able to perform several protecting group manipulations (e.g. deacetylation, silylation, and acetalation) and glycosylation reactions in good yields and with simple FSPE purification in between steps. Cleavage of the Froc group was achieved by reduction with zinc in a mixture of acetic anhydride and triethylamine that revealed an acetamido group at C-2 of the glycoside. Because the Froc group is typically a protecting group for amines, it can be installed in both the glycosyl acceptor and the glycosyl donor, and it is also suitable for peptide synthesis, as also demonstrated by the team.

15.2.5 Silyl Ether FTags

In 2009, the Liu group developed a glycosyl donor with a light fluorinated silyl protecting group at the C-6 position of a trichloroacetimidate glucosyl donor that was used for the synthesis and isolation of β-1,6-linked oligosaccharides and glycosylated amino acids [22]. The F-silyl group was introduced at the free primary hydroxy group of 2,3,4-tri-*O*-benzoyl-protected thioglycoside **27** by reaction with the corresponding silane in the presence of TfOH to yield FTagged glucoside **28** (Scheme 15.13). The thioglycoside was then transformed into FTagged trichloroacetimidate **29** using standard conditions [23]. Cleavage of this light FTag protecting group was accomplished under acidic conditions using 0.02 M HCl in a mixture of methanol/water to give non-FTagged glucosyl donor **30**. Despite having the FTag present in the glycosyl donor, the team was able to show that no significant amount of FTagged side products (e.g. FTagged hydrolyzed donor) that could interfere with the purification was formed during the exemplified reactions.

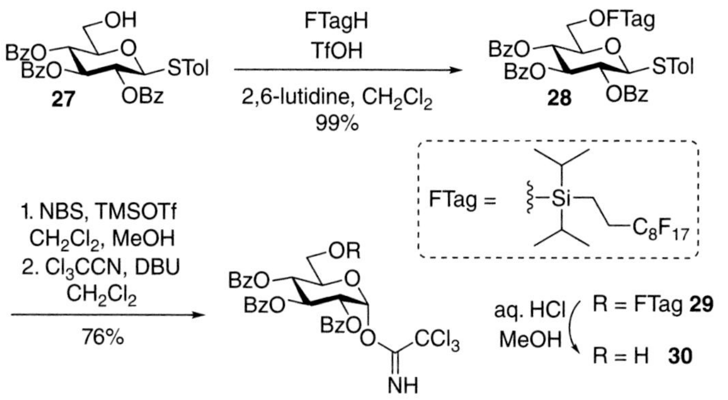

Scheme 15.13 Fluorinated silyl protecting group in oligosaccharide synthesis. *Source:* Zhang et al. 2009 [22]. Reproduced with permission of American Chemical Society.

15.2.6 Phosphate FTags

Phosphate groups are a common motif in many biologically relevant molecules, such as nucleic acids, phospholipids, proteins, and complex carbohydrates, and these motifs are important not only structurally but also in terms of their function [24]. In this context, the development of facile protocols for the synthesis of sugar phosphates for biological research is very relevant. In 2011, the Pohl group developed a novel cleavable, fluorous protecting group for phosphates that could be used as a facile label for purification and removed under mild reducing conditions [25]. In order to demonstrate the robustness of this FTag in carbohydrate chemistry, a fluorous protected phosphite, which was prepared from reaction of fluorous bromo-alcohol and benzyl phosphoramidite in the presence of

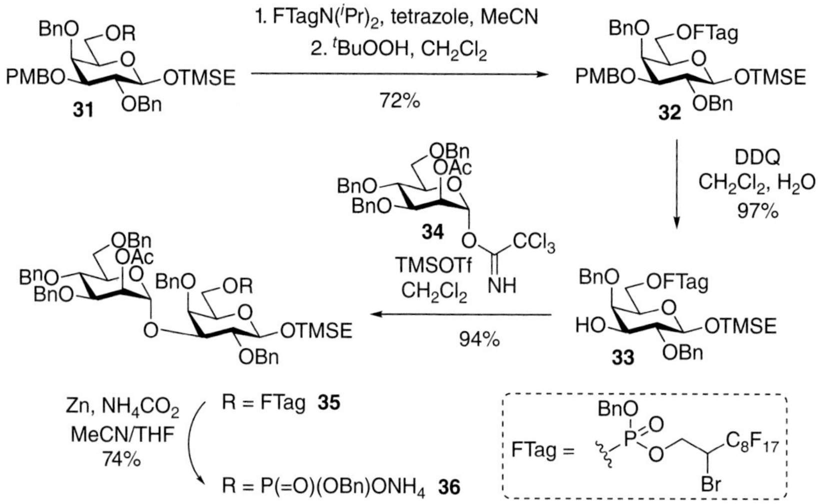

Scheme 15.14 Fluorinated phosphate protecting group in the synthesis of sugar phosphates. *Source:* Liu and Pohl 2011 [25]. Reproduced with permission of American Chemical Society.

tetrazole, was introduced at the C-6 hydroxy group of orthogonally protected galactoside **31** to give FTagged galactoside **32** (Scheme 15.14). Selective cleavage of the *para*-methoxybenzyl (PMB) ether at C-3 of **32** with DDQ afforded FTagged acceptor **33**, which was consequently glycosylated with mannosyl trichloroacetimidate donor **34** to give FTagged disaccharide **35** in an excellent yield of 94%. Once the FTag was installed, FSPE was used as the purification method after each step. The FTag was finally cleaved under mild reductive conditions with zinc and ammonium formate, to reveal phosphate-bearing disaccharide **36** as the ammonium salt in 74% yield.

15.3 Ionic Liquid Supports

Ionic liquids (ILs) have emerged as a popular class of reagents in organic chemistry as new vehicles for the immobilization of reagents because of their unique physical and chemical properties [26]. The application of ILs as protecting groups and soluble supports in oligosaccharide synthesis have only been developed in the past 10 years [27]. An ITag generally consists of a linker, to which an ionic portion, usually a quaternary amine salt, is attached. The ionic nature of the ITag renders the molecule highly polar and thus easily separable from nonpolar material present in the reaction mixture. So far, in ITag-assisted oligosaccharide synthesis, only 1-methylimidazolium-based ILs have been employed because of their facile and cost-effective synthesis by S_N2 displacement with 1-methylimidazole of a leaving group (normally a halide) present in the linker attached to the glycoside (Scheme 15.15). The halide is later on displaced by a more stable anion, such as hexafluorophosphate, tetrafluoroborate, or trifluoromethylsulfonates, that can withstand the glycosylation conditions and that will not be easily exchanged by any of the anionic species present in the reaction media.

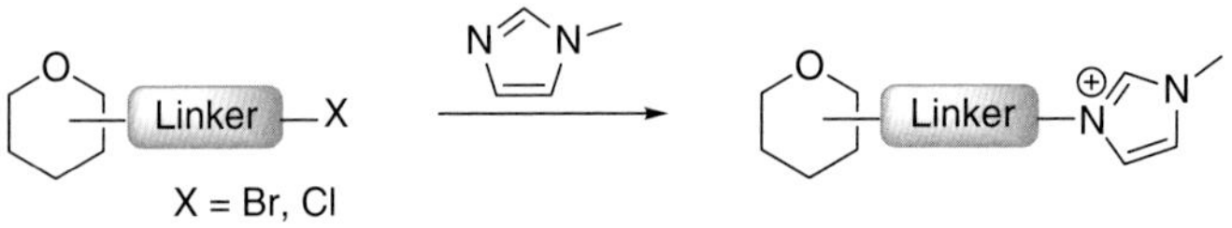

Scheme 15.15 Formation of ionic liquid tag (ITag).

Another feature of ITags is their solubility in sufficiently polar solvents, such as dichloromethane and acetonitrile, while they are insoluble in less polar solvents such as diethyl ether or hexanes. This allows for homogeneous reaction conditions in polar solvents, which facilitates *in situ* reaction monitoring, as well as limiting the need for excess reagents in order to drive the reaction to completion. Once the reaction is complete, non-ITagged materials can be separated from the ITagged product in solvents of low polarity, facilitating product purification by either liquid–liquid extraction or precipitation of the ITagged compounds in nonpolar solvents (Scheme 15.16).

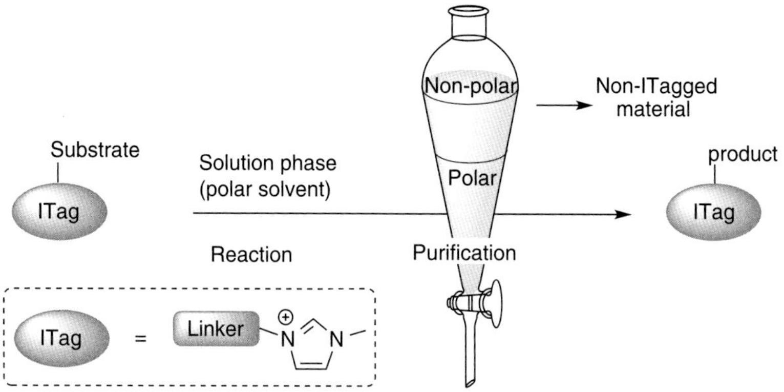

Scheme 15.16 ITag-supported oligosaccharide synthesis and product purification.

So far, only two types of linkages for the attachment of the ITag to the saccharide have been reported, ester linkages and ether linkages. The initial reports employed an ester-linked ITag that was attached to the oligosaccharide at either C-4 or C-6 (Scheme 15.17a). These strategies again raise the same issues as already described for the ester-linked FTag-assisted oligosaccharide synthesis: (i) lack of orthogonality with other ester protecting groups commonly used in oligosaccharide synthesis and (ii) unwanted formation of ITagged side products stemming from anomeric hydrolysis. Consequently, subsequent reports employed ether-linked ITags that masked the anomeric position of the reducing end of the growing oligosaccharide chain (Scheme 15.17b).

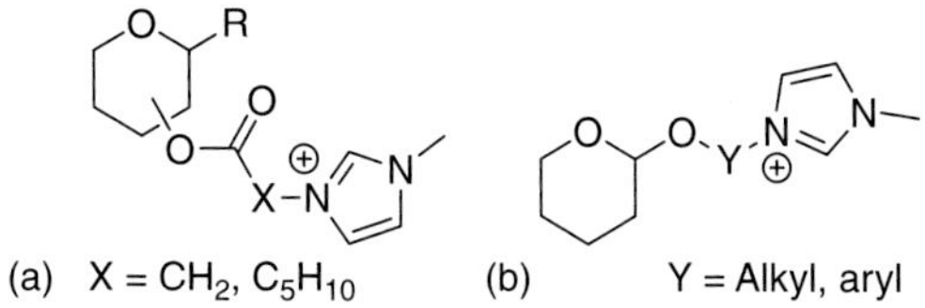

Scheme 15.17 Ester-linked ITags at C-4 or C-6 (a) and ether-linked ITags to the anomeric position (b).

15.3.1 Ester-linked ITags at the C-4 and C-6 Position

To access ester-linked ITags, typically an acyl chloride or a free acid-bearing linker is reacted with an alcohol of either the C-4 or C-6 position of a given glycoside ring. The linker contains a terminal halide (bromide or chloride), ready to be displaced in an S_N2 reaction with 1-methylimidazole to form the IL-tagged product (Scheme 15.18a). Cleavage of ester-linked ITags can be achieved through hydrolysis under basic conditions (Scheme 15.18b).

The groups of Chan [28] and Huang [29] reported, almost in parallel in 2006, the first ITag-assisted oligosaccharide synthesis. Both strategies employed the same ITag; however, there are several differences. For instance, Chan et al. attached ITag1 at the C-6 position of benzyl protected thioglucosyl donor **37**, by

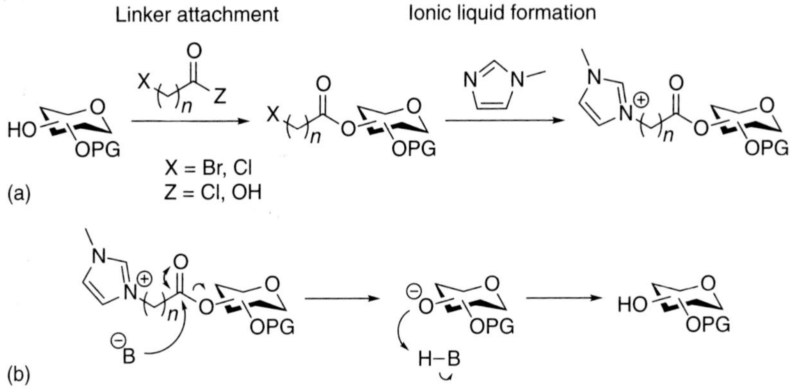

Scheme 15.18 Attachment (a) and cleavage (b) of ester-linked ITags.

first reacting the primary hydroxy group with bromoacetic acid in the presence of *N,N′*-dicyclohexylcarbodiimide (DCC), followed by displacement of the halide with 1-methylimidazole to form ITagged donor **38** in 87% yield (Scheme 15.19a). Thioglycoside activation by oxidation of sulfur with *meta*-chloroperoxybenzoic acid (mCPBA) followed by glycosylation with acceptor **37** in the presence of Tf_2O afforded the corresponding disaccharide, which was subjected to the same sequence of oxidation followed by glycosylation, to form ITagged trisaccharide **39** in 26% yield over the two cycles. In the final step, the ITag was cleaved using cesium carbonate in methanol, revealing non-ITagged trisaccharide **40** in quantitative yield. The reaction mixture was purified after each step by washing with *n*-pentane, *n*-hexane, or isopropyl ether. On the other hand, the group of Wang attached ITag1 to the C4 position of the glycosyl acceptor, as the target molecules were 1,6-linked disaccharide and trisaccharide. To that end, the C-4 hydroxy group of glucoside **41** was reacted with chloroacetic chloride, followed by treatment with 1-methylimidazole to give the imidazolium cation. Counterion exchange with potassium hexafluorophosphate formed ITagged glucoside **42** (Scheme 15.19b), which, after selective deprotection of the silyl ether at C-6 with hydrochloric acid, afforded ITagged acceptor **43** in 87% over 4 steps.

The ITagged glycoside acceptor was then glycosylated with a range of glycosyl donors, to form ITagged disaccharide and trisaccharide **45** in good yields. After the synthesis was completed, the ITag was hydrolyzed under mildly basic conditions with saturated sodium bicarbonate and tetra-*n*-butylammonium bromide (TBAB) to afford non-ITagged oligosaccharides **46** in excellent yields. Similar to the previous example, reaction products were purified between steps by simple washings with diethyl ether or ethyl acetate.

Following those initial reports, the group of Pathak developed a similar glycosylation strategy for the synthesis of an α-1,6-linked tetramannoside. In their report, they again used ITag1, which was installed at the primary position of a benzoyl-protected mannosyl fluoride donor [30]. In this example, the ITag serves the dual function as a purification support, as well as a protecting group that temporarily masks the site that gets glycosylated in the next step. After each

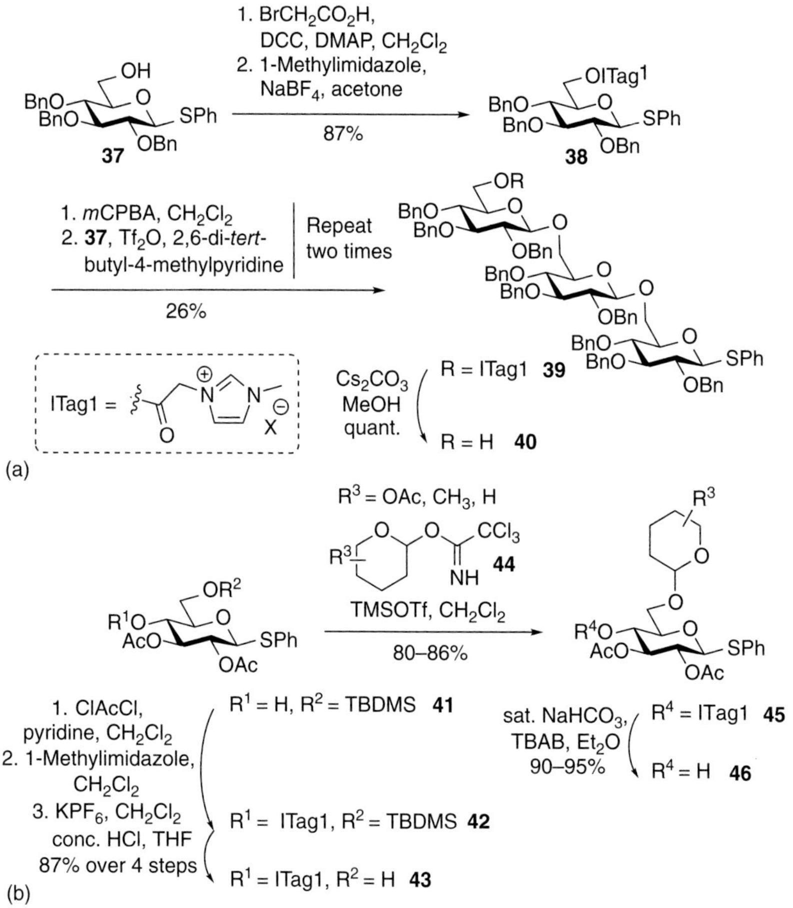

Scheme 15.19 Initial reports of ITagged oligosaccharide synthesis by the groups of Chan (a) and Wang (b).

glycosylation with the ITagged donor, using Cp_2HfCl_2 and $AgClO_4$ as the fluoro glycoside activator in CH_2Cl_2 [31], the imidazolium label is cleaved, revealing a free primary alcohol, ready to be glycosylated with the ITagged donor (Scheme 15.20). The glycosylation/cleavage cycle was repeated three times and linear α-1,6-linked tetramannoside was prepared in 31% overall yield. The ITagged intermediates were purified after each step by washing with diethyl ether, which removed all non-ITagged material from the mixture. Subsequently, the same group was able to extend their approach, with a slight modification, to the synthesis of a linear octamannoside, which demonstrated the potential of a single ITag to purify larger oligosaccharides [32].

In 2010, Gouhier et al. published an alternative approach for IL-supported glycosylation reactions, whereby the imidazolium-based ITag is connected via an ester linkage to the primary position of thioglycoside **48**, while having a free alcohol at C-4 (Scheme 15.21) [33]. The special feature of **48** is that it can act as both glycosyl acceptor and glycosyl donor. Chemoselective glycosylation of **48**

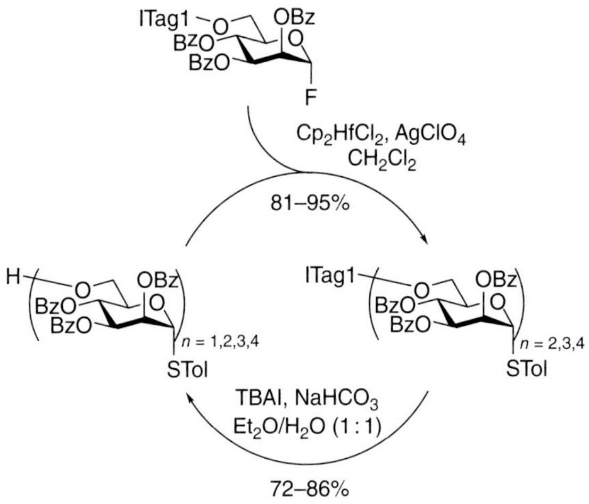

Scheme 15.20 ITag-assisted synthesis of a linear α-1,6-linked tetramannoside. *Source:* Pathak et al. 2008 [30]. Reproduced with permission of American Chemical Society.

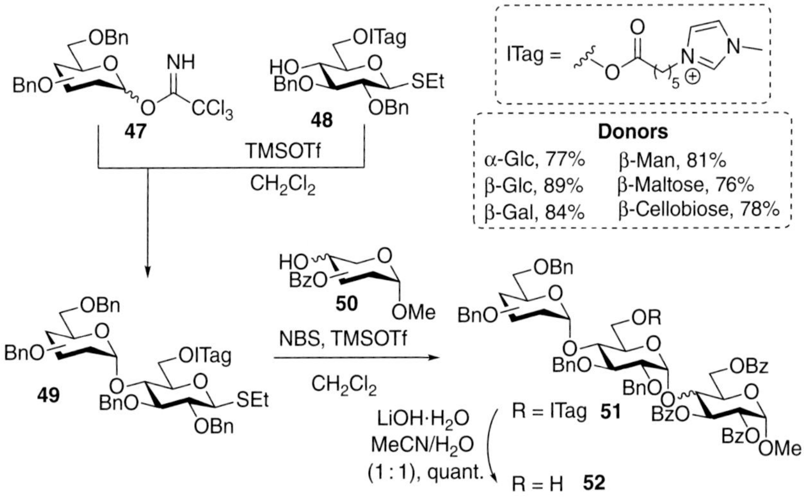

Scheme 15.21 ITag-assisted synthesis of linear α-1,4-linked oligosaccharides. *Source:* Pepin et al. 2010 [33]. Reproduced with permission of John Wiley and Sons.

with trichloroacetimidates **47** under TMSOTf activation proceeded in good yields (76–89%) to afford α-1,4-linked disaccharides **49**, which contain a thioethyl anomeric group. Subsequent activation of the thioglycoside in the presence of NBS and TMSOTf in the presence of a range of glycosyl acceptors **50** afforded a small library of ITagged trisaccharides **51** in the yields of 78–85%. At the end of the synthesis, the ITag was cleaved using lithium hydroxide to give the non-ITagged products **52**, demonstrating the versatility of the ITag in oligosaccharide synthesis.

15.3.2 Ether-linked ITags at the Anomeric Position

In order to address some of the disadvantages associated with the initial work on ITag-assisted oligosaccharide synthesis, subsequent reports used an alternative and improved strategy in which the ITag is attached to the anomeric position of the glycosyl acceptor via an ether linkage. This approach enables the use of more versatile protecting group patterns, including transient ester protecting groups as well as ensuring limited formation of ITagged side products, as the ITags are installed in the glycoside acceptor and not the glycosyl donor, which is prone to hydrolysis during glycosylation.

The ITags are typically introduced via glycosylation to the reducing end monosaccharide of the oligosaccharide target in a two-step process involving glycosylation with a linker acceptor bearing a terminal halide, followed by halide displacement with 1-methylimidazole, forming the ITag (Scheme 15.22).

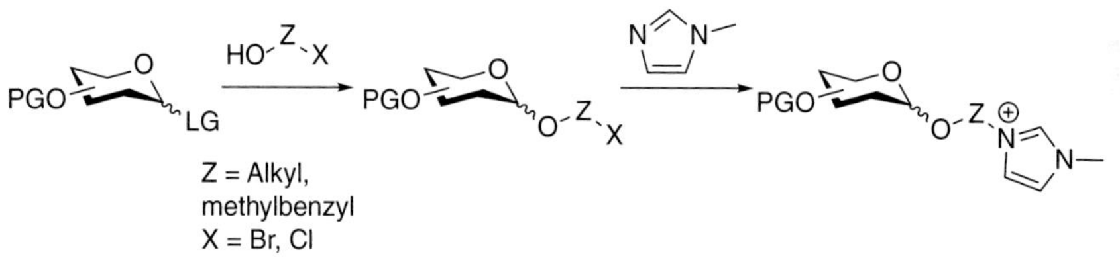

Scheme 15.22 General procedure for ITag attachment at the anomeric position via an ether linkage.

Galan et al. developed in 2011 the so-called ionic catch-and-release oligosaccharide synthesis (ICROS), in which the ITag is attached via a cleavable linker to the anomeric position of the growing oligosaccharide chain (Scheme 15.23) [34]. The team demonstrated that once the ITag is installed, protecting group

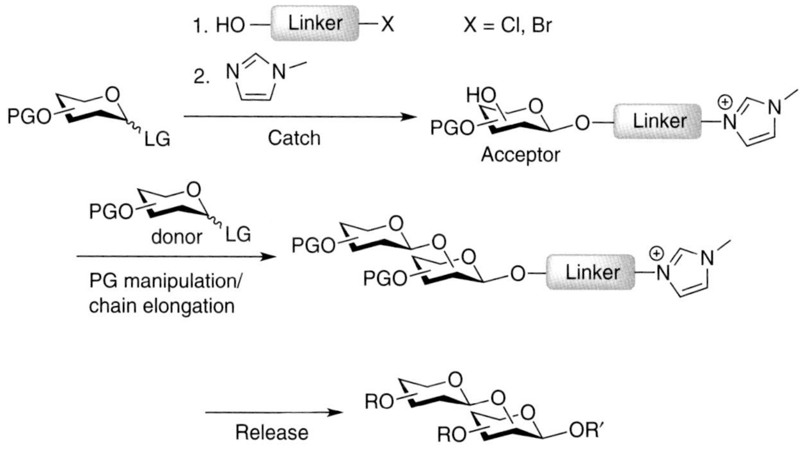

Scheme 15.23 Ionic catch-and-release oligosaccharide synthesis (ICROS). *Source:* Tran et al. 2011 [34]. Reproduced with permission of Royal Society of Chemistry.

manipulations and glycosylations can be performed and the reaction progress can be monitored *in situ* with common techniques such as NMR or mass spectrometry (MS). After the desired oligosaccharide is synthesized, the product can be released from its ITag by cleaving the linker under suitable conditions that leave the newly formed glycosidic linkages intact.

Two ITags with different cleaving mechanisms were developed for the ICROS strategy. Alkyl-linked ITag3 of glycoside **53** can be cleaved under acidic conditions using either HCl in water, releasing hemiacetal **54** or HCl in methanol, which produces methyl-substituted glycoside **55** (Scheme 15.24a). Using ITag3, Galan et al. were able to synthesize linear β-1,6-linked tetraglucan **58** in a matter of days by sequentially deprotecting the primary position of the C-6-silylated glucoside **56** with HCl in methanol, followed by glycosylation with 6-*O*-TIPS-protected trichloroacetimidate donor **57** under TMSOTf catalysis (Scheme 15.24b). Repeating the sequence of selective desilylation, followed by glycosylation with **57** for a further two cycles, afforded ITagged tetrasaccharide **58** in 69% yield. After each step, the reaction mixture was purified by washing with diethyl ether/*n*-hexane, which yielded pure ITagged material.

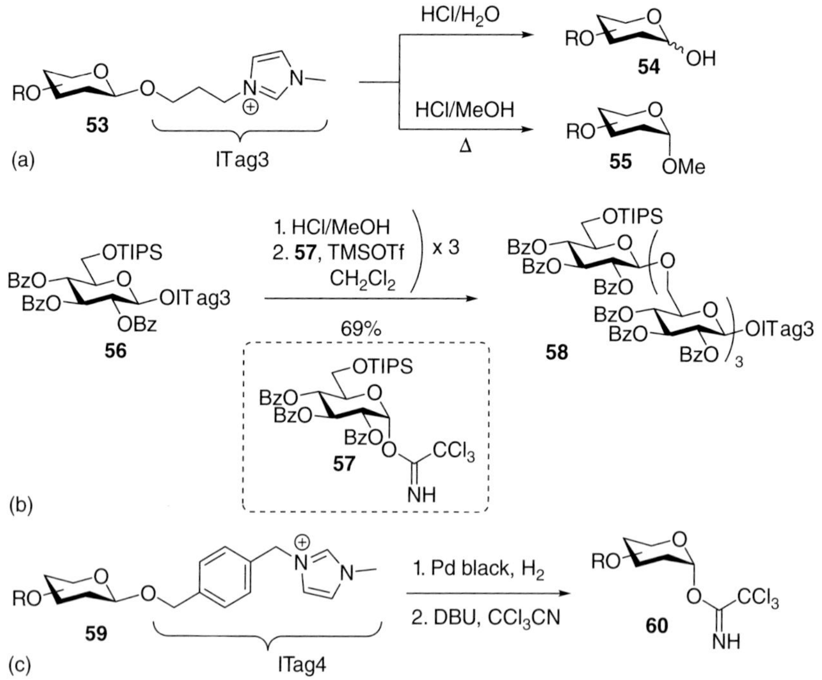

Scheme 15.24 Cleaving conditions for propyl ether-linked ITag3 (a), synthesis of a linear tetrasaccharide using the ICROS methodology (b), and benzyl ether-linked ITag4 (c).

Despite the simplicity of alkyl-linked ITag3, the harsh reaction conditions needed to remove the ITag might also hydrolyze acid-labile glycosidic linkages. Therefore, the team developed benzyl ether-linked ITag4 of glycoside **59**, which was again linked to the anomeric position. ITag4 can be cleaved by catalytic hydrogenolysis releasing a hemiacetal, which could be converted into the corresponding trichloroacetimidate donor **60** under standard conditions (Scheme 15.24c) [32]. Although Pd black is used in this example as the catalyst, Pd/C or $Pd(OH)_2$ could also be used in combination with H_2 to achieve the removal of the ITag.

Moreover, the Galan group has demonstrated the applicability of the ICROS methodology in combinatorial oligosaccharide chemistry [35]. By following the general cycle of partial glycosylation of benzyl-ITagged acceptor **62** with trichloroacetimidate glycoside donor **57** under LA catalysis, followed by partial desilylation of **61** with HCl in methanol, the group was able to synthesize a mixture of linear β-1,6-linked glucans (di- to tetrasaccharides) (Scheme 15.25). Reaction progress was monitored throughout the synthesis by HPLC and ^{1}H-NMR, which allowed the team to control the levels of glycosylation and desilylation at each step. Purification of ITagged intermediates was accomplished by washing with diethyl ether/*n*-hexane (1 : 1). Upon completion of the synthesis, size-exclusion chromatography was used to deconvolute the mixture, to afford ITagged disaccharide **63** in 10%, ITagged trisaccharide **64** in 30%, and ITagged tetrasaccharide **65** in 14% yield. Global deprotection by debenzoylation using

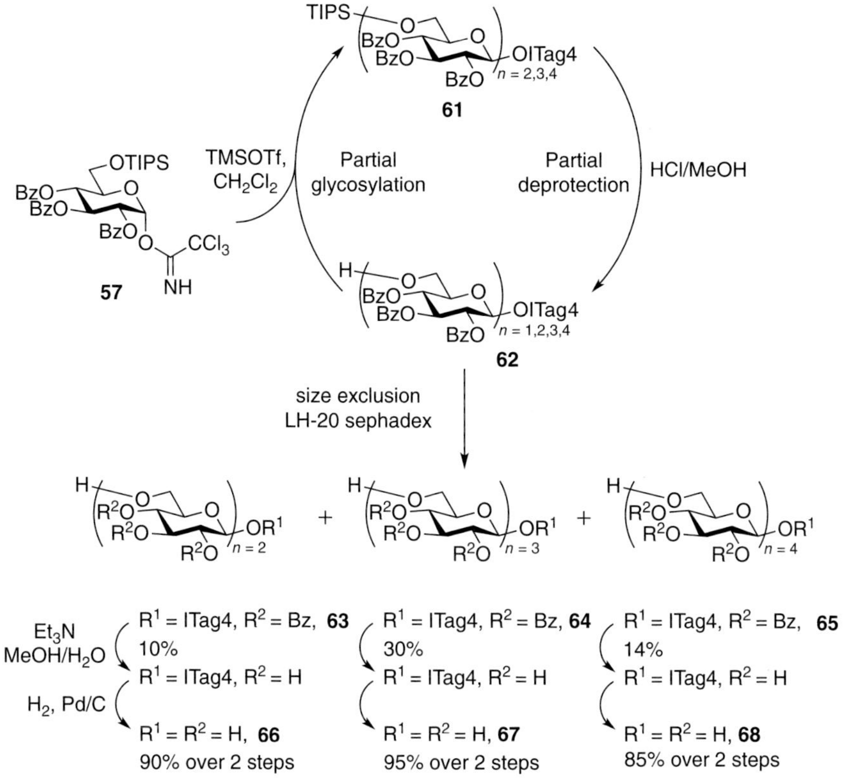

Scheme 15.25 Combinatorial ICROS. *Source:* Sittel et al. 2013 [35]. Reproduced with permission of Royal Society of Chemistry.

triethylamine in a water/methanol mixture, followed by benzyl-ITag removal via catalytic hydrogenolysis, afforded the fully unprotected oligosaccharides **66–68** in 85–95% yields.

The ICROS methodology was also subsequently employed by the group of Li, first in 2011, for the rapid synthesis of a linear α-1,2-linked nonamannoside under 30 h [36]. To that end, benzyl ether-linked ITag4 was also attached to the anomeric position of the reducing end of the growing oligosaccharide chain. Eight chemoselective deprotection/glycosylation cycles were needed for the assembly of the nonamannoside, which was obtained in an overall yield of 22% (Scheme 15.26). In order to ensure chemoselective deprotection, the C-2 hydroxy group was masked with an acetyl protecting group, while the remaining alcohols were protected as benzyl ethers. The acetate group of ITagged mannoside **70** was removed with sodium methoxide in methanol to reveal the monohydroxylated glycosyl acceptor **71**, ready to be glycosylated with C-2-acetyl-protected trichloroacetimidate donor **69**. At the end of the synthesis, global benzyl ether deprotection, with the concomitant product release from the benzyl-type ITag in 1 step by catalytic hydrogenolysis, afforded unprotected nonamannoside **72** in 22% yield over 17 steps. The reaction mixture was purified in between steps by precipitation of the ITagged products in isopropyl ether, followed by centrifugation and separation from the supernatant, which contained the non-ITagged materials.

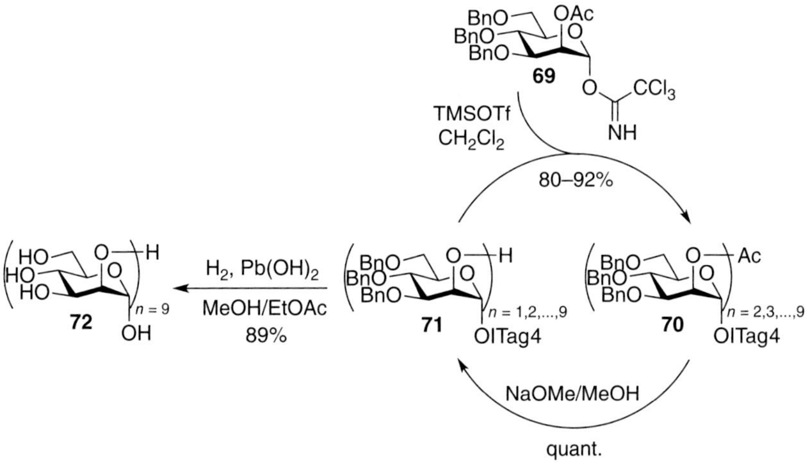

Scheme 15.26 Synthesis of linear α-1,2-linked nonamannoside. *Source:* Ma et al. 2011 [36]. Reproduced with permission of American Chemical Society.

A few years later, the same group used a similar strategy, also employing benzyl-type ITag4 for the assembly of a more complex, branched oligosaccharide, further demonstrating the utility of the approach [37].

Another recent example on the application of ICROS comes from the Boyer and Beau group [38]. An orthogonally protected β-1,4-chitotetrasaccharide, which is a key intermediate for the synthesis of lipochitooligosaccharides and

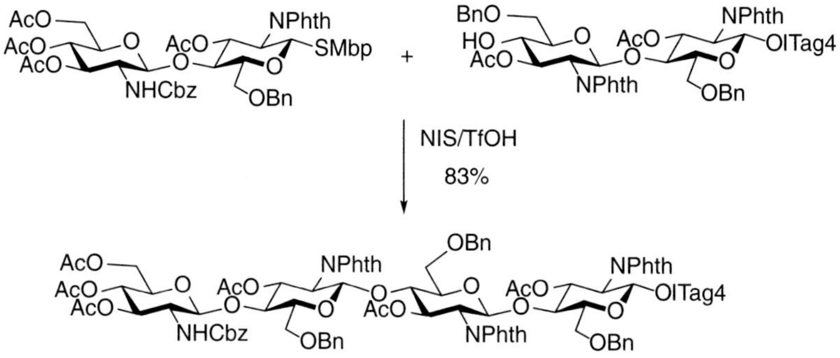

Scheme 15.27 Synthesis of protected chitotetrasaccharide. *Source:* Gillard et al. 2016 [38]. Reproduced with permission of John Wiley and Sons.

the TMG chitotriomycin, was prepared by employing anomeric benzyl ether ITag4 in an excellent yield of 83%. The target molecule was obtained by employing a 2 + 2 blockwise approach using thioglycoside-type glycosylations, whereby both the disaccharide acceptor and the donor bear different N-protecting groups (*N*-benzyloxycarbonyl (N-Cbz) and *N*-phthalimide (N-Phth)) to allow for chemoselective deprotection at these positions. The final tetrasaccharide deprotection probe to be challenging, which the authors proposed, is due to the unsuitable association of an oligo-chitin acetamido framework with the dialkylimidazolium IL.

15.3.3 ITag-assisted Chemoenzymatic Oligosaccharide Synthesis

ITags do not only facilitate purification of ITagged oligosaccharides from non-polar materials but they also improve significantly the detection limits in MS, as the ITagged material is already ionized [39]. Therefore, the imidazolium-based tags are ideal for reaction progress monitoring and more excitingly, offer a more economic and practical alternative to fluorescent or radioactive labels for the study of enzymatic glycosylations.

The Galan group also developed the first ITag-based MS-probes to monitor enzymatic glycosylations. Initial work described the use of disulfide-based ITag5 as cleavable MS probe (Scheme 15.28a) [40], and subsequently the team developed the more chemically stable ITag6, which was linked via an aryl sulfonamide

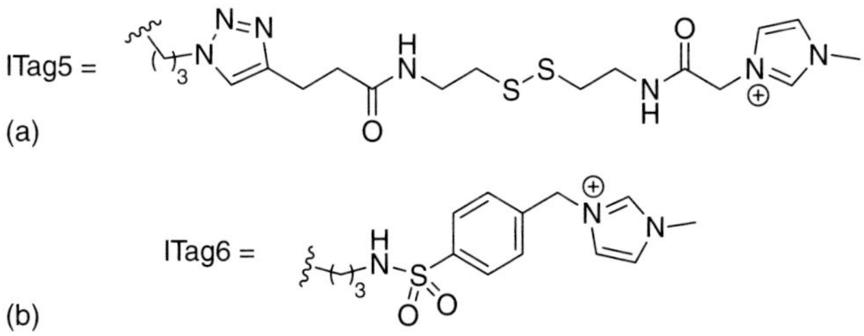

Scheme 15.28 Disulfide-based (a) and arylsulfonamide-based (b) MS probes.

functional group (Scheme 15.28b) [41]. Although cleaved under different reaction conditions (e.g. the disulfide bond can be cleaved under reductive conditions to reveal a thiol functionality, while the sulfonamide bond can be cleaved by nucleophilic substitution reaction with $KPPh_2$ [42] to reveal an amine functionality), upon product release, both ITags yield oligosaccharide targets bearing a functional group that is ready for attachment to a microarray platform for subsequent biological screening.

A range of biologically relevant oligosaccharides was prepared using this chemoenzymatic strategy, showcasing the compatibility of ITag5 and ITag6 with a variety of glycosyltransferases (Scheme 15.29). The group was also able to obtain kinetic data for the enzyme catalyzed transformations, using the ITags as the MS reporters in LC/MS-based assays.

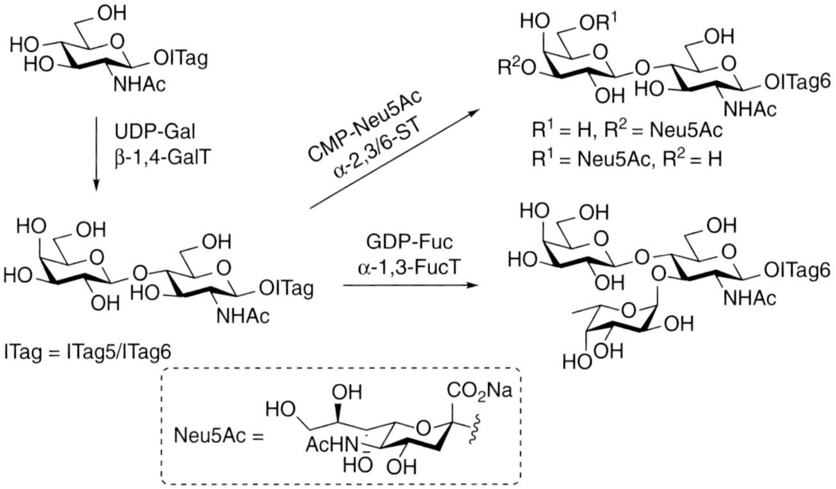

Scheme 15.29 ITags in enzymatic oligosaccharide synthesis [39, 40].

15.3.3.1 Synthesis of 3-(Perfluorooctyl)propanyloxybutenyl-4-O-benzyl-3-O-levulinyl-6-O-(4-O-benzyl-6-O-*t*-butyldiphenylsilyl-3-O-levulinyl-2-O-pivaloyl-α-D-mannopyranoside)-2-O-pivaloyl-α-D-mannopyranoside [15a]

A solution of 3-(perfluorooctyl)propanyloxybutenyl-4-*O*-benzyl-6-hydroxy-3-*O*-levunilyl-2-*O*-pivaloyl-α-D-mannopyranoside (78 mg, 0.079 mmol) and 4-*O*-benzyl-6-*O*-*t*-butyldiphenylsilyl-3-*O*-levunilyl-2-*O*-pivaloyl-α/β-D-mannopyranosyl trichloroacetimidate (86 mg, 0.103 mmol) in CH_2Cl_2 (3 ml) is cooled to 5 °C and TMSOTf (7 μl, 0.40 mmol) is added. The reaction mixture is stirred for 15 min. The reaction mixture is then quenched with triethylamine (20 μl) and then concentrated under reduced pressure. The crude product is purified by using a FSPE cartridge. Nonfluorous compounds are eluted with 80% MeOH/water and the desired product is eluted by 100% MeOH. The solvent is removed under reduced pressure to obtain mannose disaccharide (118 mg, 90%) as a yellow gel.

15.3.3.2 Synthesis of 3-(3-Methylimidazolium)-propyl 2,3,4-tri-O-benzoyl-6-O-(2,3,4-tri-O-benzoyl-6-O-triisopropylsilyl-β-D-glucopyranosyl)-β-D-glucopyranoside trifluoromethanesulfonate [34]

To a solution of 3-(3-methylimidazolium)-1-propyl 2,3,4-tri-*O*-benzoyl-β-D-glucopyranoside trifluoromethanesulfonate (140 mg, 0.23 mmol) and 2,3,4-tri-*O*-benzoyl-6-*O*-triisopropylsilyl-β-D-glucopyranosyl trichloroacetimidate (236 mg, 0.46 mmol) in CH_2Cl_2 (5 ml) were added 4 Å molecular sieves, and the mixture was stirred for 30 min at room temperature and then cooled to 0 °C. TMSOTf (13 μl, 0.07 mmol) was added at 0 °C, and the reaction was left to warm to room temperature and stirred overnight. Reaction progress was monitored by MALDI-TOF analysis. After reaction completion, the mixture was filtered and concentrated under reduced pressure. The obtained oil was then washed twice with *n*-hexane (2 ml) and *n*-hexane/Et_2O 1 : 1 (2 ml) to give the coupling product in 94% yield.

15.4 Conclusions

The development of supported methodologies in oligosaccharide synthesis that can help simplify purification has seen tremendous development over the past 20 years and has spawned new methodologies for the rapid assembly of large and complex oligosaccharides and oligosaccharide libraries using combinatorial chemistry. Among the different strategies, FTags and ITags have been devised as solution-phase supports that allow the chemistry to be carried out in homogeneous reaction conditions as an attempt to overcome the inherent problems of solid support oligosaccharide synthesis (e.g. sluggish reactions, low loadings, and need for high amounts of glycosyl building blocks). In addition, the new tags also bear the role of temporary protecting groups, and as such, orthogonal protecting group schemes have been devised to allow for the installation and selective cleavage of the labels in the presence of other functional groups. Furthermore, the FTag-strategy has been successfully adapted to automated synthesis and FTags have also been used for noncovalent attachment of oligosaccharides to microarray platforms. On the other hand, imidazolium-based ITags have demonstrated their potential in the rapid assembly of large oligosaccharides in a matter of days, using chemical and enzymatic approaches. The ITags can also be used as MS reporters to measure reaction progress and kinetics.

As new developments in the area of stereoselective glycosylation emerge, the use of solution-phase labels that can take advantage of more efficient and expedient approaches (e.g. with high yields and complete stereocontrol in any given coupling reaction) offers a promising handle toward tackling some of the obstacles that remain toward the automation of oligosaccharide synthesis.

Abbreviations

Ac	acetyl
aq.	aqueous

Bn	benzyl
BTF	benzotrifluoride
Bz	benzoyl
cat.	catalytic
CMP	cytidine monophosphate
conc.	concentrated
DTBMP	2,6-di-*tert*-butyl-4-methylpyridine
Fuc	fucose
Gal	galactose
GDP	guanosine diphosphate
GPI	glycophosphatidylinositol
Glc	glucose
HPLC	high-performance liquid chromatography
LC/MS	liquid chromatography–mass spectrometry
LG	leaving group
Man	mannose
Neu5Ac	*N*-acetylneuraminic acid
org.	organic
Piv	pivaloyl
PG	protecting group
PMB	*para*-methoxybenzyl
NPhth	*N*-phthalimide
quant.	quantitative
sat	saturated
TBAI	tetra-*n*-butylammonium iodide
TBDMS	*tert*-butyldimethylsilyl
TBDPS	*tert*-butyldiphenylsilyl
TDS	thexyldimethylsilyl
Tf	trifluoromethanesulfonyl
TFA	trifluoroacetic acid
TFT	trifluorotoluene
THF	tetrahydrofuran
TIPS	triisopropylsilyl
TMB	1,3,5-trimethoxybenzene
TMS	trimethylsilyl
TMSE	2-trimethylsilylethyl
Tol	toluyl
Ts	toluenesulfonyl
UDP	uridine diphosphate

References

1 (a) Galan, M.C., Benito-Alifonso, D., and Watt, G.M. (2011). *Org. Biomol. Chem.* 9: 3598–3610. (b) Zhu, X.M. and Schmidt, R.R. (2009). *Angew. Chem. Int. Ed.* 48: 1900–1934. c) Yasomanee, J.P. and Demchenko, A.V. (2013). *Trends Glycosci. Glycotechnol.* 25: 13–42.

2 Roychoudhury, R. and Pohl, N.L.B. (2014). *Modern Synthetic Methods in Carbohydrate Chemistry: From Monosaccharides to Complex Glycoconjugates* (ed. D.B. Werz and S. Vidal), 221–239. Weinhem: Wiley.
3 Galan, M.C., Jones, R.A., and Tran, A.T. (2013). *Carbohydr. Res.* 375: 35–46.
4 (a) Tang, S.-L., Linz, L.B., Bonning, B.C., and Pohl, N.L.B. (2015). *J. Org. Chem.* . (b) Tang, S.-L. and Pohl, N.L.B. (2015). *Org. Lett.* 17: 2642–2645.
5 (a) Chen, G.-S. and Pohl, N.L. (2008). *Org. Lett.* 10: 785–788. (b) Edwards, H.D., Nagappayya, S.K., and Pohl, N.L.B. (2012). *Chem. Commun.* 48: 510–512. (c) Jaipuri, F.A., Collet, B.Y.M., and Pohl, N.L. (2008). *Angew. Chem. Int. Ed.* 47: 1707–1710. (d) Ko, K.-S., Jaipuri, F.A., and Pohl, N.L. (2005). *J. Am. Chem. Soc.* 127: 13162–13163.
6 Dandapani, S. (2006). *QSAR Comb. Sci.* 25: 681–688.
7 (a) Manzoni, L. (2003). *Chem. Commun.* 2930–2931. (b) Kasuya, M.C.Z., Cusi, R., Ishihara, O. et al. (2004). *Biochem. Biophys. Res. Commun.* 316: 599–604. (c) Jing, Y. and Huang, X. (2004). *Tetrahedron Lett.* 45: 4615–4618. (d) Carrel, F.R., Geyer, K., Codée, J.D.C., and Seeberger, P.H. (2007). *Org. Lett.* 9: 2285–2288. (e) Kojima, M., Nakamura, Y., and Takeuchi, S. (2007). *Tetrahedron Lett.* 48: 4431–4436. (f) Tanaka, H., Tanimoto, Y., Kawai, T., and Takahashi, T. (2011). *Tetrahedron* 67: 10011–10016. (g) Tanaka, H., Tateno, Y., and Takahashi, T. (2012). *Org. Biomol. Chem.* 10: 9570–9582.
8 (a) Miura, T., Goto, K., Hosaka, D., and Inazu, T. (2003). *Angew. Chem. Int. Ed.* 42: 2047–2051. (b) Miura, T., Satoh, A., Goto, K. et al. (2005). *Tetrahedron: Asymmetry* 16: 3–6. (c) Mizuno, M., Matsumoto, H., Goto, K., and Hamasaki, K. (2006). *Tetrahedron Lett.* 47: 8831–8835. (d) Mizuno, M., Kitazawa, S., and Goto, K. (2008). *J. Fluorine Chem.* 129: 955–960.
9 Curran, D.P., Ferritto, R., and Hua, Y. (1998). *Tetrahedron Lett.* 39: 4937–4940.
10 Goto, K. and Mizuno, M. (2007). *Tetrahedron Lett.* 48: 5605–5608.
11 Carrel, F.R., Geyer, K., Codée, J.D.C., and Seeberger, P.H. (2007). *Org. Lett.* 9: 2285–2288.
12 Kim, J.-H., Yang, H., Park, J., and Boons, G.-J. (2005). *J. Am. Chem. Soc.* 127: 12090–12097.
13 Yang, B., Jing, Y., and Huang, X. (2010). *Eur. J. Org. Chem.* 2010: 1290–1298.
14 Mootoo, D.R., Konradsson, P., Udodong, U., and Fraser-Reid, B. (1988). *J. Am. Chem. Soc.* 110: 5583–5584.
15 (a) Jaipuri, F.A. and Pohl, N.L. (2008). *Org. Biomol. Chem.* 6: 2686–2691. (b) Zong, C., Venot, A., Dhamale, O., and Boons, G.-J. (2013). *Org. Lett.* 15: 342–345. (c) Huang, W., Gao, Q., and Boons, G.-J. (2015). *Chem. Eur. J.* 21: 12920–12926.
16 Miura, T., Goto, K., Hosaka, D., and Inazu, T. (2003). *Angew. Chem. Int. Ed.* 42: 2047–2051.
17 Park, G., Ko, K.-S., Zakharova, A., and Pohl, N.L. (2008). *J. Fluorine Chem.* 129: 978–982.
18 Jaipuri, F.A. and Pohl, N.L. (2008). *Org. Biomol. Chem.* 6: 2686–2691.
19 Zhang, Y., Liu, B., and Liu, G. (2013). *Mol. Diversity* 17: 613–618.
20 Tojino, M. and Mizuno, M. (2008). *Tetrahedron Lett.* 49: 5920–5923.
21 Manzoni, L. and Castelli, R. (2006). *Org. Lett.* 8: 955–957.
22 Zhang, F., Zhang, W., Zhang, Y. et al. (2009). *J. Org. Chem.* 74: 2594–2597.

23 Yu, H. and Ensley, H.E. (2003). *Tetrahedron Lett.* 44: 9363–9366.

24 (a) Gait, M.J., Michael Blackburn, G., Loakes, D., and Williams, D.M. (2006). *Nucleic Acids in Chemistry and Biology*, 3e. Cambridge, UK: The Royal Society of Chemistry. (b) Hanahan, D.J. (1997). *A Guide to Phospholipid Chemistry*. Oxford: Oxford University Press. (c) Deutscher, J., Francke, C., and Postma, P.W. (2006). *Microbiol. Mol. Biol. Rev.* 70: 939–1031.

25 Liu, L. and Pohl, N.L.B. (2011). *Org. Lett.* 13: 1824–1827.

26 Hallett, J. and Welton, T. (2011). *Chem. Rev.* 111: 3508–3576.

27 Galan, M., Jones, R., and Tran, A. (2013). *Carbohydr. Res.* 375: 35–46.

28 He, X. and Chan, T. (2006). *Synthesis* 2006: 1645–1651.

29 Huang, J., Lei, M., and Wang, Y. (2006). *Tetrahedron Lett.* 47: 3047–3050.

30 Pathak, A., Yerneni, C., Young, Z., and Pathak, V. (2008). *Org. Lett.* 10: 145–148.

31 Suzuki, K., Maeta, H., Matsumoto, T., and Tsuchihashi, G. (1988). *Tetrahedron Lett.* 29: 3571–3574.

32 Yerneni, C., Pathak, V., and Pathak, A. (2009). *J. Org. Chem.* 74: 6307–6310.

33 Pepin, M., Hubert-Roux, M., Martin, C. et al. (2010). *Eur. J. Org. Chem.* 6366–6371.

34 Tran, A., Burden, R., Racys, D., and Galan, M. (2011). *Chem. Commun.* 47: 4526–4528.

35 Sittel, I., Tran, A., Benito-Alifonso, D., and Galan, M. (2013). *Chem. Commun.* 49: 4217–4219.

36 Ma, Q., Sun, S., Meng, X. et al. (2011). *J. Org. Chem.* 76: 5652–5660.

37 Gao, Z.-S., Sun, S., Li, W. et al. (2014). *Chin. Chem. Lett.* 25: 1525–1530.

38 Gillard, L., Tran, A.T., Boyer, F.D., and Beau, J.M. (2016). *Eur. J. Org. Chem.* 1103–1109.

39 Liu, J., Joensson, J., and Jiang, G. (2005). *Trends Anal. Chem.* 24: 20–27.

40 Tran, A., Bernard, C., and Galan, M. (2010). *Chem. Commun.* 8968–8970.

41 (a) Galan, M., Tran, A., Bromfield, K. et al. (2012). *Org. Biomol. Chem.* 10: 7091–7097. (b) Sittel, I. and Galan, M.C. (2015). *Bioorg. Med. Chem. Lett.* 25: 4329–4332.

42 Yoshida, S., Igawa, K., and Tomooka, K. (2012). *J. Am. Chem. Soc.* 134: 19358–19361.

16

Orthogonally Protected Building Blocks for Automated Glycan Assembly

Fabian Pfrengle[1,2] and Peter H. Seeberger[1,2]

[1] *Max Planck Institute of Colloids and Interfaces, Department of Biomolecular Systems, Am Mühlenberg 1, 14476 Potsdam, Germany*
[2] *Freie Universität Berlin, Department of Biology, Chemistry and Physics, Institute of Chemistry and Biochemistry, Arnimallee 22, 14195 Berlin, Germany*

16.1 Introduction

Solid-phase synthesis has evolved over the past 45 years as an efficient method for the preparation of oligosaccharides [1]. As during peptide and oligonucleotide synthesis, the building blocks (BBs) are successively added to a solid phase that is equipped with a cleavable linker. The major advantage of solid-phase chemistry, when compared to solution-phase chemistry is the possibility to wash away reagents and unwanted side products by simple filtration. Therefore, a large excess glycosylating agent ensures completion of all glycosylation reactions, ideally resulting in the formation of a single product after cleavage from the solid support.

Automation of the solid-phase technology holds great potential in assembling oligosaccharides in a short space of time. Oligosaccharide synthesizers are able to carry out temperature-controlled reactions and washes in a fully automated manner, saving the operator numerous manual operations [2]. Many large and complex glycans have been produced by automated glycan assembly [3] since its initial invention in 2001 (Scheme 16.1) [4].

Among the solid supports explored for solid-phase oligosaccharide synthesis (SPOS), polystyrene resins proved most reliable for its chemical inertness, compatibility with all common organic solvents and good loading capacities. Many linkers between the solid support and oligosaccharides have been explored over the years (Figure 16.1). The linker must withstand all conditions employed in the glycosylation and deprotection reactions performed on the solid phase. An ester-based linker is an attractive option for the attachment of the oligosaccharide to the solid support [2, 5]. Following glycan assembly, the linker is readily cleaved by base. Different types of base-labile linkers have been developed that provide the final oligosaccharide products either as the free reducing sugars [5] or equipped with an aminopentyl linker [2]. Olefin metathesis allows for the cleavage of alkene-based linkers [6]. However, it precludes the use of thioglycoside BBs as the

Protecting Groups: Strategies and Applications in Carbohydrate Chemistry, First Edition.
Edited by Sébastien Vidal.

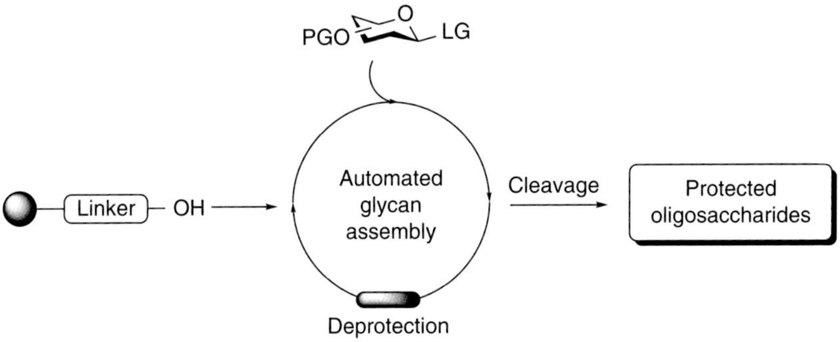

Scheme 16.1 Overview of the automated glycan assembly process.

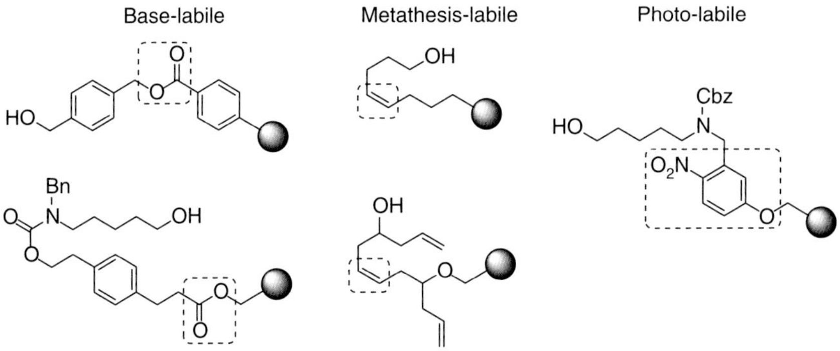

Figure 16.1 Base-labile, metathesis-labile and photo-labile linkers for SPOS.

linker is attacked by the electrophiles required for the activation of the leaving group. A photolabile linker that is cleaved in a commercial continuous flow photoreactor is currently most commonly used for automated glycan assembly [7]. This linker is the most versatile upon the available options and tolerates all kinds of reported manipulations.

Solid-phase synthesis relies on the same leaving groups used as solution-phase carbohydrate chemistry. Acetimidates [8], thioglycosides [9], and phosphates [10] are most commonly used as the glycosylating agents.

16.2 Protecting Groups

When planning the solid-phase synthesis of an oligosaccharide, the most important decision to make is the selection of suitable protecting groups (PGs) for the monosaccharide BBs. The demands for PGs in solid phase are much higher than in solution-phase chemistry. In solution, minor side products are readily removed by intermediary purifications. In SPOS, deletion sequences or unwanted stereoisomers are carried along until the end of the synthesis and complex mixtures are obtained. Thus, the PGs need to be fully orthogonal with each other and their deprotection procedures compatible with linker and solid support. Ideally, the

PGs are compatible with all three types of leaving groups commonly used in SPOS. Two general types of PGs are required: permanent and temporary PGs. While temporary PGs are removed during oligosaccharide assembly "on resin," permanent PGs remain untouched until the final oligosaccharide has been cleaved from the solid support.

16.2.1 Permanent Protecting Groups

Permanent PGs need to tolerate all reaction conditions employed for the removal of temporary PGs during the glycan assembly process. Therefore, functional groups such as simple ethers, esters, or amides are the best choice to retain specialized chemistry for temporary protection. The decision whether 1,2-*cis*- or 1,2-*trans*-linkages are formed when using a particular BB mainly depends on the type of the PG installed at the C2-position of the BB. 1,2-*cis*-Linkages are preferentially formed when nonparticipating PGs are used, whereas participating PGs, which block one face of the intermediary oxonium ion, favor the formation of 1,2-*trans*-glycosidic bonds. However, remote substituents can also control the configuration of the anomeric center. Placement of participating groups at selected positions of the glycosyl donor can promote the formation of 1,2-*cis*-linkages [11].

16.2.1.1 Participating Protecting Groups

Different esters including acetyl, pivaloyl, and benzoyl esters have been used as permanent participating PGs in SPOS (Figure 16.2) [3]. These esters are cleaved by methanolysis upon completion of the solid-phase synthesis. Over time, the benzoyl esters evolved as a superior choice for the permanent protection of hydroxyl groups because acetyl esters can be unwantedly cleaved during glycan assembly, and pivaloyl esters are hard to remove from complex glycans after completion of the synthesis. Glycosylamines were formerly protected as phthalimides [5] but more recently with trichloroacetyl (TCA) groups [3] that provide good coupling yields and are readily converted into the corresponding *N*-acetyl glycosylamines upon hydrogenation with palladium on charcoal [12].

16.2.1.2 Nonparticipating Protecting Groups

Benzyl Ethers Benzyl ethers are the most common permanent PG as they are resistant to acidic and basic conditions. When placed in the C2-position, benzyl ethers as nonparticipating groups favor the formation of α-glycosidic bonds because of the anomeric effect. SPOS mandates extraordinary selectivity in every

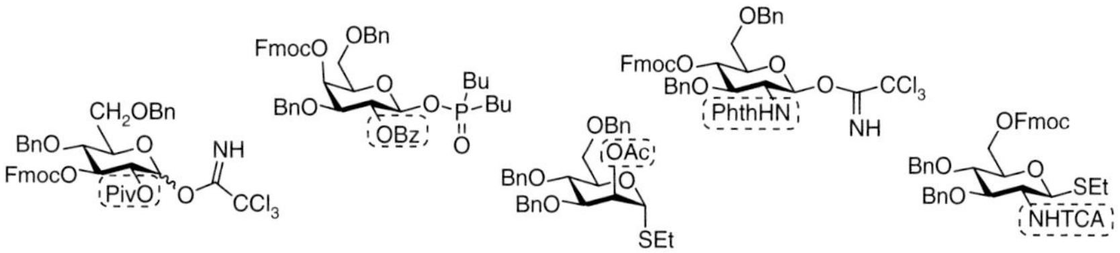

Figure 16.2 Participating PGs for the formation of 1,2-*trans* β-glycosidic bonds.

step of the synthesis because of the lack of possibilities for intermediary purifications. The selectivity of the glycosylation reactions strongly depends on the type of glycosyl donor and its specific substitution pattern. As part of the automated glycan assembly of the tumor-associated carbohydrate antigen Globo-H, the introduction of α-fucose proceeded with excellent stereoselectivity using BB **1** that is protected in the C3- and C4-positions with pivaloyl esters (Scheme 16.2) [13]. The use of esters as remote participating PGs is essential in

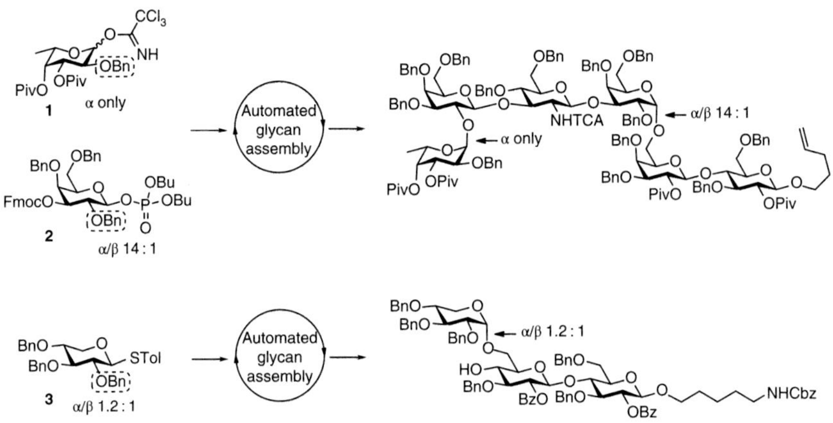

Scheme 16.2 α-Selective BBs in the automated glycan assembly of the tumor associated antigen Globo-H and a xyloglucan-related trisaccharide.

this case as a perbenzylated BB gave poor stereoselectivity in similar reactions [14]. Other glycosylations such as galactosylations required more fine-tuning of the BB PG pattern and the glycosylation conditions employed. In this Globo-H synthesis, BB **2** allowed for the formation of an α-galactosidic linkage with a selectivity of 14 : 1 (α/β). Low reaction temperatures and prolonged reaction times were required to achieve the desired selectivity. Introduction of α-xylosides proved to be particularly challenging. Glycosylation of a glucose unit in the C6-position with xylose BB **3** yielded the respective xyloside as a α/β 1.2 : 1 mixture of anomers [15].

To further improve the α-selectivity of glucosidation and galactosidation reactions, Boons et al. developed an approach where a chiral auxiliary at the C2-position promotes formation of 1,2-*cis*-linkages [16]. By employing this strategy, a glucan oligosaccharide containing four α-glucosidic bonds was obtained from BB **4** (Scheme 16.3). Usually, participating PGs in the C2-position of monosaccharide BBs promote the formation of β-glycosidic bonds through stabilization of the intermediary oxocarbenium ion by ester or carbonate groups. The Boons approach relies on a cyclic anomeric sulfonium ion that is formed after activation of the leaving group and blocks the top face of the glycosylating agent, leading to exclusive formation of the 1,2-*cis*-products in solution- and solid-phase chemistries [17]. On the solid phase, it was necessary to remove the auxiliary before subsequent glycosylation reactions because they were hampered by the high

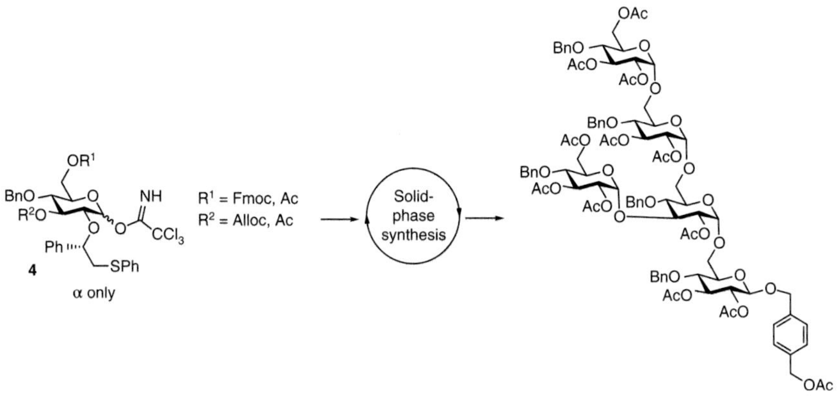

Scheme 16.3 A chiral auxiliary in the C2-position enhances the α-selectivity in glucosidation reactions on solid phase.

steric demand of the auxiliary. Because the removal of the auxiliary with borontrifluoride and acetic anhydride results in the formation of an acetylated compound rather than the free hydroxyl, the auxiliary is regarded as a permanent PG here. Further experiments to show the general applicability of this method are yet to be disclosed.

Special PG patterns are required to enable the formation of particularly difficult 1,2-*cis*-glycosidic linkages such as β-mannosides. β-Mannosylation is favored when the mannose BB is conformationally locked by a 4,6-benzylidene acetal [18]. By using mannose BB **5**, moderate selectivity toward the desired β-isomers was achieved in automated glycan assembly (Scheme 16.4) [19].

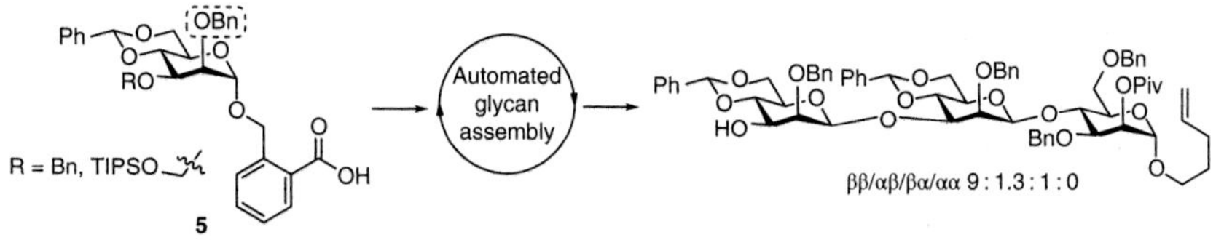

Scheme 16.4 A benzylidene protected mannose BB promotes the preferential formation of β-mannosides.

The formation of β-mannosidic linkages proceeded with perfect β-selectivity when BB **6** was used at low temperatures during the synthesis of β-mannuronic acid alginates (Scheme 16.5). Therefore, oligosaccharides containing up to 12 β-mannosidic linkages were obtained [20].

Azides α-Glycosylamines are common structural motifs in many important glycoconjugates [21]. The glycosaminoglycans heparin and heparan sulfate containing α-glucosides and α-galactosides are typically found in O-linked glycoproteins. Still, only very few examples have been reported for the installation of α-glycosylamines on solid phase. BB 7 has been successfully used

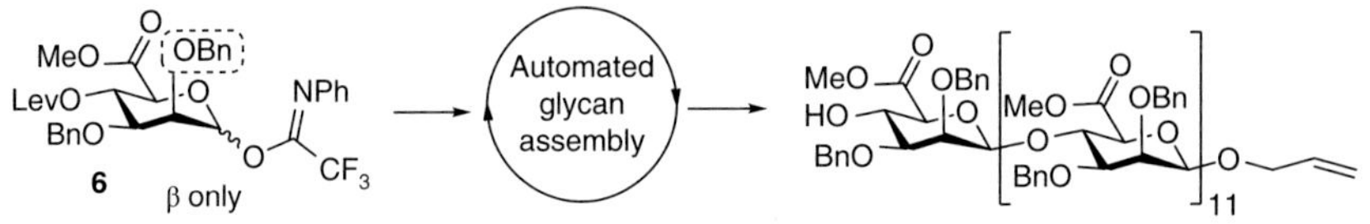

Scheme 16.5 Automated glycan assembly of β-mannuronic acid alginates.

for the solid-phase synthesis of heparin sulfate precursors (Scheme 16.6) [22]. Although high selectivity toward the desired α-isomer was observed, the conversion of the reaction was not satisfactory.

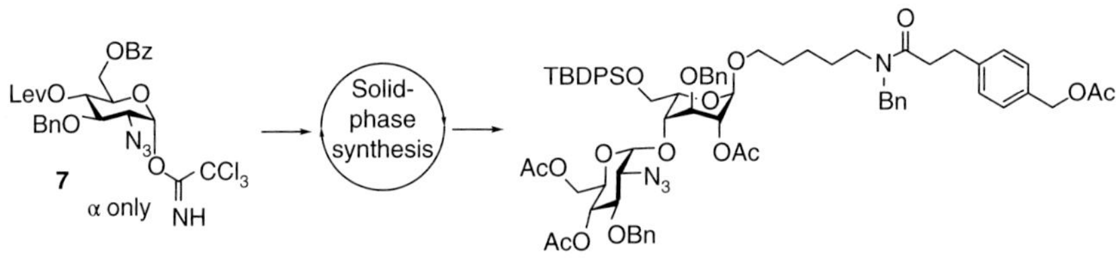

Scheme 16.6 Azide protected BB **7** in the SPOS of heparin sulfate precursors.

16.2.1.3 Cyclic Protecting Groups

Besides benzylidene, other cyclic PGs have been reported (see Chapter 11). Silyl ketals can serve as acid-stable PGs to mask the C4- and C6-hydroxyl functions of glucosamine as demonstrated in the automated glycan assembly of hyaluronan oligosaccharides using BB **8** (Figure 16.3) [23]. The C2- and C3-positions of glucosamine have been protected with a cyclic carbamate that provided a 2 : 1 selectivity in favor of the α-product when a glucose acceptor was glycosylated with BB **9** in the C6-position [24].

16.2.2 Temporary Protecting Groups

A set of temporary PGs is required for the synthesis of complex-branched oligosaccharides on solid phase. The PGs need to be fully orthogonal to each other and tolerate all common glycosylation conditions. When installed in the C2-position, temporary PGs can either serve as participating or nonparticipating groups just like the permanent PGs. Participating PGs are usually attached to the BB through an ester or carbonate moiety, rendering the BB less reactive because of their electron-withdrawing nature (disarming PGs). On the other hand, non-

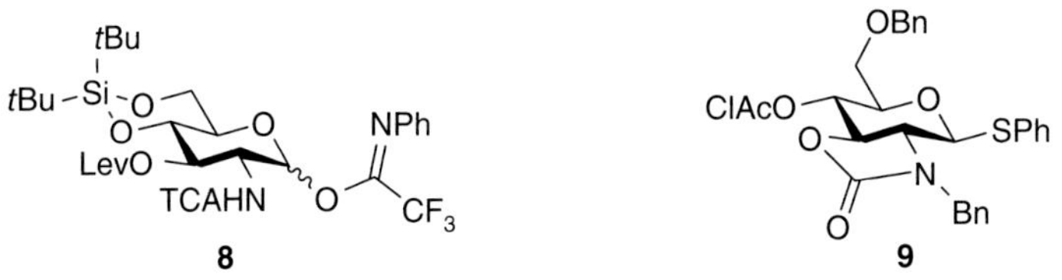

Figure 16.3 Cyclic PGs for the permanent protection of glucosamine BBs.

participating PGs usually contain an ether linkage and do not reduce activity (arming PGs) [25]. Balancing the reactivity of glycosylating agents through fine-tuning of the PGs is of utmost importance as a relatively high reactivity of the BBs is often beneficial in SPOS.

16.2.2.1 Participating Protecting Groups

Fluorenylmethoxycarbonyl (Fmoc) As in solid-phase peptide synthesis, the fluorenylmethoxycarbonyl group (Fmoc) proved to be enormously useful in SPOS. Fmoc allows for monitoring the glycosylation reactions by colorimetric determination of dibenzofulvene production during Fmoc removal. Fmoc carbonates are not very popular for solution-phase chemistry because of their high lability toward base, demanding care in the basic workup after glycosylation reactions. In SPOS, all reagents are washed away after completion of an individual reaction. The lability toward base becomes an advantage when it comes to efficient deprotection protocols. Treatment of the resin with triethylamine three times for 5 min ensured full removal of the Fmoc group in all reported cases. Triethylamine appeared to be superior to piperidine, which is typically used in peptide chemistry, as it avoids occasional migration of neighboring esters (see Chapter 8). Fmoc is the most commonly used temporary PG in automated glycan assembly and has been installed in every position of the common pyranoses as well as in arabinofuranose (Figure 16.4).

Levulinoyl (Lev) Levulinoyl esters (Lev) are the second most frequently used temporary PGs in automated glycan assembly. Lev cleavage by hydrazinium acetate renders it fully orthogonal to Fmoc carbonates. Three incubations of the resin with a hydrazinium acetate solution in pyridine/acetic acid/water for 30 min ensure complete removal of Lev groups. The water is needed to avoid precipitation of the hydrazinium acetate. Lev has been successfully used in combination with Fmoc groups in all positions of the common pyranoses except in the C2-position (Figure 16.5). Lev esters have, in principle, the potential to

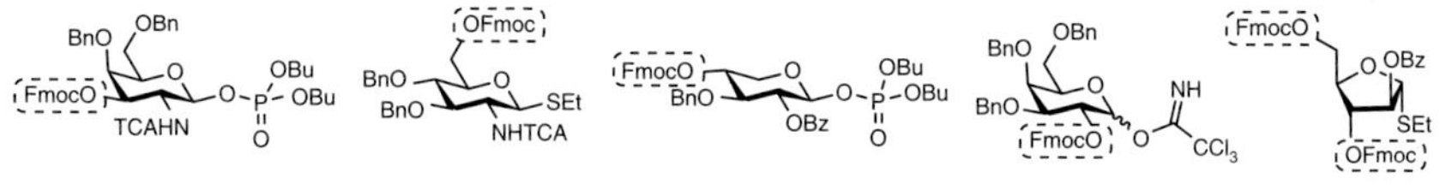

Figure 16.4 Selection of BBs bearing Fmoc as temporary PG.

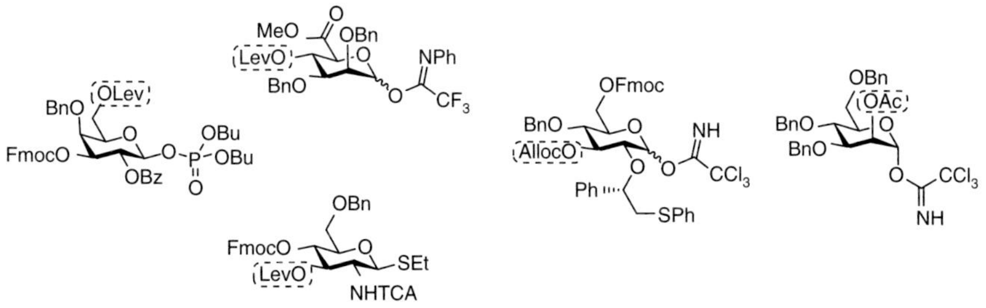

Figure 16.5 BBs equipped with Lev, Alloc and acetyl as participating temporary PGs.

promote β-selective glycosylations, but the keto-function of the Lev moiety might interfere with the intermediary oxocarbenium ion during glycosylations, preventing its routine use in the C2-position.

Allyloxycarbonyl (Alloc) Boons described the use of the allyloxycarbonyl group (Alloc) as temporary PG for SPOS. In the synthesis of a branched α-glucan, the Alloc group was selectively removed using $Pd(PPh_3)_4$ in the presence of Fmoc [17]. Incubation times of 16 h render the repeated use of Alloc in SPOS unpractical. Nevertheless, Alloc is a promising alternative to Fmoc and Lev that complements the set of orthogonal PGs available for SPOS. Experimental evidence to demonstrate the broad utility of the Alloc group for SPOS still has to be accumulated. It remains to be determined whether Alloc can be used as a participating C2-PG.

Acetyl (Ac) Early SPOS work was based on acetates as temporary PGs [4, 26]. However, the lack of selective methods to remove acetates in the presence of benzoyl esters is a severe drawback. In addition, acetates are not compatible with ester-based linkers and get occasionally cleaved during glycosylation reactions [12].

16.2.2.2 Nonparticipating Protecting Groups

(2-Naphthyl)methyl (Nap) (2-Naphthyl)methyl (Nap) was recently introduced as a new ether-based temporary PG in automated glycan assembly (Figure 16.6) [27]. As nonparticipating PG, it has the potential to promote 1,2-*cis*-selective glycosylations when installed in the C2-position of glycosylating agents. Unlike bulky groups such as silyl ethers, the size and chemical properties of Nap ethers are comparable to benzyl ethers so that similar results in glycosylations with the respective BBs are expected. Nap groups can be selectively removed in the presence of most common PGs such as Fmoc using DDQ in DCM/MeOH/H_2O. Seven reaction cycles with 10 equivalents of DDQ at 40 °C for 20 min each were required to reliably remove the Nap groups. In certain cases, benzyl ethers are cleaved under the oxidative reaction conditions applied. A new protocol for the acidic removal of Nap ethers using catalytic amounts of HCl in trifluoroethanol might solve this issue in the near future [28].

tert-Butyldimethylsilyl (TBS) As early as in 1999, *tert*-butyldimethylsilyl ethers (TBS) were reported as temporary PGs in the solid-phase synthesis of a β-1,4-linked glucan trisaccharide [6a]. Silyl ethers can be selectively cleaved by fluoride-containing reagents such as tetrabutylammonium fluoride [6a, 29]. Although

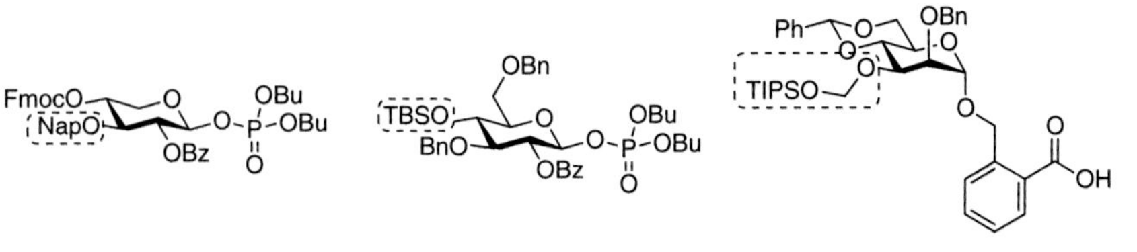

Figure 16.6 BBs equipped with Nap, TBS, and Tom as nonparticipating temporary PGs.

there have been very limited reports about silyl ethers as PGs in SPOS since the early reports, silyl ethers might be established as a valuable addition to the currently used set of temporary PGs in the future.

[(Triisopropylsilyl)oxy]methyl (Tom) [(Triisopropylsilyl)oxy]methyl ether (Tom) has been used in the synthesis of oligosaccharides such as the core structure of mammalian *N*-glycans containing β-mannosidic linkages [2]. It was installed in the C3-position of the mannose BB as the steric bulk of the alternative silyl ethers compromised the desired β-selectivity in the glycosylation reaction. The Tom group can be easily deprotected using tetrabutylammonium fluoride. Although the removal of TBS on solid support required 12 h, 20 min were sufficient for the cleavage of the Tom ether [19].

16.2.2.3 Colorimetric Protecting Groups for Monitoring Reaction Progress

The nitrophtalimidobutyric acid ester (NPB) has been proposed as an alternative PG to Fmoc that enables colorimetric monitoring of glycosylation cycles [30]. NPB can be removed under reaction conditions similar to those that affect Lev removal. Upon treatment of the oligosaccharide with hydrazinium acetate, bright orange nitrophthalhydrazide is produced and can be detected using UV/Vis spectroscopy (Scheme 16.7).

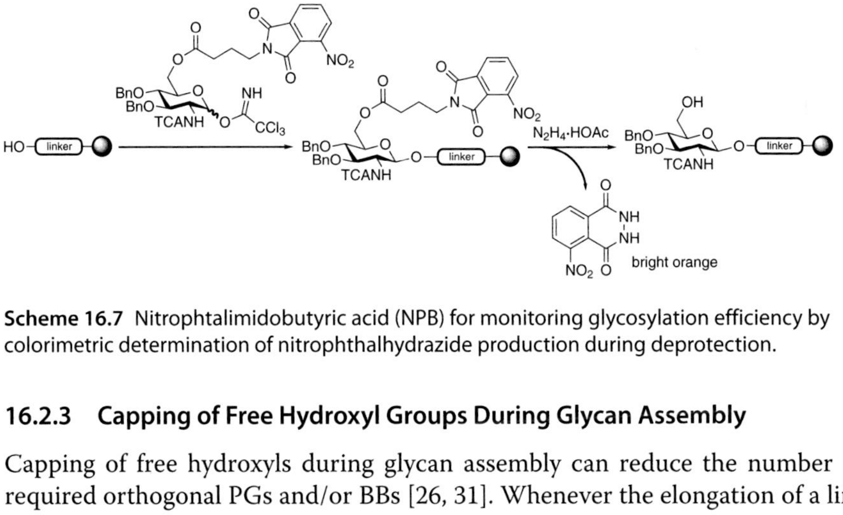

Scheme 16.7 Nitrophtalimidobutyric acid (NPB) for monitoring glycosylation efficiency by colorimetric determination of nitrophthalhydrazide production during deprotection.

16.2.3 Capping of Free Hydroxyl Groups During Glycan Assembly

Capping of free hydroxyls during glycan assembly can reduce the number of required orthogonal PGs and/or BBs [26, 31]. Whenever the elongation of a linear glycan chain is completed, a terminal temporary PG (PG_1) can be removed and replaced by an ester (Scheme 16.8). Once another PG orthogonal to PG_1 is cleaved, PG_1 can be used again with the next BB, allowing for the introduction of an additional branching point in the growing oligosaccharide chain. Acetylation of the terminal hydroxyl is performed using acetic anhydride in pyridine. However, the resulting acetyl esters are not compatible with Lev esters as they are cleaved under the same reaction conditions. A new method was devised to cap hydroxyls on solid phase as benzoyl esters [31]. Quantitative capping was achieved by incubating the resin for 30 min at 40 °C with a solution of benzoic anhydride and dimethylaminopyridine in dichloroethane.

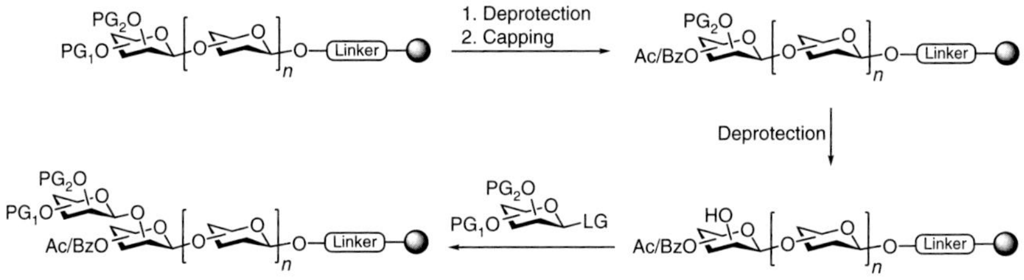

Scheme 16.8 Capping strategy to reduce the number of required temporary orthogonal PGs in automated glycan assembly.

16.3 General Strategy for the Design of Orthogonally Protected Building Blocks

We propose a general strategy for the selection of orthogonal protection groups for the synthesis of any oligosaccharide target. Table 16.1 summarizes the most promising PGs available for SPOS. Currently, benzyl ethers and benzoyl esters are used routinely as permanent PGs and Fmoc carbonates and Lev esters as temporary PGs. All other PGs we list are valuable alternatives worth to be tested.

The following point-by-point instructions provide a general strategy for the selection of PGs for automated glycan assembly:

1) Determine a suitable linear chain in the glycan target that can smoothly be prepared using Fmoc chemistry. Equip all BBs with Fmoc groups accordingly.
2) Equip all BBs that will introduce a branching point into the target glycan with an additional Lev group. Consider using Nap or TBS instead of Lev if BB synthesis is easier or high reactivity of the BB is required.
3) Consider reducing the number of required orthogonal PGs by including capping steps in the synthesis.
4) Symmetrical branches can be elongated simultaneously. Use the same set of BBs for both branches.
5) Equip all BBs that will form β-glycosidic bonds either with benzoyl esters or, in the case of glycosylamines, with trichloroacetyl (TCA) groups in the C2-position.
6) If formation of α-glycosidic bonds is required, equip the C2-position of the respective BBs with benzyl ethers or azides depending on the type of monosaccharide used. Estimate based on literature reports if your designed BBs will provide sufficient 1,2-*cis*-selectivity. If not, consider using an auxiliary in the C2-position that provides 1,2-*cis*-selective neighboring group participation. This will mandate the use of imidates as the leaving group [17].
7) Equip all remaining positions with benzyl ethers or benzoyl esters depending on the individual requirements on the BB block regarding reactivity and anomeric selectivity.
8) If the target glycan contains β-mannosides, either use the BBs described in Section 2.1.2 or a disaccharide BB having the β-mannosidic linkage preinstalled [32].

Table 16.1 Selection of PGs recommended for the design of BBs for automated glycan assembly.

Permanent protecting groups		**Deprotection**	**Neighboring group participation**
Bn		H_2, Pd/C	Nonparticipating
N_3		H_2, Pd/C	Nonparticipating
Bz		NaOMe in MeOH	Participating
NHTCA		H_2, Pd/C	Participating
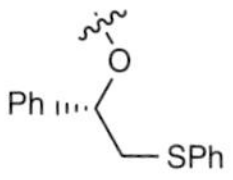		$BF_3{\cdot}OEt_2/Ac_2O$ in DCM	Participating (1,2-*cis*-selective!)
Temporary protecting groups	**Cleavable in presence of**	**Deprotection**	
Fmoc	Lev, Nap, (TBS), (Alloc), (Tom)	NEt_3 in DMF	Participating
Lev	Fmoc, Nap, (TBS), (Alloc), (Tom)	$N_2H_4{\cdot}HOAc$ in THF/H_2O	Participating
Nap	Fmoc, (Lev), (TBS), (Alloc), (Tom)	DDQ in DCM/ MeOH/H_2O[a]	Nonparticipating
TBS	(Fmoc), (Nap), (Lev), (Alloc), (Tom)	TBAF in THF	Nonparticipating
Alloc	Fmoc, (Nap), (Lev), (TBS), (Tom)	$Pd(PPh_3)_4$ in THF/ AcOH	Participating
Tom	Fmoc, (Nap), (Lev), (TBS), (Alloc)	TBAF in THF	Nonparticipating
Capping groups	**Capping conditions**	**Stable during removal of**	
Ac	Ac_2O in pyridine	Fmoc, Nap, (TBS), Alloc, (Tom)	
Bz	Bz_2O/DMAP in DCE/ pyridine	Fmoc, Lev, Nap, (TBS), (Alloc), (Tom)	

Routinely used PGs are marked in bold.
a) Cleavage of primary benzyl ethers was occasionally observed.

16.4 "Approved Building Blocks" for Automated Glycan Assembly

A key to a broad application of automated glycan assembly is the standardization of the applied BBs in order to enable the nonspecialist to perform successful syntheses. Figure 16.7 shows a selection of BBs that were reported to reliably give the respective glycosylation products with complete conversion and high stereoselectivity. Glucose, galactose, and glucosamine BBs have been used most frequently as they are the most common components of mammalian glycans. For the synthesis of plant and bacterial glycans, xylose and arabinose BBs have also

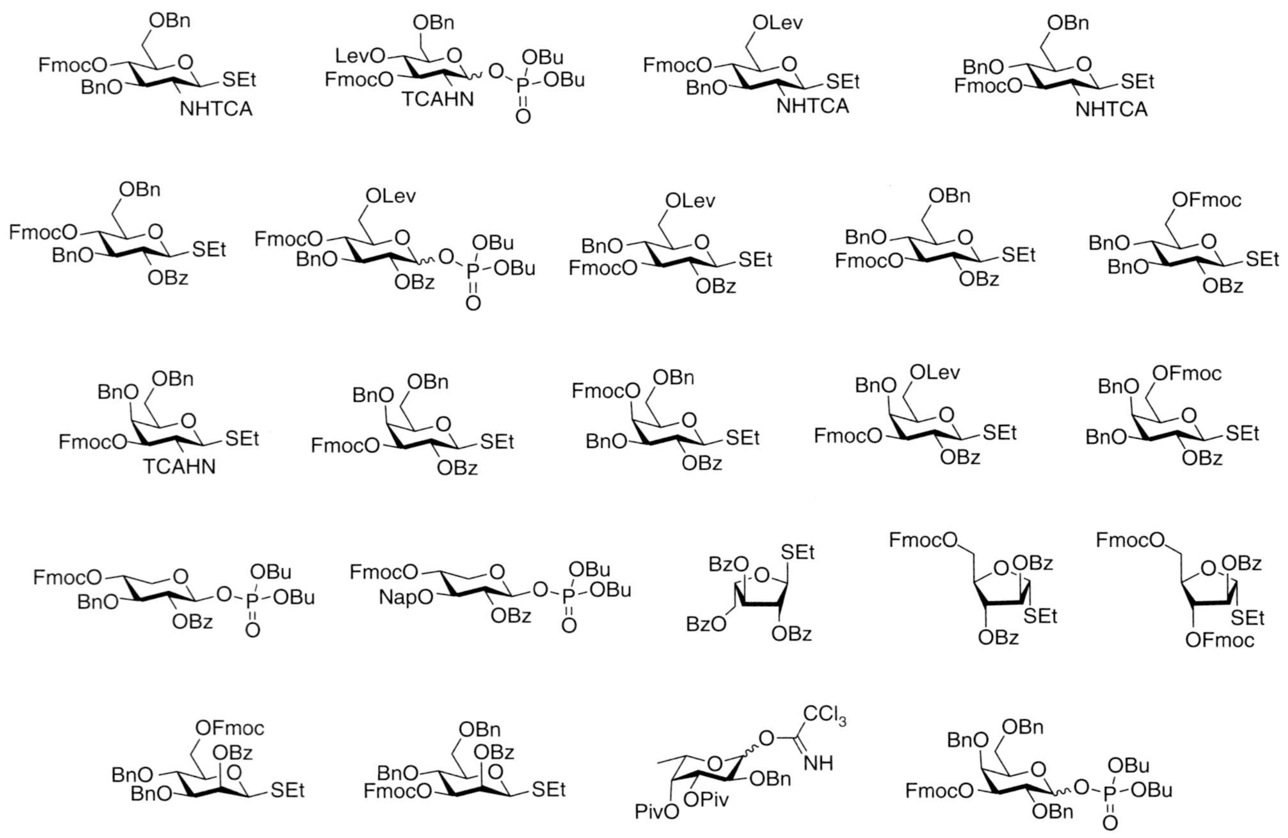

Figure 16.7 Selection of building blocks that have been successfully used in automated glycan assembly.

been used. An increasing number of "approved" BBs have been established that can be prepared on a multigram scale, are stable to storage, and reliably form the desired glycosidic linkage. Current automated glycan assembly protocols in our laboratories rely on several classes of BBs. Thioglycosides are reliable glycosylating agents because of their stability and, as they are crystalline solids, easy handling. Glycosylphosphates are powerful alternatives as has been shown for long homoligomers [7, 26, 33]. *N*-phenyltrifluoroimidates proved promising for the formation of β-mannosidic linkages [20]. Unlike thioglycosides and phosphates, imidates can be partially recycled by recovering unreacted glycosylating agent as the free reducing sugar and reinstalling the imidate leaving group [20].

Many of the described BBs are accessible by similar synthetic routes. A promising strategy for the synthesis of orthogonally protected monosaccharides is to convert the unprotected carbohydrates into thioglycosides at an early stage of the synthesis. Thioethers are stable toward the reaction conditions required for most protection and deprotection procedures and therefore act as a protecting group themselves (see Chapter 5). Another common synthetic transformation in many syntheses of glucose, galactose, and mannose BBs is the installation of a benzylidene acetal that allows for the selective opening of the acetal in either direction. A prototypical synthesis of an orthogonally protected galactose BB carrying two temporary and two permanent protecting groups is shown in Scheme 16.9 [31]. Peracetylated galactose is converted into the respective thioglycoside, the remaining acetyl groups are removed by methanolysis, and the O4- and O6-hydroxyls are selectively protected as a benzylidene acetal. The C3-position of the resulting diol **10** is selectively protected with a sterically demanding TBS group and the C2-position as a benzoyl ester. At this stage, the

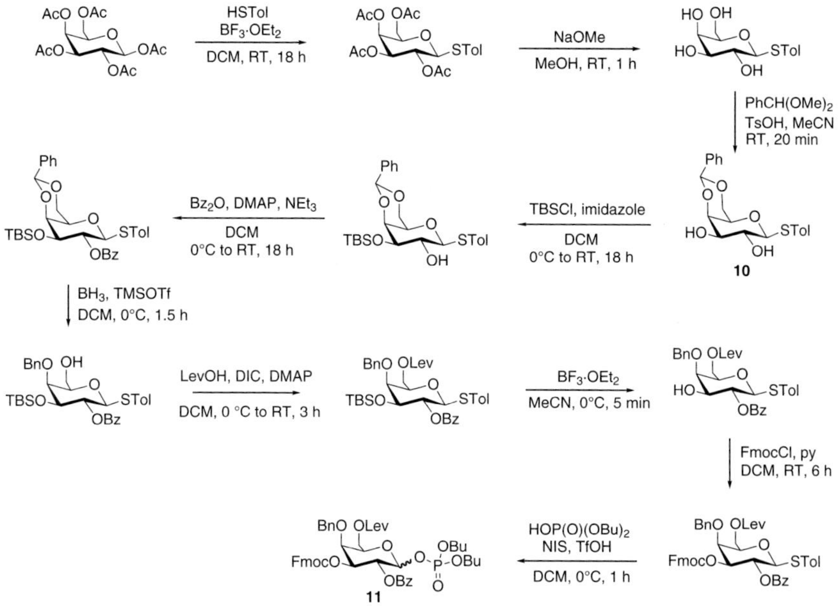

Scheme 16.9 Synthesis of orthogonally protected galactose BB **11**.

benzylidene is selectively opened using BH_3 in Lewis acidic conditions, providing a free hydroxyl group in the C6-position. Subsequent attachment of a Lev group was followed by exchange of the TBS group by an Fmoc group. Final conversion of thioglycoside into the corresponding glycosyl phosphate afforded BB **11** in 33% yield over 10 steps.

Although synthetic routes starting from the respective unprotected thioglycosides provide access to most of the depicted BBs (Figure 16.7), some BB syntheses need special considerations. The pentose xylose, for example, is particularly difficult to selectively protect as all the hydroxyl groups exhibit a similar nucleophilicity. Orthogonally protected xylose BB **14** carrying two temporary and one permanent protecting group was synthesized starting from D-xylofuranose derivative **12** (Scheme 16.10) [27]. Selective tritylation of the

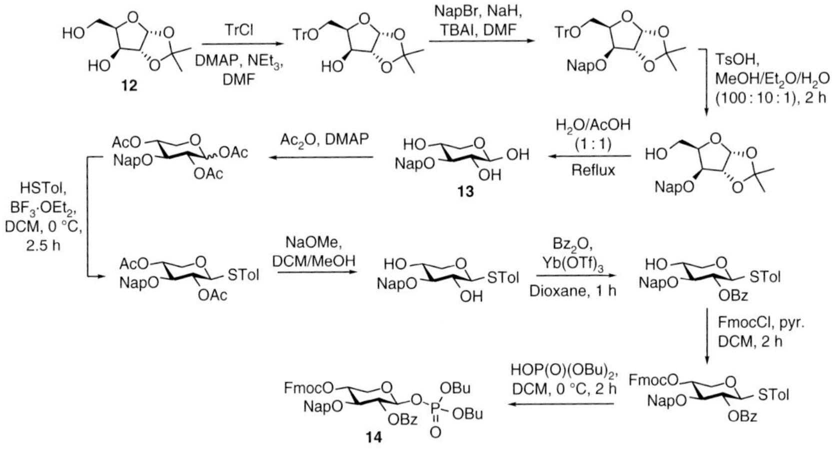

Scheme 16.10 Synthesis of orthogonally protected xylose BB **14**.

primary hydroxyl function in **12** enabled the early stage introduction of the Nap ether in the C3-position. Only after acidic removal of the trityl group and the acetonide, xylopyranose **13** was peracetylated and converted into the respective thioglycoside. Selective benzoylation of the C2-position was performed using catalytic amounts of ytterbium triflate. Installation of the second temporary protecting group using FmocCl in pyridine was followed by conversion of the thioglycoside into the corresponding glycosyl phosphate, affording BB **14** in 13% yield over 10 total steps.

16.5 Solid-phase Syntheses of Mammalian, Microbial, and Plant Oligosaccharides

We illustrate the general strategy outlined for the design of suitably protected BBs in the context of the automated glycan assembly of a mammalian, a microbial, and a plant oligosaccharide.

The first example is the automated glycan assembly of a Lewis-X-type oligosaccharide [13]. Lewis-X is a blood group oligosaccharide found in glycosphingolipids in the outer membrane of healthy and diseased mammalian cells [34]. It is a branched structure containing 1,2-*cis*- as well as 1,2-*trans*-glycosidic bonds, and the BB design is straightforward. Four BBs have been designed for the automated glycan assembly of a Lewis-X pentasaccharide (Scheme 16.11). The BBs required for the synthesis of the Glc-Gal-GlcN-Fuc chain were equipped with Fmoc groups to enable smooth chain elongation. Besides the Fmoc groups, the first two glucose and galactose BBs were equipped with pivaloyl esters in the C2-position to ensure β-selective glycosylations. The remaining positions were protected as benzyl ethers. The third BB introduces a branching point into the oligosaccharide. Therefore, in addition to the Fmoc group in the C3-position, a Lev ester was installed in the C4-position of the glucosamine BB. The amine in the C2-position was protected as a participating trichloroacetamide and the remaining C6 hydroxyl with a benzyl group. The fucose BB had to be furnished with a nonparticipating benzyl ether in the C2-position. The remaining two hydroxyls were protected as pivaloyl esters, which are required to ensure α-selectivity for the fucosylation reaction. After linear assembly of the tetrasaccharide by iterative glycosylation and Fmoc deprotection cycles, the Lev-group at the glucosamine was cleaved and another glycosylation with the galactose BB was performed. Cleavage from the solid support provided pentasaccharide **15** in 13% yield after HPLC purification. By employing the same strategy, also larger and more complex oligosaccharides such as a Lewis-Y oligosaccharide and a Le^y–Le^x dimer were prepared [13].

The second example describes the automated glycan assembly of oligoarabinofuranosides as found in arabinogalactan and lipoarabinomannan polysaccharides of mycobacterial cell envelopes [35]. Two different BBs were employed in the synthesis of a branched hexasaccharide (Scheme 16.12). Because two symmetrical branches were elongated in parallel, only one type of temporary PG was required. The BBs were equipped with either one or two Fmoc groups depending on the need for linear chain elongation or installation of a branching point. All remaining positions were protected as benzoyl esters. Furanosyl BBs are generally more reactive than pyranosyl BBs, and the use of arming benzyl ethers is unnecessary. After linear extension of the linker by glycosylation with BBs **16** and **17**, linker-bound disaccharide **18** was obtained. Simultaneous elongation of the C3- and C5-positions with two times BB **16** provided hexasaccharide **19** in 63% overall yield after cleavage of the glycan from the solid support and a final hydrogenation reaction.

The third example describes the automated glycan assembly of a plant arabinogalactan oligosaccharide [31]. Type-II arabinogalactan polysaccharides constitute the glycan portion of arabinogalactan proteins, which are important proteoglycans in plants. In contrast to mycobacterial arabinogalactans that consist of D-arabino- and D-galactofuranose, plant arabinogalactans are constructed from L-arabinofuranoses and D-galactopyranoses. Two galactose BBs and one arabinose BB were designed for automated glycan assembly (Scheme 16.13). The galactose BB for linear 1,3-galactan chain elongation was equipped with a Fmoc

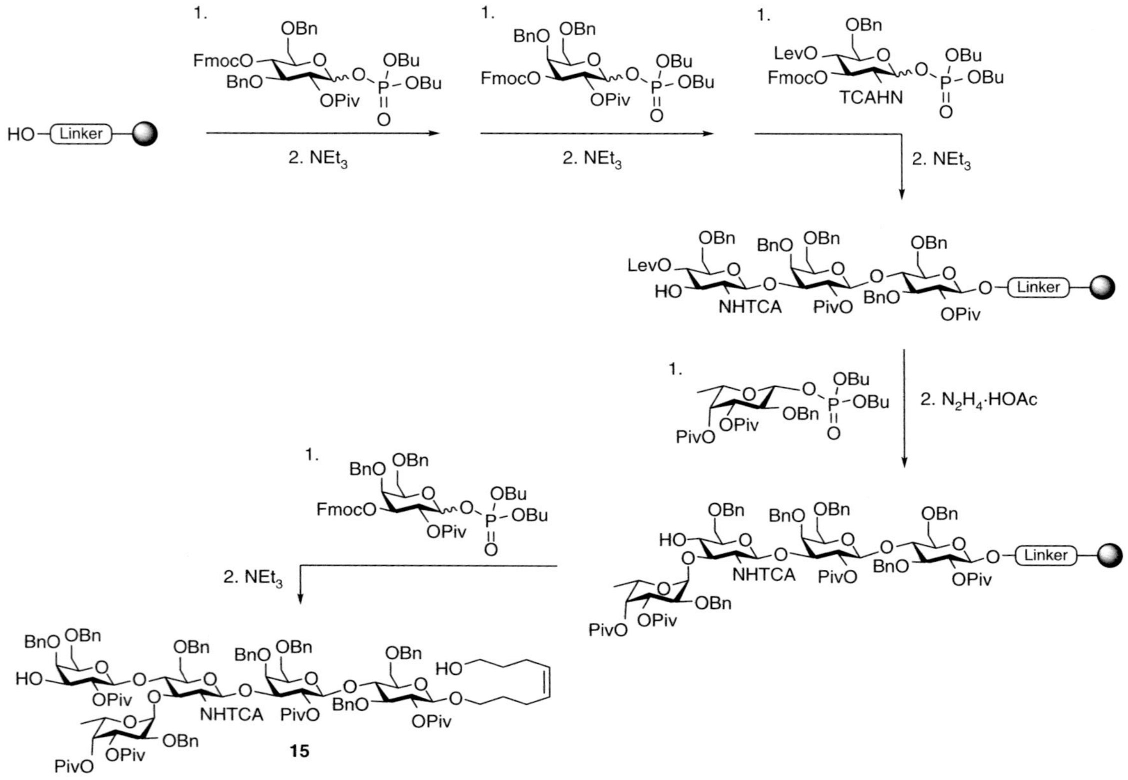

Scheme 16.11 Automated glycan assembly of a Lewis-X pentasaccharide.

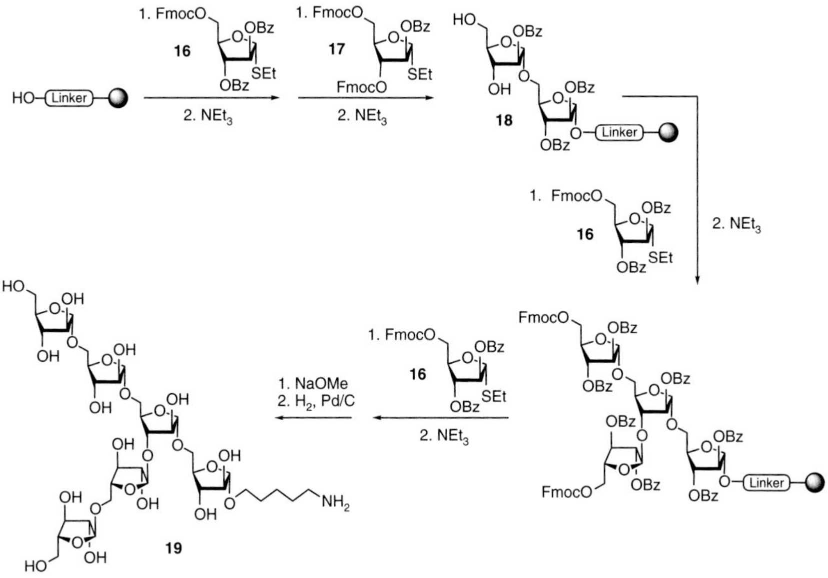

Scheme 16.12 Automated glycan assembly of a branched oligoarabinofuranoside from the mycobacterial cell envelope.

group in the C3-position. To permit branching, another galactose BB was additionally equipped with a Lev ester in the C6-position. Both BBs carry benzoyl esters in the C2-position to ensure β-selective glycosylations, and all remaining positions were protected with benzyl groups. Installation of the arabinofuranoses relied on a perbenzoylated BB. The target oligosaccharide contains two independent branching points. By using the capping strategy (see Section 16.2.3), the need for three orthogonal temporary PGs was avoided, and it was possible to draw only on Fmoc and Lev as routine temporary PGs. After assembly of the (1,3)-galactan backbone, the terminal Fmoc group was removed and replaced by a benzoyl ester. After Lev removal at the central galactose, another branching point can be introduced using BB **21**. After Fmoc removal and a glycosylation with arabinose BB **22**, the C6-position was further elongated by Lev deprotection and another glycosylation with BB **20**. Cleavage from the solid support and global deprotection provided hexasaccharide **23** as one example of a set of 14 arabinogalactan oligosaccharides.

16.6 Chances, Challenges, and Commercialization of Automated Glycan Assembly

The development of automated glycan assembly took about 15 years from the first exploratory work in the late 1990s to the proof of principle in 2001 [4] to finally yield a commercial synthesizer that was first sold in 2014 (www.glycouniverse.de). Ever more complex glycans have become tractable both as far as the length of the glycans is concerned, a 50-mer polysaccharide completed

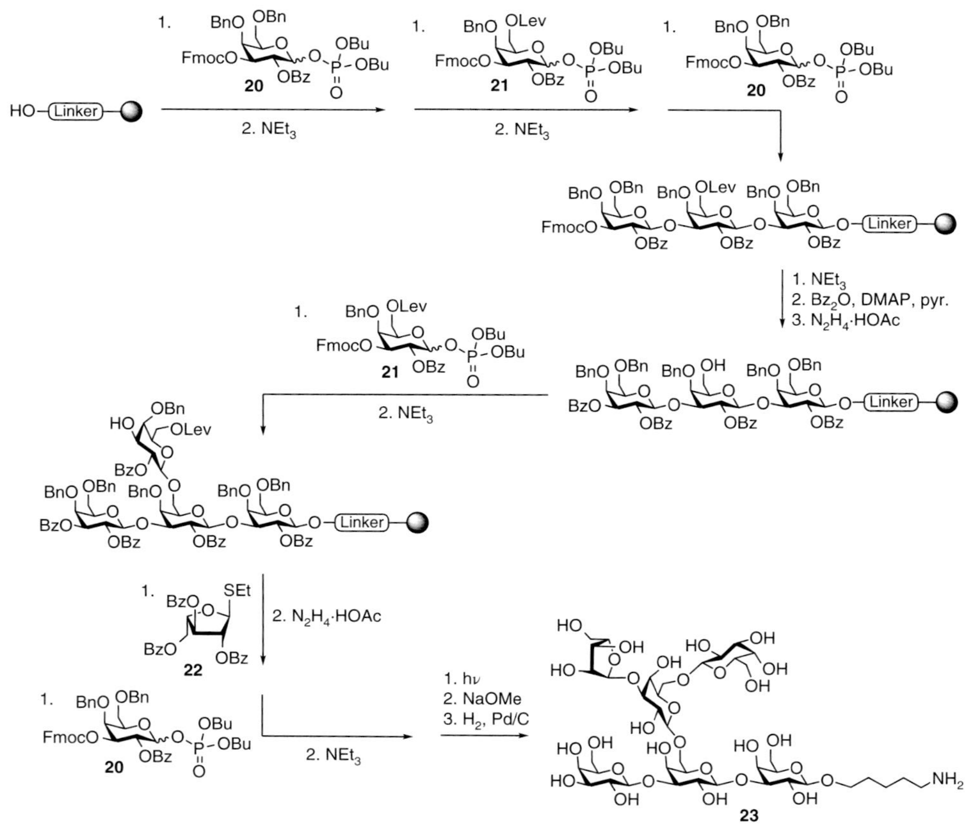

Scheme 16.13 Automated glycan assembly of a plant arabinogalactan oligosaccharide.

[27], and structural complexity such as is the case for the synthesis of glycosaminoglycans [7]. With a reliable commercial instrument in place, the bottleneck for automated glycan assembly has moved to the availability of "approved" BBs. Most recently, a spin-off company, GlycoUniverse GmbH, made these differentially protected BBs commercially available. As the number of BBs that are commercially offered increases and the number of practitioners of automated glycan assembly that utilizes these BBs increases, the cost for these BBs will decrease. A similar situation was observed for oligopeptide and oligonucleotide assembly. As more scientists enter this field, all aspects of automated glycan assembly stand to be systematically improved. New protecting groups will give rise to higher yields and/or selectivities, and the overall process will be streamlined. Further improvements regarding the automated synthesis machine are underway to allow for an even broader portfolio of chemistries to be supported.

Judging based on the development in the peptide and oligonucleotide fields, in addition to the sale of instruments and BBs, one or more custom synthesis companies using the patented automated glycan assembly process will develop to meet the needs of biologists, material scientists, and other scientists for well-defined, synthetic glycans. With all components of the process in place and commercialized, we expect a rapid growth of the field.

Abbreviations

Ac	acetyl
Alloc	allyloxycarbonyl
BB	building block
Bn	benzyl
Bz	benzoyl
DCE	1,2-dichloroethane
DCM	dichloromethane
DDQ	2,3-dichloro-5,6-dicyano-1,4-benzoquinone
DMAP	4-dimethylaminopyridine
Fmoc	fluorenylmethoxycarbonyl
Lev	levulinoyl
Nap	(2-naphthyl)methyl
LG	leaving group
PG	protecting group
Phtal	phthalimido
Piv	pivaloyl
Pyr.	pyridine
SPOS	solid-phase oligosaccharide synthesis
TBS	*tert*-butyldimethylsilyl
TCA	trichloracetyl
THF	tetrahydrofuran
TIPS	triisopropylsilyl
Tom	[(triisopropylsilyl)oxy]methyl

References

1 (a) Fréchet, J.M.J. and Schuerch, C. (1971). *J. Am. Chem. Soc.* 93: 492–496.(b) Seeberger, P.H. and Haase, W.-C. (2000). *Chem. Rev.* 100: 4349–4393.(c) Seeberger, P.H. (ed.) (2001). *Solid Support Oligosaccharide Synthesis and Combinatorial Carbohydrate Libraries*. NY, USA: Wiley-VCH.(d) Bennett, C.S. (2014). *Org. Biomol. Chem.* 12: 1686–1698.

2 Kröck, L., Esposito, D., Castagner, B. et al. (2012). *Chem. Sci.* 3: 1617–1622.

3 (a) Seeberger, P.H. (2008). *Chem. Soc. Rev.* 37: 19–28.(b) Seeberger, P.H. (2015). *Acc. Chem. Res.* 48: 1450–1463.

4 Plante, O.J., Palmacci, E.R., and Seeberger, P.H. (2001). *Science* 291: 1523–1527.

5 Wu, X., Grathwohl, M., and Schmidt, R.R. (2002). *Angew. Chem. Int. Ed.* 41: 4489–4493.

6 (a) Andrade, R.B., Plante, O.J., Melean, L.G., and Seeberger, P.H. (1999). *Org. Lett.* 1: 1811–1814.(b) de Jong, A.R., Volbeda, A.G., Hagan, B. et al. (2013). *Eur. J. Org. Chem.* 6644–6655.

7 Eller, S., Collot, M., Yin, J. et al. (2013). *Angew. Chem. Int. Ed.* 52: 5858–5861.

8 Schmidt, R.R. and Kinzy, W. (1994). *Adv. Carbohydr. Chem. Biochem.* 50: 21–123.

9 Shiao, T.C. and Roy, R. (2011). *Top. Curr. Chem.* 301: 69–108.

10 Palmacci, E.R., Plante, O.J., and Seeberger, P.H. (2002). *Eur. J. Org. Chem.* 595–606.

11 Nigudkar, S.S. and Demchenko, A.V. (2015). *Chem. Sci.* 6: 2687–2704.

12 Fair, R.J., Hahm, H.S., and Seeberger, P.H. (2015). *Chem. Commun.* 6813–6815.

13 Werz, D.B., Castagner, B., and Seeberger, P.H. (2007). *J. Am. Chem. Soc.* 129: 2770–2771.

14 Love, K.R., Andrade, R.B., and Seeberger, P.H. (2001). *J. Org. Chem.* 66: 8165–8176.

15 Dallabernardina, P., Schuhmacher, F., Seeberger, P.H., and Pfrengle, F. (2016). *Org. Biomol. Chem.* 14: 309–313.

16 Kim, J.H., Yang, H., and Boons, G.J. (2005). *Angew. Chem. Int. Ed.* 44: 947–949.

17 Boltje, T.J., Kim, J.-H., Park, J., and Boons, G.-J. (2010). *Nat. Chem.* 2: 552–557.

18 Crich, D. and Sun, S.X. (1996). *J. Org. Chem.* 61: 4506–4507.

19 Codée, J.D.C., Kröck, L., Castagner, B., and Seeberger, P.H. (2008). *Chem. Eur. J.* 14: 3987–3994.

20 Walvoort, M.T.C., van den Elst, H., Plante, O.J. et al. (2012). *Angew. Chem. Int. Ed.* 51: 4393–4396.

21 Varki, A., Cummings, R.D., Esko, J.D. et al. (ed.) (2009). *Essentials of Glycobiology*. Cold Spring Harbor, USA: Cold Spring Harbor Laboratory Press.

22 Czechura, P., Guedes, N., Kopitzki, S. et al. (2011). *Chem. Commun.* 2390–2392.

23 Walvoort, M.T.C., Volbeda, A.G., Reintjes, N.R.M. et al. (2012). *Org. Lett.* 14: 3776–3779.

24 Manabe, S., Ishii, K., and Ito, Y. (2011). *Eur. J. Org. Chem.* 3: 497–516.

25 Mootoo, D.R., Konradsson, P., Udodong, U., and Fraser-Reid, B.O. (1988). *J. Am. Chem. Soc.* 110: 5583–5584.

26 Rademann, J. and Schmidt, R.R. (1997). *J. Org. Chem.* 62: 3650–3653.

27 Schmidt, D., Schuhmacher, F., Geissner, A. et al. (2015). *Chem. Eur. J.* 21: 5709–5713.

28 Volbeda, A.G., Kistemaker, H.A.V., Overkleeft, H.S. et al. (2015). *J. Org. Chem.* 80: 8796–8806.

29 Hewitt, M.C., Snyder, D.A., and Seeberger, P.H. (2002). *J. Am. Chem. Soc.* 124: 13434–13436.

30 Ko, K.-S., Park, G., Yu, Y., and Pohl, N.L. (2008). *Org. Lett.* 10: 5381–5384.

31 Bartetzko, M.P., Schuhmacher, F., Hahm, H.S. et al. (2015). *Org. Lett.* 17: 4344–4347.

32 Ratner, D.M., Swanson, E.R., and Seeberger, P.H. (2003). *Org. Lett.* 5: 4717–4720.

33 (a) Calin, O., Eller, S., and Seeberger, P.H. (2013). *Angew. Chem. Int. Ed.* 52: 5862–5865.(b) Weishaupt, M.W., Matthies, S., and Seeberger, P.H. (2013). *Chem. Eur. J.* 19: 12497–12503.

34 Hakomori, S. and Zhang, Y. (1997). *Chem. Biol.* 4: 97–104.

35 Kandasamy, J., Hurevich, M., and Seeberger, P.H. (2013). *Chem. Commun.* 4453–4455.

17

Kilogram-scale Production of Synthetic Heparin Analogs: Some Chemical Considerations

Patrick Trouilleux[1], *Pierre Potier*[2], *and Pierre-Alexandre Driguez*[3]

[1] *Sanofi, Chemistry and Biotechnology Development, Lead Generation and Candidate Realization, 371 Rue du Professeur Blayac, 34184 Montpellier, France*
[2] *Sanofi, Chemistry and Biotechnology Development, Route d'Avignon, 30390 Aramon, France*
[3] *Sanofi R&D, Emerging Chemical Modalities, Integrated Drug Discovery, 13, Quai Jules Guesde, 94400 Vitry sur Seine, France*

17.1 Introduction

Heparin is a natural complex anionic heteropolysaccharide extracted from animal sources. Heparin backbone consists of a heteropolymer composed of alternating sugar units of 2-deoxy-2-sulfamino-α-D-glucopyranose (N-sulfonated, or N-acetylated, O-sulfonated) and O-sulfonated uronic acids (α-L-iduronic acid or β-D-glucuronic acid). Enzymatic biosynthesis of heparin generates structural variation and microheterogeneity in the sulfation pattern of disaccharidic building blocks, which are responsible for several important biological activities [1]. Despite being used in clinics as an anticoagulant for more than 70 years [2], serious side effects mainly because of activation of plasmatic proteins are sometimes observed in some patients treated with standard heparin [3]. Depolymerization of the latter through enzymatic and/or chemical methods generate smaller fragments, otherwise known as low-molecular-weight heparin (LMWH) [4]. Such fractions that display more favorable pharmacokinetic parameters than standard heparin are better tolerated and may be administered subcutaneously by the patients themselves [5].

Another option to avoid the main drawbacks of full-length heparin is the use of pure synthetic analogs that display only the chemical features necessary for antithrombin activation, a prerequisite for the inhibition of serine proteases of the blood coagulation cascade. It is well documented that the replacement of hydroxyls by methoxy groups, N-sulfonates by O-sulfonates, and modulation of the sulfation pattern give rise to optimized synthetic heparin analogs with tailor-made antithrombotic activities and pharmacokinetic properties [6]. Furthermore, it was shown that while a pentasaccharide is sufficient for the antithrombin-mediated inhibition of factor Xa, a pentadecasaccharide displays the minimal length to inhibit factor IIa as well [7]. However, depending on their requested

Protecting Groups: Strategies and Applications in Carbohydrate Chemistry, First Edition.
Edited by Sébastien Vidal.

profile (pure antifactor Xa or dual antifactors Xa and IIa, short or long acting, and bearing in mind the need or not for a biotin moiety for possible removal from the bloodstream with avidin, if required), such syntheses require 50–80 chemical steps that may become challenging at industrial scale. Optimization of the synthetic pathway, reaction conditions, and selection of appropriate temporary protecting groups are therefore mandatory.

This chapter is devoted to the disclosure of some chemical means developed for the multi-kilogram-scale preparation of key building blocks used for the generation of idrabiotaparinux (**1**) and hexadecasaccharide **2**, two "long-acting" synthetic heparin analogs, the purity of which being higher than 99.5% for clinical trials. Special attention will be devoted to protecting groups that improve reaction yields and that facilitate purification of the saccharidic intermediates. Furthermore, reaction conditions that respond to safety concerns by allowing the use of less-toxic reagents and/or solvents will be discussed.

17.2 Kilogram Synthesis of Heparin Building Blocks

17.2.1 General Considerations

17.2.1.1 Development and Production of Synthetic Heparin Analogs

Several improvements in chemoenzymatic methods have been performed over recent years to facilitate the preparation of heparinoid structures close to natural ones [8]. However, chemical synthesis still permits a higher diversity in the generation of non-natural heparin derivatives [9] with extremely safe conditions for the production of good manufacturing process (GMP)-grade batches [10].

It is clear that the process must be well designed, robust, and safe to be successfully implemented in a production plant. If such conditions are respected, it should be possible to produce kilogram or ton scale of defined intermediates up to the active pharmaceutical ingredient (API) in a reproducible way: each production of a batch will lead to the same yield and impurity profile from a qualitative and quantitative point of view. In addition, the price of the whole process needs to be acceptable when considering the cost of goods (reagents, raw materials, and overhead costs).

Therefore, the choice of the most straightforward and convergent synthetic route as well as the choice of the most relevant protecting groups are critical for the process industrialization.

As a consequence, the selection of the appropriate protecting groups has to take into account the following requirements, knowing the fact that the ultimate goal is to reach a process with a limited number of chromatographic purifications:

- High yield for both protection and deprotection to limit product losses.
- High selectivity to reduce formation of impurities.
- Atom economy to avoid ballast/waste and to limit the number of purification steps.
- Crystallization of compounds to facilitate isolation and purification (as an example, during all our chemical synthesis studies, peracetylated glycosides were generally easier to crystallize than their benzoylated analogs).

- Finally, environmental aspects, health and safety (EHS), may be illustrated with the following examples: to deprotect anomeric acetates, harmful hydrazine could be advantageously replaced by the safer morpholine; to introduce a benzoate group, benzoic anhydride has to be preferred from benzoyloxybenzotriazole because of its explosiveness; and to activate an anomeric position, thiophenol or *para*-chlorothiophenol has to be preferred from ethanethiol for hygiene consideration.

Considering the oligosaccharide synthesis, the selection of the most relevant protecting groups (and deprotecting conditions) is highly critical, considering the number of steps and the structural similarity of each glycoside units.

17.2.2 New Synthetic Oligosaccharides with Neutralizable Antithrombotic Activities

The antithrombin-binding domain of heparin is composed of a highly conserved and specific pentasaccharidic sequence, originally designated as $GlcN^{V}$–$GlcUA^{IV}$–$GlcN^{III}$–$IdoUA^{II}$–$GlcN^{I}$ (Figure 17.1) and is responsible for the potent antithrombotic properties of heparin and its derivatives [11]. Recent studies however showed that residues located at both ends of the pentasaccharide display an active role in the binding to and activation of antithrombin [12], as exemplified with the positive role of IdoUA at the nonreducing end, or GlcNS,3,6S at the reducing end. The analysis of such phenomena, which have been recently reviewed [13], has been facilitated with the development of new depolymerization procedures.

The synthetic pentasaccharidic analog fondaparinux has been available on the market in the USA and Europe since 2002.

The antithrombin-binding sequences of oligosaccharides **1** and **2** display the same chemical features required for a potent biological activity (Figure 17.2). However, compared to fondaparinux, hydroxyls are replaced by methoxy groups and N-sulfonates by O-sulfonates. Such modifications diminish structural variability and therefore considerably simplify the synthesis by reducing the overall number of steps. Furthermore, the modulation of the sulfation pattern of this region gave rise to compounds with enhanced antifactor Xa activities compared to those with the parent heparin pentasaccharidic sequence.

The potency of both oligosaccharides **1** and **2** is associated with a half-life that permits once weekly administration in humans. Although such properties appear convenient for a treatment under standard conditions, they may become an issue in cases of emergency situations such as hemorrhagic failure during surgery.

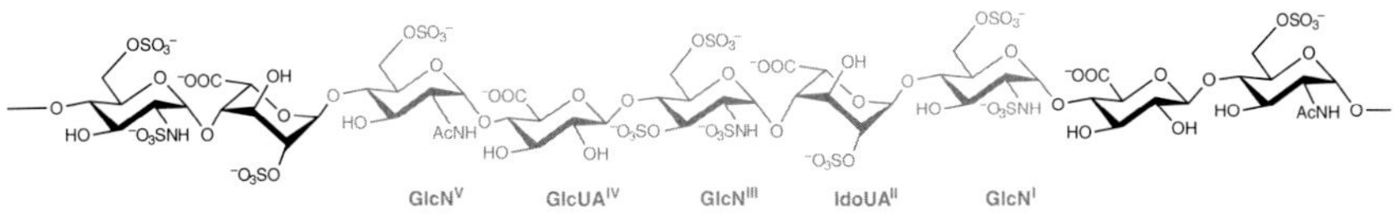

Figure 17.1 Schematic representation of the antithrombin-binding domain (grey) in a HPN/HS polysaccharide.

1: R^1 = MeO– ; 2: R^1 =

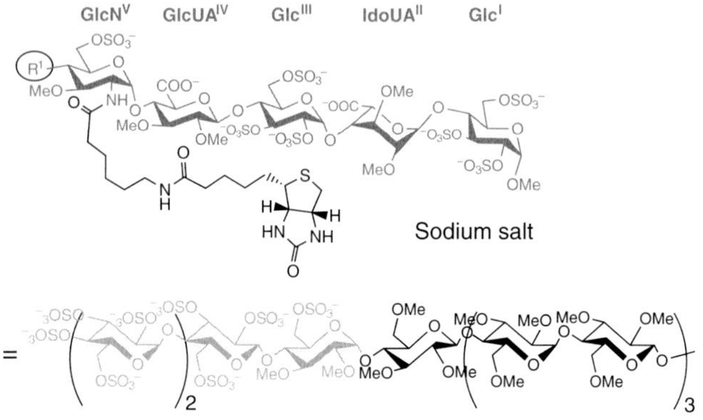

Figure 17.2 Schematic representation of the antithrombin-binding domain (dark grey) in idrabiotaparinux (**1**) and hexadecasaccharide **2**.

Therefore, for a safer use, both compounds have been designed with a covalently linked biotin moiety, thus allowing a possible neutralization with a high-affinity protein such as avidin, upon request from hospital surgeons.

In order to reduce the production cost of the batches required for clinical trials, it was decided to use the maximum common building blocks for the synthesis of both oligosaccharides **1** and **2**. Therefore, apart from the substitution on position 4 of the $GlcN^V$ unit, antithrombin-binding pentasaccharides are identical for both compounds, and the biotin is located at the same glucosamine residue.

Unlike idrabiotaparinux (**1**), which is a specific antifactor Xa compound, hexadecasaccharide **2** is a dual antifactor Xa/antifactor IIa. Such properties are obtained with the antithrombin high-affinity pentasaccharide that was coupled at its nonreducing end [14] to a partially O-sulfonated tetrasaccharide corresponding to a factor IIa-binding domain. Both sequences are linked via a neutral spacer, thus permitting an optimal antithrombin/hexadecasaccharide/factor IIa ternary complex formation (Figure 17.2). The neutral spacer, which was preferred over its partially sulfonated analog [15], is composed of a permethylated heptasaccharide displaying alternatively β- and α-linked glucose units [16]. Such a sequence proved to be efficient in preventing platelet factor IV interactions and therefore heparin-induced thrombocytopenia.

17.2.3 Retrosynthetic Analysis

The retrosynthetic analysis of the antithrombin-binding domain of oligosaccharides **1** and **2** is depicted in Scheme 17.1, where one goal was to find out the best compromise regarding convergence of the synthesis, efficiency in the chemistry, and short time cycle production. The total synthesis of oligosaccharides **1** and **2** requires about 50 or 80 chemical steps respectively, but for the reader's interest, it was decided in this chapter to focus only on the late steps of the synthesis.

The 6-(biotinamido) hexanoyl group of both oligosaccharides **1** and **2** was introduced at the final step on position 2 of the Glc^V unit through an amide bond.

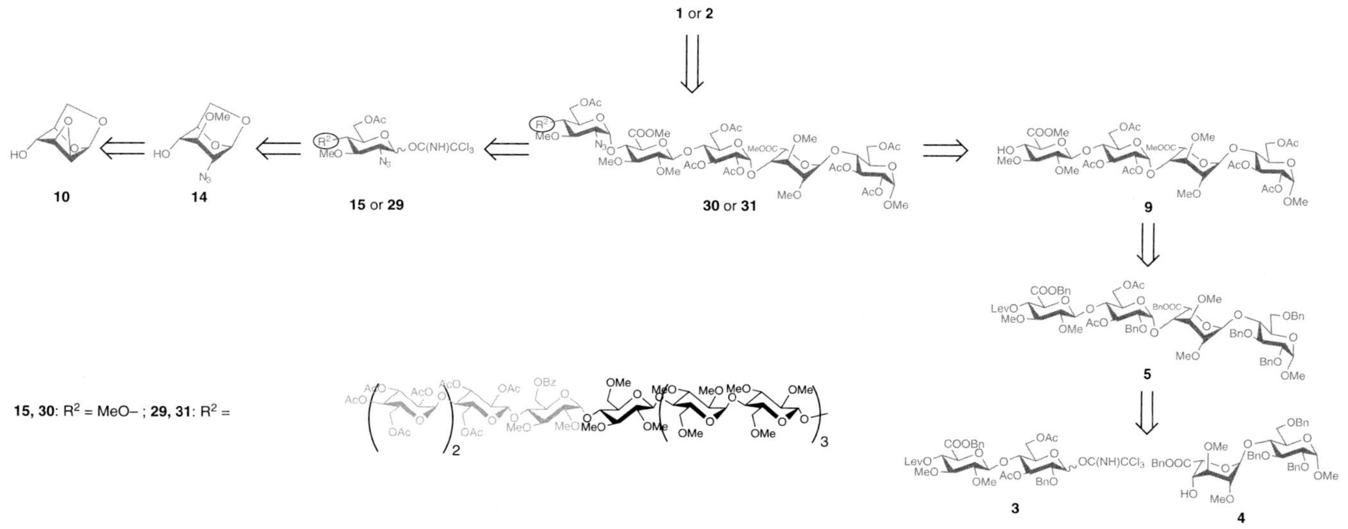

Scheme 17.1 Retrosynthesis of the antithrombin-binding domain (dark grey) of idrabiotaparinux (**1**) and hexadecasaccharide **2** (final steps).

The amino group thus required came from the reduction of the azido group of the carbohydrate moiety. The added value of this group is to first generate the desired α-D anomer during the coupling reaction of glycosyl donors **15** or **29** with tetrasaccharide **9** to generate either **30** or **31**. Moreover, the azido group is orthogonal to the other protecting groups during the remaining deprotection/functionalization steps before its hydrogenolysis. Finally, the generation of nitrogen as the only by-product during that reaction is highly valuable at the industrial scale.

The sulfonated compounds are obtained from the corresponding fully protected derivatives **30** or **31** where, in order to reduce the number of deprotection steps on a high-value pentasaccharide, all the primary and secondary hydroxyls are temporarily masked by esters (acetyl or benzoyl) or permanently methylated. Whenever possible, acetyl protecting groups were preferred over benzoates mainly for their deprotection step. This choice was motivated by environmental reasons (atom economy for 17 hydroxyls in the case of hexadecasaccharide final transesterification) and by the fact that Zemplén reaction generates methyl acetate (bp = 58 °C) which, through a simple evaporation, is easier to remove from the media than methyl benzoate (bp = 200 °C) on a kilogram scale. Saccharides **30** or **31** were obtained by the reaction of key glycosyl acceptor **9** with glycosyl imidates **15** (for **1**) or **29** (for **2**). The synthesis of tetrasaccharide **9** has been previously described [10] and was prepared from **5**, obtained from the coupling reaction between disaccharides **4** and **3**.

Among the various possible alternatives to synthesize dodecasaccharide **29** (Scheme 17.2), we chose to add tetrasaccharide donor **26** to position 4 of the monosaccharide acceptor **14** (4 + 1 strategy). The pentasaccharide obtained was processed to generate glycosyl acceptor **27** that was also coupled to **26** [4 + (4 + 1)] to give a nonasaccharide, which was further transformed as above to yield glycosyl acceptor **28**. Maltotriose **38** was transformed into derivative **39** in 2 steps. The latter was coupled to **28** {3 + [4 + (4 + 1)]} to generate, after chemical transformations at the reducing end unit, dodecasaccharide **29**. Tetrasaccharide **26** could be prepared from disaccharide **21**, which was generated from D-maltose (**16**).

17.2.4 Synthesis of Idrabiotaparinux and Hexadecasaccharide 2

17.2.4.1 Synthesis of $GlcN^{V}$–$GlcUA^{IV}$–Glc^{III}–$IdoUA^{II}$–Glc^{I} Precursors

Coupling reaction of glycosyl donor **3** and acceptor **4** yielded tetrasaccharide **5** (Scheme 17.3) [17]. In this specific coupling reaction, toluene was preferentially replaced by dichloromethane despite the classification of the latter in the carcinogenic, mutagenic, or reprotoxic (CMR) list of solvents. First, it allows the reaction to be performed in a more concentrated medium (67 g l^{-1} vs 18 g l^{-1}). More importantly, the formation of only 1% of undesired β-anomer in dichloromethane vs 4% in toluene, at the same reaction temperature, (−20 °C) was observed, thereby increasing the selectivity. The very low ratio of β-anomer coupling product detected in the reaction mixture may be explained by the axially oriented hydroxyl at C-4 because of the probable $^{1}C_{4}$ conformation of the iduronic acid acceptor **4** [18], which enhances an α-selective formation of the new glycosidic

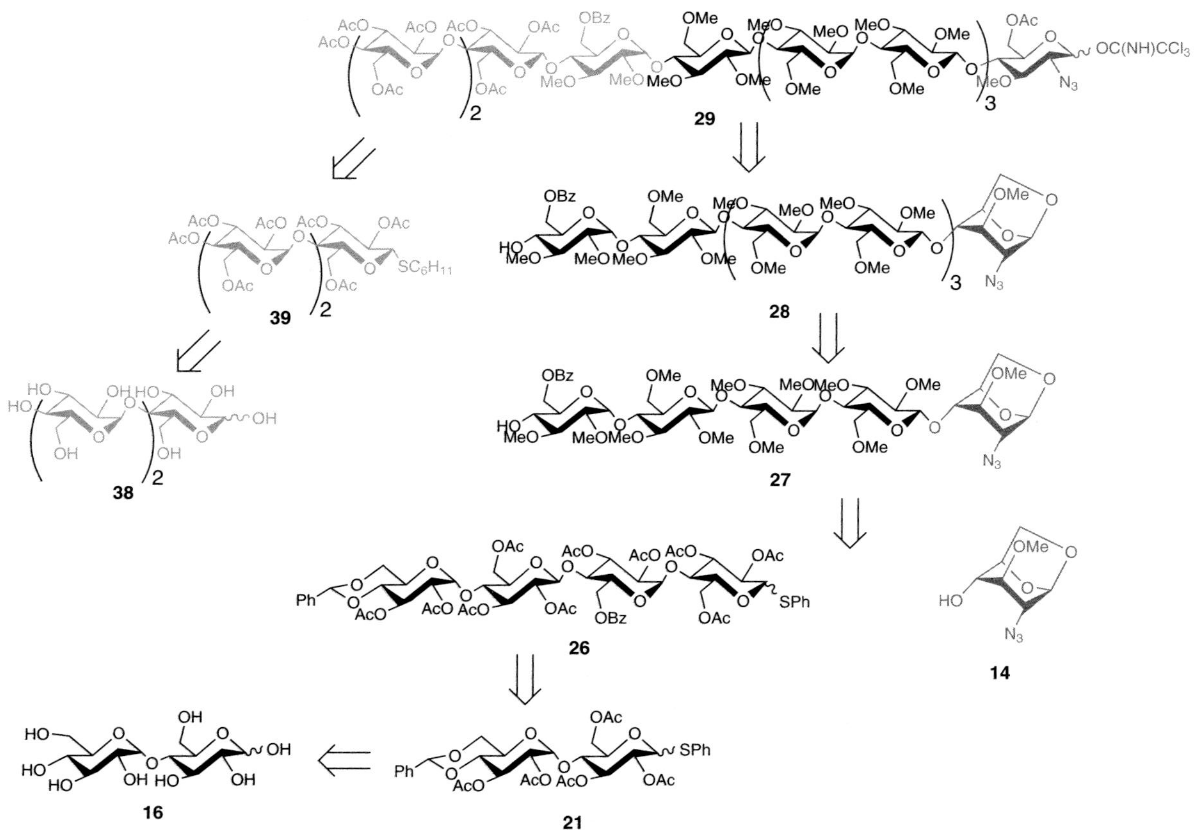

Scheme 17.2 Retrosynthesis of the factor IIa-binding domain (light grey) and the carbohydrate spacer (black) of hexadecasaccharide **2**.

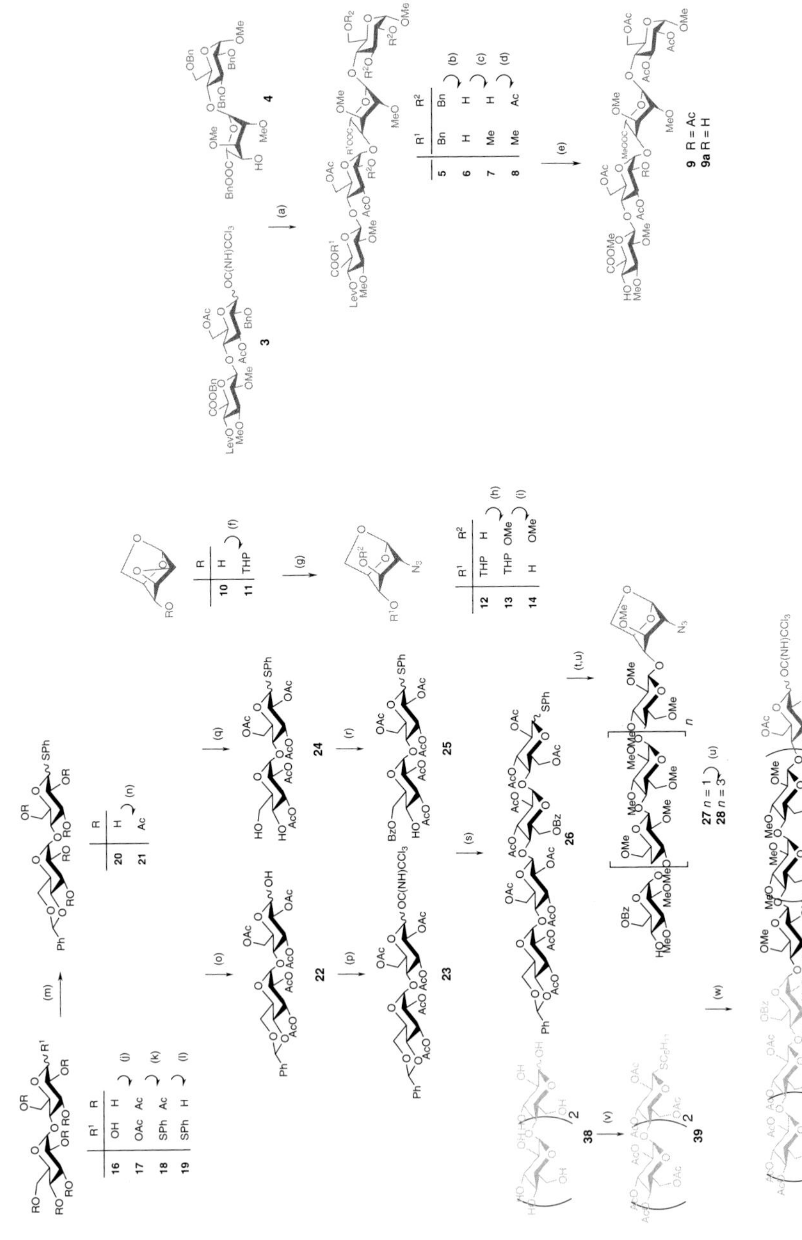

Scheme 17.3 Synthesis of key building blocks **29** and **9**. (a) TMSOTf (cat.), CH_2Cl_2, −20 °C, 1 h, crystallization *tert*-amyl alcohol, 85%; (b) H_2, Pd/C 10%, THF, 40 °C, 6 h; (c) DMS, $KHCO_3$, THF, RT, 16 h; (d) Ac_2O, Et_3N, DMAP (cat.), THF, RT, 2.5 h, crystallization *i*PrOH; (e) NH_2NH_2·AcOH, toluene/EtOH (1 : 2), 8 °C, 3.5 h, crystallization *i*PrOH, 75% (4 steps); (f) 2,3-dihydropyrane, methanesulfonic acid, CH_2Cl_2, 20 °C, 0.5 h; (g) sodium azide, AcOBu, Et_3N, NMP/H_2O, 100 °C, 24 h; (h) DMS, sodium *tert*-amylate, morpholine, 0–5 °C, 2 h; (i) HCl 36% in EtOH, RT, 6 h, 80%, 4 steps; (j) Ac_2O, DMAP, CH_2Cl_2, 0–100 °C, 6 h, α/β 1 : 4, crystallization EtOH, α/β 2:98, 78%; (k) PhSH, BF_3·OEt_2, toluene, RT, 1 h, crystallization EtOH, 90%; (l) MeONa (cat.), MeOH/CH_2Cl_2, 20 °C, 3 h, crude; (m) $PhCH(OMe)_2$, H_2SO_4, DMF, 65 °C, 3 h, crystallization CH_3CN, 78% (2 steps); (n) Ac_2O, DMAP, CH_2Cl_2, RT, 2.5 h, crystallization EtOH, 91%; (o) NBS, THF/H_2O, 3 °C, 2 h, crystallization, 90%; (p) Cl_3CCN, DBU, toluene, RT, 16 h, crude; (q) HCl, 5/1 CH_3CN/H_2O, 60 °C, 3 h; (r) Bz_2O, Et_3N, CH_2Cl_2, 20 °C, 5 h, 70% (2 steps); (s) TMSOTf (cat.), toluene, −20 °C, 1 h, 85%; (t) 1,3-dibromo-5,5-dimethylhydantoin, triflic acid, 2 : 1 toluene/ CH_2Cl_2, −7 °C, 30 min, 90%; (u) (i) MeONa, MeOH/toluene; (ii) DMS, *tert*-BuONa, NMP, RT, 1 h, quant.; (iii) HCl, 5/1 CH_3CN/H_2O, 60 °C, 5 h; (iv) Bz_2O, Et_3N, toluene/CH_2Cl_2, 20 °C, 4 h, 60% for **27**, 57% for **28**, (4 steps)); (v) (i) Ac_2O, AcONa, 100 °C, 6 h; (ii) $C_6H_{11}SH$, BF_3·OEt_2, toluene, 86% (2 steps); (w) (i) NIS, TfOH, dioxane/CH_2Cl_2, −20 °C, 1.25 h, crude; (ii) Ac_2O, TfOH, toluene, −30 °C, 5.5 h, crude; (iii) morpholine, toluene, 40 °C, 16 h, 68% (3 steps); (iv) CCl_3CN, DBU, RT, 16 h, crude.

bond [19, 20]. The use of 4 Å molecular sieves during coupling reactions usually limits glycosyl donor hydrolysis and therefore gives rise to higher yields of the coupling product. However, molecular sieves are abrasive to the reactor and should be avoided at a large scale. Water quantification is therefore mandatory before adding the catalyst, and a maximum range of 10–100 ppm of water is required. Elimination of by-products (2,2,2-trichloroacetamide, hydrolyzed glycosyl donor used in excess) was performed using an efficient crystallization in *tert*-amyl alcohol and yielded **5** (85%). The pure material exhibited a purity of 97% (HPLC-UV 210 nm) and contained manageable low amounts of the undesired β-anomer (0.3%).

The conditions of hydrogenolysis of **5** have been optimized to remove all the benzyl groups without deacetylation as a side reaction. After screening several protocols, THF appeared the best solvent allowing the solubilization of the starting material **5**, the resulting compound **6**, and the toluene generated by the benzyl hydrogenolysis with Pd/C 50% wet (0.1 $g\,g^{-1}$ of starting material).

Tetrasaccharide **6** was isolated in THF solution and directly engaged in the next step of O-methylation. As methyl iodide is a very volatile compound (bp = 42 °C) and was not easy to manage at the industrial scale, the less-volatile dimethylsulfate (bp = 188 °C) was therefore preferred for EHS considerations. Even if only one methyl group is used in dimethylsulfate, it was very reactive and 1.5 eq./COOH group is enough to have a complete reaction. The dimethylsulfate excess was quenched with morpholine, dichloromethane was added, and washings were performed to eliminate salts together with partially methylated compounds. The methylated tetrasaccharide **7** was isolated in a THF/dichloromethane mixture and engaged directly into the next step. The peracetylation of **7** was performed under standard conditions (Ac_2O, Et_3N, DMAP, RT). The resulting tetrasaccharide **8** was crystallized in isopropanol and then engaged in the delevulinoylation step. Selective removal of the levulinoyl group was challenging. An incomplete deprotection led to unreactive compound, also nonselective reactions yielded deacetylated derivatives that could generate nonlinear pentasaccharides and/or hexasaccharides in the next glycosylation step instead of the desired linear saccharide.

Thus, operating conditions were optimized to limit the formation of by-products. Using a concentrated medium (67 g of saccharide per liter) of a 1 : 2 toluene/ethanol mixture as a solvent, the stoichiometry in hydrazine acetate was reduced to 1.5 eq. and the temperature decreased to 8 °C. The main impurity corresponding to the deacetylated tetrasaccharide in position 2 of unit Glc^{III} (**9a**) was limited to 0.3% (vs 1.2% at room temperature) using these conditions. After work-up and crystallization in isopropanol, the desired tetrasaccharide **9** was isolated with a purity higher than 98% (HPLC-UV 210 nm). The overall yield for the 4 steps (hydrogenation, methylation, acetylation, and delevulinoylation) was 75%. These processes have been optimized and implemented on 5 kg batch size with a reproducible impurity profile, demonstrating their robustness.

17.2.4.2 Synthesis of $GlcN^{V}$ Unit Precursor 14

Glycosyl acceptor **14** was prepared in 4 steps from epoxide **10** (Scheme 17.3) [21]. Position 4 of the latter was first temporarily protected using 2,3-dihydropyrane in the presence of methanesulfonic acid. Although already described in dioxane [22], the reaction was performed in dichloromethane. The crude reaction mixture containing **11** as the only saccharidic product was then diluted in *N*-methyl-2-pyrrolidone (NMP). Special attention was paid to the epoxide opening of **11** with sodium azide. It should be mentioned that the use of sodium azide needs specific control of basicity of the reaction mixture to avoid formation of harmful and explosive hydrazoic acid. Furthermore, it has to be noted that compound **12**, as well as all the other 2-deoxy-2-azido compounds synthesized during the work presented in this chapter, was thermically stable as proven by differential scanning calorimetry (DSC) (especially when compared to trichloroacetimidates compounds that are highly exothermic when reaching decomposition). Indeed, their manipulation did not require special protocols compared to other carbohydrate intermediates described here. The free alcohol generated in **12** was methylated using dimethylsulfate to give **13**. Sodium *tert*-amylate was selected as a base because it is much easier to handle than sodium hydride. Although utilized very often, the use of the latter nevertheless generates hazardous hydrogen. Hydrolysis of the tetrahydropyranyl ether of **13** was performed using concentrated HCl in ethanol. Starting from the crude reaction mixture, the desired product **14** was isolated by stripping with toluene. It was possible to optimize the process in such a way that the synthesis could be performed on 35 kg of epoxide **10**.

17.2.4.3 Synthesis of the Key Disaccharidic Building Block 21

Peracetylation of maltose (**16**) (Scheme 17.3) was described in the literature on a 20-g scale [23]. However, the process developed by us does not use pyridine, which has to be avoided on a 500-kg scale for obvious EHS considerations (Table 17.1). Only DMAP was used as a catalyst, the addition of another base was not required. The reaction led to the desired peracetylated maltose **17** in a 1 : 4

Table 17.1 Maltose peracetylation conditions developed at Sanofi compared to the literature.

	Ref. [21]	Sanofi
Scale	20 g	500 kg
Solvent	CH_2Cl_2 7.9 ml g^{-1} maltose	CH_2Cl_2 10 ml g^{-1} maltose
Base	Pyridine 2.5 ml g^{-1} maltose	—
Catalyst	DMAP 0.02 eq.	DMAP 0.4 eq.
T (°C)	0 °C	0–100 °C
Ac_2O	10.6 eq. (1.3 eq./OH) freshly distilled	16 eq. (2 eq./OH) commercial grade
α/β ratio	N/A	1 : 4
Work-up	HCl 2 M 10.5 ml g^{-1} maltose Extraction CH_2Cl_2 10.5 ml g^{-1} maltose (x3), Washing: $NaHCO_3$ sat. 10.5 ml g^{-1} maltose NaCl sat. 10.5 ml g^{-1} maltose Drying over $MgSO_4$	Water 5.3 ml g^{-1} maltose NaOH 16% 4 ml g^{-1} maltose (x2), Washing: Water 3 ml g^{-1} maltose (x2)
Purification	Concentration to dryness	Cryst EtOH 15 ml g^{-1} maltose
	Amorphous colored solid	White crystals
Yield	92%	78% α/β >2 : 98

Source: Data from Cerny et al. 1965 [21].

α/β anomeric mixture. At this stage, it was necessary to remove the less-reactive α-anomer, which did not react in our hands during the next step of thiolation. The β-anomer was isolated by crystallization after stripping dichloromethane with ethanol without any concentration to dryness, which needs to be avoided whenever possible at the industrial scale. Assuming that literature conditions also generate a mixture of anomers, our process is highly efficient and yields white crystals containing mainly the β-anomer (>98%) after crystallization in ethanol.

In addition to the reasons mentioned in Section 17.2.3, protection of hydroxyls by acetylation has also been preferred to benzoylation because peracetylated compounds crystallized more easily in alcohol solvents.

Technical constraints were then associated with the use of ethanethiol on a large scale because of its high volatility. To overcome this situation, phenylthioglycoside **18** and its cyclohexyl analog (not shown) were prepared starting from β-anomer of **17**. However, optimization of the crystallization conditions could only be performed on phenylthioglycoside **18**, and this was therefore selected. Zemplén saponification of **18** gave **19** in quantitative yield.

The benzylidenation of **19** was performed under standard conditions using benzaldehyde dimethyl acetal and a Brönsted acid. Compared to the conditions described in the literature [24] in which acetonitrile was used as a solvent, DMF was preferred to obtain a homogeneous solution, leading to reproducible kinetics of the reaction from batch to batch (Table 17.2). Sulfuric acid was selected instead of camphorsulfonic acid, for the cost and availability at the industrial scale.

Table 17.2 Benzylidenation conditions of maltoside **19** developed at Sanofi compared to the literature.

	Ref. [22]	Sanofi
Scale	340 g	500 kg
Solvent	MeCN 16.8 ml g^{-1} material	DMF 6.4 ml g^{-1} material
	suspension	solution
Acid	CSA 0.1 eq., pH 3	H_2SO_4 0.05 eq.
	DMAP 0.02 eq.	DMAP 0.4 eq.
Benzylidenation agent	Benzaldehyde dimethyl acetal 3.2 eq.	Benzaldehyde dimethyl acetal 2 eq.
Work-up	Quench Et_3N, concentration to dryness	Quench NaOH 30%, concentration, dilution AcOEt, washing NaCl aq.
Purification	Not isolated	Not isolated

Source: Hermans et al. 1988 [22]. Reproduced with permission of John Wiley and Sons.

An acyl group was necessary at position 2 of the disaccharide for the anchimeric assistance in future glycosylations. The benzoylated analog of **21** proved difficult to crystallize. It was therefore decided to prepare acetylated derivative **21** starting from **20** for the production batches on 500 kg scale. The reaction was performed under standard conditions, and a yield of 91% was obtained after crystallization in EtOH.

17.2.4.4 Synthesis of Tetrasaccharide 26

Hydrolysis of thioglycoside **21** generated hemiacetal **22** [25], which was further transformed into trichloroacetimidate **23** under standard conditions (Scheme 17.3). Several acidic conditions (HCl (0.01–0.1 eq.), PTSA, PPTS, TfOH (0.05–0.2 eq.), AcOH, and IR120 H^+ resin) were tested for hydrolysis of the benzylidene of **21** to generate **24**. The temperature was also raised from 30 °C to 45 °C and up to 60 °C. The main observations were about kinetic effects, but the quality of the reaction was not improved. Transacetalization using MeOH/TfOH led to degradation of the disaccharide. Optimized conditions in our hands required the use of HCl diluted in acetonitrile/water and heating at 60 °C during 3 h.

For protection of the primary alcohol of **24**, a benzoyl group has been preferred to an acetyl group as it leads to a more selective reaction. For the synthesis of **25**, the benzoylating agent was initially hydroxybenzotriazole, but it was then replaced by benzoic anhydride, which is a more common and less harmful reactant (anhydrous hydroxybenzotriazole is explosive). Although benzoic anhydride is less selective than hydroxybenzotriazole, an equilibrium between stoichiometry and kinetics was found (2 eq./5 h) to ensure the completion of the reaction and limited the amount of 4,6-dibenzoylated by-product below 2%. It is worth noting that compound **25** was sensitive to temperature. The acetyl group at position 3 easily migrated to position 4 (see Chapter 8), generating a by-product difficult to remove. Several 40 kg batches of tetrasaccharide **26** have been prepared under standard glycosylation conditions (85% yield).

17.2.4.5 Elongation of the Spacer

The coupling reaction between glycosyl donor **26** and the glycosyl acceptor **14** was performed under standard conditions. The resulting pentasaccharide was not purified and was directly isolated in toluene solution. Then, MeOH/MeONa (0.2 eq.) was added to remove ester groups (10 acetates and 1 benzoate). The reaction performed in a homogeneous media was very clean. Saponification with NaOH instead of MeONa was tested, but precipitation occurred and the saponification was not complete. Methylation of the 11 free hydroxyl groups was performed in NMP, using dimethylsulfate as a methylating agent (1.5 eq./OH) and *tert*-BuONa as a base (1.5 eq./OH). It is important to note that dimethylsulfate is not fully stable in solvents such as NMP or DMF. Therefore, simultaneous adding of dimethylsulfate and *tert*-BuONa on the pentasaccharide in NMP solution as well as a monitoring by HPLC was required to ensure full completion of the reaction. Benzylidene hydrolysis as well as selective benzoylation on the primary position was performed as described for the preparation of **25** and yielded the pentasaccharide **27** (60%, 4 steps). The whole sequence (coupling reaction with **26**, saponification, methylation, benzylidene hydrolysis, and selective benzoylation) was repeated starting from **27** to give the nonasaccharide **28** (57%, 4 steps).

17.2.4.6 Synthesis of the Dodecasaccharide Glycosyl Donor 29

Cyclohexyl thioglycoside **39** was prepared in 2 steps from maltotriose (**38**) under standard conditions at multi-kilogram scale (Scheme 17.3). Glycosylation of **39** with glycosyl acceptor **28** and subsequent acetolysis of the 1,6-anhydro group, selective anomeric deacetylation, and formation of trichloroacetimidate **29** were also performed using usual procedures (see Section 17.3).

17.2.4.7 Final Glycosylation and Conjugation with Biotin

The coupling reaction between glycosyl acceptor **9** and donor **15** [10] was performed at −30 °C in a dichloromethane/methyl *tert*-butyl ether (MTBE) mixture (Scheme 17.4). MTBE lowered the formation of β-anomer to 5% compared to other solvents (15% with the use of Et_2O, 40% in pure dichloromethane). Toluene/CH_2Cl_2 replaced advantageously CH_2Cl_2/MTBE mixture for the coupling reaction of **29** and **9** to generate **31**. In this case, the formation of methylated impurities coming from MTBE degradation was limited and thus gave a higher yield of **31** (74% with a ratio α/β 87 : 13) at 800 g scale. Nevertheless, crystallization/repulping in methanol permits a purity enhancement from 60% to 92% with a very limited amount of the β-anomer around 0.3% (two successive crystallizations/repulping needed). The final isolated yield was 65% from imidate **29**.

Although compound **30** crystallized gently in MTBE, the needle-shaped crystals thus obtained were difficult to handle at the industrial scale. New conditions using acetone were established and led to crystals (clusters) without filtration issues. Compound **30** was isolated in 80% overall yield and was 98.5% pure (HPLC-UV 210 nm) with only 0.4% of β-anomer detected.

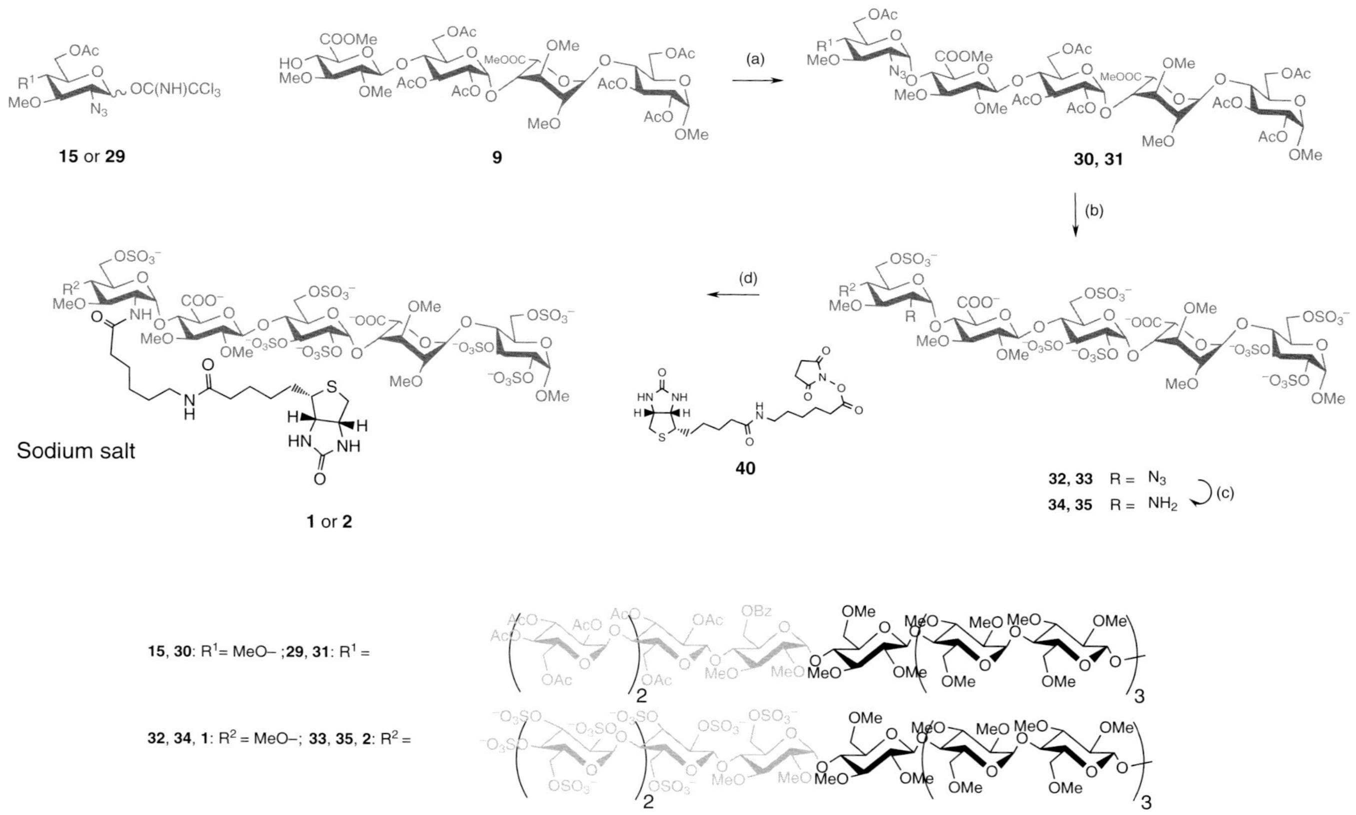

Scheme 17.4 Synthesis of idrabiotaparinux (**1**) and hexadecasaccharide **2** (final steps). (a) **15**, TBDMSOTf (cat.), CH_2Cl_2, MTBE, −30 °C, 30 min, crystallization acetone, 80% (2 steps); **29**, TBDMSOTf (cat.), toluene, CH_2Cl_2, −50 °C, 1.5 h, crystallization MeOH, 65% (2 steps); (b) (i) LiOH, *i*PrOH, H_2O, 0 °C, 19 h, crystallization *i*PrOH, 90%; (ii) SO_3·pyridine, DMF, 30 °C, 15 h, precipitation *i*PrOH, quant.; (c) **32**: H_2, 10% Pd/C, H_2O, RT, 2 h, quant.; **33**: H_2, 10% Pd/C, H_2O/*tert*-BuOH, 15 bar, RT, 3 h, quant.; (d) **40**, H_2O, acetone, 16 h, precipitation EtOH, 75% (4 steps) (**1**); **40**, DMF/H_2O, 72 h, precipitation *i*PrOH/MTBE, 67% (4 steps) (**2**).

The saponification of the ester functions was laborious because β-elimination of uronic esters may occur easily. Although the use of hydrogen peroxide/lithium hydroxide limits this side reaction, hydrogen peroxide could not be used at the industrial scale without significant safety constraints. Therefore, the best conditions were obtained using lithium hydroxide in *i*PrOH/H_2O as a solvent, at 0 °C, and about 5% of unsaturated impurities were detected. Crystallization in *i*PrOH led to the pure expected saccharides in 90% yield, eliminating salts and decreasing the unsaturated impurities from 5% to less than 0.1%, as assessed by high-field ^{1}H-NMR. The saponification product of hexadecasaccharide **31** was precipitated in MeOH/*i*PrOH and allowed an efficient elimination of lithium acetate and lithium benzoate on a 600 g scale.

The O-sulfonation of the hydroxyl groups was performed with the pyridine/sulfur trioxide complex in DMF at 30 °C for 15 h to provide compounds **32** and **33** in a quantitative yield. The salts were eliminated by precipitation in *i*PrOH as described earlier during saponification. Catalytic hydrogenolysis of the azide in the presence of 10% Pd/C in water at RT for 2 h quantitatively gave the amines **34** and **35**. Both compounds **34** and **35** reacted with commercially available *N*-hydroxysuccinimidyl biotinylated derivative **40**. The reactions were performed in water/acetone mixture as a solvent, during 16 h. After precipitation in ethanol, 3 kg of crude idrabiotaparinux (**1**) was isolated with a purity higher than 98.5% (HPLC-UV 210 nm) and 75% overall yield (over the last 4 steps). A final purification on Q-sepharose allowed isolation of up to 3 kg of **1** with a purity higher than 99.5% (HPLC-UV 210 nm). In the case of **2**, Q-Sepharose purification followed by nanofiltration (concentration) and precipitation in ethanol afforded 390 g of powdered API **2** with UV purity of 96% (global yield of 67% for purification).

17.3 Experimental Section

17.3.1 General Conditions for 1,6-Anhydro Acetolysis

To a solution of 1,6-anhydro carbohydrate in toluene (~10 ml g^{-1}) was added acetic anhydride (2.5 eq.) at −5/0 °C. Catalytic triflic acid (0.3 eq.) was then added and the reaction was monitored by HPLC. The reaction was quenched with sodium carbonate and washed with water. The crude product was used directly in the next anomeric deacetylation step.

17.3.2 General Conditions for Anomeric Deacetylation

To a solution of anomeric acetate in toluene (3 ml g^{-1}) was added morpholine (2.5 eq.). The reaction mixture was stirred at RT until reaction completion. The solution was washed successively with diluted HCl solution (pH 2.5) and water. The resulting hemiacetal was then stored as such in toluene solution.

17.3.3 General Conditions for Trichloroacetimidate Formation

Hemiacetal was dissolved in dichloromethane (10 ml g^{-1}). Water content must not be more than 50 ppm. The trichloroacetonitrile (5 eq.) and DBU (0.03 eq.)

were added. The reaction mixture was stirred at RT until completion of the reaction. The resulting trichloroacetimidate was not purified and glycosylation was initiated directly.

17.3.4 General Conditions for Glycosylation with Trichloroacetimidate

A solution of glycosyl acceptor (1.0 eq.) in anhydrous dichloromethane ($75\,g\,l^{-1}$) was stirred under nitrogen atmosphere at −20 °C. The TMSOTf (possibly TBDMSOTf) or $BF_3{\cdot}OEt_2$ (0.2 or 0.6 eq., respectively, per glycosyl donor) and then the glycosyl donor (1.2 eq.) were added to the mixture. The reaction mixture was stirred at −20 °C during 1 h and then quenched by addition of triethylamine. A $NaHCO_3$ solution (2%) was then added. The organic layer was washed several times with water. The dichloromethane was then stripped after addition of *tert*-amyl alcohol. Finally, the glycoside was isolated by crystallization in *tert*-amyl alcohol.

17.3.5 General Conditions for Glycosylation with Thioglycoside

Glycosyl donor and glycosyl acceptor (1.1 eq.) were dissolved in a 2 : 1 toluene/dichloromethane mixture ($10\,ml\,g^{-1}$). The water content was measured and must be below 100 ppm to limit hydrolysis of glycosyl donor. The reaction mixture was cooled to −7 °C and a mixture of 1,3-dibromo-5,5-dimethylhydantoin (0.6 eq.) and triflic acid (0.06 eq.) in 1 : 1 dioxane/dichloromethane ($2.4\,ml\,g^{-1}$) was added. After stirring for 30 min at −7 °C, the reaction was quenched with triethylamine (2 eq./triflic acid). The solution was washed successively with an aqueous solution of $NaHCO_3$ (2%), $Na_2S_2O_3$ (2%), and finally with water. The organic layer was then concentrated to yield the glycoside.

17.3.6 General Conditions for Delevulinoylations

To a solution of the saccharide in toluene/ethanol (1 : 2, $70\,g\,l^{-1}$) was added hydrazinium acetate (1.5 eq.) at 8 °C. After stirring for 3 h at 8 °C, dichloromethane was added ($7\,ml\,g^{-1}$ of saccharide). The organic layer was washed successively with $NaHCO_3$ (2%), HCl 1 N, and brine until pH 6–8. The dichloromethane was then stripped after addition of *i*PrOH. Finally, the glycoside was isolated by crystallization in *i*PrOH.

17.3.7 General Conditions for Transesterifications (Zemplén)

To a solution of glycoside in MeOH/toluene (2 : 1, $100\,g\,l^{-1}$) was added, at room temperature, a solution of sodium methoxide (26% in MeOH, 0.2 eq.). After stirring several hours at room temperature, the reaction mixture was quenched with acetic acid (0.2 eq.). The toluene/MeOH mixture was then replaced by stripping with NMP. Finally, the desired polyol was precipitated after addition of water.

17.3.8 General Conditions for *O*-sulfonations

To a solution, in DMF (100 $g l^{-1}$), of the saccharide was added pyridine/sulfur trioxide complex (3 eq./OH). After stirring for 14 h at 30 °C the reaction mixture was cooled to 25 °C and quenched by $NaHCO_3$ (50%). After stirring for 6 h at 25 °C, water was removed by distillation and the precipitated salts were eliminated by filtration. The O-sulfonated saccharide, in DMF solution, was finally precipitated by addition of isopropanol.

17.3.9 General Conditions for Uronate Saponification

To a solution of glycoside in isopropanol/water (5 : 1, 80 $g l^{-1}$) was added, at 0 °C, lithium hydroxide (1.4 eq./ester function). After stirring for 19 h at 0 °C and then for 5 h at room temperature, the mixture was quenched by HCl 6 N until pH 6.8. The saccharide was purified by crystallization in isopropanol (25 $g l^{-1}$).

17.3.10 General Conditions for the Catalytic Hydrogenolysis

To a toluene/THF solution (1 : 7, 40 $g l^{-1}$) of the glycoside was added 10% Pd/C (10% w/w, 50% wet). The hydrogen pressure was set at 2.5 bars. After stirring vigorously for 6 h at 40 °C, the reaction mixture was filtered. The expected polyol was kept in THF solution.

17.4 Summary and Conclusions

Chemical synthesis allows preparation of GMP-grade batches of original and non-natural complex carbohydrates in a safe and reproducible way. The 35 and 60 silica gel column chromatographies performed on the gram scale for the syntheses of compounds **1** and **2,** respectively, were efficiently replaced by crystallizations during the industrial development. This was a challenging task because the best compromise between yield, quality, and processability needed to be found for each key intermediate. It is noteworthy that the only column chromatography required along the long syntheses (51 chemical steps for **1**, 75 chemical steps for **2**) was performed at the final step. The development studies allowed the production of several batches of API **1** on a 3 kg scale, and API **2** on 390 g scale batches, with a reproducible yield and with a high quality (over 99.5% purity and without unknown impurity above 0.1%) for both APIs.

Abbreviations

API	active pharmaceutical ingredient
bp	boiling point
CMR	carcinogenic, mutagenic or reprotoxic
DBU	1,8-diazabicyclo[5.4.0]undec-7-ene
DMAP	4-dimethylaminopyridine

DMF	dimethylformamide
DMS	dimethyl sulfate
DSC	differential scanning calorimetry
EHS	environment, health and safety
GMP	good manufacturing process
HPLC	high pressure liquid chromatography
HPN	heparin
HS	heparan sulfate
LMWH	low molecular weight heparin
MTBE	methyl *tert*-butyl ether
NBS	*N*-bromosuccinimide
NIS	*N*-iodosuccinimide
NMP	*N*-methyl-2-pyrrolidone
PTSA	*para*-toluene sulfonic acid
PPTS	pyridinium *para*-toluene sulfonate
RT	room temperature
TBDMS	*tert*-butyldimethylsilyl
Tf	trifluoromethanesulfonyl
THF	tetrahydrofuran
TMS	trimethylsilyl

References

1 (a) Rosenberg, R.D., Shworak, N.W., Liu, J. et al. (1997). *Clin. Invest.* 99: 2062–2070. (b) Lindahl, U. and Li, J.P. (2009). *Int. Rev. Cell Mol. Biol.* 276: 105–157. (c) Gandhi, N.S. and Mancera, R.L. (2008). *Chem. Biol. Drug Des.* 72: 455–482. (d) Kreuger, J., Spillmann, D., Li, J.P., and Lindahl, U. (2006). *J. Cell Biol.* 174: 323–327.

2 Rodén, L. (1989). *Heparin: Chemical and Biological Properties, Clinical Applications* (ed. D.A. Lane and U. Lindahl). Boca Raton, FL: CRC Press, Inc.

3 Warkentin, T.E., Chong, B.H., and Greinacher, A. (1998). *Thromb. Haemost.* 79: 1–7.

4 Weitz, J.I. (1997). *N. Engl. J. Med.* 337: 688–698.

5 (a) Hirsh, J. and Levine, M.N. (1992). *Blood* 79: 1–17.(b) Weitz, J.I. and Linkins, L.A. (2007). *Exp. Opin. Invest. Drugs* 16: 271–282.

6 (a) van Boeckel, C.A.A. and Petitou, M. (1993). *Angew. Chem. Int. Ed.* 32: 1671–1818. (b) Petitou, M. and van Boeckel, C.A.A. (2004). *Angew. Chem. Int. Ed.* 43: 3118–3133.

7 Petitou, M., Hérault, J.P., Bernat, A. et al. (1999). *Nature* 398: 417–422.

8 (a) Chappell, E.P. and Liu, J. (2013). *Bioorg. Med. Chem.* 21: 4786–4792. (b) DeAngelis, P.L., Liu, J., and Linhardt, R.J. (2013). *Glycobiology* 23: 764–777.

9 Dulaney, S.B. and Huang, X. (2012). *Adv. Carbohydr. Chem. Biochem.* 67: 95–136.

10 Driguez, P.A., Potier, P., and Trouilleux, P. (2014). *Nat. Prod. Rep.* 31: 980–989.

11 (a) Sinaÿ, P. and Jacquinet, J.C. (1984). *Carbohydr. Res.* 132: C5–C9. (b) Jacquinet, J.C., Petitou, M., Duchaussoy, P. et al. (1984). *Carbohydr. Res.* 130: 221–241. (c) van Boeckel, C.A.A., Beetz, T., Vos, J.N. et al. (1985). *Carbohydr. Res.* 4: 293–321.

12 (a) Chen, J., Jones, C.L., and Liu, J. (2007). *Chem. Biol.* 14: 986–993. (b) Guerrini, M., Guglieri, S., Casu, B. et al. (2008). *J. Biol. Chem.* 283: 26662–26675.

13 Guerrini, M., Mourier, P.A.J., Torri, G., and Viskov, C. (2014). *Glycoconjugate J.* 31: 409–416.

14 Petitou, M., Imberty, A., Duchaussoy, P. et al. (2001). *Chem. Eur. J.* 7: 858–873.

15 Driguez, P.A., Lederman, I., Strassel, J.M. et al. (1999). *J. Org. Chem.* 64: 9512–9520.

16 Herbert, J.M., Hérault, J.P., Bernat, A. et al. (2001). *Thromb. Haemost.* 85: 852–860.

17 Westerduin, P., van Boeckel, C.A.A., Basten, J.E.M. et al. (1994). *Bioorg. Med. Chem.* 2: 1267–1280.

18 (a) Duchaussoy, P., Jaurand, G., Driguez, P.A. et al. (1999). *Carbohydr. Res.* 317: 85–99. (b) Das, S.K., Mallet, J.M., Esnault, J. et al. (2001). *Angew. Chem. Int. Ed.* 40: 1670–1673. (c) Das, S.K., Mallet, J.M., Esnault, J. et al. (2001). *Chem. Eur. J.* 7: 4821–4834.

19 (a) Lucas, R., Hamza, D., Lubineau, A., and Bonnaffé, D. (2004). *Eur. J. Org. Chem.* 2107–2117. (b) Cid, M.B., Alfonso, F., and Martin-Lomas, M. (2005). *Chem. Eur. J.* 11: 928–938. (c) Hamza, D., Lucas, R., Feizi, T. et al. (2006). *ChemBioChem* 7: 1856–1858.

20 Orgueira, H.A., Bartolozzi, A., Schell, P. et al. (2003). *Chem. Eur. J.* 9: 140–169.

21 Cerny, M., Pacak, J., and Stanek, J. (1965). *Collect. Czech. Chem. Commun.* 30: 1151–1157.

22 Hermans, J.P.G., Noort, D., van der Marel, G.A. et al. (1988). *Recl. Trav. Chim. Pays-Bas* 107: 635–640.

23 Gebbie, S.J., Gosneya, I., Harrison, P.R. et al. (1998). *Carbohydr. Res.* 308: 345–348.

24 Kamath, V.P., Yeske, R.E., Gregson, J.M. et al. (2004). *Carbohydr. Res.* 339: 1141–1146.

25 Périon, R., Lemée, L., Ferrières, V. et al. (2003). *Carbohydr. Res.* 338: 2779–2792.

Index

Protecting Groups: Strategies and Applications in Carbohydrate Chemistry, First Edition.
Edited by Sébastien Vidal.

b

c

d

e

f

g

h

u

w

x

y

z